The Theory of $\mathcal{H}(b)$ Spaces

Volume 1

An $\mathcal{H}(b)$ space is defined as a collection of analytic functions that are in the image of an operator. The theory of $\mathcal{H}(b)$ spaces bridges two classical subjects, complex analysis and operator theory, which makes it both appealing and demanding.

Volume 1 of this comprehensive treatment is devoted to the preliminary subjects required to understand the foundation of $\mathcal{H}(b)$ spaces, such as Hardy spaces, Fourier analysis, integral representation theorems, Carleson measures, Toeplitz and Hankel operators, various types of shift operators and Clark measures. Volume 2 focuses on the central theory. Both books are accessible to graduate students as well as researchers: each volume contains numerous exercises and hints, and figures are included throughout to illustrate the theory. Together, these two volumes provide everything the reader needs to understand and appreciate this beautiful branch of mathematics.

Emmanuel Fricain is Professor of Mathematics at Laboratoire Paul Painlevé, Université Lille 1, France. Part of his research focuses on the interaction between complex analysis and operator theory, which is the main content of this book. He has a wealth of experience teaching numerous graduate courses on different aspects of analytic Hilbert spaces, and he has published several papers on $\mathcal{H}(b)$ spaces in high-quality journals, making him a world specialist in this subject.

Javad Mashreghi is a Professor of Mathematics at Université Laval, Québec, Canada, where he has been selected Star Professor of the Year seven times for excellence in teaching. His main fields of interest are complex analysis, operator theory and harmonic analysis. He is the author of several mathematical textbooks, monographs and research articles. He won the G. de B. Robinson Award, the publication prize of the Canadian Mathematical Society, in 2004.

NEW MATHEMATICAL MONOGRAPHS

All the titles listed below can be obtained from good booksellers or from Cambridge University Press. For a complete series listing visit www.cambridge.org/mathematics.

1. M. Cabanes and M. Enguehard *Representation Theory of Finite Reductive Groups*
2. J. B. Garnett and D. E. Marshall *Harmonic Measure*
3. P. Cohn *Free Ideal Rings and Localization in General Rings*
4. E. Bombieri and W. Gubler *Heights in Diophantine Geometry*
5. Y. J. Ionin and M. S. Shrikhande *Combinatorics of Symmetric Designs*
6. S. Berhanu, P. D. Cordaro and J. Hounie *An Introduction to Involutive Structures*
7. A. Shlapentokh *Hilbert's Tenth Problem*
8. G. Michler *Theory of Finite Simple Groups I*
9. A. Baker and G. Wüstholz *Logarithmic Forms and Diophantine Geometry*
10. P. Kronheimer and T. Mrowka *Monopoles and Three-Manifolds*
11. B. Bekka, P. de la Harpe and A. Valette *Kazhdan's Property (T)*
12. J. Neisendorfer *Algebraic Methods in Unstable Homotopy Theory*
13. M. Grandis *Directed Algebraic Topology*
14. G. Michler *Theory of Finite Simple Groups II*
15. R. Schertz *Complex Multiplication*
16. S. Bloch *Lectures on Algebraic Cycles (2nd Edition)*
17. B. Conrad, O. Gabber and G. Prasad *Pseudo-reductive Groups*
18. T. Downarowicz *Entropy in Dynamical Systems*
19. C. Simpson *Homotopy Theory of Higher Categories*
20. E. Fricain and J. Mashreghi *The Theory of* $\mathcal{H}(b)$ *Spaces I*
21. E. Fricain and J. Mashreghi *The Theory of* $\mathcal{H}(b)$ *Spaces II*
22. J. Goubault-Larrecq *Non-Hausdorff Topology and Domain Theory*
23. J. Śniatycki *Differential Geometry of Singular Spaces and Reduction of Symmetry*
24. E. Riehl *Categorical Homotopy Theory*
25. B. A. Munson and I. Volić *Cubical Homotopy Theory*
26. B. Conrad, O. Gabber and G. Prasad *Pseudo-reductive Groups (2nd Edition)*
27. J. Heinonen, P. Koskela, N. Shanmugalingam and J. T. Tyson *Sobolev Spaces on Metric Measure Spaces*
28. Y.-G. Oh *Symplectic Topology and Floer Homology I*
29. Y.-G. Oh *Symplectic Topology and Floer Homology II*

The Theory of $\mathcal{H}(b)$ Spaces

Volume 1

EMMANUEL FRICAIN
Université Lille 1

JAVAD MASHREGHI
Université Laval, Québec

CAMBRIDGE
UNIVERSITY PRESS

CAMBRIDGE
UNIVERSITY PRESS

University Printing House, Cambridge CB2 8BS, United Kingdom

Cambridge University Press is part of the University of Cambridge.

It furthers the University's mission by disseminating knowledge in the pursuit of education, learning and research at the highest international levels of excellence.

www.cambridge.org
Information on this title: www.cambridge.org/9781107027770

First published 2016

A catalogue record for this publication is available from the British Library

Library of Congress Cataloguing in Publication data
Fricain, Emmanuel, 1971 – author.
The theory of H(b) spaces / Emmanuel Fricain, Javad Mashreghi.
2 volumes ; cm. – (New mathematical monographs ; v. 20–21)
ISBN 978-1-107-02777-0 (Hardback)
1. Hilbert space. 2. Hardy spaces. 3. Analytic functions. 4. Linear operators. I. Mashreghi, Javad, author. II. Title.
QA322.4.F73 2014
515′.733–dc23 2014005539

ISBN – 2 Volume Set 978-1-107-11941-3 Hardback
ISBN – Volume 1 978-1-107-02777-0 Hardback
ISBN – Volume 2 978-1-107-02778-7 Hardback

To our families:
Keiko & Shahzad
Hugo & Dorsa, Parisa, Golsa

Contents for Volume 1

Contents for Volume 2

Preface

In 1915, Godfrey Harold Hardy, in a famous paper published in the *Proceedings of the London Mathematical Society*, first put forward the "theory of Hardy spaces" H^p. Having a Hilbert space structure, H^2 also benefits from the rich theory of Hilbert spaces and their operators. The mutual interaction of analytic function theory, on the one hand, and operator theory, on the other, has created one of the most beautiful branches of mathematical analysis. The Hardy–Hilbert space H^2 is the glorious king of this seemingly small, but profoundly deep, territory.

In 1948, in the context of dynamics of Hilbert space operators, A. Beurling classified the closed invariant subspaces of the forward shift operator on ℓ^2. The genuine idea of Beurling was to exploit the forward shift operator S on H^2. To that end, he used some analytical tools to show that the closed subspaces of H^2 that are invariant under S are precisely of the form ΘH^2, where Θ is an inner function. Therefore, the orthogonal complement of the Beurling subspace ΘH^2, the so-called *model subspaces* K_Θ, are the closed invariant subspaces of H^2 that are invariant under the backward shift operator S^*. The model subspaces have rich algebraic and analytic structures with applications in other branches of mathematics and science, for example, control engineering and optics.

The word "model" that was used above to describe K_Θ refers to their application in recognizing the Hilbert space contractions. The main idea is to identify (via a unitary operator) a contraction as the adjoint of multiplication by z on a certain space of analytic functions on the unit disk. As Beurling's theorem says, if we restrict ourselves to closed subspaces of H^2 that are invariant under S^*, we just obtain K_Θ spaces, where Θ runs through the family of inner functions. This point of view was exploited by B. Sz.-Nagy and C. Foiaş to construct a model for Hilbert space contractions. Another way is to consider submanifolds (not necessarily closed) of H^2 that are invariant under S^*. Above half a century ago, such a modeling theory was developed by L. de Branges and J. Rovnyak. In this context, they introduced the so-called $\mathcal{H}(b)$ spaces. The de Branges–Rovnyak model is, in a certain sense, more flexible, but it causes certain difficulties. For example, the inner product in $\mathcal{H}(b)$ is not given by an explicit integral formula, contrary to the case for K_Θ, which is actually the inner product of H^2, and this makes the treatment of $\mathcal{H}(b)$ functions more difficult.

The original definition of $\mathcal{H}(b)$ spaces uses the notion of *complementary space*, which is a generalization of the orthogonal complement in a Hilbert space. But $\mathcal{H}(b)$ spaces can also be viewed as the range of a certain operator involving Toeplitz operators. This point of view was a turning point in the theory of $\mathcal{H}(b)$ spaces. Adopting the new definition, D. Sarason and several others made essential contributions to the theory. In fact, they now play a key role in many other questions of function theory (solution of the Bieberbach conjecture by de Branges, rigid functions of the unit ball of H^1, Schwarz–Pick inequalities), operator theory (invariant subspaces problem, composition operators), system theory and control theory. An excellent but very concise account of the theory of $\mathcal{H}(b)$ spaces is available in Sarason's masterpiece [460]. However, there are many results, both new and old, that are not covered there. On the other hand, despite many efforts, the structure and properties of $\mathcal{H}(b)$ spaces still remain mysterious, and numerous natural questions still remain open. However, these spaces have a beautiful structure, with numerous applications, and we hope that this work attracts more people to this domain.

In this context, we have tried to provide a rather comprehensive *introduction* to the theory of $\mathcal{H}(b)$ spaces. That is why Volume 1 is devoted to the foundation of $\mathcal{H}(b)$ spaces. In Volume 2, we discuss $\mathcal{H}(b)$ spaces and their applications. However, two facts should be kept in mind: first, we just treat the scalar case of $\mathcal{H}(b)$ spaces; and second, we do not discuss in detail the theory of model operators, because there are already two excellent monographs on this topic [388]; [508]. Nevertheless, some of the tools of model theory are implicitly exploited in certain topics. For instance, to treat some natural questions such as the inclusion between two different $\mathcal{H}(b)$ spaces, we use a geometric representation of $\mathcal{H}(b)$ spaces that comes from the relation between Sz.-Nagy–Foiaş and de Branges–Rovnyak modeling theory. Also, even if the main point of view that has been adopted in this book is based on the definition of $\mathcal{H}(b)$ via Toeplitz operators, the historical definition of de Branges and Rovnyak is also discussed and used at some points.

In the past decade, both of us have made several transatlantic trips to meet each other and work together on this book project. For these visits, we have been financially supported by Université Claude Bernard Lyon I, Université Lille 1, Université Laval, McGill University, Centre Jacques-Cartier (France), CNRS (Centre National de la Recherche Scientifique, France), ANR (Agence Nationale de la Recherche), FQRNT (Fonds Québécois de la Recherche sur la Nature et les Technologies) and NSERC (Natural Sciences and Engineering Research Council of Canada). We also benefited from the warm hospitality of CIRM (Centre International des Rencontres Mathématiques, Luminy), CRM (Centre de Recherches Mathématiques, Montréal) and the Fields Institute (Toronto). We thank them all warmly for their support.

During these past years, parallel to the writing of this book, we have also pursued our research, and some projects were directly related to $\mathcal{H}(b)$ spaces. Some of the results, mainly in collaboration with other colleagues, are contained in this monograph. Hence, we would like to thank from the bottom of our hearts our close collaborators: A. Baranov, A. Blandignères, G. Chacon, I. Chalendar, N. Chevrot, F. Gaunard, A. Hartmann, W. Ross, M. Shabankhah and D. Timotin. During the preparation of the manuscript, we also benefited from very useful discussions with P. Gorkin and D. Timotin concerning certain points of this book. We would like to warmly thank them both.

The first author would like to thank T. Ransford for inviting him to Laval University in 2014. During his six-month visit, he met the second author and the seed of mutual collaboration on $\mathcal{H}(b)$ spaces was sown. He is also profoundly influenced in his mathematical life by N. Nikolskii. He takes this opportunity to express his profound gratitude and admiration.

The preparation of the manuscript was not an easy task, and it took several years even after acceptance for publication by Cambridge University Press. We thank the CUP team, who were ultra-patient with us and on numerous occasions intervened to settle difficulties. Above all, we are grateful to Roger Astley for his unique constructive leadership during the whole procedure.

Emmanuel Fricain — Lille

Javad Mashreghi — Kashan

1

Normed linear spaces and their operators

In this chapter, we gather some basic facts about complex normed linear spaces and their operators. In particular, we discuss Banach spaces, Hilbert spaces and their bounded operators. There is no doubt that the subject is very vast and it is impossible to give a comprehensive treatment in one chapter. Our goal is to recall a few important aspects of the theory that are used in the study of $\mathcal{H}(b)$ spaces. We start by giving some examples of Banach spaces and introduce some classic operators. Then the dual space is defined and the well-known Hahn–Banach theorem is stated without proof. However, some applications of this essential result are outlined. Then we discuss the open mapping theorem (Theorem 1.14), the inverse mapping theorem (Corollary 1.15), the closed graph theorem (Corollary 1.18) and the uniform boundedness principle (Theorem 1.19). The common root of each of these theorems stems from the Baire category theorem (Theorem 1.13). Then we discuss Banach algebras and introduce the important concept of spectrum and state a simple version of the spectral mapping theorem (Theorem 1.22). At the end, we focus on Hilbert spaces, and some of their essential properties are outlined. We talk about Parseval's identity, the generalized version of the polarization identity, and Bessel's inequality. We also discuss in detail the compression of an operator to a closed subspace. Then we consider several topologies that one may face on a Hilbert space or on the space of its operators. The important concepts of adjoint and tensor product are discussed next. The chapter ends with some elementary facts about invariant subspaces and the cyclic vectors.

1.1 Banach spaces

Throughout this text we will consider only complex normed linear spaces. A complete normed linear space is called a *Banach space*. The term *linear manifold* refers to subsets of a linear space that are closed under the algebraic operations, while the term *subspace* is reserved for linear manifolds that are also closed in the norm or metric topology. Nevertheless, the combination *closed subspace* can also be found in the text. Given a subset $\mathcal{E}$ of a Banach space $\mathcal{X}$, we denote by $\text{Lin}(\mathcal{E})$ the linear manifold spanned by $\mathcal{E}$, which is the linear space whose elements are finite linear combinations of elements of $\mathcal{E}$. The closure of $\text{Lin}(\mathcal{E})$ in $\mathcal{X}$ is denoted by $\text{Span}_{\mathcal{X}}(\mathcal{E})$, or by $\text{Span}(\mathcal{E})$ if there is no ambiguity.

We start by briefly discussing some relevant families of Banach spaces that we encounter in the sequel. For a sequence $\mathfrak{z} = (z_n)_{n\geq 1}$ of complex numbers, define

$$\|\mathfrak{z}\|_p = \left(\sum_{n=1}^{\infty} |z_n|^p \right)^{1/p} \qquad (0 < p < \infty)$$

and

$$\|\mathfrak{z}\|_\infty = \sup_{n\geq 1} |z_n|.$$

Then the sequence space $\ell^p = \ell^p(\mathbb{N})$, $0 < p \leq \infty$, consists of all sequences $\mathfrak{z}$ with $\|\mathfrak{z}\|_p < \infty$. Addition and scalar multiplication are defined componentwise in ℓ^p. With this setting, $(\ell^p, \|\cdot\|_p)$, $1 \leq p \leq \infty$, is a Banach space. The sequence space

$$c_0 = c_0(\mathbb{N}) = \{\mathfrak{z} : \lim_{n\to\infty} z_n = 0\}$$

is a closed subspace of ℓ^∞. In certain applications, it is more appropriate to let the index parameter n start from zero or from $-\infty$. In the latter case, we denote our spaces by $\ell^p(\mathbb{Z})$ and $c_0(\mathbb{Z})$.

For each n, let $\mathfrak{e}_n$ be the sequence whose components are all equal to 0 except in the nth place, which is equal to 1. Clearly each $\mathfrak{e}_n$ belongs to all sequence spaces introduced above. These elements will repeatedly enter our discussion.

Let $(X, \mathcal{A}, \mu)$ be a measure space with $\mu \geq 0$. For a measurable function f, let

$$\|f\|_p = \left(\int_X |f|^p \, d\mu \right)^{1/p} \qquad (0 < p < \infty)$$

and let

$$\|f\|_\infty = \inf\{\, M : |f(x)| \leq M,\ x \in X \setminus E,\ E \in \mathcal{A},\ \mu(E) = 0 \,\}.$$

Then the family of Lebesgue spaces

$$L^p(X, \mu) = L^p(X) = L^p(\mu) = \{f : \|f\|_p < \infty\}$$

is another important example that we will need. To emphasize the role of μ, we sometimes use the more detailed notation $\|f\|_{L^p(\mu)}$ for $\|f\|_p$. For $1 \leq p \leq \infty$, $(L^p(X, \mu), \|\cdot\|_p)$ is a Banach space. In fact, the sequence spaces $\ell^p(\mathbb{N})$ can be considered as a special family in this category. It is enough to let $\mathcal{A} = \mathcal{P}(\mathbb{N})$, the power set of $\mathbb{N}$, and consider the counting measure

$$\mu(E) = \begin{cases} n, & \text{if } E \text{ is finite and has } n \text{ elements,} \\ \infty, & \text{if } E \text{ is infinite,} \end{cases}$$

on $\mathbb{N}$. Then we have $\ell^p(\mathbb{N}) = L^p(\mathbb{N}, \mu)$.

Two more special classes are vital in practice. The unit circle $\mathbb{T}$ equipped with the normalized Lebesgue measure

$$dm(e^{it}) = \frac{dt}{2\pi}$$

and the real line $\mathbb{R}$ equipped with the Lebesgue measure dt give rise to the classic Lebesgue spaces $L^p(\mathbb{T})$ and $L^p(\mathbb{R})$. If E is a Borel subset of $\mathbb{T}$, then $|E|$ will denote its length with respect to the normalized Lebesgue measure, that is

$$|E| = m(E).$$

The Banach space $(\ell^\infty, \|\cdot\|_\infty)$ is equipped with a third operation. Besides vector addition and scalar multiplication, we have vector multiplication in this space. Given $\mathfrak{x} = (x_n)_{n\geq 1}$ and $\mathfrak{y} = (y_n)_{n\geq 1}$ in ℓ^∞, let

$$\mathfrak{x}\,\mathfrak{y} = (x_n y_n)_{n\geq 1}.$$

Clearly $\mathfrak{x}\,\mathfrak{y} \in \ell^\infty$, and ℓ^∞ with this operation is an algebra that satisfies

$$\|\mathfrak{x}\,\mathfrak{y}\|_\infty \leq \|\mathfrak{x}\|_\infty \, \|\mathfrak{y}\|_\infty.$$

If a Banach space $\mathcal{B}$ is equipped with a multiplication operation that turns it into an algebra, then it is called a *Banach algebra* if it satisfies the multiplicative inequality

$$\|xy\| \leq \|x\| \, \|y\| \qquad (x \in \mathcal{B},\ y \in \mathcal{B}). \tag{1.1}$$

If, furthermore, the multiplication has a unit element, i.e. a vector $\mathfrak{e}$ such that

$$x\,\mathfrak{e} = \mathfrak{e}\,x = x \qquad (x \in \mathcal{B})$$

and

$$\|\mathfrak{e}\| = 1,$$

then it is called a *unital Banach algebra*. The sequence space ℓ^∞ is our first example of a unital commutative Banach algebra with the unit element

$$\mathfrak{e} = (1, 1, 1, \ldots).$$

Another example of a unital commutative Banach algebra is $\mathcal{C}(\mathbb{T})$, the family of continuous functions on $\mathbb{T}$ endowed with the norm

$$\|f\|_\infty = \max_{\zeta \in \mathbb{T}} |f(\zeta)|.$$

The unit element is the constant function 1. The monomials

$$\chi_n(\zeta) = \zeta^n \qquad (n \in \mathbb{Z},\ \zeta \in \mathbb{T})$$

are clearly in $\mathcal{C}(\mathbb{T})$. With this notation, the unit element is χ_0. However, for simplicity, we usually write z^n for χ_n. In particular, we denote the constant function χ_0 by 1.

A linear combination of χ_n, $n \in \mathbb{Z}$, is called a *trigonometric polynomial*, and the linear manifold of all trigonometric polynomials is denoted by $\mathcal{P}$. However, if we restrict n to be a nonnegative integer, then a linear combination of χ_n, $n \geq 0$, is called an *analytic polynomial*, and the linear manifold of all analytic polynomials is denoted by $\mathcal{P}_+$. The term "analytic" comes from the fact that each element of $\mathcal{P}_+$ extends to an analytic function on the complex plane. Similarly, $\mathcal{P}_-$ denotes the linear manifold created by χ_n, $n \leq -1$, and $\mathcal{P}_{0+}$ denotes the linear manifold created by χ_n, $n \geq 1$. In other words, $\mathcal{P}_{0+}$ is the linear manifold of analytic polynomials vanishing at 0.

The family of all complex Borel measures on the unit circle $\mathbb{T}$ is denoted by $\mathcal{M}(\mathbb{T})$. The set of positive measures in $\mathcal{M}(\mathbb{T})$ is denoted by $\mathcal{M}^+(\mathbb{T})$. Recall that all measures in $\mathcal{M}(\mathbb{T})$ are necessarily finite. The normalized *Lebesgue measure* m is a distinguished member of this class. Another interesting example is the *Dirac measure* δ_α, which attributes a unit mass to the point $\alpha \in \mathbb{T}$. For each $\mu \in \mathcal{M}(\mathbb{T})$, the smallest positive Borel measure ν that satisfies the inequality

$$|\mu(E)| \leq \nu(E)$$

for all Borel sets $E \subset \mathbb{T}$ is called the *total variation measure* of μ and is denoted by $|\mu|$. The total variation $|\mu|$ is also given by the formula

$$|\mu|(E) = \sup \sum_{k=1}^{n} |\mu(E_k)|,$$

where the supremum is taken over all possible partitions $\{E_1, E_2, \dots, E_n\}$ of E by Borel sets. The norm of an element $\mu \in \mathcal{M}(\mathbb{T})$ is defined to be its total variation on $\mathbb{T}$, i.e.

$$\|\mu\| = |\mu|(\mathbb{T}).$$

Then $\mathcal{M}(\mathbb{T})$, endowed with the above norm, is a Banach space. However, with an appropriate definition of product (the convolution of two measures) on $\mathcal{M}(\mathbb{T})$, this space becomes a commutative unital Banach algebra. But, for our applications, we do not need to treat the details of this operation. A measure $\mu \in \mathcal{M}(\mathbb{T})$ that takes only real values is called a *signed measure*. Note that we assume in the definition of signed measure that they are finite.

The nth Fourier coefficient of a measure $\mu \in \mathcal{M}(\mathbb{T})$ is defined by

$$\hat{\mu}(n) = \int_{\mathbb{T}} \chi_{-n}\, d\mu = \int_{\mathbb{T}} e^{-int}\, d\mu(e^{it}) \qquad (n \in \mathbb{Z}).$$

In the particular case $d\mu = \varphi\, dm$, where $\varphi \in L^1(\mathbb{T})$, we use $\hat{\varphi}(n)$ to denote the nth Fourier coefficient of μ, that is

$$\hat{\varphi}(n) = \int_{\mathbb{T}} \varphi\chi_{-n}\, dm = \frac{1}{2\pi}\int_0^{2\pi} \varphi(e^{it})\, e^{-int}\, dt \qquad (n \in \mathbb{Z}).$$

We refer to the sequence $(\hat{\varphi}(n))_{n\in\mathbb{Z}}$ as the *spectrum* of φ. For example, the negative part of the spectrum of any analytic polynomials is identically zero. The *uniqueness theorem* for the Fourier coefficients says that $\mu = 0$ if and only if $\hat{\mu}(n) = 0$, $n \in \mathbb{Z}$.

There is a method to connect an arbitrary Lebesgue integral to a standard Riemann integral on $\mathbb{R}$. To do so, for a measurable function f defined on a measure space (X, μ), we define the *distribution function*

$$m_{\mu,f}(t) = \mu(\{x \in X : |f(x)| > t\}).$$

For simplicity, instead of $m_{\mu,f}$, we sometimes write m_μ or m_f, or even m whenever there is no confusion. The promised connection is described in the following result.

Lemma 1.1 *Let (X, μ) be a measure space, and let $f : X \longrightarrow \mathbb{C}$ be a measurable function. Then, for any $0 < p < \infty$, we have*

$$\int_X |f|^p\, d\mu = \int_0^\infty p\, t^{p-1} m_{\mu,f}(t)\, dt.$$

Proof For $x \in X$ and $t > 0$, define

$$A_{x,t} = \{x \in X : |f(x)| > t\}$$

and let $\chi_{A_{x,t}}$ be the characteristic function of the set $A_{x,t}$. Then, by Fubini's theorem, we have

$$\begin{aligned}
\int_0^\infty p\, t^{p-1} m(t)\, dt &= \int_0^\infty p\, t^{p-1} \mu(\{x \in X : |f(x)| > t\})\, dt \\
&= \int_0^\infty p\, t^{p-1} \Big(\int_X \chi_{A_{x,t}}(x)\, d\mu(x) \Big)\, dt \\
&= \int_X \Big(\int_0^{|f(x)|} p\, t^{p-1}\, dt \Big)\, d\mu(x) \\
&= \int_X |f(x)|^p\, d\mu(x).
\end{aligned}$$

This completes the proof. □

As a special case of Lemma 1.1, we have the interesting formula

$$\int_X |f|\, d\mu = \int_0^\infty m_{\mu,f}(t)\, dt.$$

We will also need the following elementary result about convergence in $L^1(\mathbb{T})$.

Lemma 1.2 *Let $f \in L^1(\mathbb{T})$ and $(f_n)_{n\geq 1}$ be a sequence of nonnegative functions in $L^1(\mathbb{T})$ such that $\|f_n\|_1 = \|f\|_1$, $n \geq 1$, and $f_n \longrightarrow f$ a.e. on $\mathbb{T}$. Then f_n tends to f in $L^1(\mathbb{T})$, i.e.*

$$\lim_{n\to\infty} \|f_n - f\|_1 = 0.$$

Proof Define the set $E_n = \{\zeta \in \mathbb{T} : f_n(\zeta) > f(\zeta)\}$. Since $\|f_n\|_1 = \|f\|_1$, we have

$$\int_{E_n} (f_n - f)\, dm = \int_{\mathbb{T}\setminus E_n} (f - f_n)\, dm.$$

This implies that

$$\begin{aligned} \|f_n - f\|_1 &= \int_{\mathbb{T}} |f_n - f|\, dm \\ &= \int_{E_n} (f_n - f)\, dm + \int_{\mathbb{T}\setminus E_n} (f - f_n)\, dm \\ &= 2\int_{\mathbb{T}\setminus E_n} (f - f_n)\, dm. \end{aligned}$$

But $f_n - f \longrightarrow 0$ and $0 \leq f - f_n \leq f$ a.e. on $\mathbb{T} \setminus E_n$. Hence, the dominated Lebesgue convergence theorem implies that

$$\int_{\mathbb{T}\setminus E_n} (f - f_n)\, dm \longrightarrow 0 \qquad (n \longrightarrow \infty),$$

which gives the result. □

Exercises

Exercise 1.1.1 Show that, for each $0 < p < 1$, $\|\cdot\|_p$ does not fulfill the triangle inequality.
Remark: That is why ℓ^p, or more generally $L^p(X, \mu)$, $0 < p < 1$, is not a Banach space.

Exercise 1.1.2 Let $p < \infty$. Show that the linear manifold generated by $\mathfrak{e}_n$, $n \geq 1$, is dense in ℓ^p. What is the closure of this manifold in ℓ^∞?

Exercise 1.1.3 Show that ℓ^p, $0 < p < \infty$, is *separable*.
Hint: Use Exercise 1.1.2.
Remark: We recall that a space is called "separable" if it has a countable dense subset.

Exercise 1.1.4 Show that ℓ^∞ is not separable.
Hint: What is $\|\mathfrak{e}_m - \mathfrak{e}_n\|_\infty$?

Exercise 1.1.5 Consider the spaces of two-sided sequences,

$$\ell^p(\mathbb{Z}) = \left\{ (z_n)_{n\in\mathbb{Z}} : \|(z_n)_{n\in\mathbb{Z}}\|_p^p = \sum_{n=-\infty}^{\infty} |z_n|^p < \infty \right\}$$

and

$$\ell^\infty(\mathbb{Z}) = \left\{ (z_n)_{n\in\mathbb{Z}} : \|(z_n)_{n\in\mathbb{Z}}\|_\infty = \sup_{n\in\mathbb{Z}} |z_n| < \infty \right\}.$$

Let $\mathfrak{x} = (x_n)_{n\in\mathbb{Z}}$ and $\mathfrak{y} = (y_n)_{n\in\mathbb{Z}} \in \ell^1(\mathbb{Z})$. Define $\mathfrak{x} * \mathfrak{y} = (z_n)_{n\in\mathbb{Z}}$, where

$$z_n = \sum_{m=-\infty}^{\infty} x_m\, y_{n-m} \qquad (n \in \mathbb{Z}).$$

Show that the operation $*$ is well defined on $\ell^1(\mathbb{Z})$ and, moreover, that $\ell^1(\mathbb{Z})$ equipped with $*$ is a unital commutative Banach algebra.
Remark: The operation $*$ is called *convolution*.
Hint: We have

$$\sum_{n=1}^{\infty} \left(\sum_{m=-\infty}^{\infty} |x_m\, y_{n-m}| \right) = \left(\sum_{m=-\infty}^{\infty} |x_m| \right) \left(\sum_{k=-\infty}^{\infty} |y_k| \right).$$

Exercise 1.1.6 Let

$$\ell^p(\mathbb{Z}^+) = \{\, (z_n)_{n\in\mathbb{Z}} \in \ell^p(\mathbb{Z}) : z_n = 0,\ n \leq -1 \,\}.$$

Show that $\ell^p(\mathbb{Z}^+)$, $0 < p \leq \infty$, is closed in $\ell^p(\mathbb{Z})$.

Exercise 1.1.7 The family of sequences of compact support is defined by

$$c_{00} = \{(z_n)_{n\geq 1} : \exists\, N \text{ such that } z_n = 0,\ n \geq N\}.$$

Show that c_{00} is dense in c_0.

Exercise 1.1.8 Let $(X, \mathcal{A}, \mu)$ be a measure space with $\mu \geq 0$. Show that

$$\|f\|_\infty = \inf\{M : \mu(\{x \in X : |f(x)| > M\}) = 0\}.$$

Exercise 1.1.9 Let $(X, \mathcal{A}, \mu)$ be a measure space. Show that $L^\infty(X)$ endowed with the pointwise multiplication is a unital commutative Banach algebra.

Exercise 1.1.10 Show that $L^1(\mathbb{R})$ equipped with the convolution operation

$$(f * g)(x) = \int_{\mathbb{R}} f(t)\, g(x-t)\, dt$$

is a nonunital commutative Banach algebra. Why does $L^1(\mathbb{R})$ not have a unit?

Exercise 1.1.11 Show that $L^1(\mathbb{T})$ equipped with the convolution operation

$$(f * g)(e^{i\theta}) = \frac{1}{2\pi} \int_{\mathbb{T}} f(e^{it})\, g(e^{i(\theta-t)})\, dt$$

is a nonunital commutative Banach algebra. Why does $L^1(\mathbb{T})$ not have a unit?

Exercise 1.1.12 Let $(b_k)_{k\geq 1}$ be a decreasing sequence of nonnegative real numbers such that the series $\sum_{k=1}^{\infty} b_k$ converges.

(i) Using Abel's summation technique, show that

$$\sum_{k=0}^{N} k(b_k - b_{k+1}) \leq \sum_{k=1}^{N} b_k.$$

Deduce that the series $\sum_{k=0}^{N} k(b_k - b_{k+1})$ is convergent.

(ii) Show that the series $\sum_{k=1}^{\infty} (b_k - b_{k+1})$ is convergent and

$$b_n = \sum_{k=n}^{\infty} (b_k - b_{k+1}).$$

(iii) Deduce that

$$0 \leq n b_n \leq \sum_{k=n}^{\infty} k(b_k - b_{k+1}).$$

(iv) Conclude that $n b_n \longrightarrow 0$ as $n \longrightarrow \infty$.

Exercise 1.1.13 Let $\mathcal{X}$ be a Banach space, let E be a linear manifold of X and let $x \in \mathcal{X}$. Show that $x \in \text{Clos}(E)$ if and only if there exists a sequence $(x_n)_n$ of vectors of E such that $\|x_n\| \leq 1/2^n$, $n \geq 2$ and

$$x = \sum_{n=1}^{\infty} x_n.$$

Hint: If $x = \lim_{n\to\infty} u_n$ with $u_n \in E$, then consider a subsequence $(u_{\varphi(n)})_n$ such that

$$\|u_{\varphi(n+1)} - u_{\varphi(n)}\| \leq \frac{1}{2^n} \qquad (n \geq 1).$$

Then, put $x_1 = u_{\varphi(1)}$ and $x_n = u_{\varphi(n)} - u_{\varphi(n-1)}$, $n \geq 2$.

1.2 Bounded operators

Let $\mathcal{X}$, $\mathcal{X}_1$ and $\mathcal{X}_2$ be normed linear spaces. The family of all *linear and continuous maps* from $\mathcal{X}_1$ into $\mathcal{X}_2$ is denoted by $\mathcal{L}(\mathcal{X}_1, \mathcal{X}_2)$, and we will write $\mathcal{L}(\mathcal{X})$ for $\mathcal{L}(\mathcal{X}, \mathcal{X})$. Given a linear map $A : \mathcal{X}_1 \longrightarrow \mathcal{X}_2$, it is a well-known result that A is continuous if and only if

$$\|A\|_{\mathcal{L}(\mathcal{X}_1,\mathcal{X}_2)} = \sup_{\substack{x\in\mathcal{X}_1 \\ x\neq 0}} \frac{\|Ax\|_{\mathcal{X}_2}}{\|x\|_{\mathcal{X}_1}} < \infty. \tag{1.2}$$

If there is no ambiguity, we will also write $\|A\|$ for $\|A\|_{\mathcal{L}(\mathcal{X}_1,\mathcal{X}_2)}$. A linear map A belongs to $\mathcal{L}(\mathcal{X}_1, \mathcal{X}_2)$ if and only if it satisfies

$$\|Ax\|_{\mathcal{X}_2} \leq C\|x\|_{\mathcal{X}_1} \qquad (x \in \mathcal{X}_1) \tag{1.3}$$

and $\|A\|$ is the infimum of C satisfying (1.3). That is why the elements of $\mathcal{L}(\mathcal{X}_1, \mathcal{X}_2)$ are called *bounded operators*. If $\mathcal{X}_2$ is a Banach space, then the space $\mathcal{L}(\mathcal{X}_1, \mathcal{X}_2)$ endowed with the norm $\|\cdot\|_{\mathcal{L}(\mathcal{X}_1,\mathcal{X}_2)}$ is also a Banach space. In the special case where $\mathcal{X}_1 = \mathcal{X}_2 = \mathcal{X}$ is a Banach space, the space $\mathcal{L}(\mathcal{X})$, equipped with the composition of operators as its multiplication, is a unital noncommutative Banach algebra.

In the definition (1.2), the supremum is not necessarily attained. See Exercises 1.2.9 and 1.2.3. But if this is the case, any vector $x \in \mathcal{X}_1$, $x \neq 0$, for which

$$\|Ax\|_{\mathcal{X}_2} = \|A\|\,\|x\|_{\mathcal{X}_1}$$

is called a *maximizing vector* for A.

The operator $A \in \mathcal{L}(\mathcal{X}_1, \mathcal{X}_2)$ is called *lower bounded* or *bounded below* if there is a constant $c > 0$ such that

$$\|Ax\|_{\mathcal{X}_2} \geq c\|x\|_{\mathcal{X}_1} \qquad (x \in \mathcal{X}_1).$$

It is easy to see that A is lower bounded if A is left-invertible, that is, there is a bounded operator $B \in \mathcal{L}(\mathcal{X}_2, \mathcal{X}_1)$ such that $BA = I_{\mathcal{X}_1}$. The converse is true in the Hilbert space setting.

For a bounded operator $A : \mathcal{X}_1 \longrightarrow \mathcal{X}_2$, the closed subspace

$$\ker A = \{x \in \mathcal{X}_1 : Ax = 0\}$$

is called the *kernel* of A. The kernel plays a major role in our discussion. Another set that frequently appears in this subject is the *range* of A,

$$\mathcal{R}(A) = \{Ax : x \in \mathcal{X}_1\},$$

which is a linear manifold of $\mathcal{X}_2$. Note that, contrary to the kernel, the range of a bounded operator is not necessarily closed. Nevertheless, if A is lower bounded, then the kernel of A is trivial (reduced to $\{0\}$) and the range of A is closed if $\mathcal{X}_1$ is a Banach space. Conversely, if A is one-to-one and has a closed range, and if $\mathcal{X}_1$ and $\mathcal{X}_2$ are Banach spaces, then A is lower bounded. This is a profound result and will be treated after studying the open mapping theorem (Corollary 1.17).

An operator $A \in \mathcal{L}(\mathcal{X})$ is said to be *power bounded* if there exists a constant $C > 0$ such that

$$\|A^n\| \le C \qquad (n \ge 0). \tag{1.4}$$

More restrictively, A is said to be *polynomially bounded* if there exists a constant $C > 0$ such that

$$\|p(A)\| \le C\|p\|_\infty \tag{1.5}$$

for every analytic polynomial p, where $\|p\|_\infty = \sup_{|z|=1} |p(z)|$. Clearly, a polynomially bounded operator is also power bounded. But the converse is not true.

A linear map from a normed linear space $\mathcal{X}$ into the complex plane $\mathbb{C}$ is called a *functional*. The family of all continuous functionals on $\mathcal{X}$ is called the *dual space* of $\mathcal{X}$ and is denoted by $\mathcal{X}^*$. The vector space $\mathcal{X}^*$ equipped with the operator norm

$$\|f\|_{\mathcal{X}^*} = \sup_{\|x\|_{\mathcal{X}} \le 1} |f(x)| \qquad (f \in \mathcal{X}^*)$$

is a Banach space. If there is no ambiguity, we will also write $\|f\|$ for $\|f\|_{\mathcal{X}^*}$. Characterizing the elements of $\mathcal{X}^*$ is an important theme in functional analysis, and it has profound applications. Since $\mathcal{X}^*$ is a normed linear space, we can equally consider the dual of $\mathcal{X}^*$, which is called the *second dual* of $\mathcal{X}$ and is naturally denoted by $\mathcal{X}^{**}$. The mapping

$$\begin{array}{ccc} \mathcal{X} & \longrightarrow & \mathcal{X}^{**} \\ x & \longmapsto & \hat{x}, \end{array}$$

where $\hat{x}$ is defined by

$$\hat{x}(\Lambda) = \Lambda(x) \qquad (\Lambda \in \mathcal{X}^*),$$

gives an isometric injection of $\mathcal{X}$ into $\mathcal{X}^{**}$. A normed linear space $\mathcal{X}$ is called *reflexive* if this injection is also onto. First note that a reflexive space $\mathcal{X}$ is isometrically isomorphic to its second dual, and hence must be a Banach space. It is not true, however, that a Banach space $\mathcal{X}$ that is isometrically isomorphic to $\mathcal{X}^{**}$ is reflexive. The definition of reflexivity stipulates that the isometry be the natural embedding of $\mathcal{X}$ into $\mathcal{X}^{**}$. In the following, we introduce the dual of some spaces that are needed in our discussion.

For a positive number p in the interval $(1, \infty)$, there is a unique number q, in the same interval, such that

$$\frac{1}{p} + \frac{1}{q} = 1.$$

We say that q is the *conjugate exponent* of p. Note that the definition is symmetric with respect to p and q. By the same token, 1 and $+\infty$ are said to be the conjugate exponents of each other.

Theorem 1.3 *Let $1 \leq p < \infty$, and let q be the conjugate exponent of p. Define*

$$\begin{aligned} \Lambda : \quad L^q(\mathbb{T}) &\longrightarrow (L^p(\mathbb{T}))^* \\ g &\longmapsto \Lambda_g, \end{aligned}$$

where

$$\Lambda_g(f) = \int_{\mathbb{T}} fg\, dm \qquad (f \in L^p(\mathbb{T})).$$

Then Λ is an isometric isomorphism from $L^q(\mathbb{T})$ onto $\big(L^p(\mathbb{T})\big)^*$.

See [159, appdx B] for a proof of this classic result.

In light of Theorem 1.3, we abuse notation and write $(L^p(\mathbb{T}))^* = L^q(\mathbb{T})$. Hence, if $1 < p < \infty$, then $(L^p(\mathbb{T}))^{**} = L^p(\mathbb{T})$, i.e. $L^p(\mathbb{T})$ is a reflexive space. Exercises 1.2.6 and 1.2.7 show that c_0 is not reflexive.

We would like to emphasize that Λ is not the only isometric isomorphism between $L^q(\mathbb{T})$ and $(L^p(\mathbb{T}))^*$. For example, in some applications, we consider

$$\Lambda_g(f) = \int_{\mathbb{T}} \chi_1 fg\, dm \qquad (f \in L^p(\mathbb{T})).$$

In some other applications, we even consider

$$\Lambda_g(f) = \int_{\mathbb{T}} f\bar{g}\, dm \qquad (f \in L^p(\mathbb{T})).$$

However, in this case, Λ is conjugate linear.

The following theorem is another result of the above type that justifies writing $(\mathcal{C}(\mathbb{T}))^* = \mathcal{M}(\mathbb{T})$.

Theorem 1.4 *Define*

$$\begin{array}{rccc} \Lambda: & \mathcal{M}(\mathbb{T}) & \longrightarrow & (\mathcal{C}(\mathbb{T}))^* \\ & \mu & \longmapsto & \Lambda_\mu, \end{array}$$

where

$$\Lambda_\mu(f) = \int_{\mathbb{T}} f \, d\mu \qquad (f \in \mathcal{C}(\mathbb{T})). \tag{1.6}$$

Then Λ is an isometric isomorphism from $\mathcal{M}(\mathbb{T})$ onto $(\mathcal{C}(\mathbb{T}))^*$.

For a proof of this result, we may refer to [159, appdx C].

Exercises

Exercise 1.2.1 Show that the mapping

$$\begin{array}{rccc} D: & \mathcal{C}^1[0,1] & \longrightarrow & \mathcal{C}[0,1] \\ & f & \longmapsto & f' \end{array}$$

is not bounded, but

$$\begin{array}{rccc} I: & \mathcal{C}[0,1] & \longrightarrow & \mathcal{C}^1[0,1] \\ & f & \longmapsto & \displaystyle\int_0^x f(t)\, dt \end{array}$$

is a bounded operator. Both spaces $\mathcal{C}[0,1]$ and $\mathcal{C}^1[0,1]$ are endowed with the norm $\|\cdot\|_\infty$.
Hint: Consider $f(x) = \sin(n\pi x)$.

Exercise 1.2.2 Show that the mapping

$$\begin{array}{rccc} D: & \mathcal{C}^1[0,1] & \longrightarrow & \mathcal{C}[0,1] \\ & f & \longmapsto & f' \end{array}$$

is bounded operator, provided that $\mathcal{C}[0,1]$ is endowed with the norm $\|\cdot\|_\infty$ and $\mathcal{C}^1[0,1]$ is equipped with

$$\|f\|_{\mathcal{C}^1} = \|f\|_\infty + \|f'\|_\infty.$$

Remark: Compare with Exercise 1.2.1.

Exercise 1.2.3 Show that the multiplication operator

$$Af(t) = tf(t) \qquad (0 \le t \le 1)$$

on the Hilbert space $L^2(0,1)$ has no maximizing vector.
Hint: Show that $\|A\| = 1$ and then consider $\|Af\|_2^2 - \|f\|_2^2$.

Exercise 1.2.4 Let $1 < p < \infty$, and let q be the conjugate exponent of p. Show that $(\ell^p)^* = \ell^q$.
Hint: For a fixed $\mathfrak{a} = (a_n)_{n\geq 1} \in \ell^q$, consider the functional

$$\Lambda_{\mathfrak{a}}(\mathfrak{x}) = \sum_{n=1}^{\infty} a_n x_n,$$

where $\mathfrak{x} = (x_n)_{n\geq 1}$ runs through ℓ^p.

Exercise 1.2.5 Let $1 < p < \infty$. Show that ℓ^p is reflexive.
Hint: Use Exercise 1.2.4.

Exercise 1.2.6 Show that $c_0^* = \ell^1$.
Hint: For a fixed $\mathfrak{a} = (a_n)_{n\geq 1} \in \ell^1$, consider the functional

$$\Lambda_{\mathfrak{a}}(\mathfrak{x}) = \sum_{n=1}^{\infty} a_n x_n,$$

where $\mathfrak{x} = (x_n)_{n\geq 1}$ runs through c_0.

Exercise 1.2.7 Show that $(\ell^1)^* = \ell^\infty$.
Hint: For a fixed $\mathfrak{a} = (a_n)_{n\geq 1} \in \ell^\infty$, consider the functional

$$\Lambda_{\mathfrak{a}}(\mathfrak{x}) = \sum_{n=1}^{\infty} a_n x_n,$$

where $\mathfrak{x} = (x_n)_{n\geq 1}$ runs through ℓ^1.
Remark: Based on Exercise 1.2.6, we can say that $c_0^{**} = \ell^\infty$. Hence, c_0 is not reflexive.

Exercise 1.2.8 Consider the linear functional $\Lambda : \mathcal{C}[0,1] \longrightarrow \mathbb{C}$ given by

$$\Lambda f = \int_0^1 t f(t)\, dt.$$

We recall that $\mathcal{C}[0,1]$ is equipped with the norm $\|\cdot\|_\infty$. Show that $\|\Lambda\| = 1/2$ and find all its maximal vectors.

Exercise 1.2.9 Let

$$\mathcal{X} = \{f \in \mathcal{C}[0,1] : f(0) = 0\},$$

and equip $\mathcal{X}$ with the norm $\|\cdot\|_\infty$. Consider the linear functional $\Lambda : \mathcal{X} \longrightarrow \mathbb{C}$ given by

$$\Lambda f = \int_0^1 t f(t)\, dt.$$

Show that Λ has no maximal vector.
Hint: See Exercise 1.2.8.

1.3 Fourier series

Let $\varphi \in L^1 = L^1(\mathbb{T}, m)$ and let K_n be the *Fejér kernel* defined by

$$K_n(e^{it}) = \frac{1}{n+1}\left(\frac{\sin((n+1)t/2)}{\sin(t/2)}\right)^2 = \sum_{k=-n}^{n}\left(1 - \frac{|k|}{n+1}\right)e^{ikt}$$
$$(n \geq 0,\ t \in \mathbb{R}).$$

If D_n is the *Dirichlet kernel*,

$$D_n(e^{it}) = \sum_{k=-n}^{n} e^{ikt} \qquad (n \geq 0,\ t \in \mathbb{R}),$$

then we also have

$$K_n = \frac{1}{n+1}(D_0 + D_1 + \cdots + D_n).$$

Moreover, for any $\varphi \in L^1$, we have

$$(\varphi * D_n)(e^{it}) = \sum_{k=-n}^{n} \hat{f}(k)e^{ikx} = S_n(f, e^{it}),$$

which are the partial Fourier sums of φ and

$$(\varphi * K_n)(e^{it}) = \sum_{k=-n}^{n}\left(1 - \frac{|k|}{n+1}\right)\hat{\varphi}(k)e^{ikt} = \frac{1}{n+1}\sum_{m=0}^{n} S_m(f, e^{it}),$$

which are the Césaro means of the partial Fourier sum. We denote by $\sigma_n(\varphi)$ the function $\varphi * K_n$ and it is called the nth *Fejér mean* of φ.

Theorem 1.5

(i) *If $f \in C(\mathbb{T})$, then*

$$\|\sigma_n(f)\|_\infty \leq \|f\|_\infty \qquad (n \geq 1)$$

and

$$\|\sigma_n(f) - f\|_\infty \longrightarrow 0, \quad \text{as } n \longrightarrow \infty.$$

(ii) *If* $f \in L^p(\mathbb{T})$, $1 \le p < \infty$, *then* $\sigma_n(f) \longrightarrow f$ *almost everywhere on* $\mathbb{T}$, *as* $n \longrightarrow \infty$,

$$\|\sigma_n(f)\|_p \le \|f\|_p \qquad (n \ge 1)$$

and

$$\|\sigma_n(f) - f\|_p \longrightarrow 0, \quad \text{as } n \longrightarrow \infty.$$

(iii) *If* $f \in L^\infty(\mathbb{T})$, *then*

$$\|\sigma_n(f)\|_\infty \le \|f\|_\infty \qquad (n \ge 1)$$

and $\sigma_n(f) \longrightarrow f$ *in the weak-star topology of* $L^\infty(\mathbb{T})$, *as* $n \longrightarrow \infty$.

This classic result can be found in many textbooks; see e.g. [291, pp. 17–19] or [354, pp. 34 and 44].

1.4 The Hahn–Banach theorem

The Hahn–Banach theorem is one of the pillars of functional analysis. It has several variations, but we state one of its simple forms below. The spirit of all versions is that a bounded mapping on a subspace can be extended to the whole space and still keep some of its boundedness properties.

Theorem 1.6 *Let* $\mathcal{M}$ *be a linear manifold of the normed linear space* $\mathcal{X}$. *Assume that* $\lambda : \mathcal{M} \longrightarrow \mathbb{C}$ *is a bounded linear functional on* $\mathcal{M}$, *i.e.* $\lambda \in \mathcal{M}^*$. *Then there exists a bounded linear functional* $\Lambda : \mathcal{X} \longrightarrow \mathbb{C}$, *i.e.* $\Lambda \in \mathcal{X}^*$, *such that*

$$\Lambda(x) = \lambda(x) \qquad (x \in \mathcal{M})$$

and, moreover,

$$\|\Lambda\|_{\mathcal{X}^*} = \|\lambda\|_{\mathcal{M}^*}.$$

For a proof, see [159, section III.6].

The following result is one of the numerous consequences of the Hahn–Banach theorem. The second part of the theorem is probably not so well known. Hence, we give the proof.

Theorem 1.7 *Let* $\mathcal{X}$ *be a normed linear space, let* $\mathcal{M}$ *be a closed subspace of* $\mathcal{X}$, *and let* $x_0 \in \mathcal{X}$. *Then* $x_0 \notin \mathcal{M}$ *if and only if there exists a functional* $\Lambda \in \mathcal{X}^*$ *such that*

$$\Lambda(x_0) = 1$$

and

$$\Lambda(x) = 0 \qquad (x \in \mathcal{M}).$$

Furthermore, if $x_0 \notin \mathcal{M}$, then

$$\inf\{\|\Lambda\| : \Lambda \in \mathcal{X}^*,\ \Lambda(x_0) = 1,\ \Lambda \equiv 0 \text{ on } \mathcal{M}\} = \frac{1}{\text{dist}(x_0, \mathcal{M})}$$

and the infimum is attained.

Proof It is trivial that, if there exists $\Lambda_0 \in \mathcal{X}^*$ such that $\Lambda_0(x_0) = 1$ and $\Lambda_0(x) = 0$ for all $x \in \mathcal{M}$, then $x_0 \notin \mathcal{M}$. Reciprocally, assume that $x_0 \notin \mathcal{M}$. Let $\mathcal{N}$ be the subspace of $\mathcal{X}$ defined by

$$\mathcal{N} = \mathbb{C}x_0 + \mathcal{M} = \{\alpha x_0 + x : \alpha \in \mathbb{C},\ x \in \mathcal{M}\},$$

and let λ_0 be the map defined on $\mathcal{N}$ by

$$\lambda_0(\alpha x_0 + x) = \alpha \qquad (\alpha \in \mathbb{C},\ x \in \mathcal{M}).$$

Since $x_0 \notin \mathcal{M}$, one easily sees that this mapping is well defined and linear. Moreover, since $\ker(\lambda_0) = \mathcal{M}$ is a closed subspace, this linear functional is continuous on $\mathcal{N}$. This can also be proven directly. In fact, we claim that

$$\|\lambda_0\|_{\mathcal{N}^*} = \frac{1}{\text{dist}(x_0, \mathcal{M})}. \tag{1.7}$$

First, note that, for each $\alpha \in \mathbb{C}$, $\alpha \neq 0$, and each $x \in \mathcal{M}$, we have

$$\|\alpha x_0 + x\| = |\alpha| \left\| x_0 + \frac{x}{\alpha} \right\| \geq |\alpha| \, \text{dist}(x_0, \mathcal{M}),$$

which gives

$$|\lambda_0(\alpha x_0 + x)| = |\alpha| \leq \frac{\|\alpha x_0 + x\|}{\text{dist}(x_0, \mathcal{M})}.$$

Hence, $\|\lambda_0\|_{\mathcal{N}^*} \leq 1/\text{dist}(x_0, \mathcal{M})$. Moreover, for each $x \in \mathcal{M}$, we have

$$1 = |\lambda_0(x_0 - x)| \leq \|\lambda_0\|_{\mathcal{N}^*} \|x_0 - x\|,$$

whence

$$1 \leq \|\lambda_0\|_{\mathcal{N}^*} \, \text{dist}(x_0, \mathcal{M}).$$

This gives the reversed inequality, and then (1.7) follows. Now, the Hahn–Banach theorem (Theorem 1.6) enables us to extend λ_0 to a linear functional Λ_0 on $\mathcal{X}$ such that $\|\Lambda_0\| = \|\lambda_0\|_{\mathcal{N}^*}$.

If $\Lambda \in \mathcal{X}^*$ is such that $\Lambda(x_0) = 1$ and $\Lambda \equiv 0$ on $\mathcal{M}$, then, for each $x \in \mathcal{M}$, we have

$$1 = \Lambda(x_0) = \Lambda(x_0 - x) \leq \|\Lambda\| \, \|x_0 - x\|,$$

whence

$$\frac{1}{\|\Lambda\|} \leq \inf_{x \in \mathcal{M}} \|x - x_0\| = \text{dist}(x_0, \mathcal{M}) = \frac{1}{\|\Lambda_0\|}. \qquad \square$$

If in Theorem 1.7 we take $\mathcal{M} = \{0\}$, we obtain the following result.

Corollary 1.8 *Let $\mathcal{X}$ be a normed linear space, and let $x_0 \in \mathcal{X}$. Then $x_0 = 0$ if and only if for each functional $\Lambda \in \mathcal{X}^*$ we have*

$$\Lambda(x_0) = 0.$$

Furthermore, if $x_0 \neq 0$, then

$$\inf\{\|\Lambda\| : \Lambda \in \mathcal{X}^*,\ \Lambda(x_0) = 1\} = \frac{1}{\|x_0\|}$$

and the infimum is attained.

If $\mathcal{M}$ and $\mathcal{N}$ are closed subspaces of a normed linear space $\mathcal{X}$, then we define the sum

$$\mathcal{M} + \mathcal{N} = \{x + y : x \in \mathcal{M},\ y \in \mathcal{N}\}.$$

It is elementary to verify that $\mathcal{M} + \mathcal{N}$ is a linear manifold in $\mathcal{X}$. But, in the general case, $\mathcal{M} + \mathcal{N}$ is not necessarily closed. If $\mathcal{M} \cap \mathcal{N} = \{0\}$ and $\mathcal{M} + \mathcal{N}$ is a closed subspace of $\mathcal{X}$, then $\mathcal{M} + \mathcal{N}$ is called the *direct sum* of $\mathcal{M}$ and $\mathcal{N}$ and is denoted by $\mathcal{M} \oplus \mathcal{N}$. The following result asserts that, if one of the subspaces involved is finite-dimensional, then the sum is surely closed.

Theorem 1.9 *Let $\mathcal{X}$ be a normed linear space, let $\mathcal{M}$ be a closed subspace of $\mathcal{X}$, and let $\mathcal{N}$ be a finite-dimensional linear manifold in $\mathcal{X}$. Then $\mathcal{M} + \mathcal{N}$ is a closed subspace of $\mathcal{X}$. In particular, any finite-dimensional linear manifold in $\mathcal{X}$ is closed.*

For a proof, see [159, p. 71].

Let $\mathcal{X}_1$ be a closed subspace of a normed linear space $\mathcal{X}$. If there exists a closed subspace $\mathcal{X}_2$ of $\mathcal{X}$ such that

$$\mathcal{X} = \mathcal{X}_1 + \mathcal{X}_2 \quad \text{and} \quad \mathcal{X}_1 \cap \mathcal{X}_2 = \{0\},$$

then $\mathcal{X}_1$ is said to be *complemented* in $\mathcal{X}$. In other words, $\mathcal{X}$ is the direct sum of $\mathcal{X}_1$ and $\mathcal{X}_2$ and we write $\mathcal{X} = \mathcal{X}_1 \oplus \mathcal{X}_2$. In this situation, $\mathcal{X}_2$ is said to be a *complementary space* of $\mathcal{X}_1$ in $\mathcal{X}$. Two comments are in order. First, in a general normed linear space, there might be closed subspaces that do not have any complement. However, if $\mathcal{X} = \mathcal{H}$ is a Hilbert space and $\mathcal{H}_1$ is a closed subspace of $\mathcal{H}$, then $\mathcal{H}_1$ is surely complemented in $\mathcal{H}$ and a complementary space is given by $\mathcal{H}_1^\perp$ (see Section 1.7.) Second, if the complementary space exists, it is not unique.

Lemma 1.10 *Let $\mathcal{X}_1$ be a closed subspace of a normed linear space $\mathcal{X}$. If $\mathcal{X}_1$ is either of finite dimension or of finite codimension, then $\mathcal{X}_1$ is complemented in $\mathcal{X}$.*

Proof Assume first that $\dim \mathcal{X}_1 < \infty$ and let $\{e_1, \ldots, e_n\}$ be a basis for $\mathcal{X}_1$. Then, for every $x \in \mathcal{X}_1$, there are unique scalars $\alpha_1(x), \ldots, \alpha_n(x)$ such that

$$x = \sum_{i=1}^{n} \alpha_i(x)\, e_i.$$

Using the fact that norms on finite-dimensional spaces are all equivalent, it is easy to see that every linear functional $\alpha_i : x \longmapsto \alpha_i(x)$ is continuous on $\mathcal{X}_1$. By the Hahn–Banach theorem, we can extend α_i to a continuous linear functional $\widetilde{\alpha}_i$ on $\mathcal{X}$. Now let

$$\mathcal{X}_2 = \bigcap_{i=1}^{n} \ker \widetilde{\alpha}_i.$$

Since the functionals $\widetilde{\alpha}_i$ are all continuous, $\mathcal{X}_2$ is a closed subspace of $\mathcal{X}$. If $x \in \mathcal{X}_1 \cap \mathcal{X}_2$, then we have, on the one hand,

$$x = \sum_{i=1}^{n} \alpha_i(x) e_i = \sum_{i=1}^{n} \widetilde{\alpha}_i(x) e_i,$$

and, on the other, $\widetilde{\alpha}_i(x) = 0$ for every $1 \le i \le n$. Hence, $x = 0$. Moreover, for each $x \in \mathcal{X}$, if we put

$$x_1 = \sum_{i=1}^{n} \widetilde{\alpha}_i(x) e_i,$$

then $x_1 \in \mathcal{X}_1$ and, for $1 \le j \le n$, we have

$$\widetilde{\alpha}_j(x - x_1) = \widetilde{\alpha}_j(x) - \widetilde{\alpha}_j(x_1) = \widetilde{\alpha}_j(x) - \sum_{i=1}^{n} \widetilde{\alpha}_i(x) \widetilde{\alpha}_j(e_i).$$

But $\widetilde{\alpha}_j(e_i) = \delta_{ij}$, whence $\widetilde{\alpha}_j(x - x_1) = 0$. This means that $x - x_1 \in \mathcal{X}_2$ and thus $\mathcal{X} = \mathcal{X}_1 \oplus \mathcal{X}_2$.

Assume now that $\dim(\mathcal{X}/\mathcal{X}_1) < \infty$. Let $\pi : \mathcal{X} \longrightarrow \mathcal{X}/\mathcal{X}_1$ be the quotient map, and let $\{e_1, \ldots, e_n\}$ be a basis for $\mathcal{X}/\mathcal{X}_1$. For any $1 \le i \le n$, pick $x_i \in \mathcal{X}$ such that $e_i = \pi(x_i)$ and let

$$\mathcal{X}_2 = \text{Span}\{x_1, \ldots, x_n\}.$$

By Theorem 1.9, $\mathcal{X}_2$ is a closed subspace of $\mathcal{X}$. If $x \in \mathcal{X}_1 \cap \mathcal{X}_2$, we have

$$x = \sum_{i=1}^{n} \alpha_i x_i$$

and

$$0 = \pi(x) = \sum_{i=1}^{n} \alpha_i e_i.$$

Since $\{e_1, \ldots, e_n\}$ is a basis, then $\alpha_i = 0$ for all i and thus $x = 0$. Now, if $x \in \mathcal{X}$, there are scalars $\alpha_1, \ldots, \alpha_n$ such that

$$\pi(x) = \sum_{i=1}^{n} \alpha_i e_i$$

and hence

$$\pi(x) = \sum_{i=1}^{n} \alpha_i \pi(x_i) = \pi\bigg(\sum_{i=1}^{n} \alpha_i x_i \bigg).$$

Therefore, $x - \sum_{i=1}^{n} \alpha_i x_i$ belongs to $\mathcal{X}_1$, which proves that $\mathcal{X} = \mathcal{X}_1 \oplus \mathcal{X}_2$. □

If $\mathcal{M}$ is a closed subspace of a Banach space $\mathcal{X}$, then the quotient $\mathcal{X}/\mathcal{M}$ endowed with the quotient norm

$$\|x + \mathcal{M}\|_{\mathcal{X}/\mathcal{M}} = \text{dist}(x, \mathcal{M}) \qquad (x \in \mathcal{X})$$

is also a Banach space. In many applications, it is useful to have a description of the duals of $\mathcal{M}$ and $\mathcal{X}/\mathcal{M}$. To this end, we define the annihilator $\mathcal{M}$ as

$$\mathcal{M}^{\perp} = \{\Lambda \in \mathcal{X}^* : \Lambda(x) = 0, \forall\, x \in \mathcal{M}\}.$$

Theorem 1.11 *Let $\mathcal{M}$ be a closed subspace of a Banach space $\mathcal{X}$. Then there is an isometric isomorphism of $(\mathcal{X}/\mathcal{M})^*$ onto $\mathcal{M}^{\perp}$.*

Proof For simplicity, put $\mathcal{Y} = \mathcal{X}/\mathcal{M}$. Let $\pi : \mathcal{X} \longrightarrow \mathcal{Y}$ be the canonical projection of $\mathcal{X}$ onto $\mathcal{Y}$. Define

$$\begin{array}{rccc} \tau : & \mathcal{Y}^* & \longrightarrow & \mathcal{M}^{\perp} \\ & y^* & \longmapsto & y^* \circ \pi. \end{array}$$

It is now easy to verify that τ is an isometric isomorphism of $\mathcal{Y}^*$ onto $\mathcal{M}^{\perp}$. □

The following result is similar to Theorem 1.11 and it characterizes the dual of $\mathcal{M}$.

Theorem 1.12 *Let $\mathcal{M}$ be a closed subspace of a Banach space $\mathcal{X}$. Then there is an isometric isomorphism of $\mathcal{M}^*$ onto $\mathcal{X}^*/\mathcal{M}^{\perp}$.*

Proof By the Hahn–Banach theorem (Theorem 1.6) we can extend each $m^* \in \mathcal{M}^*$ to a functional $x^* \in \mathcal{X}^*$. If x_1^* and x_2^* are two such extensions, since $x_1^* = x_2^* = m^*$ on $\mathcal{M}$, we have

$$x_1^* + \mathcal{M}^{\perp} = x_2^* + \mathcal{M}^{\perp}.$$

Thus, the mapping

$$\begin{array}{rccc} \sigma : & \mathcal{M}^* & \longrightarrow & \mathcal{X}^*/\mathcal{M}^{\perp} \\ & m^* & \longmapsto & x^* + \mathcal{M}^{\perp} \end{array}$$

is well defined. Knowing this fact, it is now easy to verify that σ is an isometric isomorphism of $\mathcal{M}^*$ onto $\mathcal{X}^*/\mathcal{M}^\perp$. □

As a consequence of the characterization given in Theorem 1.12, for each $x^* \in \mathcal{X}^*$, and any closed subspace $\mathcal{M}$ of a Banach space $\mathcal{X}$, we have

$$\|x^*|_{\mathcal{M}}\|_{\mathcal{M}^*} = \operatorname{dist}(x^*, \mathcal{M}^\perp). \tag{1.8}$$

Exercises

Exercise 1.4.1 Show that $(\ell^\infty)^* \neq \ell^1$.
Hint: Use the Hahn–Banach theorem (Theorem 1.6).
Remark: Compare with Exercises 1.2.4 and 1.2.7.

Exercise 1.4.2 Show that ℓ^1 is not reflexive.
Hint: Use Exercises 1.2.7 and 1.4.1.

Exercise 1.4.3 Let

$$\mathcal{X} = \{(x_n)_{n\geq 1} \in \ell^\infty : x_n \in \mathbb{R}, \forall\, n \geq 1\}$$

be endowed with the $\|\cdot\|_\infty$ norm. Show that there is a bounded functional Λ on $\mathcal{X}$ such that

$$\liminf_{n\to\infty} x_n \leq \Lambda\mathfrak{x} \leq \limsup_{n\to\infty} x_n$$

for all $\mathfrak{x} = (x_n)_{n\geq 1} \in \mathcal{X}$.
Hint: Consider the closed subspace

$$\mathcal{M} = \left\{\mathfrak{x} \in \mathcal{X} : \lim_{n\to\infty} x_n \text{ exists}\right\}$$

and define λ on $\mathcal{M}$ by

$$\lambda\mathfrak{x} = \lim_{n\to\infty} x_n \qquad (\mathfrak{x} \in \mathcal{M}).$$

Then apply (the real version of) Theorem 1.6.

Exercise 1.4.4 Show that there is a nonzero bounded linear functional Λ on $L^\infty(\mathbb{T})$ such that

$$\Lambda f = 0 \qquad (f \in \mathcal{C}(\mathbb{T})).$$

Hint: Use Theorem 1.7.

Exercise 1.4.5 Show that there is a bounded linear functional Λ on $L^\infty(\mathbb{T})$ such that

$$\Lambda f = f(1) \qquad (f \in \mathcal{C}(\mathbb{T})).$$

Hint: Use Theorem 1.6.

Exercise 1.4.6 Let $\mathcal{X}$ be a normed linear space, and let $x_0 \in \mathcal{X}$, $x_0 \neq 0$. Show that there is $\Lambda \in \mathcal{X}^*$ such that

$$\Lambda(x_0) = \|x_0\| \quad \text{and} \quad \|\Lambda\| = 1.$$

Hint: Use Corollary 1.8.

Exercise 1.4.7 Let $\mathcal{X}$ be a real vector space, let $\mathcal{M}$ be a subspace of $\mathcal{X}$, and let K be a *convex cone* of $\mathcal{X}$ (which means that $\alpha x + \beta y$ belongs to K, for any positive scalars α, β and any x, y in K). Assume that, for each $x \in \mathcal{X}$, there exist $m_1, m_2 \in \mathcal{M}$ such that $m_1 - x \in K$ and $x - m_2 \in K$. Let φ be a linear form on $\mathcal{M}$ such that

$$x \in \mathcal{M} \cap K \quad \Longrightarrow \quad \varphi(x) \geq 0.$$

Show that there exists a linear form Φ on $\mathcal{X}$ such that $\Phi_{|\mathcal{M}} = \varphi$ and $\Phi(x) \geq 0$ for any $x \in K$.
Hint: Use the same idea usually applied in the proof of the Hahn–Banach theorem, i.e. construct an extension in $\mathcal{M} \oplus \mathbb{R}x_0$, $x_0 \notin \mathcal{M}$, and then use a transfinite induction.

1.5 The Baire category theorem and its consequences

Among other things, a Banach space is a *complete* metric space. The completeness has profound consequences. In particular, the open mapping theorem (Theorem 1.14), the uniform boundedness principle (Theorem 1.19), the closed graph theorem (Corollary 1.18) and the inverse mapping theorem (Corollary 1.15) are some celebrated properties that stem from the completeness of Banach spaces. Note that, for the Hahn–Banach theorem, completeness was not needed. The secret of the richness of complete metric spaces is hidden in the following result. We recall that in a metric space X a subset $E \subset X$ is called *nowhere dense* if its closure in X has an empty interior.

Theorem 1.13 (Baire category theorem) *Let X be a complete metric space. Suppose that $(E_n)_{n\geq 1}$ is any collection of nowhere dense subsets of X. Then*

$$X \neq \bigcup_{n\geq 1} E_n.$$

See [441, p. 43] for a proof of this classic but ultra-important result.

If $f : X \longrightarrow Y$ is a continuous function between two metric spaces X and Y, and if V is an open subset of X, then its image $f(V)$ is not necessarily an open subset of Y. In other words, f is not necessarily an open mapping. However, if the spaces X and Y are complete, then the story is different. The following result, known as the *open mapping theorem*, is another fundamental theorem of functional analysis.

Theorem 1.14 (Open mapping theorem) *Let $\mathcal{X}$ and $\mathcal{Y}$ be two Banach spaces and let $A : \mathcal{X} \longrightarrow \mathcal{Y}$ be a bounded surjective operator. Then A is an open mapping.*

A proof of this result can be found in [159, p. 90].

We now study several consequences of the open mapping theorem. If $f : X \longrightarrow Y$ is a continuous bijective function between two metric spaces X and Y, then its set theoretic inverse $f^{-1} : Y \longrightarrow X$ is not necessarily continuous. But for Banach spaces the last property automatically holds.

Corollary 1.15 (Inverse mapping theorem) *Let $\mathcal{X}$ and $\mathcal{Y}$ be two Banach spaces and let $A : \mathcal{X} \longrightarrow \mathcal{Y}$ be a bounded bijective operator. Then $A^{-1} : \mathcal{Y} \longrightarrow \mathcal{X}$ is also a bounded operator. In other words, there are positive constants c and C such that*

$$c\|x\|_{\mathcal{X}} \le \|Ax\|_{\mathcal{Y}} \le C\|x\|_{\mathcal{X}} \qquad (x \in \mathcal{X}).$$

Proof This is a direct consequence of Theorem 1.14. Since A is surjective, that results ensures that, for each open set $V \subset \mathcal{X}$, its image $A(V)$ is an open subset of $\mathcal{Y}$. But, since A is also injective, we can consider the set theoretic inverse of A and interpret the preceding statement as follows. The inverse image of each open set $V \subset \mathcal{X}$ under the mapping $A^{-1} : \mathcal{Y} \longrightarrow \mathcal{X}$ is $(A^{-1})^{-1}(V) = A(V)$, and thus it is an open subset of $\mathcal{Y}$. Hence, A^{-1} is continuous. □

There is another version of the inverse mapping theorem (Corollary 1.15) that also enters our discussion. In this version, we just assume that the operator is surjective. But it is allowed to have a nontrivial kernel and thus it is not necessarily invertible. However, a partial result holds.

Corollary 1.16 *Let $\mathcal{X}$ and $\mathcal{Y}$ be two Banach spaces and let $A : \mathcal{X} \longrightarrow \mathcal{Y}$ be a bounded surjective operator. Then there is a positive constant C such that, for every $y \in \mathcal{Y}$, we can find an $x \in \mathcal{X}$ with $y = Ax$ and satisfying the growth restriction*

$$\|x\|_{\mathcal{X}} \le C\|y\|_{\mathcal{Y}}.$$

Proof The operator

$$\begin{array}{rccc} \mathbf{A}: & \mathcal{X}/\ker A & \longrightarrow & \mathcal{Y} \\ & x + \ker A & \longmapsto & Ax \end{array}$$

is well defined, bounded and bijective. Hence, by Corollary 1.15, $\mathbf{A}$ is invertible and its inverse fulfills

$$\|\mathbf{A}^{-1}y\|_{\mathcal{X}/\ker A} \le c\|y\|_{\mathcal{Y}} \qquad (y \in \mathcal{Y}),$$

where c is a positive constant. Recall that

$$\|\mathbf{A}^{-1}y\|_{\mathcal{X}/\ker A} = \inf_{z \in \ker A} \|x_0 + z\|_{\mathcal{X}},$$

where $x_0 \in X$ is such that $y = Ax_0$. If we put $C = 2c$ (note that we are not looking for the optimal choice for C), then we can find z such that $x = x_0 + z$ satisfies the required growth restriction. □

Corollary 1.16 immediately implies the following result.

Corollary 1.17 *Let $\mathcal{X}$ and $\mathcal{Y}$ be two Banach spaces and let $A : \mathcal{X} \longrightarrow \mathcal{Y}$ be a bounded operator. Then the following are equivalent:*

(i) *A is lower bounded;*
(ii) *A is one-to-one and has a closed range.*

In that case, if furthermore the range of A is complemented into $\mathcal{Y}$, then A is left-invertible.

If $f : X \longrightarrow Y$ is a continuous bijective function between two metric spaces X and Y, then its *graph*

$$\mathcal{G}_f = \{(x, f(x)) : x \in X\}$$

is a closed subset of the product space $X \times Y$. This means that if $(x_n, y_n) \in \mathcal{G}_f$ and

$$(x_n, y_n) \longrightarrow (x, y)$$

in the space $X \times Y$ (or equivalently $x_n \longrightarrow x$ in X and $y_n \longrightarrow y$ in Y), then $(x, y) \in \mathcal{G}_f$. However, one can easily find a function $f : X \longrightarrow Y$ whose graph is a closed subset of $X \times Y$, yet f is discontinuous. Again, as another manifestation of the open mapping theorem (Theorem 1.14), we see that in Banach spaces the two concepts are equivalent.

Corollary 1.18 (Closed graph theorem) *Let $\mathcal{X}$ and $\mathcal{Y}$ be two Banach spaces and let $A : \mathcal{X} \longrightarrow \mathcal{Y}$ be a linear mapping whose graph is closed in $\mathcal{X} \times \mathcal{Y}$, i.e. the conditions*

$$x_n \longrightarrow x \ (\text{in } \mathcal{X}) \quad \text{and} \quad Ax_n \longrightarrow y \ (\text{in } \mathcal{Y})$$

imply that $y = Ax$. Then A is a bounded operator.

The reader may consult [159, p. 91] for a proof of this result.

A collection of operators $(A_\iota)_{\iota\in I}$ between the normed linear spaces $\mathcal{X}$ and $\mathcal{Y}$ is called *uniformly bounded* if there is a constant C such that

$$\|A_\iota\|_{\mathcal{L}(\mathcal{X},\mathcal{Y})} \leq C \qquad (\iota \in I).$$

In this situation, given any $x \in \mathcal{X}$, we surely have

$$\|A_\iota x\|_{\mathcal{Y}} \leq C\|x\|_{\mathcal{X}} \qquad (\iota \in I).$$

If we write $C_x = C\|x\|_{\mathcal{X}}$, the preceding inequality states that the quantity $\|A_\iota x\|_{\mathcal{Y}}$ is bounded by C_x when ι runs over the index set I. It is rather amazing that the inverse implication also holds for operators on Banach spaces.

Theorem 1.19 (Uniform boundedness principle) *Let $\mathcal{X}$ be a Banach space, let $\mathcal{Y}$ be a normed linear space, and let $A_\iota : \mathcal{X} \longrightarrow \mathcal{Y}$, $\iota \in I$, be a family of bounded operators from $\mathcal{X}$ into $\mathcal{Y}$. Assume that, for each $x \in \mathcal{X}$, there is a constant C_x such that*

$$\|A_\iota x\|_{\mathcal{Y}} \leq C_x \qquad (\iota \in I).$$

Then there is a constant C such that

$$\|A_\iota\|_{\mathcal{L}(\mathcal{X},\mathcal{Y})} \leq C \qquad (\iota \in I).$$

A proof can be found in [159, p. 95].

A typical application of the uniform boundedness principle is the following result.

Corollary 1.20 *Let $\mathcal{X}$ and $\mathcal{Y}$ be two Banach spaces and let $A_n : \mathcal{X} \longrightarrow \mathcal{Y}$, $n \geq 1$, be a sequence of bounded operators from $\mathcal{X}$ into $\mathcal{Y}$. Assume that, for each $x \in \mathcal{X}$, the limit*

$$Ax = \lim_{n\to\infty} A_n x$$

exists. Then A is a bounded operator, i.e. $A \in \mathcal{L}(\mathcal{X},\mathcal{Y})$.

Proof Clearly, the mapping A is linear. Since, for each $x \in \mathcal{X}$, the limit $\lim_{n\to\infty} A_n x$ exists, we surely have

$$\sup_{n\geq 1} \|A_n x\|_{\mathcal{Y}} < \infty.$$

Hence, by the uniform boundedness principle (Theorem 1.19),

$$C = \sup_{n\geq 1} \|A_n\|_{\mathcal{L}(\mathcal{X},\mathcal{Y})} < \infty.$$

Therefore, we can write

$$\|A_n x\|_{\mathcal{Y}} \leq C\|x\|_{\mathcal{X}} \qquad (n \geq 1,\ x \in \mathcal{X}).$$

Now, let $n \longrightarrow \infty$ to deduce that

$$\|Ax\|_{\mathcal{Y}} \leq C\|x\|_{\mathcal{X}} \qquad (x \in \mathcal{X}). \qquad \square$$

Exercises

Exercise 1.5.1 Show that the mapping

$$\begin{array}{rccc} I: & \mathcal{C}[0,1] & \longrightarrow & \mathcal{C}_0^1[0,1] \\ & f & \longmapsto & \displaystyle\int_0^x f(t)\,dt \end{array}$$

is a bounded bijective operator. However, its algebraic inverse is not bounded. Both spaces $\mathcal{C}[0,1]$ and $\mathcal{C}_0^1[0,1]$ are endowed with the norm $\|\cdot\|_\infty$.
Remark: See also Exercise 1.2.1.

Exercise 1.5.2 Let K be a continuous function on $[0,1] \times [0,1]$ and define $A : \mathcal{C}[0,1] \longrightarrow \mathcal{C}[0,1]$ by

$$(Af)(x) = \int_0^1 K(x,y) f(y)\, dy.$$

Show that A is bounded. What is $\|A\|$?

Exercise 1.5.3 Let $\mathcal{X}$ and $\mathcal{Y}$ be two Banach spaces and let $A : \mathcal{X} \longrightarrow \mathcal{Y}$ be a bounded operator with a closed range. Show that there exists a positive constant C such that, for each $y \in \mathcal{R}(A)$, there is a choice of $x \in \mathcal{X}$ such that $y = Ax$ and

$$\|x\|_{\mathcal{X}} \leq C\|y\|_{\mathcal{Y}}.$$

Hint: Use Corollary 1.16.

Exercise 1.5.4 Let $\mathcal{X}$ and $\mathcal{Y}$ be Banach spaces. Show that the set of all surjective operators from $\mathcal{X}$ to $\mathcal{Y}$ is open in $\mathcal{L}(\mathcal{X},\mathcal{Y})$.
Hint: Let $A \in \mathcal{L}(\mathcal{X},\mathcal{Y})$ be surjective. By Corollary 1.16, there exists $C > 0$ such that, for each $y \in \mathcal{Y}$, we can choose $x \in \mathcal{X}$ with $y = Ax$ and $\|x\| \leq C\|y\|$. Show that all the operators $B \in \mathcal{L}(\mathcal{X},\mathcal{Y})$ in the ball $\|B - A\| < 1/C$ are also surjective.

To verify the above claim, fix any $y_0 \in \mathcal{Y}$ with $\|y_0\| \leq 1$. For simplicity, write $r = \|B - A\|$. Then there is $x_0 \in \mathcal{X}$ with $y_0 = Ax_0$ and $\|x_0\| \leq C$. Put $y_1 = y_0 - Bx_0$. Then $\|y_1\| \leq \|Ax_0 - Bx_0\| \leq rC$. Hence, there is $x_1 \in \mathcal{X}$ with $y_1 = Ax_1$ and $\|x_1\| \leq rC^2$. Then put $y_2 = y_1 - Bx_1$. Thus, $\|y_2\| = \|Ax_1 - Bx_1\| \leq r^2C^2$. By induction, we get two series $(x_n)_{n\geq 1}$ and $(y_n)_{n\geq 1}$ such that $\|x_n\| \leq r^nC^{n+1}$, $\|y_n\| \leq r^nC^n$ and $y_{n+1} = y_n - Bx_n = (A - B)x_n$. Since $rC < 1$, the series $\sum_n x_n$ converges in norm and we can define $x = \sum_{n=0}^{\infty} x_n$. Therefore, $Bx = \sum_{n=0}^{\infty} Bx_n = \sum_{n=0}^{\infty}(y_n - y_{n+1}) = y_0$.

Exercise 1.5.5 Let $\mathcal{X}$ and $\mathcal{Y}$ be two Banach spaces and let $A : \mathcal{X} \times \mathcal{Y} \longrightarrow \mathbb{C}$ be a separately continuous bilinear mapping, i.e. for each fixed $y \in \mathcal{Y}$, the linear mapping

$$\begin{array}{ccc} \mathcal{X} & \longrightarrow & \mathbb{C} \\ x & \longmapsto & A(x,y) \end{array}$$

is continuous on $\mathcal{X}$, and, for each fixed $x \in \mathcal{X}$, the linear mapping

$$\begin{array}{ccc} \mathcal{Y} & \longrightarrow & \mathbb{C} \\ y & \longmapsto & A(x,y) \end{array}$$

is continuous on $\mathcal{Y}$. Show that A is jointly continuous, i.e. there is a constant C such that

$$|A(x,y)| \le C\|x\|_{\mathcal{X}}\, \|y\|_{\mathcal{Y}} \qquad (x \in \mathcal{X},\ y \in \mathcal{Y}).$$

Hint: Use Theorem 1.19.

Exercise 1.5.6 Let $f : \mathbb{R}^2 \longrightarrow \mathbb{R}$ be defined by $f(0,0) = 0$ and

$$f(x,y) = \frac{xy}{x^2 + y^2} \qquad \text{if}(x,y) \neq (0,0).$$

Show that f is separately continuous, but not jointly continuous, at the point $(0,0)$.
Remark: Compare with Exercise 1.5.5.

1.6 The spectrum

Let $\mathcal{B}$ be a unital Banach algebra and denote its unit by $\mathfrak{e}$. We emphasize that all Banach algebras that we consider are complex Banach algebras. Then the *spectrum* of an element $x \in \mathcal{B}$ is defined by

$$\sigma(x) = \{\lambda \in \mathbb{C} : \lambda\mathfrak{e} - x \text{ is not invertible in } \mathcal{B}\}.$$

If several nested Banach algebras are involved, we write $\sigma_{\mathcal{B}}(x)$ to distinguish between different spectra of x. The *spectral radius* of x is defined to be

$$r(x) = \sup_{\lambda \in \sigma(x)} |\lambda|.$$

Theorem 1.21 *Let $\mathcal{B}$ be a unital complex Banach algebra. Then, for each $x \in \mathcal{B}$, $\sigma(x)$ is a nonempty compact subset of $\mathbb{C}$.*

Proof Since the set of invertible elements in B is an open set, $\sigma(x)$ is a closed subset of $\mathbb{C}$. Moreover, if $|\lambda| > \|x\|$, then $\mathfrak{e} - x/\lambda$ is invertible. Hence,

$$\sigma(x) \subset \overline{D(0, \|x\|)}. \tag{1.9}$$

Therefore, $\sigma(x)$ is a compact subset of $\mathbb{C}$. That $\sigma(x)$ is always a nonempty set is a slightly deeper result.

Assume that $\sigma(x) = \emptyset$. Let $\Lambda \in \mathcal{B}^*$. Then the mapping

$$\begin{array}{rccc} f: & \mathbb{C} & \longrightarrow & \mathbb{C} \\ & \lambda & \longmapsto & \Lambda((\lambda\mathfrak{e} - x)^{-1}) \end{array}$$

is analytic on $\mathbb{C}$. Since

$$\lim_{|\lambda| \to +\infty} \|(\lambda\mathfrak{e} - x)^{-1}\| = 0,$$

f is a bounded analytic function on $\mathbb{C}$ and thus, by Liouville's theorem, $f \equiv 0$. Since this holds for all $\Lambda \in \mathcal{B}^*$, by Corollary 1.8, we must have

$$(\lambda\mathfrak{e} - x)^{-1} = 0 \qquad (\lambda \in \mathbb{C}),$$

which is absurd. □

The estimation (1.9) immediately implies that

$$r(x) \le \|x\|. \tag{1.10}$$

But we give below a precise formula for $r(x)$ (see Theorem 1.23).

The *resolvent* set of x is the set

$$\rho(x) = \mathbb{C} \setminus \sigma(x).$$

Hence, $\rho(x)$ is an open subset of $\mathbb{C}$, which at least contains $\{|\lambda| > r(x)\}$. On the set $\rho(x)$, the *resolvent operator* is defined by

$$R_\lambda(x) = (\lambda\mathfrak{e} - x)^{-1},$$

and satisfies the growth restriction

$$\|R_\lambda(x)\| \ge \frac{1}{\operatorname{dist}(\lambda, \sigma(x))} \qquad (\lambda \in \rho(x)). \tag{1.11}$$

A mapping $* : \mathcal{B} \longrightarrow \mathcal{B}$ is called an *involution* if it fulfills:

(i) $(x + y)^* = x^* + y^*$,
(ii) $(xy)^* = y^* x^*$,
(iii) $(\alpha x)^* = \bar{\alpha} x^*$,
(iv) $(x^*)^* = x$ and
(v) $\|xx^*\| = \|x\|^2$

for all $x, y \in \mathcal{B}$ and $\alpha \in \mathbb{C}$. A Banach algebra accompanied by an involution is called a *C^*-algebra*. A celebrated example of a C^*-algebra is $\mathcal{L}(\mathcal{H})$, where $\mathcal{H}$ is a Hilbert space (see Section 1.8).

Another useful tool in Banach algebras is the functional polynomial calculus. Let $x \in \mathcal{B}$ and let p be a complex (analytic) polynomial, i.e.

$$p(z) = \sum_{k=0}^{N} a_k z^k \qquad (a_k \in \mathbb{C}).$$

Then we can define the element $p(x)$ of $\mathcal{B}$ by

$$p(x) = \sum_{k=0}^{N} a_k x^k.$$

It is easy to check that, for all polynomials p and q, we have:

(i) $p(x)q(x) = q(x)p(x) = (pq)(x)$;
(ii) if $p(z) = 1$, then $p(x) = \mathfrak{e}$;
(iii) if $p(z) = z$, then $p(x) = x$; and
(iv) if $\mathcal{B}$ is a C^*-algebra, then $p(x)^* = \bar{p}(x)$, where $\bar{p}(z) = \sum_{k=0}^{N} \bar{a}_k z^k$.

The following elementary result is important in operator theory and is called the *spectral mapping theorem.*

Theorem 1.22 (Spectral mapping theorem) *Let $\mathcal{B}$ be a unital Banach algebra, let $x \in \mathcal{B}$, and let p be a complex polynomial. Then we have*

$$\sigma(p(x)) = p(\sigma(x)).$$

Proof If $p \equiv 0$, then the result is trivial. Hence, assume that p is a nonzero polynomial. For any $\lambda \in \mathbb{C}$, there is a complex polynomial q such that

$$p(z) - p(\lambda) = (z - \lambda)q(z),$$

and hence

$$p(x) - p(\lambda)\mathfrak{e} = (x - \lambda\mathfrak{e})q(x) = q(x)(x - \lambda\mathfrak{e}).$$

If $\lambda \in \sigma(x)$, then $x - \lambda\mathfrak{e}$ is not invertible. This fact implies that $p(x) - p(\lambda)\mathfrak{e}$ cannot be invertible. Thus, $p(\lambda) \in \sigma(p(x))$. In other words,

$$p(\sigma(x)) \subset \sigma(p(x)).$$

To show the reverse inclusion, let $\lambda \in \sigma(p(x))$ and factorize the polynomial $p(z) - \lambda$ as

$$p(z) - \lambda = \alpha(z - \alpha_1)(z - \alpha_2) \cdots (z - \alpha_n),$$

where the α_i are the roots of $p - \lambda$ and $\alpha \neq 0$. We thus have

$$p(x) - \lambda\mathfrak{e} = \alpha(x - \alpha_1\mathfrak{e})(x - \alpha_2\mathfrak{e}) \cdots (x - \alpha_n\mathfrak{e}).$$

If each factor $x - \alpha_i\mathfrak{e}$ $(1 \le i \le n)$ is invertible, then so is $p(x) - \lambda\mathfrak{e}$, which is a contradiction. Thus, there exists at least one index i such that $\alpha_i \in \sigma(x)$. Hence,

$$\lambda = p(\alpha_i) \in p(\sigma(x)),$$

and this proves $\sigma(p(x)) \subset p(\sigma(x))$. □

Theorem 1.23 (Spectral radius theorem) *Let $\mathcal{B}$ be a unital Banach algebra, and let $x \in \mathcal{B}$. Then*

$$r(x) = \inf_{n \ge 1} \|x^n\|^{1/n} = \lim_{n \to \infty} \|x^n\|^{1/n}.$$

Proof By Theorem 1.22, we have $r(x^n) = (r(x))^n$, $n \ge 1$, and then from (1.10) we immediately get

$$r(x) \le \|x^n\|^{1/n} \qquad (n \ge 1).$$

Hence,

$$r(x) \le \inf_{n \ge 1} \|x^n\|^{1/n} \le \liminf_{n \to \infty} \|x^n\|^{1/n}.$$

On the open set $\{|\lambda| > \|x\|\}$, the series development

$$(\lambda\mathfrak{e} - x)^{-1} = \sum_{n=0}^{\infty} \frac{x^n}{\lambda^{n+1}}$$

holds. Since $|\lambda| > \|x\|$, the series absolutely converges in $\mathcal{B}$. But the mapping $\lambda \longmapsto (\lambda\mathfrak{e} - x)^{-1}$ is well defined and analytic on $\{|\lambda| > r(x)\}$. Hence, the preceding series expansion is in fact valid on the latter open set. Since the series is divergent if

$$\limsup_{n \to \infty} \left\| \frac{x^n}{\lambda^{n+1}} \right\|^{1/n} > 1,$$

we must have

$$\limsup_{n \to \infty} \|x^n\|^{1/n} \le |\lambda|$$

for all λ in the domain $\{|\lambda| > r(x)\}$. Therefore,

$$\limsup_{n \to \infty} \|x^n\|^{1/n} \le r(x).$$

□

Since $\|x^2\| \le \|x\|^2$, we see that the sequence $\|x^{2^n}\|^{2^{-n}}$, $n \ge 0$, is a decreasing sequence, which, by Theorem 1.23, converges to $r(x)$.

Exercises

Exercise 1.6.1 Let $\mathcal{B}$ be a C^*-algebra. Show that

$$\|x^*\| = \|x\| \qquad (x \in \mathcal{B}).$$

Exercise 1.6.2 Prove the inequality (1.11).
Hint: Fix $\lambda \in \rho(x)$. Let μ be such that $|\mu - \lambda| \, \|R_\lambda(x)\| < 1$. Then $\mu \in \rho(x)$. In fact, write $\mu\mathfrak{e} - x = (\mu - \lambda)\mathfrak{e} + \lambda\mathfrak{e} - x = (\lambda\mathfrak{e} - x)(\mathfrak{e} + (\mu - \lambda)R_\lambda(x))$. Since $|\mu - \lambda| \, \|R_\lambda(x)\| < 1$, then the element $\mathfrak{e} + (\mu - \lambda)R_\lambda(x)$ is invertible, and thus so is $\mu\mathfrak{e} - x$. Hence, if $\mu \in \sigma(x)$, then $|\mu - \lambda| \, \|R_\lambda(x)\| \geq 1$. Taking the infimum with respect to $\mu \in \sigma(x)$ gives $\operatorname{dist}(\lambda, \sigma(x)) \, \|R_\lambda(x)\| \geq 1$.

1.7 Hilbert space and projections

Any inner product space is also a normed space. The norm is given via the formula

$$\|x\|^2 = \langle x, x \rangle.$$

A *Hilbert space* is an inner product space whose norm is complete. Hence, each Hilbert space is *a priori* a Banach space. But they make a very special subclass with several interesting properties.

We will only be faced with separable Hilbert spaces. Two immediate examples are in order. The space ℓ^2 endowed with the inner product

$$\langle \mathfrak{z}, \mathfrak{w} \rangle_{\ell^2} = \sum_{n=1}^{\infty} z_n \bar{w}_n \tag{1.12}$$

and $L^2(\mu)$, $\mu \in \mathcal{M}^+(\mathbb{T})$, with

$$\langle f, g \rangle_{L^2(\mu)} = \int_{\mathbb{T}} f \bar{g} \, d\mu$$

are separable Hilbert spaces. In particular, the Hilbert space $L^2(\mathbb{T})$ with the inner product

$$\langle f, g \rangle_{L^2(\mathbb{T})} = \langle f, g \rangle_2 = \frac{1}{2\pi} \int_0^{2\pi} f(e^{it}) \, \overline{g(e^{it})} \, dt = \frac{1}{2\pi i} \int_{\mathbb{T}} f(\zeta) \, \overline{g(\zeta)} \, \frac{d\zeta}{\zeta}$$

is of special interest to us.

Given the Hilbert spaces $\mathcal{H}_1$ and $\mathcal{H}_2$, the Cartesian product $\mathcal{H}_1 \times \mathcal{H}_2$ equipped with the inner product

$$\langle (x_1, x_2), (y_1, y_2) \rangle_{\mathcal{H}_1 \times \mathcal{H}_2} = \langle x_1, y_1 \rangle_{\mathcal{H}_1} + \langle x_2, y_2 \rangle_{\mathcal{H}_2} \tag{1.13}$$

is a Hilbert space, which is called the *external direct sum* of $\mathcal{H}_1$ and $\mathcal{H}_2$. We abuse notation and denote the external direct sum by $\mathcal{H}_1 \oplus \mathcal{H}_2$. Moreover, instead of (x_1, x_2), we will write $x_1 \oplus x_2$. In the case where $\mathcal{H}_1$ and $\mathcal{H}_2$ are two closed orthogonal subspaces of a Hilbert space $\mathcal{H}$, the notation $\mathcal{H}_1 \oplus \mathcal{H}_2$ could cause some ambiguity. In fact, this notation is used for the external direct sum of $\mathcal{H}_1$ and $\mathcal{H}_2$, as well as their orthogonal sum. However, this is not a serious issue because the canonical mapping

$$\begin{array}{ccc} \mathcal{H}_1 \times \mathcal{H}_2 & \longrightarrow & \mathcal{H} \\ (x_1, x_2) & \longmapsto & x_1 + x_2 \end{array}$$

provides an isometric isomorphism between the two concepts. With this understanding, if $\mathcal{H} = \mathcal{H}_1 \oplus \mathcal{H}_2$ and $\mathcal{E}_i$ is a subset of $\mathcal{H}_i$, $i = 1, 2$, then

$$\mathrm{Clos}_{\mathcal{H}}(\mathcal{E}_1 \times \mathcal{E}_2) = \mathrm{Clos}_{\mathcal{H}_1}(\mathcal{E}_1) \times \mathrm{Clos}_{\mathcal{H}_2}(\mathcal{E}_2). \tag{1.14}$$

The following result provides a complete characterization of the dual of a Hilbert space. Naively speaking and not considering the effect of complex conjugation, this result says that a Hilbert space is its own dual.

Theorem 1.24 *Let Λ be a bounded linear functional on the Hilbert space $\mathcal{H}$. Then there is a unique $y \in \mathcal{H}$ such that*

$$\Lambda x = \langle x, y \rangle \qquad (x \in \mathcal{H}).$$

Moreover, $\|\Lambda\|_{\mathcal{H}^*} = \|y\|_{\mathcal{H}}$.

For a proof see [159, p. 13].

It is easy to verify that in a Hilbert space the norm satisfies the *parallelogram law*

$$\|x + y\|^2 + \|x - y\|^2 = 2\|x\|^2 + 2\|y\|^2, \tag{1.15}$$

and the inner product is uniquely determined via the norm through the *polarization identity*

$$4\langle x, y \rangle = (\|x + y\|^2 - \|x - y\|^2) + i(\|x + iy\|^2 - \|x - iy\|^2). \tag{1.16}$$

The above identities are a direct consequence of the formula

$$\|x + y\|^2 = \|x\|^2 + \|y\|^2 + 2\,\Re\langle x, y \rangle. \tag{1.17}$$

The celebrated *Cauchy–Schwarz inequality* says that

$$|\langle x, y \rangle| \le \|x\|\,\|y\|, \tag{1.18}$$

and the equality holds in (1.18) if and only if x and y are linearly dependent. The inequality (1.18) can be derived from the identities (1.15) and (1.16).

Let $A, B \in \mathcal{L}(\mathcal{H})$ and $x, y \in \mathcal{H}$. If we develop the combinations $A(x \pm y)$ and $A(x \pm iy)$, we easily obtain

$$\begin{aligned} 4\langle Ax, By\rangle = \langle A(x+y), B(x+y)\rangle \\ - \langle A(x-y), B(x-y)\rangle \\ + i\langle A(x+iy), B(x+iy)\rangle \\ - i\langle A(x-iy), B(x-iy)\rangle. \end{aligned} \tag{1.19}$$

This is a generalization of the polarization identity (1.16). The main characteristic of the preceding formula is that on the right-hand side each term is of the form $\langle Az, Bz\rangle$ with a certain $z \in \mathcal{H}$.

Two vectors $x, y \in \mathcal{H}$ are called *orthogonal* if

$$\langle x, y\rangle = 0.$$

In this case, we write $x \perp y$. According to (1.17), for such pairs, we have the Pythagorean identity

$$\|x+y\|^2 = \|x\|^2 + \|y\|^2. \tag{1.20}$$

A subset $\mathcal{E}$ of $\mathcal{H}$ is said to be orthogonal if $x \perp y$ for each pair $x, y \in \mathcal{E}$, $x \neq y$. If, moreover, $\|x\| = 1$, for each $x \in \mathcal{E}$, we say that $\mathcal{E}$ is an *orthonormal set*.

Let $(x_n)_{n\geq 1}$ be an orthonormal sequence in $\mathcal{H}$. Put

$$y_N = x - \sum_{n=1}^{N} \langle x, x_n\rangle\, x_n.$$

It is easy to verify that $y_N \perp \{x_1, x_2, \dots, x_N\}$, and thus, in light of (1.20), from the identity

$$x = y_N + \sum_{n=1}^{N} \langle x, x_n\rangle\, x_n,$$

we deduce that

$$\|x\|^2 = \|y_N\|^2 + \sum_{n=1}^{N} |\langle x, x_n\rangle|^2.$$

Thus, for each $N \geq 1$, we have

$$\sum_{n=1}^{N} |\langle x, x_n\rangle|^2 \leq \|x\|^2 \qquad (x \in \mathcal{H}),$$

and, letting $N \longrightarrow \infty$, we obtain *Bessel's inequality*

$$\sum_{n=1}^{\infty} |\langle x, x_n\rangle|^2 \leq \|x\|^2 \qquad (x \in \mathcal{H}). \tag{1.21}$$

Naturally, we ask about the possibility of equality in (1.21) for every $x \in \mathcal{H}$. An orthonormal sequence $(x_n)_{n\geq 1}$ for which the identity

$$\|x\|^2 = \sum_{n=1}^{\infty} |\langle x, x_n \rangle|^2 \tag{1.22}$$

holds for all $x \in \mathcal{H}$ is called an *orthonormal basis*. The relation (1.22) is called *Parseval's identity*. If (1.22) holds, then a linear combination of the $(x_n)_{n\geq 1}$, with coefficients from the rational numbers, forms a countable dense subset of $\mathcal{H}$, and thus, in this case, $\mathcal{H}$ is a separable Hilbert space. We take it as granted that the sequence $(\chi_n)_{n\in\mathbb{Z}}$ is an orthonormal basis for $L^2(\mathbb{T})$. Hence, $L^2(\mathbb{T})$ is a separable Hilbert space. Theorem 1.25 gives several characterizations of an orthonormal basis.

Theorem 1.25 *Let $\mathcal{H}$ be a Hilbert space, and let $(x_n)_{n\geq 1}$ be an orthonormal sequence in $\mathcal{H}$. Then the following are equivalent:*

(i) *for each $x \in \mathcal{H}$,*

$$\|x\|^2 = \sum_{n=1}^{\infty} |\langle x, x_n \rangle|^2;$$

(ii) *for each $x, y \in \mathcal{H}$,*

$$\langle x, y \rangle = \sum_{n=1}^{\infty} \langle x, x_n \rangle \, \overline{\langle y, x_n \rangle};$$

(iii) *if $x \in \mathcal{H}$ satisfies*

$$\langle x, x_n \rangle = 0 \qquad (n \geq 1),$$

then $x = 0$.

Moreover, under the preceding equivalent conditions, we have

$$\lim_{N\to\infty} \left\| x - \sum_{n=1}^{N} \langle x, x_n \rangle \, x_n \right\| = 0 \qquad (x \in \mathcal{H}).$$

In that case, we say that $(x_n)_{n\geq 1}$ is an orthonormal basis of $\mathcal{H}$.

A proof of this classic result can be found in [159, p. 16]. Theorem 1.25 has numerous applications. We mention just one application below.

Corollary 1.26 *Any separable Hilbert space $\mathcal{H}$ is isometrically isomorphic to ℓ^2.*

If $A \in \mathcal{L}(\mathcal{H})$ and $(e_\iota)_{\iota \in I}$ is an orthonormal basis for the Hilbert space $\mathcal{H}$, then the *matrix* of A with respect to the basis $(e_\iota)_{\iota \in I}$ is the matrix whose mn entry is given by $\langle Ae_n, e_m \rangle$:

$$\begin{bmatrix} & & n\text{th column} & & \\ & & \vdots & & \\ m\text{th row} & \cdots & \langle Ae_n, e_m \rangle & \cdots \\ & & \vdots & & \end{bmatrix}.$$

The index set I is usually a finite set, $\mathbb{N}$ or $\mathbb{Z}$. Hence, the corresponding matrix can be finite, singly infinite or doubly infinite.

Given two closed subspaces $\mathcal{M}$ and $\mathcal{N}$ of $\mathcal{H}$, we say that $\mathcal{M}$ is *orthogonal* to $\mathcal{N}$ if $x \perp y$, that is

$$\langle x, y \rangle = 0$$

for all $x \in \mathcal{M}$ and $y \in \mathcal{N}$. In this case, we write $\mathcal{M} \perp \mathcal{N}$. According to (1.20), the Pythagorean identity

$$\|x + y\|^2 = \|x\|^2 + \|y\|^2$$

holds for all elements $x \in \mathcal{M}$ and $y \in \mathcal{N}$. Among other things, this identity shows that $\mathcal{M} \cap \mathcal{N} = \{0\}$. It also shows that the collection of all vectors of the form $x + y$, where $x \in \mathcal{M}$ and $y \in \mathcal{N}$, is a closed subspace of $\mathcal{H}$. We recall that this subspace was denoted by $\mathcal{M} \oplus \mathcal{N}$ and, in this situation, it is called the *orthogonal direct sum* of $\mathcal{M}$ and $\mathcal{N}$. Note that the representation $x + y$ is unique.

The *orthogonal complement* of a subspace $\mathcal{M}$ of $\mathcal{H}$ is defined by

$$\mathcal{M}^\perp = \{x \in \mathcal{H} : \langle x, y \rangle = 0 \text{ for all } y \in \mathcal{M}\}.$$

Clearly $\mathcal{M}^\perp$ is a closed subspace of $\mathcal{H}$ and $\mathcal{M} \perp \mathcal{M}^\perp$. The orthogonal complement $\mathcal{M}^\perp$ is also denoted by $\mathcal{H} \ominus \mathcal{M}$.

The most intrinsic geometrical property of a Hilbert space $\mathcal{H}$ is that, for each closed convex subset $\mathcal{E}$ of $\mathcal{H}$ and every $x \in \mathcal{H}$, there is a unique $y \in \mathcal{E}$ such that

$$\text{dist}(x, \mathcal{E}) = \|x - y\|. \tag{1.23}$$

In the particular case, when $\mathcal{E} = \mathcal{M}$ is a closed subspace of $\mathcal{H}$, the vector y can also be characterized as the unique element $y \in \mathcal{M}$ such that $x - y \perp z$, for all vectors $z \in \mathcal{E}$. That is why the vector y is called the *orthogonal projection* of x onto $\mathcal{M}$ and is denoted by $P_\mathcal{M}(x)$. The linear map $P_\mathcal{M} : \mathcal{H} \longrightarrow \mathcal{H}$ is called the *orthogonal projection* of $\mathcal{H}$ onto $\mathcal{M}$. The representation

$$x = P_\mathcal{M}(x) + (x - P_\mathcal{M}(x)) \quad \text{with property} \quad P_\mathcal{M}(x) \perp (x - P_\mathcal{M}(x))$$

shows that

$$\mathcal{H} = \mathcal{M} \oplus \mathcal{M}^\perp.$$

It is straightforward to verify that $P_\mathcal{M}$ satisfies the following properties:

$$P_\mathcal{M}^2 = P_\mathcal{M}, \tag{1.24}$$

$$\|x\|^2 = \|P_\mathcal{M} x\|^2 + \|P_{\mathcal{M}^\perp} x\|^2 \qquad (x \in \mathcal{H}), \tag{1.25}$$

$$\langle P_\mathcal{M} x, y\rangle = \langle x, P_\mathcal{M} y\rangle = \langle P_\mathcal{M} x, P_\mathcal{M} y\rangle \qquad (x, y \in \mathcal{H}), \tag{1.26}$$

$$\langle P_\mathcal{M} x, x\rangle = \|P_\mathcal{M} x\|^2 \qquad (x \in \mathcal{H}). \tag{1.27}$$

If $\mathcal{M}$ and $\mathcal{N}$ are two closed subspaces of $\mathcal{H}$, then

$$\mathcal{M}^{\perp\perp} = \mathcal{M} \tag{1.28}$$

and

$$(\mathcal{M} \cap \mathcal{N})^\perp = \mathcal{M}^\perp + \mathcal{N}^\perp. \tag{1.29}$$

Relation (1.25) reveals that $P_\mathcal{M}$ is a bounded operator on $\mathcal{H}$ with $\|P_\mathcal{M}\| \leq 1$. In fact, if $\mathcal{M} \neq \{0\}$, we even have $\|P_\mathcal{M}\| = 1$.

In a Banach space $\mathcal{X}$, besides the norm topology, there is another important topology, namely the *weak topology*, which is defined as the weakest topology on $\mathcal{X}$ for which all the applications

$$\begin{array}{ccc} \mathcal{X} & \longrightarrow & \mathbb{C} \\ x & \longmapsto & \Lambda x \end{array}$$

are continuous. The parameter Λ runs through $\mathcal{X}^*$. Of course, the norm topology on $\mathcal{X}$ is stronger than the weak topology.

In a Banach space $\mathcal{X}$, a subset $E \subset \mathcal{X}$ is called *weakly bounded* if for each $\Lambda \in \mathcal{X}^*$ there is a constant C_Λ such that

$$|\Lambda x| \leq C_\Lambda \qquad (x \in E).$$

A sequence $(x_n)_{n\geq 1} \subset \mathcal{X}$ is said to be *weakly convergent* to $x \in \mathcal{X}$ if it converges to x for the weak topology, which means that

$$\lim_{n\to\infty} \Lambda x_n = \Lambda x$$

for all $\Lambda \in \mathcal{X}^*$. We will sometimes use the notation

$$x_n \xrightarrow{w} x$$

for the weak convergence, whereas as usual the notation $x_n \longrightarrow x$ means that $(x_n)_{n\geq 1}$ converges to x in the norm. In light of Theorem 1.24, in the context of Hilbert spaces, a subset $E \subset \mathcal{H}$ is weakly bounded if for each $y \in \mathcal{H}$ there is a constant C_y such that

$$|\langle x, y\rangle_\mathcal{H}| \leq C_y \qquad (x \in E).$$

Similarly, the sequence $(x_n)_{n\geq 1} \subset \mathcal{H}$ weakly converges to $x \in \mathcal{H}$ if

$$\lim_{n\to\infty} \langle x_n, y\rangle = \langle x, y\rangle$$

for all fixed $y \in \mathcal{H}$.

A standard example of a weakly convergent sequence is obtained via Bessel's inequality. Let $(e_n)_{n\geq 1}$ be any orthonormal sequence in a Hilbert space $\mathcal{H}$. Then, by (1.21), we have

$$\sum_{n=1}^{\infty} |\langle e_n, y\rangle|^2 \leq \|y\|^2 < \infty \qquad (y \in \mathcal{H}).$$

Hence,

$$\lim_{n\to\infty} \langle e_n, y\rangle = 0 \qquad (y \in \mathcal{H}).$$

This means that $(e_n)_{n\geq 1}$ weakly converges to zero. The following result is a variation of the Banach–Steinhaus theorem (Theorem 1.19) and the Banach–Alaoglu theorem on the weak-star compactness of the closed unit ball of the dual space. However, for a separable Hilbert space, other direct proofs are available.

Theorem 1.27 *A weakly bounded set is (norm) bounded. In particular, a weakly convergent sequence is bounded. On the other hand, each bounded sequence has a weakly convergent subsequence.*

Proof The first assertion follows immediately from the Banach–Steinhaus theorem. For the second assertion, we should imagine that $(x_n)_{n\geq 1}$ and x live in the dual space and apply the Banach–Alaoglu theorem. □

If $(x_n)_{n\geq 1}$ weakly converges to x, since

$$|\langle x_n, y\rangle| \leq \|x_n\|\,\|y\| \qquad (y \in \mathcal{H}),$$

we have

$$|\langle x, y\rangle| \leq \left(\liminf_{n\to\infty} \|x_n\|\right) \|y\| \qquad (y \in \mathcal{H}).$$

Therefore,

$$\|x\| \leq \liminf_{n\to\infty} \|x_n\|. \tag{1.30}$$

We saw that each orthonormal sequence weakly converges to zero. In this case, strict inequality holds in (1.30).

Given two Hilbert spaces $\mathcal{H}_1$ and $\mathcal{H}_2$, the space $\mathcal{L}(\mathcal{H}_1, \mathcal{H}_2)$ endowed with the operator norm is a Banach space (see Section 1.2). Besides the *norm operator topology*, there are two other important topologies on $\mathcal{L}(\mathcal{H}_1, \mathcal{H}_2)$, i.e. the strong operator topology and the weak operator topology. The *strong operator topology* is the weakest topology on $\mathcal{L}(\mathcal{H}_1, \mathcal{H}_2)$ for which all the applications

$$\begin{array}{rcl} \mathcal{L}(\mathcal{H}_1, \mathcal{H}_2) & \longrightarrow & \mathcal{H}_2 \\ A & \longmapsto & Ax \end{array}$$

are continuous. The parameter x runs through $\mathcal{H}_1$. The *weak operator topology* is the weakest topology on $\mathcal{L}(\mathcal{H}_1, \mathcal{H}_2)$ for which all the applications

$$\begin{array}{rcl} \mathcal{L}(\mathcal{H}_1, \mathcal{H}_2) & \longrightarrow & \mathbb{C} \\ A & \longmapsto & \langle Ax, y \rangle \end{array}$$

are continuous. The parameters x and y run respectively through $\mathcal{H}_1$ and $\mathcal{H}_2$. It is easy to see that the norm operator topology in $\mathcal{L}(\mathcal{H}_1, \mathcal{H}_2)$ is stronger than the strong operator topology, which is itself stronger than the weak operator topology.

Based on the above definitions, if $(A_n)_{n\geq 1}$ is a sequence in $\mathcal{L}(\mathcal{H}_1, \mathcal{H}_2)$ and $A \in \mathcal{L}(\mathcal{H}_1, \mathcal{H}_2)$, then the following assertions hold:

(i) $(A_n)_{n\geq 1}$ converges to A in the norm operator topology if and only if

$$\lim_{n\to\infty} \left(\sup_{x\in\mathcal{H}_1, \|x\|\leq 1} \|A_n x - Ax\| \right) = 0;$$

(ii) $(A_n)_{n\geq 1}$ converges to A in the strong operator topology if and only if

$$\lim_{n\to\infty} \|A_n x - Ax\| = 0$$

for all $x \in \mathcal{H}_1$; and

(iii) $(A_n)_{n\geq 1}$ converges to A in the weak operator topology if and only if

$$\lim_{n\to\infty} \langle (A_n - A)x, y \rangle = 0$$

for all $x \in \mathcal{H}_1$ and $y \in \mathcal{H}_2$.

In the following, if we do not specify the topology on $\mathcal{L}(\mathcal{H}_1, \mathcal{H}_2)$, e.g. if we write "$(A_n)_{n\geq 1}$ converges to A in $\mathcal{L}(\mathcal{H}_1, \mathcal{H}_2)$", it means with respect to the topology of the norm. Whenever we need another topology, we clearly mention it.

If (E, τ) is a topological space, then a subset K is called *sequentially compact* if any sequence $(x_n)_{n\geq 1}$ in K has a converging subsequence. Note that we do not assume that the limit stays in K. Recall also that a subset of a topological space is said to be *relatively compact* if its closure is compact. In a metrizable topological space, it is well known that the two notions of relative compactness and sequential compactness are equivalent. But, in general, these two properties are independent. We mention another situation in which the two concepts coincide.

Theorem 1.28 *Let K be a subset of a Banach space $\mathcal{X}$. Then the following are equivalent:*

(i) *K is relatively weakly compact;*
(ii) *K is sequentially weakly compact.*

This is a standard result in functional analysis, and a proof can be found for instance in [12], [186] or [557].

Let $(X, \mathcal{A}, \mu)$ be a measurable space with μ a finite positive measure. The following result gives a nice characterization of a weakly sequentially compact subset of $L^1(X, \mu)$.

Theorem 1.29 *Let K be a subset of $L^1(X, \mu)$. Then K is sequentially weakly compact if and only if it satisfies the following two conditions:*

(i) *K is norm-bounded; and*
(ii) *given $\varepsilon > 0$, there exists $\delta > 0$ such that for any measurable set A satisfying $\mu(A) \leq \delta$ we have*

$$\left| \int_A f(x)\, d\mu(x) \right| \leq \varepsilon \qquad (f \in K).$$

For a proof of this result, see [186, p. 292].

Exercises

Exercise 1.7.1 Let Λ be a bounded linear functional on the Hilbert space $\mathcal{H}$. Suppose that $\Lambda \not\equiv 0$, and put

$$\mathcal{M} = \{x \in \mathcal{H} : \Lambda x = 0\}.$$

Show that $\mathcal{M}^\perp$ is one-dimensional.

Exercise 1.7.2 Let $\omega = e^{i2\pi/n}$, $n \geq 3$. Let $(\mathcal{H}, \langle \cdot, \cdot \rangle)$ be a complex inner product space. Show that

$$\langle x, y \rangle = \frac{1}{n} \sum_{k=0}^{n-1} \omega^k \, \|x + \omega^k y\|^2 \qquad (x, y \in \mathcal{H}).$$

Hint: Use the fact that

$$\sum_{k=0}^{n-1} \omega^{2k} = 0.$$

Remark: The polarization identity (1.16) is the special case $n = 4$.

Exercise 1.7.3 Let $A \in \mathcal{L}(\mathcal{H})$ be such that

$$\langle Ax, x\rangle_{\mathcal{H}} = 0 \qquad (x \in \mathcal{H}).$$

Show that $A = 0$.
Hint: Use (1.19).

Exercise 1.7.4 Construct a bounded operator A on a *real* Hilbert space $\mathcal{H}$ such that

$$\langle Ax, x\rangle_{\mathcal{H}} = 0 \qquad (x \in H),$$

but still $A \neq 0$.
Hint: $\mathbb{R}^2$ is a real Hilbert space and consider the operator

$$\begin{aligned} A : \mathbb{R}^2 &\longrightarrow \mathbb{R}^2 \\ (x, y) &\longmapsto (-y, x). \end{aligned}$$

Remark: Compare with Exercise 1.7.3.

Exercise 1.7.5 Show that $(\mathfrak{e}_n)_{n\geq 1}$ is an orthonormal basis for ℓ^2.
Hint: Appeal to the definition (1.12).

Exercise 1.7.6 Show that ℓ^2 is a separable Hilbert space.
Hint: Consider (rational) linear combinations of $\mathfrak{e}_n$, $n \geq 1$.

Exercise 1.7.7 Let $\mathcal{E}$ be a subset of the Hilbert space $\mathcal{H}$. Show that

$$\mathcal{E}^{\perp\perp} = \operatorname{Span}\mathcal{E} = \operatorname{Clos}_{\mathcal{H}}\{\alpha_1 x_1 + \cdots + \alpha_n x_n : \alpha_i \in \mathbb{C},\ x_i \in \mathcal{E}\}.$$

Hint: Use (1.28).

Exercise 1.7.8 Let $\mathcal{H}_1$ and $\mathcal{H}_2$ be Hilbert spaces, and let $A \in \mathcal{L}(\mathcal{H}_1, \mathcal{H}_2)$, with $A \neq 0$. Suppose that $x \in \mathcal{H}_1$ is a maximizing vector for A. Show that $x \perp \ker A$.
Hint: Note that $Ax = APx$, where P is the projection of $\mathcal{H}_1$ onto the closed subspace $(\ker A)^{\perp}$.

Exercise 1.7.9 Let X be a topological space, let $x \in X$, and let $(x_n)_{n\geq 1}$ be a sequence in X. Assume that every subsequence of $(x_n)_{n\geq 1}$ has a subsequence that converges to x. Show that $(x_n)_{n\geq 1}$ converges to x.
Hint: Assume on the contrary that $(x_n)_{n\geq 1}$ does not converge to x. Hence we can find a neighborhood of x and a subsequence $(x_{\varphi(n)})_{n\geq 1}$ such that $x_{\varphi(n)} \notin V$, for all $n \geq 1$. But $(x_{\varphi(n)})_{n\geq 1}$ has a subsequence, say $(x_{\varphi(\theta(n))})_{n\geq 1}$, that converges to x.

1.8 The adjoint operator

If $A \in \mathcal{L}(\mathcal{H}_1, \mathcal{H}_2)$, then an application of Theorem 1.24 shows that there is a unique operator $A^* \in \mathcal{L}(\mathcal{H}_2, \mathcal{H}_1)$ such that

$$\langle Ax, y\rangle_{\mathcal{H}_2} = \langle x, A^*y\rangle_{\mathcal{H}_1} \tag{1.31}$$

for all $x \in \mathcal{H}_1$ and $y \in \mathcal{H}_2$. Moreover,

$$\|A\|_{\mathcal{L}(\mathcal{H}_1,\mathcal{H}_2)} = \|A^*\|_{\mathcal{L}(\mathcal{H}_2,\mathcal{H}_1)}.$$

The operator A^* is called the *adjoint* of A. This concept can also be defined for operators between Banach spaces. See [159, chap. VI] for precise definitions in the Banach space setting. For further reference, we list some elementary properties of the adjoint operator.

Theorem 1.30 *Let $A, B \in \mathcal{L}(\mathcal{H}_1, \mathcal{H}_2)$ and $C \in \mathcal{L}(\mathcal{H}_2, \mathcal{H}_3)$, and let $\alpha, \beta \in \mathbb{C}$. Then the following hold.*

(i) $(\alpha A + \beta B)^* = \bar{\alpha}A^* + \bar{\beta}B^*$.
(ii) $A^{**} = A$.
(iii) $(CA)^* = A^*C^*$.
(iv) *If A is invertible, then so is A^* and* $(A^*)^{-1} = (A^{-1})^*$.
(v) $\|A^*A\| = \|AA^*\| = \|A\|^2$.
(vi) $\ker A^*A = \ker A$.
(vii) $\ker A = \mathcal{R}(A^*)^\perp$ *and* $(\ker A)^\perp = \mathrm{Clos}_{\mathcal{H}_1}\mathcal{R}(A^*)$.
(viii) $\ker A^* = \mathcal{R}(A)^\perp$ *and* $(\ker A^*)^\perp = \mathrm{Clos}_{\mathcal{H}_2}\mathcal{R}(A)$.
(ix) *If $A \in \mathcal{L}(\mathcal{H})$, then*

$$\sigma(A^*) = \overline{\sigma(A)} = \{\bar{\lambda} : \lambda \in \sigma(A)\}.$$

If, moreover, A is invertible, then

$$\sigma(A^{-1}) = \sigma(A)^{-1} = \{1/\lambda : \lambda \in \sigma(A)\}.$$

Most of the properties stated in Theorem 1.30 are standard and straightforward, and we refer to [159] for more details.

By taking $y = Ax$ in (1.31), we obtain

$$\|Ax\|^2_{\mathcal{H}_2} = \langle x, A^*Ax\rangle_{\mathcal{H}_1} \qquad (x \in \mathcal{H}_1). \tag{1.32}$$

Using Theorem 1.30 and the fact that an operator is bounded below if and only if it is one-to-one and has a closed range (Corollary 1.17), we easily get the following result.

Corollary 1.31 *Let $A \in \mathcal{L}(\mathcal{H}_1, \mathcal{H}_2)$. Then the following are equivalent.*

(i) *A is invertible.*
(ii) *A is bounded below and its range is dense in $\mathcal{H}$.*
(iii) *A and A^* are bounded below.*

If $A \in \mathcal{L}(\mathcal{H})$, then, by Theorem 1.30, we have

$$\mathcal{H} = \ker A \oplus \mathrm{Clos}_{\mathcal{H}}\mathcal{R}(A^*) = \ker A^* \oplus \mathrm{Clos}_{\mathcal{H}}\mathcal{R}(A). \tag{1.33}$$

If, moreover, $\mathcal{H}$ is finite-dimensional, each subspace of $\mathcal{H}$ is automatically closed. Hence, in this case, we can say more about the relations between A and A^*.

Lemma 1.32 *Let $\mathcal{H}$ be a finite-dimensional Hilbert space, and let $A \in \mathcal{L}(\mathcal{H})$. Then the following hold:*

(i) $\dim \mathcal{H} = \dim \mathcal{R}(A) + \dim \ker(A)$;
(ii) $\dim \mathcal{R}(A) = \dim \mathcal{R}(A^*)$;
(iii) $\dim \ker A = \dim \ker A^*$.

In the infinite-dimensional case, the ranges of A and A^* are not necessarily closed, but we have the following result, which has many important consequences.

Theorem 1.33 *Let $A \in \mathcal{L}(\mathcal{H}_1, \mathcal{H}_2)$. Then $\mathcal{R}(A)$ is a closed subspace of $\mathcal{H}_2$ if and only if $\mathcal{R}(A^*)$ is a closed subspace of $\mathcal{H}_1$.*

Proof First note that, according to Theorem 1.30, it is sufficient to prove one side of the equivalence. Hence, let us assume that $\mathcal{R}(A)$ is a closed subspace of $\mathcal{H}_2$, and then we prove that $\mathcal{R}(A^*)$ is a closed subspace of $\mathcal{H}_1$.

By Theorem 1.30, we have $\mathrm{Clos}_{\mathcal{H}_1}(\mathcal{R}(A^*)) = (\ker A)^\perp$ and thus it is sufficient to show that $(\ker A)^\perp \subset \mathcal{R}(A^*)$. To do so, fix $x_0 \in (\ker A)^\perp$ and define

$$\begin{array}{rccc} \Lambda: & \mathcal{R}(A) & \longrightarrow & \mathbb{C} \\ & y & \longmapsto & \langle x, x_0\rangle_{\mathcal{H}_1}, \end{array}$$

where x is such that $y = Ax$. Note that, if $Ax_1 = Ax_2$, then $x_1 - x_2 \in \ker A$ and thus $\langle x_1 - x_2,\ x_0\rangle = 0$, which shows that Λ is well defined. Moreover, as a consequence of the open mapping theorem (Theorem 1.14), we know that there exists a finite constant C such that, for each $y \in \mathcal{R}(A)$, there is a choice of $x \in \mathcal{H}_1$ such that $y = Ax$ and

$$\|x\|_{\mathcal{H}_1} \le C\|y\|_{\mathcal{H}_2}$$

(see Exercise 1.5.3). Thus, exploiting the above representative of y, we get

$$|\Lambda(y)| = |\langle x, x_0\rangle_{\mathcal{H}_1}| \le \|x\|_{\mathcal{H}_1}\,\|x_0\|_{\mathcal{H}_1} \le C\|x_0\|_{\mathcal{H}_1}\,\|y\|_{\mathcal{H}_2}.$$

Therefore, Λ is a continuous linear functional on the Hilbert space $\mathcal{R}(A)$. (Here we used the fact that $\mathcal{R}(A)$ is a closed subspace of $\mathcal{H}_2$.) Then, by the Riesz representation theorem (Theorem 1.24), there exists $y_0 \in \mathcal{R}(A)$ such that

$$\Lambda(y) = \langle y, y_0\rangle_{\mathcal{H}_2} \qquad (y \in \mathcal{R}(A)).$$

This relation can be rewritten as

$$\Lambda(Ax) = \langle Ax, y_0\rangle_{\mathcal{H}_2} \qquad (x \in \mathcal{H}_1).$$

Therefore,

$$\langle x, x_0\rangle_{\mathcal{H}_1} = \langle x, A^*y_0\rangle_{\mathcal{H}_1} \qquad (x \in \mathcal{H}_1),$$

which gives $x_0 = A^*y_0 \in \mathcal{R}(A^*)$. □

Corollary 1.34 *Let $A \in \mathcal{L}(\mathcal{H}_1, \mathcal{H}_2)$. Then the following are equivalent.*

(i) *A is right-invertible.*
(ii) *A^* is left-invertible.*
(iii) *A is onto.*
(iv) *A^* is bounded below.*

Proof The equivalence (i) $\Longleftrightarrow$ (ii) and the implication (i) $\Longrightarrow$ (iii) are trivial.

(iii) $\Longrightarrow$ (i) By Corollary 1.16, there exists a constant $c > 0$ such that, for any $y \in \mathcal{H}_2$, there exists $x \in \mathcal{H}_1$ such that $y = Ax$ and $\|x\| \le c\|y\|$. We can uniquely choose the vector x by assuming $x \perp \ker A$. Define $By = x$. Then we have $ABy = Ax = y$ and $\|By\| = \|x\| \le c\|y\|$, which proves that B is a continuous linear map from $\mathcal{H}_2$ into $\mathcal{H}_1$. Thus A is right-invertible.

(ii) $\Longrightarrow$ (iv) Since A^* is left-invertible, there exists an operator $C \in \mathcal{L}(\mathcal{H}_1, \mathcal{H}_2)$ such that $CA^* = I_{\mathcal{H}_2}$. Hence, for any vector $y \in \mathcal{H}_2$, we have

$$\|y\| = \|CA^*y\| \le \|C\|\,\|A^*y\|,$$

whence $\|A^*y\| \ge \|C\|^{-1}\|y\|$. This shows that A^* is bounded below.

(iv) $\Longrightarrow$ (iii) Since A^* is bounded below, in particular, it is one-to-one and its range is closed. Then, by Theorem 1.33, we know that the range of A is also closed, and since $\ker(A^*) = \mathcal{R}(A)^\perp$, we get that $\mathcal{R}(A) = \mathcal{H}_2$. □

The following simple surjectivity criterion will also be needed in applications.

Corollary 1.35 *Let $A \in \mathcal{L}(\mathcal{H}_1, \mathcal{H}_2)$ and assume that A^* is one-to-one and A^*A has a closed range. Then A is onto.*

Proof Let $y \in \mathrm{Clos}_{\mathcal{H}_2}(\mathcal{R}(A))$. Then there exists a sequence $(x_n)_{n\ge1}$ in $\mathcal{H}_1$ such that

$$y = \lim_{n\to\infty} Ax_n.$$

Hence, $A^*y = \lim_{n\to\infty} A^*Ax_n$, which implies $A^*y \in \mathrm{Clos}_{\mathcal{H}_1}(\mathcal{R}(A^*A)) = \mathcal{R}(A^*A)$. Thus, there exists $x \in \mathcal{H}_1$ such that $A^*y = A^*Ax$ and, since A^* is injective, we get $y = Ax$. This proves that A has a closed range. But, by Theorem 1.30, we also have $(\mathcal{R}(A))^\perp = \ker A^* = \{0\}$, whence A has a dense range. Therefore, $\mathcal{R}(A) = \mathcal{H}_2$. □

Let $\mathcal{M}$ be a closed subspace of $\mathcal{H}$. We recall that $P_{\mathcal{M}}$ denotes the orthogonal projection of $\mathcal{H}$ onto $\mathcal{M}$. A closely related operator is the *restricted projection*

$$\begin{array}{rccc} \mathbf{P}_{\mathcal{M}}: & \mathcal{H} & \longrightarrow & \mathcal{M} \\ & x & \longmapsto & P_{\mathcal{M}}x. \end{array}$$

Note that $P_{\mathcal{M}} \in \mathcal{L}(\mathcal{H})$, while $\mathbf{P}_{\mathcal{M}} \in \mathcal{L}(\mathcal{H}, \mathcal{M})$. By the same token, we consider the identity operator

$$\begin{array}{rccc} i: & \mathcal{H} & \longrightarrow & \mathcal{H} \\ & x & \longmapsto & x \end{array}$$

and then define the *inclusion*

$$\begin{array}{rccc} \mathbf{i}_{\mathcal{M}}: & \mathcal{M} & \longrightarrow & \mathcal{H} \\ & x & \longmapsto & x. \end{array}$$

Then, for each $x \in \mathcal{H}$, $y \in \mathcal{M}$, we have

$$\begin{aligned} \langle x, \mathbf{P}^*_{\mathcal{M}} y \rangle_{\mathcal{H}} &= \langle \mathbf{P}_{\mathcal{M}} x, y \rangle_{\mathcal{M}} \\ &= \langle x, y \rangle_{\mathcal{H}} \\ &= \langle x, \mathbf{i}_{\mathcal{M}} y \rangle_{\mathcal{H}}. \end{aligned}$$

Therefore,

$$\mathbf{P}^*_{\mathcal{M}} = \mathbf{i}_{\mathcal{M}}. \tag{1.34}$$

In other words, $\mathbf{i}_{\mathcal{M}}$ is the adjoint of $\mathbf{P}_{\mathcal{M}}$.

Now, consider the bounded operator $A : \mathcal{H}_1 \longrightarrow \mathcal{H}_2$. Let $\mathcal{M}_1 \subset \mathcal{H}_1$ and $\mathcal{M}_2 \subset \mathcal{H}_2$ be two closed subspaces. Then the restricted mapping

$$\begin{array}{rccc} A_{\mathcal{M}_1 \to \mathcal{M}_2}: & \mathcal{M}_1 & \longrightarrow & \mathcal{M}_2 \\ & x & \longmapsto & \mathbf{P}_{\mathcal{M}_2} A x \end{array}$$

defines a bounded operator in $\mathcal{L}(\mathcal{M}_1, \mathcal{M}_2)$, and we can write

$$A_{\mathcal{M}_1 \to \mathcal{M}_2} = \mathbf{P}_{\mathcal{M}_2} A \, \mathbf{i}_{\mathcal{M}_1}. \tag{1.35}$$

It is natural and important in many applications to know the adjoint of $A_{\mathcal{M}_1 \to \mathcal{M}_2}$. In fact, using (1.35), (1.34) and Theorem 1.30(ii), we immediately get

$$A^*_{\mathcal{M}_1 \to \mathcal{M}_2} = \mathbf{P}_{\mathcal{M}_1} A^* \, \mathbf{i}_{\mathcal{M}_2}. \tag{1.36}$$

In particular, given $A \in \mathcal{L}(\mathcal{H}_1, \mathcal{H}_2)$, we define the restricted operator

$$\begin{array}{rccc} A_{|(\ker A)^\perp}: & (\ker A)^\perp & \longrightarrow & \mathcal{H}_2 \\ & x & \longmapsto & Ax. \end{array}$$

Using the above notation, we have

$$A_{|(\ker A)^\perp} = A_{(\ker A)^\perp \to \mathcal{H}_2} = A \, \mathbf{i}_{(\ker A)^\perp}.$$

Clearly $A_{|(\ker A)^\perp}$ is injective and (1.36) immediately gives

$$(A_{|(\ker A)^\perp})^* = \mathbf{P}_{(\ker A)^\perp} A^*. \tag{1.37}$$

Given two closed subspaces $\mathcal{M}$ and $\mathcal{N}$ of a Hilbert space $\mathcal{H}$, let

$$\begin{array}{rccl} \mathbf{P}_{\mathcal{N}\to\mathcal{M}}: & \mathcal{N} & \longrightarrow & \mathcal{M} \\ & x & \longmapsto & P_{\mathcal{M}}x. \end{array}$$

In other words,

$$\mathbf{P}_{\mathcal{N}\to\mathcal{M}} = \mathbf{P}_{\mathcal{M}}\mathbf{i}_{\mathcal{N}}.$$

Hence, by (1.34),

$$(\mathbf{P}_{\mathcal{N}\to\mathcal{M}})^* = \mathbf{i}_{\mathcal{N}}^*\mathbf{P}_{\mathcal{M}}^* = \mathbf{P}_{\mathcal{N}}\mathbf{i}_{\mathcal{M}} = \mathbf{P}_{\mathcal{M}\to\mathcal{N}}. \tag{1.38}$$

Some properties of this operator are given in the following lemma.

Lemma 1.36 *Let $\mathcal{H}$ be a Hilbert space and suppose we have two orthogonal decompositions*

$$\mathcal{H} = \mathcal{M}_1 \oplus \mathcal{M}_2 = \mathcal{N}_1 \oplus \mathcal{N}_2.$$

Then the following statements are equivalent.

(i) $\mathbf{P}_{\mathcal{M}_1\to\mathcal{N}_1}$ *is surjective.*
(ii) $\mathbf{P}_{\mathcal{N}_1\to\mathcal{M}_1}$ *is bounded below.*
(iii) $\|\mathbf{P}_{\mathcal{N}_1\to\mathcal{M}_2}\| < 1$.
(iv) $\|\mathbf{P}_{\mathcal{M}_2\to\mathcal{N}_1}\| < 1$.
(v) $\mathbf{P}_{\mathcal{M}_2\to\mathcal{N}_2}$ *is bounded below.*
(vi) $\mathbf{P}_{\mathcal{N}_2\to\mathcal{M}_2}$ *is surjective.*

Proof Equivalences (i) $\Longleftrightarrow$ (ii) and (v) $\Longleftrightarrow$ (vi) follow from (1.38) and Corollary 1.34.

(iii) $\Longleftrightarrow$ (iv) This follows immediately from (1.38) and Theorem 1.30.

(ii) $\Longleftrightarrow$ (iii) By definition, $\mathbf{P}_{\mathcal{N}_1\to\mathcal{M}_1}$ is bounded below if and only if there is a constant $\delta > 0$ such that

$$\delta\|y_1\|^2 \le \|P_{\mathcal{M}_1}y_1\|^2 \qquad (y_1 \in \mathcal{N}_1).$$

Note that necessarily $\delta \le 1$ (because $\|P_{\mathcal{M}_1}\| \le 1$). Using the identity

$$\|y_1\|^2 = \|P_{\mathcal{M}_1}y_1\|^2 + \|P_{\mathcal{M}_2}y_1\|^2,$$

we see that $\mathbf{P}_{\mathcal{N}_1\to\mathcal{M}_1}$ is bounded below if and only if there is a constant $\delta \in (0,1]$ such that

$$\|P_{\mathcal{M}_2}y_1\|^2 \le (1-\delta)\|y_1\|^2 \qquad (y_1 \in \mathcal{N}_1),$$

which means that $\|\mathbf{P}_{\mathcal{N}_1\to\mathcal{M}_2}\| < 1$.

(iv) $\Longleftrightarrow$ (v) This follows from the previously established equivalence (ii) $\Longleftrightarrow$ (iii) by interchanging the roles of $\mathcal{M}_2$ and $\mathcal{N}_2$ with $\mathcal{M}_1$ and $\mathcal{N}_1$. □

Exercises

Exercise 1.8.1 If $A \in \mathcal{L}(\mathcal{H})$ has a finite rank, show that $A^* \in \mathcal{L}(\mathcal{H})$ also has a finite rank.
Hint: If $\mathcal{R}(A) = \operatorname{Span}(y_1, \ldots, y_n)$, then $\mathcal{R}(A^*) = \operatorname{Span}(A^*y_1, \ldots, A^*y_n)$. Indeed, let $z \in \mathcal{H}$ and assume that $z \perp A^*y_j$, $1 \leq j \leq n$. By Theorem 1.30, $\ker A^* = \mathcal{R}(A)^\perp$. Then, show that $z \in \ker(A^*A)$. Hence, again by Theorem 1.30, $z \in \ker(A) = (\mathcal{R}(A^*))^\perp$.

Exercise 1.8.2 Let $P \in \mathcal{L}(\mathcal{H})$ be an orthogonal projection on the Hilbert space $\mathcal{H}$. Show that

$$\langle Px,\, y - Py\rangle_{\mathcal{H}} = 0 \qquad (x, y \in \mathcal{H}).$$

Hint: Use $P^* = P^2 = P$.

Exercise 1.8.3 Let $A \in \mathcal{L}(\mathcal{H})$, and let $x \in \mathcal{H}$ be such that $x \perp \ker(I - AA^*)$. Show that $A^*x \perp \ker(I - A^*A)$.

Exercise 1.8.4 Let $\mathcal{H}_1$ and $\mathcal{H}_2$ be Hilbert spaces, and let $A \in \mathcal{L}(\mathcal{H}_1, \mathcal{H}_2)$, $A \neq 0$. Show that any maximizing vector for A^*A is also a maximizing vector for A.
Hint: Use $\|A^*Ax\| \leq \|A^*\|\,\|Ax\|$ and that $\|A^*A\| = \|A\|^2 = \|A^*\|^2$.

Exercise 1.8.5 Let $\mathcal{H}_1$ and $\mathcal{H}_2$ be Hilbert spaces, and let $A \in \mathcal{L}(\mathcal{H}_1, \mathcal{H}_2)$, $A \neq 0$. Show that, if x is a maximizing vector for A, then Ax is a maximizing vector for A^*.
Hint: Consider $\langle A^*Ax, x\rangle$.

1.9 Tensor product and algebraic direct sum

Let $x \in \mathcal{H}_1$ and $y \in \mathcal{H}_2$. Then the *tensor product* $x \otimes y$ is the rank-one operator in $\mathcal{L}(\mathcal{H}_2, \mathcal{H}_1)$ that is defined by

$$(x \otimes y)z = \langle z, y\rangle_{\mathcal{H}_2} x \qquad (z \in \mathcal{H}_2). \tag{1.39}$$

Directly from the definition, we see that

$$(\alpha_1 x_1 + \alpha_2 x_2) \otimes y = \alpha_1(x_2 \otimes y) + \alpha_2(x_2 \otimes y)$$

and

$$x \otimes (\alpha_1 y_1 + \alpha_2 y_2) = \bar{\alpha}_1(x \otimes y_1) + \bar{\alpha}_2(x \otimes y_2).$$

We gather some further properties of the tensor product in the following theorem.

Theorem 1.37 *Let $A \in \mathcal{L}(\mathcal{H}_1, \mathcal{H}_2)$. Then, for each $x \in \mathcal{H}_1$ and $y \in \mathcal{H}_2$, we have*

(i) $A(x \otimes y) = Ax \otimes y$,
(ii) $(x \otimes y)A = x \otimes A^*y$,
(iii) $(x \otimes y)^* = y \otimes x$,
(iv) $\|x \otimes y\|_{\mathcal{L}(\mathcal{H}_2,\mathcal{H}_1)} = \|x\|_{\mathcal{H}_1} \times \|y\|_{\mathcal{H}_2}$.

Proof (i) For each $z \in \mathcal{H}_2$, we have

$$\begin{aligned}(A(x \otimes y))(z) &= A((x \otimes y)(z))\\ &= A(\langle z, y\rangle_{\mathcal{H}_2} x)\\ &= \langle z, y\rangle_{\mathcal{H}_2} Ax\\ &= (Ax \otimes y)(z).\end{aligned}$$

(iii) For each $w \in \mathcal{H}_1$ and $z \in \mathcal{H}_2$, we have

$$\begin{aligned}\langle (x \otimes y)^* w, z\rangle_{\mathcal{H}_2} &= \langle w, (x \otimes y)z\rangle_{\mathcal{H}_1}\\ &= \langle w, \langle z, y\rangle_{\mathcal{H}_2} x\rangle_{\mathcal{H}_1}\\ &= \overline{\langle z, y\rangle}_{\mathcal{H}_2} \times \langle w, x\rangle_{\mathcal{H}_1}\\ &= \langle y, z\rangle_{\mathcal{H}_2} \times \langle w, x\rangle_{\mathcal{H}_1}\\ &= \langle \langle w, x\rangle_{\mathcal{H}_1} y, z\rangle_{\mathcal{H}_2}\\ &= \langle (y \otimes x)w, z\rangle_{\mathcal{H}_2}.\end{aligned}$$

(ii) According to (i) and (iii),

$$\begin{aligned}((x \otimes y)A)^* &= A^*(x \otimes y)^*\\ &= A^*(y \otimes x)\\ &= A^*y \otimes x.\end{aligned}$$

Hence, again by (iii),

$$(x \otimes y)A = (A^*y \otimes x)^* = x \otimes A^*y.$$

(iv) By the definition of norm,

$$\begin{aligned}\|x \otimes y\|_{\mathcal{L}(\mathcal{H}_2,\mathcal{H}_1)} &= \sup_{\|z\|_{\mathcal{H}_2}=1} \|(x \otimes y)z\|_{\mathcal{H}_1}\\ &= \sup_{\|z\|_{\mathcal{H}_2}=1} \|\langle z, y\rangle_{\mathcal{H}_2} x\|_{\mathcal{H}_1}\\ &= \|x\|_{\mathcal{H}_1} \sup_{\|z\|_{\mathcal{H}_2}=1} |\langle z, y\rangle_{\mathcal{H}_2}|\\ &= \|x\|_{\mathcal{H}_1} \|y\|_{\mathcal{H}_2}.\end{aligned}$$

This completes the proof. □

Consider a bounded operator

$$A : \mathcal{H}_1 \oplus \mathcal{H}_2 \longrightarrow \mathcal{H}'_1 \oplus \mathcal{H}'_2,$$

which has the matrix representation

$$A = \begin{bmatrix} A_{11} & A_{12} \\ A_{21} & A_{22} \end{bmatrix},$$

where $A_{ij} \in \mathcal{L}(\mathcal{H}_i, \mathcal{H}'_j)$, $i, j = 1, 2$. Direct verification shows that the adjoint

$$A^* : \mathcal{H}'_1 \oplus \mathcal{H}'_2 \longrightarrow \mathcal{H}_1 \oplus \mathcal{H}_2$$

has the matrix representation

$$A^* = \begin{bmatrix} A^*_{11} & A^*_{21} \\ A^*_{12} & A^*_{22} \end{bmatrix}. \tag{1.40}$$

Now, we consider a special case of this representation.

Lemma 1.38 *Let* $\mathcal{H} = \mathcal{H}_1 \oplus \mathcal{H}_2$, $\mathcal{H}' = \mathcal{H}'_1 \oplus \mathcal{H}'_2$, $A_1 \in \mathcal{L}(\mathcal{H}_1, \mathcal{H}'_1)$, $A_2 \in \mathcal{L}(\mathcal{H}_2, \mathcal{H}'_2)$ *and* $A \in \mathcal{L}(\mathcal{H}, \mathcal{H}')$. *Assume that, relative to the preceding decompositions,* A *has the matrix representation*

$$A = \begin{bmatrix} A_1 & 0 \\ 0 & A_2 \end{bmatrix}.$$

Then the operator A *has a closed range if and only if both operators* A_1 *and* A_2 *have closed ranges.*

Proof The result easily follows from the relation

$$\mathcal{R}(A) = \mathcal{R}(A_1) \times \mathcal{R}(A_2),$$

which, with (1.14), gives

$$\mathrm{Clos}_{\mathcal{H}'}(\mathcal{R}(A)) = \mathrm{Clos}_{\mathcal{H}'_1}(\mathcal{R}(A_1)) \times \mathrm{Clos}_{\mathcal{H}'_2}(\mathcal{R}(A_2)). \qquad \square$$

We will need another special case where the focus is on the decomposition of the domain of A. In other words, we consider A as an operator from $\mathcal{H}_1 \oplus \mathcal{H}_2$ into $\mathcal{H}$, i.e.

$$A : \mathcal{H}_1 \oplus \mathcal{H}_2 \longrightarrow \mathcal{H}.$$

Then A has the matrix representation $A = [A_1 \;\; A_2]$, where $A_i \in \mathcal{L}(\mathcal{H}_i, \mathcal{H})$, $i = 1, 2$. Hence, we can write

$$A(x_1 \oplus x_2) = A_1 x_1 + A_2 x_2 \qquad (x_1 \in \mathcal{H}_1,\; x_2 \in \mathcal{H}_2). \tag{1.41}$$

Since

$$\begin{aligned}\langle A(x_1 \oplus x_2),\, x\rangle_{\mathcal{H}} &= \langle A_1x_1 + A_2x_2,\, x\rangle_{\mathcal{H}}\\ &= \langle A_1x_1, x\rangle_{\mathcal{H}} + \langle A_2x_2, x\rangle_{\mathcal{H}}\\ &= \langle x_1, A_1^*x\rangle_{\mathcal{H}_1} + \langle x_2, A_2^*x\rangle_{\mathcal{H}_2}\\ &= \langle x_1 \oplus x_2,\, A_1^*x \oplus A_2^*x\rangle_{\mathcal{H}_1\oplus\mathcal{H}_2},\end{aligned}$$

we deduce that

$$A^*x = A_1^*x \oplus A_2^*x \qquad (x \in \mathcal{H}). \tag{1.42}$$

Using (1.41) and (1.42), we obtain the identity

$$AA^* = A_1A_1^* + A_2A_2^*. \tag{1.43}$$

According to (1.41), it is clear that

$$\ker A_1 \oplus \ker A_2 \subset \ker A. \tag{1.44}$$

However, in the general case, the inclusion (1.44) is proper. In any case, this relation implies that

$$(\ker A_1)^{\perp} \oplus (\ker A_2)^{\perp} \supset (\ker A)^{\perp}, \tag{1.45}$$

where the three orthogonal complements are respectively taken in $\mathcal{H}_1$, $\mathcal{H}_2$ and $\mathcal{H}_1 \oplus \mathcal{H}_2$. From the definitions of A_1 and A_2, we see that

$$\mathcal{R}(A) = \mathcal{R}(A_1) + \mathcal{R}(A_2).$$

If $\mathcal{R}(A_1) \cap \mathcal{R}(A_2) = \{0\}$, then $\mathcal{R}(A)$ is the *algebraic direct sum* of $\mathcal{R}(A_1)$ and $\mathcal{R}(A_2)$. An easy calculation shows that

$$\mathcal{R}(A_1) \cap \mathcal{R}(A_2) = \{0\} \quad \Longleftrightarrow \quad \ker A = \ker A_1 \oplus \ker A_2.$$

In this situation, (1.45) becomes

$$(\ker A)^{\perp} = (\ker A_1)^{\perp} \oplus (\ker A_2)^{\perp}, \tag{1.46}$$

and thus, for each $x_1 \in \mathcal{H}_1$ and $x_2 \in \mathcal{H}_2$, we have

$$P_{(\ker A)^{\perp}}(x_1 \oplus x_2) = P_{(\ker A_1)^{\perp}}x_1 \oplus P_{(\ker A_2)^{\perp}}x_2 \tag{1.47}$$

(see Exercise 1.9.4).

Exercises

Exercise 1.9.1 Show that

$$(x \otimes y)\,(x' \otimes y') = \langle x', y\rangle\,(x \otimes y').$$

Exercise 1.9.2 Show that

$$(x \otimes y)^* \, (x \otimes y) = \|x\|^2 \, (y \otimes y).$$

Hint: Use Theorem 1.37 and Exercise 1.9.1.

Exercise 1.9.3 Construct an example for which the inclusion (1.44) is proper. Hint: Take $A_1 = -A_2 = id : \mathbb{C} \longrightarrow \mathbb{C}$.

Exercise 1.9.4 Let $\mathcal{H}_1$ and $\mathcal{H}_2$ be Hilbert spaces, and let $\mathcal{M}_1$ and $\mathcal{M}_2$ be respectively some closed subspaces of $\mathcal{H}_1$ and $\mathcal{H}_2$. Show that

$$P_{\mathcal{M}_1 \oplus \mathcal{M}_2}(x_1 \oplus x_2) = P_{\mathcal{M}_1}(x_1) \oplus P_{\mathcal{M}_2}(x_2) \qquad (x_1 \in \mathcal{H}_1,\ x_2 \in \mathcal{H}_2).$$

1.10 Invariant subspaces and cyclic vectors

Let A be a bounded operator on a Hilbert space $\mathcal{H}$. A closed subspace $\mathcal{M} \subset \mathcal{H}$ is said to be *invariant* under A if

$$x \in \mathcal{M} \quad \Longrightarrow \quad Ax \in \mathcal{M}.$$

In other words, $A\mathcal{M} \subset \mathcal{M}$. Furthermore, if $\mathcal{M} \neq \mathcal{H}$, then $\mathcal{M}$ is called a *proper invariant subspace* of A; and if $\{0\} \neq \mathcal{M} \neq \mathcal{H}$, then it is called a *nontrivial invariant subspace*. The collection of closed subspaces of $\mathcal{X}$ that are invariant under A is denoted by $\text{Lat}(A)$. It is clear that $\{0\}$ and $\mathcal{X}$ are always in $\text{Lat}(A)$.

The study of invariant subspaces of an operator is one of the main goals of operator theory. A simple family of closed invariant subspaces of A is obtained in the following way. Fix $x_0 \in \mathcal{H}$, and put

$$\langle x_0 \rangle = \text{Span}\{A^n x_0 : n \geq 0\}.$$

It is easy to verify that $\langle x_0 \rangle$ is the smallest closed subspace of $\mathcal{H}$ that contains x_0 and is invariant under A. It may happen that $\langle x_0 \rangle$ coincides with the whole space $\mathcal{H}$, and, in this case, we say that x_0 is a *cyclic vector* for A. More explicitly, a vector $x_0 \in \mathcal{H}$ is cyclic for A if

$$\text{Span}\{A^n x_0 : n \geq 0\} = \mathcal{H}.$$

By the same token, an invariant subspace $\mathcal{M}$ is said to be cyclic for A if there is a vector $x_0 \in \mathcal{M}$ such that $\langle x_0 \rangle = \mathcal{M}$. Thus, based on the latter definition, we can say that A has a cyclic vector if and only if the space $\mathcal{H}$ is cyclic for A. For an *invertible* operator A, cyclic vectors are sometimes called 1-*cyclic vectors*, to distinguish them from 2-*cyclic vectors* defined by

$$\text{Span}\{A^n x_0 : n \in \mathbb{Z}\} = \mathcal{H}.$$

An invariant subspace of A is not necessarily invariant under A^*. The following result clarifies the connection between the invariant closed subspaces of A and A^*.

Lemma 1.39 *Let $A \in \mathcal{L}(\mathcal{H})$. Let $\mathcal{M}$ be a closed subspace of $\mathcal{H}$. Then $\mathcal{M}$ is invariant under A if and only if $\mathcal{M}^\perp$ is invariant under A^*.*

Proof Since $\mathcal{M}^{\perp\perp} = \mathcal{M}$ and $(A^*)^* = A$, it is sufficient to prove that

$$A\mathcal{M} \subset \mathcal{M} \quad \Longrightarrow \quad A^*\mathcal{M}^\perp \subset \mathcal{M}^\perp.$$

But this implication follows immediately from

$$\langle A^*y, x\rangle = \langle y, Ax\rangle = 0 \qquad (x \in \mathcal{M},\ y \in \mathcal{M}^\perp). \qquad \square$$

In spectral theory, it is important to distinguish the closed subspaces that are invariant only under A from those that are invariant under both A and A^*. In this context, a closed subspace $\mathcal{M}$ is called *simply invariant* if $\mathcal{M}$ is invariant under A, but not under A^*. In contrast, $\mathcal{M}$ is called *doubly invariant* or *reducing* if it is invariant under A and A^*. Note that, according to Lemma 1.39, a closed subspace $\mathcal{M}$ is doubly invariant if and only if $\mathcal{M}$ and $\mathcal{M}^\perp$ are both invariant under A. In Chapter 8, we will describe explicitly the invariant subspaces and the cyclic vectors of the bilateral and unilateral shift operators and their adjoints.

The following result concerns block-diagonal operators and their invariant subspaces.

Lemma 1.40 *Let $\mathcal{H} = \mathcal{H}_1 \oplus \mathcal{H}_2$, let $A \in \mathcal{L}(\mathcal{H})$, and assume that A admits the matrix representation*

$$A = \begin{pmatrix} A_1 & 0 \\ 0 & A_2 \end{pmatrix},$$

where $A_1 \in \mathcal{L}(\mathcal{H}_1)$ and $A_2 \in \mathcal{L}(\mathcal{H}_2)$, with respect to the decomposition $\mathcal{H} = \mathcal{H}_1 \oplus \mathcal{H}_2$. Let $\mathcal{M} = \mathcal{M}_1 \oplus \mathcal{M}_2$, where $\mathcal{M}_1 \subset \mathcal{H}_1$ and $\mathcal{M}_2 \subset \mathcal{H}_2$. Then the following assertions hold.

(i) *$\mathcal{M}$ is invariant under A if and only if $\mathcal{M}_1$ is invariant under A_1 and $\mathcal{M}_2$ is invariant under A_2.*
(ii) *$\mathcal{M}$ is simply invariant under A if and only if either $\mathcal{M}_1$ is simply invariant under A_1 or $\mathcal{M}_2$ is simply invariant under A_2.*
(iii) *$\mathcal{M}$ is reducing for A if and only if $\mathcal{M}_1$ is reducing for A_1 and $\mathcal{M}_2$ is reducing for A_2.*

Proof We start by showing that

$$AP_{\mathcal{H}_1} = P_{\mathcal{H}_1}A \quad \text{and} \quad AP_{\mathcal{H}_2} = P_{\mathcal{H}_2}A. \tag{1.48}$$

Indeed, for each $x \in \mathcal{H}$, we have

$$x = P_{\mathcal{H}_1}x + P_{\mathcal{H}_2}x$$

and thus

$$Ax = AP_{\mathcal{H}_1}x + AP_{\mathcal{H}_2}x.$$

But, we also have

$$Ax = P_{\mathcal{H}_1}Ax + P_{\mathcal{H}_2}Ax,$$

whence

$$AP_{\mathcal{H}_1}x + AP_{\mathcal{H}_2}x = P_{\mathcal{H}_1}Ax + P_{\mathcal{H}_2}Ax.$$

Since $A\mathcal{H}_1 \subset \mathcal{H}_1$ and $A\mathcal{H}_2 \subset \mathcal{H}_2$, we have $AP_{\mathcal{H}_1}x \in \mathcal{H}_1$ and $AP_{\mathcal{H}_2}x \in \mathcal{H}_2$, and because $\mathcal{H}_1 \cap \mathcal{H}_2 = \{0\}$, we get (1.48). Another method to derive (1.48) is to observe the matrix representations

$$AP_{\mathcal{H}_1} = P_{\mathcal{H}_1}A = \begin{pmatrix} A_1 & 0 \\ 0 & 0 \end{pmatrix} \quad \text{and} \quad AP_{\mathcal{H}_2} = P_{\mathcal{H}_2}A = \begin{pmatrix} 0 & 0 \\ 0 & A_2 \end{pmatrix}.$$

(i) If $\mathcal{M}$ is invariant under A, then using relations (1.48), we have

$$A_i\mathcal{M}_i = AP_{\mathcal{H}_i}\mathcal{M} = P_{\mathcal{H}_i}A\mathcal{M} \subset P_{\mathcal{H}_i}\mathcal{M} = \mathcal{M}_i,$$

for $i = 1, 2$. Conversely, if $\mathcal{M}_1$ is invariant under A_1 and $\mathcal{M}_2$ is invariant under A_2, take a vector x in $\mathcal{M}$, and write $x = x_1 \oplus x_2$, with $x_i \in \mathcal{M}_i$, $i = 1, 2$. Then

$$Ax = A_1x_1 \oplus A_2x_2 \in \mathcal{M}_1 \oplus \mathcal{M}_2 = \mathcal{M}.$$

(iii) Note that

$$A^* = \begin{pmatrix} A_1^* & 0 \\ 0 & A_2^* \end{pmatrix}.$$

Assume that $\mathcal{M}$ is a reducing subspace of A, i.e. $A\mathcal{M} \subset \mathcal{M}$ and $A^*\mathcal{M} \subset \mathcal{M}$. The assertion (i) also applies to A^*, and thus we deduce that $\mathcal{M}_1$ is reducing for A_1 and $\mathcal{M}_2$ is reducing for A_2. Reciprocally, assume that each subspace $\mathcal{M}_i$ is reducing for A_i, $i = 1, 2$. Thus, we have $A_i^*\mathcal{M}_i \subset \mathcal{M}_i$ and

$$A^*\mathcal{M} = A_1^*\mathcal{M}_1 \oplus A_2^*\mathcal{M}_2 \subset \mathcal{M}_1 \oplus \mathcal{M}_2 = \mathcal{M}.$$

Therefore, $\mathcal{M}$ is a reducing subspace for A, which proves the assertion (iii).

(ii) This follows easily from (iii), because an invariant subspace is either simply invariant or reducing. □

Exercises

Exercise 1.10.1 Let $A : \mathcal{H} \longrightarrow \mathcal{H}$ be a bounded operator, and let $\mathcal{M}$ be an invariant subspace of A. Define

$$\begin{array}{rccl} \tilde{A}: & \mathcal{M} & \longrightarrow & \mathcal{M} \\ & x & \longmapsto & Ax. \end{array}$$

In other words,

$$\tilde{A} = A_{\mathcal{M}\to\mathcal{M}} = \mathbf{P}_{\mathcal{M}} A\, \mathbf{i}_{\mathcal{M}}.$$

(i) Let $\mathcal{N}$ be a closed subspace of $\mathcal{M}$ and assume that $\mathcal{N}$ is reducing for A. Show that $\mathcal{N}$ is also reducing for $\tilde{A}$.
Hint: Use (1.36) to get

$$\tilde{A}^*\mathcal{N} = A^*\mathcal{N} \subset \mathcal{N}.$$

(ii) Let $\mathcal{N}$ be a closed subspace of $\mathcal{H}$ that is reducing for A. If $\mathcal{N} \cap A^*\mathcal{M} \subset \mathcal{M}$, then show that $\mathcal{N} \cap \mathcal{M}$ is a reducing subspace of $\tilde{A}$.
Hint: Put $\mathcal{N}' = \mathcal{M} \cap \mathcal{N}$ and use part (i).

(iii) Let $\mathcal{N}$ be a closed subspace of $\mathcal{M}$ and assume that $\mathcal{M}$ is a reducing subspace of A. Then show that $\mathcal{N}$ is reducing for A if and only if it is reducing for $\tilde{A}$.

Exercise 1.10.2 Let $A : \mathcal{H} \longrightarrow \mathcal{H}$ be a bounded operator, and let $\mathcal{M}$ be a closed subspace of $\mathcal{H}$. Show that the matrix representation of A with respect to the decomposition $\mathcal{H} = \mathcal{M} \oplus \mathcal{M}^\perp$ is block-diagonal if and only if the subspace $\mathcal{M}$ is reducing for A.
Hint: Use Lemma 1.39.

1.11 Compressions and dilations

Let $A : \mathcal{H} \longrightarrow \mathcal{H}$ be a bounded operator on the Hilbert space $\mathcal{H}$, and let $\mathcal{M}$ be a closed subspace of $\mathcal{H}$. Then consider the bounded operator $B : \mathcal{M} \longrightarrow \mathcal{M}$ defined by the formula

$$B = \mathbf{P}_{\mathcal{M}} A\, \mathbf{i}_{\mathcal{M}}.$$

In the general case, we cannot claim that $B^n = \mathbf{P}_{\mathcal{M}} A^n\, \mathbf{i}_{\mathcal{M}}$, for all integers $n \geq 0$. If it happens that

$$p(B) = \mathbf{P}_{\mathcal{M}} p(A)\, \mathbf{i}_{\mathcal{M}}$$

for all (analytic) polynomials p, we say that B is a *compression* of A on $\mathcal{M}$. We also say that A is a *dilation* of B. The following result provides a useful characterization of compressions.

Theorem 1.41 *Let $A : \mathcal{H} \longrightarrow \mathcal{H}$ be a bounded operator on the Hilbert space $\mathcal{H}$, let $\mathcal{M}$ be a closed subspace of $\mathcal{H}$, and let $B = \mathbf{P}_{\mathcal{M}} A \mathbf{i}_{\mathcal{M}}$. Then the following are equivalent.*

(i) *B is a compression of A (or A is a dilation of B).*
(ii) *There is $\mathcal{N} \in \operatorname{Lat}(A)$, $\mathcal{N} \perp \mathcal{M}$, such that*

$$\mathcal{N}' = \mathcal{M} \oplus \mathcal{N} \in \operatorname{Lat}(A).$$

(iii) *There are $\mathcal{N} \in \operatorname{Lat}(A)$ and $\mathcal{K} \in \operatorname{Lat}(A^*)$ such that*

$$\mathcal{H} = \mathcal{M} \oplus \mathcal{N} \oplus \mathcal{K}.$$

Proof (i) $\Longrightarrow$ (ii) Put

$$\mathcal{N}' = \operatorname{Span}_{\mathcal{H}}\{A^n \mathcal{M} : n \ge 0\}.$$

Surely, we have $A\mathcal{N}' \subset \mathcal{N}'$, i.e. $\mathcal{N}' \in \operatorname{Lat}(A)$. Moreover, for $x \in \mathcal{M}$ and $n \ge 0$,

$$\begin{aligned}
\mathbf{P}_{\mathcal{M}} A(A^n x) &= \mathbf{P}_{\mathcal{M}} A^{n+1} \mathbf{i}_{\mathcal{M}} x \\
&= B^{n+1} x \\
&= B B^n x \\
&= B \mathbf{P}_{\mathcal{M}} A^n \mathbf{i}_{\mathcal{M}} x \\
&= B \mathbf{P}_{\mathcal{M}} (A^n x).
\end{aligned}$$

In other words, we have the intertwining identity

$$\mathbf{P}_{\mathcal{M}} A y = B \mathbf{P}_{\mathcal{M}} y \qquad (y \in \mathcal{N}'). \tag{1.49}$$

Now, define $\mathcal{N} = \mathcal{N}' \ominus \mathcal{M}$. In particular, if we pick $y \in \mathcal{N}$ in (1.49), then we get $\mathbf{P}_{\mathcal{M}} A y = B \mathbf{P}_{\mathcal{M}} y = 0$, which implies that $Ay \in \mathcal{N}' \ominus \mathcal{M} = \mathcal{N}$. Therefore, $\mathcal{N} \in \operatorname{Lat}(A)$.

(ii) $\Longrightarrow$ (iii) Using the notation of the preceding paragraph, put $\mathcal{K} = \mathcal{N}'^{\perp}$. Since $\mathcal{N}' \in \operatorname{Lat}(A)$, we also have $\mathcal{K} \in \operatorname{Lat}(A^*)$. Moreover, the relation $\mathcal{N}' = \mathcal{N} \oplus \mathcal{M}$ implies that

$$\mathcal{H} = \mathcal{N}' \oplus \mathcal{N}'^{\perp} = \mathcal{M} \oplus \mathcal{N} \oplus \mathcal{K}.$$

(iii) $\Longrightarrow$ (i) With respect to the decomposition $\mathcal{H} = \mathcal{M} \oplus \mathcal{N} \oplus \mathcal{K}$, the operator A has the matrix representation

$$[A] = \begin{pmatrix} B & 0 & * \\ * & * & * \\ 0 & 0 & * \end{pmatrix}.$$

This is a consequence of the assumptions: $\mathcal{N} \in \text{Lat}(A)$ gives the second column; $\mathcal{K}^\perp = \mathcal{M} \oplus \mathcal{N} \in \text{Lat}(A)$ gives the third row; and $\mathbf{P}_\mathcal{M} A \mathbf{i}_\mathcal{M} = B$ gives the first column. Hence, for all $n \geq 1$, we get

$$[A^n] = \begin{pmatrix} B^n & 0 & * \\ * & * & * \\ 0 & 0 & * \end{pmatrix},$$

which gives $B^n = \mathbf{P}_\mathcal{M} A^n \mathbf{i}_\mathcal{M}$ $(n \geq 0)$. Thus, by linearity, for each analytic polynomial $p \in \mathcal{P}_+$, we have $p(B) = \mathbf{P}_\mathcal{M} p(A) \mathbf{i}_\mathcal{M}$. □

The following special case of Theorem 1.41 is important in our applications.

Corollary 1.42 *Let $A : \mathcal{H} \longrightarrow \mathcal{H}$ be a bounded operator on the Hilbert space $\mathcal{H}$, and let $\mathcal{M}$ be a closed subspace of $\mathcal{H}$ such that $\mathcal{M}^\perp \in \text{Lat}(A)$. Then $B = \mathbf{P}_\mathcal{M} A \, \mathbf{i}_\mathcal{M}$ is a compression of A.*

1.12 Angle between two subspaces

In this section, we introduce some geometric concepts regarding two linear manifolds in a Banach space. In particular, we generalize the notion of orthogonal projection, which was introduced in Section 1.7.

Let $\mathcal{M}$ and $\mathcal{N}$ be two linear manifolds of a linear space $\mathcal{X}$ with $\mathcal{M} \cap \mathcal{N} = \{0\}$. We define on the algebraic sum $\mathcal{M} + \mathcal{N}$ the linear mapping

$$\begin{array}{rccc} P_{\mathcal{M}\|\mathcal{N}} : & \mathcal{M} + \mathcal{N} & \longrightarrow & \mathcal{M} \\ & x + y & \longmapsto & x. \end{array}$$

Abusing notation, it is clear that this mapping fulfills the following properties:

(i) $P^2_{\mathcal{M}\|\mathcal{N}} = P_{\mathcal{M}\|\mathcal{N}}$;
(ii) $P_{\mathcal{M}\|\mathcal{N}}|_\mathcal{M} = I$;
(iii) $P_{\mathcal{M}\|\mathcal{N}}|_\mathcal{N} = 0$.

This mapping is called the *skew projection* onto $\mathcal{M}$ parallel to $\mathcal{N}$.

Lemma 1.43 *Let $\mathcal{M}$ and $\mathcal{N}$ be two linear manifolds in a Banach space $\mathcal{X}$ with $\mathcal{M} \cap \mathcal{N} = \{0\}$. Denote by $\bar{\mathcal{M}}$ (respectively $\bar{\mathcal{N}}$) the closure of $\mathcal{M}$ (respectively $\mathcal{N}$) in $\mathcal{X}$. Then the following statements hold.*

(i) *$P_{\mathcal{M}\|\mathcal{N}}$ is continuous if and only if so is $P_{\bar{\mathcal{M}}\|\bar{\mathcal{N}}}$.*
(ii) *If $\mathcal{M}$ and $\mathcal{N}$ are closed, then $P_{\mathcal{M}\|\mathcal{N}}$ is continuous if and only if $\mathcal{M} + \mathcal{N}$ is closed.*

Proof (i) If $P_{\bar{\mathcal{M}}\|\bar{\mathcal{N}}}$ is continuous, then trivially so is $P_{\mathcal{M}\|\mathcal{N}}$. For the converse, a simple limiting argument gives the result.

(ii) If $\mathcal{M}+\mathcal{N}$ is closed, by the closed graph theorem (Corollary 1.18), $P_{\mathcal{M}\|\mathcal{N}}$ is continuous. For the converse, if $P_{\mathcal{M}\|\mathcal{N}}$ is continuous, then

$$\|x\| \le C\|x+y\| \qquad (x \in \mathcal{M},\ y \in \mathcal{N}).$$

If $z \in \operatorname{Clos}(\mathcal{M}+\mathcal{N})$, then there is a sequence $(x_n+y_n)_{n\ge 1} \subset \mathcal{M}+\mathcal{N}$ such that $x_n+y_n \longrightarrow z$ in $\mathcal{X}$. Hence, $(x_n+y_n)_{n\ge1}$ is a Cauchy sequence. The above inequality shows that $(x_n)_{n\ge1}$ is a Cauchy sequence in $\mathcal{M}$, and thus it converges to $x \in \mathcal{M}$. In return, this implies that $(y_n)_{n\ge1}$ also converges to $y \in \mathcal{N}$. Therefore, $z = x+y \in \mathcal{M}+\mathcal{N}$. □

Assume that $\mathcal{M}$ and $\mathcal{N}$ are linear manifolds in a Hilbert space $\mathcal{H}$. Then the *angle* between $\mathcal{M}$ and $\mathcal{N}$ is the unique real number $\alpha \in [0,\pi/2]$ that satisfies the equation

$$\cos\alpha = \sup \frac{|\langle x,y\rangle|}{\|x\|\,\|y\|},$$

where x and y run respectively through $\mathcal{M}\setminus\{0\}$ and $\mathcal{N}\setminus\{0\}$. This angle is denoted by $\langle \mathcal{M},\mathcal{N}\rangle$. To specify that all the actions are taken in the Hilbert space $\mathcal{H}$, we also denote this angle by $\langle \mathcal{M},\mathcal{N}\rangle_{\mathcal{H}}$. Note that the definition is symmetric with respect to $\mathcal{M}$ and $\mathcal{N}$, i.e.

$$\langle \mathcal{M},\mathcal{N}\rangle_{\mathcal{H}} = \langle \mathcal{N},\mathcal{M}\rangle_{\mathcal{H}}.$$

In the following result, we use the following convention: if an operator T is not bounded, we take $\|T\| = \infty$.

Lemma 1.44 *Let $\mathcal{M}$ and $\mathcal{N}$ be linear manifolds in a Hilbert space $\mathcal{H}$, and let $\bar{\mathcal{M}} = \operatorname{Clos}(\mathcal{M})$ and $\bar{\mathcal{N}} = \operatorname{Clos}(\mathcal{N})$. Then the following statements hold.*

(i) $\displaystyle \cos\langle \mathcal{M},\mathcal{N}\rangle = \sup_{x\in\mathcal{M}\setminus\{0\}} \frac{\|P_{\bar{\mathcal{N}}}x\|}{\|x\|} = \sup_{x\in\mathcal{N}\setminus\{0\}} \frac{\|P_{\bar{\mathcal{M}}}x\|}{\|x\|}.$

(ii) $\cos\langle \mathcal{M},\mathcal{N}\rangle = \|P_{\bar{\mathcal{N}}}P_{\bar{\mathcal{M}}}\| = \|P_{\bar{\mathcal{M}}}P_{\bar{\mathcal{N}}}\|.$

(iii) $\displaystyle \sin\langle \mathcal{M},\mathcal{N}\rangle = \inf_{x\in\mathcal{M}\setminus\{0\}} \frac{\|(I-P_{\bar{\mathcal{N}}})x\|}{\|x\|} = \inf_{x\in\mathcal{N}\setminus\{0\}} \frac{\|(I-P_{\bar{\mathcal{M}}})x\|}{\|x\|}.$

(iv) $\sin\langle \mathcal{M},\mathcal{N}\rangle = \|P_{\mathcal{M}\|\mathcal{N}}\|^{-1} = \|P_{\mathcal{N}\|\mathcal{M}}\|^{-1}.$

Proof Since $\langle \mathcal{M},\mathcal{N}\rangle = \langle \mathcal{N},\mathcal{M}\rangle$, we prove the first identity in each part.

(i) We have

$$\begin{aligned}\cos\langle\mathcal{M},\mathcal{N}\rangle &= \sup_{x\in\mathcal{M}\setminus\{0\},\, y\in\mathcal{N}\setminus\{0\}} \frac{|\langle x,y\rangle|}{\|x\|\,\|y\|}\\ &= \sup_{x\in\mathcal{M}\setminus\{0\}} \frac{1}{\|x\|} \sup_{y\in\mathcal{N}\setminus\{0\}} \frac{|\langle x,y\rangle|}{\|y\|}\\ &= \sup_{x\in\mathcal{M}\setminus\{0\}} \frac{1}{\|x\|} \sup_{y\in\mathcal{N}\setminus\{0\}} \frac{|\langle P_{\bar{\mathcal{N}}}x,y\rangle|}{\|y\|}\\ &= \sup_{x\in\mathcal{M}\setminus\{0\}} \frac{\|P_{\bar{\mathcal{N}}}x\|}{\|x\|}.\end{aligned}$$

(ii) On the one hand, we have

$$\frac{\|P_{\bar{\mathcal{N}}}x\|}{\|x\|} = \frac{\|P_{\bar{\mathcal{N}}}P_{\bar{\mathcal{M}}}x\|}{\|x\|} \leq \|P_{\bar{\mathcal{N}}}P_{\bar{\mathcal{M}}}\| \qquad (x\in\mathcal{M}\setminus\{0\}).$$

On the other, for each $y\in\mathcal{H}$ with $y\notin\mathcal{M}^{\perp}$, we have

$$\frac{\|P_{\bar{\mathcal{N}}}P_{\bar{\mathcal{M}}}y\|}{\|y\|} \leq \frac{\|P_{\bar{\mathcal{N}}}P_{\bar{\mathcal{M}}}y\|}{\|P_{\bar{\mathcal{M}}}y\|} \leq \sup_{x\in\mathcal{M}\setminus\{0\}} \frac{\|P_{\bar{\mathcal{N}}}x\|}{\|x\|}.$$

If $y\in\mathcal{M}^{\perp}$, then the above estimation is trivial. Hence,

$$\|P_{\bar{\mathcal{N}}}P_{\bar{\mathcal{M}}}\| = \sup_{y\in\mathcal{H}\setminus\{0\}} \frac{\|P_{\bar{\mathcal{N}}}P_{\bar{\mathcal{M}}}y\|}{\|y\|} \leq \sup_{x\in\mathcal{M}\setminus\{0\}} \frac{\|P_{\bar{\mathcal{N}}}x\|}{\|x\|}.$$

(iii) Using the trigonometric identity $\sin^2\alpha+\cos^2\alpha=1$, write

$$\begin{aligned}\sin^2\langle\mathcal{M},\mathcal{N}\rangle &= 1-\cos^2\langle\mathcal{M},\mathcal{N}\rangle\\ &= 1-\sup_{x\in\mathcal{M}\setminus\{0\}} \frac{\|P_{\bar{\mathcal{N}}}x\|^2}{\|x\|^2}\\ &= \inf_{x\in\mathcal{M}\setminus\{0\}} \left(1-\frac{\|P_{\bar{\mathcal{N}}}x\|^2}{\|x\|^2}\right)\\ &= \inf_{x\in\mathcal{M}\setminus\{0\}} \frac{\|(I-P_{\bar{\mathcal{N}}})x\|^2}{\|x\|^2}.\end{aligned}$$

(iv) We have

$$\begin{aligned}\|P_{\mathcal{M}\|\mathcal{N}}\| &= \sup_{x\in\mathcal{M},\, y\in\mathcal{N},\, x+y\neq 0} \frac{\|x\|}{\|x+y\|}\\ &= \sup_{x\in\mathcal{M}\setminus\{0\},\, y\in\mathcal{N}} \frac{\|x\|}{\|x+y\|}\\ &= \sup_{x\in\mathcal{M}\setminus\{0\}} \frac{\|x\|}{\inf_{y\in\mathcal{N}}\|x+y\|}\\ &= \sup_{x\in\mathcal{M}\setminus\{0\}} \frac{\|x\|}{\inf_{y\in\bar{\mathcal{N}}}\|(x-P_{\bar{\mathcal{N}}}x)+y\|}\\ &= \sup_{x\in\mathcal{M}\setminus\{0\}} \frac{\|x\|}{\|(I-P_{\bar{\mathcal{N}}})x\|} = \frac{1}{\sin\langle\mathcal{M},\mathcal{N}\rangle}.\end{aligned}$$

This completes the proof. □

Note that a byproduct of Lemma 1.44 is

$$\|P_{\mathcal{M}\|\mathcal{N}}\| = \|P_{\mathcal{N}\|\mathcal{M}}\|. \tag{1.50}$$

Corollary 1.45 *Let $\mathcal{M}$ and $\mathcal{N}$ be linear manifolds in a Hilbert space $\mathcal{H}$, and let $\bar{\mathcal{M}} = \text{Clos}(\mathcal{M})$ and $\bar{\mathcal{N}} = \text{Clos}(\mathcal{N})$. Then the following are equivalent.*

(i) $P_{\mathcal{M}\|\mathcal{N}}$ *is continuous.*
(ii) $P_{\mathcal{N}\|\mathcal{M}}$ *is continuous.*
(iii) $\|P_{\bar{\mathcal{M}}}P_{\bar{\mathcal{N}}}\| < 1$.
(iv) $\|P_{\bar{\mathcal{N}}}P_{\bar{\mathcal{M}}}\| < 1$.
(v) $\langle\mathcal{M},\mathcal{N}\rangle_{\mathcal{H}} > 0$.

Notes on Chapter 1

For a thorough discussion of normed linear spaces see [11, 565].

Section 1.1

The name *Banach space* honors the profound contribution of Banach to normed linear spaces. His achievements were presented in his book [67].

Section 1.2

Theorems 1.3 and 1.4 are due to F. Riesz.

Section 1.3

Theorem 1.5 is due to Fejér [211].

Section 1.4

The Hahn–Banach theorem (Theorem 1.6) has its roots in the work of Helly [278]. The presented version is due to Hahn [247] and Banach [66]. The extension of positive linear forms presented in Exercise 1.4.7 is due to M. Riesz [432] in his study of a moment problem. See the book of Akhiezer [10, theorem 2.6.2, p. 69].

Section 1.5

The Baire category theorem (Theorem 1.13) was first given in [56]. For a proof see [421, p. 80]. The general version was given by Kuratowski [332] and Banach [333]. Theorem 1.19 is also known as the Banach–Steinhaus theorem. It was first presented in [68].

Section 1.6

The term "C^*-algebra" was introduced by Segal to describe norm-closed subalgebras of $\mathcal{L}(\mathcal{H})$ [474]. The letter C stands for closed. The fact that a bounded operator on an arbitrary complex Banach space has nonvoid spectrum was proved by A. E. Taylor [515]. A special case of the fact that $r(T) = \lim_{n\to+\infty} \|T^n\|^{1/n}$ was proved by Beurling [96] and the general case by Gelfand [234]. Theorem 1.22 is proved in [249] in the context of Hilbert space operators. Then it was extended in many directions for various classes of operators and functions.

Section 1.7

Corollary 1.26 is known as the Riesz–Fischer theorem. Theorem 1.24 is due to F. Riesz. The strong and weak operator topologies for bounded operators on Hilbert spaces were introduced and employed systematically by von Neumann [536]. However, the notion of strong and weak convergence of sequences of operators had been used earlier by Hilbert [285] and F. Riesz [424]. The norm operator topology was introduced by von Neumann [538]. Theorem 1.28 is due to Eberlein [205] and Šmulian [491]. Theorem 1.29 was proved in the case of finite measure by Dunford [184] and it was generalized in the σ-finite case by Dunford and Pettis [185].

Section 1.8

The concept of adjoint goes back to matrix theory and to the theories of differential and integral equations. In this context, if $\mathcal{H}_1$ and $\mathcal{H}_2$ are both finite-dimensional Hilbert spaces, every $A \in \mathcal{L}(\mathcal{H}_1, \mathcal{H}_2)$ can be represented by a matrix $[A]$, and in that case $[A^*]$ is the transpose of $[A]$. The notion of adjoint for some concrete operators on L^2 spaces was introduced by Hilbert [287] and F. Riesz [424]. The general definition was given by Schauder [468], Hildebrandt [289] and Banach [67]. Theorems 1.30 and 1.33 are due to Banach and appear in [66, 67].

Section 1.9

Lemma 1.36 is taken from [135].

Section 1.10

The study of invariant subspaces and cyclic vectors is of particular interest and gives rise to a whole direction of research. In particular, there is a famous problem, named the *invariant subspace problem*, which asks whether any bounded operator A on a separable, complex, infinite-dimensional Banach space has a nontrivial invariant subspace. This problem was solved by Enflo [207], Read [419, 420] and Beauzamy [84] in the context of Banach spaces. More precisely, these authors have constructed examples of operators acting on some Banach spaces without nontrivial invariant subspaces. However, for the time being, this problem is still open in the context of Hilbert spaces. See [130, 134] for an account on this famous problem.

Section 1.11

Theorem 1.41 is due to Sarason [445]. See Section 7.10 for more details about dilations.

Section 1.12

The concept of the angle between two subspaces of a Hilbert space has different definitions depending on the authors. The definition and the results presented here are taken from [387].

2

Some families of operators

In this chapter, we introduce several important classes of bounded operators between Hilbert spaces. More precisely, we define finite-rank, compact, self-adjoint, positive, normal and unitary operators. A particularly interesting class for us is the contractions. We also define some subsets of the spectrum of an operator, e.g. point spectrum, approximate spectrum, continuous spectrum, essential spectrum and essential left-spectrum, and discuss their relations. Then we treat several important examples, i.e. the backward and forward shift operators on the sequence space ℓ^2, the multiplication operator on $L^2(\mu)$, and the flip operator on $L^2(\mathbb{T})$. In each case, we determine the norm, the spectrum and some of its subsets. The chapter ends with two sections devoted to doubly infinite Toeplitz and Hankel matrices.

2.1 Finite-rank operators

An operator $A \in \mathcal{L}(\mathcal{H}_1, \mathcal{H}_2)$ whose range is a finite-dimensional subspace of $\mathcal{H}_2$ is called a *finite-rank operator*. The set of all finite-rank operators in $\mathcal{L}(\mathcal{H}_1, \mathcal{H}_2)$ is denoted by $\mathcal{F}(\mathcal{H}_1, \mathcal{H}_2)$. Whenever $\mathcal{H}_1 = \mathcal{H}_2 = \mathcal{H}$, we write $\mathcal{F}(\mathcal{H}) = \mathcal{F}(\mathcal{H}, \mathcal{H})$. Clearly, the rank-one operator $y \otimes x$, where $x \in \mathcal{H}_1$ and $y \in \mathcal{H}_2$ with $x \neq 0$ and $y \neq 0$, which was defined in Section 1.9, is of rank one and thus belongs to $\mathcal{F}(\mathcal{H}_1, \mathcal{H}_2)$. Therefore, the combination

$$\sum_{n=1}^{N} y_n \otimes x_n$$

gives a finite-rank operator (note that its rank is not necessarily N; it might be less than N). But, more interestingly, the inverse is also true.

Theorem 2.1 *Let $A \in \mathcal{F}(\mathcal{H}_1, \mathcal{H}_2)$. Then there are $(x_n)_{1 \leq n \leq N} \subset \mathcal{H}_1$ and $(y_n)_{1 \leq n \leq N} \subset \mathcal{H}_2$ such that*

$$A = \sum_{n=1}^{N} y_n \otimes x_n.$$

In other words, we have

$$\mathcal{F}(\mathcal{H}_1, \mathcal{H}_2) = \mathrm{Lin}\{y \otimes x : x \in \mathcal{H}_1,\ y \in \mathcal{H}_2\}. \tag{2.1}$$

Proof Since $\mathcal{R}(A)$ is finite-dimensional, Theorem 1.9 implies that $\mathcal{R}(A)$ is closed and then A establishes an isomorphism between $(\ker A)^\perp$ and $\mathcal{R}(A)$. Hence, $(\ker A)^\perp$ is finite-dimensional, say

$$(\ker A)^\perp = \mathrm{Lin}\{x_1, x_2, \dots, x_N\}.$$

Without loss of generality, assume that $\|x_k\| = 1$ $(1 \le k \le N)$. Then, for each $x \in \mathcal{H}_1$, put

$$x' = x - \sum_{n=1}^{N} \langle x, x_n \rangle_{\mathcal{H}_1} x_n = (I - P_{(\ker A)^\perp})x.$$

Therefore, $x' \in \ker A$ and we have

$$Ax = \sum_{n=1}^{N} \langle x, x_n \rangle_{\mathcal{H}_1} A x_n.$$

This relation can be rewritten as

$$A = \sum_{n=1}^{N} A x_n \otimes x_n. \qquad \square$$

Corollary 2.2 *Let $A \in \mathcal{L}(\mathcal{H}_1, \mathcal{H}_2)$. Then A is a finite-rank operator if and only if so is A^*.*

Proof This is a direct consequence of the representation given in Theorems 2.1 and 1.37. $\square$

It is rare that strong convergence leads to norm convergence. The following result is of this type and will be exploited in studying compact operators.

Theorem 2.3 *Let $A \in \mathcal{F}(\mathcal{H}_1, \mathcal{H}_2)$ and let $B \in \mathcal{L}(\mathcal{H}_2)$. Assume that $(B^n)_{n \ge 1}$ strongly tends to zero, i.e.*

$$\lim_{n \to \infty} \|B^n y\|_{\mathcal{H}_2} = 0 \qquad (y \in \mathcal{H}_2).$$

Then

$$\lim_{n \to \infty} \|B^n A\|_{\mathcal{L}(\mathcal{H}_1, \mathcal{H}_2)} = 0.$$

Proof According to Theorem 2.1, there are vectors $(x_k)_{1\le k\le K} \subset \mathcal{H}_1$ and $(y_k)_{1\le k\le K} \subset \mathcal{H}_2$ such that

$$A = \sum_{k=1}^{K} y_k \otimes x_k.$$

Hence, by Theorem 1.37,

$$B^n A = \sum_{k=1}^{K} B^n(y_k \otimes x_k) = \sum_{k=1}^{K} B^n y_k \otimes x_k,$$

and thus

$$\|B^n A\| \le \sum_{k=1}^{K} \|B^n y_k\| \times \|x_k\|.$$

By assumption, the right-hand side tends to zero as $n \longrightarrow \infty$. □

We need an equivalent variant of Theorem 2.3, which is presented below.

Corollary 2.4 *Let $A \in \mathcal{F}(\mathcal{H}_1, \mathcal{H}_2)$ and let $B \in \mathcal{L}(\mathcal{H}_1)$ be such that $(B^{*n})_{n\ge1}$ strongly tends to zero. Then*

$$\lim_{n\to\infty} \|AB^n\|_{\mathcal{L}(\mathcal{H}_1,\mathcal{H}_2)} = 0.$$

Proof Apply Theorem 2.3 to A^* and B^*. Note that $A^* \in \mathcal{F}(\mathcal{H}_2, \mathcal{H}_1)$ and

$$\|AB^n\|_{\mathcal{L}(\mathcal{H}_1,\mathcal{H}_2)} = \|B^{*n}A^*\|_{\mathcal{L}(\mathcal{H}_2,\mathcal{H}_1)}.$$ □

2.2 Compact operators

The closure of $\mathcal{F}(\mathcal{H}_1, \mathcal{H}_2)$, the family of finite-rank operators between $\mathcal{H}_1$ and $\mathcal{H}_2$, in $\mathcal{L}(\mathcal{H}_1, \mathcal{H}_2)$ is denoted by $\mathcal{K}(\mathcal{H}_1, \mathcal{H}_2)$ and its elements are called *compact operators*. In other words, an operator $K \in \mathcal{L}(\mathcal{H}_1, \mathcal{H}_2)$ is compact if and only if there exists a sequence $(A_n)_{n\ge1}$ of finite-rank operators such that

$$\lim_{n\to\infty} \|K - A_n\|_{\mathcal{L}(\mathcal{H}_1,\mathcal{H}_2)} = 0.$$

As usual, for simplicity, whenever $\mathcal{H}_1 = \mathcal{H}_2 = \mathcal{H}$, we write $\mathcal{K}(\mathcal{H}) = \mathcal{K}(\mathcal{H}, \mathcal{H})$.

Corollary 2.2 implies that $A \in \mathcal{K}(\mathcal{H}_1, \mathcal{H}_2)$ if and only if $A^* \in \mathcal{K}(\mathcal{H}_2, \mathcal{H}_1)$. Based on the above definition, it is straightforward to see that, if $A \in \mathcal{L}(\mathcal{H}_1, \mathcal{H}_2)$ and $B \in \mathcal{L}(\mathcal{H}_2, \mathcal{H}_3)$, and at least one of them is compact, then BA is also compact. Therefore, in technical language, $\mathcal{K}(\mathcal{H})$ is a norm-closed two-sided $*$-invariant ideal in $\mathcal{L}(\mathcal{H})$. Knowing this, we consider the quotient algebra $\mathcal{L}(\mathcal{H})/\mathcal{K}(\mathcal{H})$, endowed with the quotient norm

$$\|A + \mathcal{K}(\mathcal{H})\| = \inf_{K\in\mathcal{K}(\mathcal{H})} \|A + K\|_{\mathcal{L}(\mathcal{H})}.$$

In the literature, $\mathcal{L}(\mathcal{H})/\mathcal{K}(\mathcal{H})$ is called the *Calkin algebra*, which is a unital Banach algebra. We use π for the canonical projection

$$\begin{array}{rrcl}\pi : & \mathcal{L}(\mathcal{H}) & \longrightarrow & \mathcal{L}(\mathcal{H})/\mathcal{K}(\mathcal{H}) \\ & A & \longmapsto & A+\mathcal{K}(\mathcal{H}).\end{array}$$

One may wonder why elements of $\mathcal{K}(\mathcal{H}_1,\mathcal{H}_2)$ are called compact operators. The following useful characterization explains the connection with compactness.

Theorem 2.5 *Let $A \in \mathcal{L}(\mathcal{H}_1,\mathcal{H}_2)$. Then the following assertions are equivalent.*

(i) *A is compact.*
(ii) *A maps the closure of any bounded subset of $\mathcal{H}_1$ into a relatively compact subset of $\mathcal{H}_2$.*
(iii) *A maps the closed unit ball of $\mathcal{H}_1$ into a relatively compact subset of $\mathcal{H}_2$.*
(iv) *A transforms each weakly convergent sequence to a strongly convergent sequence, i.e.*

$$x_n \xrightarrow{w} x \ (in\ \mathcal{H}_1) \quad \Longrightarrow \quad Ax_n \longrightarrow Ax \ (in\ \mathcal{H}_2).$$

(v) *For any orthonormal sequence $(e_n)_{n\geq 1}$ in $\mathcal{H}_1$, we have*

$$\lim_{n\to\infty} \|Ae_n\|_{\mathcal{H}_2} = 0.$$

The spectral theory of compact operators is now well understood and goes back to the work of F. Riesz around 1910. The key point of the theory is the result of F. Riesz that asserts that the closed unit ball of a normed linear space $\mathcal{X}$ is compact if and only if $\mathcal{X}$ is of finite dimension. Using this result, one can prove what is now called the *Fredholm alternative*.

Theorem 2.6 *Let $\mathcal{H}$ be a Hilbert space and let $K \in \mathcal{K}(\mathcal{H})$. Then the following hold:*

(i) *$\mathcal{R}(I-K)$ is closed; and*
(ii) $\dim\ker(I-K) = \dim\ker(I-K)^* < \infty$.

Using the Fredholm alternative, one can give a useful description of the spectrum of a compact operator $K \in \mathcal{K}(\mathcal{H})$: either $\sigma(K)$ is finite, or $\sigma(K) = \{0\}\bigcup\{\lambda_n : n \geq 1\}$, where λ_n is a sequence of distinct complex numbers tending to zero. Moreover, if $\lambda \in \sigma(K)\setminus\{0\}$, then λ is an eigenvalue of K of finite multiplicity, that is

$$0 < \dim\ker(K-\lambda I) < \infty.$$

All the properties on compact operators mentioned above are classic and can be found in any textbook on functional analysis. We refer the reader to [118], [159], [176] and [421].

We now give a generalization of Corollary 2.4, which will be exploited in studying Hankel operators.

Theorem 2.7 *Let $A \in \mathcal{K}(\mathcal{H}_1, \mathcal{H}_2)$, and let $B \in \mathcal{L}(\mathcal{H}_1)$ be a contraction (i.e. $\|B\| \leq 1$) such that*

$$\lim_{n\to\infty} \|B^{*n}x\|_{\mathcal{H}_1} = 0 \qquad (x \in \mathcal{H}_1).$$

Then we have

$$\lim_{n\to\infty} \|AB^n\|_{\mathcal{L}(\mathcal{H}_1,\mathcal{H}_2)} = 0.$$

Proof Given $\varepsilon > 0$, there is a finite-rank operator $C \in \mathcal{F}(\mathcal{H}_1, \mathcal{H}_2)$ such that

$$\|A - C\| < \varepsilon.$$

Hence,

$$\begin{aligned} \|AB^n\| &\leq \|(A - C)B^n\| + \|CB^n\| \\ &\leq \|A - C\| \times \|B^n\| + \|CB^n\| \\ &\leq \varepsilon + \|CB^n\|. \end{aligned}$$

Therefore, by Corollary 2.4,

$$\limsup_{n\to\infty} \|AB^n\| \leq \varepsilon. \qquad \square$$

Exercises

Exercise 2.2.1 Let $\mathcal{H}$ be a Hilbert space and let $(e_n)_{n\geq1}$ be an orthonormal basis of $\mathcal{H}$. Let $(\lambda_n)_{n\geq1}$ be a bounded sequence of complex numbers and let A be the linear map defined on the linear span of $(e_n)_{n\geq1}$ by

$$Ae_n = \lambda_n e_n \qquad (n \geq 1).$$

(i) Show that A extends into a bounded operator on $\mathcal{H}$.
(ii) Show that A is compact on $\mathcal{H}$ if and only if $\lim_{n\to\infty} \lambda_n = 0$.

Hint: Use Theorem 2.5.

Exercise 2.2.2 Let $\mathcal{H}$ be a Hilbert space and let $K \in \mathcal{K}(\mathcal{H})$. Show that if $\mathcal{R}(K)$ is closed then K is a finite-rank operator.
Hint: The restricted operator $K_{(\ker K)^\perp \to \mathcal{R}(K)}$ is invertible and compact. Hence, the identity operator on $\mathcal{R}(K)$ is compact.

Exercise 2.2.3 Let $\mathcal{H}$ be a separable Hilbert space and let $(e_n)_{n\geq 1}$ be an orthonormal basis of $\mathcal{H}$. We say that $A \in \mathcal{L}(\mathcal{H})$ is a *Hilbert–Schmidt operator* if

$$\sum_{n=1}^{\infty} \|Ae_n\|^2 < \infty.$$

We denote the set of Hilbert–Schmidt operators on $\mathcal{H}$ by $\mathcal{C}_2(\mathcal{H})$. Show that the following assertions hold.

(i) The quantity $\sum_{n=1}^{\infty} \|Ae_n\|^2$ is independent of the choice of the basis $(e_n)_{n\geq 1}$.
Knowing this fact, put

$$\|A\|_{\mathcal{C}_2} = \left(\sum_{n=1}^{\infty} \|Ae_n\|^2 \right)^{1/2}.$$

(ii) $\mathcal{C}_2(\mathcal{H})$, equipped with $\|\cdot\|_{\mathcal{C}_2}$, is a Banach space.
(iii) $\mathcal{C}_2(\mathcal{H})$ is a subset of $\mathcal{K}(\mathcal{H})$.
(iv) $\mathcal{C}_2(\mathcal{H})$ is a two-sided ideal of $\mathcal{L}(\mathcal{H})$.

Exercise 2.2.4 Let $k \in L^2([0,1] \times [0,1])$ and let

$$Af(x) = \int_0^1 k(x,y) f(y)\, dy \qquad (x \in (0,1),\ f \in L^2(0,1)).$$

First, show that A is in $\mathcal{L}(L^2(0,1))$. Second, verify that A is a Hilbert–Schmidt operator.

Exercise 2.2.5 Let $\mathcal{H}_1$ and $\mathcal{H}_2$ be Hilbert spaces, and let $A \in \mathcal{L}(\mathcal{H}_1, \mathcal{H}_2)$, $A \neq 0$, be a compact operator. Show that A has a maximizing vector, i.e. there exists a vector $x \in \mathcal{H}_1$, $\|x\|_1 = 1$, such that $\|Ax\|_2 = \|A\|$.
Hint: Use Theorem 2.5.

2.3 Subdivisions of spectrum

In the finite-dimensional case, an operator (or a matrix) is invertible if and only if it is injective. This is also equivalent to saying that it is onto. In the infinite-dimensional case, this is no longer true, and there are several (different) reasons that an operator might fail to be invertible. Some such possibilities are listed below.

(i) The operator is not injective. This is equivalent to saying that the kernel is not equal to $\{0\}$.

(ii) The operator is not lower bounded. This is equivalent to saying that the kernel is not equal to $\{0\}$ or the range of A is not closed (see Corollary 1.17).

(iii) Its projection under the mapping $\pi : \mathcal{L}(\mathcal{H}) \longrightarrow \mathcal{L}(\mathcal{H})/\mathcal{K}(\mathcal{H})$ is not invertible.

Based on these cases, we respectively define some components of the spectrum as follows. All these components are in general distinct from the whole spectrum and they are also distinct from each other.

(i) The *point spectrum* $\sigma_p(A)$ is the set of all $\lambda \in \mathbb{C}$ such that $\lambda I - A$ is not injective. Equivalently, this means that there is an $x \in \mathcal{H}$, $x \neq 0$, such that

$$Ax = \lambda x.$$

In this case, λ is called an *eigenvalue* and x is its corresponding *eigenvector*. Hence, $\sigma_p(A)$ is precisely the set of all eigenvalues of A.

(ii) The *approximate spectrum* $\sigma_a(A)$ is the set of all $\lambda \in \mathbb{C}$ such that $\lambda I - A$ is not lower bounded, i.e.

$$\inf_{\substack{x \in \mathcal{H} \\ \|x\|=1}} \|(\lambda I - A)x\| = 0.$$

Clearly, this is equivalent to saying that there is a sequence $(x_n)_{n\geq 1} \subset \mathcal{H}$, with $\|x_n\| = 1$, such that

$$\lim_{n\to\infty} \|(\lambda I - A)x_n\| = 0.$$

A sequence is called *noncompact* if it has no convergent subsequences. As a part of $\sigma_a(A)$, the *continuous spectrum* $\sigma_c(A)$ is the set of all $\lambda \in \mathbb{C}$ such that there is a noncompact sequence $(x_n)_{n\geq 1}$, with $\|x_n\| = 1$, such that

$$\lim_{n\to\infty} \|(\lambda I - A)x_n\| = 0.$$

(iii) The *essential spectrum* $\sigma_{ess}(A)$ is the set of all $\lambda \in \mathbb{C}$ such that $\pi(\lambda I - A)$ is not invertible in $\mathcal{L}(\mathcal{H})/\mathcal{K}(\mathcal{H})$. Hence, we have

$$\sigma_{ess}(A) = \sigma(\pi(A)),$$

where the right-hand side spectrum is for an operator in $\mathcal{L}(\mathcal{H})/\mathcal{K}(\mathcal{H})$. Naively speaking, $\sigma_{ess}(A)$ is the set of all the λ such that, modulo compact operators, $\lambda I - A$ is not invertible.

In the above definition, the main concentration was on the "invertibility" of $\lambda I - A$. We can have parallel definitions if we replace "invertible" by "left-invertible". As an example, we define the left-spectrum $\sigma^{\ell}(A)$ as the set of all

$\lambda \in \mathbb{C}$ such that $\lambda I - A$ is not left-invertible in $\mathcal{L}(\mathcal{H})$. But, since an operator is bounded below if and only if it is left-invertible (see Corollary 1.34), we deduce that

$$\sigma^{\ell}(A) = \sigma_a(A). \tag{2.2}$$

An important consequence of this formula is that the approximate spectrum $\sigma_a(A)$ is a closed subset of $\mathbb{C}$, because the set of left-invertible operators forms an open subset in $\mathcal{L}(\mathcal{H})$. Moreover, we certainly have

$$\sigma_p(A) \subset \sigma_a(A) \subset \sigma(A). \tag{2.3}$$

But, in general, the above inclusions are proper. Similarly, parallel to Theorem 1.30, simple computations show that

$$\sigma_{ess}(A^*) = \overline{\sigma_{ess}(A)}, \tag{2.4}$$

where the bar stands for complex conjugation.

As another important situation, the *essential left-spectrum* $\sigma^{\ell}_{ess}(A)$ is the set of all $\lambda \in \mathbb{C}$ such that the projection $\pi(\lambda I - A)$ is not left-invertible in $\mathcal{L}(\mathcal{H})/\mathcal{K}(\mathcal{H})$.

The following result reveals the relation between different parts of the spectrum. As usual, the boundary of a subset $E \subset \mathbb{C}$ is denoted by ∂E.

Theorem 2.8 *Let $A \in \mathcal{L}(\mathcal{H})$. Then*

$$\partial\sigma(A) \subset \sigma_a(A)$$

and

$$\sigma_a(A) = \sigma_p(A) \cup \sigma_c(A),$$

with $\sigma_p(A) \cap \sigma_c(A) = \emptyset$ and

$$\sigma_c(A) \subset \sigma^{\ell}_{ess}(A) \subset \sigma_{ess}(A).$$

Proof To show that $\partial\sigma(A) \subset \sigma_a(A)$, take any $\lambda \in \partial\sigma(A)$. Hence, there is a sequence $(\lambda_n)_{n\geq 1} \subset \rho(A)$ that converges to λ. In particular, $\text{dist}(\lambda_n, \sigma(A)) \longrightarrow 0$, as $n \longrightarrow \infty$. But, by (1.11), we have

$$\|R_{\lambda_n}(A)\| \geq \frac{1}{\text{dist}(\lambda_n, \sigma(A))},$$

which implies that $\|R_{\lambda_n}(A)\| \longrightarrow \infty$. Therefore, there is a sequence $(x_n)_{n\geq 1} \subset \mathcal{H}$, $\|x_n\| = 1$, such that

$$\lim_{n\to\infty} \|R_{\lambda_n}(A)x_n\| = \infty.$$

Put

$$y_n = \frac{R_{\lambda_n}(A)x_n}{\|R_{\lambda_n}(A)x_n\|}.$$

We have $\|y_n\| = 1$ and

$$(A - \lambda_n I)y_n = \frac{x_n}{\|R_{\lambda_n}(A)x_n\|},$$

which implies that

$$\lim_{n\to\infty} \|(A - \lambda_n I)y_n\| = 0.$$

Now, note that

$$\begin{aligned}\|(A - \lambda I)y_n\| &= \|(A - \lambda_n I)y_n + (\lambda_n - \lambda)y_n\| \\ &\leq \|(A - \lambda_n I)y_n\| + |\lambda_n - \lambda|,\end{aligned}$$

and we get $\lim_{n\to\infty} \|(\lambda I - A)y_n\| = 0$. This proves $\lambda \in \sigma_a(A)$.

To show $\sigma_a(A) = \sigma_p(A) \cup \sigma_c(A)$, first note that the inclusion $\sigma_p(A) \cup \sigma_c(A) \subset \sigma_a(A)$ follows immediately from the definitions of these objects. To prove the inverse inclusion, let $\lambda \in \sigma_a(A)$. Then there exists a sequence $(x_n)_{n\geq 1}$ in $\mathcal{H}$, $\|x_n\| = 1$, such that $\|(\lambda I - A)x_n\| \longrightarrow 0$, as $n \longrightarrow \infty$. If $(x_n)_{n\geq 1}$ has no convergent subsequence, then, again by definition, $\lambda \in \sigma_c(A)$. Otherwise, there is a subsequence $(x_{n_k})_{k\geq 1}$ that converges, say, to $x \in \mathcal{H}$. In particular, we have

$$0 = \lim_{k\to\infty} ((\lambda I - A)x_{n_k}) = (\lambda I - A)x,$$

which means that $x \in \ker(\lambda I - A)$. Moreover,

$$\|x\| = \lim_{k\to\infty} \|x_{n_k}\| = 1.$$

Thus, $x \neq 0$ and we conclude that $\lambda \in \sigma_p(A)$. That $\sigma_p(A) \cap \sigma_c(A) = \emptyset$ follows from the definitions of $\sigma_p(A)$ and $\sigma_c(A)$.

The inclusion $\sigma_{ess}^{\ell}(A) \subset \sigma_{ess}(A)$ is trivial. To show that $\sigma_c(A) \subset \sigma_{ess}^{\ell}(A)$, take $\lambda \in \sigma_c(A)$ and assume that $\lambda \notin \sigma_{ess}^{\ell}(A)$. Hence, on the one hand, there exists an operator $B \in \mathcal{L}(\mathcal{H})$ such that $\pi(B)\,\pi(\lambda I - A) = \pi(I)$. This is equivalent to saying that the operator

$$K = B(\lambda I - A) - I \tag{2.5}$$

is compact. On the other hand, there exists a noncompact sequence $(x_n)_{n\geq 1}$ such that $\|x_n\| = 1$ and $\lim_{n\to\infty} \|(\lambda I - A)x_n\| = 0$. Thus, by (2.5), we get

$$\lim_{n\to\infty} \|x_n + Kx_n\| = 0.$$

By Theorem 1.27, $(x_n)_{n\geq 1}$ has a weakly convergent subsequence. Thus, since K is compact, there is a subsequence $(x_{n_j})_{j\geq 1}$ such that $(Kx_{n_j})_{j\geq 1}$ is norm-convergent, say, to $y \in \mathcal{H}$. Then

$$\lim_{j\to\infty} x_{n_j} = \lim_{j\to\infty} ((x_{n_j} + Kx_{n_j}) - Kx_{n_j}) = -y,$$

which is a contradiction. Therefore, $\sigma_c(A) \subset \sigma_{ess}^{\ell}(A)$. □

The following result is an immediate consequence of Theorem 2.8.

Corollary 2.9 *Let $A \in \mathcal{L}(\mathcal{H})$ be such that $\sigma_p(A) = \emptyset$. Then*

$$\partial\sigma(A) \subset \sigma_a(A) = \sigma_c(A) \subset \sigma_{ess}(A).$$

Another useful relation between different parts of the spectrum is the following result.

Theorem 2.10 *Let $A \in \mathcal{L}(\mathcal{H})$. Then*

$$\sigma(A) = \sigma_a(A) \cup \overline{\sigma_p(A^*)}.$$

(The bar stands for complex conjugation.)

Proof Let $\lambda \in \sigma(A) \setminus \sigma_a(A)$. Then the operator $A - \lambda I$ is bounded below but is not invertible. In particular, its range cannot be dense in $\mathcal{H}$. Hence,

$$\ker(A^* - \bar{\lambda} I) = (\mathcal{R}(A - \lambda I))^{\perp} \neq \{0\}.$$

In other words, $\bar{\lambda} \in \sigma_p(A^*)$. For the converse, it is sufficient to note that

$$\overline{\sigma_p(A^*)} \subset \overline{\sigma(A^*)} = \sigma(A). \qquad \square$$

We will encounter the decomposition of a spectrum and a point spectrum for operators that are block-diagonal. To be more precise, let $\mathcal{H} = \mathcal{H}_1 \oplus \mathcal{H}_2$, $\mathcal{H}' = \mathcal{H}_1' \oplus \mathcal{H}_2'$, $A_1 \in \mathcal{L}(\mathcal{H}_1, \mathcal{H}_1')$ and $A_2 \in \mathcal{L}(H_2, \mathcal{H}_2')$. Let $A \in \mathcal{L}(\mathcal{H}, \mathcal{H}')$ be such that, relative to the preceding decompositions, it admits the matrix representation

$$A = \begin{pmatrix} A_1 & 0 \\ 0 & A_2 \end{pmatrix}.$$

Then, for each $\lambda \in \mathbb{C}$, we have

$$\ker(A - \lambda I) = \ker(A_1 - \lambda I) \oplus \ker(A_2 - \lambda I), \tag{2.6}$$

whence we get

$$\sigma_p(A) = \sigma_p(A_1) \cup \sigma_p(A_2). \tag{2.7}$$

Moreover,

$$\sigma(A) = \sigma(A_1) \cup \sigma(A_2) \quad \text{and} \quad \sigma_a(A) = \sigma_a(A_1) \cup \sigma_a(A_2). \tag{2.8}$$

Exercise

Exercise 2.3.1 Show that the set difference $\sigma(A) \setminus \sigma_a(A)$, if not empty, precisely contains all $\lambda \in \mathbb{C}$ such that $\mathcal{R}(\lambda I - A)$ is a proper closed subspace of $\mathcal{H}$, and that $\lambda I - A$, as an operator between $\mathcal{H}$ and $\mathcal{R}(\lambda I - A)$, is invertible.

2.4 Self-adjoint operators

An operator $A \in \mathcal{L}(\mathcal{H})$ is called *self-adjoint* if $A = A^*$. To a certain extent, the role played by real numbers among complex numbers is similarly played by self-adjoint operators in the family of all bounded operators on $\mathcal{H}$. Using the polarization identity, it is easy to see that

$$A = A^* \quad \Longleftrightarrow \quad \langle Ax, x\rangle \in \mathbb{R} \qquad (x \in \mathcal{H}). \tag{2.9}$$

In other words, A is self-adjoint if and only if $\langle Ax, x\rangle$ is real for all $x \in \mathcal{H}$. If A is self-adjoint, then it follows easily from Corollary 1.31 that A is invertible if and only if it is bounded below. This fact is exploited below in the proof of Theorem 2.13.

Lemma 2.11 *Let $A \in \mathcal{L}(\mathcal{H})$. If A is self-adjoint, then*

$$\sigma(A) \subset \mathbb{R}.$$

Proof Let $\lambda = \alpha + i\beta \in \mathbb{C}$ with $\alpha, \beta \in \mathbb{R}$ and $\beta \neq 0$. Then

$$\begin{aligned}|\langle (A-\lambda I)x,\ x\rangle|^2 &= |\langle (A-\alpha I)x,\ x\rangle - i\beta\|x\|^2|^2\\ &= |\langle (A-\alpha I)x,\ x\rangle|^2 + \beta^2\|x\|^4,\end{aligned}$$

which implies that

$$|\langle (A-\lambda I)x,\ x\rangle| \geq |\beta|\,\|x\|^2.$$

Thus, by the Cauchy–Schwarz inequality, $\|(A-\lambda I)x\| \geq |\beta|\,\|x\|$, for all $x \in \mathcal{H}$. Hence, $A-\lambda I$ is bounded below. But, since $A = A^*$, by the same reasoning the operator $(A-\lambda I)^*$ is also bounded below. Hence, by Corollary 1.31, $A-\lambda I$ is invertible. Therefore, we conclude that $\sigma(A) \subset \mathbb{R}$. □

An operator $A \in \mathcal{L}(\mathcal{H})$ is called *positive* if

$$\langle Ax, x\rangle \geq 0 \qquad (x \in \mathcal{H}).$$

By (1.27), an orthogonal projection is a positive operator. By (2.9), a positive operator is self-adjoint.

Lemma 2.12 *Let $A \in \mathcal{L}(\mathcal{H})$ be positive. Then*

$$|\langle Ax, y\rangle|^2 \leq \langle Ax, x\rangle\,\langle Ay, y\rangle \qquad (x, y \in \mathcal{H})$$

and

$$\|Ax\|^2 \leq \|A\|\,\langle Ax, x\rangle \qquad (x \in \mathcal{H}).$$

In particular,

$$Ax \perp x \quad \Longrightarrow \quad Ax = 0 \qquad (x \in \mathcal{H}).$$

Proof Define

$$\phi(x, y) = \langle Ax, y\rangle \qquad (x, y \in \mathcal{H}).$$

Since A is positive, ϕ is a positive bilinear form and thus the Cauchy–Schwarz inequality holds for ϕ, i.e.

$$|\phi(x, y)|^2 \leq \phi(x, x)\,\phi(y, y) \qquad (x, y \in \mathcal{H}).$$

This is exactly the first inequality. Moreover, $\phi(y, y) = \langle Ay, y\rangle \leq \|A\|\,\|y\|^2$. Hence

$$|\langle Ax, y\rangle|^2 \leq \|A\|\,\langle Ax, x\rangle\,\|y\|^2 \qquad (x, y \in \mathcal{H}).$$

Take $y = Ax$ to obtain the second inequality. The last assertion is an immediate consequence of this inequality. □

For a self-adjoint operator A, we define

$$m(A) = \inf_{\substack{x\in\mathcal{H}\\ \|x\|=1}} \langle Ax, x\rangle$$

and

$$M(A) = \sup_{\substack{x\in\mathcal{H}\\ \|x\|=1}} \langle Ax, x\rangle.$$

Based on the above two notions, we can improve Lemma 2.11.

Theorem 2.13 *Let $A \in \mathcal{L}(\mathcal{H})$ be self-adjoint. Then*

$$\sigma(A) \subset [m(A), M(A)].$$

Moreover,

$$m(A) \in \sigma(A) \quad \textit{and} \quad M(A) \in \sigma(A).$$

In other words, $[m(A), M(A)]$ is the smallest closed interval in $\mathbb{R}$ that contains $\sigma(A)$.

Proof By Lemma 2.11, $\sigma(A) \subset \mathbb{R}$. Suppose that $\lambda \in \mathbb{R} \setminus [m(A), M(A)]$. Hence,

$$d = \inf_{m(A)\leq t\leq M(A)} |\lambda - t| > 0.$$

Therefore, for each $x \in \mathcal{H}$, $x \neq 0$,

$$\begin{aligned}|\langle(A - \lambda I)x,\ x\rangle| &= |\langle Ax, x\rangle - \lambda\langle x, x\rangle|\\ &= \|x\|^2 \left|\lambda - \left\langle A\frac{x}{\|x\|}, \frac{x}{\|x\|}\right\rangle\right| \geq d\,\|x\|^2.\end{aligned}$$

Therefore, by the Cauchy–Schwarz inequality,

$$\|(A - \lambda I)x\| \geq d\,\|x\| \qquad (x \in \mathcal{H}).$$

But, since λ is real, the operator $A - \lambda I$ is self-adjoint and bounded below. Hence, we know that it must be invertible, which means that $\lambda \notin \sigma(A)$. We have thus proved that $\sigma(A) \subset [m(A), M(A)]$.

Now, according to the definition of $m(A)$, the operator $B = A - m(A)I$ is positive and there is a sequence $(x_n)_{n\geq 1} \subset \mathcal{H}$ such that $\|x_n\| = 1$ and $\langle Bx_n, x_n\rangle \longrightarrow 0$. But, by Lemma 2.12,

$$\|Bx_n\|^2 \leq \|B\| \, \langle Bx_n, x_n\rangle,$$

and thus $\|Bx_n\| \longrightarrow 0$. This fact implies that B is not invertible, since otherwise we would have $x_n = B^{-1}Bx_n \longrightarrow 0$. Therefore, $m(A) \in \sigma(A)$. Similarly, by considering $M(A)I - A$, we conclude that $M(A) \in \sigma(A)$. □

The preceding theorem shows that $[m(A), M(A)]$ is the smallest closed interval that contains $\sigma(A)$. But, in general, the spectrum is a proper subset of $[m(A), M(A)]$, which always contains $m(A)$ and $M(A)$. It is indeed easy to construct a self-adjoint operator A for which $\sigma(A) = \{m(A), M(A)\}$.

Corollary 2.14 *If $A \in \mathcal{L}(\mathcal{H})$ is self-adjoint, then*

$$r(A) = \|A\| = \max\{\, |m(A)|, \, |M(A)| \,\} = \sup_{\substack{x\in\mathcal{H} \\ \|x\|=1}} |\langle Ax, x\rangle|.$$

Proof The identity

$$r(A) = \max\{\, |m(A)|, \, |M(A)| \,\}$$

is an immediate consequence of Theorem 2.13. To prove $r(A) = \|A\|$, note that, by Theorem 1.30, $\|A^2\| = \|AA^*\| = \|A\|^2$. Hence, by induction,

$$\|A^{2^n}\| = \|A\|^{2^n} \qquad (n \geq 1).$$

But, by the spectral radius theorem (Theorem 1.23), $\|A^{2^n}\|^{1/2^n} \longrightarrow r(A)$. Therefore, $r(A) = \|A\|$. The last identity is a consequence of the definitions of $m(A)$ and $M(A)$. □

According to Theorem 2.13, a positive operator A satisfies

$$\sigma(A) \subset [0, +\infty).$$

In fact, we can explore further and obtain the following useful characterization of positive operators.

Corollary 2.15 *Let $A \in \mathcal{L}(\mathcal{H})$. Then the following are equivalent.*

(i) *A is a positive operator.*
(ii) *A is self-adjoint and $\sigma(A) \subset [0, +\infty)$.*

Proof (i) $\Longrightarrow$ (ii) This follows from (2.9) and Theorem 2.13.

(ii) $\Longrightarrow$ (i) By Theorem 2.13, we must have $m(A) \geq 0$. This means that

$$\langle Ax, x\rangle \geq m(A) \geq 0$$

for all $x \in \mathcal{H}$. Hence, A is positive. □

One should be careful in applying the above result. Generally speaking, a mere assumption on the spectrum is not enough to make the conclusion. For example, there is an operator A with $\sigma(A) = \{0\}$, but A is not even self-adjoint, and hence not positive.

The notion of positivity enables us to define a partial ordering on the set of self-adjoint operators. Given two self-adjoint operators A and B, we write $A \leq B$ (or $B \geq A$) if $B - A$ is a positive operator, that is

$$\langle Ax, x\rangle \leq \langle Bx, x\rangle \qquad (z \in \mathcal{H}).$$

In particular, for a positive operator A, we write $A \geq 0$. The following properties are easy to establish:

(i) If $\alpha \geq 0$ and $A \leq B$, then $\alpha A \leq \alpha B$.
(ii) If $A \leq B$ and $B \leq C$, then $A \leq C$.
(iii) If $A \leq B$ and C is self-adjoint, then $A + C \leq B + C$.
(iv) If $A \leq B$ and $B \leq A$, then $A = B$.
(v) If $A \leq B$ and $C \in \mathcal{L}(\mathcal{H})$, then $C^*AC \leq C^*BC$. In particular, if $B \geq 0$, then for any $C \in \mathcal{L}(\mathcal{H})$, we have $C^*BC \geq 0$.

For each $A \in \mathcal{L}(\mathcal{H})$, clearly $A^*A \geq 0$. Moreover, A is bounded below if and only if there is $\delta > 0$ such that $A^*A \geq \delta I$. The simpler relation $A \geq \alpha I$ means that

$$\langle Ax, x\rangle \geq \langle \alpha Ix, x\rangle = \alpha\|x\|^2 \qquad (x \in \mathcal{H}).$$

By the Cauchy–Schwarz inequality, $|\langle Ax, x\rangle| \leq \|Ax\|\,\|x\|$, and thus we obtain

$$A \geq \alpha I \quad \Longrightarrow \quad \|Ax\| \geq \alpha\|x\| \qquad (x \in \mathcal{H}). \tag{2.10}$$

In particular, the condition $A \geq \alpha I$ implies that A is bounded below. We will see that the converse is also true for positive operators (see Corollary 7.9).

We end this section by an interesting property of positive semidefinite matrices. First recall that if $A = (a_{i,j})_{1\leq i,j\leq n}$ and $B = (b_{i,j})_{1\leq i,j\leq n}$ are two $n \times n$ complex matrices, the *Hadamard product* of A and B, denoted by $A \circ B$, is defined by

$$A \circ B = (a_{i,j}b_{i,j})_{1\leq i,j\leq n}.$$

Then, the Schur product theorem says that

$$A \geq 0, B \geq 0 \quad \Longrightarrow \quad A \circ B \geq 0.$$

For a proof of this classic result, see [292].

Exercises

Exercise 2.4.1 Let $\mathcal{H}$ be a *real* Hilbert space, and let $A \in \mathcal{L}(\mathcal{H})$ be positive. Suppose that $\langle Ax, x \rangle = 0$ for all $x \in \mathcal{H}$. Show that $A = 0$.
Hint: Lemma 2.12 is also valid for real Hilbert spaces.
Remark: Compare with Exercise 1.7.4.

Exercise 2.4.2 Let $A, B \in \mathcal{L}(\mathcal{H})$. Suppose that $A \le B$ and $B \le A$. Show that $A = B$.
Hint: Lemma 2.12 can be useful.

Exercise 2.4.3 Let $A, B, C \in \mathcal{L}(\mathcal{H})$ be self-adjoint and $A \le B \le C$. Show that

$$\|B\| \le \max\{\, \|A\|, \|C\| \,\}.$$

Hint: Use Corollary 2.14.

Exercise 2.4.4 The *numerical range* of an operator $A \in \mathcal{L}(\mathcal{H})$ is defined by

$$W(A) = \mathrm{Clos}_{\mathbb{C}}\{\, \langle Ax, x \rangle : x \in \mathcal{H},\ \|x\| = 1 \,\}.$$

Show that $W(A)$ is a convex subset of $\mathbb{C}$.
Hint: Study the action of A on a two-dimensional subspace of $\mathcal{H}$.

Exercise 2.4.5 Let $A \in \mathcal{L}(\mathcal{H})$ be self-adjoint. Show that the following are equivalent:

(i) $W(A) \subset [-1, 1]$;
(ii) $\sigma(A) \subset [-1, 1]$;
(iii) $\|A\| \le 1$;
(iv) $-I \le A \le I$;
(v) $A^2 \le I$.

Exercise 2.4.6 Let $A_1, A_2, \dots, B \in \mathcal{L}(\mathcal{H})$ be self-adjoint operators. Suppose that

$$A_n \le A_{n+1} \le B \qquad (n \ge 1).$$

Our goal is to show that there is a self-adjoint operator $A \in \mathcal{L}(\mathcal{H})$ such that

$$Ax = \lim_{n\to\infty} A_n x \qquad (x \in \mathcal{H}).$$

(i) Show that we can assume that $A_1 \ge 0$ and $B = I$.
For $m < n$, put $A_{m,n} = A_n - A_m$.

(ii) Show that $0 \leq A_{m,n} \leq I$.

(iii) Deduce that $\|A_{m,n}\| \leq 1$.
Hint: Use Corollary 2.14.

(iv) Show that

$$\|A_{m,n}x\|^4 \leq \langle A_{m,n}x, x\rangle \langle A_{m,n}^2 x, A_{m,n}x\rangle \qquad (x \in \mathcal{H}).$$

Hint: Use Lemma 2.12.

(v) Prove that

$$\|A_m x - A_n x\|^4 \leq (\langle A_n x, x\rangle - \langle A_m x, x\rangle)\|x\|^2.$$

(vi) Use (v) to deduce that $A_n x$ is norm convergent.

(vii) Put $Ax = \lim_{n\to\infty} A_n x$. Show that A is a self-adjoint operator.

Exercise 2.4.7 Let $A, B \in \mathcal{L}(\mathcal{H})$ be positive and $AB = BA$. We want to show that AB is positive.

(i) Check that, if $TS = ST$, T and S positive, then TS^2 is also positive.

(ii) Show that we can assume that $\|B\| \leq 1$.
Let $B_0 = B$ and

$$B_{n+1} = B_n - B_n^2 \qquad (n \geq 0).$$

(iii) Show that

$$B_{n+1} = (I - B_n)B_n^2 + B_n(I - B_n)^2$$

and

$$I - B_{n+1} = (I - B_n) + B_n^2.$$

(iv) Deduce that $0 \leq B_n \leq I$, $n \geq 0$.

(v) Show that

$$\sum_{k=0}^{n} B_k^2 \leq B \qquad (n \geq 0).$$

(vi) Deduce that $\lim_{n\to\infty} B_n x = 0$ and

$$\lim_{n\to\infty} \left(\sum_{k=0}^{n} B_k^2\right) x = Bx.$$

(vii) Conclude that AB is positive.

Exercise 2.4.8 Let A be a positive operator. We want to show that there is a positive operator B such that $B^2 = A$. First, show that we can assume that $A \leq I$. Then put $C = I - A$ and $X = I - Y$. Show that

$$X^2 = A \iff Y = \tfrac{1}{2}(C + Y^2).$$

Define $Y_0 = 0$, $Y_1 = \frac{1}{2}C$ and

$$Y_{n+1} = \tfrac{1}{2}(C + Y_n^2) \qquad (n \geq 0).$$

(i) Check that

$$Y_{n+1} - Y_n = \tfrac{1}{2}(Y_n + Y_{n-1})(Y_n - Y_{n-1}).$$

Hint: Observe that Y_n is a polynomial in C and thus $Y_n Y_m = Y_m Y_n$.

(ii) Show that $Y_n \geq 0$ and $Y_{n+1} - Y_n \geq 0$.
Hint: Use Exercise 2.4.7.

(iii) Show that $\|Y_n\| \leq 1$.

(iv) Show that there is a positive operator Y satisfying $Y = \frac{1}{2}(C + Y^2)$.
Hint: Use Exercise 2.4.6.

(v) Check that $X = I - Y$ is a positive square root of A and observe that X is the limit (in the strong operator topology) of a sequence of polynomials on A.

Remark: In Section 7.2, we give another construction of the square root of a positive operator based on the functional calculus (see Theorem 7.5).

Exercise 2.4.9 Let A be a positive operator. We want to show that the positive operator X such that $X^2 = A$, which was obtained in Exercise 2.4.8, is unique. Assume that there is $X' \in \mathcal{L}(\mathcal{H})$, $X' \geq 0$, such that $X'^2 = A$.

(i) Show that $AX' = X'A$ and $p(A)X' = X'p(A)$, for any polynomial p.

(ii) Deduce that $XX' = X'X$.
Hint: Use Exercise 2.4.8(v).

(iii) Let $x \in \mathcal{H}$ and $y = (X - X')x$. Show that

$$\langle (X + X')y, y\rangle = 0,$$

and deduce that

$$\langle Xy, y\rangle = \langle X'y, y\rangle = 0.$$

(iv) Show that $Xy = X'y = 0$.
Hint: Use Lemma 2.12.

(v) Conclude that $X' = X$.
Hint: Compute $\|(X - X')x\|^2$.

Remark: The unique positive operator $X \in \mathcal{L}(\mathcal{H})$ such that $X^2 = A$ is called the *positive square root* of the positive operator A and is denoted by $\sqrt{A}$ or $A^{1/2}$.

2.5 Contractions

An operator $A \in \mathcal{L}(\mathcal{H}_1, \mathcal{H}_2)$ is called a *contraction* if $\|A\| \leq 1$. In other words, contractions are the elements of the closed unit ball of $\mathcal{L}(\mathcal{H}_1, \mathcal{H}_2)$. It is a rather well-known fact that the elements of the closed unit ball of a Banach space play a special role in the theory. That is why contractions are so important in operator theory. If A is a contraction, since $r(A) \leq \|A\| \leq 1$, we have

$$\sigma(A) \subset \bar{\mathbb{D}}. \tag{2.11}$$

Moreover, given an operator $A \in \mathcal{L}(\mathcal{H}_1, \mathcal{H}_2)$, then it is immediate to see that A is a contraction if and only if

$$I - AA^* \geq 0. \tag{2.12}$$

Since $\|A\| = \|A^*\|$, (2.12) is equivalent to

$$I - A^*A \geq 0. \tag{2.13}$$

In Chapter 7, we will study more specific properties of contractions. In particular, we will prove that any contraction has an isometric dilation. This profound result has many consequences and is the cornerstone of the Sz.-Nagy–Foiaş model theory. At this point, we just mention an elementary result that describes the manifold on which a contraction is an isometry.

Lemma 2.16 *Let $A \in \mathcal{L}(\mathcal{H}_1, \mathcal{H}_2)$ be a contraction. Then*

$$\ker(I - A^*A) = \{x \in \mathcal{H}_1 : \|Ax\|_{\mathcal{H}_2} = \|x\|_{\mathcal{H}_1}\}.$$

Proof First note that $\|Ax\|_{\mathcal{H}_2} = \|x\|_{\mathcal{H}_1}$ if and only if $\langle (I - A^*A)x,\, x\rangle_{\mathcal{H}_1} = 0$. Now, knowing (2.13), it remains to apply Lemma 2.12. □

Exercise

Exercise 2.5.1 Let $T \in \mathcal{L}(\mathcal{H})$ be a contraction, and let

$$K_{r,t}(T) = (I - re^{it}T^*)^{-1} + (I - re^{-it}T)^{-1} - I$$

for all $0 \leq r < 1$ and $t \in \mathbb{R}$.

(i) Check that $(K_{r,t}(T))^* = K_{r,t}(T)$.
(ii) Show that $K_{r,t}(T) = (I - re^{-it}T)^{-1}(I - r^2TT^*)(I - re^{it}T^*)^{-1}$.
(iii) Show that

$$K_{r,t}(T) = \sum_{n=0}^{\infty} r^n e^{int} T^{*n} + \sum_{n=0}^{\infty} r^n e^{-int} T^n - I.$$

(iv) Show that we have

$$p(rT) = \frac{1}{2\pi} \int_0^{2\pi} p(e^{it}) K_{r,t}(T)\, dt$$

for all $0 \leq r < 1$ and for all analytic polynomials p.

(v) Show that $K_{r,t}(T) \geq 0$.
Hint: Use (ii).

(vi) Show that, for any $x, y \in \mathcal{H}$ and for any $0 \leq r < 1$, we have

$$|\langle p(rT)x, y\rangle_{\mathcal{H}}| \\ \leq \|p\|_\infty \left(\frac{1}{2\pi} \int_0^{2\pi} \|x_{r,t}\|_{\mathcal{H}}^2\, dt \right)^{1/2} \left(\frac{1}{2\pi} \int_0^{2\pi} \|y_{r,t}\|_{\mathcal{H}}^2\, dt \right)^{1/2},$$

where $x_{r,t} = (K_{r,t}(T))^{1/2} x$ and $y_{r,t} = (K_{r,t}(T))^{1/2} y$.

(vii) Show that

$$\frac{1}{2\pi} \int_0^{2\pi} \|x_{r,t}\|_{\mathcal{H}}^2\, dt = \|x\|_{\mathcal{H}}^2 \quad \text{and} \quad \frac{1}{2\pi} \int_0^{2\pi} \|y_{r,t}\|_{\mathcal{H}}^2\, dt = \|y\|_{\mathcal{H}}^2.$$

Hint: Apply (iv) with p identically equal to 1.

(viii) Show that $\lim_{r \to 1} \|p(rT)\| = \|p(T)\|$ and deduce that

$$\|p(T)\| \leq \|p\|_\infty. \tag{2.14}$$

Remark: Inequality (2.14) is known as the *von Neumann inequality*. See Theorem 7.48 for another proof based on the theory of dilation.

2.6 Normal and unitary operators

An operator $A \in \mathcal{L}(\mathcal{H})$ is called *normal* if $AA^* = A^*A$. In particular, a self-adjoint operator is normal. If A is any normal operator on a Hilbert space $\mathcal{H}$, then, for each $\lambda \in \mathbb{C}$, we have

$$\|(A - \lambda I)x\| = \|(A^* - \bar{\lambda} I)x\| \qquad (x \in \mathcal{H}). \tag{2.15}$$

This simple relation has important consequences. For example, (2.15) implies that

$$\ker(A - \lambda I) = \ker(A^* - \bar{\lambda} I) \tag{2.16}$$

and

$$\sigma_p(A^*) = \overline{\sigma_p(A)} \quad \text{and} \quad \sigma_a(A^*) = \overline{\sigma_a(A)}. \tag{2.17}$$

In Corollary 2.14, we saw that, for any self-adjoint operator A, we have $r(A) = \|A\|$. This formula can be generalized to the larger class of normal operators.

Theorem 2.17 *Let $A \in \mathcal{L}(\mathcal{H})$ be normal. Then $r(A) = \|A\|$.*

Proof Since A^*A is a self-adjoint operator, Corollary 2.14 and Theorem 1.30 imply

$$r(A^*A) = \|A^*A\| = \|A\|^2.$$

We claim now that if $A_1, A_2 \in \mathcal{L}(\mathcal{H})$ are commuting, i.e. $A_1A_2 = A_2A_1$, then

$$r(A_1A_2) \leq r(A_1)r(A_2). \tag{2.18}$$

Indeed, since $A_1A_2 = A_2A_1$, we have

$$(A_1A_2)^n = A_1^n A_2^n \qquad (n \geq 1),$$

and thus

$$\|(A_1A_2)^n\|^{1/n} \leq \|A_1^n\|^{1/n}\|A_2^n\|^{1/n} \qquad (n \geq 1).$$

Now, letting n tend to $+\infty$ and using the spectral radius formula, we get (2.18).

Since A is normal, we can apply (2.18) to A and A^*, which gives

$$\|A\|^2 = r(A^*A) \leq r(A^*)r(A) = r(A)^2.$$

Thus $\|A\| \leq r(A)$. The opposite inequality is true for any operator. □

An operator $A \in \mathcal{L}(\mathcal{H}_1, \mathcal{H}_2)$ is called an *isometry* if

$$\|Ax\|_{\mathcal{H}_2} = \|x\|_{\mathcal{H}_1} \qquad (x \in \mathcal{H}_1).$$

It follows easily from the above definition that A is an isometry if and only if $A^*A = I_{\mathcal{H}_1}$. Therefore, if A is an isometry, then it is one-to-one and has a closed range.

An operator $A \in \mathcal{L}(\mathcal{H}_1, \mathcal{H}_2)$ is called *unitary* if $AA^* = I_{\mathcal{H}_2}$ and $A^*A = I_{\mathcal{H}_1}$. In other words, A is a unitary operator if and only if it is invertible and $A^{-1} = A^*$. There are other simple ways to characterize unitary operators. For example, A is unitary if and only if A and A^* are isometries. We can also say that A is unitary if and only if A is a surjective isometry.

If $\mathcal{H}$ is finite-dimensional and $A \in \mathcal{L}(\mathcal{H})$ is an isometry, then A is necessarily a unitary operator, i.e. $AA^* = A^*A = I_{\mathcal{H}}$. This is no longer true for infinite-dimensional Hilbert spaces. For example, as we will see in Section 2.7, the forward shift operator $U \in \mathcal{L}(\ell^2)$ is an isometry that is not surjective, and thus it is not a unitary operator. While the spectrum of self-adjoint and positive operators rests on the real line, the following result shows that the spectrum of unitary operators is located on the unit circle.

Theorem 2.18 *Let $A \in \mathcal{L}(\mathcal{H})$ be a unitary operator. Then*

$$\sigma(A) \subset \mathbb{T}.$$

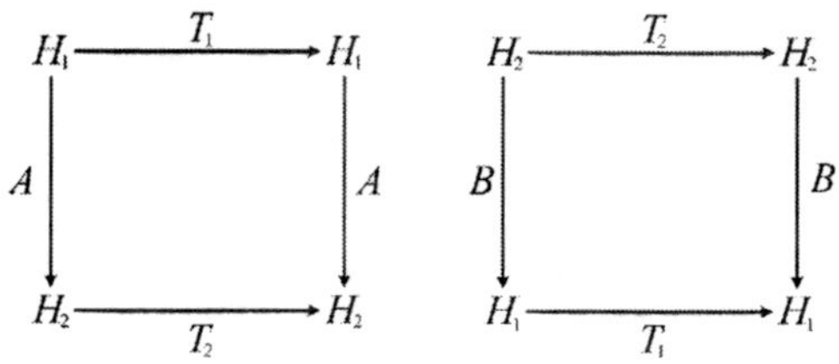

Figure 2.1 T_1 and T_2 are quasi-similar.

Proof By Theorem 1.30,

$$\|A\|^2 = \|AA^*\|^2 = \|I\|^2 = 1.$$

Hence, $\sigma(A) \subset \bar{\mathbb{D}}$. But, according to Theorem 1.30, and the fact that $A^{-1} = A^*$, we must have $\sigma(A)^{-1} = \overline{\sigma(A)}$. Thus, no point of $\mathbb{D}$ can be in the spectrum. □

Let $\mathcal{H}_1, \mathcal{H}_2$ be two Hilbert spaces. We say that $T_1 \in \mathcal{L}(\mathcal{H}_1)$ is *unitarily* equivalent to $T_2 \in \mathcal{L}(\mathcal{H}_2)$ if there exists a unitary operator $U \in \mathcal{L}(\mathcal{H}_1, \mathcal{H}_2)$ such that $A = U^*BU$. We say that $T_1 \in \mathcal{L}(\mathcal{H}_1)$ is *similar* to $T_2 \in \mathcal{L}(\mathcal{H}_2)$ if there exists an invertible operator $V \in \mathcal{L}(\mathcal{H}_1, \mathcal{H}_2)$ such that $A = V^{-1}BV$. Clearly, if A and B are unitarily equivalent, then they are similar too. But the inverse is not true. Moreover, we can easily check that both relations are equivalence relations.

In our discussion, we need a generalized version of the equivalence relations that were introduced above. We say that $A \in \mathcal{L}(\mathcal{H}_1, \mathcal{H}_2)$ is *quasi-affine* (or a *quasi-affinity*) if A is injective and has a dense range. In particular, any invertible operator from $\mathcal{H}_1$ onto $\mathcal{H}_2$ is a quasi-affinity. But the family of quasi-affine operators is generally much larger than the invertible one. We say that $T_1 \in \mathcal{L}(\mathcal{H}_1)$ and $T_2 \in \mathcal{L}(\mathcal{H}_2)$ are *quasi-similar* if there exist two quasi-affinities $A \in \mathcal{L}(\mathcal{H}_1, \mathcal{H}_2)$ and $B \in \mathcal{L}(\mathcal{H}_2, \mathcal{H}_1)$ such that $AT_1 = T_2A$ and $T_1B = BT_2$. In other words, T_1 and T_2 are quasi-similar if there exist two quasi-affinities A and B such that the diagrams in Figure 2.1 commute.

2.7 Forward and backward shift operators on ℓ^2

The most important operator that naturally enters our discussion is the unilateral forward shift operator. In this section, we define it as an operator on ℓ^2. Later on, in Section 8.2, we will study a similar object on the Hardy space H^2. The mapping

$$\begin{array}{rrcl} U: & \ell^2 & \longrightarrow & \ell^2 \\ & (z_1, z_2, \dots) & \longmapsto & (0, z_1, z_2, \dots) \end{array}$$

is called the *unilateral forward shift operator* on ℓ^2. Clearly

$$\|U\mathfrak{z}\|_{\ell^2} = \|\mathfrak{z}\|_{\ell^2} \qquad (\mathfrak{z} \in \ell^2). \tag{2.19}$$

Hence, U is an isometry on $\mathcal{L}(\ell^2)$. In particular, $\|U\|_{\mathcal{L}(\ell^2)} = 1$. It is not difficult to find the adjoint of U. In fact, for each $\mathfrak{z}, \mathfrak{w} \in \ell^2$, we have

$$\langle U\mathfrak{z},\, \mathfrak{w}\rangle_{\ell^2} = \sum_{n=1}^{\infty} z_n\, \bar{w}_{n+1} = \langle \mathfrak{z},\, U^*\mathfrak{w}\rangle_{\ell^2},$$

where U^* is given by

$$\begin{array}{rccc} U^* : & \ell^2 & \longrightarrow & \ell^2 \\ & (w_1, w_2, \dots) & \longmapsto & (w_2, w_3, \dots). \end{array}$$

The operator U^* is called the *unilateral backward shift operator*. Note that U^* is not injective and

$$\ker U^* = \mathbb{C}\,\mathfrak{e}_1 = \{(\alpha, 0, 0, \dots) : \alpha \in \mathbb{C}\}.$$

Theorem 2.19 *We have*

$$\sigma_p(U) = \emptyset \quad \textit{and} \quad \sigma_p(U^*) = \mathbb{D}.$$

However,

$$\sigma(U) = \sigma(U^*) = \bar{\mathbb{D}}.$$

Proof Since $\|U^*\| = 1$, then $r(U^*) \leq 1$, and this implies that

$$\sigma(U^*) \subset \bar{\mathbb{D}}. \tag{2.20}$$

Fix $\lambda \in \mathbb{D}$, and consider

$$\mathfrak{z} = (1, \lambda, \lambda^2, \dots) \in \ell^2.$$

This nonzero vector satisfies

$$U^*\mathfrak{z} = \lambda\,\mathfrak{z}.$$

Hence, each $\lambda \in \mathbb{D}$ is an eigenvalue of U^* and thus

$$\mathbb{D} \subset \sigma_p(U^*).$$

But $\sigma_p(U^*) \subset \sigma(U^*)$ and $\sigma(U^*)$ is a closed set. Hence, by (2.20), we necessarily have $\sigma(U^*) = \bar{\mathbb{D}}$. Knowing this, Theorem 1.30 now ensures that $\sigma(U) = \bar{\mathbb{D}}$.

To show that U^* has no eigenvalues on $\mathbb{T}$, suppose that, on the contrary, there is a $\lambda \in \mathbb{T}$ and a nonzero vector $\mathfrak{z} \in \ell^2$ such that $U^*\mathfrak{z} = \lambda\,\mathfrak{z}$. The identity $U^*\mathfrak{z} = \lambda\,\mathfrak{z}$, in terms of the components of $\mathfrak{z}$, is equivalent to

$$z_{n+1} = \lambda z_n \qquad (n \geq 1).$$

Hence, by induction, we necessarily have $\mathfrak{z} = z_1(1, \lambda, \lambda^2, \dots)$. This vector is in ℓ^2 if and only if $z_1 = 0$, which implies $\mathfrak{z} = 0$, and that is a contradiction. Hence $\sigma_p(U^*) = \mathbb{D}$.

By a similar reasoning, we can show that U has no eigenvalue. Indeed, suppose that the identity $U\mathfrak{z} = \lambda\mathfrak{z}$ holds for a $\lambda \in \mathbb{C}$ and a nonzero vector $\mathfrak{z} \in \ell^2$. This identity is equivalent to

$$0 = \lambda z_1$$

and

$$z_n = \lambda z_{n+1} \qquad (n \geq 1).$$

If $\lambda \neq 0$, we immediately see that $\mathfrak{z} = 0$. If $\lambda = 0$, the second relation implies $\mathfrak{z} = 0$. In both cases, we arrive at a contradiction. Therefore, $\sigma_p(U) = \emptyset$. $\square$

To find the matrix representations of U and U^*, equip the Hilbert space ℓ^2 with the standard orthonormal basis $(\mathfrak{e}_n)_{n\geq 1}$. Then, according to the definition of U, we have

$$\langle U\mathfrak{e}_j, \mathfrak{e}_i\rangle = \begin{cases} 1 & \text{if } i = j+1, \\ 0 & \text{if } i \neq j+1. \end{cases}$$

Hence, the matrix representation of U is

$$\begin{bmatrix} 0 & 0 & 0 & 0 & 0 & \cdots \\ 1 & 0 & 0 & 0 & 0 & \cdots \\ 0 & 1 & 0 & 0 & 0 & \cdots \\ 0 & 0 & 1 & 0 & 0 & \cdots \\ 0 & 0 & 0 & 1 & 0 & \cdots \\ \vdots & \vdots & \vdots & \vdots & \vdots & \ddots \end{bmatrix},$$

and the matrix representation of U^* is

$$\begin{bmatrix} 0 & 1 & 0 & 0 & 0 & \cdots \\ 0 & 0 & 1 & 0 & 0 & \cdots \\ 0 & 0 & 0 & 1 & 0 & \cdots \\ 0 & 0 & 0 & 0 & 1 & \cdots \\ 0 & 0 & 0 & 0 & 0 & \cdots \\ \vdots & \vdots & \vdots & \vdots & \vdots & \ddots \end{bmatrix}.$$

2.8 The multiplication operator on $L^2(\mu)$

To define another important operator, fix $\mu \in \mathcal{M}^+(\mathbb{T})$ and let $L^p(\mu) = L^p(\mathbb{T}, \mu)$, $1 \le p \le \infty$. For $\varphi \in L^\infty(\mu)$, define

$$\begin{array}{rccc} M_\varphi : & L^2(\mu) & \longrightarrow & L^2(\mu) \\ & f & \longmapsto & \varphi f. \end{array}$$

Since $\|\varphi f\|_2 \le \|\varphi\|_\infty \|f\|_2$, M_φ is a bounded operator on $L^2(\mu)$ and we have $\|M_\varphi\| \le \|\varphi\|_\infty$. In the following, we show that $\|M_\varphi\| = \|\varphi\|_\infty$. As the notation suggests, this operator is called the *multiplication* operator M_φ. We highlight that the notation M_φ could be misleading because it carries no indication about the space on which the multiplication is performed. Indeed, in other parts of the book, we will also consider multiplication operators on Hilbert spaces other than Lebesgue space $L^2(\mu)$. Hence, to avoid ambiguity, we will be precise about the space on which our multiplication operator acts.

Theorem 2.20 *Let $\varphi \in L^\infty(\mu)$. Then*

$$\|M_\varphi\|_{\mathcal{L}(L^2(\mu))} = \|\varphi\|_{L^\infty(\mu)}.$$

Proof Fix $\varepsilon > 0$, and let

$$E = \{\zeta \in \mathbb{T} : |\varphi(\zeta)| \ge \|\varphi\|_\infty - \varepsilon\}.$$

Consider $f = \chi_E$, the characteristic function of E. Since $\mu(E) > 0$, we have $f \in L^2(\mu)$ with $\|f\|_2 \neq 0$. We use the inequality

$$\|M_\varphi f\|_2 \le \|M_\varphi\| \, \|f\|_2$$

to obtain a lower bound for $\|M_\varphi\|$. From the estimation

$$\begin{aligned} \|M_\varphi f\|_2^2 &= \int_{\mathbb{T}} |(M_\varphi f)(\zeta)|^2 \, d\mu(\zeta) \\ &= \int_{\mathbb{T}} |\varphi(\zeta)|^2 \, |f(\zeta)|^2 \, d\mu(\zeta) \\ &= \int_E |\varphi(\zeta)|^2 \, d\mu(\zeta) \\ &\ge (\|\varphi\|_\infty - \varepsilon)^2 \int_E d\mu(\zeta) \\ &= (\|\varphi\|_\infty - \varepsilon)^2 \, \|f\|_2^2, \end{aligned}$$

and the fact that $\|f\|_2 \neq 0$, we get $\|M_\varphi\| \ge \|\varphi\|_\infty - \varepsilon$. Let $\varepsilon \longrightarrow 0$ to get the result. □

We also address the following inverse question. If a measurable function φ on $\mathbb{T}$ is such that, for each $f \in L^2(\mu)$, the product φf belongs to $L^2(\mu)$, then what can we say about φ? In light of Theorem 2.20, does it have to be an essentially bounded function? This is answered affirmatively below.

Theorem 2.21 *Let φ be a measurable function on $\mathbb{T}$ that has the following property:*

$$f \in L^2(\mu) \quad \Longrightarrow \quad \varphi f \in L^2(\mu).$$

Then $\varphi \in L^\infty(\mu)$.

Proof Assume that $\varphi \notin L^\infty(\mu)$. Put

$$E_n = \{\zeta \in \mathbb{T} : n \le |\varphi(\zeta)| < n+1\} \qquad (n \ge 1)$$

and

$$\Lambda = \{n \ge 1 : \mu(E_n) \ne 0\}.$$

By assumption, there are infinitely many n values such that $\mu(E_n) \ne 0$. In other words, Λ is an infinite subset of $\mathbb{N}$. Also note that the measurable sets E_n are pairwise disjoint. Now define

$$f = \sum_{n\in\Lambda} \frac{1}{n(\mu(E_n))^{1/2}}\, \chi_{E_n}. \tag{2.21}$$

In the first place, we have

$$\|f\|_{L^2(\mu)}^2 = \int_{\mathbb{T}} |f|^2\, d\mu = \sum_{n\in\Lambda} \frac{1}{n^2} < \infty,$$

whence, the series in (2.21) defines an $L^2(\mu)$ function. But, second, we have

$$\|\varphi f\|_{L^2(\mu)}^2 = \int_{\mathbb{T}} |\varphi f|^2\, d\mu = \sum_{n\in\Lambda} \frac{1}{|\mu(E_n)|} \int_{E_n} \left|\frac{\varphi}{n}\right|^2 d\mu \ge \sum_{n\in\Lambda} 1 = \infty.$$

This is a contradiction. □

It is easy to determine the adjoint of M_φ. According to the definition of inner product, for each $f, g \in L^2(\mu)$, we have

$$\begin{aligned}
\langle M_\varphi f, g\rangle_2 &= \int_{\mathbb{T}} (M_\varphi f)(\zeta)\, \overline{g(\zeta)}\, d\mu(\zeta) \\
&= \int_{\mathbb{T}} \varphi(\zeta) f(\zeta)\, \overline{g(\zeta)}\, d\mu(\zeta) \\
&= \int_{\mathbb{T}} f(\zeta)\, \overline{\overline{\varphi(\zeta)}\, g(\zeta)}\, d\mu(\zeta) \\
&= \int_{\mathbb{T}} f(\zeta)\, \overline{(M_{\bar\varphi} g)(\zeta)}\, d\mu(\zeta) \\
&= \langle f, M_{\bar\varphi} g\rangle_2.
\end{aligned}$$

Therefore, for each $\varphi \in L^\infty(\mu)$,

$$M_\varphi^* = M_{\bar\varphi}. \tag{2.22}$$

Based on Theorem 2.20, the elementary properties

$$\alpha M_\varphi + \beta M_\psi = M_{\alpha\varphi+\beta\psi}, \qquad M_\varphi M_\psi = M_\psi M_\varphi = M_{\varphi\psi} \tag{2.23}$$

and (2.22), we can say that the mapping

$$\begin{array}{ccc} L^\infty(\mu) & \longrightarrow & \mathcal{L}(L^2(\mu)) \\ \varphi & \longmapsto & M_\varphi \end{array}$$

is an isometric $*$-homomorphism of Banach algebras. Naively speaking, this means that this mapping puts a copy of $L^\infty(\mu)$ into $\mathcal{L}(L^2(\mu))$.

To find the spectrum of M_φ on $L^2(\mu)$, we start with the following preliminary result.

Lemma 2.22 *Let $\varphi \in L^\infty(\mu)$. Then the following assertions are equivalent.*

(i) *M_φ is invertible in $\mathcal{L}(L^2(\mu))$;*
(ii) *M_φ is lower bounded on $L^2(\mu)$;*
(iii) *φ is invertible in $L^\infty(\mu)$.*

Proof (i) $\Longrightarrow$ (ii) This is trivial.

(ii) $\Longrightarrow$ (iii) Suppose that M_φ is lower bounded on $L^2(\mu)$. Then there is a constant $c > 0$ such that

$$\|M_\varphi f\|_2 \geq c\|f\|_2 \qquad (f \in L^2(\mu)). \tag{2.24}$$

Put $E_n = \{\zeta \in \mathbb{T} : |\varphi(\zeta)| < nc/(n+1)\}$, $n \geq 1$, and let $f = \chi_{E_n}$. Hence, (2.24) implies that

$$\frac{nc}{n+1}\,(\mu(E_n))^{1/2} \geq c\,(\mu(E_n))^{1/2}.$$

Therefore, we must have $\mu(E_n) = 0$, $n \geq 1$. Considering the union of all such sets, we see that $|\varphi| \geq c$, μ-almost everywhere on $\mathbb{T}$. Hence, we have $1/\varphi \in L^\infty(\mu)$.

(iii) $\Longrightarrow$ (i) Suppose that φ is invertible in $L^\infty(\mu)$, i.e. $1/\varphi \in L^\infty(\mu)$. Then, by (2.23),

$$M_\varphi M_{1/\varphi} = M_{1/\varphi} M_\varphi = I.$$

Thus, M_φ is invertible in $\mathcal{L}(L^2(\mu))$ and, moreover,

$$M_\varphi^{-1} = M_{1/\varphi}.$$

□

Considering the preceding lemma, we naturally may wonder when a function $\varphi \in L^\infty(\mu)$ is invertible in the Banach algebra $L^\infty(\mu)$. To answer this question, we need the notion of *essential range*. The essential range of φ, with respect to measure μ, is the set

$$\mathcal{R}_e^\mu(\varphi) = \{\lambda \in \mathbb{C} : \forall\, r > 0,\ \mu(\varphi^{-1}(D(\lambda, r))) > 0\},$$

where $D(\lambda, r) = \{z \in \mathbb{C} : |z - \lambda| < r\}$. If μ is the Lebesgue measure m, instead of $\mathcal{R}_e^m(\varphi)$ we simply write $\mathcal{R}_e(\varphi)$. Based on this notion, it is easy to see that φ is invertible in $L^\infty(\mu)$ if and only if $0 \notin \mathcal{R}_e^\mu(\varphi)$.

Theorem 2.23 *Let $\varphi \in L^\infty(\mu)$. Then*

$$\sigma_a(M_\varphi) = \sigma(M_\varphi) = \mathcal{R}_e^\mu(\varphi).$$

Proof Note that $\lambda I - M_\varphi = M_{\lambda-\varphi}$. Hence, the equality $\sigma_a(M_\varphi) = \sigma(M_\varphi)$ follows easily from Lemma 2.22 and the fact that the approximate spectrum of an operator T corresponds to the points $\lambda \in \mathbb{C}$ such that $\lambda I - T$ is not bounded below. Furthermore, using Lemma 2.22 again, we see that $\lambda \notin \sigma(M_\varphi)$ if and only if $\lambda - \varphi$ is invertible in $L^\infty(\mathbb{T})$. This is equivalent to saying that $0 \notin \mathcal{R}_e^\mu(\lambda - \varphi) = \lambda - \mathcal{R}_e^\mu(\varphi)$. Thus, $\sigma(M_\varphi) = \mathcal{R}_e^\mu(\varphi)$. □

2.9 Doubly infinite Toeplitz and Hankel matrices

We first introduce an operator whose properties are easy to determine. However, it will be needed to make the bridge between the Toeplitz and Hankel type operators. The *reversion* or *flip operator* R on $L^2(\mathbb{T})$ is defined by the formula

$$\begin{array}{rccc} R: & L^2(\mathbb{T}) & \longrightarrow & L^2(\mathbb{T}) \\ & \chi_n & \longmapsto & \chi_{-n} \end{array}$$

for all $n \in \mathbb{Z}$. Another way to describe R is to write

$$(Rf)(\zeta) = f(\bar{\zeta}) \qquad (\zeta \in \mathbb{T}).$$

It is clear that

$$R^2 = I. \tag{2.25}$$

Moreover, the chain of identities

$$\begin{aligned} \langle Rf, g\rangle_2 &= \int_{\mathbb{T}} (Rf)(\zeta)\, \overline{g(\zeta)}\, dm(\zeta) \\ &= \int_{\mathbb{T}} f(\bar{\zeta})\, \overline{g(\zeta)}\, dm(\zeta) \\ &= \int_{\mathbb{T}} f(\zeta)\, \overline{g(\bar{\zeta})}\, dm(\zeta) \\ &= \int_{\mathbb{T}} f(\zeta)\, \overline{(Rg)(\zeta)}\, dm(\zeta) \\ &= \langle f, Rg\rangle_2 \end{aligned}$$

show that

$$R = R^*. \tag{2.26}$$

Hence, R is a self-adjoint unitary operator. Thus, by Theorems 2.13 and 2.18, we have

$$\sigma(R) \subset \{-1, 1\}.$$

The relation

$$R\chi_0 = \chi_0$$

even shows that 1 is an eigenvalue of R. Moreover, if we write the identity (2.25) as $(R - I)(R + I) = 0$, since $R \neq \pm I$, we conclude that

$$\sigma(R) = \{-1, 1\}. \tag{2.27}$$

To obtain a matrix representation of $R \in \mathcal{L}(L^2(\mathbb{T}))$, as usual equip $L^2(\mathbb{T})$ with the orthonormal basis $(\chi_n)_{n\in\mathbb{Z}}$. Then, for each $m, n \in \mathbb{Z}$,

$$\langle R\chi_n, \chi_m \rangle_2 = \langle \chi_{-n}, \chi_m \rangle_2 = \delta_{m+n}.$$

Hence, the matrix of R is

$$\begin{bmatrix} \ddots & \ddots & \ddots & \ddots & \ddots & \ddots & \ddots \\ \ddots & 0 & 0 & 0 & 0 & 1 & \ddots \\ \ddots & 0 & 0 & 0 & 1 & 0 & \ddots \\ \ddots & 0 & 0 & \mathbf{1} & 0 & 0 & \ddots \\ \ddots & 0 & 1 & 0 & 0 & 0 & \ddots \\ \ddots & 1 & 0 & 0 & 0 & 0 & \ddots \\ \ddots & \ddots & \ddots & \ddots & \ddots & \ddots & \ddots \end{bmatrix}, \tag{2.28}$$

where the boldface marks the 00 position.

A (finite, doubly infinite or singly infinite) matrix $A = [a_{ij}]$ that is constant on the main diagonals is called a *Toeplitz matrix*. Hence, its components fulfill the defining property

$$i - j = i' - j' \quad \Longrightarrow \quad a_{ij} = a_{i'j'},$$

or equivalently, e.g. for doubly infinite Toeplitz matrices,

$$a_{ij} = a_{(i+k)(j+k)} \qquad (k \in \mathbb{Z}).$$

A doubly infinite Toeplitz matrix has the form

$$\begin{bmatrix} \ddots & \ddots & \ddots & \ddots & \ddots & \ddots & \ddots \\ \ddots & \alpha_0 & \alpha_{-1} & \alpha_{-2} & \alpha_{-3} & \alpha_{-4} & \ddots \\ \ddots & \alpha_1 & \alpha_0 & \alpha_{-1} & \alpha_{-2} & \alpha_{-3} & \ddots \\ \ddots & \alpha_2 & \alpha_1 & \boldsymbol{\alpha_0} & \alpha_{-1} & \alpha_{-2} & \ddots \\ \ddots & \alpha_3 & \alpha_2 & \alpha_1 & \alpha_0 & \alpha_{-1} & \ddots \\ \ddots & \alpha_4 & \alpha_3 & \alpha_2 & \alpha_1 & \alpha_0 & \ddots \\ \ddots & \ddots & \ddots & \ddots & \ddots & \ddots & \ddots \end{bmatrix}, \tag{2.29}$$

where again the boldface marks the 00 position. The identity matrix is the simplest Toeplitz matrix. In this section, we characterize all doubly infinite Toeplitz matrices that give a bounded operator of $\ell^2(\mathbb{Z})$. Briefly speaking, we can say that, in a sense, they are the same as the multiplication operators on $L^2(\mathbb{T})$. The classification of bounded singly infinite Toeplitz matrices is more difficult and will be done in Chapter 12.

To start the discussion, fix $\varphi \in L^\infty(\mathbb{T}) = L^\infty(\mathbb{T}, m)$ and consider the multiplication operator M_φ on $L^2(\mathbb{T}) = L^2(\mathbb{T}, m)$, which was introduced in Section 2.8. As usual, we equip $L^2(\mathbb{T})$ with the orthonormal basis $(\chi_n)_{n\in\mathbb{Z}}$. Then, for each $m, n \in \mathbb{Z}$,

$$\begin{aligned}\langle M_\varphi \chi_n, \chi_m \rangle &= \int_{\mathbb{T}} \varphi(\zeta)\, \chi_n(\zeta)\, \overline{\chi_m(\zeta)}\, dm(\zeta) \\ &= \int_{\mathbb{T}} \varphi(\zeta)\, \chi_{n-m}(\zeta)\, dm(\zeta) \\ &= \hat{\varphi}(m-n).\end{aligned}$$

Hence, the matrix of M_φ is

$$\begin{bmatrix} \ddots & \ddots & \ddots & \ddots & \ddots & \ddots & \ddots \\ \ddots & \hat{\varphi}(0) & \hat{\varphi}(-1) & \hat{\varphi}(-2) & \hat{\varphi}(-3) & \hat{\varphi}(-4) & \ddots \\ \ddots & \hat{\varphi}(1) & \hat{\varphi}(0) & \hat{\varphi}(-1) & \hat{\varphi}(-2) & \hat{\varphi}(-3) & \ddots \\ \ddots & \hat{\varphi}(2) & \hat{\varphi}(1) & \boldsymbol{\hat{\varphi}(0)} & \hat{\varphi}(-1) & \hat{\varphi}(-2) & \ddots \\ \ddots & \hat{\varphi}(3) & \hat{\varphi}(2) & \hat{\varphi}(1) & \hat{\varphi}(0) & \hat{\varphi}(-1) & \ddots \\ \ddots & \hat{\varphi}(4) & \hat{\varphi}(3) & \hat{\varphi}(2) & \hat{\varphi}(1) & \hat{\varphi}(0) & \ddots \\ \ddots & \ddots & \ddots & \ddots & \ddots & \ddots & \ddots \end{bmatrix}, \tag{2.30}$$

where the boldface marks the 00 position. In other words, the matrix of the multiplication operator M_φ with respect to the standard orthonormal basis of $L^2(\mathbb{T})$ is a doubly infinite Toeplitz matrix. More interestingly, the converse is also true.

Theorem 2.24 *Let the doubly infinite Toeplitz matrix*

$$A = \begin{bmatrix} \ddots & \ddots & \ddots & \ddots & \ddots & \ddots & \ddots \\ \ddots & \alpha_0 & \alpha_{-1} & \alpha_{-2} & \alpha_{-3} & \alpha_{-4} & \ddots \\ \ddots & \alpha_1 & \alpha_0 & \alpha_{-1} & \alpha_{-2} & \alpha_{-3} & \ddots \\ \ddots & \alpha_2 & \alpha_1 & \boldsymbol{\alpha_0} & \alpha_{-1} & \alpha_{-2} & \ddots \\ \ddots & \alpha_3 & \alpha_2 & \alpha_1 & \alpha_0 & \alpha_{-1} & \ddots \\ \ddots & \alpha_4 & \alpha_3 & \alpha_2 & \alpha_1 & \alpha_0 & \ddots \\ \ddots & \ddots & \ddots & \ddots & \ddots & \ddots & \ddots \end{bmatrix},$$

where $\alpha_n \in \mathbb{C}$, give a bounded operator on $\ell^2(\mathbb{Z})$. Then there is a function $\varphi \in L^\infty(\mathbb{T})$ such that

$$\hat{\varphi}(n) = \alpha_n \qquad (n \in \mathbb{Z})$$

and, moreover,

$$\|A\|_{\mathcal{L}(\ell^2(\mathbb{Z}))} = \|\varphi\|_\infty.$$

Proof By assumption, we have

$$A\mathfrak{e}_0 = \sum_{n\in\mathbb{Z}} \alpha_n \mathfrak{e}_n \in \ell^2(\mathbb{Z}).$$

This means that

$$\sum_{n\in\mathbb{Z}} |\alpha_n|^2 < \infty,$$

and thus the function

$$\varphi = \sum_{n\in\mathbb{Z}} \alpha_n \, \chi_n$$

is a well-defined element of $L^2(\mathbb{T})$. But we can say more about φ.

One the one hand, if $f \in L^2(\mathbb{T})$, then $\varphi f \in L^1(\mathbb{T})$ and, by Parseval's identity,

$$\widehat{\varphi f}(n) = \sum_{m\in\mathbb{Z}} \hat{\varphi}(n-m)\hat{f}(m) = \sum_{m\in\mathbb{Z}} \alpha_{n-m}\hat{f}(m) \qquad (n \in \mathbb{Z}).$$

On the other, if the vector

$$\sum_{n\in\mathbb{Z}} \beta_n \mathfrak{e}_n \in \ell^2(\mathbb{Z})$$

is given, then

$$A\bigg(\sum_{n\in\mathbb{Z}} \beta_n \mathfrak{e}_n\bigg) = \sum_{n\in\mathbb{Z}} \eta_n \mathfrak{e}_n \in \ell^2(\mathbb{Z}),$$

where the coefficients η_n, $n \in \mathbb{Z}$, are given by

$$\eta_n = \sum_{m\in\mathbb{Z}} \alpha_{n-m}\beta_m.$$

This observation reveals that the sequence $(\widehat{\varphi f}(n))_{n\in\mathbb{Z}}$ is in $\ell^2(\mathbb{Z})$. Therefore, by the Riesz–Fischer theorem, $\varphi f \in L^2(\mathbb{T})$. Theorem 2.21 now ensures that $\varphi \in L^\infty(\mathbb{T})$, and the action of A on ℓ^2 is unitarily equivalent to the action of M_φ on $L^2(\mathbb{T})$. Finally, Theorem 2.20 shows that

$$\|A\|_{\mathcal{L}(\ell^2(\mathbb{Z}))} = \|M_\varphi\|_{\mathcal{L}(L^2(\mathbb{T}))} = \|\varphi\|_\infty. \qquad \square$$

A (finite, doubly infinite or singly infinite) matrix $A = [a_{ij}]$ that is constant on the skew diagonals is called a *Hankel matrix*. Hence, its components fulfill the defining property

$$i + j = i' + j' \quad \Longrightarrow \quad a_{ij} = a_{i'j'},$$

or equivalently, e.g. for doubly infinite Toeplitz matrices,

$$a_{ij} = a_{(i+k)(j-k)} \qquad (k \in \mathbb{Z}).$$

A doubly infinite Hankel matrix has the form

$$\begin{bmatrix} \ddots & \ddots & \ddots & \ddots & \ddots & \ddots & \ddots \\ \ddots & \alpha_4 & \alpha_3 & \alpha_2 & \alpha_1 & \alpha_0 & \ddots \\ \ddots & \alpha_3 & \alpha_2 & \alpha_1 & \alpha_0 & \alpha_{-1} & \ddots \\ \ddots & \alpha_2 & \alpha_1 & \boldsymbol{\alpha_0} & \alpha_{-1} & \alpha_{-2} & \ddots \\ \ddots & \alpha_1 & \alpha_0 & \alpha_{-1} & \alpha_{-2} & \alpha_{-3} & \ddots \\ \ddots & \alpha_0 & \alpha_{-1} & \alpha_{-2} & \alpha_{-3} & \alpha_{-4} & \ddots \\ \ddots & \ddots & \ddots & \ddots & \ddots & \ddots & \ddots \end{bmatrix}, \tag{2.31}$$

with the boldface again marking the 00 position. According to (2.28), the flip operator has a doubly infinite Hankel matrix representation. In fact, this matrix makes the connection between the Hankel and Toeplitz matrices. It is easy to verify that left-multiplication by R transforms one type to another. Hence, we immediately get the following result.

Theorem 2.25 *Let the doubly infinite Hankel matrix*

$$A = \begin{bmatrix} \ddots & \ddots & \ddots & \ddots & \ddots & \ddots & \ddots \\ \ddots & \alpha_4 & \alpha_3 & \alpha_2 & \alpha_1 & \alpha_0 & \ddots \\ \ddots & \alpha_3 & \alpha_2 & \alpha_1 & \alpha_0 & \alpha_{-1} & \ddots \\ \ddots & \alpha_2 & \alpha_1 & \boldsymbol{\alpha_0} & \alpha_{-1} & \alpha_{-2} & \ddots \\ \ddots & \alpha_1 & \alpha_0 & \alpha_{-1} & \alpha_{-2} & \alpha_{-3} & \ddots \\ \ddots & \alpha_0 & \alpha_{-1} & \alpha_{-2} & \alpha_{-3} & \alpha_{-4} & \ddots \\ \ddots & \ddots & \ddots & \ddots & \ddots & \ddots & \ddots \end{bmatrix},$$

where $\alpha_n \in \mathbb{C}$, give a bounded operator on $\ell^2(\mathbb{Z})$. Then there is a function $\varphi \in L^\infty(\mathbb{T})$ such that

$$\hat{\varphi}(n) = \alpha_n \qquad (n \in \mathbb{Z})$$

and, moreover,

$$\|A\|_{\mathcal{L}(\ell^2(\mathbb{Z}))} = \|\varphi\|_\infty.$$

Proof Use R also to denote the doubly infinite matrix of the reversion operator R. Hence, RA gives a bounded doubly infinite Toeplitz matrix on $\ell^2(\mathbb{Z})$. The result now follows from Theorem 2.24. Note that $\|A\| = \|RA\|$. □

The classification of bounded singly infinite Hankel matrices is more difficult and will be done in Chapter 11.

Exercises

Exercise 2.9.1 Improve the relation (2.27) by showing that

$$\sigma_p(R) = \{-1, 1\}.$$

Hint: To show $-1 \in \sigma_p(R)$, consider

$$f(\zeta) = \begin{cases} 1 & \text{if } \Im(\zeta) > 0, \\ 0 & \text{if } \zeta = \pm 1, \\ -1 & \text{if } \Im(\zeta) < 0. \end{cases}$$

Exercise 2.9.2 Let $\varphi \in L^\infty(\mathbb{T})$. Show that M_φ is self-adjoint if and only if $\mathcal{R}_e(\varphi) \subset \mathbb{R}$.
Hint: Use (2.22). Note that Theorem 2.20 ensures that $M_\varphi = 0 \Longleftrightarrow \varphi = 0$.

Exercise 2.9.3 Let $\varphi \in L^\infty(\mathbb{T})$. Show that M_φ is positive if and only if $\mathcal{R}_e(\varphi) \subset [0, \infty)$.
Hint: Use Exercise 2.9.2.

Exercise 2.9.4 The purpose of this exercise is to introduce a discrete multiplication operator. Equip $\ell^2 = \ell^2(\mathbb{N})$ with the standard orthonormal basis $(\mathfrak{e}_n)_{n\geq 1}$. Let $(\lambda_n)_{n\geq 1}$ be a bounded sequence in $\mathbb{C}$. Define Λ by

$$\Lambda\Big(\sum_{n=1}^{\infty} c_n \mathfrak{e}_n\Big) = \sum_{n=1}^{\infty} c_n \lambda_n \mathfrak{e}_n \qquad (n \geq 1).$$

(i) Verify that Λ is a well-defined bounded operator on ℓ^2 whose norm is

$$\|\Lambda\|_{\mathcal{L}(\ell^2)} = \sup_{n\geq 1} |\lambda_n|.$$

(ii) Show that Λ has the matrix representation

$$\begin{bmatrix} \lambda_1 & 0 & 0 & 0 & 0 & \cdots \\ 0 & \lambda_2 & 0 & 0 & 0 & \cdots \\ 0 & 0 & \lambda_3 & 0 & 0 & \cdots \\ 0 & 0 & 0 & \lambda_4 & 0 & \cdots \\ 0 & 0 & 0 & 0 & \lambda_5 & \cdots \\ \vdots & \vdots & \vdots & \vdots & \vdots & \ddots \end{bmatrix}.$$

That is why we sometimes write $\Lambda = \text{diag}(\lambda_1, \lambda_2, \dots)$.

(iii) Show that $\sigma_p(\Lambda) = \{\lambda_1, \lambda_2, \dots\}$ and that $\sigma(\Lambda) = \text{Clos}_{\mathbb{C}}\{\lambda_1, \lambda_2, \dots\}$.

(iv) Show that $\Lambda^* = \text{diag}(\bar{\lambda}_1, \bar{\lambda}_2, \dots)$.

(v) When is Λ self-adjoint? When is it positive?

Exercise 2.9.5 The goal of this exercise is to highlight the effect of using different orthonormal bases for the domain and codomain of an operator. In deriving the matrix representation (2.30), we put the same orthonormal basis (with the same order) on the domain and codomain of the multiplication operator $M_\varphi \in \mathcal{L}(L^2(\mathbb{T}))$.

Equip $L^2(\mathbb{T})$ on the domain with the orthonormal basis $(\chi_n)_{n\in\mathbb{Z}}$, and on the codomain with $(\chi_{-n})_{n\in\mathbb{Z}}$. The second basis has the same element as the first one, but the ordering is different. However, this causes certain changes in the representation of the matrix. Show that the matrix of M_φ in the new setting is

$$\begin{bmatrix} \ddots & \ddots & \ddots & \ddots & \ddots & \ddots & \ddots \\ \ddots & \hat{\varphi}(4) & \hat{\varphi}(3) & \hat{\varphi}(2) & \hat{\varphi}(1) & \hat{\varphi}(0) & \ddots \\ \ddots & \hat{\varphi}(3) & \hat{\varphi}(2) & \hat{\varphi}(1) & \hat{\varphi}(0) & \hat{\varphi}(-1) & \ddots \\ \ddots & \hat{\varphi}(2) & \hat{\varphi}(1) & \boldsymbol{\hat{\varphi}(0)} & \hat{\varphi}(-1) & \hat{\varphi}(-2) & \ddots \\ \ddots & \hat{\varphi}(1) & \hat{\varphi}(0) & \hat{\varphi}(-1) & \hat{\varphi}(-2) & \hat{\varphi}(-3) & \ddots \\ \ddots & \hat{\varphi}(0) & \hat{\varphi}(-1) & \hat{\varphi}(-2) & \hat{\varphi}(-3) & \hat{\varphi}(-4) & \ddots \\ \ddots & \ddots & \ddots & \ddots & \ddots & \ddots & \ddots \end{bmatrix}, \tag{2.32}$$

where the boldface marks the 00 position.

Hint: The element in the position mn is

$$\langle M_\varphi \chi_n, \chi_{-m} \rangle = \hat{\varphi}(-m-n).$$

Notes on Chapter 2

The essential content of the theory of Hilbert spaces is the study of linear operators on them. The concept of a Hilbert space itself was formulated in the works of Hilbert [287] and Schmidt [469] on the theory of integral equations, while the abstract definition of a Hilbert space was given by von Neumann [535], F. Riesz [429] and Stone [496] in their studies of adjoint operators.

Section 2.2

The earliest results on compact operators are implicit in the studies of Volterra and Fredholm on integral equations. However, the concept itself is essentially due to Hilbert [286], where it was defined for bilinear forms in ℓ^2. In terms of operators, Hilbert requires the operator to map weakly convergent sequences into strongly convergent sequences. In Hilbert spaces (even in reflexive Banach spaces) this is equivalent to requiring that the operator maps the closed unit ball into a relatively compact set. This definition was introduced by F. Riesz [426], who adopted a more abstract point of view and formulated the so-called *Fredholm alternative* (Theorem 2.6). Some results of this section can be generalized in the context of operators acting on Banach spaces. But, in this context, we should adopt the definition of Riesz. One of the famous and classic problems of functional analysis, the so-called *approximation problem*, is to determine whether every compact operator T from a Banach space $\mathcal{X}$ into a Banach space $\mathcal{Y}$ is the uniform limit (i.e. in the uniform operator topology) of a sequence of finite-rank operators. In [206], Enflo settles this problem in the negative by a deep and beautiful counterexample. Schauder was the first to remark that, if $T \in \mathcal{L}(\mathcal{X}, \mathcal{Y})$, then T is compact if and only if $T^* \in \mathcal{L}(\mathcal{X}^*, \mathcal{Y}^*)$ is compact. Compact operators are discussed in many excellent books on operator theory. We mention here the work of Brezis [118], Conway [159], Douglas [176], Dunford and Schwartz [186], Gohberg and Kreĭn [242], Reed and Simon [421], Ringrose [434] and Zhu [570]. In particular, the proofs of Theorems 2.5 and 2.6 can be found in these books.

Section 2.3

The essential spectrum was originally defined by Weyl [546] for a certain differential operator.

Section 2.4

The results concerning self-adjoint operators presented in this section have been established by Hilbert [287] and Schmidt [469] in the case of integral operators in L^2 and by von Neumann [535] in the general case. Corollary 2.14 is proved in [249]. In this book, we do not discuss the important theory of spectral decomposition for bounded self-adjoint operators founded by Hilbert [286]. Nowadays, several approaches to the spectral theory of self-adjoint and normal operators are available. One of the most profound ones is given by the theory of Banach algebras. The spectral decomposition of an unbounded self-adjoint operator was found by von Neumann [535]. We refer the interested reader to [187] for a presentation of this theory and more references.

Study of positive operators started at the beginning of the nineteenth century. They were tied to integral operators (whose study triggered the birth of functional analysis) and to matrices with nonnegative entries. However, positive operators were investigated in a systematic manner much later. Their study began with the works of Riesz [429], Wecken [544], Stone [496] and von Neumann [535]. One of the main results concerning positive operators is the existence (and uniqueness) of their positive square root. This result is proved in Chapter 7 using the functional calculus. The construction of the square root presented in Exercise 2.4.8 is due to Visser [531] and the uniqueness (presented in Exercise 2.4.9) comes from Sz.-Nagy [503]. Another construction of the square root was established by Wecken [544]. The result presented in Exercise 2.4.7 is due to Riesz [429]. The result of Exercise 2.4.6 was proved by Sz.-Nagy [504] with the extra assumption that the A_n are commuting. The result in its whole generality is due to Vigier [528]. See [431] for an account of positive operators and more references.

Section 2.5

The famous inequality proved in Exercise 2.5.1 is due to von Neumann. The proof presented here was discovered by Heinz [276]. This proof is reproduced by Riesz and Sz.-Nagy [431]. For a more recent account, see [134]. Note that Foiaş [218] also proved that von Neumann's inequality characterizes the Hilbert spaces among the complex Banach spaces.

Section 2.7

The shift operator on ℓ^2 is one of the most important operators in mathematics. In Chapter 8, we will introduce its analytic counterpart, which will be one of the main objects of this book. The book of Nikolskii [386] is entirely devoted to the study of the shift operator on the Hardy space.

Section 2.8

The multiplication operator on L^2 is also an important example in the theory of operators. Its importance can be manifested through the spectral theorem, which says that any normal operator with simple spectrum is unitarily equivalent to the operator M_z on $L^2(\mu)$, where μ is a compactly supported Borel measure in the complex plane. See Rudin [441].

Section 2.9

The study of Toeplitz operators originated with Toeplitz [523]. See also the paper of Carathéodory and Fejér [124] where Toeplitz matrices appeared in the solution of the following interpolation problem. Given n complex numbers c_k, $1 \le k \le n$, there is a function f analytic and bounded by 1 in $\mathbb{D}$ such that

$$\hat{f}(k) = c_k \qquad (1 \le k \le n)$$

if and only if the Toeplitz matrix associated with the sequence $c_0, c_1, \dots, c_n$ forms a contraction. For a modern presentation of the Carathéodory–Fejér theorem, see [386, p. 179]. The relationship between the Toeplitz matrices and symbols of Laurent is due to Brown and Halmos [119].

Hankel matrices are named after the nineteenth-century mathematician Hankel, whose doctoral dissertation [256] included the study of the determinant of finite Hankel matrices. These matrices are frequently encountered in applications. For instance, as a special case, the Hilbert matrix defined by $H = ((i+j-1)^{-1})_{i,j\ge 1}$ plays an important role in the study of spectral properties of integral operators of Carleman type. See [298] for the theoretical properties of Hankel matrices and [98, 120, 228] for numerous relations between Hankel matrices and polynomial computations.

3

Harmonic functions on the open unit disk

Harmonic functions, i.e. the solutions of Laplace's equation, play a crucial role in many area of mathematics. The purpose of this chapter is not to develop the whole theory but just to gather some important results that will be important for us in the theory of $\mathcal{H}(b)$ spaces. Most of the results are presented without proofs. However, these topics are standard and can be found in many textbooks, e.g. [354, 442]. We start with the definition of nontangential boundary values and a classic result on angular derivatives, which paves the road to the definition of angular derivative in the sense of Carathéodory. Then, after recalling some facts about a measure μ, e.g. its support, carrier and some decompositions, we study the boundary behavior of the Poisson integral $P\mu$ and its conjugate $Q\mu$. We also discuss the integral means of these two objects. Subharmonic functions are generalizations of harmonic functions. We introduce a particular subharmonic function that is needed later on in the study of the corona problem. The chapter ends with an application of Green's formula to the harmonic function $\log |z|$.

3.1 Nontangential boundary values

Let f be a function defined on the open unit disk $\mathbb{D} = \{z \in \mathbb{C} : |z| < 1\}$. Then, in many situations, we need to know what happens to $f(z)$ when z approaches the boundary $\mathbb{T} = \partial\mathbb{D}$. A particular, but very important, case is when z approaches nontangentially to boundary points. To explain this concept more precisely, let ζ_0 be a point on the unit circle $\mathbb{T}$, and let $C \geq 1$. A region of the form

$$S_C(\zeta_0) = \{z \in \mathbb{D} : |z - \zeta_0| \leq C(1 - |z|)\}$$

is called a *Stolz's domain* anchored at the point ζ_0 (see Figure 3.1).

We say that f has the *nontangential limit* L at ζ_0 provided that, for every fixed value of the parameter C,

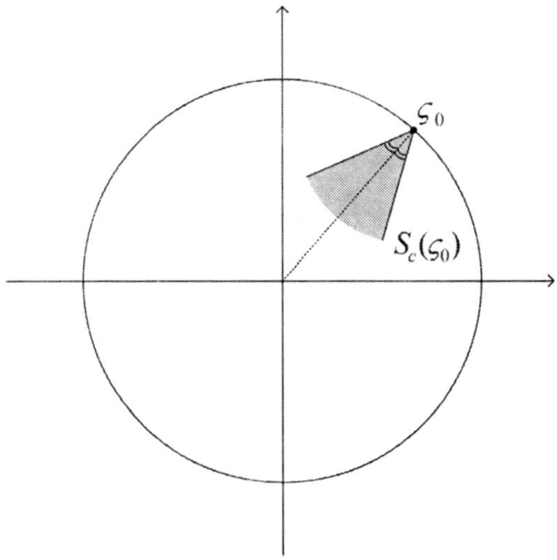

Figure 3.1 A Stolz's domain.

$$\lim_{\substack{z\to\zeta_0\\ z\in S_C(\zeta_0)}} f(z) = L.$$

If so, we define $f(\zeta_0) = L$ and write

$$\lim_{\substack{z\to\zeta_0\\ \sphericalangle}} f(z) = f(\zeta_0).$$

The quantity $f(\zeta_0)$ is referred to as the *boundary value* of f at ζ_0.

We also say that f has the *radial limit* L at $\zeta_0 \in \mathbb{T}$ provided that

$$\lim_{r\to 1} f(r\zeta_0) = L.$$

Clearly, if f has the nontangential limit L at ζ_0, then it has the radial limit L at ζ_0. For bounded analytic functions, a result of Lindelöf asserts that the inverse implication is also true, i.e. the existence of a radial limit implies the existence of a nontangential limit. But, in general, this is no longer true.

We need the following simple geometrical property of Stolz's domains in studying the boundary behavior of analytic functions. Let $z \in S_C(\zeta_0)$ and consider the circle

$$\Gamma_z = \{w : |w - z| = (1 - |z|)/2\}.$$

Then, for each $w \in \Gamma_z$, we have $1 - |z| = 2\,|w - z| \geq 2|w| - 2|z|$, which gives $1 + |z| \geq 2|w|$, and this inequality is equivalent to

$$1 - |z| \leq 2(1 - |w|).$$

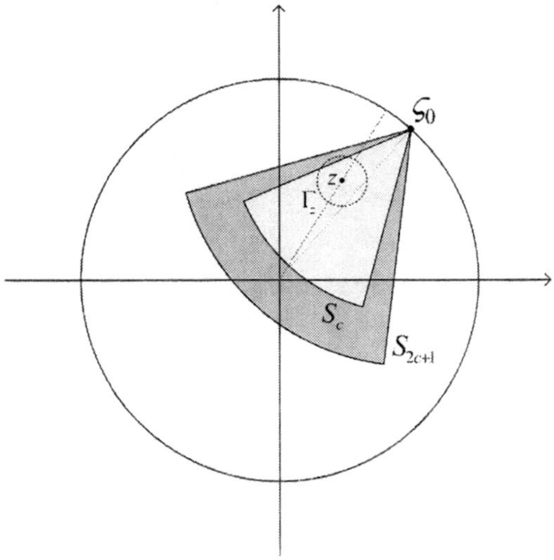

Figure 3.2 The circle Γ_z.

Thus,

$$\begin{aligned} |w-\zeta_0| &\le |w-z| + |z-\zeta_0| \\ &\le \frac{1-|z|}{2} + C\,(1-|z|) \\ &\le (2C+1)\,(1-|w|). \end{aligned} \tag{3.1}$$

In other words, as shown in Figure 3.2, we have

$$\Gamma_z \subset S_{2C+1}(\zeta_0). \tag{3.2}$$

Therefore, all the disks $\{w : |w-z| \le (1-|z|)/2\}$, where the parameter z runs through $S_C(\zeta_0)$, rest entirely in a larger Stolz's domain whose constant does not depend on the location of the disk.

3.2 Angular derivatives

Suppose that a real function $h : [a,b] \longrightarrow \mathbb{R}$ is differentiable on $[a,b]$. In particular, this means that at the point b both limits

$$h(b) = \lim_{t\to b} h(t) \quad \text{and} \quad h'(b) = \lim_{t\to b} \frac{h(t)-h(b)}{t-b}$$

exist. However, we cannot deduce that the limit of $h'(t)$, as t approaches b from the right side, also exists. But for analytic functions a partial result of this type holds. The following result reveals the connection between the nontangential values of $f'(z)$, on the one hand, and that of the quotient $(f(z) - f(\zeta_0))/(z - \zeta_0)$, on the other, as the variable z tends to ζ_0 from within a Stolz's domain.

Theorem 3.1 *Let f be analytic on the open unit disk $\mathbb{D}$, and let $\zeta_0 \in \mathbb{T}$. Then the following are equivalent.*

(i) *Both nontangential limits*

$$f(\zeta_0) = \lim_{\substack{z \to \zeta_0 \\ \sphericalangle}} f(z)$$

and

$$\lim_{\substack{z \to \zeta_0 \\ \sphericalangle}} \frac{f(z) - f(\zeta_0)}{z - \zeta_0}$$

exist.

(ii) *There is a complex number λ such that the nontangential limit*

$$\lim_{\substack{z \to \zeta_0 \\ \sphericalangle}} \frac{f(z) - \lambda}{z - \zeta_0}$$

exists.

(iii) *The function f' has a nontangential limit at ζ_0, i.e.*

$$\lim_{\substack{z \to \zeta_0 \\ \sphericalangle}} f'(z)$$

exists.

Moreover, under the preceding equivalent conditions, we have

$$\lim_{\substack{z \to \zeta_0 \\ \sphericalangle}} \frac{f(z) - f(\zeta_0)}{z - \zeta_0} = \lim_{\substack{z \to \zeta_0 \\ \sphericalangle}} f'(z).$$

Proof (i) $\Longrightarrow$ (ii) This is trivial.

(ii) $\Longrightarrow$ (iii) Put

$$L = \lim_{\substack{z \to \zeta_0 \\ \sphericalangle}} \frac{f(z) - \lambda}{z - \zeta_0}$$

and define

$$g(z) = \frac{f(z) - \lambda}{z - \zeta_0} - L \qquad (z \in \mathbb{D}).$$

Fix a Stolz's domain $S_C(\zeta_0)$. Then, according to our assumption, for any given $\varepsilon > 0$, there is $\delta = \delta(\varepsilon, C)$ such that

$$|g(w)| < \varepsilon \tag{3.3}$$

provided that $w \in S_{2C+1}(\zeta_0)$ and $|w - \zeta_0| < \delta$ (see Figure 3.3).

Let $z \in S_C(\zeta_0)$, and let Γ_z denote the circle of radius $(1 - |z|)/2$ and center z (see Figure 3.2). Hence, by Cauchy's integral formula,

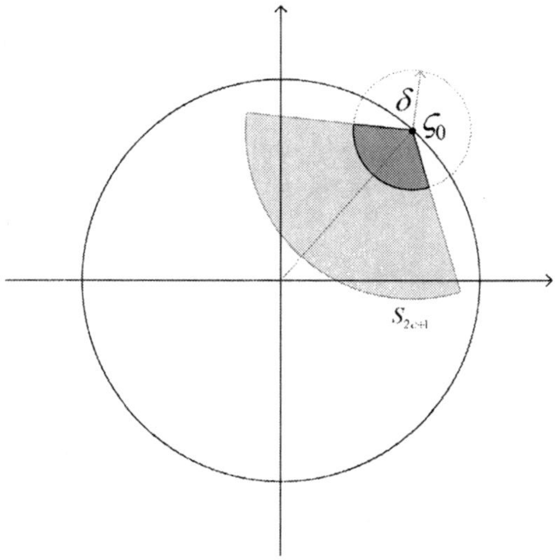

Figure 3.3 A nontangential neighborhood of ζ_0.

$$\begin{aligned} f'(z) &= \frac{1}{2\pi i} \int_{\Gamma_z} \frac{f(w)}{(w-z)^2}\, dw \\ &= L + \frac{1}{2\pi i} \int_{\Gamma_z} \frac{g(w)\,(w-\zeta_0)}{(w-z)^2}\, dw. \end{aligned}$$

If we further assume that

$$1 - |z| < \delta' = \frac{2\delta}{1+2C},$$

then, by (3.1), for each $w \in \Gamma_z$,

$$|w - \zeta_0| < \delta$$

and thus, by (3.2) and (3.3), we have

$$\left| \frac{g(w)\,(w-\zeta_0)}{(w-z)^2} \right| \le \frac{2\varepsilon(1+2C)}{1-|z|}$$

for each $w \in \Gamma_z$. Since the length of Γ_z is $\pi(1-|z|)$, we obtain the estimate

$$|f'(z) - L| \le \varepsilon(1+2C)$$

for each $z \in S_C(\zeta_0)$ with $1 - |z| < 2\delta/(1+2C)$. This means that

$$\lim_{\substack{z \to \zeta_0 \\ \sphericalangle}} f'(z) = L.$$

(iii) $\Longrightarrow$ (i) The assumption implies that the integral

$$\int_{[0,\zeta_0]} f'(w)\, dw$$

is well defined and thus we exploit this fact to define the quantity

$$\lambda = f(0) + \int_{[0,\zeta_0]} f'(w)\,dw.$$

Moreover, by Cauchy's theorem, we have

$$\lambda = f(z) + \int_{[z,\zeta_0]} f'(w)\,dw$$

for each $z \in \mathbb{D}$. On the Stolz's domain $S_C(\zeta_0)$,

$$\left| \int_{[z,\zeta_0]} f'(w)\,dw \right| \leq \left(\sup_{w \in S_C(\zeta_0)} |f'(w)| \right) |z - \zeta_0|,$$

and thus

$$\lim_{\substack{z \to \zeta_0 \\ \sphericalangle}} f(z) = \lambda.$$

As usual, write $\lambda = f(\zeta_0)$. Hence,

$$\begin{aligned} \frac{f(z) - f(\zeta_0)}{z - \zeta_0} &= -\frac{1}{z - \zeta_0} \int_{[z,\zeta_0]} f'(w)\,dw \\ &= f'(\zeta_0) + \frac{1}{z - \zeta_0} \int_{[z,\zeta_0]} (f'(\zeta_0) - f'(w))\,dw. \end{aligned}$$

In each Stolz's domain, the last integral tends to zero az $z \longrightarrow \zeta_0$. Therefore,

$$\lim_{\substack{z \to \zeta_0 \\ \sphericalangle}} \frac{f(z) - f(\zeta_0)}{z - \zeta_0} = f'(\zeta_0). \qquad \square$$

Under the conditions of Theorem 3.1, the quantity

$$f'(\zeta_0) = \lim_{\substack{z \to \zeta_0 \\ \sphericalangle}} \frac{f(z) - f(\zeta_0)}{z - \zeta_0} = \lim_{\substack{z \to \zeta_0 \\ \sphericalangle}} f'(z)$$

is called the *angular derivative* of f at ζ_0. In Chapter 20, using the theory of $\mathcal{H}(b)$ spaces, we will further study this subject for functions that map $\mathbb{D}$ into itself.

3.3 Some well-known facts in measure theory

Through some integral formulas, we will see that there exists a deep connection between harmonic functions on the unit disk and Borel measures on $\mathbb{T}$. In this direction, the study of the boundary behavior of harmonic functions on $\mathbb{D}$ is closely connected to measure theory. That is why we first need to recall some properties and concepts related to Borel measures. These are gathered in this section. All the facts stated in this section are standard and can be found, for example, in [442] or [208].

Let $\zeta \in \mathbb{T}$ and for each $\epsilon > 0$ define

$$I(\zeta, \epsilon) = \{\zeta e^{it} : -\epsilon < t < \epsilon\}.$$

This is an open arc of length 2ϵ centered around ζ. Then the *support* of a positive measure $\mu \in \mathcal{M}(\mathbb{T})$ is the set of points $\zeta \in \mathbb{T}$ such that

$$\mu(I(\zeta, \epsilon)) > 0$$

for all $\epsilon > 0$. The support of μ is a compact subset of $\mathbb{T}$ and is denoted by $\text{supp}(\mu)$. In fact, $U = \mathbb{T} \setminus \text{supp}(\mu)$ is the largest open subset of $\mathbb{T}$ such that $\mu(U) = 0$. A *carrier* of μ is any Borel set E such that $\mu(\mathbb{T} \setminus E) = 0$. Note that the carrier is not uniquely attributed to a measure. For example, for the measure

$$\mu = \sum_{n=1}^{\infty} \frac{1}{2^n} \, \delta_{\zeta_n},$$

where $(\zeta_n)_{n \geq 1}$ is a dense subset of $\mathbb{T}$, we have $\text{supp}(\mu) = \mathbb{T}$, while the countable set $\{\zeta_n : n \geq 1\}$ is a carrier for μ. However, any other Borel set between $\{\zeta_n : n \geq 1\}$ and $\mathbb{T}$ can be equally considered as a carrier for μ. The support and carrier of a complex Borel measure are defined correspondingly to be the carrier and support of its total variation.

According to the *Jordan decomposition theorem*, each $\mu \in \mathcal{M}(\mathbb{T})$ can be written as

$$\mu = (\mu_1 - \mu_2) + i(\mu_3 - \mu_4),$$

where μ_k, $1 \leq k \leq 4$, are positive Borel measures on $\mathbb{T}$. If, furthermore, we ask that the carriers of μ_1 and μ_2, and the carriers of μ_3 and μ_4, be disjoint, then μ_k, $1 \leq k \leq 4$, are unique. When we refer to the Jordan decomposition of μ, we mean this unique representation. In this case, we have

$$\frac{1}{\sqrt{2}} \sum_{k=1}^{4} \mu_k(E) \leq |\mu|(E) \leq \sum_{k=1}^{4} \mu_k(E)$$

for each Borel set $E \subset \mathbb{T}$.

The measure μ is said to be *absolutely continuous* with respect to ν, and is denoted by $\mu \ll \nu$, if the condition $|\nu|(E) = 0$, where E is a Borel subset of $\mathbb{T}$, implies $\mu(E) = 0$. The following assertions are equivalent:

(i) $\mu \ll \nu$;
(ii) $|\mu| \ll \nu$;
(iii) $\mu \ll |\nu|$;

(iv) $|\mu| \ll |\nu|$;
(v) $\mu_k \ll \nu, 1 \leq k \leq 4$;

where $\mu = (\mu_1 - \mu_2) + i(\mu_3 - \mu_4)$ is the Jordan decomposition of μ.

The measure μ and ν are mutually *singular*, which is denoted by $\mu \perp \nu$, if they have disjoint carriers. More explicitly, we can write $\mathbb{T} = A \cup B$, where A and B are disjoint Borel subsets of $\mathbb{T}$ and

$$\mu(E) = \mu(E \cap A) \quad \text{and} \quad \nu(E) = \nu(E \cap B)$$

for all Borel sets $E \subset \mathbb{T}$. Naively speaking, we can say that $\mu \perp \nu$ if they live on disjoint subsets of $\mathbb{T}$. The following assertions are equivalent:

(i) $\mu \perp \nu$;
(ii) $|\mu| \perp \nu$;
(iii) $\mu \perp |\nu|$;
(iv) $|\mu| \perp |\nu|$;
(v) $\mu_k \perp \nu_\ell,\ 1 \leq k \leq 4,\ 1 \leq \ell \leq 4$;

where $\mu = (\mu_1 - \mu_2) + i(\mu_3 - \mu_4)$ and $\nu = (\nu_1 - \nu_2) + i(\nu_3 - \nu_4)$ are the Jordan decompositions of μ and ν.

In practice, ν is usually the Lebesgue measure m and, in this situation, we simply say that μ is absolutely continuous, or singular, without referring to the Lebesgue measure.

Given $\nu \in \mathcal{M}(\mathbb{T})$, the *Lebesgue decomposition theorem* says that each $\mu \in \mathcal{M}(\mathbb{T})$ is uniquely decomposed as

$$\mu = \mu_a + \mu_s, \tag{3.4}$$

where μ_a is absolutely continuous and μ_s is singular, both with respect to ν. In this case, we have

$$\|\mu\| = \|\mu_a\| + \|\mu_s\|. \tag{3.5}$$

If $\nu = m$, then we refer to the above decomposition as the Lebesgue decomposition of μ. According to the *Radon–Nikodym theorem*, μ_a is given by

$$\mu_a(E) = \int_E f \, d\nu,$$

where $f \in L^1(\nu)$ and E is any Borel set of $\mathbb{T}$. The preceding identity is also written as $d\mu_a = f\, d\nu$. Hence, putting together the last two results, we see that, given ν, each $\mu \in \mathcal{M}(\mathbb{T})$ is uniquely decomposed as

$$d\mu = f \, d\nu + d\mu_s, \tag{3.6}$$

where $f \in L^1(\nu)$ and μ_s is singular with respect to ν.

Since an element $\mu \in \mathcal{M}(\mathbb{T})$ has a finite total variation, the Borel set $E_n = \{\zeta \in \mathbb{T} : |\mu|(\{\zeta\}) > 1/n\}$ has to be finite. As a consequence, the set

$$E = \{\zeta \in \mathbb{T} : |\mu|(\{\zeta\}) > 0\}$$

is at most countable. But the set E is precisely the collection of all points of $\mathbb{T}$ at which μ has an atom. Hence, we can define the measure

$$\mu_d = \sum_{\zeta \in E} \mu(\{\zeta\})\, \delta_\zeta.$$

Such a singular measure is called a *discrete measure*. In other words, a discrete measure is a measure that is carried on a countable set. Then put

$$\mu_c = \mu_s - \mu_d.$$

The measure μ_c is singular, but it does not have any atom, i.e. we have $\mu_c(\{\zeta\}) = 0$ for all $\zeta \in \mathbb{T}$. Such a measure is called a *continuous measure*. Based on these new definitions, the Lebesgue decomposition (3.4) becomes

$$\mu = \mu_a + \mu_c + \mu_d, \tag{3.7}$$

where $\mu_a \ll m$, $\mu_d \perp m$, $\mu_c \perp m$, $\mu_c \perp \mu_d$, μ_c is continuous and μ_d is discrete.

There is a practical way to find the function f in the decomposition (3.6). Fix $\zeta \in \operatorname{supp}(\nu)$, and define

$$\underline{d\mu/d\nu}(\zeta) = \liminf_{\epsilon \to 0+} \frac{\mu(I(\zeta,\epsilon))}{\nu(I(\zeta,\epsilon))},$$
$$\overline{d\mu/d\nu}(\zeta) = \limsup_{\epsilon \to 0+} \frac{\mu(I(\zeta,\epsilon))}{\nu(I(\zeta,\epsilon))}.$$

If the above two quantities coincide, i.e. $\underline{d\mu/d\nu}(\zeta) = \overline{d\mu/d\nu}(\zeta)$, we say that the derivative of μ with respect to ν at ζ exists and we denote this common value by $d\mu/d\nu(\zeta)$. Hence,

$$\frac{d\mu}{d\nu}(\zeta) = \lim_{\epsilon \to 0+} \frac{\mu(I(\zeta,\epsilon))}{\nu(I(\zeta,\epsilon))}.$$

For points outside the support of ν, since $\nu(I(\zeta,\epsilon)) = 0$ for small values of ϵ, there is a technical difficulty. But we define $d\mu/d\nu$ to be zero at these points. The *Lebesgue differentiation theorem* says that

$$\frac{d\mu}{d\nu}(\zeta) = f(\zeta)$$

for ν-almost all $\zeta \in \mathbb{T}$. In most applications, we take $\nu = m$ and use the abbreviated notation

$$\underline{D}\mu = \underline{d\mu/dm}, \qquad \overline{D}\mu = \overline{d\mu/dm}, \qquad D\mu = d\mu/dm.$$

Hence, by (3.6), for each measure $\mu \in \mathcal{M}(\mathbb{T})$, we have the decomposition

$$d\mu = f\, dm + d\sigma, \tag{3.8}$$

where σ is a singular measure and $f \in L^1(\mathbb{T})$ is given by

$$f(\zeta) = D\mu(\zeta) = \lim_{\epsilon \to 0} \frac{\mu(I(\zeta,\epsilon))}{m(I(\zeta,\epsilon))} \tag{3.9}$$

and

$$D\sigma(\zeta) = \lim_{\epsilon \to 0} \frac{\sigma(I(\zeta,\epsilon))}{m(I(\zeta,\epsilon))} = 0 \tag{3.10}$$

for almost all $\zeta \in \mathbb{T}$. Note that $m(I(\zeta,\epsilon)) = \epsilon/\pi$. If, furthermore, we assume that $\mu \geq 0$, then, in the decomposition (3.8), we have $f \geq 0$ and $\sigma \geq 0$, and

$$D\sigma(\zeta) = \lim_{\epsilon \to 0} \frac{\sigma(I(\zeta,\epsilon))}{m(I(\zeta,\epsilon))} = +\infty \tag{3.11}$$

for σ-almost all $\zeta \in \mathbb{T}$. Moreover, some possible carriers for $d\mu_a = f\,dm$ are

$$\{\zeta \in \mathbb{T} : 0 < f(\zeta) < +\infty\}$$

or

$$\{\zeta \in \mathbb{T} : 0 < \underline{D}\mu(\zeta) < +\infty\}. \tag{3.12}$$

A carrier for σ is

$$\{\zeta \in \mathbb{T} : D\sigma(\zeta) = +\infty\} \tag{3.13}$$

and, moreover,

$$\mathrm{supp}(\sigma) = \mathrm{Clos}_{\mathbb{T}}\{\zeta \in \mathbb{T} : D\sigma(\zeta) = +\infty\}.$$

Note that

$$\{\zeta \in \mathbb{T} : D\sigma(\zeta) = +\infty\} \subset \{\zeta \in \mathbb{T} : D\mu(\zeta) = +\infty\}.$$

Hence, the set $\{\zeta \in \mathbb{T} : D\mu(\zeta) = +\infty\}$ is also a carrier for σ.

Let μ be a finite positive Borel measure on $\mathbb{T}$, and let $\mu = \mu_a + \mu_s$ be its Lebesgue decomposition, where $\mu_a \ll m$ and $\mu_s \perp m$. Hence, for all $f \in L^p(\mu)$, $1 \leq p < +\infty$, we have

$$\int_{\mathbb{T}} |f|^p\,d\mu = \int_{\mathbb{T}} |f|^p\,d\mu_a + \int_{\mathbb{T}} |f|^p\,d\mu_s. \tag{3.14}$$

This identity enables us to construct a direct decomposition of $L^p(\mu)$, which is orthogonal in the case $p = 2$. To this end, let E be a Borel subset of $\mathbb{T}$ that is a carrier for μ_s. Furthermore, we choose E to be of Lebesgue measure zero. Then we have

$$f = f\chi_E \qquad (f \in L^p(\mu_s))$$

and

$$f = f\chi_{\mathbb{T}\setminus E} \qquad (f \in L^p(\mu_a)).$$

Therefore, $L^p(\mu_a) \cap L^p(\mu_s) = \{0\}$. Moreover, for a given $f \in L^p(\mu)$, write

$$f = f\chi_{\mathbb{T}\setminus E} + f\chi_E = f_a + f_s.$$

Clearly, $f_a \in L^p(\mu_a)$ and $f_s \in L^p(\mu_s)$. Thus,

$$L^p(\mu) = L^p(\mu_a) \oplus L^p(\mu_s),$$

where, by (3.14), the sum is direct when $1 \le p < +\infty$ and orthogonal when $p = 2$. With the above notation, we can rewrite (3.14) as

$$\|f\|^p_{L^p(\mu)} = \|f_a\|^p_{L^p(\mu_a)} + \|f_s\|^p_{L^p(\mu_s)}.$$

Exercises

Exercise 3.3.1 Put

$$\mathcal{M}_a = \{\mu \in \mathcal{M}(\mathbb{T}) : \mu \ll m\} \quad \text{and} \quad \mathcal{M}_s = \{\mu \in \mathcal{M}(\mathbb{T}) : \mu \perp m\}.$$

Show that $\mathcal{M}_a$ and $\mathcal{M}_s$ are closed subspaces of $\mathcal{M}(\mathbb{T})$ and moreover that

$$\mathcal{M}(\mathbb{T}) = \mathcal{M}_a \oplus \mathcal{M}_s.$$

Hint: Use (3.5).

Exercise 3.3.2 Let μ be a real Borel measure on $\mathbb{T}$, and let

$$\mu = \mu_1 - \mu_2$$

be its Jordan decomposition. Show that

$$|\mu| = \mu_1 + \mu_2.$$

3.4 Boundary behavior of $P\mu$

The operator

$$\Delta = \frac{\partial^2}{\partial x^2} + \frac{\partial^2}{\partial y^2}$$

is called the *Laplacian*. It is an easy exercise to check that $\Delta = 4\partial\bar{\partial}$, where

$$\partial = \frac{1}{2}\left(\frac{\partial}{\partial x} - i\frac{\partial}{\partial y}\right) \quad \text{and} \quad \bar{\partial} = \frac{1}{2}\left(\frac{\partial}{\partial x} + i\frac{\partial}{\partial y}\right).$$

Let Ω be an open set of the complex plane and let $f : \Omega \longrightarrow \mathbb{C}$ be a continuous function such that $\partial^2 f/\partial x^2$ and $\partial^2 f/\partial y^2$ exist at every point of Ω. Then f is said to be *harmonic* in Ω if

$$\Delta f = 0$$

at every point of Ω. Three excellent references for harmonic functions are [188], [291] and [442].

The kernel

$$\zeta \longmapsto \frac{\zeta + z}{\zeta - z} \qquad (\zeta \in \mathbb{T},\ z \in \mathbb{D})$$

plays a central role in our studies. Hence, it is no wonder that its real and imaginary parts are also essential. We exploit them to introduce two important functions. The functions

$$P_z(\zeta) = \Re\left(\frac{\zeta + z}{\zeta - z}\right) = \frac{1 - |z|^2}{|\zeta - z|^2}$$

and

$$Q_z(\zeta) = \Im\left(\frac{\zeta + z}{\zeta - z}\right) = \frac{2\,\Im(\bar{\zeta} z)}{|\zeta - z|^2}$$

are respectively called the *Poisson kernel* and the *conjugate Poisson kernel*. Then the *Poisson integral* and the *conjugate Poisson integral* of the measure $\mu \in \mathcal{M}(\mathbb{T})$ are respectively defined by

$$P\mu(z) = \int_{\mathbb{T}} \frac{1 - |z|^2}{|\zeta - z|^2}\, d\mu(\zeta)$$

and

$$Q\mu(z) = \int_{\mathbb{T}} \frac{2\,\Im(\bar{\zeta} z)}{|\zeta - z|^2}\, d\mu(\zeta).$$

If $d\mu = \varphi\, dm$, where $\varphi \in L^1(\mathbb{T})$, we will denote the Poisson integrals of μ by $P\varphi$ and $Q\varphi$. The above formulas create harmonic functions on $\mathbb{D}$. We easily check that

$$P_z(\zeta) = \frac{1 - |z|^2}{|z - \zeta|^2} = 1 + \sum_{n=1}^{\infty} \bar{z}^n \zeta^n + \sum_{n=1}^{\infty} \bar{\zeta}^n z^n \qquad (\zeta \in \mathbb{T},\ z \in \mathbb{D})$$

and

$$Q_z(\zeta) = \frac{2\,\Im(\bar{\zeta} z)}{|z - \zeta|^2} = i\sum_{n=1}^{\infty} \bar{z}^n \zeta^n - i\sum_{n=1}^{\infty} \bar{\zeta}^n z^n \qquad (\zeta \in \mathbb{T},\ z \in \mathbb{D}).$$

Thus, for any measure $\mu \in \mathcal{M}(\mathbb{T})$, we have

$$(P\mu)(z) = \hat{\mu}(0) + \sum_{n=1}^{\infty} \hat{\mu}(-n)\bar{z}^n + \sum_{n=1}^{\infty} \hat{\mu}(n) z^n \qquad (z \in \mathbb{D}) \tag{3.15}$$

and

$$(Q\mu)(z) = i\sum_{n=1}^{\infty} \hat{\mu}(-n)\bar{z}^n - i\sum_{n=1}^{\infty} \hat{\mu}(n) z^n \qquad (z \in \mathbb{D}). \tag{3.16}$$

Since $|\hat{\mu}(n)| \leq \|\mu\|$, the functions g_1 and g_2 defined by

$$g_1(z) = \sum_{n=1}^{\infty} \overline{\hat{\mu}(-n)}\, z^n \quad \text{and} \quad g_2(z) = \sum_{n=1}^{\infty} \hat{\mu}(n)\, z^n \qquad (z \in \mathbb{D})$$

are analytic on $\mathbb{D}$, and (3.15) and (3.16) can be rewritten as

$$P\mu = \hat{\mu}(0) + \bar{g}_1 + g_2 \quad \text{and} \quad Q\mu = i\bar{g}_1 - ig_2.$$

Moreover, we see that

$$(P\mu)(z) + i(Q\mu)(z) = \hat{\mu}(0) + 2g_2(z) = \hat{\mu}(0) + 2\sum_{n=1}^{\infty} \hat{\mu}(n) z^n.$$

In particular, we see that $P\mu + iQ\mu$ is an analytic function on $\mathbb{D}$. Moreover, we also easily see, using (3.15) and the uniqueness theorem for Fourier coefficients, that, for $\mu_1, \mu_2 \in \mathcal{M}(\mathbb{T})$,

$$P\mu_1 = P\mu_2 \quad \Longrightarrow \quad \mu_1 = \mu_2.$$

This means that the Poisson integral $P\mu$ uniquely determines the measure μ.

One of the oldest results about the boundary behavior of $P\mu$ is due to Fatou.

Theorem 3.2 *If μ is a real Borel measure in $\mathcal{M}(\mathbb{T})$, then, for each $\zeta \in \mathbb{T}$, we have*

$$\underline{D}\mu(\zeta) \leq \liminf_{r \to 1} P\mu(r\zeta) \leq \limsup_{r \to 1} P\mu(r\zeta) \leq \overline{D}\mu(\zeta). \tag{3.17}$$

In particular, if $\zeta \in \mathbb{T}$ is such that $D\mu(\zeta)$ exists, then the radial limit of $P\mu$ at ζ exists and, moreover,

$$\lim_{r \to 1} P\mu(r\zeta) = D\mu(\zeta). \tag{3.18}$$

This classic result can be found in many textbooks, e.g. see [188], [291] and [354].

Note that, by considering the real part and the imaginary part of μ, the formula (3.18) is still valid for complex measure $\mu \in \mathcal{M}(\mathbb{T})$. Several special cases of Fatou's result are frequently used in applications. According to the Lebesgue differentiation theorem, we can decompose $\mu \in \mathcal{M}(\mathbb{T})$ as

$$d\mu = \varphi\, dm + d\sigma,$$

where $\varphi \in L^1(\mathbb{T})$ and σ is singular and, moreover, $D\mu(\zeta)$ exists for almost all $\zeta \in \mathbb{T}$ and is equal to $\varphi(\zeta)$ (see (3.8) and (3.9)). Hence, the following result is immediate.

Corollary 3.3 *If*

$$d\mu = \varphi\, dm + d\sigma,$$

where $\varphi \in L^1(\mathbb{T})$ *and* σ *is singular with respect to the Lebesgue measure, then*

$$\lim_{r \to 1} P\mu(r\zeta) = \varphi(\zeta)$$

for all $\zeta \in \mathbb{T}$ *except possibly on a set of Lebesgue measure zero.*

As a special case of Corollary 3.3, we obtain the following result (see also (3.10)).

Corollary 3.4 *If* $\sigma \in \mathcal{M}(\mathbb{T})$ *is singular, then*

$$\lim_{r \to 1} P\sigma(r\zeta) = 0$$

for all $\zeta \in \mathbb{T} \setminus E$*, where* E *is a Borel set with* $m(E) = 0$.

If the singular measure σ is positive, then, because of (3.11), we can shed more light on the radial behavior of $P\sigma$.

Corollary 3.5 *If* $\sigma \in \mathcal{M}(\mathbb{T})$ *is a positive singular measure, then*

$$\lim_{r \to 1} P\sigma(r\zeta) = +\infty$$

for all $\zeta \in \mathbb{T} \setminus E$*, where* E *is a Borel set with* $\sigma(E) = 0$.

In fact, using Theorem 3.2, (3.10) and (3.11), we can give a more detailed version of Corollaries 3.4 and 3.5. Put

$$\begin{aligned} E_0 &= \{\zeta \in \mathbb{T} : \overline{D}\sigma(\zeta) = 0\}, \\ E_\infty &= \{\zeta \in \mathbb{T} : \underline{D}\sigma(\zeta) = +\infty\} \end{aligned}$$

and $E = \mathbb{T} \setminus (E_0 \cup E_\infty)$. Then the following hold:

(i) E, E_0 and E_∞ are disjoint Borel sets and $\mathbb{T} = E \cup E_0 \cup E_\infty$;
(ii) $\lim_{r\to 1} P\sigma(r\zeta) = 0$ for all $\zeta \in E_0$;
(iii) $\lim_{r\to 1} P\sigma(r\zeta) = +\infty$ for all $\zeta \in E_\infty$;
(iv) $\sigma(E_0) = 0$ and $m(E_0) = m(\mathbb{T}) = 1$;
(v) $m(E_\infty) = 0$ and $\sigma(E_\infty) = \sigma(\mathbb{T}) = \|\sigma\|$;
(vi) $m(E) = \sigma(E) = 0$.

In all cases, $E_0 \neq \emptyset$. We have $E_\infty \neq \emptyset$ if and only if $\sigma \neq 0$. The exceptional set E is more delicate. First, it is a small set in the sense that $m(E) = \sigma(E) = 0$. Second, the Poisson integral might fail to have a radial limit at these points, or, if the limit exists, it is not necessarily equal to 0 or to $+\infty$.

By (3.13), σ is carried on the set E_∞ and $\operatorname{supp}(\sigma) = \operatorname{Clos}_{\mathbb{T}} E_\infty$. But, by Theorem 3.2, we have

$$E_\infty \subset S = \left\{\zeta \in \mathbb{T} : \lim_{r \to 1} P\sigma(r\zeta) = +\infty\right\},$$

and the Poisson integral has harmonic continuation across $\mathbb{T} \setminus \operatorname{supp}(\sigma)$. Thus, we can also use S as a carrier for σ. This in turn implies $\operatorname{supp}(\sigma) = \operatorname{Clos}_{\mathbb{T}} S$.

In most of the situations above in which we studied the existence of a radial limit, the result still holds if we replace the "radial limit" by the "nontangential limit".

Exercise

Exercise 3.4.1 Let μ be a positive measure in $\mathcal{M}(\mathbb{T})$, and let

$$d\mu = f\,dm + d\sigma$$

be its Lebesgue decomposition. Show that the singular measure σ is carried on the set

$$\{\zeta \in \mathbb{T} : \lim_{r\to 1} P\mu(r\zeta) = +\infty\}.$$

Hint: Use (3.13) and Theorem 3.2.

3.5 Integral means of $P\mu$

The identity

$$\int_{\mathbb{T}} \frac{1-|z|^2}{|\zeta - z|^2}\, dm(\zeta) = 1 \qquad (|z| < 1) \tag{3.19}$$

is easy to verify. Then, using Fubini's theorem and (3.19), we see that $h = P\mu$ satisfies

$$\int_{\mathbb{T}} |h(r\zeta)|\, dm(\zeta) \le \int_{\mathbb{T}} \left(\int_{\mathbb{T}} \frac{1-r^2}{|r\zeta - \omega|^2}\, dm(\zeta) \right) d|\mu|(\omega) = \|\mu\| \qquad (0 \le r < 1).$$

Hence, h fulfills the growth restriction

$$\sup_{0\le r<1} \|h_r\|_1 < \infty, \tag{3.20}$$

where $h_r(\zeta) = h(r\zeta)$, $\zeta \in \mathbb{T}$. In fact, the converse to this assertion is also true, and we have the following complete characterization.

Theorem 3.6 *Let h be a function defined on $\mathbb{D}$. Then the following are equivalent.*

(i) *Function h is a harmonic function on $\mathbb{D}$ that satisfies*

$$\sup_{0\le r<1} \|h_r\|_1 < \infty.$$

(ii) *There exists a unique complex Borel measure* μ *on* $\mathbb{T}$ *such that* $h = P\mu$.

Moreover, in this case, we have $\|h_r\|_1 \leq \|\mu\|$ *and*

$$\lim_{r\to 1} \int_{\mathbb{T}} \psi(\zeta) h_r(\zeta)\, dm(\zeta) = \int_{\mathbb{T}} \psi(\zeta)\, d\mu(\zeta) \tag{3.21}$$

for each $\psi \in \mathcal{C}(\mathbb{T})$. *In particular, if* $h = P\varphi$, *where* $\varphi \in L^1(\mathbb{T})$, *then*

$$\lim_{r\to 1} \|h_r - \varphi\|_1 = 0.$$

Note that (3.21) can be reformulated by saying that $h_r\, dm$ tends to $d\mu$, as $r \longrightarrow 1$, in the weak-star topology of $\mathcal{M}(\mathbb{T}) = \mathcal{C}(\mathbb{T})^*$.

We can apply Theorem 3.6 to positive harmonic functions. On the one hand, if μ is a positive and finite Borel measure, then $h = P\mu$ is a positive harmonic function on $\mathbb{D}$ that satisfies (3.20). On the other hand, if h is a given positive harmonic function, then, by the mean value property for harmonic functions, it satisfies

$$\|h_r\|_1 = \int_{\mathbb{T}} |h_r(\zeta)|\, dm(\zeta) = \frac{1}{2\pi} \int_0^{2\pi} h(re^{i\theta})\, d\theta = h(0) \qquad (0 \leq r < 1).$$

Thus, by Theorem 3.6, there is a positive measure μ such that $h = P\mu$. Therefore, Theorem 3.6 can be rewritten as follows.

Corollary 3.7 *Let* h *be a function defined on* $\mathbb{D}$. *Then the following are equivalent.*

(i) *Function* h *is a positive harmonic function on* $\mathbb{D}$.
(ii) *There exists a unique finite positive Borel measure* μ *on* $\mathbb{T}$ *such that* $h = P\mu$.

We can improve Theorem 3.6 if we know more about the behavior of h_r, or about the function φ in the Poisson representation $h = P\varphi$. Two such results are stated below.

Theorem 3.8 *Let* h *be a function defined on* $\mathbb{D}$, *and let* $1 < p < \infty$. *Then the following are equivalent.*

(i) *Function* h *is a harmonic function on* $\mathbb{D}$ *that satisfies*

$$\sup_{0\leq r<1} \|h_r\|_p < \infty.$$

(ii) *There exists a unique function* $\varphi \in L^p(\mathbb{T})$ *such that* $h = P\varphi$.

Moreover, in this case, we have $\|h_r\|_p \leq \|\varphi\|_p$ *and*

$$\lim_{r\to 1} \|h_r - \varphi\|_p = 0.$$

Theorem 3.8 does not hold either for $p = 1$ or for $p = \infty$. The case $p = 1$ is in fact treated in Theorem 3.6. The case $p = \infty$ is discussed below.

Theorem 3.9 *Let h be a function defined on $\mathbb{D}$. Then the following are equivalent.*

(i) *Function h is a harmonic function on $\mathbb{D}$ that satisfies*

$$\sup_{z\in\mathbb{D}} |h(z)| = \sup_{0\le r<1} \|h_r\|_\infty < \infty.$$

(ii) *There exists a unique function $\varphi \in L^\infty(\mathbb{T})$ such that $h = P\varphi$.*

Moreover, in this case, the following hold:

(a) $\|h_r\|_\infty \le \|\varphi\|_\infty$;

(b) *we have*

$$\lim_{r\to 1} \int_{\mathbb{T}} \psi(\zeta) h_r(\zeta)\, dm(\zeta) = \int_{\mathbb{T}} \psi(\zeta)\varphi(\zeta)\, dm(\zeta)$$

for each $\psi \in L^1(\mathbb{T})$;

(c) *but if the stronger condition $\varphi \in \mathcal{C}(\mathbb{T})$ is fulfilled, then*

$$\lim_{r\to 1} \|h_r - \varphi\|_\infty = 0.$$

The results above (Theorems 3.6, 3.8 and 3.9 and Corollary 3.7) are standard and can be found in many textbooks of complex and harmonic analysis (see e.g. [442] or [291]).

Exercise

Exercise 3.5.1 Show that the linear span of the Poisson kernels P_z, where $z \in \mathbb{D}$, is uniformly dense in $\mathcal{C}(\mathbb{T})$.

Hint: Use the Riesz representation theorem (1.6), and the uniqueness part of Theorem 3.6.

3.6 Boundary behavior of $Q\mu$

The conjugate Poisson integral of the measure $\mu \in \mathcal{M}(\mathbb{T})$ was defined by the formula

$$Q\mu(z) = \int_{\mathbb{T}} \Im\left(\frac{\zeta + z}{\zeta - z}\right) d\mu(\zeta) = \int_{\mathbb{T}} \frac{2\,\Im(\bar{\zeta} z)}{|\zeta - z|^2}\, d\mu(\zeta).$$

Generally speaking, the study of the radial behavior of $Q\mu$ is more delicate than that of $P\mu$ and leads to the theory of singular integrals. One of the oldest results in this theory says that the limit

$$\lim_{\varepsilon\to 0+} \int_{|\omega-\zeta|\geq\varepsilon} \frac{d\mu(\omega)}{1-\bar{\omega}\zeta}$$

exists and is finite for almost all $\zeta \in \mathbb{T}$. This is called the *principal value* of the integral and is written as

$$P.V. \int_{\mathbb{T}} \frac{d\mu(\omega)}{1-\bar{\omega}\zeta} = \lim_{\varepsilon\to 0+} \int_{|\omega-\zeta|\geq\varepsilon} \frac{d\mu(\omega)}{1-\bar{\omega}\zeta}.$$

Since

$$1 + \frac{\omega+\zeta}{\omega-\zeta} = \frac{2}{1-\bar{\omega}\zeta} \tag{3.22}$$

and, with $\omega = e^{it}$ and $\zeta = e^{i\theta}$,

$$\frac{\omega+\zeta}{\omega-\zeta} = i\cot\left(\frac{\theta-t}{2}\right),$$

we deduce that

$$P.V. \int_{\mathbb{T}} \cot\left(\frac{\theta-t}{2}\right) d\mu(e^{it}) = \lim_{\varepsilon\to 0+} \int_{\pi\geq|\theta-t|\geq\varepsilon} \cot\left(\frac{\theta-t}{2}\right) d\mu(e^{it})$$

also exists and is finite for almost all $e^{i\theta} \in \mathbb{T}$. This result can be exploited to show that the radial limit of $Q\mu$ also exists and is finite for almost all $e^{i\theta} \in \mathbb{T}$, and, moreover,

$$\lim_{r\to 1} Q\mu(re^{i\theta}) = P.V. \int_{\mathbb{T}} \cot\left(\frac{\theta-t}{2}\right) d\mu(e^{it}).$$

Owing to the importance of this quantity, it is also denoted by $\tilde{\mu}(\zeta)$ and is called the *Hilbert transform* of μ at ζ. Hence, for the record,

$$\tilde{\mu}(\zeta) = \lim_{r\to 1} Q\mu(r\zeta). \tag{3.23}$$

Note that, according to (3.22), we have

$$\mu(\mathbb{T}) + i\tilde{\mu}(\zeta) = P.V. \int_{\mathbb{T}} \frac{2\,d\mu(\omega)}{1-\bar{\omega}\zeta}. \tag{3.24}$$

As usual, if $d\mu = \varphi\,dm$, then instead of $\tilde{\mu}$ we will write $\tilde{\varphi}$.

The properties above concerning the radial behavior of the conjugate Poisson integral are standard and can be found, for example, in [354, chap. V] or [320, chap. I].

3.7 Integral means of $Q\mu$

It should come as no surprise that the integral means of $Q\mu$ are also difficult to study. Some relevant results are stated below.

Theorem 3.10 *Let $\varphi \in L^p(\mathbb{T})$, $1 < p < \infty$. Then the following assertions hold:*

(i) *$\tilde{\varphi} \in L^p(\mathbb{T})$ and there is a constant c_p such that*

$$\|(Q\varphi)_r\|_p \le \|\tilde{\varphi}\|_p \le c_p\|\varphi\|_p;$$

(ii) *$Q\varphi = P\tilde{\varphi}$ and*

$$\lim_{r\to 1} \|(Q\varphi)_r - \tilde{\varphi}\|_p = 0.$$

A proof of this classic result can be found in many textbooks; see e.g. [141, p. 65], [188, p. 54], [233, p. 108] or [572, p. 253].

The following result is an easy application of Theorem 3.10 and gives the relation between the Fourier coefficients of an L^p function and the Fourier coefficient of its Hilbert transform. It will be useful to study the Riesz projection on $L^p(\mathbb{T})$.

Lemma 3.11 *Let $\varphi \in L^p(\mathbb{T})$, $1 < p < \infty$. Then*

$$\widehat{\tilde{\varphi}}(n) = -i\,\mathrm{sgn}(n)\hat{\varphi}(n) \qquad (n \in \mathbb{Z}),$$

where

$$\mathrm{sgn}(n) = \begin{cases} 1 & \text{if } n > 0, \\ 0 & \text{if } n = 0, \\ -1 & \text{if } n < 0. \end{cases}$$

Proof By Theorem 3.10, we have $Q\varphi = P\tilde{\varphi}$. Using (3.15) and (3.16), we have

$$i\sum_{n=1}^{+\infty} \hat{\varphi}(-n)\bar{z}^n - i\sum_{n=1}^{+\infty} \hat{\varphi}(n)z^n = \widehat{\tilde{\varphi}}(0) + \sum_{n=1}^{+\infty} \widehat{\tilde{\varphi}}(-n)\bar{z}^n + \sum_{n=1}^{+\infty} \widehat{\tilde{\varphi}}(n)z^n,$$

which implies that

$$\begin{cases} \widehat{\tilde{\varphi}}(0) = 0, & \\ \widehat{\tilde{\varphi}}(-n) = i\hat{\varphi}(-n), & n \ge 1, \\ \widehat{\tilde{\varphi}}(n) = -i\hat{\varphi}(n), & n \ge 1. \end{cases}$$

This completes the proof. □

By Theorem 3.6, the condition $\varphi \in L^1(\mathbb{T})$ is enough to ensure that

$$\sup_{0\le r<1} \|(P\varphi)_r\|_1 < \infty.$$

However, this condition is not strong enough to imply $\sup_{0\le r<1} \|(Q\varphi)_r\|_1 < \infty$ and $\tilde{\varphi} \in L^1(\mathbb{T})$. To obtain this property, a slightly stronger assumption is needed. To state the precise result, we need to introduce some notation. Let

$$\log^+ t = \max(\log t, 0) \quad \text{and} \quad \log^- t = \max(-\log t, 0), \quad t > 0. \tag{3.25}$$

Hence, we have

$$\log = \log^+ - \log^- \quad \text{and} \quad |\log| = \log^+ + \log^- . \tag{3.26}$$

Theorem 3.12 *Suppose that φ satisfies*

$$\int_{\mathbb{T}} |\varphi| \log^+ |\varphi| \, dm < \infty.$$

Then

$$\sup_{0 \le r < 1} \|(Q\varphi)_r\|_1 < \infty.$$

Moreover, in this case, we have $\tilde{\varphi} \in L^1(\mathbb{T})$, $Q\varphi = P\tilde{\varphi}$ and

$$\lim_{r \to 1} \|(Q\varphi)_r - \tilde{\varphi}\|_1 = 0.$$

For a proof of this result, see [188, p. 58] or [572, p. 254].

But if we only know that $\varphi \in L^1(\mathbb{T})$, the best available result is due to Kolmogorov, which says that $\tilde{\varphi}$ is in weak-L^1.

Theorem 3.13 *Let $\mu \in \mathcal{M}(\mathbb{T})$. Then*

$$m(|\tilde{\mu}| > t) \le C \frac{\|\mu\|}{t} \qquad (t > 0),$$

where C is a universal constant. In particular, for each $\varphi \in L^1(\mathbb{T})$, we have

$$m(|\tilde{\varphi}| > t) \le C \frac{\|\varphi\|_1}{t} \qquad (t > 0).$$

A proof of this classic result can also be found in many textbooks, see e.g. [320, p. 94].

Under a mild condition, the Hilbert transform is reversible. This is made precise below.

Theorem 3.14 *Let $\varphi \in L^1(\mathbb{T})$ be such that $\tilde{\varphi} \in L^1(\mathbb{T})$. Then*

$$\tilde{\tilde{\varphi}} = \hat{\varphi}(0) - \varphi.$$

By the same token, the condition $\varphi \in L^\infty(\mathbb{T})$ is not enough to ensure that $\sup_{0 \le r < 1} \|(Q\varphi)_r\|_\infty < \infty$ and $\tilde{\varphi} \in L^\infty(\mathbb{T})$. One of the best available results in this case is the following.

Theorem 3.15 *Let $\varphi \in L^\infty(\mathbb{T})$ with $\|\varphi\|_\infty < \pi/2$. Then $\exp(\tilde{\varphi}) \in L^1(\mathbb{T})$.*

For a proof see [233, p. 110] or [320, p. 100].

Corollary 3.16 *Let $\varphi \in L^\infty(\mathbb{T})$. Then*

$$\tilde{\varphi} \in \bigcap_{0 < p < \infty} L^p(\mathbb{T}).$$

To explain better the Hilbert transform of $L^\infty(\mathbb{T})$ functions, we need to introduce a new space. A function $\varphi \in L^1(\mathbb{T})$ is of *bounded mean oscillation*, in short $\varphi \in BMO$, if

$$\|\varphi\|_* = \sup_{I \text{ subarc of } \mathbb{T}} \left(\frac{1}{m(I)} \int_I |\varphi - \varphi_I| \, dm \right) < \infty,$$

where

$$\varphi_I = \frac{1}{m(I)} \int_I \varphi \, dm.$$

Theorem 3.17 *Let $\varphi \in L^\infty(\mathbb{T})$. Then $\tilde{\varphi} \in BMO$. Moreover, there is a universal constant C such that*

$$\|\tilde{\varphi}\|_* \le C \|\varphi\|_\infty.$$

This standard result can also be found in many textbooks, see [141, p. 69] or [233, p. 220].

Let us emphasize that $\|\cdot\|_*$ is a seminorm on BMO. To turn it into a norm, we can define

$$\|\varphi\|_{BMO} = \|\varphi\|_* + \left| \int_{\mathbb{T}} \varphi \, dm \right|.$$

Exercises

Exercise 3.7.1 Let $\mu \in \mathcal{M}(\mathbb{T})$, and define

$$f(z) = \int_{\mathbb{T}} \frac{\zeta + z}{\zeta - z} \, d\mu(\zeta) \qquad (z \in \mathbb{D}).$$

Show that

$$\lim_{r \to 1} f(r\zeta) = D\mu(\zeta) + i\tilde{\mu}(\zeta)$$

for almost all $\zeta \in \mathbb{T}$.

Hint: Use Theorem 3.2 and the definition (3.23).

Exercise 3.7.2 Let $0 < a \le 1$ and $b > 0$. Show that

$$\log a + \log^+ b \le \log^+(ab) \le \log^+ b.$$

3.8 Subharmonic functions

A function $\varphi : \Omega \longrightarrow [-\infty, +\infty)$ is called *subharmonic* on Ω if it is upper semicontinuous on Ω, i.e. $f^{-1}[-\infty, c)$ is an open set for each c, and fulfills the submean property

$$\varphi(z) \le \frac{1}{2\pi}\int_0^{2\pi} \varphi(z+re^{it})\,dt$$

whenever $\overline{D(z,r)} \subset \Omega$. If $\varphi \in \mathcal{C}^2$, i.e. twice continuously differentiable, then φ is subharmonic on Ω if and only if

$$\Delta\varphi \ge 0 \qquad (\text{on } \Omega).$$

Two good references for subharmonic functions are [233] and [274].

The following result gives a criterion for subharmonicity for a particular family of smooth functions.

Theorem 3.18 *Let $\varphi : \Omega \longrightarrow \mathbb{R}$ be a function of class $\mathcal{C}^2$, and let f be an analytic function on Ω. Then we have*

$$\Delta(e^{\varphi}|f|^2) = e^{\varphi}|f|^2\Delta\varphi + 4e^{\varphi}|f\partial\varphi + \partial f|^2.$$

Moreover, if φ is subharmonic, then so is $e^{\varphi}|f|^2$.

Proof Without loss of generality, we may assume that $f \ne 0$. For simplicity, put $u = e^{\varphi}|f|^2$. Then, using the fact that $\bar\partial f = 0$ and that $\bar\partial \bar f = \overline{\partial f}$, we have

$$\begin{aligned}
\Delta u &= 4\partial\bar\partial(e^{\varphi}|f|^2)\\
&= 4\partial(e^{\varphi}(|f|^2\bar\partial\varphi + f\bar\partial\bar f))\\
&= 4e^{\varphi}(|f|^2\bar\partial\varphi + f\overline{\partial f})\partial\varphi + 4e^{\varphi}\partial(|f|^2\bar\partial\varphi + f\overline{\partial f})\\
&= 4e^{\varphi}((|f|^2\bar\partial\varphi + f\overline{\partial f})\partial\varphi + |f|^2\partial\bar\partial\varphi + \bar f\,\bar\partial\varphi\partial f + \partial f\overline{\partial f}).
\end{aligned}$$

Since φ is real-valued, $\bar\partial\varphi = \overline{\partial\varphi}$ and we obtain

$$\begin{aligned}
\Delta u &= e^{\varphi}|f|^2\Delta\varphi + 4e^{\varphi}(|f|^2|\partial\varphi|^2 + f\overline{\partial f}\partial\varphi + \bar f\,\overline{\partial\varphi}\partial f + |\partial f|^2)\\
&= e^{\varphi}|f|^2\Delta\varphi + 4e^{\varphi}|\partial f + f\partial\varphi|^2.
\end{aligned}$$

In particular, we get $\Delta u \ge e^{\varphi}|f|^2\Delta\varphi$. Therefore, if φ is subharmonic, i.e. $\Delta\varphi \ge 0$, then we surely have $\Delta u \ge 0$ and thus the function u is also subharmonic. □

3.9 Some applications of Green's formula

If u and v are of class $\mathcal{C}^2$ in a neighborhood of a compact set $K \subset \mathbb{R}^2$ with a rectifiable boundary, the classic *Green's formula* says that

$$\int_K (u\Delta v - v\Delta u)\,dA = \int_{\partial K}\left(u\frac{\partial v}{\partial n} - v\frac{\partial u}{\partial n}\right)ds, \tag{3.27}$$

where $\partial/\partial n$ stands for differentiation in the direction of the outer normal n_K, ds is the arc length on ∂K, and $dA(z) = dxdy$ is the area measure. As a particular case, if K is the corona $\Omega = \Omega_{r_1,r_2} = \{z \in \mathbb{C} : r_1 \leq |z| \leq r_2\}$, then

$$n_\Omega(r_2 e^{i\vartheta}) = e^{i\vartheta} \quad \text{and} \quad n_\Omega(r_1 e^{i\vartheta}) = -e^{i\vartheta} \qquad (\vartheta \in [0, 2\pi]).$$

Thus, for a function f of class $\mathcal{C}^1$ in a neighborhood of Ω, we get

$$\frac{\partial f}{\partial n}(\zeta) = \nabla f(\zeta) \cdot n_\Omega(\zeta) = \begin{cases} \dfrac{\partial f}{\partial r}(r_2 e^{i\vartheta}) & \text{if} \quad \zeta = r_2 e^{i\vartheta}, \\ -\dfrac{\partial f}{\partial r}(r_1 e^{i\vartheta}) & \text{if} \quad \zeta = r_1 e^{i\vartheta}. \end{cases}$$

Therefore, Green's formula can be rewritten as

$$\begin{aligned} &\int_{\Omega_{r_1,r_2}} (u\Delta v - v\Delta u)\, dxdy \\ &\quad = -r_1 \int_0^{2\pi} \left(u(r_1 e^{i\vartheta}) \frac{\partial v}{\partial r}(r_1 e^{i\vartheta}) - v(r_1 e^{i\vartheta}) \frac{\partial u}{\partial r}(r_1 e^{i\vartheta}) \right) d\vartheta \\ &\qquad + r_2 \int_0^{2\pi} \left(u(r_2 e^{i\vartheta}) \frac{\partial v}{\partial r}(r_2 e^{i\vartheta}) - v(r_2 e^{i\vartheta}) \frac{\partial u}{\partial r}(r_2 e^{i\vartheta}) \right) d\vartheta. \end{aligned}$$

Green's formula is classic and can be found in many textbooks of complex and harmonic analysis, e.g. [42]. In fact, we only need this formula in the case where K is a corona (see the above formula). In this particular case, the proof is very simple and can be reduced to an integration by parts; see e.g. [136, p. 298].

We exploit Green's formula to obtain the following result.

Theorem 3.19 *Let w be a $\mathcal{C}^2$ function on a neighborhood of $\bar{\mathbb{D}}$. Then*

$$\frac{1}{2\pi} \int_{\mathbb{D}} \log \frac{1}{|z|}\, \Delta w(z)\, dA(z) = \int_{\mathbb{T}} w(\zeta)\, dm(\zeta) - w(0).$$

Proof Let $K = K_\varepsilon = \{z \in \mathbb{C} : \varepsilon \leq |z| \leq 1\}$, where $0 < \varepsilon < 1$ is fixed. Then we apply Green's formula to the functions

$$u(z) = \log \frac{1}{|z|} \quad \text{and} \quad v(z) = w(z).$$

It is straightforward to verify that $\Delta u \equiv 0$ on K_ε, $u \equiv 0$ on $\mathbb{T}$, $(\partial u/\partial r)(\varepsilon e^{i\vartheta}) = -1/\varepsilon$ and $(\partial u/\partial r)(e^{i\vartheta}) = -1$. Hence, we get

$$\begin{aligned} &\int_{K_\varepsilon} \log \frac{1}{|z|}\, \Delta w(z)\, dA(z) \\ &\quad = \varepsilon \log \varepsilon \int_0^{2\pi} \frac{\partial w}{\partial r}(\varepsilon e^{i\vartheta})\, d\vartheta - \int_0^{2\pi} w(\varepsilon e^{i\vartheta})\, d\vartheta + \int_0^{2\pi} w(e^{i\vartheta})\, d\vartheta. \end{aligned}$$

Using the fact that $\log |z|$ is integrable on $\bar{\mathbb{D}}$, we easily see that the left-hand side tends to

$$\int_{\mathbb{D}} \log \frac{1}{|z|}\, \Delta w(z)\, dA(z)$$

as ε tends to 0. Moreover, $\partial w/\partial r$ is bounded on the compact set $\bar{\mathbb{D}}$, whence

$$\int_0^{2\pi} \frac{\partial w}{\partial r}(\varepsilon e^{i\vartheta})\, d\vartheta = O(1), \qquad \varepsilon \longrightarrow 0.$$

Thus,

$$\lim_{\varepsilon\to 0} \left(\varepsilon \log \varepsilon \int_0^{2\pi} \frac{\partial w}{\partial r}(\varepsilon e^{i\vartheta})\, d\vartheta \right) = 0.$$

Finally, by continuity of w,

$$\lim_{\varepsilon\to 0} \left(\int_0^{2\pi} w(\varepsilon e^{i\vartheta}) \right) = 2\pi w(0).$$

Therefore, we deduce that

$$\int_{\mathbb{D}} \log \frac{1}{|z|} \Delta w(z)\, dA(z) = -2\pi w(0) + 2\pi \int_{\mathbb{T}} w(\zeta)\, dm(\zeta),$$

which gives the desired identity. □

As another application, let f be a function of class $\mathcal{C}^1$ in an open neighborhood of Ω_{r_1,r_2}. Then by Green's formula

$$\int_{\Omega_{r_1,r_2}} \bar{\partial} f\, dA = \frac{1}{2} \left(r_2 \int_0^{2\pi} f(r_2 e^{i\vartheta}) e^{i\vartheta}\, d\vartheta - r_1 \int_0^{2\pi} f(r_1 e^{i\vartheta}) e^{i\vartheta}\, d\vartheta \right). \tag{3.28}$$

See [136, p. 297] for a proof of this formula. We use it to obtain the following result.

Theorem 3.20 *Let φ be a compactly supported function of class $\mathcal{C}^\infty$ on $\mathbb{C}$, and define*

$$\psi(a) = \frac{1}{\pi} \iint_{\mathbb{C}} \frac{\varphi(z)}{a - z}\, dA(z) \qquad (a \in \mathbb{C}).$$

Then ψ is of class $\mathcal{C}^\infty$ and satisfies

$$\bar{\partial}\psi = \varphi.$$

Proof First, note that ψ is well defined because φ is compactly supported and the function $z \longmapsto 1/(z-a)$ is integrable on any ball centered at a. Second, we have

$$\psi(a) = \frac{1}{\pi} \iint_{\mathbb{C}} \frac{\varphi(a - z)}{z}\, dA(z).$$

This formula ensures that ψ is of class $\mathcal{C}^\infty$ and

$$\bar{\partial}\psi(a) = \frac{1}{\pi} \iint_{\mathbb{C}} \frac{\bar{\partial}\varphi(a - z)}{z}\, dA(z).$$

Assume that $a = 0$. Then the above identity shows that

$$\begin{aligned}\bar{\partial}\psi(0) &= -\frac{1}{\pi}\iint_{\mathbb{C}} \frac{\bar{\partial}\varphi(z)}{z}\, dA(z)\\ &= -\lim_{\substack{\varepsilon\to 0\\ R\to+\infty}} \left(\frac{1}{\pi}\iint_{\Omega_{\varepsilon,R}} \frac{\bar{\partial}\varphi(z)}{z}\, dA(z)\right)\\ &= -\lim_{\substack{\varepsilon\to 0\\ R\to+\infty}} \left(\frac{1}{\pi}\iint_{\Omega_{\varepsilon,R}} \bar{\partial}\left(\frac{\varphi(z)}{z}\right) dA(z)\right).\end{aligned}$$

If R is sufficiently large so that $\varphi \equiv 0$ on $\{z \in \mathbb{C} : |z| = R\}$, then the formula (3.28) implies that

$$\frac{1}{\pi}\iint_{\Omega_{\varepsilon,R}} \bar{\partial}\left(\frac{\varphi(z)}{z}\right) dA(z) = -\frac{1}{2\pi}\int_0^{2\pi} \varphi(\varepsilon e^{i\vartheta})\, d\vartheta,$$

and this gives

$$\bar{\partial}\psi(0) = \lim_{\varepsilon\to 0} \frac{1}{2\pi}\int_0^{2\pi} \varphi(\varepsilon e^{i\vartheta})\, d\vartheta = \varphi(0).$$

For the general case, i.e. $a \neq 0$, we introduce the function $\phi(z) = \varphi(z+a)$. Then

$$\Psi(b) = \psi(b+a) = \frac{1}{\pi}\iint_{\mathbb{C}} \frac{\varphi(z)}{b+a-z}\, dA(z) = \frac{1}{\pi}\iint_{\mathbb{C}} \frac{\phi(z)}{b-z}\, dA(z).$$

Hence, we get

$$\bar{\partial}\psi(a) = \bar{\partial}\Psi(0) = \phi(0) = \varphi(a). \qquad \square$$

Notes on Chapter 3

The results presented in this chapter are standard and can be found in [53, 188, 233, 291, 354].

Section 3.3

All the results presented in this section can be found in [442]. The Lebesgue decomposition (3.8) is proved by von Neumann in a very elegant way in [539]. See also [442, pp. 117–118]. The Lebesgue differentiation theorem is proved in [442, pp. 148–149].

Section 3.4

Theorem 3.2 is due to Fatou [210]. Modern proofs can be found in [188, p. 39] or [291, p. 34].

Section 3.5

Corollary 3.7 is due to Herglotz. Reference [284] contains the original proof, while [291, p. 34] and [188, p. 2] have more modern proofs. Helly [277] also gave a proof of this result. See also [115] and [288]. An analogous representation for positive harmonic functions in a half-plane is proved by Cauer [129]. From the Herglotz representation, one can easily derive a characterization of an analytic function on the unit disk with positive real part. Such functions had already been characterized by Carathéodory [122] in terms of the positive definiteness of their Taylor coefficients. The theory of analytic functions on $\mathbb{D}$ with positive real part goes back to this paper of Carathéodory in 1907, with an explosion of follow-up papers in 1911 by Carathéodory himself [123], Toeplitz [522], Carathéodory and Fejér [124], Fischer [216], Herglotz [284] and F. Riesz [423]. See also the papers of Fejér [212], Riesz [425] and Schur [471, 472].

Section 3.7

Theorem 3.10 is a celebrated result of M. Riesz [433]. Theorem 3.12 is known as Zygmund's $L \log L$ theorem [571]. Theorem 3.13 was discovered by Kolmogorov [319]. Finally, Theorem 3.17, which is a reminiscent of a theorem of M. Riesz, was obtained by Stein [495].

Section 3.9

Green's formula, which is a particular case of Stokes' formula, is one of the most important formulas in real and complex analysis. See [91, pp. 28 and 32] for a proof of these two formulas. The identity given by Theorem 3.19 is known as the *Green–Riesz formula*.

4

Hardy spaces

This chapter is devoted to a rapid and concise treatment of Hardy spaces. Since our favorite objects, $\mathcal{H}(b)$ spaces, live inside the Hardy space H^2, understanding Hardy spaces is essential for the theory of $\mathcal{H}(b)$ spaces. We discuss the main topics in this chapter. Despite a few standard results that are stated without proof but for which we give precise references, all the results are proved. For further details consult other sources [188, 233, 291, 320, 354, 442]. After a brief review of hyperbolic geometry, we start with the definition of H^p spaces and find several integral representations. Then, we introduce the Riesz projection and study its properties, evaluate the duals and preduals of H^p, and discuss various topologies and convergence types in H^p. The canonical factorization theorem (Theorem 4.19) is one of the fundamental results that is stated without a complete proof. However, several corollaries of this theorem are proved. Especially, several important properties of outer functions are obtained. We finish with the Littlewood–Paley formula and some of its consequences.

4.1 Hyperbolic geometry

Let $f : \mathbb{D} \longrightarrow \bar{\mathbb{D}}$ be a nonconstant analytic function. The generalized form of Schwarz's lemma says

$$\left| \frac{f(z) - f(w)}{1 - \overline{f(w)}f(z)} \right| \leq \left| \frac{z - w}{1 - \bar{w}z} \right| \qquad (z, w \in \mathbb{D}). \tag{4.1}$$

The *hyperbolic distance* between two points z and w in $\mathbb{D}$ is defined by

$$\rho(z, w) = \left| \frac{z - w}{1 - \bar{w}z} \right|.$$

According to this definition, Schwarz's lemma can be rewritten as

$$\rho(f(z), f(w)) \leq \rho(z, w) \qquad (z, w \in \mathbb{D}). \tag{4.2}$$

In other words, each function $f : \mathbb{D} \longrightarrow \mathbb{D}$ is in fact a contraction (Lipschitz) on $\mathbb{D}$ whenever $\mathbb{D}$ is equipped with the hyperbolic metric. As an interesting special case, if f is the Möbius transformation

$$f(z) = \gamma \frac{z - z_0}{1 - \bar{z}_0 z},$$

where $\gamma \in \mathbb{T}$ and $z_0 \in \mathbb{D}$, then equality holds in (4.2). In fact, this is a characterization of having equality in (4.2).

The *hyperbolic disk* with center z_0 and radius r_0 is defined by

$$D_{hyp}(z_0, r_0) = \{z \in \mathbb{D} : \rho(z, z_0) < r_0\}.$$

However, a hyperbolic disk is in fact a Euclidean disk. More precisely, we have $D_{hyp}(z_0, r_0) = D(c, r)$, where

$$c = \frac{1 - r_0^2}{1 - r_0^2 |z_0|^2} z_0 \quad \text{and} \quad r = \frac{1 - |z_0|^2}{1 - r_0^2 |z_0|^2} r_0.$$

The above relations immediately imply that

$$\frac{|z_0| - r_0}{1 - r_0 |z_0|} \leq |z| \leq \frac{|z_0| + r_0}{1 + r_0 |z_0|} \tag{4.3}$$

for all $z \in D_{hyp}(z_0, r_0)$. The lower bound is useful whenever $r_0 < |z_0|$. Otherwise, if the origin is in $D_{hyp}(z_0, r_0)$ or on its frontier, then the lower bound is zero.

Exercises

Exercise 4.1.1 Let $f \in H^\infty$, $\|f\|_\infty \leq 1$. Show that

$$\frac{|f'(z)|}{1 - |f(z)|^2} \leq \frac{1}{1 - |z|^2} \qquad (z \in \mathbb{D}).$$

Hint: Use (4.1).

Exercise 4.1.2 Show that, for each $z_1, z_2, z_3 \in \mathbb{D}$, we have

$$\frac{\rho(z_1, z_3) - \rho(z_3, z_2)}{1 - \rho(z_1, z_3)\rho(z_3, z_2)} \leq \rho(z_1, z_2) \leq \frac{\rho(z_1, z_3) + \rho(z_3, z_2)}{1 + \rho(z_1, z_3)\rho(z_3, z_2)}. \tag{4.4}$$

Use these inequalities to verify that ρ is a distance on $\mathbb{D}$.

4.2 Classic Hardy spaces H^p

For an analytic function f on the open unit disk $\mathbb{D}$ and $0 \le r < 1$, we denote by f_r the function defined on $\mathbb{T}$ by $f_r(\zeta) = f(r\zeta)$. Then we set

$$\|f\|_p = \sup_{0 \le r < 1} \|f_r\|_p = \sup_{0 \le r < 1} \left(\int_{\mathbb{T}} |f(r\zeta)|^p \, dm(\zeta) \right)^{1/p},$$

if $p \in (0, \infty)$, and

$$\|f\|_\infty = \sup_{0 \le r < 1} \|f_r\|_\infty = \sup_{z \in \mathbb{D}} |f(z)|.$$

The *Hardy space* $H^p(\mathbb{D})$ is the family of all analytic functions f that satisfy $\|f\|_p < \infty$. The subspace $\{f \in H^p(\mathbb{D}) : f(0) = 0\}$ is denoted by $H_0^p(\mathbb{D})$. It turns out that, for $1 \le p \le +\infty$, the space $H^p(\mathbb{D})$ is a Banach space and $H_0^p(\mathbb{D})$ is a closed subspace of $H^p(\mathbb{D})$.

In the following, we are mainly concerned with $H^1(\mathbb{D})$, $H^2(\mathbb{D})$ and $H^\infty(\mathbb{D})$. A simple application of Hölder's inequality shows that

$$H^\infty(\mathbb{D}) \subset H^2(\mathbb{D}) \subset H^1(\mathbb{D}).$$

In light of Theorems 3.2, 3.6, 3.8 and 3.9, for each p, we expect to have a relevant family of functions on the unit circle $\mathbb{T}$. These facts are gathered in the following theorem.

Theorem 4.1 *Let $f \in H^p(\mathbb{D})$, $0 < p \le \infty$. Then*

$$f^*(\zeta) = \lim_{r \to 1} f(r\zeta)$$

exists for almost all $\zeta \in \mathbb{T}$ and $f^* \in L^p(\mathbb{T})$. Moreover

$$\lim_{r \to 1} \|f_r - f^*\|_p = 0 \qquad (0 < p < \infty) \tag{4.5}$$

and

$$\lim_{r \to 1} \int_{\mathbb{T}} \psi(\zeta) f_r(\zeta) \, dm(\zeta) = \int_{\mathbb{T}} \psi(\zeta) f^*(\zeta) \, dm(\zeta) \qquad (p = \infty) \tag{4.6}$$

for all $\psi \in L^1(\mathbb{T})$. Finally, if $f \in H^1(\mathbb{D})$, then we have

$$\widehat{f^*}(n) = \begin{cases} \dfrac{f^{(n)}(0)}{n!} & \text{if } \ n \ge 0, \\ 0 & \text{if } \ n < 0, \end{cases} \tag{4.7}$$

and

$$f(z) = \int_{\mathbb{T}} \frac{1 - |z|^2}{|\zeta - z|^2} f^*(\zeta) \, dm(\zeta) = Pf^*(z) \qquad (z \in \mathbb{D}). \tag{4.8}$$

A proof of this classic result can be found in [188, chap. 2], [320, chap. IV], [354, chap. 7] and [442, chap. 17].

Note that (4.5) and (4.6) imply that

$$\|f^*\|_{L^p(\mathbb{T})} = \|f\|_{H^p(\mathbb{D})} \qquad (0 < p \le \infty). \tag{4.9}$$

The relation (4.9) establishes a norm-preserving correspondence between $H^p(\mathbb{D})$ and a closed subspace of $L^p(\mathbb{T})$, which we denote by $H^p(\mathbb{T})$. Note that by (4.5) and (4.9) we have

$$\lim_{r\to 1} \|f_r\|_p = \|f\|_p \qquad (0 < p \le \infty).$$

We will now give two useful characterizations of $H^p(\mathbb{T})$. It is clear that $H^p(\mathbb{T})$ contains all the analytic polynomials, i.e. elements of $\mathcal{P}_+$. Since $H^p(\mathbb{T})$ is a closed subspace of $L^p(\mathbb{T})$, we have

$$\mathrm{Span}_{L^p(\mathbb{T})}(\mathcal{P}_+) \subset H^p(\mathbb{T}) \qquad (0 < p < \infty).$$

But the converse is also true.

Theorem 4.2 *For* $0 < p < \infty$,

$$H^p(\mathbb{T}) = \mathrm{Span}_{L^p(\mathbb{T})}(\mathcal{P}_+).$$

Proof Let $f \in H^p(\mathbb{D})$ and let $\varepsilon > 0$. Denote by $f_r(z) = f(rz)$. Then, by (4.5), there exists $r < 1$ such that $\|f_r - f^*\|_p < \varepsilon/2$. Now, let $S_n(z)$ denote the nth sum of a Taylor series of f at the origin,

$$S_n(z) = \sum_{k=0}^{n} a_k z^k,$$

where $a_k = f^{(k)}(0)/k!$. Since $S_n \longrightarrow f$ uniformly on the disk $|z| = r$, we have

$$\|(S_n)_r - f_r\|_p < \varepsilon/2$$

for n sufficiently large. Then, using Minkowski's inequality in the case $p \ge 1$, we find

$$\|(S_n)_r - f^*\|_p < \varepsilon.$$

But $(S_n)_r$ belongs to $\mathcal{P}_+$, which shows that $f^* \in \mathrm{Span}_{L^p(\mathbb{T})}(\mathcal{P}_+)$. For $p < 1$, the inequality

$$(a+b)^p \le 2^p(a^p + b^p)$$

gives the same result. □

Theorem 4.2 is false for $p = \infty$ since $H^\infty(\mathbb{T})$ contains functions that do not coincide almost everywhere with continuous functions. One example is the function $\exp((z+1)/(z-1))$ whose boundary function is $\exp(-i\cot(\theta/2))$, $\theta \neq 0$.

There is another approach, however, that leads to a description of $H^p(\mathbb{T})$, $1 \leq p \leq \infty$.

Theorem 4.3 *For* $1 \leq p \leq \infty$,

$$H^p(\mathbb{T}) = \{f \in L^p(\mathbb{T}) : \hat{f}(n) = 0,\ n < 0\}.$$

Proof One inclusion follows from (4.7). Now let $\varphi \in L^p(\mathbb{T})$, $\hat{\varphi}(n) = 0$, $n < 0$, and let

$$f(z) = \int_{\mathbb{T}} \frac{1-|z|^2}{|\zeta - z|^2} \varphi(\zeta)\, dm(\zeta) \qquad (z \in \mathbb{D})$$

be the Poisson integral of φ. Since $\hat{\varphi}(n) = 0$, $n < 0$, then (3.15) implies that f is analytic on $\mathbb{D}$. Moreover, we know from Theorems 3.8 and Theorem 3.9 that

$$\|f_r\|_p \leq \|\varphi\|_p,$$

which means that $f \in H^p(\mathbb{D})$. But, by Corollary 3.3,

$$f^* = \varphi \qquad \text{a.e. on } \mathbb{T},$$

which gives that $\varphi \in H^p(\mathbb{T})$. □

The correspondence between $H^p_0(\mathbb{D})$ and $H^p_0(\mathbb{T})$ is defined similarly, and using the same arguments we can show that

$$H^p_0(\mathbb{T}) = \{f \in L^p(\mathbb{T}) : \hat{f}(n) = 0,\ n \leq 0\} \qquad (1 \leq p \leq \infty) \tag{4.10}$$

and

$$\begin{aligned} H^p_0(\mathbb{T}) &= \operatorname{Span}_{L^p(\mathbb{T})}(\chi_n : n \geq 1) \\ &= \operatorname{Span}_{L^p(\mathbb{T})}(\mathfrak{p} \in \mathcal{P}_+ : \mathfrak{p}(0) = 0) \qquad (0 < p < \infty). \end{aligned}$$

Considering the isometric isomorphism between $H^p(\mathbb{D})$ and $H^p(\mathbb{T})$, from now on, we will also write f for f^*. This will not cause any serious difficulty. We will also mostly write H^p to refer to these Hardy spaces, either as a collection of analytic functions living on the unit disk $\mathbb{D}$, or as a family of measurable functions on the unit circle $\mathbb{T}$.

The representation (4.8) for the case $p = 1$ is, in a sense, equivalent to a deep theorem of F. and M. Riesz. This fundamental result exhibits the main difference between the space of all harmonic functions satisfying the growth condition (3.20) and $H^1(\mathbb{D})$. A harmonic function satisfying this growth condition is represented by the Poisson integral of a Borel measure on $\mathbb{T}$, which is not necessarily absolutely continuous with respect to the Lebesgue measure. A simple example is

$$h(z) = P\delta_1(z) = \frac{1-|z|^2}{|1-z|^2} \qquad (z \in \mathbb{D}).$$

But the elements of $H^1(\mathbb{D})$ are the Poisson integral of integrable functions with a vanishing negative spectrum. To see why this happens, let us call a complex Borel measure $\mu \in \mathcal{M}(\mathbb{T})$ satisfying

$$\hat{\mu}(-n) = \int_{\mathbb{T}} \zeta^n \, d\mu(\zeta) = 0 \qquad (n \geq 1)$$

an *analytic measure*. If $f \in H^1(\mathbb{T})$ and we define $d\mu = f\, dm$, then, by the definition of $H^1(\mathbb{T})$, μ is an analytic measure. The fundamental result of the Riesz brothers affirms that the converse is also true.

Theorem 4.4 *Let μ be an analytic measure on $\mathbb{T}$. Then there is an $f \in H^1(\mathbb{T})$ such that $d\mu = f\, dm$.*

This result is classic and can be found in many textbooks of complex analysis, e.g. in [188, p. 41], [354, p. 118] and [442, p. 341]. Since

$$\frac{\bar{\zeta}+z}{\bar{\zeta}-z} = 1 + 2\sum_{n=1}^{\infty} \zeta^n z^n,$$

Theorem 4.4 can be rewritten as

$$\int_{\mathbb{T}} \frac{\bar{\zeta}+z}{\bar{\zeta}-z}\, d\mu(\zeta) = 0 \quad (z \in \mathbb{D}) \quad \Longrightarrow \quad d\mu = f\, dm \quad (f \in H^1_0(\mathbb{T})). \tag{4.11}$$

We will need this version in the following.

The representation (4.8) has several important consequences. For example, if f is a real-valued function on $\mathbb{T}$, since the Poisson kernel is also real, the Poisson integral of f creates a real-valued function on $\mathbb{D}$. But the only analytic functions on a domain that assume only real values are real constant functions. Hence, we deduce that

$$f \in H^1, \quad f(\mathbb{T}) \subset \mathbb{R} \quad \Longrightarrow \quad f \text{ is constant.} \tag{4.12}$$

To get further results, fix $f \in H^p(\mathbb{D})$, $1 \leq p \leq \infty$. Let $z \in \mathbb{D}$, and let $|z| < r < 1$. Then, by Cauchy's integral formula,

$$f(z) = \frac{1}{2\pi i}\int_{\{|w|=r\}} \frac{f(w)}{w-z}\, dw = \int_{\mathbb{T}} \frac{r f_r(\zeta)}{r - \bar{\zeta} z}\, dm(\zeta).$$

Let $r \longrightarrow 1$. Then, by (4.5) and (4.6), we obtain

$$f(z) = \int_{\mathbb{T}} \frac{f(\zeta)}{1-\bar{\zeta} z}\, dm(\zeta) \qquad (z \in \mathbb{D}). \tag{4.13}$$

Now, if $1 < p < \infty$, by Hölder's inequality, we get

$$|f(z)| \leq \|f\|_p \left(\int_{\mathbb{T}} \frac{dm(\zeta)}{|1-\bar{\zeta}z|^q} \right)^{1/q}.$$

Thus, for any $1 < p < \infty$, there is a constant C_p such that

$$|f(z)| \leq C_p \frac{\|f\|_p}{(1-|z|)^{1/p}} \qquad (z \in \mathbb{D}), \tag{4.14}$$

for all $f \in H^p(\mathbb{D})$. We emphasize that the constant C_p depends neither on f nor on z (see Exercise 4.2.1). Moreover, we immediately see that (4.14) is also valid for $p = 1$ and $p = +\infty$ with $C_1 = C_\infty = 1$. Thus, for any $1 \leq p \leq \infty$,

$$|f(z)| \leq C_p \frac{\|f\|_p}{(1-|z|)^{1/p}} \qquad (z \in \mathbb{D}), \tag{4.15}$$

for all $f \in H^p(\mathbb{D})$. As a special case, we have

$$|f(0)| \leq \|f\|_p. \tag{4.16}$$

An immediate but important consequence of (4.15) is that the evaluation functionals $E_z : f \longmapsto f(z)$ are continuous on $H^p(\mathbb{D})$, for any point z in $\mathbb{D}$ and any $1 \leq p \leq \infty$.

In the family of Lebesgue spaces $L^p(\mathbb{T})$, a special role is played by $L^2(\mathbb{T})$ and its closed subspace $H^2(\mathbb{T})$. This is mainly because $L^2(\mathbb{T})$ and $H^2(\mathbb{T})$ are Hilbert spaces endowed with the inner product

$$\langle f, g \rangle_2 = \int_{\mathbb{T}} f(\zeta)\, \overline{g(\zeta)}\, dm(\zeta) = \sum_{n=-\infty}^{\infty} \hat{f}(n)\, \overline{\hat{g}(n)}. \tag{4.17}$$

Note that if $f, g \in H^2$, the index n in fact starts from 0. In H^2, the relation (4.13) can be rewritten as

$$f(z) = \langle f, k_z \rangle_2 \qquad (z \in \mathbb{D}), \tag{4.18}$$

where

$$k_z(w) = \frac{1}{1-\bar{z}w} \qquad (z, w \in \mathbb{D}). \tag{4.19}$$

We refer to k_z as the *Cauchy kernel*. Note that k_z belongs to each space $H^p(\mathbb{D})$, $1 \leq p \leq +\infty$. Moreover, (4.18) implies that

$$\|k_z\|_2^2 = k_z(z) = \frac{1}{1-|z|^2} \qquad (z \in \mathbb{D}). \tag{4.20}$$

Theorem 4.5 *Let K be a compact set in the complex plane such that $\mathbb{C} \setminus K$ is connected. Let $f \in H^\infty$ with*

$$f(\zeta) \in K \qquad (\textit{a.e. on } \mathbb{T}).$$

Then

$$f(z) \in K \qquad (z \in \mathbb{D}).$$

Proof Suppose that there is $z_0 \in \mathbb{D}$ such that $w_0 = f(z_0) \notin K$. Then, by Mergelyan's theorem, there is an analytic polynomial p such that

$$|p(w_0) - 1| < \frac{1}{4} \quad \text{and} \quad |p(z)| < \frac{1}{4} \quad (z \in K).$$

Consider $g = p \circ f$. Clearly, $g \in H^\infty$. On the one hand, we have $|g(z_0) - 1| < 1/4$, which gives $|g(z_0)| > 3/4$. On the other, $|g(\zeta)| < 1/4$ a.e. on $\mathbb{T}$, which in light of Poisson's formula implies $|g(z)| < 1/4$, $z \in \mathbb{D}$. This is a contradiction. □

Exercises

Exercise 4.2.1 Let $1 < p < \infty$, and let q be its conjugate exponent. Show that there is a constant C_p such that

$$\left(\int_{\mathbb{T}} \frac{1}{|1 - \bar{\zeta} z|^q} \, dm(\zeta) \right)^{1/q} \leq \frac{C_p}{(1 - |z|)^{1/p}} \qquad (z \in \mathbb{D}).$$

Hint: Use

$$|1 - re^{i\theta}| \asymp (1 - r) + |\theta|.$$

Exercise 4.2.2 Let $f : \mathbb{D} \longrightarrow \mathbb{D}$ be analytic. Show that the set of points $\zeta \in \mathbb{T}$ for which the radial limit

$$\lim_{r \to 1} f(r\zeta)$$

exists is an $F_{\sigma\delta}$ set.

Remark: We recall that, in a topological space, a set E is $F_{\sigma\delta}$ if there are closed sets $E(m, n)$ such that

$$E = \bigcap_{m=1}^{\infty} \bigcup_{n=1}^{\infty} E(m, n).$$

Hint: Consider

$$E(m, n) = \{\zeta \in \mathbb{T} : |f(r\zeta) - f(\rho\zeta)| \leq 1/m \text{ for all } r, \rho \geq 1 - 1/n\}.$$

Exercise 4.2.3 Let

$$f(z) = \frac{1}{1 - z} \log\left(\frac{1}{1 - z}\right) \qquad (z \in \mathbb{D}).$$

Show that

$$f \in \bigcap_{0<p<1} H^p.$$

Also verify that

$$f(z) = \sum_{n=1}^{\infty} \left(\sum_{k=1}^{n} \frac{1}{k} \right) z^n \qquad (z \in \mathbb{D}).$$

Use this representation to deduce that $f \notin H^1$.
Hint: For any $f \in H^1$, and for any integer n, $|\hat{f}(n)| \le \|f\|_1$.

Exercise 4.2.4 Let q be any polynomial of degree $n \ge 1$ with zeros inside $\mathbb{D}$, and let p be any polynomial of degree $n-1$. Show that there are polynomials r and s, with zeros of s outside $\bar{\mathbb{D}}$, such that

$$\langle f, p/q \rangle_{L^2(\mathbb{T})} = \frac{1}{2\pi i} \int_{\mathbb{T}} f(z) \frac{r(z)}{s(z)} \, dz \qquad (f \in L^2(\mathbb{T})).$$

Hint: Use the fact that $z^n \overline{q(z)}$, $z \in \mathbb{T}$, is a polynomial of degree at most n, with zeros outside $\bar{\mathbb{D}}$.

4.3 The Riesz projection P_+

Let $f \in L^2(\mathbb{T})$. Then, by Parseval's identity,

$$\|f\|_2 = \left(\int_{\mathbb{T}} |f(\zeta)|^2 \, dm(\zeta) \right)^{1/2} = \left(\sum_{n=-\infty}^{\infty} |\hat{f}(n)|^2 \right)^{1/2}.$$

Therefore, by the Riesz–Fischer theorem and Theorem 4.2, the function

$$P_+ f = \sum_{n=0}^{\infty} \hat{f}(n) \chi_n \tag{4.21}$$

is a well-defined element of $H^2(\mathbb{T})$. Again by Parseval's identity, we have

$$\|P_+ f\|_2 = \left(\sum_{n=0}^{\infty} |\hat{f}(n)|^2 \right)^{1/2}$$

and thus

$$\|P_+ f\|_2 \le \|f\|_2 \qquad (f \in L^2(\mathbb{T})). \tag{4.22}$$

Moreover,

$$P_+ f = f \iff f \in H^2(\mathbb{T}). \tag{4.23}$$

The operator

$$\begin{array}{rccl} P_+: & L^2(\mathbb{T}) & \longrightarrow & H^2(\mathbb{T}) \\ & f & \longmapsto & \sum_{n=0}^{\infty} \hat{f}(n)\, \chi_n \end{array}$$

is called the *Riesz projection*. Note that, by (4.22) and (4.23), $\|P_+\| = 1$.

Since P_+ is well defined on trigonometric polynomials, we can regard P_+ as a densely defined operator on $L^p(\mathbb{T})$, for $1 \le p < \infty$. It turns out that, for $1 < p < \infty$, P_+ is a bounded operator on $L^p(\mathbb{T})$.

Theorem 4.6 *Let $1 < p < +\infty$. Then P_+ extends into a bounded operator from $L^p(\mathbb{T})$ into $H^p(\mathbb{T})$. Moreover, for any $f \in L^p(\mathbb{T})$, we have*

$$\widehat{P_+ f}(n) = \hat{f}(n) \qquad (n \ge 0).$$

Proof First let

$$f = \sum_{k=-M}^{N} \hat{f}(k) \chi_k$$

be a trigonometric polynomial. Then

$$P_+ f = \sum_{k=0}^{N} \hat{f}(k) \chi_k.$$

Put $g := \hat{f}(0) + f + i\tilde{f}$. By Theorem 3.10, $g \in L^p(\mathbb{T})$ and

$$\|g\|_p \le c_p \|f\|_p. \tag{4.24}$$

Moreover, according to Lemma 3.11, for any $\ell \in \mathbb{Z}$, we have

$$\begin{aligned} \hat{g}(\ell) &= \hat{f}(0)\hat{\chi}_0(\ell) + \hat{f}(\ell) + i\hat{\tilde{f}}(\ell) \\ &= \hat{f}(0)\hat{\chi}_0(\ell) + \hat{f}(\ell) + \operatorname{sgn}(\ell)\hat{f}(\ell). \end{aligned}$$

Hence

$$\hat{g}(\ell) = \begin{cases} 2\hat{f}(\ell) & \text{if } \ell \ge 0, \\ 0 & \text{if } \ell < 0. \end{cases}$$

That proves that $2P_+ f = g$ and (4.24) implies that

$$\|P_+ f\|_p \le \frac{c_p}{2} \|f\|_p.$$

The boundedness of P_+ on $L^p(\mathbb{T})$ follows now from the density of $\mathcal{P}$ in $L^p(\mathbb{T})$.

Moreover, given $f \in L^p(\mathbb{T})$, let f_n be a sequence of trigonometric polynomials such that $\|f_n - f\|_p \longrightarrow 0$ as $n \longrightarrow +\infty$. Hence, $P_+ f_n$ converges to $P_+ f$ in $L^p(\mathbb{T})$. Fix an integer $\ell \geq 0$. Then,

$$\widehat{P_+ f}(\ell) = \lim_{n \to +\infty} \widehat{P_+ f_n}(\ell) = \lim_{n \to +\infty} \hat{f}_n(\ell) = \hat{f}(\ell),$$

which proves the result. □

Corollary 4.7 *Let* $1 < p < +\infty$. *Then*

$$L^p(\mathbb{T}) = H^p(\mathbb{T}) \oplus \overline{H_0^p(\mathbb{T})},$$

and for $p = 2$ *the sum is orthogonal.*

Proof First let $f \in H^p(\mathbb{T}) \cap \overline{H_0^p(\mathbb{T})}$. On the one hand, according to Theorem 4.3, we have $\hat{f}(n) = 0$, $n < 0$. On the other, write $f = \bar{g}$, with $g \in H_0^p(\mathbb{T})$. Then, by (4.10), we have $\hat{g}(n) = 0$, $n \leq 0$. Hence $\hat{f}(n) = \overline{\hat{g}(-n)} = 0$ for $n \geq 0$. Thus $\hat{f}(n) = 0$, $n \in \mathbb{Z}$, which implies that $f \equiv 0$.

Now, let us prove that we can decompose any function $f \in L^p(\mathbb{T})$ as the sum of an $H^p(\mathbb{T})$ function plus an $\overline{H_0^p(\mathbb{T})}$ function. According to Theorem 4.6, the function $g := P_+ f \in H^p(\mathbb{T})$. Put $h = f - g = f - P_+ f$. Using Theorem 4.6 one more time, we have

$$\hat{h}(n) = \hat{f}(n) - \widehat{P_+ f}(n) = 0 \qquad (n \geq 0).$$

Hence $\bar{h} \in H_0^p(\mathbb{T})$, that is, $h \in \overline{H_0^p(\mathbb{T})}$ and $f = h + g$ is the desired decomposition.

It remains to use Parseval's identity, (4.10) and Theorem 4.3 to conclude that, in the case when $p = 2$, the decomposition is orthogonal. □

We also denote by $H_-^2 = H_-^2(\mathbb{T}) = (H^2(\mathbb{T}))^\perp = \overline{H_0^2(\mathbb{T})}$.

By the same token, we also define

$$\begin{array}{rccc} P_- : & L^2(\mathbb{T}) & \longrightarrow & H^2(\mathbb{T})^\perp \\ & f & \longmapsto & \displaystyle\sum_{n=-\infty}^{-1} \hat{f}(n)\, \chi_n, \end{array}$$

where $H^2(\mathbb{T})^\perp = \overline{H_0^2(\mathbb{T})}$. Hence,

$$\|P_- f\|_2 = \left(\sum_{n=-\infty}^{-1} |\hat{f}(n)|^2 \right)^{1/2}$$

and thus

$$\|P_- f\|_2 \leq \|f\|_2 \qquad (f \in L^2(\mathbb{T})). \tag{4.25}$$

Moreover,

$$P_- f = f \iff f \in H^2(\mathbb{T})^\perp. \tag{4.26}$$

The definitions of P_+ and P_- immediately imply that

$$f = P_+ f + P_- f \qquad (f \in L^2(\mathbb{T})).$$

The following lemma is a direct consequence of Parseval's identity. This result will appear frequently in our future discussions.

Lemma 4.8 *Let $f \in L^2(\mathbb{T})$ and $g \in H^2(\mathbb{T})$. Then*

$$\langle P_+ f, g\rangle_{H^2(\mathbb{T})} = \langle f, g\rangle_{L^2(\mathbb{T})}.$$

Proof Parseval's identity says that

$$\langle f, h\rangle_{L^2(\mathbb{T})} = \sum_{n=-\infty}^{\infty} \hat{f}(n)\,\overline{\hat{h}(n)} \qquad (f, h \in L^2(\mathbb{T})).$$

Based on this formula and that

$$\hat{g}(n) = 0 \qquad (n \le -1)$$

and

$$\widehat{P_+ f}(n) = \hat{f}(n) \qquad (n \ge 0),$$

we obtain

$$\langle P_+ f, g\rangle_{H^2(\mathbb{T})} = \langle f, g\rangle_{L^2(\mathbb{T})} = \sum_{n=0}^{\infty} \hat{f}(n)\,\overline{\hat{g}(n)}. \qquad \square$$

Similar to Lemma 4.8, we have the following useful identity.

Lemma 4.9 *Let $f \in L^2(\mathbb{T})$ and $g \in H^2(\mathbb{T})^\perp$. Then*

$$\langle P_- f, g\rangle_{H^2(\mathbb{T})^\perp} = \langle f, g\rangle_{L^2(\mathbb{T})}.$$

Proof By Parseval's identity and that

$$\hat{g}(n) = 0 \qquad (n \ge 0)$$

and

$$\widehat{P_- f}(n) = \hat{f}(n) \qquad (n \le -1),$$

we have

$$\langle P_- f, g\rangle_{H^2(\mathbb{T})^\perp} = \langle f, g\rangle_{L^2(\mathbb{T})} = \sum_{n=-\infty}^{-1} \hat{f}(n)\,\overline{\hat{g}(n)}. \qquad \square$$

Lemma 4.8 has an interesting interpretation. It shows that the adjoint of the Riesz projection $P_+ \in \mathcal{L}(L^2, H^2)$ is the inclusion map $i_+ \in \mathcal{L}(H^2, L^2)$ given by

$$\begin{array}{rccc} i_+ : & H^2(\mathbb{T}) & \longrightarrow & L^2(\mathbb{T}) \\ & f & \longmapsto & f. \end{array}$$

This is another manifestation of the fact that P_+ is not an orthogonal projection in the sense defined in Section 1.7. While an orthogonal projection is self-adjoint, we have

$$P_+^* = i_+. \tag{4.27}$$

This identity implies that

$$P_+ P_+^* = P_+ i_+ = I_{H^2(\mathbb{T})}, \tag{4.28}$$

while

$$P_+^* P_+ = i_+ P_+ = P_{H^2(\mathbb{T})}, \tag{4.29}$$

where $P_{H^2(\mathbb{T})}$ is the orthogonal projection of $L^2(\mathbb{T})$ onto its closed subspace $H^2(\mathbb{T})$.

By the same token, Lemma 4.9 shows that the adjoint of the Riesz projection P_- is the inclusion map $i_- \in \mathcal{L}(H^{2\perp}, L^2)$ given by

$$\begin{array}{rccc} i_- : & H^2(\mathbb{T})^\perp & \longrightarrow & L^2(\mathbb{T}) \\ & f & \longmapsto & f. \end{array}$$

For further reference, we write

$$P_-^* = i_-. \tag{4.30}$$

Exercises

Exercise 4.3.1 Let

$$f(e^{it}) = -i(\pi - t) \qquad (0 < t < 2\pi).$$

Show that

$$(P_+ f)(z) = \log(1 - z).$$

Hint: Verify that the Fourier coefficients of f are given by

$$\hat{f}(n) = \begin{cases} -\dfrac{1}{n} & \text{if} \quad n \neq 0, \\ 0 & \text{if} \quad n = 0. \end{cases}$$

Moreover, remember that $-\log(1 - z) = \sum_{n=1}^{\infty} z^n / n$.

Exercise 4.3.2 Let

$$f(e^{it}) = \begin{cases} 1 & \text{if} \quad 0 < t < \pi, \\ -1 & \text{if} \quad -\pi < t < 0. \end{cases}$$

Find $P_+ f$.
Hint: Apply the same method as in Exercise 4.3.1.

Exercise 4.3.3 Show that

$$H^2(\mathbb{T})^\perp = \overline{H_0^2(\mathbb{T})} = \{f \in L^2(\mathbb{T}) : \hat{f}(n) = 0,\ n \geq 0\}.$$

Exercise 4.3.4 Show that the mapping

$$\begin{array}{rccc} A: & H^2(\mathbb{T}) & \longrightarrow & H^2(\mathbb{T})^\perp \\ & \displaystyle\sum_{n=0}^{\infty} a_n \chi_n & \longmapsto & \displaystyle\sum_{n=1}^{\infty} a_{n-1} \chi_{-n} \end{array}$$

is a Hilbert space isomorphism.

Remark: Note that A is also given by the formula

$$(Af)(\zeta) = \frac{1}{\zeta} f\left(\frac{1}{\zeta}\right) \qquad (\zeta \in \mathbb{T}).$$

Exercise 4.3.5 Let f and $(f_n)_{n\geq 1}$ be in H^1. Show that the following are equivalent:

(i) f_n uniformly converges to f on compact subsets of $\mathbb{D}$;
(ii) we have

$$\lim_{n\to\infty} \hat{f}_n(k) = \hat{f}(k) \qquad (k \in \mathbb{Z}).$$

Exercise 4.3.6 Show that, if $p \geq 1$, then $H^p(\mathbb{T}) \cap \overline{H^p(\mathbb{T})}$ contains only the constant functions. What about the case $0 < p < 1$?
Hint: For the second part, consider the function $f(z) = (z-1)^{-1}$.

4.4 Kernels of P_+ and P_-

Using the original definition of P_+ and P_-, it is not difficult to find the kernels of P_+ and P_-. This is done in the following lemma. However, these characterizations lead to an important result (Corollary 4.11), which will be used in many results afterwards.

Lemma 4.10 *We have*

$$\ker P_+ = \overline{H_0^2}$$

and

$$\ker P_- = H^2.$$

Proof Let

$$f = \sum_{n=-\infty}^{\infty} \hat{f}(n)\chi_n$$

be a function in $L^2(\mathbb{T})$. Then, according to (4.21) and the uniqueness theorem, f belongs to $\ker P_+$ if and only if $\hat{f}(n) = 0$ for all $n \geq 0$. Therefore, the elements of the kernel are precisely $L^2(\mathbb{T})$ functions, which have the representation

$$f = \sum_{n=-\infty}^{-1} \hat{f}(n)\chi_n.$$

But this means that

$$\bar{f} = \sum_{n=1}^{\infty} \overline{\hat{f}(-n)}\chi_n \in H_0^2,$$

which is equivalent to $f \in \overline{H_0^2}$. The proof of $\ker P_- = H^2$ is similar. □

Corollary 4.11 *Let $\varphi \in H^\infty$. Then*

$$P_+(\bar{\varphi}f) = P_+(\bar{\varphi}P_+f)$$

for each $f \in L^2(\mathbb{T})$.

Proof Put

$$g = \bar{\varphi}(f - P_+f) = \bar{\varphi}P_-f.$$

Then, by Parseval's identity,

$$\hat{g}(n) = 0 \qquad (n \geq 0).$$

Hence, by Lemma 4.10, $g \in \ker P_+$, i.e. $P_+g = 0$. But the relation $P_+g = 0$ is equivalent to the required identity. □

Corollary 4.12 *Let $f \in L^2(\mathbb{T})$. Then the following are equivalent.*

(i) $f \in H^2(\mathbb{T})$.
(ii) $P_+f = f$.
(iii) $\|P_+f\|_2 = \|f\|_2$.

Proof The implications (i) $\Longrightarrow$ (ii) $\Longrightarrow$ (iii) are trivial.
(iii) $\Longrightarrow$ (i) Since

$$f = P_+f + P_-f \qquad (f \in L^2(\mathbb{T}))$$

and $P_+f \in H^2(\mathbb{T})$ while $P_-f \in H^2(\mathbb{T})^\perp$, we have

$$\|f\|_2^2 = \|P_+f\|_2^2 + \|P_-f\|_2^2.$$

Therefore, $\|P_+f\|_2 = \|f\|_2$ if and only if $P_-f = 0$, i.e. $f \in \ker P_-$, which, by Lemma 4.10, is equivalent to saying that $f \in H^2(\mathbb{T})$. □

A slight modification of the above result leads to the following result for P_-.

Corollary 4.13 *Let $f \in L^2(\mathbb{T})$. Then the following are equivalent.*

(i) $f \in H^2(\mathbb{T})^\perp$.
(ii) $P_-f = f$.
(iii) $\|P_-f\|_2 = \|f\|_2$.

Exercises

Exercise 4.4.1 Let $\lambda \in \mathbb{D}$, let $n \geq 1$ be an integer, and put

$$f(z) = \frac{1}{(z-\lambda)^n} \qquad (z \in \mathbb{D}).$$

Show that $f \in \ker P_+$.
Hint: We have

$$\frac{1}{z-p} = \sum_{k=1}^{\infty} \frac{p^{k-1}}{z^k}$$

and the series converges uniformly on $\mathbb{T}$.

Exercise 4.4.2 Let f be a rational function such that $f \in L^2(\mathbb{T})$. Find P_+f. Hint: Do the partial fraction expansion, and then apply Exercise 4.4.1. Note that, by Exercise 5.3.3, f has no pole on $\mathbb{T}$. Another proof can be based on Exercise 4.2.4.

4.5 Dual and predual of H^p spaces

We recall that the dual of $L^p(\mathbb{T})$, $1 \leq p < \infty$, is isometrically isomorphic to $L^q(\mathbb{T})$, where $1/p + 1/q = 1$. The correspondence is given by

$$L_g(f) = \int_{\mathbb{T}} f(\zeta)g(\zeta)\,dm(\zeta) \qquad (f \in L^p(\mathbb{T}),\ g \in L^q(\mathbb{T})) \tag{4.31}$$

(see Section 1.2). By the same token, the dual of H^p, $1 \leq p < \infty$, can be identified with L^q/H_0^q. Indeed, by Theorem 1.12, we have

$$(H^p)^* = (L^p)^*/(H^p)^\perp = L^q/(H^p)^\perp,$$

where

$$(H^p)^\perp = \{g \in L^q(\mathbb{T}) : L_g | H^p = 0\}.$$

But, using Theorem 4.3, an easy computation shows that

$$(H^p)^\perp = H^q_0. \tag{4.32}$$

Similarly, we have

$$(H^p_0)^\perp = H^q. \tag{4.33}$$

Hence

$$(H^p)^* = L^q / H^q_0.$$

Actually, we may as well replace L^q/H^q_0 by L^q/H^q, since the map

$$\begin{array}{rccc} \pi : & L^q/H^q & \longrightarrow & L^q/H^q_0 \\ & f + H^q & \longmapsto & \chi_1 f + H^q_0 \end{array}$$

is an isometric isomorphism. Then

$$(H^p)^* = L^q/H^q \qquad (1 \le p < +\infty). \tag{4.34}$$

It is even possible to give a canonical representation of the bounded linear functionals on H^p, for $1 < p < +\infty$. Indeed, any $\phi \in (H^p)^*$ can be extended (by the Hahn–Banach theorem) to a functional on L^p, and hence may be represented in the form

$$\phi(f) = \int_{\mathbb{T}} fg \, dm \qquad (f \in H^p), \tag{4.35}$$

for some $g \in L^q$. By Corollary 4.7, we can decompose g as $g = g_1 + \overline{g_2}$, where $g_1 \in H^q$ and $g_2 \in H^q_0$ (observe that, since $1 < p < \infty$, then $1 < q < \infty$). Write $g = g_3 + \overline{g_4}$, where $g_3 = g_1 - g_1(0) \in H^q_0$ and $g_4 = g_2 + \overline{g_1(0)} \in H^q$. Thanks to (4.35) and (4.32), we obtain, for any $f \in H^p$,

$$\phi(f) = \int_{\mathbb{T}} f g_3 \, dm + \int_{\mathbb{T}} f \overline{g_4} \, dm = \int_{\mathbb{T}} f \overline{g_4} \, dm.$$

Moreover, assume that there exists another $\varphi \in H^q$ such that

$$\phi(f) = \int_{\mathbb{T}} f \bar{\varphi} \, dm \qquad (f \in H^p).$$

Then

$$\int_{\mathbb{T}} f \overline{(\varphi - g_4)} \, dm = 0$$

for any $f \in H^p$. That means that $\overline{(\varphi - g_4)} \in (H^p)^\perp = H^q_0$. Thus, we get that

$$\varphi - g_4 \in \overline{H^q_0} \cap H^q = \{0\},$$

and that proves that $\varphi = g_4$. In summary, we have proved the following result.

Theorem 4.14 *For $1 \le p < +\infty$, the space $(H^p)^*$ is isometrically isomorphic to L^q/H^q, where q is the conjugate exponent of p. Furthermore, if $1 < p < +\infty$, each $\phi \in (H^p)^*$ is representable in the form* (4.35) *by a unique $g \in H^q$, while each $\phi \in (H^1)^*$ is representable in the form* (4.35) *by some $g \in L^\infty$.*

In this context, for $f \in H^p$ and $g \in H^q$, identifying g with $L_{\bar{g}}$, we write

$$\langle f, g \rangle_{H^p,(H^p)^*} = \int_{\mathbb{T}} f(\zeta)\, \overline{g(\zeta)}\, dm(\zeta).$$

Therefore, the relation (4.13) can be rewritten as

$$f(z) = \langle f, k_z \rangle_{H^p,(H^p)^*} \tag{4.36}$$

for each $f \in H^p$, $z \in \mathbb{D}$, $1 \le p < \infty$.

Another basic fact about H^p spaces is that they have preduals. More precisely, we mean that, for $1 < p \le +\infty$, the space H^p is isomorphically isometric to $(L^q(\mathbb{T})/H^q_0)^*$, where q is the conjugate exponent of p. This follows immediately from Theorem 1.11 and (4.33). More importantly, by appealing to Theorems 1.11 and 4.4, we see that H^1 is the dual of $\mathcal{C}(\mathbb{T})/\mathcal{A}_0$, where

$$\mathcal{A}_0 = \mathrm{Span}_{\mathcal{C}(\mathbb{T})}(\chi_n : n \ge 1).$$

In Section 5.3, we will introduce the disk algebra $\mathcal{A}$ and then $\mathcal{A}_0 = \{f \in \mathcal{A} : f(0) = 0\}$. Thus, we can state the following theorem.

Theorem 4.15 *For $1 < p \le +\infty$, the space H^p is isomorphically isometric to $(L^q(\mathbb{T})/H^q_0)^*$, where q is the conjugate exponent of p, while the space H^1 is isomorphically isometric to $(\mathcal{C}(\mathbb{T})/\mathcal{A}_0)^*$.*

Hence, if $(f_n)_{n\ge1}$ is a sequence in H^p, $1 < p \le +\infty$, then saying that $(f_n)_{n\ge1}$ converges to $f \in H^p$ in the weak-star topology means that

$$\lim_{n\to\infty} \int_{\mathbb{T}} f_n(\zeta) g(\zeta)\, dm(\zeta) = \int_{\mathbb{T}} f(\zeta) g(\zeta)\, dm(\zeta), \tag{4.37}$$

for every $g \in L^q(\mathbb{T})$. Similarly, if $(f_n)_{n\ge1}$ is a sequence in H^1, then saying that $(f_n)_{n\ge1}$ converges to $f \in H^1$ in the weak-star topology means that (4.37) holds for every $g \in \mathcal{C}(\mathbb{T})$.

The next result gives a useful characterization of the weak-star convergence of sequences in H^p, $1 \le p \le \infty$.

Theorem 4.16 *Let $1 \leq p \leq \infty$. A sequence $(f_n)_{n\geq 1}$ in H^p converges to $f \in H^p$ in the weak-star topology if and only if*

$$\sup_{n\geq 1} \|f_n\|_p < \infty$$

and

$$\lim_{n\to\infty} f_n(z) = f(z) \qquad (z \in \mathbb{D}).$$

Moreover, in that situation, f_n converges uniformly to f on compact subsets of $\mathbb{D}$.

Proof Assume that $(f_n)_{n\geq 1}$ converges to $f \in H^p$ in the weak-star topology. First, the uniform boundedness principle (Theorem 1.19) ensures that $\sup_{n\geq 1} \|f_n\|_p < \infty$. Second, by (4.13) and (4.37), we have

$$\lim_{n\to\infty} f_n(z) = \lim_{n\to\infty} \int_{\mathbb{T}} f_n(\zeta)\,\overline{k_z(\zeta)}\,dm(\zeta) = \int_{\mathbb{T}} f(\zeta)\,\overline{k_z(\zeta)}\,dm(\zeta) = f(z)$$

for every $z \in \mathbb{D}$.

Conversely, assume that $\sup_n \|f_n\|_p < \infty$ and $\lim_{n\to\infty} f_n(z) = f(z)$ for every $z \in \mathbb{D}$. Then, according to the Banach–Alaoglu theorem, there exists a subsequence $(f_{n_k})_{k\geq 1}$ of $(f_n)_{n\geq 1}$ that converges to an element $g \in H^p$ in the weak-star topology. But, since, for each $z \in \mathbb{D}$, $f_n(z) \longrightarrow f(z)$ as $n \longrightarrow \infty$, we must have $f = g$. Therefore, $(f_n)_{n\geq 1}$ converges to f in the weak-star topology of H^p.

Now, fix a compact subset K of $\mathbb{D}$. Note that (4.15) implies that

$$\sup_{n\geq 1, z\in K} |f_n(z)| \leq C_K < \infty.$$

The uniform convergence on compact subsets of $\mathbb{D}$ is thus an immediate consequence of a normal family argument. □

It is well known that the set of trigonometric polynomials is dense in $L^p(\mathbb{T})$, $1 \leq p < \infty$. This is not true for $p = +\infty$ because, by the Stone–Weierstrass theorem, we know that the uniform closure of trigonometric polynomials is precisely $\mathcal{C}(\mathbb{T})$, which is a strictly smaller subspace of $L^\infty(\mathbb{T})$. It is natural to ask about the closure of analytic polynomials in $L^p(\mathbb{T})$, $1 \leq p \leq \infty$. As we have already seen in Theorem 4.2, the set of analytic polynomials is dense in $H^p(\mathbb{T})$ when $1 \leq p < \infty$. We already noted that, for $p = \infty$, this cannot be true. But we have the following result.

Theorem 4.17 *The space $H^\infty(\mathbb{T})$ is precisely the weak-star closure of analytic polynomials in $L^\infty(\mathbb{T})$.*

Proof It is sufficient to prove that, if $f \in H^\infty$, then we can find a sequence of analytic polynomials $(\mathfrak{p}_n)_{n\geq 1}$ such that $\mathfrak{p}_n \longrightarrow f$ in the weak-star topology of H^∞. But, by Theorem 1.5, $\sigma_n(f) \longrightarrow f$ in the weak-star topology of L^∞, as $n \longrightarrow +\infty$, where $\sigma_n(f)$ is the nth Fejér mean of f. It remains to note that, since $f \in H^\infty$, then $\sigma_n(f)$ is an analytic polynomial. □

4.6 The canonical factorization

If f is analytic on a domain containing the closed disk $\overline{D}_r = \{z \in \mathbb{C} : |z| \le r\}$ with $f(0) \neq 0$ and $z_1, z_2, \dots, z_n$ denote the zeros of f inside D_r, repeated according to their multiplicities, then *Jensen's formula* says

$$\frac{1}{2\pi}\int_0^{2\pi} \log|f(re^{it})|\,dt = \log|f(0)| + \sum_{k=1}^{n} \log\left(\frac{r}{|z_k|}\right).$$

This simple-looking identity can be found in many textbooks, see e.g. [442, p. 307], and it has several essential applications. The hypothesis $f(0) \neq 0$ causes no harm in applications, because, if f has a zero of order k at 0, the formula can be applied to $f(z)/z^k$. Note that Jensen's formula implies that $\int_{\mathbb{T}} \log|f(r\zeta)|\,dm(\zeta)$ increases with r.

We will also need a little generalization of this formula. Let f be analytic on a domain containing the closed disk $\overline{D}_r = \{z \in \mathbb{C} : |z| \le r\}$ and let $z_1, z_2, \dots, z_n$ denote the zeros of f inside D_r repeated according to their multiplicities. Then, for each $z \in D_r$, we have

$$\log|f(z)| = -\sum_{k=1}^{n} \log\left|\frac{r^2 - \bar{z}_k z}{r(z - z_k)}\right| + \frac{1}{2\pi}\int_0^{2\pi} \frac{r^2 - |z|^2}{|re^{it} - z|^2} \log|f(re^{it})|^2\,dt.$$

This formula is called the *Poisson–Jensen formula*. For $z = 0$ we recover Jensen's formula. We can state this formula slightly differently. There exists a constant of modulus one, such that, for each $z \in D_r$, we have

$$f(z) = \gamma \prod_{k=1}^{n} \frac{r(z - z_k)}{r^2 - \bar{z}_k z} \exp\left(\frac{1}{2\pi}\int_0^{2\pi} \frac{re^{it} + z}{r^{it} - z} \log|f(re^{it})|\,dt\right) \tag{4.38}$$

(see e.g. [354, p. 163]).

Let $f \in H^p(\mathbb{D})$, $0 < p \le \infty$, $f \not\equiv 0$, and let $(z_n)_{n\ge1}$ be the zero sequence of f in $\mathbb{D}$, each zero being repeated according to its multiplicity. Then Jensen's formula implies that

$$\sum_{n\ge1} (1 - |z_n|) < \infty. \tag{4.39}$$

This is the so-called *Blaschke condition* and it ensures that the infinite product defined by

$$B(z) = \prod_{n\ge1} \frac{|z_n|}{z_n} \frac{z_n - z}{1 - \bar{z}_n z} \qquad (z \in \mathbb{D}) \tag{4.40}$$

converges uniformly on compact subsets of $\mathbb{D}$. Here we adopt the convention that, when $z_n = 0$, then $|z_n|/z_n = 1$, so that the corresponding factor appearing in the product is z. Since we have uniform convergence on compact subsets of $\mathbb{D}$, the function B is analytic on $\mathbb{D}$ and it is easy to see that

$$|B(z)| \le 1 \qquad (z \in \mathbb{D}).$$

In particular, $B \in H^\infty$ and we can check that

$$|B(e^{i\theta})| = 1 \quad \text{a.e. on } \mathbb{T}.$$

An infinite product of the form (4.40) is called a *Blaschke product*. Let $\mathfrak{Z}$ denote the set of all accumulation points of $(z_n)_{n\ge1}$. Since $|z_n| \longrightarrow 1$ as $n \longrightarrow +\infty$, $\mathfrak{Z}$ is a closed nonempty subset of $\mathbb{T}$. Moreover, $\mathfrak{Z}$ also coincides with the accumulation points of $(1/\bar{z}_n)_{n\ge1}$ and the set

$$\Omega = \mathbb{C} \setminus \{\mathfrak{Z} \cup \{1/\bar{z}_n : n \ge 1\}\}$$

is an open subset. Moreover, the infinite product (4.40) converges uniformly on compact subsets of Ω, so that B indeed defines an analytic function on Ω. In particular, if an arc of $\mathbb{T}$ is free of the accumulation points of the zeros of a Blaschke product B, then B is a well-defined analytic function on this arc and $|B| \equiv 1$ there.

In connection with Blaschke products, we will also need the following two formulas. If $\alpha, z \in \mathbb{D}$, then

$$\left|\frac{\alpha - z}{1 - \bar{\alpha} z}\right|^2 = 1 - \frac{(1-|\alpha|^2)(1-|z|^2)}{|1-\bar{\alpha}z|^2}. \tag{4.41}$$

The denominator on the right-hand side is also given by

$$|1-\bar{\alpha}z|^2 = (1-|\alpha z|)^2 + 4|\alpha z| \sin^2(\tfrac{1}{2}\theta), \tag{4.42}$$

where $\theta \in (-\pi, \pi]$ is the argument of $\bar{\alpha}z$. We will also denote by b_α the simple Blaschke factor given by

$$b_\alpha(z) = \frac{\alpha - z}{1 - \bar{\alpha}z}. \tag{4.43}$$

The formula (4.40) can thus be rewritten as

$$B(z) = \prod_{n=1} \frac{|z_n|}{z_n} b_{z_n}(z) \qquad (z \in \mathbb{D}).$$

Note also that the hyperbolic distance introduced in Section 4.1 satisfies

$$\rho(z, w) = |b_z(w)| = |b_w(z)| \qquad (z, w \in \mathbb{D}).$$

Using (4.18), (4.19) and (4.20), we can also rewrite (4.41) as

$$|b_\alpha(z)|^2 = 1 - |\langle \hat{k}_\alpha, \hat{k}_z \rangle_2|^2,$$

where $\hat{k}_\alpha = k_\alpha / \|k_\alpha\|_2$ is the normalized reproducing kernel of H^2.

Two further profound consequences of Jensen's formula are

$$\log|f| \in L^1(\mathbb{T}) \tag{4.44}$$

and

$$\log|f(z)| \le \int_{\mathbb{T}} \frac{1-|z|^2}{|\zeta - z|^2} \log|f(\zeta)|\, dm(\zeta) \qquad (z \in \mathbb{D}), \tag{4.45}$$

for any $f \in H^p$, $f \not\equiv 0$. In particular, (4.44) shows that $f \neq 0$ almost everywhere on $\mathbb{T}$. The property (4.44) enables us to define the nonvanishing analytic function

$$[f](z) = \exp\left(\int_{\mathbb{T}} \frac{\zeta + z}{\zeta - z} \log|f(\zeta)|\, dm(\zeta) \right) \qquad (z \in \mathbb{D}). \tag{4.46}$$

Note that this function belongs to H^p. Indeed, applying the arithmetic–geometric mean inequality (see e.g. [442, p. 63]), we have

$$\begin{aligned} |[f](re^{i\theta})|^p &= \exp\left(\int_{\mathbb{T}} \frac{1-r^2}{|re^{i\theta} - \zeta|^2} \log|f(\zeta)|^p\, dm(\zeta) \right) \\ &\le \int_{\mathbb{T}} \frac{1-r^2}{|re^{i\theta} - \zeta|^2} |f(\zeta)|^p\, dm(\zeta). \end{aligned}$$

Thus

$$\int_0^{2\pi} |[f](re^{i\theta})|^p \, \frac{d\theta}{2\pi} \le \int_{\mathbb{T}} |f(\zeta)|^p\, dm(\zeta) = \|f\|_p^p.$$

Hence $[f] \in H^p$ and $\|[f]\|_p \le \|f\|_p$. Moreover, we easily see from (4.45) and (4.46) that

$$|f(z)| \le |[f](z)| \qquad (z \in \mathbb{D}),$$

whence $\|f\|_p \le \|[f]\|_p$, and we finally obtain that

$$\|[f]\|_p = \|f\|_p.$$

Note also that Corollary 3.3 and Theorem 4.1 imply that

$$|[f](\zeta)| = |f(\zeta)| \qquad \text{a.e. on } \mathbb{T}.$$

As we will see, the function defined in (4.46) plays a vital role in the theory of Hardy spaces.

The first step in the canonical factorization theorem (Theorem 4.19) is the following result, which shows that the zeros of any $f \in H^p$ can be divided out without increasing the norm.

Theorem 4.18 *Let $f \in H^p$, $0 < p \leq \infty$, $f \not\equiv 0$. Then f can be factored in the form*

$$f(z) = B(z)g(z) \qquad (z \in \mathbb{D}),$$

where B is the Blaschke product formed by the zeros of f, and g is a function in H^p that does not vanish in $\mathbb{D}$. Furthermore, we have

$$|f(\zeta)| = |g(\zeta)| \qquad \text{a.e. on } \mathbb{T}$$

and $\|f\|_p = \|g\|_p$.

This classic result can be found in many textbooks, see e.g. [188, p. 20] or [442, p. 338]. Now let $f \in H^p(\mathbb{D})$, $p \geq 0$, $f \not\equiv 0$. Let us consider the factorization $f = Bg$ given by Theorem 4.18. Since $|f| = |g|$ a.e. on $\mathbb{T}$, we easily see that $[f](z) = [g](z)$, $z \in \mathbb{D}$, and

$$|g(z)| \leq |[f](z)| \qquad (z \in \mathbb{D}).$$

Also, from the discussion above, we have

$$|g(\zeta)| = |[f](\zeta)| \qquad \text{a.e. on } \mathbb{T}.$$

Hence, if $e^{i\gamma} = g(0)/|g(0)|$, the function

$$S(z) = e^{-i\gamma} \frac{g(z)}{[f](z)} \qquad (z \in \mathbb{D})$$

is analytic on $\mathbb{D}$ and has the properties

$$0 < |S(z)| \leq 1, \quad |S(\zeta)| = 1 \text{ a.e. on } \mathbb{T}, \quad S(0) > 0.$$

This shows that $-\log |S(z)|$ is a positive harmonic function that vanishes almost everywhere on $\mathbb{T}$. Thus by Corollary 3.7 and Theorem 3.2,

$$-\log |S(z)| = P(\sigma)(z) \qquad (z \in \mathbb{D}),$$

where σ is a finite and positive Borel measure on $\mathbb{T}$, singular with respect to the Lebesgue measure. Since $S(0) > 0$, analytic completion gives

$$S(z) = S_\sigma(z) = \exp\left(- \int_{\mathbb{T}} \frac{\zeta + z}{\zeta - z} \, d\sigma(\zeta) \right) \qquad (z \in \mathbb{D}). \tag{4.47}$$

Putting all this together, we have the factorization

$$f(z) = e^{i\gamma} B(z) S_\sigma(z) [f](z) \qquad (z \in \mathbb{D}).$$

Thus we have obtained the following theorem.

Theorem 4.19 (Canonical factorization theorem) *Let $f \in H^p(\mathbb{D})$, $0 < p \leq \infty$, $f \not\equiv 0$. Then we have the unique factorization*

$$f = \lambda B S_\sigma [f],$$

where λ is a unimodular constant, B is the Blaschke product formed with $(z_n)_{n\geq 1}$, the zeros of f in $\mathbb{D}$, and

$$S_\sigma(z) = \exp\left(- \int_{\mathbb{T}} \frac{\zeta + z}{\zeta - z} \, d\sigma(\zeta) \right) \qquad (z \in \mathbb{D}),$$

with σ a finite and positive Borel measure on $\mathbb{T}$, singular with respect to the Lebesgue measure. Moreover, $[f] \in H^p(\mathbb{D})$, and

$$\|[f]\|_p = \|f\|_p.$$

In the canonical factorization, the function $[f]$ is called the *outer part* of f, the function S_σ is called its *singular inner part*, whereas BS_σ is called its *inner part*. The multiplicative factor λ can be merged into either the inner or outer factors. Hence, a function $f \in H^p$ is called *outer* if $f = \lambda[f]$, and *inner* if $f = \lambda BS_\sigma$. In other words, f is outer if and only if it has no zeros on $\mathbb{D}$ and the harmonic function $\log|f|$ is given by the Poisson integral of its boundary values. By the same token, f is inner if and only if f is bounded and $|f| = 1$ almost everywhere on $\mathbb{T}$. In the following, a prototype inner function will usually be denoted by Θ.

For further reference, some of the properties of B, S_σ and $[f]$ are recalled next:

(i)

$$|B(z)| \leq 1 \quad \text{and} \quad |S_\sigma(z)| \leq 1 \qquad (z \in \mathbb{D}),$$

(ii)

$$|B| = |S_\sigma| = 1 \quad \text{(a.e. on } \mathbb{T}),$$

(iii)

$$-\log|S_\sigma(z)| = \int_{\mathbb{T}} \frac{1 - |z|^2}{|\zeta - z|^2} \, d\sigma(\zeta) \qquad (z \in \mathbb{D}), \tag{4.48}$$

(iv)

$$|f(z)| \leq |[f](z)| \qquad (z \in \mathbb{D}),$$

(v)

$$|f| = |[f]| \qquad \text{(a.e. on } \mathbb{T}). \tag{4.49}$$

If $f \in H^p$, $0 < p \leq \infty$ and Θ is any inner function, then clearly $\Theta f \in H^p$. Moreover, by the canonical factorization and (4.49), we have

$$\|\Theta f\|_p = \|f\|_p = \|[f]\|_p. \tag{4.50}$$

Given $w \in L^p(\mathbb{T})$ such that $\log|w| \in L^1(\mathbb{T})$, we can use the idea in (4.46) to define

$$[w](z) = \exp\left(\int_{\mathbb{T}} \frac{\zeta+z}{\zeta-z} \log|w(\zeta)|\, dm(\zeta) \right) \qquad (z \in \mathbb{D}). \tag{4.51}$$

Based on the above observation, it is immediate to see that $[w]$ is an outer function in H^p such that $|[w]| = |w|$ a.e. on $\mathbb{T}$. We can take one further step in generalizing this concept and just consider a measurable function h on $\mathbb{T}$ such that $\log|h| \in L^1(\mathbb{T})$. Then we define the analytic function h on $\mathbb{D}$ by

$$[h](z) = \gamma \exp\left(\int_{\mathbb{T}} \frac{\zeta+z}{\zeta-z} \log|h(\zeta)|\, dm(\zeta) \right) \qquad (z \in \mathbb{D}), \tag{4.52}$$

where γ is a unimodular constant. Since

$$\frac{\zeta+z}{\zeta-z} = P_z(\zeta) + iQ_z(\zeta),$$

according to Corollary 3.3 and (3.23), we have

$$[h] = \gamma|h| \exp(i\, \widetilde{\log|h|}) \qquad \text{(a.e. on } \mathbb{T}). \tag{4.53}$$

Moreover, since $\log|[h](z)| = P(\log|h|)(z)$, $z \in \mathbb{D}$, Theorem 3.6 implies that

$$\sup_{0 \le r < 1} \int_{\mathbb{T}} \big|\log|[h](r\zeta)|\big|\, dm(\zeta) < \infty.$$

In particular,

$$\sup_{0 \le r < 1} \int_{\mathbb{T}} \log^+ |[h](r\zeta)|\, dm(\zeta) < \infty.$$

That means that the function $[h]$ defined by (4.52) is in the Nevanlinna class $\mathcal{N}$ (see Section 5.1). Such a function is called an *outer function of* $\mathcal{N}$. Note that, if f_1 and f_2 are in some (maybe even different) Hardy spaces, with $f_1, f_2 \not\equiv 0$, then $\log|f_1/f_2|$ is integrable on $\mathbb{T}$ and

$$\frac{[f_1]}{[f_2]} = \left[\frac{f_1}{f_2}\right]. \tag{4.54}$$

In the same spirit, we can also generalize a little bit the concept of singular inner function. Let $\sigma \in \mathcal{M}(\mathbb{T})$ be a signed measure on $\mathbb{T}$ and define S_σ by

$$S_\sigma(z) = \exp\left(-\int_{\mathbb{T}} \frac{\zeta+z}{\zeta-z}\, d\sigma(\zeta) \right) \qquad (z \in \mathbb{D}).$$

It is immediate to see that S_σ is analytic on $\mathbb{D}$ and we have

$$\log|S_\sigma(z)| = -P(\sigma)(z). \tag{4.55}$$

Hence, Theorem 3.6 implies that

$$\sup_{0\le r<1}\int_{\mathbb{T}}\big|\log|S_\sigma(r\zeta)|\big|\,dm(\zeta)<\infty.$$

In particular, we deduce that S_σ belongs to the Nevanlinna class. Now, on the one hand, according to Corollary 3.7 and (4.55), we immediately see that $|S_\sigma(z)|\le 1$ if and only if $\sigma\ge 0$. On the other hand, we get from Corollary 3.3 that $|S_\sigma|=1$ a.e. on $\mathbb{T}$ if and only if σ is singular with respect to the Lebesgue measure. Hence, we deduce that S_σ is an inner function if and only if σ is positive and singular with respect to the Lebesgue measure.

The following nice characteristic property of Blaschke products is a consequence of Jensen's formula.

Lemma 4.20 *Let B be a Blaschke product in the open unit disk. Then*

$$\lim_{r\to 1}\int_0^{2\pi}\log|B(re^{it})|\,dt=0. \tag{4.56}$$

Proof This is obvious if B has only a finite number of factors because in that case B is continuous on the closed unit disk $\bar{\mathbb{D}}$, and $|B|\equiv 1$ on $\mathbb{T}$. So we may assume that

$$B(z)=\prod_{n=1}^{\infty}\frac{|a_n|}{a_n}\frac{a_n-z}{1-\bar{a}_n z},\qquad 0<|a_n|\le|a_{n+1}|<1.$$

By Jensen's formula, we have

$$\frac{1}{2\pi}\int_0^{2\pi}\log|B(re^{it})|\,dt=\log|B(0)|+\sum_{\{n:|a_n|<r\}}\log\left(\frac{r}{|a_n|}\right).$$

Fix an integer N. Thus, for all $r>|a_N|$, we have

$$\begin{aligned}\frac{1}{2\pi}\int_0^{2\pi}\log|B(re^{it})|\,dt&\ge\log|B(0)|+\sum_{n=1}^{N}\log\left(\frac{r}{|a_n|}\right)\\&=\sum_{n=1}^{\infty}\log|a_n|+\sum_{n=1}^{N}\log\left(\frac{r}{|a_n|}\right)\\&=N\log r+\sum_{n=N+1}^{\infty}\log|a_n|.\end{aligned}$$

Consequently,

$$\sum_{n=N+1}^{\infty}\log|a_n|\le\lim_{r\to 1}\frac{1}{2\pi}\int_0^{2\pi}\log|B(re^{it})|\,dt\le 0.$$

The result now follows by letting $N\longrightarrow\infty$. □

All the facts stated above are very standard and can be found in [188, 233, 320, 442].

4.7 The Schwarz reflection principle for H^1 functions

Before giving a condition that ensures that an H^1 function can be analytically extended through an open subarc of $\mathbb{T}$, we recall the classic Schwarz reflection principle for analytic functions.

Theorem 4.21 *Let I be an open subarc of $\mathbb{T}$ and let f be an analytic function on $\mathbb{D}$. Suppose that*

$$\lim_{n\to\infty} \Im(f(z_n)) = 0$$

for every sequence $(z_n)_{n\geq 1}$ in $\mathbb{D}$ that converges to a point of I. Then the function f can be analytically extended through I.

For a function in H^1, we can slightly relax the condition of the above theorem. Briefly speaking, we can replace "every" by "almost all" in this situation. Moreover, it is implicit in the theorem that we restrict ourselves to sequences converging nontangentially to the boundary points.

Theorem 4.22 *Let $f \in H^1$ and let I be an open subarc of $\mathbb{T}$. Assume that*

$$\Im(f(\zeta)) = 0$$

for almost all $\zeta \in I$. Then f can be analytically extended through I.

Proof According to (4.8), we have

$$\begin{aligned}\Im(f(z)) &= \int_{\mathbb{T}} \frac{1-|z|^2}{|\zeta - z|^2}\, \Im(f(\zeta))\, dm(\zeta)\\ &= \int_{\mathbb{T}\setminus I} \frac{1-|z|^2}{|\zeta - z|^2}\, \Im(f(\zeta))\, dm(\zeta).\end{aligned}$$

Let $\zeta_0 \in I$ and $\delta = \operatorname{dist}(\zeta_0, \mathbb{T}\setminus I)$. Clearly $\delta > 0$. Then, for each $z \in \mathbb{D}$, $|z - \zeta_0| \leq \delta/2$ and $\zeta \in \mathbb{T}\setminus I$, we have

$$|z-\zeta| \geq |\zeta_0 - \zeta| - |z - \zeta_0| \geq \delta - \frac{\delta}{2} = \frac{\delta}{2}.$$

Hence, for $|z - \zeta_0| \leq \delta/2$,

$$|\Im(f(z))| \leq \frac{4}{\delta^2}(1-|z|^2)\|f\|_1,$$

which proves that $\Im(f(z)) \longrightarrow 0$ as $z \longrightarrow \zeta_0$. We can thus apply Theorem 4.21 and deduce that f can be analytically extended through I. □

We now state a well-known result that is an easy consequence of the Littlewood subordination principle. This classic result says that, if φ is an analytic map of the unit disk into itself and if $f \in H^p$, $1 \leq p$, then $f \circ \varphi \in H^p$. See Section 19.6 for a discussion of this result using the theory of reproducing kernels. For a further application, we need the following technical corollary.

Corollary 4.23 *Let I be an open subarc of $\mathbb{T}$, let $f \in H^1$ and let g be an outer function that is locally bounded below on I. Suppose that $h = f/g$ satisfies*

$$\Im(h(\zeta)) = 0$$

for almost all $\zeta \in I$. Then h can be analytically extended through I.

Proof Let $\zeta_0 \in I$. By hypothesis, there exists $r_0 > 0$ satisfying

$$\inf\{|g(z)| : z \in \mathbb{D},\ |z - \zeta_0| < r_0\} > 0.$$

Put $\Omega_0 = \{z \in \mathbb{D} : |z - \zeta_0| < r_0\}$, $J_0 = \partial\Omega_0 \cap \mathbb{T}$ and denote by I_0 the interior of J_0. By taking r_0 as small as needed, we also assume that $I_0 \subset I$. Since Ω_0 is a bounded simply connected domain and $\partial\Omega_0$ is a Jordan curve, the Riemann–Carathéodory theorem implies that there exists a conformal map φ from Ω_0 onto $\mathbb{D}$ that extends into a homeomorphism of $\bar{\Omega}_0$ onto $\bar{\mathbb{D}}$. Put $h = f/g$ and consider $\tilde{h} = h \circ \varphi^{-1}$. By the Littlewood subordination principle $f \circ \varphi^{-1} \in H^1$, and clearly $1/g \circ \varphi^{-1} \in H^\infty$. Hence, $\tilde{h} \in H^1$. Moreover,

$$\Im(\tilde{h}(\zeta)) = 0$$

for almost all $\zeta \in \tilde{I}_0 = \varphi(I_0)$. According to Theorem 4.22, we deduce that $\tilde{h}$ extends analytically through $\tilde{I}_0$ and thus h can be analytically extended through I_0. Since ζ_0 is an arbitrary point of I, we get the result. □

4.8 Properties of outer functions

In this section, we study several further consequences of Theorem 4.19. Naively speaking, the divisors of any outer function have to be outer, and similarly for inner functions.

Corollary 4.24 *Let $p, q, r \in (0, \infty]$. Then the following hold.*

(i) *If $f \in H^p$ is outer and $g \in H^q$ is such that $f/g \in H^r$, then g and f/g are both outer functions.*
(ii) *If $f \in H^p$ and $g \in H^q$, then the product fg is an outer function if and only if f and g are both outer functions.*
(iii) *If $f \in H^p$ and $1/f \in H^q$, then f is outer.*
(iv) *If f and g are inner functions such that $f/g \in H^p$, then f/g is also inner.*

Proof Take $s = \min\{p, q, r\}$. Then all the functions considered above belong to H^s.

(i) This is an immediate consequence of the uniqueness part of the canonical factorization theorem (Theorem 4.19). Put $h = f/g$. Then the identity $f = gh$ holds in H^s. Since f is outer, then, by the uniqueness, g and h also have to be outer.

(ii) Assume that fg is outer. Let $f = \lambda_1 B_1 S_1 [f]$ and $g = \lambda_2 B_2 S_2 [g]$ be, respectively, the canonical factorizations of f and g. Then

$$fg = \lambda_1 \lambda_2 B_1 B_2 S_1 S_2 [fg].$$

Since fg is outer, according to the uniqueness of the canonical factorization, the product $B_1 B_2 S_1 S_2$ must be a unimodular constant. Thus, B_1, B_2, S_1 and S_2 are all unimodular constants. Hence, we have $f = \gamma_1 [f]$ and $g = \gamma_2 [g]$, which means that f and g are outer functions. Since $[f][g] = [fg]$, the reverse implication is obvious.

(iii) This follows from (ii).

(iv) As in (i), put $h = f/g$. Then $f = hg$, and the uniqueness of the canonical factorization reveals that h cannot have an outer factor. □

We can now apply Corollary 4.24 to obtain a vast class of outer functions, which will be exploited in our discussions.

Corollary 4.25 *Let f be an analytic function in $\mathbb{D}$, $f \not\equiv 0$, such that $\Re(f(z)) \geq 0$, for all $z \in \mathbb{D}$. Then f is an outer function in H^p, for each $0 < p < 1$. Moreover,*

$$\|f\|_p \leq c_p |f(0)|,$$

where c_p is a constant satisfying

$$c_p = O\left(\frac{1}{1-p}\right) \qquad (\textit{as } p \longrightarrow 1).$$

Proof First of all, note that $\Re(f(z)) > 0$, for all $z \in \mathbb{D}$. Indeed, if there exists a point $z_0 \in \mathbb{D}$ such that $\Re(f(z_0)) = 0$, then, by the maximum principle for harmonic functions, $\Re(f) \equiv 0$ on $\mathbb{D}$ and so f is constant, identically equal to 0, which is a contradiction.

Let $\log z$ be the main branch of the logarithm with $\Im(\log z) \in]-\pi, \pi]$. Then the mapping $z \longmapsto f(z)^p = \exp(p \log f(z))$ is analytic on $\mathbb{D}$. Since $0 < p < 1$ and $\Re(f(z)) > 0$, we have

$$|f(z)|^p \leq k_p \, \Re(f(z)^p) \qquad (z \in \mathbb{D}),$$

where $k_p = 1/\cos(p\pi/2)$. Then, if we apply the mean value theorem to the harmonic function $\Re(f(z)^p)$, we get

$$\frac{1}{2\pi}\int_0^{2\pi} |f(re^{i\theta})|^p \, d\theta \leq \frac{k_p}{2\pi}\int_0^{2\pi} \Re(f(re^{i\theta})^p)\, d\theta = k_p\, \Re(f(0)^p)$$

for each $0 \leq r < 1$. This estimation ensures that $f \in H^p$ with

$$\|f\|_p \leq k_p^{1/p}|f(0)|.$$

That $k_p^{1/p} = O(1/(1-p))$, as $p \longrightarrow 1$, is elementary to establish. Moreover, since $1/f$ is also an analytic function on $\mathbb{D}$ and

$$\Re\left(\frac{1}{f(z)}\right) = \frac{\Re(f(z))}{|f(z)|^2} > 0 \qquad (z \in \mathbb{D}),$$

we also have $1/f \in H^p$. Thus, by Corollary 4.24, we conclude that f is outer. □

There are several applications of Corollary 4.25. As an interesting example, if $f \in H^\infty$, $f \not\equiv 0$, then, for each constant $C \geq \|f\|_\infty$, the function $C + f$ is outer. Two more applications are stated below.

Corollary 4.26 *Let b be a nonconstant function in the closed unit ball of H^∞. Put*

$$f = \frac{1}{1-b}.$$

Then f is an outer function in H^p, for each $0 < p < 1$.

Proof By the maximum principle, $|b(z)| < \|b\|_\infty \leq 1$ for each $z \in \mathbb{D}$. Hence, f is analytic on $\mathbb{D}$. Moreover, we have

$$\Re\left(\frac{1}{1-b(z)}\right) = \frac{1-\Re(b(z))}{|1-b(z)|^2} \geq \frac{1-|b(z)|}{|1-b(z)|^2} > 0 \qquad (z \in \mathbb{D}).$$

Hence, Corollary 4.25 implies that f is an outer function in H^p. □

Corollary 4.27 *Let $\mathfrak{p}$ be an analytic polynomial, whose zeros are all located on the unit circle. Then $\mathfrak{p}$ is an outer function in H^∞.*

Proof Write

$$\mathfrak{p}(z) = \lambda \prod_{i=1}^{n} (z - z_i)^{\alpha_i},$$

with $\lambda \in \mathbb{C}$, $\alpha_i \in \mathbb{N}$ and $z_i \in \mathbb{T}$. But $z - z_i = -z_i(1 - \bar{z}_i z)$ and by Corollary 4.25, the function $1 - \bar{z}_i z$ is an H^∞ outer function. Now the result follows immediately from Corollary 4.24. □

In the family of Hardy spaces, dividing by an analytic function, even if it does not have any zero, is a delicate process and the result could be a function that does not belong to any Hardy space. For example, if S is a singular inner function, then $1/S$ does not belong to any Hardy space (see Corollary 4.24). However, at the same time, its boundary values are unimodular and one is (wrongly) tempted to say that $1/S$ is an inner function. The following result says that dividing by an outer function is legitimate as long as the boundary values remain in a Lebesgue space.

Corollary 4.28 *Let $p, q, r \in (0, \infty]$, and let $f \in H^p$. Then the following are equivalent.*

(i) *The function f is outer.*
(ii) *If $g \in H^q$ and $g/f \in L^r(\mathbb{T})$, then $g/f \in H^r$.*

In particular, if $g \in H^q$ and $g \in L^r(\mathbb{T})$, then $g \in H^r$.

Proof (i) $\Longrightarrow$ (ii) Let $g = \lambda_1 BS[g]$ be the canonical factorization of g. By assumption, $f = \lambda_2[f]$. Hence,

$$\frac{g}{f} = \frac{\lambda_1 BS[g]}{\lambda_2[f]} = \frac{\lambda_1}{\lambda_2} BS \left[\frac{g}{f}\right].$$

But the condition $g/f \in L^r$ implies $[g/f] \in H^r$. Thus, $g/f \in H^r$.

(ii) $\Longrightarrow$ (i) By the assumption, we certainly have $f \not\equiv 0$. Put

$$g(\zeta) = \min\{|f(\zeta)|, 1\} \qquad (\zeta \in \mathbb{T}).$$

Since $g \in L^\infty(\mathbb{T}) \subset L^q(\mathbb{T})$, the outer function $[g]$ is in H^q and, by (4.49), we have

$$|[g]/f| = |g/f| \leq 1 \qquad \text{(a.e. on } \mathbb{T}).$$

Hence, by assumption, we must have $[g]/f \in H^r$, i.e. $[g] = fh$, where $h \in H^r$. The uniqueness part of the canonical factorization theorem (Theorem 4.19) now implies that both f and h have to be outer functions. □

As another application of the canonical factorization, we slightly generalize (4.12). It is no surprise that, owing to the lack of a Cauchy (or Poisson) integral representation for H^p functions when $0 < p < 1$, the proof of the following result calls for a different approach.

Theorem 4.29 *Let $f \in H^{1/2}$. Assume that $f \geq 0$ almost everywhere on $\mathbb{T}$. Then f is constant.*

Proof If $f \equiv 0$, we are done. Hence, assume that $f \not\equiv 0$. By the canonical factorization, we have $f = Bg$, where B is the Blaschke product associated with f and g belongs to $H^{1/2}$ and has no zeros on $\mathbb{D}$. That is why we can define $h = g^{1/2}$, and the function h belongs to H^1, with

$$\|h\|_1 = \|g\|_{1/2}.$$

Clearly, $f = Bh^2$.

The condition $f \geq 0$ ensures that $f = |f|$ almost everywhere on $\mathbb{T}$. Hence, since B is unimodular on $\mathbb{T}$, we have

$$Bh^2 = |h|^2 \qquad \text{(a.e. on } \mathbb{T}).$$

Since $f \not\equiv 0$ on $\mathbb{D}$, (4.44) implies that $f \neq 0$ almost everywhere on $\mathbb{T}$, and thus $h \neq 0$ and the previous identity implies that

$$Bh = \bar{h} \qquad \text{(a.e. on } \mathbb{T}).$$

Now, on the one hand, we have $Bh \in H^1$, and, on the other, $\bar{h} \in \overline{H^1}$. We know that $H^1 \cap \overline{H^1}$ contains only the constant functions. Therefore, Bh is a constant function. By the uniqueness of the canonical factorization, this happens precisely when B is a unimodular constant and h is a constant. Thus, eventually f is constant. □

Consider the function

$$f(z) = -\left(\frac{1+z}{1-z}\right)^2 \qquad (z \in \mathbb{D}).$$

By a direct analysis, one can simply show that

$$f \in \bigcap_{0<p<1/2} H^p.$$

This fact can also be deduced from Corollary 4.26. We also have

$$f(e^{i\theta}) = \cot^2(\tfrac{1}{2}\theta) \geq 0.$$

This example shows that the exponent $p = 1/2$ in Theorem 4.29 is optimal.

Exercises

Exercise 4.8.1 Let $\varphi \in H^\infty$ with $\|\varphi\|_\infty \leq \pi/2$. Show that e^φ is an outer function.
Hint: Use Corollary 4.25.

Exercise 4.8.2 The following result is referred as the *maximum principle for outer functions*. Let $f, g \in H^p$ and assume that g is outer. Show that

$$|f| \leq |g| \text{ on } \mathbb{T} \quad \Longrightarrow \quad |f| \leq |g| \text{ on } \mathbb{D}.$$

4.9 A uniqueness theorem

The idea of considering $h = g^{1/2}$, which was exploited in the proof of Theorem 4.29, is fruitful and has many other interesting consequences. For example, see Section 4.10 for an interpretation of $H^1 = H^2H^2$. It can also be used for studying the boundary values of functions in H^p spaces with $p < 1$.

The property (4.44) can be interpreted as a uniqueness theorem for H^1 functions. This result is so important that we mention it again as a lemma. This result will be exploited in the theorem that follows.

Lemma 4.30 *Let $f \in H^1$ be such that*

$$\int_0^{2\pi} \big|\log|f(e^{i\theta})|\big|\, d\theta = +\infty.$$

Then $f \equiv 0$.

This classic result can be strengthened in the following sense. Remember that, if $f \in H^1$, then

$$\hat{f}(-n) = 0 \qquad (n \geq 1).$$

Theorem 4.31 *Let $f \in L^1(\mathbb{T})$ fulfill the following properties:*

(i) *there is a constant $c > 0$ such that*

$$\hat{f}(-n) = O(e^{-cn}) \qquad (n \longrightarrow \infty);$$

(ii) *one has*

$$\int_0^{2\pi} \big|\log|f(e^{i\theta})|\big|\, d\theta = +\infty.$$

Then $f \equiv 0$.

Proof The first assumption is equivalent to saying that there are constants $a > 0$ and $0 < b < 1$ such that

$$|\hat{f}(-n)| \leq ab^n \qquad (n \geq 1). \tag{4.57}$$

This is the version that is used in the following proof. Since $f \in L^1(\mathbb{T})$, the second assumption is also equivalent to

$$\int_0^{2\pi} \log^- |f(e^{i\theta})|\, d\theta = +\infty.$$

We assume that $f \not\equiv 0$ on $\mathbb{T}$ and we obtain a contradiction. In light of the uniqueness theorem for the Fourier coefficients of an integrable function, this assumption ensures the existence of an $n \in \mathbb{Z}$ such that $\hat{f}(n) \neq 0$. The condition (4.57) implies that the functions

$$f_1(z) = \sum_{n=1}^{\infty} \frac{\hat{f}(-n)}{z^n} = \sum_{n=-\infty}^{-1} \hat{f}(n) z^n$$

and

$$f_2(z) = \sum_{n=1}^{\infty} \overline{\hat{f}(-n)}\, z^n$$

are analytic on $|z| > b$ and $|z| < 1/b$, respectively, such that $f_1 = \bar{f}_2$ on $\mathbb{T}$. We also clearly have

$$|\hat{f}(n)| \leq \|f\|_1 \qquad (n \in \mathbb{Z}).$$

This condition implies that the function

$$f_3(z) = \sum_{n=0}^{\infty} \hat{f}(n) z^n$$

is analytic on $\mathbb{D}$. Now, we exploit the functions f_1, f_2 and f_3. Naively speaking, we can say that we undertake surgery to replace the conjugate-analytic part of the Poisson extension of f by an analytic function and still keep some properties. This is clarified below.

Recall that f is initially defined on $\mathbb{T}$. First, extend f on the annulus $b < |z| < 1$ by the formula

$$f(z) = f_1(z) + f_3(z) = \sum_{n=-\infty}^{\infty} \hat{f}(n) z^n.$$

This function is analytic on the annulus $b < |z| < 1$, and thus, for each fixed $b < r < 1$, one has

$$\int_0^{2\pi} |f(re^{i\theta})|\, d\theta < \infty. \tag{4.58}$$

Moreover, $f \not\equiv 0$ on this annulus. Indeed, for any $n \in \mathbb{Z}$, we have

$$\hat{f}(n) = r^{-n} \int_{\mathbb{T}} f(r\zeta)\bar{\zeta}^n\, dm(\zeta),$$

where r is any fixed number with $b < r < 1$. Thus, the condition $f \equiv 0$ on the annulus would imply $\hat{f}(n) = 0$ for any $n \in \mathbb{Z}$, which contradicts the assumption.

Second, Pf, the Poisson integral of f, can be written as

$$Pf(z) = \overline{f_2(z)} + f_3(z) \qquad (z \in \mathbb{D})$$

(see (3.15)). According to Theorem 3.6, the assumption $f \in L^1(\mathbb{T})$ ensures that the L^1-integral means of Pf are uniformly bounded and, moreover, that $\|(Pf)_r - f\|_1 \longrightarrow 0$ as $r \longrightarrow 1^-$. Then, since $f = f_1 + f_3$ and $Pf = \bar{f}_2 + f_3$, we have

$$\begin{aligned}\int_0^{2\pi} |f(re^{i\theta}) - f(e^{i\theta})|\, d\theta &\le \int_0^{2\pi} |(Pf)(re^{i\theta}) - f(e^{i\theta})|\, d\theta \\ &\quad + \int_0^{2\pi} |f(re^{i\theta}) - (Pf)(re^{i\theta})|\, d\theta \\ &= \int_0^{2\pi} |(Pf)(re^{i\theta}) - f(e^{i\theta})|\, d\theta \\ &\quad + \int_0^{2\pi} |f_1(re^{i\theta}) - \overline{f_2(re^{i\theta})}|\, d\theta.\end{aligned}$$

We already know that $\|(Pf)_r - f\|_1$ tends to 0, as $r \longrightarrow 1$. As for the last integral, using the fact that f_1 is analytic on $|z| > b$ and f_2 is analytic on $|z| < 1/b$, we easily get

$$\lim_{r\to 1} \int_0^{2\pi} \left|f_1(re^{i\theta}) - \overline{f_2(e^{i\theta})}\right| d\theta = \int_0^{2\pi} \left|f_1(e^{i\theta}) - \overline{f_2(e^{i\theta})}\right| d\theta = 0.$$

Remember, $f_1 = \bar{f}_2$ on $\mathbb{T}$. Hence,

$$\lim_{r\to 1} \int_0^{2\pi} |f(re^{i\theta}) - f(e^{i\theta})|\, d\theta = 0. \tag{4.59}$$

The identity $f = Pf + f_1 - \bar{f}_2$ on the corona $b < |z| < 1$ also ensures that

$$\lim_{r\to 1} f(re^{i\theta}) = f(e^{i\theta}) \qquad \text{(a.e. on } \mathbb{T}).$$

Since $|\log^+ \alpha - \log^+ \beta| \le |\alpha - \beta|$, (4.59) implies that

$$\lim_{r\to 1} \int_0^{2\pi} \log^+ |f(re^{i\theta})|\, d\theta = \int_0^{2\pi} \log^+ |f(e^{i\theta})|\, d\theta.$$

Hence, using the relation $\log t = \log^+ t - \log^- t$, Fatou's lemma and part (ii) of the theorem, we obtain

$$\lim_{r\to 1} \int_0^{2\pi} \log |f(re^{i\theta})|\, d\theta = -\infty. \tag{4.60}$$

Now let

$$L(r) = \int_0^{2\pi} \log |f(re^{i\theta})|\, d\theta \qquad (b < r \le 1).$$

For $b < r < 1$, by (4.58), $L(r)$ is a well-defined quantity in $[-\infty, \infty)$. Then, since $f \not\equiv 0$ on the annulus $b < |z| < 1$, by a well-known result of subharmonic functions, $L(r)$ is a finite and convex function of $\log r$ on the interval $(b, 1)$, i.e.

$$L(r) \leq \frac{\log(r_2) - \log(r)}{\log(r_2) - \log(r_1)} L(r_1) + \frac{\log(r) - \log(r_1)}{\log(r_2) - \log(r_1)} L(r_2)$$

for $b < r_1 < r < r_2 < 1$. Let $r_2 \longrightarrow 1$. Hence, by (4.60), we obtain $L(r) = -\infty$ on $(b, 1)$. This is absurd. □

We also recall the following famous uniqueness result, to which we will appeal on some occasions.

Theorem 4.32 *If f is meromorphic on $\mathbb{D}$ and has vanishing nontangential limits on a set of positive Lebesgue measure, then f is the zero function.*

This theorem says that, if two meromorphic functions on $\mathbb{D}$ have the same (nontangential) boundary values on a set of positive measure, then they are indeed equal everywhere. A proof of this classic result can be found, for example, in [320, p. 325].

4.10 More on the norm in H^p

In any Banach space, the norm convergence implies the weak convergence, and the inverse is not necessarily true. However, under some mild extra conditions, we can pass from the weak convergence to the norm convergence. It is easy to see that, if $(f_n)_{n\geq 1}$ is a sequence of functions in H^2 such that $\|f_n\|_2 \longrightarrow \|f\|_2$ and $(f_n)_{n\geq 1}$ tends weakly to f, then $f_n \longrightarrow f$ in H^2 norm. This is in fact true in any Hilbert space. It is remarkable that this result still holds in H^1. To prove this, we need a preliminary lemma, which is interesting in its own right.

Lemma 4.33 *Let $f \in H^1$. Then there are functions $g, h \in H^2$ such that*

$$f = gh$$

and, moreover,

$$\|f\|_1 = \|g\|_2^2 = \|h\|_2^2.$$

Proof By the canonical factorization theorem (Theorem 4.19), we can write $f = \Theta[f]$, where Θ is an inner function and $[f]$ is the outer function defined by (4.46). Recall that $[f]$ is an H^1 function and $\|f\|_1 = \|[f]\|_1$ (see (4.50)). Since $[f]$ is zero-free, we can define

$$g = \Theta[f]^{1/2} \quad \text{and} \quad h = [f]^{1/2}.$$

We easily see that g^2 and h^2 belong to H^1; in other words, $g, h \in H^2$. Moreover,

$$|g|^2 = |h|^2 = |f| \qquad \text{a.e. on } \mathbb{T}.$$

Hence we also have $\|g\|_2^2 = \|h\|_2^2 = \|f\|_1$. Clearly, g and h are defined such that $f = gh$. □

Theorem 4.34 *Let f and $(f_n)_{n\geq 1}$ be in H^1. Suppose that f_n uniformly converges to f on compact subsets of $\mathbb{D}$, and*

$$\lim_{n\to\infty} \|f_n\|_1 = \|f\|_1.$$

Then f_n converges to f in H^1 norm, i.e.

$$\lim_{n\to\infty} \|f_n - f\|_1 = 0.$$

Proof Without loss of generality, we may assume that $\|f_n\|_1 = \|f\|_1 = 1$. Assume now that $\|f_n - f\|_1$ does not converge to 0. Hence, by passing to a subsequence if needed, we can assume that there is an $\varepsilon_0 > 0$ such that $\|f_n - f\|_1 \geq \varepsilon_0$ for all $n \geq 1$. Now, we seek a contradiction.

By Lemma 4.33, each f_n has a factorization $f_n = g_n h_n$, where $g_n, h_n \in H^2$ and $|g_n|^2 = |h_n|^2 = |f_n|$ almost everywhere on $\mathbb{T}$. In particular, we have

$$\|g_n\|_2 = \|h_n\|_2 = \|f_n\|_1^{1/2} = 1.$$

Since the unit ball of H^2 is weakly compact, there exist $g, h \in H^2$ such that $\|g\|_2 \leq 1$ and $\|h\|_2 \leq 1$, and subsequences $(g_{n_k})_{k\geq 1}$ and $(h_{n_k})_{k\geq 1}$ such that g_{n_k} and h_{n_k} weakly converge to g and h, respectively. Hence, according to Theorem 4.16, g_{n_k} and h_{n_k} converge uniformly on compact subsets of $\mathbb{D}$ to g and h, respectively. Therefore, $f_{n_k} = g_{n_k} h_{n_k}$ converges uniformly on compact subsets of $\mathbb{D}$ to gh. By assumption, we must have $f = gh$. Moreover, by the Cauchy–Schwarz inequality,

$$1 = \|f\|_1 = \|gh\|_1 \leq \|g\|_2 \times \|h\|_2 \leq 1,$$

and thus $\|g\|_2 = \|h\|_2 = 1$. Now, we appeal to the fact that we mentioned about H^2 before the theorem to deduce that $g_{n_k} \longrightarrow g$ and $h_{n_k} \longrightarrow h$ in H^2. Hence, $f_{n_k} = g_{n_k} h_{n_k} \longrightarrow gh = f$ in H^1, i.e. $\|f_{n_k} - f\|_1 \longrightarrow 0$. This is absurd. □

The norm of a function in the Hardy space H^p is defined via a line integral. The following result, due to Littlewood and Paley, shows that we may also get the same result via a double integral. We prove the result just for $p = 2$.

Theorem 4.35 *Let $f \in H^2$. Then*

$$\|f\|_2^2 = |f(0)|^2 + \int_{\mathbb{D}} |f'(z)|^2 \, d\lambda(z),$$

where the measure λ *is given by*

$$d\lambda(z) = \frac{2}{\pi}\log(1/|z|)\,dA(z) = \frac{2}{\pi}\log(1/|z|)\,dx\,dy.$$

Proof We consider two cases in the proof.

Case I: Assume that f is analytic on an open neighborhood of $\bar{\mathbb{D}}$. Then $u = |f|^2$ is a $\mathcal{C}^2$ function and we have $\Delta u = 4|f'|^2$. Theorem 3.19 implies that

$$\int_{\mathbb{T}} |f(\zeta)|^2\,dm(\zeta) = |f(0)|^2 + \int_{\mathbb{D}} |f'(z)|^2\,d\lambda(z),$$

which gives the desired identity.

Case II: For the general situation, we introduce the function $f_r(z) = f(rz)$, $0 < r < 1$. Then we have

$$\|f_r\|_2^2 = |f(0)|^2 + r^2 \int_{\mathbb{D}} |f'(rz)|^2\,d\lambda(z).$$

By the change of variable $u = rz$, we get

$$\|f_r\|_2^2 = |f(0)|^2 + \frac{2}{\pi}\int_{r\mathbb{D}} |f'(u)|^2 \log\frac{r}{|u|}\,dA(u).$$

Letting r tend to 1, the monotone convergence theorem implies that

$$\|f\|_2^2 = |f(0)|^2 + \frac{2}{\pi}\int_{\mathbb{D}} |f'(u)|^2 \log\frac{1}{|u|}\,dA(u).$$ □

Corollary 4.36 *Let* $f \in H^2$. *Then*

$$\left(\int_{\mathbb{D}} |f'|^2\,d\lambda\right)^{1/2} \le \|f\|_2.$$

Note that, by taking $f(z) = z$ in Theorem 4.35, we deduce that λ is a probability measure. We now give two more applications of the Littlewood–Paley formula.

Corollary 4.37 *Let* $f \in H^\infty$ *and* $g \in H^1$. *Then*

$$\int_{\mathbb{D}} |f'g'|\,d\lambda \le 4\|f\|_\infty \|g\|_1.$$

Proof Without loss of generality, we can assume that $\|f\|_\infty = 1$. By Lemma 4.33, there exist $g_1, g_2 \in H^2$ such that $g = g_1g_2$ with $\|g_1\|_2^2 = \|g_2\|_2^2 = \|g\|_1$. Then

$$\int_{\mathbb{D}} |f'g'|\,d\lambda \le \int_{\mathbb{D}} |f'g_1'g_2|\,d\lambda + \int_{\mathbb{D}} |f'g_1g_2'|\,d\lambda.$$

By the Cauchy–Schwarz inequality, we have

$$\int_{\mathbb{D}} |f'g_1'g_2|\,d\lambda \le \left(\int_{\mathbb{D}} |g_1'|^2\,d\lambda\right)^{1/2} \left(\int_{\mathbb{D}} |f'g_2|^2\,d\lambda\right)^{1/2}. \tag{4.61}$$

But, by Corollary 4.36,

$$\left(\int_{\mathbb{D}} |g_1'|^2\, d\lambda\right)^{1/2} \leq \|g_1\|_2 = \|g\|_1^{1/2}. \tag{4.62}$$

For the second integral on the right-hand side of (4.61), we write $f'g_2 = (fg_2)' - fg_2'$, which, again by Corollary 4.36, gives

$$\begin{aligned}\left(\int_{\mathbb{D}} |f'g_2|^2\, d\lambda\right)^{1/2} &\leq \left(\int_{\mathbb{D}} |(fg_2)'|^2\, d\lambda\right)^{1/2} + \left(\int_{\mathbb{D}} |fg_2'|^2\, d\lambda\right)^{1/2}\\ &\leq \|fg_2\|_2 + \left(\int_{\mathbb{D}} |g_2'|^2\, d\lambda\right)^{1/2}\\ &\leq 2\|g_2\|_2.\end{aligned}$$

Hence,

$$\left(\int_{\mathbb{D}} |f'g_2|^2\, d\lambda\right)^{1/2} \leq 2\|g\|_1^{1/2}. \tag{4.63}$$

Thus, using (4.61), (4.62) and (4.63), we obtain

$$\int_{\mathbb{D}} |f'g_2g_1'|\, d\lambda \leq 2\|g\|_1.$$

By symmetry, we also have

$$\int_{\mathbb{D}} |f'g_1g_2'|\, d\lambda \leq 2\|g\|_1,$$

which gives the desired estimate. □

Corollary 4.38 *Let $f_1, f_2 \in H^\infty$, and let $g \in H^1$. Then*

$$\int_{\mathbb{D}} |f_1'f_2'g|\, d\lambda \leq 4\|f_1\|_\infty \|f_2\|_\infty \|g\|_1.$$

Proof Without loss of generality, we can assume that $\|f_1\|_\infty = \|f_2\|_\infty = 1$. Then, as in Corollary 4.37, we can write $g = g_1g_2$, where $g_1, g_2 \in H^2$ and $\|g_1\|_2^2 = \|g_2\|_2^2 = \|g\|_1$. Then the Cauchy–Schwarz inequality implies that

$$\int_{\mathbb{D}} |f_1'f_2'g|\, d\lambda \leq \left(\int_{\mathbb{D}} |f_1'g_1|^2\, d\lambda\right)^{1/2} \left(\int_{\mathbb{D}} |f_2'g_2|^2\, d\lambda\right)^{1/2},$$

and the conclusion follows immediately from (4.63). □

We end this section with the construction of a sequence of H^∞ functions whose properties will be used in the proof of some characterization of exposed points of the closed unit ball of H^1.

Theorem 4.39 *Let $(E_k)_{k\geq 1}$ be a sequence of measurable subsets of $\mathbb{T}$ such that $|E_k| \longrightarrow 0$, $k \longrightarrow \infty$. Then there is a sequence $(g_k)_{k\geq 1}$ of outer functions in H^∞ and a sequence $(\varepsilon_k)_{k\geq 1}$ of positive real numbers such that*

(i) $\sup_{E_k} |g_k| \longrightarrow 0$, $k \longrightarrow \infty$,
(ii) $g_k(0) \longrightarrow 1$, $k \longrightarrow \infty$,
(iii) $|g_k(z)| + |1 - g_k(z)| \le 1 + \varepsilon_k$, $z \in \mathbb{D}$ *and* $k \ge 1$,
(iv) $\varepsilon_k \longrightarrow 0$, $k \longrightarrow \infty$.

Proof Choose a sequence of positive numbers $(t_k)_{k\ge 1}$ such that $t_k \longrightarrow \infty$ and $t_k|E_k| \longrightarrow 0$ (for instance, take $t_k = |E_k|^{-1/2}$). Put

$$f_k(z) = t_k \int_{E_k} \frac{\zeta + z}{\zeta - z}\, dm(\zeta) \qquad (z \in \mathbb{D},\ k \ge 1).$$

We have

$$\begin{aligned} f_k(z)/t_k &= \int_{E_k} \frac{\zeta + z}{\zeta - z}\, dm(\zeta) \\ &= \int_{\mathbb{T}} \frac{\zeta + z}{\zeta - z} \chi_{E_k}(\zeta)\, dm(\zeta) \\ &= P(\chi_{E_k})(z) + iQ(\chi_{E_k})(z). \end{aligned}$$

Thus, f_k is an analytic function on $\mathbb{D}$ with values in the right half-plane (see Section 3.4). Also $f_k(0) = t_k|E_k|$, and, according to Corollary 3.3, we have $\Re(f_k) = t_k$ almost everywhere on E_k.

Now, let

$$h_k = \frac{1}{1 + f_k} \qquad (k \ge 1).$$

Since $\Re(f_k(z)) > 0$, $z \in \mathbb{D}$, we have

$$\left| \frac{1 - f_k(z)}{1 + f_k(z)} \right| < 1 \qquad (z \in \mathbb{D}),$$

and we obtain

$$\left| h_k(z) - \frac{1}{2} \right| = \frac{1}{2} \left| \frac{1 - f_k(z)}{1 + f_k(z)} \right| < \frac{1}{2} \qquad (z \in \mathbb{D}).$$

In other words, h_k maps $\mathbb{D}$ into the disk $|w - \frac{1}{2}| < \frac{1}{2}$. Furthermore, we have

$$h_k(0) = \frac{1}{1 + f_k(0)} = \frac{1}{1 + t_k|E_k|} \longrightarrow 1$$

while

$$\sup_{E_k} |h_k| \le \frac{1}{1 + t_k} \longrightarrow 0$$

as $k \longrightarrow \infty$. Note that, by Corollary 4.25, f_k and h_k are outer functions. Pick a sequence $\delta_k > 0$ that tends to zero so slowly that still

$$\sup_{E_k} |h_k^{\delta_k}| \longrightarrow 0.$$

The reason for introducing δ_k is clarified below.

The mapping $z \longmapsto z^\delta$ sends the disk $|w - \frac{1}{2}| < \frac{1}{2}$ into the cone $\Delta_\delta = \{z \in \mathbb{C} : |z| \le 1 \text{ and } |\arg z| \le \delta\pi/2\}$. Furthermore, the cone itself is contained in the interior of the ellipse $\{z \in \mathbb{C} : \operatorname{dist}(z,0) + \operatorname{dist}(z,1) = 1 + 2\sin(\delta\pi/4)\}$. In other words, we have

$$|w - \tfrac{1}{2}| < \tfrac{1}{2} \quad \Longrightarrow \quad |w^\delta| + |1 - w^\delta| \le 1 + 2\sin(\delta\pi/4).$$

With δ_k chosen as above, $g_k = h_k^{\delta_k}$ satisfies conditions (i), (ii) and (iii) with $\varepsilon_k = 2\sin(\delta_k\pi/4)$, and it is obvious that $\varepsilon_k \longrightarrow 0$, $k \longrightarrow \infty$. □

Exercises

Exercise 4.10.1 Let $f \in H^1$. Show that there are functions $g, h \in H^1$ such that

$$f = g + h$$

and, moreover, g and h are outer (and hence zero-free) functions.
Hint: Use a similar idea to that exploited in Lemma 4.33. But this time consider $g = \frac{1}{2}(\Theta + 1)[f]$ and $h = \frac{1}{2}(\Theta - 1)[f]$.

Exercise 4.10.2 Let f and $(f_n)_{n\ge1}$ be in H^1. Suppose that f_n converges pointwise to f, and

$$\lim_{n\to\infty} \|f_n\|_1 = \|f\|_1.$$

Show that f_n converges to f in H^1 norm.
Hint: Use the theorem of normal families (e.g. see [442, theorem 14.6]) and apply Theorem 4.34.

Exercise 4.10.3 Let $g \in H^1$. Show that

$$\int_{\mathbb{D}} |g'| \, d\lambda \le 4\|g\|_1$$

and

$$\int_{\mathbb{D}} |g| \, d\lambda \le 4\|g\|_1.$$

Hint: Use Corollaries 4.37 and Theorem 4.38.

Exercise 4.10.4 Let Θ be an inner function such that $\Theta(0) = 0$. The purpose of this exercise is to prove the following result: *for any Borel subset E of $\mathbb{T}$, we have*

$$m(\Theta^{-1}(E)) = m(E). \tag{4.64}$$

We say that the map $\zeta \longmapsto \Theta(\zeta)$ is *measure-preserving*.

(i) Show that

$$\int_{\mathbb{T}} \Theta(\zeta)^n \, dm(\zeta) = \begin{cases} 1 & \text{if} \quad n = 0, \\ 0 & \text{if} \quad n \neq 0. \end{cases}$$

(ii) Deduce that

$$\int_{\mathbb{T}} p(\Theta(\zeta)) \, dm(\zeta) = \int_{\mathbb{T}} p(\zeta) \, dm(\zeta)$$

for any trigonometric polynomial p.

(iii) Prove that

$$\int_{\mathbb{T}} h(\Theta(\zeta)) \, dm(\zeta) = \int_{\mathbb{T}} h(\zeta) \, dm(\zeta)$$

for any bounded Borel function h.
Hint: Approximate h by its Fejér means in the weak-star topology of $L^\infty(\mathbb{T})$ and use the bounded convergence theorem.

(iv) Obtain the claimed result.
Hint: For a Borel subset E of $\mathbb{T}$, apply (iii) to $h = \chi_E$, the characteristic function of E.

Notes on Chapter 4

The theory of Hardy spaces of the open unit disk is a classic and rich theory that is presently very well developed. It has its origin in discoveries made at the beginning of the twentieth century by mathematicians such as Hardy, Littlewood, Privalov, the Riesz brothers, Smirnov and Szegő. Some standard results in this section have not been proved. The reader can find proofs of these and a thorough discussion of Hardy spaces in [188, 233, 291, 354, 442].

Section 4.1

The classic Schwarz's lemma says that, if $f : \mathbb{D} \longrightarrow \bar{\mathbb{D}}$ and $f(0) = 0$, then $|f(z)| \leq |z|$, $z \in \mathbb{D}$. This is a simple consequence of the maximum principle (see [442, p. 232]). Then the inequality (4.1) is just an invariant form of Schwarz's lemma, which is due to Pick [412]. See also [233, p. 2]. The fact that ρ is a distance, and in particular satisfies the triangle inequality, is an easy consequence of the following inequality:

$$\frac{\rho(z_0, z_2) - \rho(z_2, z_1)}{1 - \rho(z_0, z_2)\rho(z_2, z_1)} \leq \rho(z_0, z_1) \leq \frac{\rho(z_0, z_2) + \rho(z_2, z_1)}{1 + \rho(z_0, z_2)\rho(z_2, z_1)}.$$

For a proof of this last inequality, see [233, p. 4]. That a hyperbolic disk is a Euclidean disk can be proved by direct calculations, but in [233] it is derived by geometric considerations.

Section 4.2

Theorem 4.1 gathers some standard facts due to different authors. Fatou [210] showed in 1906 that each $f \in H^\infty$ has a nontangential limit $f(e^{i\vartheta})$ almost everywhere and that $f(e^{i\vartheta})$ cannot vanish on an open arc unless $f(z) \equiv 0$. In 1916, M. and F. Riesz [430] improved this to H^1 and showed that $f(z) \equiv 0$ if its boundary function vanishes on a set of positive measure. In 1923, F. Riesz [427] proved that, if $f \in H^p$, $p > 0$, $f \not\equiv 0$, then f can be factored in the form $f = Bg$, where B is a Blaschke product and g is an H^p function that does not vanish on $\mathbb{D}$. Then he was able to prove that each $f \in H^p$, $p > 0$, has a radial limit almost everywhere. The mean convergence (4.5) is also due to F. Riesz [427]. The representation (4.8) appeared in the famous paper of F. and M. Riesz [430], where they also obtained Theorem 4.4, probably one of the most fundamental facts of one-dimensional harmonic analysis.

Section 4.3

Theorem 4.6 is due to M. Riesz [433] and has now become a classic result. It can be found in many textbooks, see e.g. [188, p. 54], [141, p. 65] or [572, p. 253]. It should be noted that, when $0 < p < 1$, we still have the decomposition

$$L^p(\mathbb{T}) = H^p(\mathbb{T}) + \overline{H_0^p(\mathbb{T})},$$

but

$$H^p(\mathbb{T}) \cap \overline{H_0^p(\mathbb{T})} \neq \{0\}.$$

In fact, $H^p(\mathbb{T}) \cap \overline{H_0^p(\mathbb{T})}$ is the closed linear span in L^p of the Cauchy transform of Dirac measures δ_ζ, $\zeta \in \mathbb{T}$. See [13] or [142, p. 116].

Section 4.5

A. E. Taylor [516–518] was among the first to study H^p, $1 \le p \le \infty$, as a Banach space. In particular, he represented the linear functionals of H^p. There also exists a representation for $0 < p < 1$ obtained by Walters [542, 543], Romberg [436] and Duren, Romberg and Shields [189]. See also [188, p. 115].

Section 4.6

For a proof of Jensen's formula, see [354, p. 162]. The canonical factorization theorem (Theorem 4.19) is due to Smirnov [489] (as mentioned above, in [427] F. Riesz had a weaker form). That is why this result is known as *Riesz–Smirnov canonical factorization*.

Blaschke products were introduced by Blaschke in [100]. The condition (4.56) in Lemma 4.20 actually characterizes Blaschke products among all analytic functions $f : \mathbb{D} \longrightarrow \bar{\mathbb{D}}$. This result is due to M. Riesz, but was first published in Frostman [224]. See also Zygmund [572, vol. I, p. 281].

The terms *inner* and *outer* were introduced by Beurling in [97] without explanation concerning the choice of the terminology. Smirnov [488–490] used no specific name for inner functions, and referred to the outer ones as *maximal functions*, being motivated by a maximum principle discovered in [490]. See Exercise 4.8.2.

Section 4.7

There are a lot of versions and generalizations of the Schwarz reflection principle in the literature. That presented in Theorem 4.21 is classic and can be found for instance in [442, pp. 237–238].

Section 4.8

Corollaries 4.25 and 4.28 are due to Smirnov [489, 490]. Theorem 4.29 is in a paper of Neuwirth and Newman [374] and implicitly in a paper of Helson and Sarason [282].

Section 4.9

Lemma 4.30 was first noticed by Szegő [512] for $p = 2$ and by F. Riesz [427] for other values of p. Theorem 4.31 is due to Sarason [444]. Theorem 4.32 is due to Lusin and Privalov [349]. We refer to [320] for a proof of this result.

Section 4.10

Theorem 4.34 is due to Newman [377]. The proof given is by Kellogg [315]. Theorem 4.35 was obtained by Littlewood and Paley; see [233, Lemma 6.3.1]. Lemma 4.39 is taken from [233].

5

More function spaces

In this chapter, we continue our discussion on classic Banach spaces of analytic functions on the open unit disk. In particular, we introduce the Nevanlinna class $\mathcal{N}$ and its subclass $\mathcal{N}^+$, and give a different characterization of each space. We also discuss the disk algebra $\mathcal{A}$ and the algebra $\mathcal{C}+H^\infty$. Generalized Hardy spaces $H^2(\mu)$ also naturally enter our discussion. We pay special attention to the characterization of the spectrum of b, where b is an element in the closed unit ball of H^∞. This subject is evidently important in the study of $\mathcal{H}(b)$ spaces. The chapter ends with a thorough discussion of Carleson measures for Hardy spaces, their characterization, and their application in the famous corona problem.

5.1 The Nevanlinna class $\mathcal{N}$

An important property of Hardy spaces is that any function $f \in H^p$, $0 < p \le \infty$, can be written as the quotient of two H^∞ functions. The proof is based on the canonical factorization.

Lemma 5.1 *Let $f \in H^p$, $0 < p \le \infty$. Then there exist $f_1, f_2 \in H^\infty$, f_2 outer, such that*

$$f = \frac{f_1}{f_2} \qquad (a.e.\ on\ \mathbb{T}).$$

Proof If $f \equiv 0$, the result is trivial. Hence, assume that $f \not\equiv 0$. Therefore, the outer functions $[\min(|f|, 1)]$ and $[\min(|f|^{-1}, 1)]$ are well defined. Let $f = \lambda[f]BS$ be the canonical decomposition of f and define

$$f_1 = \lambda[\min(|f|, 1)]BS \quad \text{and} \quad f_2 = [\min(|f|^{-1}, 1)].$$

Since $\min(|f|^{\pm 1}, 1) \in L^\infty(\mathbb{T})$, the outer functions $[\min(|f|^{\pm 1}, 1)]$ belong to H^∞. Thus, f_1 and f_2 also belong to H^∞ and we have

$$\frac{f_1}{f_2} = \lambda BS \frac{[\min(|f|, 1)]}{[\min(|f|^{-1}, 1)]} = f \frac{[\min(|f|, 1)]}{[\min(|f|^{-1}, 1)][f]}.$$

But since

$$\frac{\min(|f|, 1)}{\min(|f|^{-1}, 1)\,|f|} = 1 \qquad \text{(a.e. on } \mathbb{T}\text{)},$$

we have

$$\frac{[\min(|f|, 1)]}{[\min(|f|^{-1}, 1)][f]} = 1.$$

Hence, $f_1/f_2 = f$. □

Lemma 5.1 also inspires the following definition. The *Nevanlinna class* $\mathcal{N}$ consists of all analytic functions of the form g/h, where g and h are in H^∞ and h is zero-free. According to this lemma, we surely have

$$\bigcup_{p>0} H^p \subset \mathcal{N}.$$

But the inclusion is proper. We easily see from the properties of H^∞ functions that, if $f \in \mathcal{N}$, $f \not\equiv 0$, then

$$f^*(\zeta) := \lim_{r \to 1} f(r\zeta)$$

exists for almost all $\zeta \in \mathbb{T}$ and $\log|f^*| \in L^1(\mathbb{T})$. In particular, $f^* \neq 0$ a.e. on $\mathbb{T}$. The following result gives a characterization of $\mathcal{N}$, which resembles the definition of Hardy spaces.

Theorem 5.2 *Let f be an analytic function on $\mathbb{D}$. Then $f \in \mathcal{N}$ if and only if*

$$\sup_{0 \le r < 1} \int_{\mathbb{T}} \log^+ |f(r\zeta)|\, dm(\zeta) < \infty.$$

Proof If $f \in \mathcal{N}$, then $f = g/h$, where $g, h \in H^\infty$ and h is zero-free. Without loss of generality, we assume that $\|g\|_\infty \le 1$ and $\|h\|_\infty \le 1$. Hence,

$$\log^+ |f(r\zeta)| \le -\log|h(r\zeta)|$$

and thus, by Jensen's formula (or harmonicity of the function $\log|h|$),

$$\int_{\mathbb{T}} \log^+ |f(r\zeta)|\, dm(\zeta) \le -\int_{\mathbb{T}} \log|h(r\zeta)|\, dm(\zeta) = -\log|h(0)|.$$

The inverse is more delicate. Fix $\rho < 1$. Then, for each $z \in D_\rho = \{w \in \mathbb{C} : |w| < \rho\}$, we have

$$f(z) = \gamma \prod_{k=1}^{n} \frac{\rho(z - z_k)}{\rho^2 - \bar{z}_k z} \exp\left\{ \frac{1}{2\pi} \int_0^{2\pi} \frac{\rho e^{it} + z}{\rho e^{it} - z} \log|f(\rho e^{it})|\, dt \right\},$$

where $\gamma \in \mathbb{T}$ and $z_1, \ldots, z_n$ are the zeros of f in D_ρ, repeated according to their multiplicities. Hence, by (4.38), for each $z \in \mathbb{D}$,

$$f(\rho z) = \gamma \prod_{k=1}^{n} \frac{z - z_k/\rho}{1 - z\bar{z}_k/\rho} \exp\left\{ \frac{1}{2\pi} \int_0^{2\pi} \frac{e^{it} + z}{e^{it} - z} \log|f(\rho e^{it})| \, dt \right\}.$$

Put

$$g_\rho(z) = \gamma \prod_{k=1}^{n} \frac{z - z_k/\rho}{1 - z\bar{z}_k/\rho} \exp\left\{ -\frac{1}{2\pi} \int_0^{2\pi} \frac{e^{it} + z}{e^{it} - z} \log^- |f(\rho e^{it})| \, dt \right\},$$

$$h_\rho(z) = \exp\left\{ -\frac{1}{2\pi} \int_0^{2\pi} \frac{e^{it} + z}{e^{it} - z} \log^+ |f(\rho e^{it})| \, dt \right\}.$$

Then g_ρ and h_ρ are analytic functions in the closed unit ball of $H^\infty(\mathbb{D})$ and

$$f(\rho z) = \frac{g_\rho(z)}{h_\rho(z)} \qquad (z \in \mathbb{D}).$$

By assumption, there is a constant $C > 0$ such that

$$h_\rho(0) = \exp\left\{ -\frac{1}{2\pi} \int_0^{2\pi} \log^+ |f(\rho e^{it})| \, dt \right\} \geq C \tag{5.1}$$

for all $\rho < 1$. Now consider a sequence $(\rho_n)_{n \geq 1}$ that tends to 1. Hence, by Montel's theorem, there is a subsequence $(n_k)_{k \geq 1}$ and functions g and h in the unit ball of $H^\infty(\mathbb{D})$ such that

$$g_{\rho_{n_k}}(z) \longrightarrow g(z) \quad \text{and} \quad h_{\rho_{n_k}}(z) \longrightarrow h(z)$$

uniformly on compact subsets of $\mathbb{D}$. Therefore,

$$f(z)h(z) = g(z) \qquad (z \in \mathbb{D}). \tag{5.2}$$

Moreover, by (5.1) we have $h(0) \geq C$ and we deduce that $h \not\equiv 0$. The relation (5.2) implies that, if $h(z) = 0$ for some $z \in \mathbb{D}$, then necessarily $g(z) = 0$. Thus using Theorem 4.18, we can factorize $h = Bh_1$, $g = Bg_1$, with $g_1, h_1 \in H^\infty$ and $h_1(z) \neq 0$, $z \in \mathbb{D}$. Hence we get that $f = g_1/h_1$, which proves that $f \in \mathcal{N}$. □

For an $f \in \mathcal{N}$, by appealing to Theorem 5.2 and the canonical factorization theorem (Theorem 4.19), we easily see that f has the decomposition $f = BS_\sigma[f]$, where σ is a signed Borel measure, which is singular with respect to the Lebesgue measure. Hence, we define the subclass

$$\mathcal{N}^+ = \{f \in \mathcal{N} : f = BS_\sigma[f] \text{ with } \sigma \geq 0\}.$$

The following result provides a nice characterization of the elements of $\mathcal{N}^+$.

Theorem 5.3 *Let $f \in \mathcal{N}$. Then $f \in \mathcal{N}^+$ if and only if*

$$\lim_{r\to 1} \int_0^{2\pi} \log^+ |f(re^{i\theta})|\, d\theta = \int_0^{2\pi} \log^+ |f(e^{i\theta})|\, d\theta. \tag{5.3}$$

Proof Suppose that $f \in \mathcal{N}^+$. Hence, f has the decomposition $f = BS_\sigma[f]$, where σ is a positive singular measure. In particular, since $\sigma \geq 0$, we have $|B(z)S_\sigma(z)| \leq 1$, and then $\log^+ |f(z)| \leq \log^+ |[f](z)|$, $z \in \mathbb{D}$. Therefore, by (4.52),

$$\log^+ |f(re^{i\theta})| \leq \frac{1}{2\pi} \int_0^{2\pi} \frac{1-r^2}{1+r^2-2r\cos(\theta - t)} \log^+ |f(e^{it})|\, dt$$
$$(re^{i\theta} \in \mathbb{D}),$$

and an application of Fubini's theorem gives

$$\int_0^{2\pi} \log^+ |f(re^{i\theta})|\, d\theta \leq \int_0^{2\pi} \log^+ |f(e^{it})|\, dt \qquad (0 \leq r < 1).$$

By Fatou's lemma,

$$\int_0^{2\pi} \log^+ |f(e^{i\theta})|\, d\theta \leq \liminf_{r\to 1} \int_0^{2\pi} \log^+ |f(re^{i\theta})|\, d\theta.$$

Hence, (5.3) follows.

Now, to prove the reverse implication, suppose that (5.3) holds. Let $f = BS_\sigma[f]$ be the canonical decomposition of $f \in \mathcal{N}$. Put $g = S_\sigma[f]$. Since $f = Bg$, it is enough to prove that $g \in \mathcal{N}^+$.

We have $|f| = |g|$ a.e. on $\mathbb{T}$, and

$$\log |B(z)| + \log^+ |g(z)| \leq \log^+ |f(z)| \leq \log^+ |g(z)|.$$

Thus, by (5.3) and Lemma 4.20,

$$\lim_{r\to 1} \int_0^{2\pi} \log^+ |g(re^{i\theta})|\, d\theta = \int_0^{2\pi} \log^+ |g(e^{i\theta})|\, d\theta. \tag{5.4}$$

The main characteristic of g is that it has the representation

$$\log |g(re^{i\theta})| = \frac{1}{2\pi} \int_{\mathbb{T}} \frac{1-r^2}{1+r^2-2r\cos(\theta - t)}\, d\lambda(e^{it}) \qquad (re^{i\theta} \in \mathbb{D}),$$

where

$$d\lambda(e^{it}) = \log |g(e^{it})|\, dt - d\sigma(e^{it}).$$

Hence, by Theorem 3.6, the measures $\log |g(re^{it})|\, dt$ converge in the weak-star topology to $d\lambda(e^{it})$.

Take any sequence $r_n > 0$, $n \geq 1$, such that $r_n \longrightarrow 1$. Since the sequences $(\log^+ |g(r_n e^{it})|\, dt)_{n\geq 1}$ and $(\log^- |g(r_n e^{it})|\, dt)_{n\geq 1}$ are uniformly bounded in

$\mathcal{M}(\mathbb{T})$, by the Banach–Alaoglu theorem, there is a subsequence $(n_k)_{k\geq 1}$ and two positive measures $\lambda_1, \lambda_2 \in \mathcal{M}(\mathbb{T})$ such that

$$\log^+ |g(r_{n_k} e^{it})|\, dt \longrightarrow d\lambda_1(e^{it}) \quad \text{and} \quad \log^- |g(r_{n_k} e^{it})|\, dt \longrightarrow d\lambda_2(e^{it})$$

in the weak-star topology. Hence, $\lambda = \lambda_1 - \lambda_2$. We now show that λ_1 is absolutely continuous with respect to the Lebesgue measure. This fact is equivalent to saying that $\sigma \geq 0$.

Let E be any Borel subset of $\mathbb{T}$. Then, by Fatou's lemma,

$$\int_E \log^+ |g(e^{it})|\, dt \leq \liminf_{r\to 1} \int_E \log^+ |g(re^{it})|\, dt$$

and

$$\int_{\mathbb{T}\setminus E} \log^+ |g(e^{it})|\, dt \leq \liminf_{r\to 1} \int_{\mathbb{T}\setminus E} \log^+ |g(re^{it})|\, dt.$$

In both cases, equality holds. In fact, if in any one of the last two inequalities the strict inequality holds, then we add them up and obtain a strict inequality for integrals over $\mathbb{T}$, and this contradicts (5.4). Thus,

$$\liminf_{r\to 1} \int_E \log^+ |g(re^{it})|\, dt = \int_E \log^+ |g(e^{it})|\, dt$$

for all Borel subsets E of $\mathbb{T}$. In particular, if we take $r = r_{n_k}$, $k \longrightarrow \infty$, then, by (5.4), the measures $\log^+ |g(r_{n_k} e^{it})|\, dt$ converge to $\log^+ |g(e^{it})|\, dt$ in the weak-star topology. Therefore,

$$d\lambda_1(e^{it}) = \log^+ |g(e^{it})|\, dt.$$

□

Exercises

Exercise 5.1.1 Show that

$$\bigcup_{p>0} H^p(\mathbb{D}) \subsetneq \mathcal{N}^+ \subsetneq \mathcal{N}.$$

Exercise 5.1.2 Let $f \in \mathcal{N}^+$ and let $0 < p \leq \infty$. Show that the assumption $f \in L^p(\mathbb{T})$ implies that $f \in H^p(\mathbb{D})$. If we replace $f \in \mathcal{N}^+$ by $f \in \mathcal{N}$, is the result still true?

Exercise 5.1.3 Let $f \in \mathcal{N}$. Show that there is a Blaschke product B and singular inner functions S_1, S_2 and an outer function O such that

$$f = BS_1O/S_2.$$

Exercise 5.1.4 Let g and h be two measurable functions on $\mathbb{T}$ such that $\log|g|, \log|h| \in L^1(\mathbb{T})$ and $g/h \in L^p(\mathbb{T})$ for some $0 < p \le \infty$. Show that

$$\frac{[g]}{[h]} \in H^p(\mathbb{D}).$$

5.2 The spectrum of b

The main object of this book, the $\mathcal{H}(b)$ spaces, is a family of Hilbert spaces that is parameterized by an element b of the closed unit ball of H^∞. That is why the closed unit ball of H^∞ plays an important role in this context. To avoid some trivial explanations or certain technical dilemmas, we usually assume that b is not a constant function. As we will see later on, e.g. in Section 19.5, the boundary behavior of functions in $\mathcal{H}(b)$ spaces intrinsically depends on the boundary behavior of the symbol b. To explore the latter, we introduce the notion of regular and singular points for b.

A point $z \in \bar{\mathbb{D}}$ is said to be *regular* for b if either $z \in \mathbb{D}$ and $b(z) \neq 0$, or $z \in \mathbb{T}$ and b admits an analytic continuation across a neighborhood $D(z,\varepsilon) = \{w : |w-z| < \varepsilon\}$ of z with $|b| \equiv 1$ on the arc $D(z,\varepsilon) \cap \mathbb{T}$. The *spectrum* of b, denoted by $\sigma(b)$, is then defined as the complement in $\bar{\mathbb{D}}$ of all regular points of b. The points of $\sigma(b)$ are also referred to as the *singular points* of b.

An inner function Θ is already unimodular almost everywhere on $\mathbb{T}$. Hence, a point $z \in \mathbb{T}$ is regular for Θ provided that Θ admits an analytic continuation on some neighborhood $D(z,\varepsilon)$. We do not need to emphasize that $|\Theta|$ must be equal to 1 on $D(z,\varepsilon) \cap \mathbb{T}$. In fact, if Θ is analytic on an arc of $\mathbb{T}$, then, by continuity, it has to be unimodular there. Knowing this, naively speaking, we can say that $\sigma(\Theta)$ consists of all zeros of Θ on $\bar{\mathbb{D}}$. This statement is explained correctly below.

Theorem 5.4 *Let b be a function in the closed unit ball of H^∞ and let $\lambda \in \bar{\mathbb{D}}$. Then the following assertions are equivalent.*

(i) *Point λ is a regular point for b.*
(ii) *Either $\lambda \in \mathbb{D}$ and $b(\lambda) \neq 0$, or $\lambda \in \mathbb{T}$ and there exists $\varepsilon > 0$ such that*

$$|b(\zeta)| = 1, \qquad \textit{a.e. } \zeta \in \mathbb{T} \cap D(\lambda,\varepsilon)$$

and

$$\inf\{|b(z)| : z \in \mathbb{D} \cap D(\lambda,\varepsilon)\} > 0.$$

In particular, if $b = \Theta$ is a nonconstant inner function, then

$$\sigma(\Theta) = \left\{ z \in \bar{\mathbb{D}} : \liminf_{\substack{w \to z \\ w \in \mathbb{D}}} |\Theta(w)| = 0 \right\} = \mathrm{Clos}(Z(\Theta)) \cup \mathrm{supp}(\nu),$$

where $Z(\Theta) = \{\lambda \in \mathbb{D} : \Theta(\lambda) = 0\}$ *and* ν *is the measure representing the singular part of* Θ.

Proof (i) $\Longrightarrow$ (ii) If $\lambda \in \mathbb{T}$ is a regular point of b, then, by definition, there exists $\varepsilon > 0$ such that b admits an analytic continuation across $D(\lambda, \varepsilon)$, with $|b(\zeta)| = 1$ for any $\zeta \in D(\lambda, \varepsilon) \cap \mathbb{T}$. Thus, by continuity, since $|b(\lambda)| = 1$, we could find an $0 < \varepsilon' \le \varepsilon$ such that

$$\inf\{|b(z)| : z \in \mathbb{D} \cap D(\lambda, \varepsilon')\} > 0.$$

(ii) $\Longrightarrow$ (i) If $\lambda \in \mathbb{D}$ and $b(\lambda) \neq 0$, then, by definition, λ is regular. Now suppose that $\lambda \in \mathbb{T}$, and b fulfills the local behavior described in (ii). To prove (i), we need the canonical factorization of b and some properties of Blaschke products and singular inner functions. Assume that

$$b = BS[b],$$

where B is the Blaschke product formed with the zeros of b and $S = S_\nu$ is a singular inner function that is produced by the singular measure ν. Put

$$I = I_{\varepsilon,\lambda} = \mathbb{T} \cap D(\lambda, \varepsilon).$$

Outer part $[b]$. The assumption $|b(\zeta)| = 1$, a.e. on $\zeta \in I$, implies $\log|b| \equiv 0$ a.e. on the arc I, and thus, just by looking at the formula for an outer function, we see that the outer function $[b]$ has an analytic continuation across I, and $|[b]| \equiv 1$ on I.

Inner part B. The zeros of B cannot cluster at any point of I, since otherwise the assumption $\inf\{|b(z)| : z \in \mathbb{D} \cap D(\lambda, \varepsilon)\} > 0$ would have been violated. Hence, B also has an analytic continuation across the arc I, and $|B| \equiv 1$ on I (see Section 4.6).

Inner part S. The support of ν must be on $\mathbb{T} \setminus I$. In fact, by Corollary 3.5 and (4.48), we have

$$\liminf_{r \to 1} |S(re^{i\theta})| = 0 \tag{5.5}$$

for ν-almost all $e^{i\theta} \in \mathbb{T}$. Denote by

$$E = \left\{\zeta \in \mathbb{T} : \liminf_{\substack{w \to \zeta \\ w \in \mathbb{D}}} |S(w)| = 0\right\}$$

and observe that this set is a closed subset of $\mathbb{T}$. We claim that $\operatorname{supp}(\nu) \subset E$. Indeed, assume that there is a point $\zeta_0 \in \operatorname{supp}(\nu) \cap \mathbb{T} \setminus E$. Hence, since E is closed, we can find $\varepsilon_0 > 0$ such that $I_{\zeta_0,\varepsilon_0} \subset \mathbb{T} \setminus E$ and $\nu(I_{\zeta_0,\varepsilon_0}) > 0$, which contradicts not fulfilling (5.5). That implies

$$\liminf_{\substack{w \to \zeta \\ w \in \mathbb{D}}} |S(w)| = 0$$

at *all* points ζ in the support of ν. Thus, the support of ν must be disjoint from I. Otherwise, again, the condition $\inf\{|b(z)| : z \in \mathbb{D} \cap D(\lambda, \varepsilon)\} > 0$ will not hold. In this case, as in the case of the outer part, the formula for S shows that S has an analytic continuation across the arc I, and $|S| \equiv 1$ on I.

Summing up the above discussion, we see that all three functions in the decomposition $b = BS[b]$ extend analytically across I, and thus so does b, and $|b| \equiv 1$ on I. This means that λ is regular.

If $b = \Theta$ is an inner function, since Θ is unimodular on $\mathbb{T}$, the above characterization of regular points of Θ is equivalent to saying

$$\sigma(\Theta) = \left\{ z \in \bar{\mathbb{D}} : \liminf_{\substack{w \to z \\ w \in \mathbb{D}}} |\Theta(w)| = 0 \right\}.$$

That the last set is precisely equal to $\mathrm{Clos}(Z(\Theta)) \cup \mathrm{supp}(\nu)$ was also discussed above. □

In the above proof, we saw that S extends analytically across $\mathbb{T} \setminus \mathrm{supp}(\nu)$ and, moreover, $|S| \equiv 1$ on this open set. However, without demanding the property $|S| = 1$, we may wonder if it is possible to have analytic continuation across any point of $\mathrm{supp}(\nu)$. The answer is negative. In fact, a direct calculation shows that we have

$$\liminf_{r \to 1} |S^{(n)}(re^{i\theta})| = 0 \qquad (n \geq 0),$$

for ν-almost all $e^{i\theta} \in \mathbb{T}$. Hence, on the support of ν, S cannot have any analytic continuation. A similar fact holds for the Blaschke product. According to the uniqueness theorem for analytic functions, a Blaschke product B cannot be analytically continued across any open arc that contains a cluster point of the zero sets of B.

5.3 The disk algebra $\mathcal{A}$

According to Theorem 4.2, $H^p(\mathbb{T})$, $0 < p < \infty$, is the closure of analytic polynomials in $L^p(\mathbb{T})$. But this is not true for $p = \infty$. In other words, H^∞ is not the uniform closure of analytic polynomials. Indeed, the uniform limit of a sequence of continuous functions has to be continuous. But there are plenty of elements of $H^\infty(\mathbb{T})$ that are not continuous on $\mathbb{T}$, and thus such elements cannot be uniformly approximated by a sequence of analytic polynomials.

The uniform closure of analytic polynomials in $L^\infty(\mathbb{T})$ is called the *disk algebra* and is denoted by $\mathcal{A}(\mathbb{T})$. Note that, since the analytic polynomials belong to $H^\infty(\mathbb{T})$, we have

$$\mathcal{A}(\mathbb{T}) \subsetneqq H^\infty(\mathbb{T}).$$

Clearly, each analytic polynomial is also in $\mathcal{C}(\mathbb{T})$ and we can say that $\mathcal{A}(\mathbb{T})$ is the uniform closure of analytic polynomials in $\mathcal{C}(\mathbb{T})$. It is easy to check that $\mathcal{A}(\mathbb{T})$ is a closed subalgebra of $\mathcal{C}(\mathbb{T})$ and we have

$$\mathcal{A}(\mathbb{T}) \subset H^\infty(\mathbb{T}) \cap \mathcal{C}(\mathbb{T}).$$

The last inclusion is in fact an equality. Indeed, let $f \in H^\infty(\mathbb{T}) \cap \mathcal{C}(\mathbb{T})$. By Fejér's theorem,

$$\|\sigma_n(f) - f\|_\infty \longrightarrow 0, \qquad \text{as } n \longrightarrow \infty,$$

where $\sigma_n(f)$ are the Fejér means of f. But, since $f \in H^\infty(\mathbb{T})$, we have $\hat{f}(k) = 0$, $k < 0$, and thus $\sigma_n(f)$ is an analytic polynomial. Thus, $f \in \mathcal{A}(\mathbb{T})$, which proves that

$$\mathcal{A}(\mathbb{T}) = H^\infty(\mathbb{T}) \cap \mathcal{C}(\mathbb{T}). \tag{5.6}$$

By Theorem 4.3 we can also write

$$\mathcal{A}(\mathbb{T}) = \{f \in \mathcal{C}(\mathbb{T}) : \hat{f}(k) = 0,\ k < 0\}. \tag{5.7}$$

We may naturally ask where the word "disk" comes from. The situation is similar to the classic Hardy spaces $H^p(\mathbb{T})$, as a subclass of $L^p(\mathbb{T})$, where, using the Poisson integral formula, we can extend their elements to $\mathbb{D}$ and obtain the corresponding space $H^p(\mathbb{D})$. In the same manner, each element of $\mathcal{A}(\mathbb{T})$ extends to an analytic function on the open unit disk and the extended function is in fact continuous on $\bar{\mathbb{D}}$. Indeed, let $f \in \mathcal{A}(\mathbb{T})$ and let

$$(Uf)(z) = \begin{cases} (Pf)(z), & \text{if } z \in \mathbb{D}, \\ f(z), & \text{if } z \in \mathbb{T}. \end{cases}$$

Since $f \in \mathcal{A}(\mathbb{T})$, then $\hat{f}(k) = 0$, $k < 0$ and thus Uf is analytic on $\mathbb{D}$ (see (3.15)). Moreover, using (3.19), we easily see that

$$|(Pf)(z)| \le \|f\|_\infty = \sup_{z\in\mathbb{T}} |f(z)|,$$

whence

$$\sup_{z\in\bar{\mathbb{D}}} |(Uf)(z)| = \|f\|_\infty. \tag{5.8}$$

Now, since $f \in \mathcal{A}(\mathbb{T})$, we know that there are analytic polynomials $(p_k)_k$ such that

$$\|p_k - f\|_\infty \longrightarrow 0, \qquad \text{as } k \longrightarrow \infty.$$

By (5.8), it follows that

$$\sup_{z\in\bar{\mathbb{D}}} |(Up_k - Uf)(z)| \longrightarrow 0, \qquad \text{as } k \longrightarrow \infty. \tag{5.9}$$

But if $p(\zeta) = \sum_{k=0}^{N} a_k \zeta^k$ is an analytic polynomial, it follows from (3.15) that

$$(Up_k)(z) = \sum_{k=0}^{N} a_k z^k \qquad (z \in \mathbb{D}),$$

so that $Up_k \in \mathcal{C}(\bar{\mathbb{D}})$. Moreover, (5.9) says that Up_k converges uniformly on $\bar{\mathbb{D}}$ to Uf. Hence $Uf \in \mathcal{C}(\mathbb{D})$. If we denote by Hol($\mathbb{D}$) the space of analytic functions on $\mathbb{D}$ and

$$\mathcal{A}(\mathbb{D}) = \mathrm{Hol}(\mathbb{D}) \cap \mathcal{C}(\bar{\mathbb{D}}),$$

then we have proved that the operator U is an isometry from $\mathcal{A}(\mathbb{T})$ into $\mathcal{A}(\mathbb{D})$. It turns out that this operator is also onto. Indeed, let $F \in \mathcal{A}(\mathbb{D})$. Consider its Taylor series,

$$F(z) = \sum_{n=0}^{\infty} a_n z^n \qquad (z \in \mathbb{D}),$$

and put $f := F_{|\mathbb{T}}$. Using the fact that F is continuous on the closed unit disk, an easy computation gives

$$\hat{f}(n) = a_n,\ n \geq 0 \quad \text{and} \quad \hat{f}(n) = 0,\ n < 0.$$

It follows from (5.7) that $f \in \mathcal{A}(\mathbb{T})$. Moreover, formula (3.15) gives

$$(Pf)(z) = \sum_{n=0}^{\infty} \hat{f}(n) z^n = F(z) \qquad (z \in \mathbb{D}).$$

Hence $Uf = F$ and U is onto. Finally, we get that $\mathcal{A}(\mathbb{T})$ and $\mathcal{A}(\mathbb{D})$ are isometrically isomorphic, and, as we have done for the classic Hardy spaces, in the following we simply write $\mathcal{A}$ for the disk algebra.

The following result shows that $\mathcal{A}$ can also be defined as the closure of Cauchy kernels k_z, $z \in \mathbb{D}$, in $\mathcal{C}(\mathbb{T})$.

Theorem 5.5 *Let*

$$k_z(w) = \frac{1}{1 - \bar{z}w} \qquad (z \in \mathbb{D},\ w \in \bar{\mathbb{D}}).$$

Then the following hold.

(i) *For each $z \in \mathbb{D}$, $k_z \in \mathcal{A}$. More specifically, we have*

$$\lim_{N \to \infty} \left\| \sum_{n=0}^{N} \bar{z}^n \chi_n - k_z \right\|_\infty = 0.$$

(ii) *The linear manifold created by* $\{k_z,\ z \in \mathbb{D}\}$ *is uniformly dense in* $\mathcal{A}$. *More specifically,* $\chi_0 = k_0$ *and, for a fixed* $n \geq 1$, *we have*

$$\lim_{r \to 0} \left\| \frac{k_r + k_{r\zeta} + \cdots + k_{r\zeta^{n-1}} - nk_0}{nr^n} - \chi_n \right\|_\infty = 0,$$

where $\zeta = e^{i2\pi/n}$.

(iii) *The collection* $\{k_z : z \in \mathbb{D}\}$ *is a finitely linearly independent subset of* $\mathcal{A}$, *which means that*

$$\sum_{i \in I} \lambda_i k_{z_i} = 0 \quad \Longrightarrow \quad \lambda_i = 0,\ i \in I,$$

for any finite sequence of complex numbers $(\lambda_i)_{i \in I}$ *and any finite sequence of distinct points* $(z_i)_{i \in I}$ *of* $\mathbb{D}$.

Proof (i) Fix $z \in \mathbb{D}$. Then we have

$$\left| k_z(w) - \sum_{n=0}^{N} \bar{z}^n w^n \right| \leq \frac{|z|^{N+1}}{1 - |z|} \tag{5.10}$$

for all $w \in \mathbb{T}$. The above estimation reveals that each k_z can be uniformly approximated by analytic polynomials $\sum_{n=0}^{N} \bar{z}^n \chi_n$ on $\mathbb{T}$. Thus, $k_z \in \mathcal{A}$.

(ii) To show that the linear manifold of finite linear combinations of Cauchy kernels is uniformly dense in $\mathcal{A}$, it is enough to verify that each monomial $\chi_n(w) = w^n$, $n \geq 0$, can be uniformly approximated on $\mathbb{T}$ by a sequence of linear combinations of Cauchy kernels. For $n = 0$, just note that $k_0 = \chi_0 \equiv 1$. For $n \geq 1$, put $\zeta = e^{i2\pi/n}$. For each integer $\ell \geq 0$, this number satisfies

$$\frac{1 + \zeta^\ell + \zeta^{2\ell} + \cdots + \zeta^{(n-1)\ell}}{n} = \begin{cases} 0 & \text{if} \quad n \nmid \ell, \\ 1 & \text{if} \quad n | \ell. \end{cases}$$

Therefore, for each $0 < r < 1$, we have

$$\frac{k_r(w) + k_{r\zeta}(w) + \cdots + k_{r\zeta^{n-1}}(w)}{n} = \sum_{k=0}^{\infty} r^{kn} w^{kn}.$$

From this formula we obtain

$$\left| \frac{k_r(w) + k_{r\zeta}(w) + \cdots + k_{r\zeta^{n-1}}(w) - nk_0(w)}{nr^n} - \chi_n(w) \right| \leq \frac{r^n}{1 - r^n} \tag{5.11}$$

for all $w \in \mathbb{T}$. Hence,

$$\frac{k_r + k_{r\zeta} + \cdots + k_{r\zeta^{n-1}} - nk_0}{nr^n} \longrightarrow \chi_n$$

uniformly on $\mathbb{T}$, as $r \longrightarrow 0$.

(iii) Let $(\lambda_i)_{i\in I}$ be a finite sequence of complex numbers and $(z_i)_{i\in I}$ be a finite sequence of distinct points of $\mathbb{D}$ such that

$$\sum_{i\in I} \lambda_i k_{z_i} = 0.$$

That means that

$$\sum_{i\in I} \frac{\lambda_i}{1-\bar{z}_i z} = 0 \tag{5.12}$$

for all $z \in \bar{\mathbb{D}}$. But, by the analytic continuation principle, the relation should be true in $\mathbb{C} \setminus \{1/\bar{z}_i : i \in I\}$. Then multiplying this relation by $(1-\bar{z}_i z)$ and letting z tend to $1/\bar{z}_i$ gives that $\lambda_i = 0$, for all $i \in I$. We thus get that $\{k_{\lambda_i} : i \in I\}$ is a finitely linearly independent set. □

In light of Theorems 4.2 and Theorem 5.5, we can now say that

$$H^p(\mathbb{T}) = \text{Span}_{L^p(\mathbb{T})}\{k_z : z \in \mathbb{D}\} \qquad (0 < p < \infty)$$

and

$$\mathcal{A}(\mathbb{T}) = \text{Span}_{L^\infty(\mathbb{T})}\{k_z : z \in \mathbb{D}\}.$$

We saw that $\mathcal{A}$ is a closed subalgebra of $\mathcal{C}(\mathbb{T})$. We may naturally wonder if there is any other closed algebra in between $\mathcal{A}$ and $\mathcal{C}(\mathbb{T})$. Surely, there is no closed algebra $\mathcal{B}$ such that

$$\mathcal{A} \subsetneqq \mathcal{B} \subsetneqq \mathcal{C}(\mathbb{T}) \quad \text{and} \quad \chi_{-1} \in \mathcal{B}.$$

This is because the condition $\chi_{-1} \in \mathcal{B}$ implies $\chi_{-n} \in \mathcal{B}$, for all $n \geq 1$, and, since $\mathcal{A} \subset \mathcal{B}$, we also have $\chi_n \in \mathcal{B}$, for all $n \geq 0$. Therefore, we must have $\mathcal{B} = \mathcal{C}(\mathbb{T})$. The following result, known as *Wermer's maximality theorem*, says that in fact there is no proper closed algebra between $\mathcal{A}$ and $\mathcal{C}(\mathbb{T})$. It is rather amazing that, in the proof, we use the above observation.

Theorem 5.6 *Suppose that $\mathcal{B}$ is a closed subalgebra of $\mathcal{C}(\mathbb{T})$ such that*

$$\mathcal{A} \subsetneqq \mathcal{B} \subset \mathcal{C}(\mathbb{T}).$$

Then $\mathcal{B} = \mathcal{C}(\mathbb{T})$.

Proof It is enough to show that $\chi_{-1} \in \mathcal{B}$. Since $\mathcal{A} \subsetneqq \mathcal{B}$, by (5.7), there is an element $f \in \mathcal{B}$ such that $\hat{f}(n) \neq 0$ at least for one $n \leq -1$. Replacing f by $\chi_{-n-1} f/\hat{f}(n)$, without loss of generality, we can assume that

$$\hat{f}(-1) = 1.$$

This identity can be rewritten as

$$\hat{\varphi}(0) = 1,$$

where $\varphi = \chi_1 f$. Now, Fejér's approximation theorem ensures that

$$\left\| \varphi - \sum_{k=-n}^{n} \left(1 - \frac{|k|}{n}\right) \hat{\varphi}(k) \chi_k \right\|_\infty < \frac{1}{2},$$

for a sufficiently large n. Write

$$\sum_{k=-n}^{n} \left(1 - \frac{|k|}{n}\right) \hat{\varphi}(k) \chi_k = 1 + \chi_1 g + \overline{\chi_1 h},$$

where g and h are analytic polynomials. Thus, we have

$$\chi_1 f = 1 + \chi_1 g + \overline{\chi_1 h} + k,$$

with $k \in \mathcal{C}(\mathbb{T})$ and $\|k\|_\infty < 1/2$. The main trick is to write this identity as

$$\chi_1 (f - g - h) = 1 + \overline{\chi_1 h} - \chi_1 h + k.$$

Note that $f - g - h \in \mathcal{B}$. Another essential observation is that, since $\overline{\chi_1 h} - \chi_1 h$ is purely imaginary,

$$\|1 + t(\overline{\chi_1 h} - \chi_1 h)\|_\infty^2 \le 1 + t^2 \|\overline{\chi_1 h} - \chi_1 h\|_\infty^2$$

for all $t \in \mathbb{R}$. Hence, for each $t > 0$,

$$\begin{aligned} \|1 + t - t\chi_1 (f - g - h)\|_\infty &= \|1 - t(\overline{\chi_1 h} - \chi_1 h) - tk\|_\infty \\ &\le \|1 - t(\overline{\chi_1 h} - \chi_1 h)\|_\infty + \|tk\|_\infty \\ &\le (1 + t^2 \|\overline{\chi_1 h} - \chi_1 h\|_\infty^2)^{1/2} + \frac{t}{2}. \end{aligned}$$

We choose $t > 0$ small enough so that

$$(1 + t^2 \|\overline{\chi_1 h} - \chi_1 h\|_\infty^2)^{1/2} < 1 + \frac{t}{2}.$$

With this fixed choice of t, we have

$$\|1 + t - t\chi_1 (f - g - h)\|_\infty < 1 + t.$$

Therefore, the element $\chi_1 (f - g - h)$ is invertible in $\mathcal{B}$. Since

$$\chi_{-1} = (f - g - h) \times (\chi_1 (f - g - h))^{-1},$$

we deduce that $\chi_{-1} \in \mathcal{B}$. □

The family $\mathrm{Hol}(\bar{\mathbb{D}})$ consists of all functions that are analytic on a domain containing $\bar{\mathbb{D}}$. A standard argument of complex analysis shows that elements of $\mathrm{Hol}(\bar{\mathbb{D}})$ are defined and analytic on some disks $\mathbb{D}_R = \{z \in \mathbb{C} : |z| < R\}$ with $R > 1$ (note that R is not a universal constant and depends on the particular function in $\mathrm{Hol}(\bar{\mathbb{D}})$). Certainly this class is properly included in $\mathcal{A}$. In the following, we present two conditions that ensure that a function primarily defined and analytic on $\mathbb{D}$ is actually an element of $\mathrm{Hol}(\bar{\mathbb{D}})$.

Theorem 5.7 *Let f be analytic on $\mathbb{D}$ and write*

$$f(z) = \sum_{n=0}^{\infty} a_n z^n \qquad (z \in \mathbb{D}).$$

Then $f \in \mathrm{Hol}(\bar{\mathbb{D}})$ if and only if there is a constant $c = c(f) > 0$ such that

$$a_n = O(e^{-cn}) \qquad (n \longrightarrow \infty).$$

Proof According to Hadamard's formula, the radius of convergence R of the power series is given by the formula

$$\frac{1}{R} = \limsup_{n \to \infty} |a_n|^{1/n}.$$

Assume first that $f \in \mathrm{Hol}(\bar{\mathbb{D}})$. Thus, by definition, we surely have $R > 1$. Fix any δ such that

$$\frac{1}{R} < \delta < 1.$$

Hence, there is an integer n_0 such that

$$|a_n|^{1/n} < \delta$$

for all $n \geq n_0$. Therefore, with $c = \log(1/\delta) > 0$, we have

$$|a_n| < \delta^n = e^{-cn} \qquad (n \geq n_0).$$

Conversely, assume that

$$|a_n| \leq K e^{-cn} \qquad (n \geq 0),$$

where c and K are some positive constants. Therefore, we have

$$\limsup_{n \to \infty} |a_n|^{1/n} \leq e^{-c},$$

which shows that $R \geq e^c > 1$. This means that $f \in \mathrm{Hol}(\bar{\mathbb{D}})$. □

A rational function P/Q, where P and Q are analytic polynomials with no common factors, is in $\mathrm{Hol}(\bar{\mathbb{D}})$ if and only if all the zeros of Q are outside the closed unit disk. In this case, we clearly have $P/Q \in H^1$. The following result shows that the condition $P/Q \in H^1$ is also sufficient to conclude that $P/Q \in \mathrm{Hol}(\bar{\mathbb{D}})$.

Theorem 5.8 *Let f be a rational function in H^1. Then $f \in \mathrm{Hol}(\bar{\mathbb{D}})$.*

Proof Let $f = P/Q$, where P and Q are analytic polynomials with no common factors. Since f is analytic on $\mathbb{D}$, the zeros of Q are contained in $\mathbb{C} \setminus \mathbb{D}$ and the only thing we need to prove is that Q has no zeros on $\mathbb{T}$. Assume on the contrary that there is $\lambda_1 \in \mathbb{T}$ such that $Q(\lambda_1) = 0$. Put $Q_1(z) = Q(z)/(z - \lambda_1)$, and define

$$g(z) = Q_1(z)f(z) = P_1(z) + \frac{c_1}{z - \lambda_1},$$

where P_1 is an analytic polynomial and c_1 is a constant. Since P and Q have no common factors, $c_1 \neq 0$. The first equality reveals that $g \in H^1$. Using the second one, we obtain

$$\frac{1}{z - \lambda_1} \in H^1.$$

Therefore, we must have

$$\int_0^{2\pi} \frac{d\theta}{|e^{i\theta} - \lambda_1|} < +\infty.$$

But, after doing a simple change of variable, we get

$$\int_0^{2\pi} \frac{d\theta}{|e^{i\theta} - \lambda_1|} = 2\int_0^{\pi} \frac{dt}{|e^{it} - 1|} = \int_0^{\pi} \frac{dt}{\sin(t/2)} = +\infty.$$

Therefore, all the zeros of Q must be outside $\bar{\mathbb{D}}$. □

Exercises

Exercise 5.3.1 Let $f \in \mathcal{C}(\mathbb{T})$. Show that the following are equivalent:

(i) $f \in A(\mathbb{T})$;
(ii) $\hat{f}(k) = 0$, for all $k < 0$.

Hint: Use Fejér's result, which says that, if $f \in \mathcal{C}(\mathbb{T})$,

$$s_n(e^{i\theta}) = \sum_{k=-n}^{n} \hat{f}(k)e^{ik\theta}$$

and

$$\sigma_n(f) = \frac{1}{n+1}(s_0 + s_1 + \cdots + s_n),$$

then $\|\sigma_n(f) - f\|_\infty \longrightarrow 0$ as $n \longrightarrow \infty$.

Exercise 5.3.2 Let $f : \bar{\mathbb{D}} \longrightarrow \mathbb{C}$ and assume that f is analytic on $\mathbb{D}$ and continuous on $\bar{\mathbb{D}}$. Show that $f_{|\mathbb{T}}$ belongs to $\mathcal{A}(\mathbb{T})$.
Hint : Use Fejér's theorem and show that $\hat{f}(k) = 0$, $k < 0$.

Exercise 5.3.3 Let f be a rational function such that $f \in L^2(\mathbb{T})$. Show that f has no pole on $\mathbb{T}$.
Hint: This is similar to the proof of Lemma 5.8.

5.4 The algebra $\mathcal{C}(\mathbb{T}) + H^\infty$

The spaces $\mathcal{C}(\mathbb{T})$ and H^∞ are two Banach algebras that live inside $L^\infty(\mathbb{T})$. As we have seen, their intersection is precisely the disk algebra, i.e.

$$\mathcal{A} = \mathcal{C}(\mathbb{T}) \cap H^\infty.$$

We naturally wonder about the smallest algebra that contains both $\mathcal{C}(\mathbb{T})$ and H^∞. Surely, this algebra must contain

$$\mathcal{C}(\mathbb{T}) + H^\infty = \{f + g : f \in \mathcal{C}(\mathbb{T}) \text{ and } g \in H^\infty\}.$$

What is less trivial is that $\mathcal{C}(\mathbb{T}) + H^\infty$ is itself a closed subalgebra of $L^\infty(\mathbb{T})$. We establish this result below as a consequence of a more general theorem.

Theorem 5.9 *Let $\mathcal{X}$ be a Banach space, and let $\mathcal{M}$ and $\mathcal{N}$ be two closed subspaces of $\mathcal{X}$. Assume that $(A_\iota)_\iota \subset \mathcal{L}(\mathcal{X})$ is a family of contractions on $\mathcal{X}$ that fulfill the following properties:*

(i) *for each ι, $A_\iota \mathcal{X} \subset \mathcal{M}$;*
(ii) *for each ι, $A_\iota \mathcal{N} \subset \mathcal{N}$; and*
(iii) *for each $x \in \mathcal{M}$ and each $\varepsilon > 0$, there is an index $\iota = \iota(x, \varepsilon)$ such that*

$$\|A_\iota x - x\| \le \varepsilon.$$

Then $\mathcal{M} + \mathcal{N}$ is a closed subspace of $\mathcal{X}$.

Proof Assume that x is in the closure of $\mathcal{M} + \mathcal{N}$. Hence, there is a sequence $(x_k)_{k\ge1} \subset \mathcal{M}$ and a sequence $(y_k)_{k\ge1} \subset \mathcal{N}$ such that

$$\|x_k + y_k\| \le \frac{1}{2^k} \qquad (k \ge 2) \tag{5.13}$$

and

$$x = \sum_{k=1}^{\infty} (x_k + y_k)$$

(see Exercise 1.1.13). Now, we choose $A_k \in \mathcal{L}(\mathcal{X})$ such that

$$\|A_k x_k - x_k\| \le \frac{1}{2^k} \qquad (k \ge 1). \tag{5.14}$$

Since the A_k are contractions, and by (5.13) and (5.14) and assumption (i) of the theorem, we immediately see that the series

$$\sum_{k=1}^{\infty}(A_k x_k - x_k) \quad \text{and} \quad \sum_{k=1}^{\infty} A_k(x_k + y_k)$$

converge in $\mathcal{M}$. Then, by (ii), the formula

$$y_k - A_k y_k = (x_k + y_k) + (A_k x_k - x_k) - A_k(x_k + y_k) \qquad (k \geq 1)$$

shows that the series

$$\sum_{k=1}^{\infty}(y_k - A_k y_k)$$

converges to an element of $\mathcal{N}$. If we rewrite the last identity as

$$x_k + y_k = (A_k(x_k + y_k) - (A_k x_k - x_k)) + (y_k - A_k y_k) \qquad (k \geq 1)$$

and take the summation to get

$$\begin{aligned} x &= \sum_{k=1}^{\infty}(x_k + y_k) \\ &= \sum_{k=1}^{\infty}(A_k(x_k + y_k) - (A_k x_k - x_k)) + \sum_{k=1}^{\infty}(y_k - A_k y_k), \end{aligned}$$

the result follows. □

Now we apply Theorem 5.9 in a special case to show that $\mathcal{C}(\mathbb{T}) + H^\infty$ is a Banach algebra.

Theorem 5.10 *The algebra $\mathcal{C}(\mathbb{T}) + H^\infty$ is a closed subalgebra of $L^\infty(\mathbb{T})$.*

Proof The only (nontrivial) things to prove are that $\mathcal{C}(\mathbb{T}) + H^\infty$ is closed with respect to the topology and with respect to multiplication. We first show that $\mathcal{C}(\mathbb{T}) + H^\infty$ is a closed subspace of $L^\infty(\mathbb{T})$. Then this property helps to prove that $\mathcal{C}(\mathbb{T}) + H^\infty$ is in fact also closed under multiplication.

To apply Theorem 5.9, we put $\mathcal{X} = L^\infty(\mathbb{T})$, $\mathcal{M} = \mathcal{C}(\mathbb{T})$ and $\mathcal{N} = H^\infty(\mathbb{T})$. We also consider the sequence of operators $A_n \in \mathcal{L}(L^\infty(\mathbb{T}))$ defined by

$$A_n f = \sigma_n(f) = \sum_{k=-n}^{n}\left(1 - \frac{|k|}{n}\right)\hat{f}(k)\chi_k \qquad (n \geq 1).$$

This is precisely the nth Fejér mean of f. Then the assumptions of Theorem 5.9 hold and thus $\mathcal{C}(\mathbb{T}) + H^\infty$ is a closed subspace of $L^\infty(\mathbb{T})$.

To show that $\mathcal{C}(\mathbb{T}) + H^\infty$ is an algebra, since $\mathcal{C}(\mathbb{T}) \times \mathcal{C}(\mathbb{T}) = \mathcal{C}(\mathbb{T})$ and $H^\infty \times H^\infty = H^\infty$, we only need to check that

$$\mathcal{C}(\mathbb{T}) \times H^\infty \subset \mathcal{C}(\mathbb{T}) + H^\infty.$$

Hence, fix $f \in \mathcal{C}(\mathbb{T})$ and $g \in H^\infty$. Remember that

$$\lim_{n\to\infty} \|A_n f - f\|_\infty = 0.$$

Write

$$fg = (f - A_n f)g + gA_n f. \tag{5.15}$$

Since $A_n f$ is a trigonometric polynomial of order (at most) n and we can write g as

$$g = \sum_{k=0}^{n-1} \hat{g}(k)\chi_k + \chi_n h,$$

where $h \in H^\infty$, we deduce that

$$gA_n f = \left(\sum_{k=0}^{n-1} \hat{g}(k)\chi_k\right) A_n f + (\chi_n A_n f)h \in \mathcal{C}(\mathbb{T}) + H^\infty.$$

Note that $\chi_n A_n f$ is an analytic polynomial and thus it belongs to H^∞. Therefore, by (5.15),

$$\operatorname{dist}(fg,\, \mathcal{C}(\mathbb{T}) + H^\infty) \le \|(f - A_n f)g\|_\infty \le \|f - A_n f\|_\infty \, \|g\|_\infty.$$

Let $n \longrightarrow \infty$ to conclude that fg belongs to the closure of $\mathcal{C}(\mathbb{T}) + H^\infty$. But we have already seen that $\mathcal{C}(\mathbb{T}) + H^\infty$ is closed. Thus, $fg \in \mathcal{C}(\mathbb{T}) + H^\infty$. □

5.5 Generalized Hardy spaces $H^p(\nu)$

As we have seen in Theorem 4.2, the Hardy space $H^p(\mathbb{T})$, $0 < p < \infty$, is the closure of analytic polynomials in $L^p(\mathbb{T}) = L^p(m)$. This interpretation provides the motivation to define $H^p(\nu)$, where $\nu \in \mathcal{M}^+(\mathbb{T})$, as the closure of analytic polynomials in $L^p(\nu)$. Note that, since each $\nu \in \mathcal{M}^+(\mathbb{T})$ is a finite measure, analytic polynomials are in fact elements of $L^p(\nu)$. Thus, the above definition makes sense. In a similar manner, $H_0^p(\nu)$ is defined to be the closure in $L^p(\nu)$ of $\mathcal{P}_{0+}$, the family of analytic polynomials p with $p(0) = 0$. Hence,

$$H^p(\nu) = \mathrm{Clos}_{L^p(\nu)}\mathcal{P}_+ \quad \text{and} \quad H_0^p(\nu) = \mathrm{Clos}_{L^p(\nu)}\mathcal{P}_{0+},$$

and it is easy to see that

$$H_0^p(\nu) = \chi_1 H^p(\nu) \quad \text{and} \quad H^p(\nu) = \chi_{-1} H_0^p(\nu).$$

It is obvious from the definition that $H^p(\nu)$ and $H_0^p(\nu)$ are invariant under multiplication by χ_n, $n \ge 0$. One should keep in mind that the elements of $H^p(\nu)$ are objects that live on $\mathbb{T}$. If ν does not have good properties, many results of the classic H^p spaces fail to hold for the general case. In particular, one should pay attention to the fact that elements of $H^p(\nu)$ cannot be necessarily extended to analytic functions on the open unit disk (see Exercise 5.5.1).

Theorem 5.11 *Let $\nu \in \mathcal{M}^+(\mathbb{T})$, and let $0 < p < \infty$. Then $\mathcal{A} \subset H^p(\nu)$ and*

$$\|f\|_{H^p(\nu)} \leq \|\nu\|^{1/p}\, \|f\|_\infty \qquad (f \in \mathcal{A}).$$

In particular, for each $z \in \mathbb{D}$, $k_z \in H^p(\nu)$, and the linear manifold generated by $\{k_z, z \in \mathbb{D}\}$ is dense in $H^p(\nu)$.

Proof For every function $f \in \mathcal{C}(\mathbb{T})$, we have

$$\|f\|_{L^p(\nu)} \leq \|\nu\|^{1/p}\, \|f\|_\infty. \tag{5.16}$$

If $f \in \mathcal{A}$, then, by definition, there exists a sequence $(p_n)_{n\geq 1}$ of analytic polynomials such that $\|p_n - f\|_\infty \longrightarrow 0$, as $n \longrightarrow \infty$. Thus (5.16) implies that $\|p_n - f\|_{L^p(\nu)} \longrightarrow 0$. This means that $f \in H^p(\nu)$ and, moreover, (5.16) holds for such elements. The rest follows from Theorem 5.5. □

We can derive several results directly from the original definition of $H^p(\nu)$. We mention some of them below. If μ_1 and μ_2 are positive and finite Borel measures that are supported on disjoint subsets of $\mathbb{T}$, then

$$H^p(\mu_1 + \mu_2) = H^p(\mu_1) \oplus H^p(\mu_2),$$

where the sum is direct when $0 < p < \infty$ and orthogonal when $p = 2$. In particular, for the Lebesgue decomposition $\mu = \mu_a + \mu_s$, we have

$$H^p(\mu) = H^p(\mu_a) \oplus H^p(\mu_s). \tag{5.17}$$

As a consequence of the original definition, for the case $p = 2$, we can say

$$H^2(\nu)^\perp = L^2(\nu) \ominus H^2(\nu) = \{f \in L^2(\nu) : \langle f, \chi_n\rangle_\nu = 0,\ n \geq 0\},$$

where $\chi_n(z) = z^n$. Knowing that the family of Cauchy kernels also generates the Hardy space $H^2(\nu)$, we immediately see that

$$H^2(\nu)^\perp = \{f \in L^2(\nu) : \langle f, k_z\rangle_\nu = 0 \text{ for all } z \in \mathbb{D}\}. \tag{5.18}$$

If $d\nu = \varphi\, dm$, where $\varphi \in L^\infty(\mathbb{T})$, $\varphi \geq 0$ a.e. on $\mathbb{T}$, since

$$\langle f, \chi_n\rangle_\nu = 0 \quad \Longleftrightarrow \quad \langle f\varphi, \chi_n\rangle_2 = 0 \qquad (n \geq 0),$$

then we have the slightly more explicit description of the orthogonal of $H^2(\varphi)$:

$$f \in H^2(\varphi)^\perp \quad \Longleftrightarrow \quad \varphi f \in \overline{H^2_0}. \tag{5.19}$$

The subspace $H^2(\mathbb{T})^\perp$ is infinite-dimensional. However, in certain situations, $H^2(\nu)$ occupies the whole Lebesgue space $L^2(\nu)$, and thus its orthogonal complement reduces to $\{0\}$. See Exercise 5.5.1 and Corollary 8.23.

The following result is a characterization of the situation $H^2(\nu)^\perp = \{0\}$. However, later on, we will obtain a more intrinsic characterization in terms of the Radon–Nikodym derivative of ν.

Lemma 5.12 *Let $\nu \in \mathcal{M}^+(\mathbb{T})$, and let $0 < p < \infty$. Then the following are equivalent.*

(i) $H^p(\nu) = L^p(\nu)$.
(ii) $\chi_{-1} \in H^p(\nu)$.
(iii) $1 \in H^p_0(\nu)$.

Proof (i) $\Longrightarrow$ (ii) This is obvious.

(ii) $\Longrightarrow$ (iii) This is seen by multiplying both sides by χ_1.

(iii) $\Longrightarrow$ (ii) This is seen by multiplying both sides by χ_{-1}.

(ii) $\Longrightarrow$ (i) The idea is to show that $\chi_{-1} \in H^p(\nu)$ implies $\chi_{-n} \in H^p(\nu)$ for all $n \geq 1$ (see also Exercise 5.5.3). For each analytic polynomial $\mathfrak{p}(z) = a_0 + a_1 z + \cdots + a_n z^n$, we have

$$\chi_{-1}(z)\mathfrak{p}(z) = a_0\chi_{-1}(z) + a_1 + \cdots + a_n z^{n-1}$$

and thus $\chi_{-1}\mathfrak{p} \in H^p(\nu)$. By assumption, there is a sequence of analytic polynomials $(\mathfrak{p}_n)_{n\geq 1}$ such that $\mathfrak{p}_n \longrightarrow \chi_{-1}$ in $L^p(\nu)$ norm. Since χ_{-1} is bounded on $\mathbb{T}$, we see that $\chi_{-1}\mathfrak{p}_n \longrightarrow \chi_{-2}$ in $L^p(\nu)$ norm. Thus $\chi_{-2} \in H^p(\nu)$. By induction, we see that $\chi_{-n} \in H^p(\nu)$, $n \geq 1$. Therefore, all trigonometric polynomials are in $H^p(\nu)$, which in turn implies that $H^p(\nu) = L^p(\nu)$. □

We can generalize the concepts discussed in Section 4.3 for the Hilbert space $H^2(\varphi)$, as a closed subspace of $L^2(\varphi)$, where φ is a fixed element of $L^\infty(\mathbb{T})$. It is no surprise that the generalization is done in such a way that, for $\varphi \equiv 1$, we obtain the preceding classic results. Let

$$\begin{array}{rccc} i_\varphi : & H^2(\varphi) & \longrightarrow & L^2(\varphi) \\ & f & \longmapsto & f \end{array}$$

denote the inclusion mapping from $H^2(\varphi)$ into $L^2(\varphi)$. In the same manner that the Riesz projection P_+ was defined, let

$$\begin{array}{rccc} P_\varphi : & L^2(\varphi) & \longrightarrow & H^2(\varphi) \\ & f & \longmapsto & P_{H^2(\varphi)} f \end{array}$$

be the projection of $L^2(\varphi)$ onto $H^2(\varphi)$. As usual, $P_{H^2(\varphi)} \in \mathcal{L}(L^2(\varphi))$ denotes the orthogonal projection of $L^2(\varphi)$ onto its closed subspace $H^2(\varphi)$. Thus, we have

$$P_\varphi i_\varphi = I_{H^2(\varphi)} \tag{5.20}$$

while

$$i_\varphi P_\varphi = P_{H^2(\varphi)}. \tag{5.21}$$

Since, for each $f \in L^2(\varphi)$ and $g \in H^2(\varphi)$,

$$\begin{aligned} \langle i_\varphi^* f, g\rangle_{H^2(\varphi)} &= \langle f, i_\varphi g\rangle_{L^2(\varphi)} \\ &= \langle f, g\rangle_{L^2(\varphi)} \\ &= \langle P_{H^2(\varphi)} f, g\rangle_{L^2(\varphi)} \\ &= \langle P_\varphi f, g\rangle_{L^2(\varphi)} \\ &= \langle P_\varphi f, g\rangle_{H^2(\varphi)}, \end{aligned}$$

we deduce that

$$i_\varphi^* = P_\varphi. \tag{5.22}$$

Exercises

Exercise 5.5.1 Let $\nu = \sum_{i=0}^N a_i \delta_{\zeta_i}$, where $a_i > 0$, $\zeta_i \in \mathbb{T}$, $1 \le i \le N$.

(i) Show that $H^2(\nu)$ is a finite-dimensional space.
(ii) Deduce that $\{k_z : z \in \mathbb{D}\}$ is not a linearly independent set in $H^2(\nu)$.
(iii) Compare $H^2(\nu)$ and $L^2(\nu)$.

Exercise 5.5.2 Let ν be a positive Borel measure on $\mathbb{T}$ and let $\nu = \nu_a + \nu_s$ be its Lebesgue decomposition, with $\nu_a \ll m$ and $\nu_s \perp m$. Suppose that $\nu_a \neq 0$. Show that $\{k_z : z \in \mathbb{D}\}$ is a linearly independent set in $H^2(\nu)$.
Hint: If $c_1 k_{z_1} + \cdots + c_n k_{z_n} = 0$ in $H^2(\nu)$, at how many $\zeta \in \mathbb{T}$ must we have

$$c_1 k_{z_1}(\zeta) + \cdots + c_n k_{z_n}(\zeta) = 0?$$

Exercise 5.5.3 Let $f, g \in H^2(\nu)$, and assume that

$$|g(\zeta)| \le 1 \qquad (\zeta \in \mathbb{T}).$$

Show that $fg \in H^2(\nu)$.
Hint: Consider the identity

$$fg - PQ = (f - P)g + P(g - Q),$$

where P and Q are appropriate polynomials.

5.6 Carleson measures

Let μ be a positive Borel measure on the unit disk $\mathbb{D}$. We say that μ is a *Carleson measure* for H^2 if

$$f \in H^2 \quad \Longrightarrow \quad f \in L^2(\mu).$$

Before stating a characterization of Carleson measures, we give an important class of measures that are Carleson.

Lemma 5.13 *Let $\varphi : \mathbb{D} \longrightarrow \mathbb{R}$ be a C^2 subharmonic function such that $\varphi \le 0$ on $\mathbb{D}$. Then the measure*

$$d\mu(z) = \frac{1}{2\pi} e^{\varphi(z)} \Delta\varphi(z) \log \frac{1}{|z|}\, dA(z)$$

is a Carleson measure for H^2. More precisely, for each $f \in H^2$, we have

$$\int_{\mathbb{D}} |f(z)|^2\, d\mu(z) \le \|f\|_2^2.$$

Proof Let $f \in H^2$ and let $u = e^{\varphi}|f|^2$. According to Theorem 3.18, we have $\Delta u \ge e^{\varphi}|f|^2 \Delta\varphi$, which implies that

$$\int_{\mathbb{D}} |f(z)|^2\, d\mu(z) \le \frac{1}{2\pi} \int_{\mathbb{D}} \Delta u(z) \log \frac{1}{|z|}\, dA(z). \tag{5.23}$$

Denote by $u_r(z) = u(rz) = e^{\varphi(rz)}|f(rz)|^2$, $0 < r < 1$ and $|z| < 1/r$. The function u_r is C^2 on a neighborhood of $\bar{\mathbb{D}}$ and we can apply Theorem 3.19 with $w = u_r$ to get

$$\begin{aligned} \frac{1}{2\pi} \int_{\mathbb{D}} (\Delta u_r)(z) \log \frac{1}{|z|}\, dA(z) &= \int_{\mathbb{T}} u_r(\zeta)\, dm(\zeta) - u_r(0) \\ &\le \int_{\mathbb{T}} u_r(\zeta)\, dm(\zeta). \end{aligned}$$

Since $\varphi(r\zeta) \le 0$, we surely have

$$\frac{1}{2\pi} \int_{\mathbb{D}} (\Delta u_r)(z) \log \frac{1}{|z|}\, dA(z) \le \int_{\mathbb{T}} |f(r\zeta)|^2\, dm(\zeta).$$

But $(\Delta u_r)(z) = r^2 (\Delta u)(rz)$, and thus

$$\frac{r^2}{2\pi} \int_{\mathbb{D}} (\Delta u)(rz) \log \frac{1}{|z|}\, dA(z) \le \int_{\mathbb{T}} |f(r\zeta)|^2\, dm(\zeta).$$

Letting r tend to 1, we obtain, with Fatou's lemma,

$$\frac{1}{2\pi} \int_{\mathbb{D}} (\Delta u)(z) \log \frac{1}{|z|}\, dA(z) \le \int_{\mathbb{T}} |f(\zeta)|^2\, dm(\zeta),$$

which combined with (5.23) gives the desired inequality. □

Corollary 5.14 *Let $\varphi : \mathbb{D} \longrightarrow \mathbb{R}$ be a $\mathcal{C}^2$ subharmonic and bounded function on $\mathbb{D}$ and assume that φ is either ≤ 0 or ≥ 0 on $\mathbb{D}$. Then the measure*

$$d\mu(z) = \frac{1}{2\pi} \Delta\varphi(z) \log \frac{1}{|z|}\, dA(z)$$

is a Carleson measure for H^2. More precisely, for each $f \in H^2$, we have

$$\frac{1}{2\pi} \int_{\mathbb{D}} |f(z)|^2 \Delta\varphi(z) \log \frac{1}{|z|}\, dA(z) \leq e\, \|\varphi\|_\infty\, \|f\|_2^2.$$

Proof We can assume that φ is nonconstant. We consider the two cases of φ separately.

Case I. Assume that $\varphi \leq 0$ on $\mathbb{D}$. Let $r = \|\varphi\|_\infty > 0$. Thus,

$$\varphi(z) \geq -r \qquad (z \in \mathbb{D}),$$

and an application of Lemma 5.13 gives

$$e^{-r} \int_{\mathbb{D}} |f(z)|^2\, d\mu(z) \leq \|f\|_2^2 \qquad (f \in H^2).$$

By applying this result to $t\varphi$, where t is a positive real number, we obtain

$$te^{-tr} \int_{\mathbb{D}} |f(z)|^2\, d\mu(z) \leq \|f\|_2^2 \qquad (f \in H^2).$$

But the function $t \longmapsto te^{-tr}$ attains its maximum at $t = 1/r$. This choice of t implies that

$$\int_{\mathbb{D}} |f(z)|^2\, d\mu(z) \leq e\, \|\varphi\|_\infty\, \|f\|_2^2.$$

Case II. Assume that $\varphi \geq 0$ on $\mathbb{D}$. Take a function $f \in H^2$ and $t > 0$. According to Theorem 3.18, we have

$$\Delta(|f|^2 e^{t\varphi}) = te^{t\varphi}|f|^2\Delta(\varphi) + 4e^{t\varphi}t^2|f\partial(\varphi) + \varphi f|^2 \geq t|f|^2\Delta(\varphi),$$

where we have used that $\varphi \geq 0$ on $\mathbb{D}$. Therefore, for any $t > 0$, we have

$$\begin{aligned} \frac{1}{2\pi} &\int_{\mathbb{D}} |f(z)|^2 \Delta\varphi(z) \log \frac{1}{|z|}\, dA(z) \\ &\leq \frac{t^{-1}}{2\pi} \int_{\mathbb{D}} \Delta(|f(z)|^2 e^{t\varphi(z)}) \log \frac{1}{|z|}\, dA(z). \end{aligned}$$

Let $u(z) = |f(z)|^2 e^{t\varphi(z)}$ and $u_r(z) = u(rz)$, $0 < r < 1$. Now, apply Theorem 3.19 with $w = u_r$ to get

$$\begin{aligned} \frac{1}{2\pi} \int_{\mathbb{D}} \Delta(u_r)(z) \log \frac{1}{|z|}\, dA(z) &= \int_{\mathbb{T}} |f(r\zeta)|^2 e^{t\varphi(r\zeta)}\, dm(\zeta) - |f(0)|^2 e^{t\varphi(0)} \\ &\leq e^{t\|\varphi\|_\infty} \int_{\mathbb{T}} |f(r\zeta)|^2\, dm(\zeta). \end{aligned}$$

Using that $(\Delta u_r)(z) = r^2(\Delta u)(rz)$ and Fatou's lemma implies that

$$\frac{1}{2\pi}\int_{\mathbb{D}} (\Delta u)(z) \log \frac{1}{|z|}\, dA(z) \le e^{t\|\varphi\|_\infty} \|f\|_2^2,$$

whence

$$\frac{1}{2\pi}\int_{\mathbb{D}} |f(z)|^2 \Delta\varphi(z) \log \frac{1}{|z|}\, dA(z) \le t^{-1} e^{t\|\varphi\|_\infty} \|f\|_2^2.$$

Using the same argument of optimization as in the first case, we get

$$\frac{1}{2\pi}\int_{\mathbb{D}} |f(z)|^2 \Delta\varphi(z) \log \frac{1}{|z|}\, dA(z) \le e\|\varphi\|_\infty \|p\|_2^2. \qquad \square$$

If μ is a Carleson measure, then, in particular, we have $1 \in L^2(\mu)$, which implies that μ is finite. To characterize Carleson measures, we can assume (without loss of generality) that the measures are finite. The characterization of Carleson measures for H^2 will involve some particular subsets of the open unit disk. For $\zeta_0 \in \mathbb{T}$ and $0 < h < 1$, let

$$S(\zeta_0, h) = \{z \in \mathbb{D} : |z - \zeta_0| < h\}.$$

This set is the intersection of the open unit disk with the open disk of radius h centered at ζ_0 (see Figure 5.1).

Theorem 5.15 *Let μ be a finite positive Borel measure on $\mathbb{D}$. The following are equivalent.*

(i) *Measure μ is a Carleson measure for H^2.*

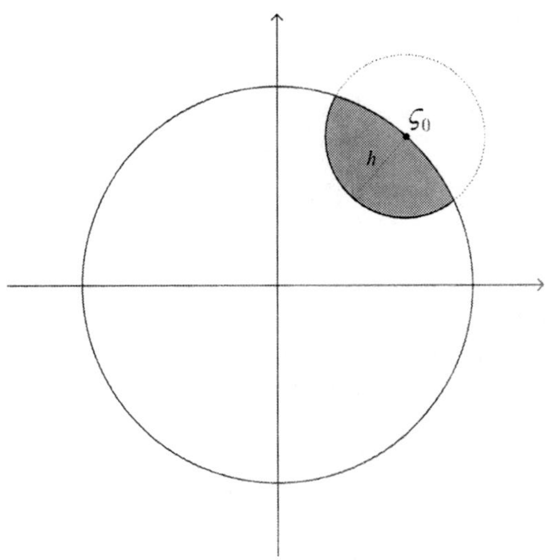

Figure 5.1 A neighborhood of ζ in $\mathbb{D}$.

(ii) *There exists a constant $K > 0$ such that*

$$\int_{\mathbb{D}} \frac{1}{|1-\bar{\lambda}z|^2}\, d\mu(z) \le \frac{K}{1-|\lambda|^2}, \tag{5.24}$$

for all $\lambda \in \mathbb{D}$.

(iii) *There exists a constant $K' > 0$ such that*

$$\int_{\mathbb{D}} \frac{1}{|1-\bar{\lambda}z|^2}\, d\mu(z) \le \frac{K'}{1-|\lambda|^2}, \tag{5.25}$$

for all $\lambda \in \operatorname{supp}(\mu)$.

(iv) *There exists a constant $C > 0$ such that*

$$\mu(S(\zeta,h)) \le Ch, \tag{5.26}$$

for all $h > 0$ and $\zeta \in \mathbb{T}$.

Moreover, if we put

$$\begin{aligned} J(\mu) &= \sup_{f\in H^2,\, f\neq 0} \frac{\|f\|^2_{L^2(\mu)}}{\|f\|_2^2}, \\ K(\mu) &= \sup_{\lambda\in\mathbb{D}} \left(\int_{\mathbb{D}} \frac{1-|\lambda|^2}{|1-\bar{\lambda}z|^2}\, d\mu(z) \right), \\ K'(\mu) &= \sup_{\lambda\in\operatorname{supp}(\mu)} \left(\int_{\mathbb{D}} \frac{1-|\lambda|^2}{|1-\bar{\lambda}z|^2}\, d\mu(z) \right), \\ C(\mu) &= \sup_{\zeta\in\mathbb{T},\, 0<h<1} \frac{\mu(S(\zeta,h))}{h}, \end{aligned}$$

then we have

$$K'(\mu) \le K(\mu) \le J(\mu) \le 2eK'(\mu)$$

and

$$\tfrac{1}{6}K(\mu) \le C(\mu) \le 4K(\mu).$$

Proof The strategy is to prove (i) $\Longrightarrow$ (ii) $\Longrightarrow$ (iii) $\Longrightarrow$ (i) and then prove (ii) $\Longrightarrow$ (iv) $\Longrightarrow$ (ii). The proof is quite long.

(i) $\Longrightarrow$ (ii) Assume that μ is a Carleson measure. By an application of the closed graph theorem (Corollary 1.18), the inclusion operator

$$\begin{array}{ccc} H^2 & \longrightarrow & L^2(\mu) \\ f & \longmapsto & f \end{array}$$

is bounded. Hence, $J(\mu) < \infty$ and we have

$$\|f\|^2_{L^2(\mu)} \le J(\mu)\|f\|_2^2 \qquad (f \in H^2).$$

In particular, for each $f = k_\lambda$, $\lambda \in \mathbb{D}$, we have

$$\int_{\mathbb{D}} \frac{1}{|1-\bar{\lambda}z|^2}\, d\mu(z) = \|k_\lambda\|^2_{L^2(\mu)} \le J(\mu)\|k_\lambda\|_2^2,$$

and since $\|k_\lambda\|_2^2 = (1-|\lambda|^2)^{-1}$, we get the desired inequality. Moreover, we have $K(\mu) \le J(\mu)$.

(ii) $\Longrightarrow$ (iii) This is trivial, and we have $K'(\mu) \le K(\mu)$.

(iii) $\Longrightarrow$ (i) Let

$$\varphi(z) = -\int_{\mathbb{D}} \frac{1-|z|^2}{|1-\bar{z}\lambda|^2}\, d\mu(\lambda) \qquad (z \in \mathbb{D}).$$

(For the sake of the differential operator ∇, which is applied with respect to the variable z, we have interchanged the roles of z and λ in the previous integral.) By assumption,

$$\varphi(z) \le 0 \qquad (z \in \mathbb{D})$$

and

$$\varphi(z) \ge -K'(\mu) \qquad (z \in \text{supp}(\mu)).$$

Since

$$\begin{aligned}\Delta\left(\frac{1-|z|^2}{|1-\bar{z}\lambda|^2}\right) &= 4\frac{\partial}{\partial z}\frac{\partial}{\partial \bar{z}}\left(\frac{1-|z|^2}{|1-\bar{z}\lambda|^2}\right)\\ &= 4\frac{\partial}{\partial z}\left(\frac{\lambda - z}{(1-\bar{z}\lambda)^2(1-z\bar{\lambda})}\right)\\ &= 4\frac{|\lambda|^2-1}{|1-\bar{z}\lambda|^4},\end{aligned}$$

we have

$$\Delta\varphi(z) = 4\int_{\mathbb{D}} \frac{1-|\lambda|^2}{|1-\bar{z}\lambda|^4}\, d\mu(\lambda) \qquad (z \in \mathbb{D}).$$

In particular, we see that φ is subharmonic. We now appeal to Lemma 5.13. According to this lemma, we have

$$\int_{\mathbb{D}} |f(z)|^2\, d\nu(z) \le \|f\|_2^2 \tag{5.27}$$

for all $f \in H^2$, where

$$d\nu(z) = \frac{1}{2\pi} e^{\varphi(z)} \Delta\varphi(z) \log\frac{1}{|z|}\, dA(z).$$

Hence, in order to obtain part (i) of the theorem, it is sufficient to prove the estimate

$$\int_{\mathbb{D}} |f(\lambda)|^2\, d\mu(\lambda) \le 2e^{K'(\mu)} \int_{\mathbb{D}} |f(z)|^2\, d\nu(z) \qquad (f \in H^2). \tag{5.28}$$

This is because, using (5.28) and (5.27), we immediately get

$$\int_{\mathbb{D}} |f(\lambda)|^2 \, d\mu(\lambda) \le 2e^{K'(\mu)} \|f\|_2^2 \qquad (f \in H^2).$$

By applying this inequality to the measure $d\mu_t = t\, d\mu$, where $t > 0$, we obtain

$$\int_{\mathbb{D}} |f(\lambda)|^2 \, d\mu(\lambda) \le 2\frac{e^{tK'(\mu)}}{t} \|f\|_2^2 \qquad (f \in H^2).$$

But the function $t \longmapsto t^{-1}e^{tK'(\mu)}$ attains its minimum at $t = K'(\mu)^{-1}$. Thus,

$$\int_{\mathbb{D}} |f(\lambda)|^2 \, d\mu(\lambda) \le 2eK'(\mu) \|f\|_2^2 \qquad (f \in H^2).$$

This proves (i) and $J(\mu) \le 2eK'(\mu)$.

It remains to prove (5.28). To do so, note that

$$\begin{aligned}\int_{\mathbb{D}} |f(z)|^2 \, d\nu(z) &= \frac{1}{2\pi} \int_{\mathbb{D}} |f(z)|^2 e^{\varphi(z)} \log \frac{1}{|z|} \Delta\varphi(z) \, dA(z) \\ &= \frac{4}{2\pi} \int_{\mathbb{D}} \int_{\mathbb{D}} |f(z)|^2 e^{\varphi(z)} \frac{1 - |\lambda|^2}{|1 - \bar{\lambda} z|^4} \log \frac{1}{|z|} \, dA(z) \, d\mu(\lambda).\end{aligned}$$

Using the estimate

$$\frac{1}{2}(1 - t^2) \le \log \frac{1}{t} \qquad (t > 0)$$

we get

$$\int_{\mathbb{D}} |f(z)|^2 \, d\nu(z) \ge \frac{1}{\pi} \int_{\mathbb{D}} \int_{\mathbb{D}} |f(z)|^2 e^{\varphi(z)} \frac{(1 - |\lambda|^2)(1 - |z|^2)}{|1 - \bar{\lambda} z|^4} \, dA(z) \, d\mu(\lambda). \tag{5.29}$$

We claim that, for any $\lambda \in \mathbb{D}$,

$$\frac{1}{\pi} \int_{\mathbb{D}} |f(z)|^2 e^{\varphi(z)} \frac{(1 - |\lambda|^2)(1 - |z|^2)}{|1 - \bar{\lambda} z|^4} \, dA(z) \ge \frac{1}{2} e^{\varphi(\lambda)} |f(\lambda)|^2. \tag{5.30}$$

To establish the claim, let $w = b_\lambda(z) = (\lambda - z)/(1 - \bar{\lambda} z)$. Thus, $z = b_\lambda(w)$ and we have $dA(w) = |b_\lambda'(z)|^2 \, dA(z)$. An easy computation shows that

$$|b_\lambda'(z)| = \frac{1 - |\lambda|^2}{|1 - \bar{\lambda} z|^2},$$

whence

$$dA(w) = \left(\frac{1 - |\lambda|^2}{|1 - \bar{\lambda} z|^2} \right)^2 dA(z).$$

Therefore,

$$\begin{aligned}&\frac{1}{\pi} \int_{\mathbb{D}} |f(z)|^2 e^{\varphi(z)} \frac{(1 - |\lambda|^2)(1 - |z|^2)}{|1 - \bar{\lambda} z|^4} \, dA(z) \\ &\quad = \frac{1}{\pi} \int_{\mathbb{D}} |\tilde{f}(w)|^2 e^{\tilde{\varphi}(w)} \frac{1 - |b_\lambda(w)|^2}{1 - |\lambda|^2} \, dA(w),\end{aligned}$$

where $\tilde{f}(w) = (f \circ b_\lambda)(w)$ and $\tilde{\varphi}(w) = (\varphi \circ b_\lambda)(w)$. But using the equality

$$1 - |b_\lambda(w)|^2 = \frac{(1 - |w|^2)(1 - |\lambda|^2)}{|1 - \bar{\lambda} w|^2},$$

we get

$$\begin{aligned}\frac{1}{\pi} \int_{\mathbb{D}} |f(z)|^2 e^{\varphi(z)} \frac{(1 - |\lambda|^2)(1 - |z|^2)}{|1 - \bar{\lambda} z|^4} \, dA(z) \\ = \frac{1}{\pi} \int_{\mathbb{D}} e^{\tilde{\varphi}(w)} \left| \frac{\tilde{f}(w)}{1 - \bar{\lambda} w} \right|^2 (1 - |w|^2) \, dA(w).\end{aligned}$$

The function $w \longmapsto \tilde{f}(w)/(1 - \bar{\lambda} w)$ is analytic, and $\tilde{\varphi}$ is subharmonic. Hence, according to Theorem 3.18, the function

$$u(w) = e^{\tilde{\varphi}(w)} \left| \frac{\tilde{f}(w)}{1 - \bar{\lambda} w} \right|^2$$

is subharmonic. Therefore, integrating in polar coordinates and using the mean value property for subharmonic functions, we obtain

$$\begin{aligned}\int_{\mathbb{D}} u(w)(1 - |w|^2) \, dA(w) &= \int_0^1 \int_0^{2\pi} u(re^{i\vartheta})(1 - r^2) r \, dr \, d\vartheta \\ &= \int_0^1 r(1 - r^2) \left(\int_0^{2\pi} u(re^{i\vartheta}) \, d\vartheta \right) dr \\ &\geq 2\pi u(0) \int_0^1 r(1 - r^2) \, dr = \frac{\pi}{2} u(0).\end{aligned}$$

Thus,

$$\frac{1}{\pi} \int_{\mathbb{D}} |f(z)|^2 e^{\varphi(z)} \frac{(1 - |\lambda|^2)(1 - |z|^2)}{|1 - \bar{\lambda} z|^4} \, dA(z) \geq \frac{1}{2} u(0).$$

But $u(0) = e^{\varphi(\lambda)} |f(\lambda)|^2$, and then

$$\frac{1}{\pi} \int_{\mathbb{D}} |f(z)|^2 e^{\varphi(z)} \frac{(1 - |\lambda|^2)(1 - |z|^2)}{|1 - \bar{\lambda} z|^4} \, dA(z) \geq \frac{1}{2} e^{\varphi(\lambda)} |f(\lambda)|^2,$$

which proves the claim. Now, integrating (5.30) with respect to $d\mu(\lambda)$ and using (5.29), we obtain

$$\int_{\mathbb{D}} |f(z)|^2 \, d\nu(z) \geq \frac{1}{2} \int_{\text{supp}(\mu)} e^{\varphi(\lambda)} |f(\lambda)|^2 \, d\mu(\lambda).$$

Remembering that $\varphi(\lambda) \geq -K'(\mu)$, we get the estimate (5.28), which ends the proof of (iii) $\Longrightarrow$ (i).

(ii) $\Longrightarrow$ (iv) Let $\zeta \in \mathbb{T}$ and $0 < h < 1$. Set $\lambda = (1-h)\zeta$ and let $z \in S(\zeta, h)$. Then we have

$$|1 - \bar{\lambda} z| = |1 - (1 - h)\bar{\zeta} z| = |\zeta - (1 - h)z| \leq |\zeta - z| + h < 2h.$$

Hence,

$$|k_\lambda(z)|^2 = \frac{1}{|1-\bar{\lambda}z|^2} \geq \frac{1}{4h^2}$$

for each $z \in S(\zeta, h)$. Since $1-|\lambda|^2 = 1-(1-h)^2 = 2h-h^2$, the inequality (5.25) implies that

$$\frac{1}{4h^2}\mu(S(\zeta,h)) \leq \int_{S(\zeta,h)} |k_\lambda(z)|^2\, d\mu(z) \leq \int_{\mathbb{D}} |k_\lambda(z)|^2\, d\mu(z) \leq \frac{K(\mu)}{2h-h^2}.$$

Thus,

$$\mu(S(\zeta,h)) \leq \frac{4K(\mu)h^2}{2h-h^2} = \frac{4K(\mu)h}{2-h} \leq 4K(\mu)h,$$

which proves part (iii) of the theorem and $C(\mu) \leq 4K(\mu)$.

(iv) $\Longrightarrow$ (ii) Let $\zeta \in \mathbb{T}$ and $0 \leq r < 1$ be such that $\lambda = r\zeta$. Put

$$\Omega_{\varepsilon,\lambda} = \left\{ z \in \mathbb{D} : \frac{1-|\lambda|^2}{|1-\bar{\lambda}z|^2} > \varepsilon \right\} \qquad (\varepsilon > 0).$$

Let us check that

$$\Omega_{\varepsilon,\lambda} \subset S\left(\zeta,\ \sqrt{\frac{1-r^2}{\varepsilon}} + (1-r)\right). \tag{5.31}$$

If $z \in \Omega_{\varepsilon,\lambda}$, then we have $|\zeta - rz|^2 = |1-\bar{\lambda}z|^2 < (1-r^2)/\varepsilon$, which implies that

$$|z-\zeta| = |rz - \zeta + z(1-r)| \leq |\zeta - rz| + (1-r) \leq \sqrt{\frac{1-r^2}{\varepsilon}} + (1-r).$$

This proves (5.31). Moreover, since

$$\frac{1-|\lambda|^2}{|1-\bar{\lambda}z|^2} \leq \frac{1-r^2}{(1-r)^2} = \frac{1+r}{1-r},$$

we have $\Omega_{\varepsilon,\lambda} = \emptyset$ if $(1+r)(1-r)^{-1} \leq \varepsilon$. Now, according to Lemma 1.1, we have

$$\int_{\mathbb{D}} \frac{1-|\lambda|^2}{|1-\bar{\lambda}z|^2}\, d\mu(z) = \int_0^\infty \mu(\Omega_{\varepsilon,\lambda})\, d\varepsilon = \int_0^{(1+r)(1-r)^{-1}} \mu(\Omega_{\varepsilon,\lambda})\, d\varepsilon.$$

But, taking into account (5.31) and (5.26), we have

$$\begin{aligned}
\int_{\mathbb{D}} \frac{1-|\lambda|^2}{|1-\bar{\lambda}z|^2}\, d\mu(z) &\leq C(\mu) \int_0^{(1+r)(1-r)^{-1}} \left(\sqrt{\frac{1-r^2}{\varepsilon}} + (1-r) \right) d\varepsilon \\
&= C(\mu)\left(2\sqrt{1-r^2}\sqrt{\frac{1+r}{1-r}} + (1+r) \right) \\
&= 3C(\mu)(1+r) \leq 6C(\mu),
\end{aligned}$$

which proves part (ii) of the theorem and $K(\mu) \leq 6C(\mu)$.

The proof of Theorem 5.15 is now complete. □

The set $S(\zeta, h)$ appearing in condition (iii) of Theorem 5.15 can be replaced by some other equivalent sets, which, in specific applications, may be easier to use. In particular, one sometimes uses the *Carleson boxes* defined for $\zeta \in \mathbb{T}$ and $0 < h < 1$ by

$$W(\zeta, h) = \left\{ z \in \mathbb{D} : 1 - h < |z| < 1 \text{ and } \left| \frac{z}{|z|} - \zeta \right| < h \right\}$$

(see Figure 5.2).

We have

$$W(\zeta, h/2) \subset S(\zeta, h) \subset W(\zeta, 2h). \tag{5.32}$$

Indeed, to establish the first inclusion, let $z \in W(\zeta, h/2)$. Then $1 - h/2 < |z|$ and $|z/|z| - \zeta| < h/2$. Thus,

$$\begin{aligned} |z - \zeta| &\leq \left| z - \frac{z}{|z|} \right| + \left| \frac{z}{|z|} - \zeta \right| \\ &= (1 - |z|) + \left| \frac{z}{|z|} - \zeta \right| \\ &< \frac{h}{2} + \frac{h}{2} = h, \end{aligned}$$

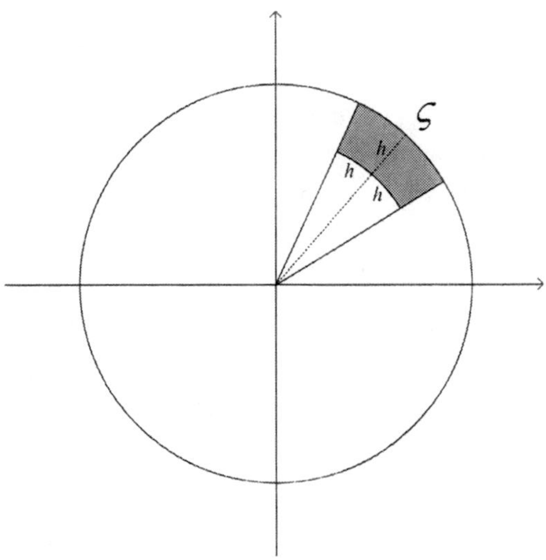

Figure 5.2 A Carleson box.

which shows that $W(\zeta, h/2) \subset S(\zeta, h)$. For the second inclusion, let $z \in S(\zeta, h)$. Then

$$|z| \geq |\zeta| - |z - \zeta| = 1 - |z - \zeta| > 1 - h > 1 - 2h.$$

Moreover,

$$\left| \frac{z}{|z|} - \zeta \right| \leq \left| \frac{z}{|z|} - z \right| + |z - \zeta| = (1 - |z|) + |z - \zeta| < h + h = 2h,$$

which proves that $S(\zeta, h) \subset W(\zeta, 2h)$. The two inclusions (5.32) ensure that, for a finite positive Borel measure on the unit disk $\mathbb{D}$, the condition (5.26) is satisfied for sets $S(\zeta, h)$ if and only if it is satisfied for sets $W(\zeta, h)$ (with eventually a different constant C).

Let $I = (e^{i\theta_1}, e^{i\theta_2})$, $0 \leq \theta_1 < \theta_2 \leq 2\pi$, be any arc of $\mathbb{T}$ and let $\zeta_0 = e^{i(\theta_1+\theta_2)/2}$ be the center of I. We recall that $|I|$ denotes the length of I, that is, $|I| = m(I) = (\Theta_2 - \Theta_1)/2\pi$. Then, we define

$$\begin{aligned} W_I &= \left\{ z \in \mathbb{D} : 1 - |I| < |z| < 1 \text{ and } \frac{z}{|z|} \in I \right\} \\ &= \{ z = re^{i\theta} : 1 - |I| < r < 1 \text{ and } \theta_1 < \theta < \theta_2 \}. \end{aligned}$$

We check that $W_I \subset W(\zeta_0, \pi|I|)$ and $W_0(\zeta, |I|) \subset W_I$. Hence, the condition (5.26) can also be tested using W_I rather than $S(\zeta, h)$. To summarize, the following are equivalent:

(i) μ is a Carleson measure;
(ii) $\mu(S(\zeta, h)) \lesssim h$ for all $h > 0$ and $\zeta \in \mathbb{T}$;
(iii) $\mu(W(\zeta, h)) \lesssim h$ for all $h > 0$ and $\zeta \in \mathbb{T}$;
(iv) $\mu(W_I) \lesssim |I|$ for all arcs I of $\mathbb{T}$.

Note also that it is sufficient to check the condition (5.26) for small values of $h > 0$. Indeed, assume that (5.26) is fulfilled whenever $h < \delta$ for some $\delta \in (0, 1)$. Then, for $h \geq \delta$, we have

$$\mu(S(\zeta, h)) \leq \mu(\mathbb{D}) = \frac{\mu(\mathbb{D})}{\delta} \delta \leq \frac{\mu(\mathbb{D})}{\delta} h,$$

and thus (5.26) is satisfied for all $0 < h < 1$.

It is natural to wonder what happens if μ is a measure defined on the closed unit disk. Of course, since functions in H^2 are defined almost everywhere on $\mathbb{T}$ (with respect to m), we should restrict the embedding inequality

$$\int_{\bar{\mathbb{D}}} |f|^2 \, d\mu \leq J \|f\|_2^2 \tag{5.33}$$

to a subclass of H^2 functions f that are at least μ-measurable. Since $H^2 \cap \mathcal{C}(\bar{\mathbb{D}})$ is dense in H^2 (for instance, it contains the set of analytic polynomials), it is natural to ask when we have (5.33) for any $f \in H^2 \cap \mathcal{C}(\bar{\mathbb{D}})$. In the following, $S^-(\zeta, h)$ denotes the closure of $S(\zeta, h)$.

Theorem 5.16 *Let μ be a finite and positive Borel measure on $\bar{\mathbb{D}}$. The following are equivalent.*

(i) *There exists a constant $J > 0$ such that*

$$\int_{\bar{\mathbb{D}}} |f|^2 \, d\mu \leq J \|f\|_2^2, \tag{5.34}$$

for any $f \in H^2 \cap \mathcal{C}(\bar{\mathbb{D}})$.

(ii) *There exists a constant $K > 0$ such that*

$$\int_{\bar{\mathbb{D}}} \frac{1}{|1 - \bar{\lambda} z|^2} \, d\mu(z) \leq \frac{K}{1 - |\lambda|^2},$$

for any $\lambda \in \mathbb{D}$.

(iii) *There exists a constant $C > 0$ such that*

$$\mu(S^-(\zeta, h)) \leq Ch,$$

for all $h > 0$ and $\zeta \in \mathbb{T}$.

Proof The implication (i) $\Longrightarrow$ (ii) is trivial since $k_\lambda \in H^2 \cap \mathcal{C}(\bar{\mathbb{D}})$.

(ii) $\Longrightarrow$ (iii) We argue as in the proof of the implication (iii) $\Longrightarrow$ (iv) in Theorem 5.15. So let $\zeta \in \mathbb{T}$ and $0 < h < 1$. Set $\lambda = (1 - h)\zeta$ and let $z \in S^-(\zeta, h)$. Then we have $|1 - \bar{\lambda} z| \leq 2h$. Hence

$$\frac{1}{|1 - \bar{\lambda} z|^2} \geq \frac{1}{4h^2},$$

for each $z \in S^-(\zeta, h)$. Thus

$$\frac{1}{4h^2} \mu(S^-(\zeta, h)) \leq \int_{\bar{\mathbb{D}}} \frac{1}{|1 - \bar{\lambda} z|^2} \, d\mu(z) \leq \frac{K}{1 - |\lambda|^2} \leq \frac{K}{h},$$

whence $\mu(S^-(\zeta, h)) \leq 4Kh$.

(iii) $\Longrightarrow$ (i) Let I be any arc of $\mathbb{T}$ and choose $\zeta \in \mathbb{T}$ and $h > 0$ such that $\bar{I} = S^-(\zeta, h) \cap \mathbb{T}$. Note that $|I|$ is comparable to h and by hypothesis we have

$$\mu(I) \leq \mu(S^-(\zeta, h)) \leq Ch \lesssim |I|.$$

Then, the Radon–Nikodym theorem implies that $\mu_{|\mathbb{T}}$ is absolutely continuous with respect to m and its Radon–Nikodym derivative g is in $L^\infty(\mathbb{T})$ and satisfies $\|g\|_\infty \leq C$. Thus we have

$$d\mu = d\mu_{|\mathbb{T}} + d\mu_0 = g \, dm + d\mu_0,$$

where μ_0 is supported on $\mathbb{D}$ and satisfies $\mu_0(S(\zeta,h)) \le Ch$. By Theorem 5.15, there exists a constant $K > 0$ such that

$$\int_{\mathbb{D}} |f|^2\,d\mu \le K\|f\|_2^2,$$

for any $f \in H^2$. Moreover, since $\mu_{|\mathbb{T}}$ is absolutely continuous with respect to m, any $f \in H^2$ is μ-measurable and we have

$$\int_{\mathbb{T}} |f|^2\,d\mu = \int_{\mathbb{T}} |f|^2 g\,dm \le C\|f\|_2^2.$$

Thus

$$\int_{\bar{\mathbb{D}}} |f|^2\,d\mu \le (C+K)\|f\|_2^2,$$

for any $f \in H^2$. □

5.7 Equivalent norms on H^2

In this section, we are interested in measures μ on the closed unit disk that produce an equivalent norm on H^2, which means that there are two constants $c, C > 0$ such that

$$c\|f\|_2^2 \le \int_{\bar{\mathbb{D}}} |f|^2\,d\mu \le C\|f\|_2^2$$

for all $f \in H^2 \cap \mathcal{C}(\bar{\mathbb{D}})$. We first begin with an isometric embedding.

Lemma 5.17 *Let μ be a positive and finite Borel measure on $\bar{\mathbb{D}}$. Assume that*

$$\int_{\bar{\mathbb{D}}} |f|^2\,d\mu = \|f\|_2^2 \tag{5.35}$$

for all $f \in H^2 \cap \mathcal{C}(\bar{\mathbb{D}})$. Then μ is the normalized Lebesgue measure on $\mathbb{T}$.

Proof Apply (5.35) to $f = \chi^n$, $n \ge 0$. Then we get

$$\int_{\mathbb{D}} |z|^{2n}\,d\mu(z) + \mu(\mathbb{T}) = 1 \qquad (n \ge 1).$$

The first term on the right-hand side goes to zero as $n \longrightarrow \infty$, and so $\mu(\mathbb{T}) = 1$. But, by setting $n = 0$ in the previous equation, this means that $\mu(\mathbb{D}) = 0$. This implies that μ is supported on $\mathbb{T}$. Applying now (5.35) to $f = k_\lambda$, $\lambda \in \mathbb{T}$, gives

$$\int_{\mathbb{T}} \frac{1-|\lambda|^2}{|\zeta-\lambda|^2}\,d\mu(\zeta) = 1,$$

which can be rewritten as $(P\mu)(\lambda) = 1$, $\lambda \in \mathbb{D}$. Thus $P\mu \equiv Pm$ and by uniqueness we get that $\mu = m$. □

The next step is to characterize reverse Carleson inequality, i.e.

$$\|f\|_2^2 \lesssim \int_{\bar{\mathbb{D}}} |f|^2 \, d\mu,$$

for all $f \in H^2 \cap \mathcal{C}(\bar{\mathbb{D}})$. To state the characterization, we need to introduce a slight modification of the set W_I introduced in Section 5.6. Given $I = (e^{i\theta_1}, e^{i\theta_2})$, $0 \le \theta_1 < \theta_2 \le 2\pi$, we set

$$\mathcal{W}_I = \{re^{i\theta} : 1 - |I| \le r \le 1 \text{ and } \theta_1 < \theta < \theta_2\}.$$

Theorem 5.18 *Let μ be a positive and finite Borel measure on $\bar{\mathbb{D}}$ and denote by $h = d\mu_{|\mathbb{T}}/dm$ the Radon–Nikodym derivative of $\mu_{|\mathbb{T}}$ with respect to m. Then the following are equivalent.*

(i) *There exists $c_1 > 0$ such that, for every $f \in H^2 \cap \mathcal{C}(\bar{\mathbb{D}})$,*

$$\int_{\bar{\mathbb{D}}} |f|^2 \, d\mu \ge c_1 \|f\|_2^2. \tag{5.36}$$

(ii) *There exists $c_2 > 0$ such that, for every $\lambda \in \mathbb{D}$,*

$$\int_{\bar{\mathbb{D}}} \frac{1}{|1 - \bar{\lambda} z|^2} \, d\mu(z) \ge c_2 \frac{1}{1 - |\lambda|^2}. \tag{5.37}$$

(iii) *There exists $c_3 > 0$ such that, for any arc $I \subset \mathbb{T}$,*

$$\mu(\mathcal{W}_I) \ge c_3 |I|.$$

(iv) *We have* $\operatorname{ess\,inf}_{\zeta \in \mathbb{T}} h(\zeta) > 0$.

Proof (i) $\Longrightarrow$ (ii) This is trivial.

(iii) $\Longrightarrow$ (iv) Let $I = (e^{i\theta_1}, e^{i\theta_2})$, $0 \le \theta_1 < \theta_2 \le 2\pi$, be any open arc of $\mathbb{T}$, and, for $h > 0$ sufficiently small, put

$$\mathcal{W}_{I,h} = \{re^{i\theta} \in \bar{\mathbb{D}} : 1 - h \le r \le 1 \text{ and } \theta_1 < \theta < \theta_2\}.$$

Note that $\mathcal{W}_{I,|I|} = \mathcal{W}_I$. Then, for every open set $\mathcal{O}$ in $\bar{\mathbb{D}}$ for which $\bar{I} \subset \mathcal{O}$, there exists $h > 0$ such that $\mathcal{W}_{I,h} \subset \mathcal{O}$.

Since the measure μ is outer regular, we thus have

$$\mu(\bar{I}) = \inf_{\substack{\bar{I} \subset \mathcal{O} \\ \mathcal{O} \text{ open subset in } \bar{\mathbb{D}}}} \mu(\mathcal{O}) \ge \inf_{h>0} \mu(\mathcal{W}_{I,h}).$$

Since $h \longmapsto \mu(\mathcal{W}_{I,h})$ is an increasing function, we have

$$\inf_{h>0} \mu(\mathcal{W}_{I,h}) = \lim_{n \to \infty} \mu(\mathcal{W}_{I,h_n}),$$

where $h_n = |I|/n$, $n \ge 1$. For each $n \ge 1$, split I into n subarcs I_k such that $|I_k| = h_n = |I|/n$. Then $\mathcal{W}_{I_k,h_n} = \mathcal{W}_{I_k}$ and by hypothesis we have

$$\mu(\mathcal{W}_{I_k,h_n}) = \mu(\mathcal{W}_{I_k}) \ge c_3 |I_k| = c_3 \frac{|I|}{n},$$

which gives

$$\mu(\mathcal{W}_{I,h_n}) \geq \mu\bigg(\bigcup_{k=1}^{n} \mathcal{W}_{I_k,h_n}\bigg) \geq \sum_{k=1}^{n} \mu(\mathcal{W}_{I_k,h_n}) \geq c_3|I|.$$

Hence,

$$\mu(\bar{I}) \geq c_3|I|,$$

for any arc I of $\mathbb{T}$. Now, let I be a fixed open arc of $\mathbb{T}$ and consider a sequence I_n of closed arcs of $\mathbb{T}$ such that $I_n \subset I_{n+1}$ and $I = \bigcup_{n\geq 1} I_n$. The previous estimate gives

$$\mu(I_n) \geq c_3|I_n| \qquad (n \geq 1),$$

and letting $n \longrightarrow \infty$ implies that

$$\mu(I) \geq c_3|I|.$$

Finally, Lebesgue's theorem gives $h(\zeta) \geq c_3$ a.e. on $\mathbb{T}$.

(iv) $\Longrightarrow$ (i) For $f \in H^2 \cap \mathcal{C}(\bar{\mathbb{D}})$, we have

$$\begin{aligned}
\int_{\bar{\mathbb{D}}} |f|^2 \, d\mu &\geq \int_{\mathbb{T}} |f|^2 \, d\mu \\
&\geq \int_{\mathbb{T}} |f|^2 h \, dm \\
&\geq \operatorname*{ess\,inf}_{\zeta\in\mathbb{T}} h(\zeta) \|f\|_2^2.
\end{aligned}$$

(ii) $\Longrightarrow$ (iii) Let $I = (e^{i\theta_1}, e^{i\theta_2})$ be an open arc of $\mathbb{T}$ and $h > 0$ be appropriately small. Integrating (5.37) over $\mathcal{W}_{I,h}$ with respect to area measure dA on the closed unit disk and using Fubini's theorem, we get

$$c_2|I|h \leq \int_{\bar{\mathbb{D}}} \bigg(\int_{\mathcal{W}_{I,h}} \frac{1 - |\lambda|^2}{|1 - \bar{\lambda} z|^2} \, dA(\lambda) \bigg) d\mu(z).$$

Set

$$\varphi_h(z) = \frac{1}{h} \int_{\mathcal{W}_{I,h}} \frac{1 - |\lambda|^2}{|1 - \bar{\lambda} z|^2} \, dA(\lambda).$$

Then, we can rewrite the previous estimate as

$$c_2|I| \leq \int_{\bar{\mathbb{D}}} \varphi_h(z) \, d\mu(z). \tag{5.38}$$

Write

$$\int_{\bar{\mathbb{D}}} \varphi_h(z) \, d\mu(z) = \int_{\bar{I}} \varphi_h(z) \, d\mu(z) + \int_{\bar{\mathbb{D}}\setminus \bar{I}} \varphi_h(z) \, d\mu(z). \tag{5.39}$$

Note that, for any $z \in \bar{\mathbb{D}}$, we have

$$\begin{aligned}\varphi_h(z) &= \frac{1}{h}\int_{1-h}^{1}\left(\int_I \frac{1-r^2}{|1-re^{-i\theta}z|^2}\,d\theta\right) r\,dr\\ &\le \frac{1}{h}\int_{1-h}^{1}\left(\int_0^{2\pi} \frac{1-r^2|z|^2}{|1-re^{-i\theta}z|^2}\,d\theta\right) r\,dr\\ &= \frac{2\pi}{h}\int_{1-h}^{1} r\,dr \le 2\pi.\end{aligned}$$

In particular, we get

$$\int_{\bar{I}} \varphi_h(z)\,d\mu(z) \le 2\pi\mu(\bar{I}).$$

To estimate the second integral on the right-hand side of (5.39), we remark that, if $z \notin \bar{I}$, then $\delta := \operatorname{dist}(z, \bar{I}) > 0$. Thus, for any $e^{i\theta} \in \bar{I}$ and $1 - \frac{1}{2}\delta < r < 1$, we have

$$|1 - re^{-i\theta}z| = |e^{i\theta} - rz| \ge |e^{i\theta} - z| - (1-r)|z| \ge \delta - \frac{\delta}{2} = \frac{\delta}{2}.$$

Hence

$$\begin{aligned}\varphi_h(z) &\le \frac{4}{h\delta^2}\int_{1-h} 1(1-r^2)r\,dr\\ &= \frac{4}{\delta^2}\left(h - h^2 + \frac{h^3}{4}\right),\end{aligned}$$

which proves that $\varphi(z) \longrightarrow 0$, as $h \longrightarrow 0$ and $z \notin \bar{I}$. By the dominated convergence theorem, we conclude that

$$\int_{\bar{\mathbb{D}}\setminus\bar{I}} \varphi_h(z)\,d\mu(z) \longrightarrow 0, \qquad \text{as } h \longrightarrow 0.$$

Therefore,

$$\limsup_{h\to 0} \int_{\bar{\mathbb{D}}} \varphi_h(z)\,d\mu(z) \le 2\pi\mu(\bar{I}).$$

By (5.38) we get

$$\frac{c_2}{2\pi}|I| \le \mu(\bar{I}),$$

for any open arc $I \subset \mathbb{T}$. As above, this estimate implies that

$$\frac{c_2}{2\pi}|I| \le \mu(I) \le \mu(\mathcal{W}_I),$$

for any open arc $I \subset \mathbb{T}$. □

Combining Theorems 5.16 and Theorem 5.18, we immediately obtain the following result.

Corollary 5.19 *Let μ be a positive and finite Borel measure on $\bar{\mathbb{D}}$. Then the following are equivalent.*

(i) *There exist two constants $c_1, c_2 > 0$ such that, for any $f \in H^2 \cap \mathcal{C}(\bar{\mathbb{D}})$, we have*

$$c_1 \|f\|_2^2 \le \int_{\bar{\mathbb{D}}} |f|^2 \, d\mu \le c_2 \|f\|_2^2. \tag{5.40}$$

(ii) *There exist two constants $\kappa_1, \kappa_2 > 0$ such that, for any open arc I of $\mathbb{T}$, we have*

$$\kappa_1 |I| \le \mu(\mathcal{W}_I) \le \kappa_2 |I|. \tag{5.41}$$

Note that, if μ satisfies (5.41), then $\mu_{|\mathbb{T}} = h \, dm$ and h is essentially bounded below and above on $\mathbb{T}$. In that case, it is difficult to see that (5.40) is in fact satisfied for all functions $f \in H^2$.

5.8 The corona problem

Let $f_1, f_2 \in H^\infty$ and assume that there exist two functions $g_1, g_2 \in H^\infty$ such that

$$f_1 g_1 + f_2 g_2 \equiv 1 \qquad (\text{on } \mathbb{D}).$$

Then putting $M = \max(\|g_1\|_\infty, \|g_2\|_\infty)$, we get

$$0 < \frac{1}{M} \le |f_1(z)| + |f_2(z)|$$

for every $z \in \mathbb{D}$. The next result shows that the converse, which is a deep result, is also true.

Theorem 5.20 *Let f_1 and f_2 be two functions in H^∞. The following are equivalent.*

(i) *There exist two function $g_1, g_2 \in H^\infty$ such that*

$$f_1 g_1 + f_2 g_2 \equiv 1 \qquad (\text{on } \mathbb{D}). \tag{5.42}$$

(ii) *There exists $\delta > 0$ such that*

$$|f_1(z)| + |f_2(z)| \ge \delta,$$

for every z in $\mathbb{D}$.

Proof (i) $\Longrightarrow$ (ii) This simple implication has already been discussed.

(ii) $\Longrightarrow$ (i) Now, we proceed to prove the reverse implication. Without loss of generality, we can assume that there exists $\delta \in \,]0, 1]$ such that

$$\delta \le (|f_1(z)|^2 + |f_2(z)|^2)^{1/2} \le 1, \tag{5.43}$$

for every z in $\mathbb{D}$. We will show that there exist $g_1, g_2 \in H^\infty$ satisfying $f_1 g_1 + f_2 g_2 \equiv 1$ and

$$(|g_1(z)|^2 + |g_2(z)|^2)^{1/2} \le c(\delta) \qquad (z \in \mathbb{D}),$$

where

$$c(\delta) = \delta^{-1} + \delta^{-2}\big(2\sqrt{\pi e \log \delta^{-1}} + 3\sqrt{3}e \log \delta^{-1}\big).$$

The proof is broken down into a series of steps.

Step I: We assume that f_1 and f_2 are analytic on a neighborhood of $\bar{\mathbb{D}}$.

We show that there exist g_1 and g_2 analytic on $\mathbb{D}$ satisfying (5.42). Define

$$h_1 = \frac{\overline{f_1}}{|f_1|^2 + |f_2|^2} \quad \text{and} \quad h_2 = \frac{\overline{f_2}}{|f_1|^2 + |f_2|^2}.$$

The functions h_1 and h_2 are C^∞ in a neighborhood of $\bar{\mathbb{D}}$ (note that, according to (5.43), the function $|f_1|^2+|f_2|^2$ does not vanish on $\bar{\mathbb{D}}$) and $h_1 f_1 + h_2 f_2 \equiv 1$. The problem is that h_1 and h_2 are not analytic functions. We now modify h_1 and h_2 to obtain analytic functions. Assume, first, that the problem is solved. Since we want $g_1 f_1 + g_2 f_2 = 1$, we get $f_1(g_1 - h_1) + f_2(g_2 - h_2) = 0$, that is (without taking into account the problem of zeros)

$$\frac{g_1 - h_1}{f_2} = -\frac{g_2 - h_2}{f_1} = w.$$

This gives

$$\begin{cases} g_1 = h_1 + w f_2, \\ g_2 = h_2 - w f_1. \end{cases}$$

Since we want g_1 and g_2 to be analytic, we must have

$$\begin{cases} \bar\partial g_1 = \bar\partial h_1 + f_2 \bar\partial w = 0, \\ \bar\partial g_2 = \bar\partial h_2 - f_1 \bar\partial w = 0. \end{cases}$$

Multiplying the first equation by $\overline{f_2}$ and the second by $-\overline{f_1}$, and then adding the two equations, we get $\psi \bar\partial w = \overline{f_1} \bar\partial h_2 - \overline{f_2} \bar\partial h_1$, where

$$\psi = |f_1|^2 + |f_2|^2.$$

Hence, $\bar\partial w = h_1 \bar\partial h_2 - h_2 \bar\partial h_1$. Finally, putting $h_{1,2} = h_1 \bar\partial h_2 - h_2 \bar\partial h_1$, we choose g_1 and g_2 such that

$$\begin{cases} g_1 = h_1 + w f_2, \\ g_2 = h_2 - w f_1, \\ \bar\partial w = h_{1,2}. \end{cases} \tag{5.44}$$

Conversely, we can easily check that any pair of functions (g_1, g_2), defined by (5.44), gives analytic functions satisfying (5.42). It remains to verify that the equation $\bar{\partial} w = h_{1,2}$ has a solution. But, since the function $h_{1,2}$ is $\mathcal{C}^\infty$ in a neighborhood of $\bar{\mathbb{D}}$, we can (multiplying it by a function $\mathcal{C}^\infty$, compactly supported on $\mathbb{C}$ and identically equal to 1 in a neighborhood of $\bar{\mathbb{D}}$) assume that $h_{1,2}$ is also compactly supported on $\mathbb{C}$. Now, Theorem 3.20 implies that the equation $\bar{\partial} w = h_{1,2}$ has a solution $w_{1,2}$ that is $\mathcal{C}^\infty$ in a neighborhood of $\bar{\mathbb{D}}$.

The rest of the proof consists of choosing a solution w so that we can control $\|g_1\|_\infty$ and $\|g_2\|_\infty$.

Step II: We have

$$\inf_{\bar{\partial} w = h_{1,2}} \|w\|_\infty = \sup_{\substack{F \in \mathcal{P}_{0+} \\ \|F\|_1 \le 1}} \left| \int_{\mathbb{D}} F' h_{1,2} \, d\lambda + \int_{\mathbb{D}} F \partial h_{1,2} \, d\lambda \right|,$$

where we recall that

$$d\lambda(z) = \frac{2}{\pi} \log \frac{1}{|z|} \, dA(z) \qquad (z \in \mathbb{D}).$$

We have

$$\inf_{\bar{\partial} w = h_{1,2}} \|w\|_\infty = \inf_{h \in H^\infty} \|w_{1,2} + h\|_\infty = \|w_{1,2}\|_{L^\infty / H^\infty},$$

where $w_{1,2}$ is the solution to the equation $\bar{\partial} w = h_{1,2}$ given by Theorem 3.20. To estimate the quotient norm, we use the characterization (4.33), which gives $(H_0^1)^\perp = H^\infty$. Thus, by (1.8),

$$\inf_{\bar{\partial} w = h_{1,2}} \|w\|_\infty = \sup_{\substack{F \in H_0^1 \\ \|F\|_1 \le 1}} \left| \int_{\mathbb{T}} w_{1,2} F \, dm \right|.$$

Now, remembering that the set of analytic polynomials vanishing at 0 is dense in H_0^1, we can write

$$\inf_{\bar{\partial} w = h_{1,2}} \|w\|_\infty = \sup_{\substack{F \in \mathcal{P}_{0+} \\ \|F\|_1 \le 1}} \left| \int_{\mathbb{T}} w_{1,2} F \, dm \right|.$$

For any $F \in \mathcal{P}_{0+}$, since $F w_{1,2}$ is $\mathcal{C}^\infty$ in a neighborhood of $\bar{\mathbb{D}}$, we can apply Theorem 3.19, which gives

$$\int_{\mathbb{T}} w_{1,2} F \, dm = w_{1,2}(0) F(0) + \frac{1}{4} \int_{\mathbb{D}} \Delta(w_{1,2} F) \, d\lambda = \frac{1}{4} \int_{\mathbb{D}} \Delta(w_{1,2} F) \, d\lambda,$$

because $F(0) = 0$. To conclude the proof of this step, it remains to note that

$$\bar{\partial}(w_{1,2} F) = F \bar{\partial} w_{1,2} = F h_{1,2},$$

and then

$$\Delta(w_{1,2}F) = 4\partial(Fh_{1,2}) = 4(h_{1,2}F' + F\partial h_{1,2}).$$

Step III: We have

$$|h_{1,2}|^2 = \tfrac{1}{4}\psi^{-2}\Delta(\log\psi), \tag{5.45}$$

$$\partial h_{1,2} = -2\psi^{-1}(f_1'\overline{f_1} + f_2'\overline{f_2})h_{1,2} \tag{5.46}$$

and

$$\Delta(a) = 4\psi^{-1}(1+\alpha)^{-2}(1+\alpha-\beta)(|f_1'|^2 + |f_2'|^2), \tag{5.47}$$

where

$$a = \log\left(\frac{2\psi + \delta^2}{3\delta^2}\right), \qquad \alpha = \frac{\delta^2}{2\psi}$$

and

$$\beta = |f_1'\overline{f_1} + f_2'\overline{f_2}|^2\psi^{-1}(|f_1'|^2 + |f_2'|^2)^{-1}.$$

We first claim that

$$h_{1,2} = \overline{f_1 f_2' - f_2 f_1'}\,\psi^{-2}. \tag{5.48}$$

Indeed, for $i = 1, 2$, we have $h_i = \overline{f_i}\psi^{-1}$, and then $\bar\partial h_i = \overline{f_i'}\psi^{-1} - \overline{f_i}\psi^{-2}\bar\partial\psi$. Thus,

$$\begin{aligned} h_{1,2} &= h_1\bar\partial h_2 - h_2\bar\partial h_1 \\ &= \overline{f_1}\psi^{-1}(\overline{f_2'}\psi^{-1} - \overline{f_2}\psi^{-2}\bar\partial\psi) - \overline{f_2}\psi^{-1}(\overline{f_1'}\psi^{-1} - \overline{f_1}\psi^{-2}\bar\partial\psi) \\ &= \overline{f_1 f_2'}\psi^{-2} - \overline{f_2 f_1'}\psi^{-2}. \end{aligned}$$

To verify (5.45), we have

$$\Delta(\log\psi) = 4\frac{\partial\bar\partial\psi}{\psi} - 4\frac{\partial\psi\bar\partial\psi}{\psi^2},$$

and since ψ is real, the identity $\bar\partial\psi = \overline{\partial\psi}$ holds, and this gives

$$\Delta(\log\psi) = 4\frac{\partial\bar\partial\psi}{\psi} - 4\frac{|\bar\partial\psi|^2}{\psi^2}.$$

But $\bar\partial\psi = f_1\overline{f_1'} + f_2\overline{f_2'}$ and $\partial\bar\partial\psi = |f_1'|^2 + |f_2'|^2$. Hence,

$$\Delta(\log\psi) = 4\frac{|f_1'|^2 + |f_2'|^2}{\psi} - 4\frac{|f_1\overline{f_1'} + f_2\overline{f_2'}|^2}{\psi^2}.$$

Then

$$\begin{aligned}\tfrac{1}{4}\psi^{-2}\Delta(\log\psi) &= \psi^{-3}(|f_1'|^2+|f_2'|^2) - \psi^{-4}|f_1\overline{f_1'}+f_2\overline{f_2'}|^2\\ &= \psi^{-4}((|f_1|^2+|f_2|^2)(|f_1'|^2+|f_2'|^2) - |f_1\overline{f_1'}+f_2\overline{f_2'}|^2)\\ &= \psi^{-4}|f_1f_2'-f_2f_1'|^2,\end{aligned}$$

which, according to (5.48), gives (5.45).

To prove (5.46), since $f_1f_2'-f_2f_1'$ is analytic, from (5.48) we obtain

$$\begin{aligned}\partial h_{1,2} &= -2\overline{f_1f_2'-f_2f_1'}\,\psi^{-3}\partial\psi = -2\overline{f_1f_2'-f_2f_1'}\,\psi^{-3}\big(f_1'\overline{f_1}+f_2'\overline{f_2}\big)\\ &= -2\psi^{-1}\big(f_1'\overline{f_1}+f_2'\overline{f_2}\big)h_{1,2}.\end{aligned}$$

It remains to prove (5.47). It is trivial that

$$\Delta a = \frac{2(2\psi+\delta^2)\Delta\psi - 16|\partial\psi|^2}{(2\psi+\delta^2)^2}.$$

But $2\psi+\delta^2 = 2\psi(1+\alpha)$, $\Delta\psi = 4(|f_1'|^2+|f_2'|^2)$ and

$$|\partial\psi|^2 = \big|f_1\overline{f_1'}+f_2\overline{f_2'}\big|^2 = \beta\psi(|f_1'|^2+|f_2'|^2).$$

Hence,

$$\begin{aligned}\Delta a &= \frac{16\psi(1+\alpha)(|f_1'|^2+|f_2'|^2) - 16\beta\psi(|f_1'|^2+|f_2'|^2)}{4\psi^2(1+\alpha)^2}\\ &= \frac{16\psi(|f_1'|^2+|f_2'|^2)(1+\alpha-\beta)}{4\psi^2(1+\alpha)^2},\end{aligned}$$

which gives (5.47).

Step IV: We have

$$|\partial h_{1,2}| \le \frac{3\sqrt{3}}{8}\delta^{-2}\Delta a. \tag{5.49}$$

According to (5.48), we have

$$|h_{1,2}|^2 = \psi^{-4}|f_1f_2'-f_2f_1'|^2 = \psi^{-4}\big(\psi(|f_1'|^2+|f_2'|^2) - \big|f_1\overline{f_1'}+f_2\overline{f_2'}\big|^2\big).$$

Taking into account (5.46), we get

$$\begin{aligned}|\partial h_{1,2}| &= 2\psi^{-3}\big|f_1'\overline{f_1}+f_2'\overline{f_2}\big|\big(\psi(|f_1'|^2+|f_2'|^2) - \big|f_1\overline{f_1'}+f_2\overline{f_2'}\big|^2\big)^{1/2}\\ &= 2\psi^{-2}(|f_1'|^2+|f_2'|^2)\sqrt{\beta(1-\beta)}.\end{aligned}$$

Note that by the Cauchy–Schwarz inequality, we have $0\le\beta\le1$ and (5.43) implies $0\le\alpha\le1/2$. Apply (5.47) to obtain

$$|\partial h_{1,2}| = \tfrac{1}{2}\psi^{-1}\omega(\alpha,\beta)\Delta a,$$

where $\omega(\alpha,\beta) = (1+\alpha)^2(1+\alpha-\beta)^{-1}\sqrt{\beta(1-\beta)}$. Moreover, by a standard method of differential calculus, we can verify that the function ω has a maximum at $(\alpha,\beta) = (\frac{1}{2},\frac{3}{4})$ and we have $\omega(\frac{1}{2},\frac{3}{4}) = \frac{3}{4}\sqrt{3}$. Therefore, we obtain

$$|\partial h_{1,2}| \leq \frac{3\sqrt{3}}{8}\delta^{-2}\Delta a.$$

Step V: We have

$$\int_{\mathbb{D}} |F' h_{1,2}|\, d\lambda \leq 2\delta^{-2}\sqrt{\pi e \log \delta^{-1}}.$$

Let φ be the outer function in H^2 such that $|\varphi(\zeta)| = |F(\zeta)|^{1/2}$, $\zeta \in \mathbb{T}$. Note that, for each $z \in \mathbb{D}$, we have

$$|F(z)| \leq |\varphi(z)|^2 \quad \text{and} \quad \|\varphi\|_2 = \|F\|_1^{1/2} \leq 1. \tag{5.50}$$

Then, by the Cauchy–Schwarz inequality, we get

$$\int_{\mathbb{D}} |F'||h_{1,2}|\, d\lambda \leq \left(\int_{\mathbb{D}} \frac{|F'|^2}{|\varphi|^2}\, d\lambda\right)^{1/2} \left(\int_{\mathbb{D}} |h_{1,2}|^2|\varphi|^2\, d\lambda\right)^{1/2}. \tag{5.51}$$

Write

$$I_1 = \int_{\mathbb{D}} \frac{|F'|^2}{|\varphi|^2}\, d\lambda, \qquad I_2 = \int_{\mathbb{D}} |h_{1,2}|^2|\varphi|^2\, d\lambda$$

and $r = F\varphi^{-1}$. According to (5.50), we have

$$|r(z)| \leq |\varphi(z)| \qquad (z \in \mathbb{D}),$$

whence, in particular, the function r belongs to H^2 and we have $\|r\|_2 \leq 1$. Hence, for each $z \in \mathbb{D}$, we obtain

$$\begin{aligned} |F'(z)|^2 = |r'(z)\varphi(z) + r(z)\varphi'(z)|^2 &\leq 2(|r'(z)|^2|\varphi(z)|^2 + |r(z)|^2|\varphi'(z)|^2) \\ &\leq 2|\varphi(z)|^2(|r'(z)|^2 + |\varphi'(z)|^2). \end{aligned}$$

Therefore,

$$I_1 \leq 2\int_{\mathbb{D}} (|r'|^2 + |\varphi'|^2)\, d\lambda.$$

Corollary 4.36 now implies that

$$\int_{\mathbb{D}} |r'|^2\, d\lambda \leq \|r\|_2^2 \leq 1$$

and

$$\int_{\mathbb{D}} |\varphi'|^2\, d\lambda \leq \|\varphi\|_2^2 \leq 1.$$

Thus $I_1 \le 4$. To estimate I_2, use (5.45) and (5.43) to get

$$I_2 = \frac{1}{4}\int_{\mathbb{D}} \psi^{-2}\Delta(\log\psi)|\varphi|^2\,d\lambda \le \frac{1}{4}\delta^{-4}\int_{\mathbb{D}} |\varphi|^2\Delta(\log\psi)\,d\lambda.$$

Note that the function $\log\psi$ is subharmonic and $\log\psi \le 0$. Moreover, this function is bounded and we have $\|\log\psi\|_\infty \le 2\log\delta^{-1}$. Hence, according to Corollary 5.14, we have

$$I_2 \le \pi\delta^{-4}e\log\delta^{-1}\|\varphi\|_2^2 \le \pi\delta^{-4}e\log\delta^{-1}.$$

Plugging the estimates for I_1 and I_2 into (5.51) concludes the proof of this step.

Step VI: We have

$$\int_{\mathbb{D}} |F\partial h_{1,2}|\,d\lambda \le 3\sqrt{3}\,e\delta^{-2}\log\delta^{-1}.$$

Using (5.49),

$$\int_{\mathbb{D}} |F\partial h_{1,2}|\,d\lambda \le \frac{3\sqrt{3}}{8}\delta^{-2}\int_{\mathbb{D}} |F|\Delta a\,d\lambda.$$

Since $|F| \le |\varphi|^2$ on $\mathbb{D}$, we get

$$\int_{\mathbb{D}} |F\partial h_{1,2}|\,d\lambda \le \frac{3\sqrt{3}}{4\pi}\delta^{-2}\int_{\mathbb{D}} |\varphi(z)|^2(\Delta a)(z)\log\frac{1}{|z|}\,dA(z).$$

The estimation $0 \le a \le 2\log(1/\delta)$ holds on $\mathbb{D}$. Applying Corollary 5.14 gives

$$\int_{\mathbb{D}} |F\partial h_{1,2}|\,d\lambda \le 3\sqrt{3}\,e\delta^{-2}\log\delta^{-1}\|\varphi\|_2^2,$$

which gives the desired result.

Step VII: Let f_1 and f_2 be two analytic functions on a neighborhood of $\bar{\mathbb{D}}$ and satisfying (5.43)*. Then, there exist $g_1, g_2 \in H^\infty$ satisfying $f_1g_1 + f_2g_2 \equiv 1$ and*

$$|g_1(z)|^2 + |g_2(z)|^2 \le c(\delta),$$

with

$$c(\delta) = \delta^{-1} + \delta^{-2}\big(2\sqrt{\pi e\log\delta^{-1}} + 3\sqrt{3}\,e\log\delta^{-1}\big).$$

From the previous steps, there exists $w \in L^\infty(\mathbb{T})$ such that $\bar\partial w = h_{1,2}$ with

$$\sup_{|z|=1} |w(z)| \le 2\delta^{-2}\sqrt{\pi e\log\delta^{-1}} + 3\sqrt{3}\,e\delta^{-2}\log\delta^{-1}.$$

Since, according to (5.44), we have $(g_1, g_2) = (h_1, h_2) + w(f_2, -f_1)$, we get

$$(|g_1|^2 + |g_2|^2)^{1/2} \le (|h_1|^2 + |h_2|^2)^{1/2} + |w|(|f_1|^2 + |f_2|^2)^{1/2},$$

which yields

$$(|g_1|^2 + |g_2|^2)^{1/2} \le \delta^{-1} + \delta^{-2}\big(2\sqrt{\pi e \log \delta^{-1}} + 3\sqrt{3}\, e \log \delta^{-1}\big).$$

Step VIII: Assume that f_1 and f_2 are analytic on $\mathbb{D}$.

For $i = 1, 2$ and $0 < r < 1$, consider $f_i^{(r)}$ defined by $f_i^{(r)}(z) = f_i(rz)$. The function $f_i^{(r)}$ is analytic on an open neighborhood of $\bar{\mathbb{D}}$ and we have

$$|f_1^{(r)}(z)|^2 + |f_2^{(r)}(z)|^2 = |f_1(rz)|^2 + |f_2(rz)|^2 \ge \delta^2.$$

The hypothesis implies the existence of $g_1^{(r)}$ and $g_2^{(r)}$ in H^∞ satisfying

$$f_1^{(r)}(z)g_1^{(r)}(z) + f_2^{(r)}(z)g_2^{(r)}(z) = 1 \tag{5.52}$$

and

$$(|g_1^{(r)}(z)|^2 + |g_2^{(r)}(z)|^2)^{1/2} \le c(\delta) \tag{5.53}$$

for any $z \in \mathbb{D}$. By Montel's theorem, there are two functions $g_1, g_2 \in H^\infty$ such that, up to a subsequence, we have $g_i^{(r)} \longrightarrow g_i$, as $r \longrightarrow 1$, uniformly on compact subsets of $\mathbb{D}$. Thus, letting r tend to 1 in (5.52) and (5.53) gives

$$f_1(z)g_1(z) + f_2(z)g_2(z) = 1 \quad \text{and} \quad (|g_1(z)|^2 + |g_2(z)|^2)^{1/2} \le c(\delta)$$

for all $z \in \mathbb{D}$.

This completes the proof of Theorem 5.20. □

Using some small modifications in the proof of Theorem 5.20, we can obtain the following generalized result.

Theorem 5.21 *Let $f_1, f_2, \ldots, f_n$ be n functions in H^∞. The following are equivalent.*

(i) *There exist n functions $g_1, g_2, \ldots, g_n \in H^\infty$ such that*

$$f_1 g_1 + f_2 g_2 + \cdots + f_n g_n \equiv 1 \qquad (on\ \mathbb{D}).$$

(ii) *There exists a constant $\delta > 0$ such that*

$$|f_1(z)| + |f_2(z)| + \cdots + |f_n(z)| \ge \delta$$

for every z in $\mathbb{D}$.

Theorems 5.20 and 5.21 are known as *corona theorems*. A few technical words are needed to explain this terminology. In fact, it stems from the Banach algebra theory. Let $\mathcal{B}$ be a commutative Banach algebra with unit. Then it is well known that there is a one-to-one correspondence $\phi \longleftrightarrow M$ between the characters of $\mathcal{B}$ (i.e. the nontrivial algebra homomorphism of $\mathcal{B}$ into $\mathbb{C}$) and the maximal ideals M of $\mathcal{B}$. The correspondence is defined by $M = \ker \phi$. Moreover, we have

$$|\phi(x)| \leq \|x\| \qquad (x \in \mathcal{B}),$$

and, in particular, each character ϕ is continuous. Hence, if $\mathfrak{Car}(\mathcal{B})$ denotes the set of characters of $\mathcal{B}$, then $\mathfrak{Car}(\mathcal{B})$ is a subset of the closed unit ball of $\mathcal{B}^*$, the dual space of $\mathcal{B}$. Since $\mathfrak{Car}(\mathcal{B})$ is closed in the weak-star topology of $\mathcal{B}^*$, by the Banach–Alaoglu theorem, $\mathfrak{Car}(\mathcal{B})$ is a compact set. A basic neighborhood of an element $\phi_0 \in \mathfrak{Car}(\mathcal{B})$ has the form

$$\mathcal{U} = \{\phi \in \mathfrak{Car}(\mathcal{B}) : |\phi(x_k) - \phi_0(x_k)| < \varepsilon,\ k = 1, 2, \dots, n\},$$

where $\varepsilon > 0$ and the x_k are arbitrary elements of $\mathcal{B}$.

Now, associated with each point $a \in \mathbb{D}$, the evaluations $\delta_a : f \longmapsto f(a)$ are elements of $\mathfrak{Car}(H^\infty)$. The problem arises to describe the closure of the set $\{\delta_a : a \in \mathbb{D}\}$ under the weak-star topology. In other words, if we identify $\mathbb{D}$ with the set $\{\delta_a : a \in \mathbb{D}\}$, what can we say about

$$\mathfrak{Car}(H^\infty) \setminus \mathbb{D}^-,$$

where $\mathbb{D}^-$ here stands for the closure of $\mathbb{D}$ (or more precisely of $\{\delta_a : a \in \mathbb{D}\}$) in the weak-star topology? The corona theorem now says that

$$\mathfrak{Car}(H^\infty) \setminus \mathbb{D}^- = \emptyset. \tag{5.54}$$

In other words, the corona theorem says that there is no corona! To derive (5.54) from Theorem 5.21, suppose that the set $\{\delta_a : a \in \mathbb{D}\}$ is not dense in $\mathfrak{Car}(H^\infty)$. Then there is a character ϕ_0 on H^∞ and a neighborhood of ϕ_0,

$$\mathcal{U} = \{\phi \in \mathfrak{Car}(H^\infty) : |\phi(f_k) - \phi_0(f_k)| < \varepsilon,\ k = 1, 2, \dots, n\},$$

where $\varepsilon > 0$ and $f_k \in H^\infty$, $k = 1, \dots, n$, such that $\delta_a \notin \mathcal{U}$, for any $a \in \mathbb{D}$. In other words, to each point $a \in \mathbb{D}$ there corresponds an integer $k \in \{1, 2, \dots, n\}$ such that $|f_k(a) - c_k| \geq \varepsilon$, where $c_k = \phi_0(f_k)$. Hence,

$$\sum_{k=1}^{n} |f_k(a) - c_k| \geq \varepsilon \qquad (a \in \mathbb{D}).$$

According to Theorem 5.21, there exist n functions $g_1, g_2, \dots, g_n \in H^\infty$ such that

$$\sum_{k=1}^{n} (f_k - c_k) g_k \equiv 1.$$

Taking the image under ϕ_0 gives

$$1 = \phi_0(1) = \sum_{k=1}^{n} (\phi_0(f_k) - c_k)\phi_0(g_k) = 0,$$

which is a contradiction.

In light of Theorem 5.20, we say that f_1 and f_2 form a *corona pair*, and we write

$$(f_1, f_2) \in (HCR)$$

if there exists a constant $\delta > 0$ such that

$$|f_1(z)| + |f_2(z)| \geq \delta$$

for every z in $\mathbb{D}$.

Notes on Chapter 5

Section 5.1

Theorem 5.2 is due to F. and R. Nevanlinna [375]. The class $\mathcal{N}^+$, called the *Smirnov class*, is introduced in [490]. In that paper, Smirnov cited and attributed Theorem 5.3 to Madame P. Kotchkine. Smirnov's main application of the class $\mathcal{N}^+$ was to the theory of conformal mappings (Smirnov domains; see Duren [188]). Later on, it turned out that the class $\mathcal{N}^+$ plays a crucial role in many problems of complex analysis, from polynomial and rational approximation to the description of ideals in the algebra of holomorphic functions. The result presented in Exercise 5.1.2 is due to Smirnov [489], who also noted a weaker form of this property: if $f \in H^p$ and $f \in L^q(\mathbb{T})$ for some $q > p$, then $f \in H^q$.

Section 5.2

The spectrum of inner functions was studied rather long ago, e.g. by Hoffman [291] and Helson [279]. Our presentation follows [386, pp. 62–63] and [388, p. 103]. See also [508].

Section 5.3

The disk algebra $\mathcal{A}$ appears prominently in many parts of analysis. It is the central example of a uniform algebra, i.e. a Banach algebra for which the Gelfand transform is an isometry. For further results on $\mathcal{A}$, we refer the interested reader to [291, chap. 6], [557, chap. III.E] and [233]. Theorem 5.6 is due to Wermer [545], but the proof given in the text is due to Cohen [149].

Section 5.4

Theorem 5.9 is due to Rudin. However, it is a generalization of an idea of Zalcman. The statement of the theorem and its proof are taken from Koosis [320, p. 149]. Theorem 5.10 and the definition of the class $H^\infty + \mathcal{C}(\mathbb{T})$ are due to Sarason [447]. There is also an approach to prove Theorem 5.10 through Hankel operators and Hartman's theorem. This approach is proposed by Coburn; see [449] or [387, corollary B.2.3.1].

Section 5.5

The generalized Hardy space $H^p(\nu)$ appeared naturally in the context of the weighted polynomial approximation. Clearly, for every finite and positive measure ν and $1 \le p \le \infty$, the set of all trigonometric polynomials $\mathcal{P}$ is dense in $L^p(\nu)$. However, if we replace $\mathcal{P}$ by the set of analytic polynomials, the situation is not so simple, and it is natural and interesting to study the space $H^p(\nu)$. In Section 8.5, we give a characterization of measures ν for which we have the equality $H^p(\nu) = L^p(\nu)$. See [387, part A, chap. 4] for further results and a discussion about the history and references concerning the weighted polynomial approximation problem.

Section 5.6

Theorem 5.15 first appeared in Carleson [126] but it contains only the equivalence (i) $\Longleftrightarrow$ (iv). This fundamental fact of complex and harmonic analysis is a crucial point of several theories, such as free interpolation, BMO functions, Hankel and singular integral operators. The original proof of the theorem by Carleson, based on delicate dyadic techniques, is not easy. About 1974, S. Vinogradov transformed it into a very simple reasoning using an elementary test of V. Senichkin for boundedness of integral operators. More precisely, Vinogradov proved the implication (ii) $\Longrightarrow$ (i), which says that it is sufficient to check the embedding on reproducing kernels k_λ, $\lambda \in \mathbb{D}$. This is one of the first and most explicit manifestations of what is called by Havin and Nikolskii [268] the *reproducing kernel thesis*. See also [387, p. 131]. The proof of Vinogradov was first published in [386]. Here we present a proof due to Petermichl, Treil and Wick [410], which is based on Green's formula. Not only is this beautiful proof rather elementary but also it gives the best constant, $2e$, known to date in the inequality

$$J(\mu) \le 2eK'(\mu).$$

Lemma 5.13 and Corollary 5.14 are taken from [410]. Lemma 5.13 is attributed there to Uchiyama.

Section 5.7

Theorem 5.18 is from [262]. A previous version appeared in [336] under the assumption that the measure μ is a Carleson measure. The authors were motivated by applications in operator theory and they obtained a characterization for a composition operator on H^2 to be one-to-one and to have a closed range.

Section 5.8

Theorem 5.21 is due to Carleson [126]. He got an estimate

$$\|g\|_\infty = \sup_{z\in\mathbb{D}} \left(\sum_{k=1}^{n} |g_k(z)|^2 \right)^{1/2} \leq \delta^{-A(n)}$$

for the solution $g = (g_1, g_2, \ldots, g_n)$ that depends on n. In 1979, T. Wolff gave a remarkable new proof of Carleson's theorem, which (with some simplifications due to Varopoulos and Garnett) can be found in an appendix of Koosis's book [320]. The estimate for the solution has the form

$$\|g\|_\infty \leq 2\delta^{-1} + 2(n-1)M,$$

where M depends only on δ. Based on this proof (improved by Gamelin [229]), Tolokonnikov [524] finally obtained an estimate for the solution that does not depend on n:

$$\|g\|_\infty \leq \delta^{-1} + 42\delta^{-2}(\log\delta^{-1})^{1/2}(\tfrac{1}{4} + \log\delta^{-1}).$$

This independence of n provides the possibility to solve the corona problem for ℓ^2-valued functions. The proof presented in this book follows the idea of Wolff. The presentation is taken from [136]. See also [386] for more comments on this problem.

6

Extreme and exposed points

This chapter is devoted to the study of extreme and exposed points of the closed unit ball of a Banach space. The extreme points are usually addressed in a standard course on functional analysis. But the treatment of exposed points and strongly exposed points is left to the reader or to more advanced courses. We completely characterize the extreme points of the closed unit balls of $L^p(\mathbb{T})$ and H^p spaces for $1 \leq p \leq \infty$. In particular, the extreme points of the closed unit balls of H^1 are precisely the outer functions on the unit sphere, and extreme points of the closed unit balls of H^∞ are characterized via the identity

$$\int_{\mathbb{T}} \log(1 - |f|)\, dm = -\infty.$$

Then the notion of strict convexity and it usefulness in studying extreme points is discussed. In particular, this topic leads to the concept of exposing functionals. We define rigid functions and study their properties. Among other things, we see that the exposed points of the closed unit ball of H^1 are precisely the rigid functions on the unit sphere. We also provide some sufficient conditions that ensure that a function is an exposed, or a strongly exposed, point for H^1.

6.1 Extreme points

Let $\mathcal{X}$ be a normed linear space. For each $a, b \in \mathcal{X}$, $a \neq b$, the set of all convex combinations $\lambda a + (1 - \lambda)b$, where $\lambda \in [0, 1]$, is called the interval $[a, b]$. A set $\Omega \subset \mathcal{X}$ is *convex* if, for each pair of points a and b in Ω, the interval $[a, b]$ is entirely in Ω. We say that the point $p \in \Omega$ is an *extreme point* of Ω if it is not an interior point of any interval in Ω. There are some other, but equivalent, ways to define an extreme point. We mention two such characterizations. First, p is an extreme point of Ω if and only if

$$p \in [a, b] \text{ with } a, b \in \Omega \quad \Longrightarrow \quad p = a \text{ or } p = b. \tag{6.1}$$

Second, p is an extreme point of Ω if and only if the subset $\Omega \setminus \{p\}$ is convex. Using our geometric intuition, it is not hard to convince ourselves that, in the complex plane $\mathbb{C}$, the unit circle $\mathbb{T}$ is the set of extreme points of the closed unit disk $\bar{\mathbb{D}}$, while the extreme points of the rectangle $[-1,1] \times [-1,1]$ are just its four vertices $(\pm 1, \pm 1)$.

To verify (6.1), we can assume that the interval $[a, b]$ is symmetric around the point p. More precisely, we can say that $p \in \Omega$ is an extreme point of Ω if and only if

$$p = \frac{a+b}{2} \text{ with } a, b \in \Omega \quad \Longrightarrow \quad p = a = b. \tag{6.2}$$

On the one hand, if p is an extreme point of Ω and $p = (a+b)/2$, $a, b \in \Omega$, then surely $p \in [a, b]$ and thus, by the original definition of extreme points, either $p = a$ or $p = b$. If $p = a$, then $(a+b)/2 = a$, which implies $p = a = b$. If $p = b$, then $(a+b)/2 = b$, which similarly implies $p = a = b$. Therefore, p satisfies (6.2). On the other hand, suppose that p is not an extreme point of Ω. Hence, there are two points $a, b \in \Omega$ such that $p \in [a, b]$ and that $p \neq a$ and $p \neq b$. Surely $a \neq b$, since otherwise we would have $p = a = b$. Therefore, there is a λ, with $0 < \lambda < 1$, such that $p = \lambda a + (1-\lambda)b$. We construct $a', b' \in \Omega$, $a' \neq b'$, such that $p = (a'+b')/2$. To this end, choose r satisfying

$$0 < r < \min\{(1-\lambda)\|a-b\|, \lambda\|a-b\|\}$$

and define

$$t_1 = \lambda + \frac{r}{\|a-b\|}, \qquad t_2 = \lambda - \frac{r}{\|a-b\|}$$

and

$$a' = t_1 a + (1-t_1)b, \qquad b' = t_2 a + (1-t_2)b.$$

Geometrically, this means that we consider a small symmetric interval around the point p that rests entirely inside the interval $[a, b]$. Then it is easy to check that $t_1, t_2 \in (0,1)$, $t_1 \neq t_2$, and thus $a', b' \in [a, b] \subset \Omega$ with $a' \neq b'$. Moreover,

$$\begin{aligned} a' + b' &= (t_1 + t_2)a + (1 - t_1 + 1 - t_2)b \\ &= 2(\lambda a + (1-\lambda)b) = 2p. \end{aligned}$$

Thus, p does not fulfill (6.2).

In the following, for a normed linear space $\mathcal{X}$, we denote the closed unit ball of $\mathcal{X}$ by $\mathfrak{B}(\mathcal{X})$, i.e.

$$\mathfrak{B}(\mathcal{X}) = \{x \in \mathcal{X} : \|x\| \leq 1\},$$

and $Ext(\mathcal{X})$ denotes the set of all extreme points of $\mathfrak{B}(\mathcal{X})$. We recall some elementary properties of extreme points of the closed unit ball.

Lemma 6.1 *Let $\mathcal{X}$ be a normed linear space, and let $p \in \mathfrak{B}(\mathcal{X})$. Then $p \in Ext(\mathcal{X})$ if and only if p fulfills the property*

$$\|p - x\| \le 1,\ \|p + x\| \le 1,\ x \in \mathcal{X} \quad \Longrightarrow \quad x = 0. \tag{6.3}$$

Moreover,

$$Ext(\mathcal{X}) \subset \{x \in \mathcal{X} : \|x\| = 1\}.$$

Proof First, assume that $p \in Ext(\mathcal{X})$ and let $x \in \mathcal{X}$ be such that $\|p-x\| \le 1$ and $\|p + x\| \le 1$. Since $p = ((p - x) + (p + x))/2$ and both points $p + x$ and $p - x$ are in $\mathfrak{B}(\mathcal{X})$, by (6.2), we have $p = p - x = p + x$, or equivalently $x = 0$.

Second, let p satisfy (6.3) and assume that there are two points $a, b \in \mathfrak{B}(\mathcal{X})$ such that $p = (a + b)/2$. Put $x = (a - b)/2$. Then $p + x = a \in \mathfrak{B}(\mathcal{X})$ and $p - x = b \in \mathfrak{B}(\mathcal{X})$. Therefore, by assumption, we must have $x = 0$, and thus $p = a = b$. Hence, by (6.2), $p \in Ext(\mathcal{X})$.

To show that the extreme points rest on the unit sphere, let $x \in \mathcal{X}$, $\|x\| < 1$. Then there is $r > 0$ such that

$$\{z \in \mathcal{X} : \|z - x\| \le r\} \subset \mathfrak{B}(\mathcal{X}), \tag{6.4}$$

and choose any $z \in \mathcal{X}$ such that $\|z - x\| = r$. Now put $y = 2x - z$. Then we have $\|x - y\| = \|x - z\| = r$, whence by (6.4), we see that $y, z \in \mathfrak{B}(\mathcal{X})$. Moreover, by definition $x = (y + z)/2$ and $y \ne z$, since otherwise $x = y = z$, which implies $r = 0$. Therefore, we conclude that x is not an extreme point of $\mathfrak{B}(\mathcal{X})$. Hence, $Ext(\mathcal{X})$ is a subset of the unit sphere $\{x \in \mathcal{X} : \|x\| = 1\}$. □

Since a point $p \in Ext(\mathcal{X})$ fulfills $\|p\| = 1$, the conditions of Lemma 6.1 imply that

$$2 = 2\|p\| = \|2p\| \le \|p - x\| + \|p + x\| \le 2.$$

Hence, we necessarily have $\|p-x\| = \|p+x\| = 1$. Based on this observation, Lemma 6.1 can be rewritten as follows.

Corollary 6.2 *Let $\mathcal{X}$ be a normed linear space, and let $p \in \mathfrak{B}(\mathcal{X})$. Then $p \in Ext(\mathcal{X})$ if and only if $\|p\| = 1$ and it fulfills the property*

$$\|p - x\| = \|p + x\| = 1,\ x \in \mathcal{X} \quad \Longrightarrow \quad x = 0. \tag{6.5}$$

The following result provides a sufficient condition for extreme points, and at the same time gives the motivation for the definition of exposed points (see Section 6.5).

Lemma 6.3 *Let Ω be a convex subset of a normed linear space $\mathcal{X}$, and let $p \in \Omega$. Suppose that there is a linear functional $\Lambda \in \mathcal{X}^*$ such that*

$$\Re(\Lambda x) < \Re(\Lambda p)$$

for all $x \in \Omega \setminus \{p\}$. Then p is an extreme point of Ω.

Proof Suppose, on the contrary, that p is not an extreme point of Ω. Hence, there are $a, b \in \Omega$ such that $p \in [a, b]$ and $p \neq a$, $p \neq b$. Thus, we can write $p = ta + (1-t)b$, with $0 < t < 1$. But then

$$\begin{aligned}\Re(\Lambda p) &= \Re(\Lambda(ta + (1-t)b)) = t\Re(\Lambda a) + (1-t)\Re(\Lambda b)\\ &< t\Re(\Lambda p) + (1-t)\Re(\Lambda p) = \Re(\Lambda p),\end{aligned}$$

which is absurd. □

It is clear from the proof that the hypotheses of the continuity of Λ and that $\mathcal{X}$ be a normed space are not necessary. Lemma 6.3 remains true if $\mathcal{X}$ is only assumed to be a linear space with Λ a linear functional on $\mathcal{X}$.

Exercises

Exercise 6.1.1 Let $\mathcal{X}$ be a normed linear space, and let $\mathcal{Y}$ be a subspace of $\mathcal{X}$. Show that

$$Ext(\mathcal{X}) \cap \mathcal{Y} \subset Ext(\mathcal{Y}).$$

Remark: We will see examples where strict inclusion holds (see Theorems 6.5 and 6.8 below). As another case, compare Theorems 6.6 and 6.7.

Exercise 6.1.2 Let $\mathcal{H}$ be any Hilbert space. Show that

$$Ext(\mathcal{H}) = \{x \in \mathcal{H} : \|x\| = 1\}.$$

Hint: Use Lemma 6.1 and the parallelogram identity.

Exercise 6.1.3 Let K be a compact set. Show that the extreme points of the unit ball of $\mathcal{C}(K)$ are the continuous functions f such that $|f(x)| = 1$ for all $x \in K$.

6.2 Extreme points of $L^p(\mathbb{T})$

In this section, we classify the extreme points of Lebesgue spaces $L^p(\mathbb{T})$. As is usually the case in studying Lebesgue spaces, we have to consider three cases corresponding to $1 < p < \infty$, $p = 1$ and $p = \infty$.

By Lemma 6.1, we have the inclusion

$$Ext(L^p(\mathbb{T})) \subset \{f \in L^p(\mathbb{T}) : \|f\|_p = 1\}$$

for each $1 \le p \le \infty$. We first show that, if $1 < p < \infty$, then equality holds.

Theorem 6.4 *Let* $1 < p < \infty$. *Then*

$$Ext(L^p(\mathbb{T})) = \{f \in L^p(\mathbb{T}) : \|f\|_p = 1\}.$$

Proof Let $f \in L^p(\mathbb{T})$ with $\|f\|_p = 1$. Assume that $g \in L^p(\mathbb{T})$ is such that

$$\|f + g\|_p = \|f - g\|_p = 1.$$

Up to here, the proof works for all $1 \le p \le \infty$. However, if $1 < p < \infty$, then the above equality holds if and only if $g = 0$ (explore the case of equality in Minkowski's inequality). Therefore, by Corollary 6.2, f is an extreme point of $L^p(\mathbb{T})$. □

Contrary to the preceding case, which reveals that extreme points of $L^p(\mathbb{T})$, $1 < p < \infty$, form the largest possible set, the following result shows that for $L^1(\mathbb{T})$ we are faced with the smallest possible situation, i.e. there is no extreme point.

Theorem 6.5 *We have*

$$Ext(\, L^1(\mathbb{T})) = \emptyset.$$

Proof Fix $f \in L^1(\mathbb{T})$, with $\|f\|_1 = 1$. We explicitly construct two other distinct elements on the unit sphere of $L^1(\mathbb{T})$ such that f is precisely the middle point between them.

Define the function $\varphi : [0, 2\pi] \longrightarrow [0, 1]$ by the formula

$$\varphi(x) = \frac{1}{2\pi} \int_0^x |f(e^{it})|\, dt.$$

Since φ is continuous on $[0, 2\pi]$, $\varphi(0) = 0$ and $\varphi(2\pi) = 1$, by the mean value theorem, there is a point t_0 in the interval $(0, 2\pi)$ such that $\varphi(t_0) = \frac{1}{2}$. Put

$$g(e^{it}) = \begin{cases} 2f(e^{it}) & \text{if} \quad 0 \le t \le t_0, \\ 0 & \text{if} \quad t_0 < t < 2\pi, \end{cases}$$

and

$$h(e^{it}) = \begin{cases} 0 & \text{if} \quad 0 \le t \le t_0, \\ 2f(e^{it}) & \text{if} \quad t_0 < t < 2\pi. \end{cases}$$

Then $g, h \in L^1(\mathbb{T})$, $\|g\|_1 = \|h\|_1 = 1$, $f = (g + h)/2$, and moreover $f \ne g$ and $f \ne h$. Therefore, f is not an extreme point of the closed unit ball of $L^1(\mathbb{T})$. □

Our last result says that the extreme points of $L^\infty(\mathbb{T})$ are precisely unimodular elements of the closed unit ball of $L^\infty(\mathbb{T})$.

Theorem 6.6 *We have*

$$Ext(L^\infty(\mathbb{T})) = \{f \in L^\infty(\mathbb{T}) : |f| = 1 \text{ a.e. on } \mathbb{T}\}.$$

Proof First, let $f \in Ext(L^\infty(\mathbb{T}))$ and put $E = \{\zeta \in \mathbb{T} : |f(\zeta)| < 1\}$,

$$g = \begin{cases} f + \frac{1}{2}(1 - |f|) & \text{on } E, \\ f & \text{on } \mathbb{T} \setminus E, \end{cases}$$

and

$$h = \begin{cases} f - \frac{1}{2}(1 - |f|) & \text{on } E, \\ f & \text{on } \mathbb{T} \setminus E. \end{cases}$$

Then $g, h \in L^\infty(\mathbb{T})$, $\|g\|_\infty \le 1$, $\|h\|_\infty \le 1$, and moreover $f = (g + h)/2$. Therefore, we must have $f = g = h$, which is equivalent to $m(E) = 0$. In other words, $|f| = 1$ almost everywhere on $\mathbb{T}$.

Second, let $f \in L^\infty(\mathbb{T})$ with $|f| = 1$ almost everywhere on $\mathbb{T}$, and let g and h be two elements on the closed unit ball of $L^\infty(\mathbb{T})$ such that $f = (g + h)/2$. Put $E = \{\zeta \in \mathbb{T} : |g(\zeta)| < 1\}$ and $F = \{\zeta \in \mathbb{T} : |h(\zeta)| < 1\}$. On $E \cup F$, we have $|f| < 1$. Therefore, $m(E) = m(F) = 0$. A simple geometrical observation is needed now. The identity $\gamma = (\alpha + \beta)/2$, with $\alpha, \beta, \gamma \in \mathbb{T}$, holds if and only if $\alpha = \beta = \gamma$. Hence, $f = g = h$, a.e. on $\mathbb{T}$, and we conclude that $f \in Ext(L^\infty(\mathbb{T}))$. □

Exercise

Exercise 6.2.1 Show that $L^1(\mathbb{T})$ has no predual. In other words, there is no Banach space $\mathcal{X}$ such that $\mathcal{X}^* = L^1(\mathbb{T})$.
Hint: Use Theorem 6.5 and then appeal to the Kreĭn–Milman theorem.

6.3 Extreme points of H^p

The extreme points of the unit ball of H^p, $1 < p < \infty$, are easy to determine. We simply have

$$Ext(H^p) = \{f \in H^p : \|f\|_p = 1\}.$$

The situation was similar in $L^p(\mathbb{T})$. However, the story is completely different for $p = 1$ and $p = +\infty$. We know that $Ext(L^1(\mathbb{T})) = \emptyset$ (Theorem 6.5). But, in this regard, H^1 is very different from $L^1(\mathbb{T})$. The roots of this phenomenon stem from the fact that H^1 is the dual of a Banach space. More precisely, we have the Banach space isometric isomorphism

$$H^1 \simeq (\mathcal{C}(\mathbb{T})/\mathcal{A}_0)^*,$$

where $\mathcal{A}_0$ is the subspace of disk algebra $\mathcal{A}$ generated by polynomials z^n, $n \geq 1$ (see Section 4.2). But there is no Banach space $\mathcal{Y}$ such that $L^1(\mathbb{T}) \simeq \mathcal{Y}^*$ (see Exercise 6.2.1). Moreover, according to the Kreĭn–Milman theorem, if a Banach space $\mathcal{X}$ is isometrically isomorphic to the dual of a Banach space $\mathcal{Y}$, in short if $\mathcal{X} \simeq \mathcal{Y}^*$, then the unit ball of $\mathcal{X}$ not only has extreme points, but also, in a sense, it has a lot of these points. More precisely, the closed unit ball of $\mathcal{X}$ coincides with the weak-star closed convex hull of its extreme points. We emphasize that the closure is taken relative to the weak-star topology in $\mathcal{X}$. Therefore, we see that the situation for H^1 is dramatically different from the situation for $L^1(\mathbb{T})$. While the closed unit ball of $L^1(\mathbb{T})$ has no extreme point at all, the closed unit ball of H^1 has enough extreme points that the weak-star closed convex hull of them generates the whole unit ball of H^1. However, the Kreĭn–Milman theorem is a theorem about existence. It ensures the existence of many extreme points. But to decide whether a particular point is extreme or not is a quite different phenomenon and could be very difficult to do. It is remarkable that extreme points of the closed unit ball of H^1 and H^∞ could be described explicitly.

We start with the extreme points of H^∞. As motivation, we saw in Theorem 6.6 that, if Θ is an inner function, then it is an extreme point of the unit ball of $L^\infty(\mathbb{T})$. Hence, it is also automatically an extreme point of the unit ball of H^∞ (see Exercise 6.1.1). However, the set of extreme points of H^∞ is much larger than the family of inner functions. The following result gives the complete picture.

Theorem 6.7 *Let $f \in H^\infty$ with $\|f\|_\infty \leq 1$. Then f is an extreme point of the closed unit ball of H^∞ if and only if*

$$\int_0^{2\pi} \log(1 - |f(e^{it})|)\, dt = -\infty. \tag{6.6}$$

In particular, any inner function is an extreme point of the closed unit ball of H^∞.

Proof First assume that (6.6) is satisfied. In order to verify that f is an extreme point, in light of Lemma 6.1, let $g \in H^\infty$ be such that $\|f + g\|_\infty \leq 1$ and $\|f - g\|_\infty \leq 1$. Then we need to show that $g \equiv 0$. Using the parallelogram identity, we have

$$|f(z)|^2 + |g(z)|^2 = \frac{|f(z) + g(z)|^2 + |f(z) - g(z)|^2}{2} \leq 1 \qquad (z \in \mathbb{D}).$$

Hence, for almost all $e^{it} \in \mathbb{T}$,

$$|g(e^{it})|^2 \leq 1 - |f(e^{it})|^2.$$

In particular,

$$|g(e^{it})|^2 \leq (1+|f(e^{it})|)(1-|f(e^{it})|) \leq 2(1-|f(e^{it})|),$$

which gives

$$2\int_0^{2\pi} \log|g(e^{it})|\,dt \leq 2\pi\log 2 + \int_0^{2\pi} \log(1-|f(e^{it})|)\,dt.$$

Therefore, by (6.6),

$$\int_0^{2\pi} \log|g(e^{it})|\,dt = -\infty,$$

and (4.44) implies that $g \equiv 0$. Hence, f is an extreme point of the unit ball of H^∞.

For the converse implication, assume that

$$\int_0^{2\pi} \log(1-|f(e^{it})|)\,dt \neq -\infty. \tag{6.7}$$

Since $\log(1-|f(e^{it})|) \leq 0$, for almost all $e^{it} \in \mathbb{T}$, the condition (6.7) indeed means that $\log(1-|f|) \in L^1(\mathbb{T})$. This assumption enables us to construct the outer function

$$g(z) = \exp\left(\int_{\mathbb{T}} \frac{\zeta+z}{\zeta-z} \log(1-|f(\zeta)|)\,dm(\zeta)\right) \qquad (z \in \mathbb{D}).$$

Since we have $|g| = 1-|f|$ (a.e. on $\mathbb{T}$), then g is an outer function in H^∞. Moreover, $f+g \in H^\infty$, $f-g \in H^\infty$ and

$$\begin{aligned}\|f \pm g\|_\infty &= \operatorname*{ess\,sup}_{t\in[0,2\pi]} |f(e^{it}) \pm g(e^{it})| \\ &\leq \operatorname*{ess\,sup}_{t\in[0,2\pi]} (|f(e^{it})| + |g(e^{it})|) = 1.\end{aligned}$$

Therefore, by Lemma 6.1, f is not an extreme point of the unit ball of H^∞. □

Contrary to the case of $L^1(\mathbb{T})$, we now show that the closed unit ball of H^1 has a rich resource of extreme points.

Theorem 6.8 *Let $f \in H^1$. Then f is an extreme point of the closed unit ball of H^1 if and only if f is an outer function with $\|f\|_1 = 1$.*

Proof Suppose that f is an outer function in H^1 with $\|f\|_1 = 1$. Let $g \in H^1$ be such that $\|f \pm g\|_1 = 1$. By Corollary 6.2, we need to show that $g \equiv 0$. Define

$$\phi(z) = \frac{g(z)}{f(z)} \qquad (z \in \mathbb{D}).$$

Remember that, since f is an outer function, $f(z) \neq 0$ for any $z \in \mathbb{D}$, and, by (4.44), that $f \neq 0$ almost everywhere on $\mathbb{T}$. Hence, ϕ is an analytic function on $\mathbb{D}$. Furthermore, since $f, g \in H^1$ and $f \neq 0$ almost everywhere on $\mathbb{T}$, the quantity $\phi(e^{it})$, the boundary value of ϕ, exists almost everywhere on $\mathbb{T}$. Since $\|f \pm g\|_1 = \|f\|_1 = 1$, we have

$$\int_0^{2\pi} [|1 + \phi(e^{it})| + |1 - \phi(e^{it})| - 2]|f(e^{it})| \, dt = 0. \tag{6.8}$$

We have the triangle inequality

$$|1 + \phi(e^{it})| + |1 - \phi(e^{it})| \geq |1 + \phi(e^{it}) + 1 - \phi(e^{it})| = 2,$$

and the equality happens if and only if $\phi(e^{it}) \in [-1, 1]$. But, since again $f \neq 0$ almost everywhere on $\mathbb{T}$, (6.8) implies that

$$|1 + \phi(e^{it})| + |1 - \phi(e^{it})| = 2,$$

and thus $-1 \leq \phi(e^{it}) \leq 1$ for almost all $e^{it} \in \mathbb{T}$. Since f is outer, $f, g \in H^1$ and $\phi = g/f \in L^\infty(\mathbb{T})$, then, by Corollary 4.28, $\phi \in H^\infty$. However, we saw that ϕ is real-valued on $\mathbb{T}$ and thus, by (4.12), ϕ is a constant function with its value in the closed interval $[-1, 1]$. Finally,

$$\begin{aligned} 1 - \phi &= (1 - \phi)\|f\|_1 \\ &= \|(1 - \phi)f\|_1 \\ &= \|f - g\|_1 = 1, \end{aligned}$$

and thus $\phi = 0$, or equivalently $g = 0$. Therefore, f is an extreme point of the unit ball of H^1.

To prove the converse, suppose that $\|f\|_1 = 1$, but f is not outer. Thus, according to Theorem 4.19, we have $f = \Theta h$, where $h \in H^1$ is the outer part of f, and Θ is its inner part. By assumption, Θ is not constant. Consider

$$\Lambda(s) = \int_{-\pi}^{\pi} |f(e^{it})| \, \Re(e^{is}\Theta(e^{it})) \, dt \qquad (0 \leq s \leq \pi).$$

Since f is in $L^1(\mathbb{T})$ and Θ is unimodular, Λ is continuous on $[0, \pi]$. Moreover, we have

$$\Lambda(0) = \int_{-\pi}^{\pi} |f(e^{it})| \, \Re(\Theta(e^{it})) \, dt$$

and

$$\Lambda(\pi) = \int_{-\pi}^{\pi} |f(e^{it})| \, \Re(-\Theta(e^{it})) \, dt = -\Lambda(0).$$

Hence, by the mean value theorem, there is a point $s_0 \in [0, \pi]$ such that $\Lambda(s_0) = 0$. Put

$$g = \frac{e^{-is_0}}{2}(1 + e^{i2s_0}\Theta^2)h.$$

Certainly $g \in H^1$ and, since Θ is not constant, $g \not\equiv 0$. Moreover, g is defined such that

$$\begin{aligned} g(e^{it}) &= \frac{e^{-is_0}}{2}[1 + e^{i2s_0}\Theta^2(e^{it})]h(e^{it}) \\ &= \frac{e^{-is_0}\overline{\Theta(e^{it})} + e^{is_0}\Theta(e^{it})}{2}\Theta(e^{it})h(e^{it}) \\ &= f(e^{it})\,\Re(e^{is_0}\Theta(e^{it})). \end{aligned}$$

We can thus write

$$|f(e^{it}) \pm g(e^{it})| = |f(e^{it})|\,[1 \pm \Re(e^{is_0}\Theta(e^{it}))].$$

But, since

$$\int_{-\pi}^{\pi} |f(e^{it})|\,\Re(e^{is_0}\Theta(e^{it}))\,dt = \Lambda(s_0) = 0,$$

we conclude that

$$\|f \pm g\|_1 = \|f\|_1 = 1.$$

Since $g \not\equiv 0$, by Corollary 6.2, f is not extreme. □

Exercises

Exercise 6.3.1 Let $1 < p < \infty$. Show that

$$Ext(H^p) = \{f \in H^p : \|f\|_p = 1\}.$$

Hint: Use Theorem 6.4 and Exercise 6.1.1.

Exercise 6.3.2 Let $f \in H^1$, $\|f\|_1 = 1$, and assume that f is not an extreme point of the unit ball of H^1. Then, according to Theorem 6.8, we have $f = \Theta h$, where Θ is a nonconstant inner factor and h is outer. As in the proof of that theorem, we put

$$g = \frac{e^{-is_0}}{2}(1 + e^{i2s_0}\Theta^2)h.$$

(i) Show that

$$f + g = \tfrac{1}{2}e^{-is_0}(1 + e^{is_0}\Theta)^2 h$$

and

$$f - g = -\tfrac{1}{2}e^{-is_0}(1 - e^{is_0}\Theta)^2 h.$$

(ii) Deduce that $f + g$ and $f - g$ are outer functions.

(iii) Since $\|f \pm g\|_1 = 1$, conclude that there are two extreme points f_1 and f_2 such that

$$f = \tfrac{1}{2}(f_1 + f_2).$$

6.4 Strict convexity

A normed linear space $\mathcal{X}$ is said to be *strictly convex* if extreme points of its closed unit ball are precisely all the points of its unit sphere. This can also be rephrased as follows:

$$x, y \in \mathcal{X},\ x \neq y,\ \|x\| = \|y\| = 1 \quad \Longrightarrow \quad \left\| \frac{x+y}{2} \right\| < 1.$$

Based on Theorems 6.4, 6.5 and 6.6, the Lebesgue spaces $L^p(\mathbb{T})$ are strictly convex for $1 < p < \infty$, but $L^1(\mathbb{T})$ and $L^\infty(\mathbb{T})$ are not strictly convex. Similarly, the Hardy spaces H^p are strictly convex for $1 < p < \infty$, but H^1 and H^∞ are not strictly convex (see Theorems 6.7 and 6.8).

To start the discussion, we study the following simple result, which reveals a consequence of strict convexity on the Hahn–Banach separation theorem.

Lemma 6.9 *Let $\mathcal{X}$ be a normed linear space and assume that the dual space $\mathcal{X}^*$ is strictly convex. Then, for each point $x \in \mathcal{X}$ with $\|x\| = 1$, there is a unique functional $\Lambda \in \mathcal{X}^*$ such that $\Lambda(x) = \|\Lambda\| = 1$.*

Proof By the Hahn–Banach theorem, there exists a $\Lambda \in \mathcal{X}^*$ such that $\Lambda(x) = 1$ and $\|\Lambda\| = 1$. For the existence of Λ, the strict convexity of $\mathcal{X}^*$ is not needed. Our goal is to show that, under this extra condition, this linear functional is unique. To do so, assume that there is another $\Lambda' \in \mathcal{X}^*$ satisfying $\|\Lambda'\| = 1$ and $\Lambda'(x) = 1$. Then consider the functional $\Lambda'' = (\Lambda + \Lambda')/2$. We clearly have $\|\Lambda''\| \leq 1$ and

$$\Lambda''(x) = \frac{\Lambda(x) + \Lambda'(x)}{2} = 1.$$

Thus, $\|\Lambda''\| = 1$, i.e. Λ'' is on the unit sphere of $\mathcal{X}^*$. Now, the strict convexity of $\mathcal{X}^*$ forces $\Lambda = \Lambda' = \Lambda''$. □

In the case of $L^p(\mathbb{T})$ spaces, $1 < p < \infty$, we can give an explicit formula for the unique linear continuous functional whose existence was ensured in Lemma 6.9. More precisely, given $f \in L^p(\mathbb{T})$, with $\|f\|_p = 1$, put

$$h(e^{it}) = \begin{cases} |f(e^{it})|^{p-2}\overline{f(e^{it})} & \text{if} \quad f(e^{it}) \neq 0, \\ 0 & \text{if} \quad f(e^{it}) = 0. \end{cases}$$

It is easy to verify that $h \in L^q(\mathbb{T})$ and $\|h\|_q = 1$. Put

$$\Lambda(g) = \frac{1}{2\pi}\int_0^{2\pi} g(e^{it})h(e^{it})\,dt \qquad (g \in L^p(\mathbb{T})).$$

Then $\Lambda \in (L^p(\mathbb{T}))^*$, with representing symbol $h \in L^q(\mathbb{T})$, such that $\|\Lambda\| = \|h\|_q = 1$ and $\Lambda(f) = 1$. Hence, Λ is the unique linear continuous functional in the dual of $L^p(\mathbb{T})$ that we were looking for.

The space $L^1(\mathbb{T})$ is not strictly convex, and thus, for a given element $f \in L^1(\mathbb{T})$ with $\|f\|_1 = 1$, we should not expect to have a unique linear functional Λ such that $\|\Lambda\| = 1$ and $\Lambda(f) = 1$ (see Exercise 6.4.1). Nevertheless, if we add an extra hypothesis on f, then we can establish the uniqueness. The motivation for the following result comes from the above observation if we put $p = 1$ in the definition of Λ.

Theorem 6.10 *Let $f \in L^1(\mathbb{T})$, with $\|f\|_1 = 1$, be such that*

$$m(\{\zeta \in \mathbb{T} : f(\zeta) = 0\}) = 0.$$

Then there is a unique Λ in the dual of $L^1(\mathbb{T})$ such that $\Lambda(f) = \|\Lambda\| = 1$. This functional is given by the explicit formula

$$\Lambda(g) = \frac{1}{2\pi}\int_0^{2\pi} g(e^{it})\frac{\overline{f(e^{it})}}{|f(e^{it})|}\,dt \qquad (g \in L^1(\mathbb{T})).$$

Proof The existence of a continuous linear functional Λ such that $\Lambda(f) = 1$ and $\|\Lambda\| = 1$ follows from the Hahn–Banach theorem. Moreover, for any given Λ, the Riesz representation theorem gives a unique function $h \in L^\infty(\mathbb{T})$ such that $\|h\|_\infty = \|\Lambda\| = 1$ and

$$\Lambda(g) = \frac{1}{2\pi}\int_0^{2\pi} g(e^{it})h(e^{it})\,dt \qquad (g \in L^1(\mathbb{T})).$$

To establish the uniqueness part of the theorem, it remains to show that h is necessarily given by the formula $h = \bar{f}/|f|$ almost everywhere on $\mathbb{T}$.

The relations

$$\begin{aligned} 1 = \Lambda(f) &= \frac{1}{2\pi}\int_0^{2\pi} f(e^{it})h(e^{it})\,dt \\ &= \left|\frac{1}{2\pi}\int_0^{2\pi} f(e^{it})h(e^{it})\,dt\right| \\ &\le \frac{1}{2\pi}\int_0^{2\pi} |f(e^{it})h(e^{it})|\,dt \le \|f\|_1\,\|h\|_\infty = 1 \end{aligned}$$

imply that

$$f(e^{it})h(e^{it}) = |f(e^{it})h(e^{it})| \ge 0 \qquad \text{(a.e. on } \mathbb{T}) \tag{6.9}$$

and

$$|h(e^{it})| = 1 \qquad \text{(a.e. on } \mathbb{T}).$$

This is explained in more detail below.

Put

$$E = \{e^{it} : h(e^{it}) = 0\}.$$

Then, since

$$\begin{aligned} 1 &= \frac{1}{2\pi} \int_0^{2\pi} |f(e^{it})h(e^{it})| \, dt \\ &= \frac{1}{2\pi} \int_{\mathbb{T}\setminus E} |f(e^{it})h(e^{it})| \, dt \\ &\leq \frac{1}{2\pi} \int_{\mathbb{T}\setminus E} |f(e^{it})| \, dt \\ &\leq \frac{1}{2\pi} \int_{\mathbb{T}} |f(e^{it})| \, dt = 1, \end{aligned}$$

we deduce that

$$\int_E |f(e^{it})| \, dt = 0.$$

Since, by assumption, $f \neq 0$ almost everywhere on $\mathbb{T}$, we must have $m(E) = 0$, or equivalently $h \neq 0$ almost everywhere on $\mathbb{T}$. Knowing this fact, the identity (6.9) becomes equivalent to

$$\arg(h) = \arg(\bar{f}) \qquad \text{(a.e. on } \mathbb{T}). \tag{6.10}$$

With a similar technique, we now show that h is unimodular on $\mathbb{T}$. To do so, put

$$F = \{e^{it} : |h(e^{it})| < 1\},$$

and assume that $m(F) > 0$. Then we have

$$\begin{aligned} 1 &= \frac{1}{2\pi} \int_0^{2\pi} |f(e^{it})h(e^{it})| \, dt \\ &= \frac{1}{2\pi} \int_{\mathbb{T}\setminus F} |f(e^{it})h(e^{it})| \, dt + \frac{1}{2\pi} \int_F |f(e^{it})h(e^{it})| \, dt \\ &< \frac{1}{2\pi} \int_{\mathbb{T}\setminus F} |f(e^{it})| \, dt + \frac{1}{2\pi} \int_F |f(e^{it})| \, dt = 1, \end{aligned}$$

which is a contradiction (here we have used that $|h| < 1$ and $f \neq 0$ almost everywhere on F). Therefore, $m(F) = 0$, which means that $|h| = 1$ almost everywhere on $\mathbb{T}$. Therefore, (6.10) is rewritten as

$$h(e^{it}) = \frac{\overline{f(e^{it})}}{|f(e^{it})|} \qquad \text{(a.e. on } \mathbb{T}). \qquad \square$$

Exercises

Exercise 6.4.1 Let

$$f(e^{it}) = \begin{cases} 2e^{it} & \text{if} \quad 0 \le t \le \pi, \\ 0 & \text{if} \quad \pi < t < 2\pi. \end{cases}$$

Show that $f \in L^1(\mathbb{T})$, $\|f\|_1 = 1$, and construct two linear functionals Λ_1 and Λ_2 in the dual of $L^1(\mathbb{T})$, with $\Lambda_1 \neq \Lambda_2$, such that $\|\Lambda_1\| = \|\Lambda_2\| = 1$ and $\Lambda_1(f) = \Lambda_2(f) = 1$.
Hint: Consider the linear functionals corresponding to the bounded functions

$$g_1(e^{it}) = \begin{cases} e^{-it} & \text{if} \quad 0 \le t \le \pi, \\ 0 & \text{if} \quad \pi < t < 2\pi, \end{cases}$$

and $g_2(e^{it}) = e^{-it}$ for each $e^{it} \in \mathbb{T}$.

Exercise 6.4.2 Show that $L^1(\mathbb{T})$ is not strictly convex.
Hint: Either use Theorem 6.5 and do it directly, or use Lemma 6.9 and Exercise 6.4.1.

Exercise 6.4.3 Let $f \in L^1(\mathbb{T})$, $\|f\|_1 = 1$. Show that the sufficient condition

$$m(\{\zeta \in \mathbb{T} : f(\zeta) = 0\}) = 0$$

given in Theorem 6.10 is also necessary to obtain the uniqueness of $\Lambda \in (L^1(\mathbb{T}))^*$ such that $\Lambda f = \|\Lambda\| = 1$.
Hint: The idea used in Exercise 6.4.1 might be useful.

Exercise 6.4.4 Show that $L^\infty(\mathbb{T})$ is not strictly convex.
Hint: Use Theorem 6.6.

Exercise 6.4.5 Let $1 < p < \infty$. Show that $L^p(\mathbb{T})$ is strictly convex.
Hint: Use Theorem 6.4.

6.5 Exposed points of $\mathfrak{B}(\mathcal{X})$

Let Ω be a convex subset of a normed linear space $\mathcal{X}$. We say that a point $p \in \Omega$ is an *exposed point* of Ω if there exists a continuous linear functional $\Lambda \in \mathcal{X}^*$ such that

$$\Re(\Lambda x) < \Re(\Lambda p)$$

for all $x \in \Omega \setminus \{p\}$. This is equivalent to saying that there exists a continuous linear functional $\Lambda \in \mathcal{X}^*$ such that

$$\sup\{\Re(\Lambda x) : x \in \Omega\} = \Re(\Lambda p) \tag{6.11}$$

and

$$\Re(\Lambda x) = \Re(\Lambda p),\ x \in \Omega \quad \Longrightarrow \quad x = p. \tag{6.12}$$

By Lemma 6.3, each exposed point of Ω is an extreme point of Ω. But the reverse of this assertion is not true (see Exercise 6.5.2). The functional that fulfills (6.11) and (6.12) is not unique. For example, any such functional can be multiplied by a positive constant. When $\Omega = \mathfrak{B}(\mathcal{X})$, the closed unit ball of $\mathcal{X}$, we can pick a useful normalized representing functional.

Lemma 6.11 *Let $\mathcal{X}$ be a normed linear space, and let $p \in \mathcal{X}$ with $\|p\| = 1$. Then the following are equivalent.*

(i) *Point p is an exposed point of $\mathfrak{B}(\mathcal{X})$.*
(ii) *There exists a functional $\Lambda \in \mathcal{X}^*$ such that*

$$\Lambda p = \|\Lambda\| = 1$$

and, for $x \in \mathfrak{B}(\mathcal{X})$,

$$\Lambda x = \Lambda p = 1 \quad \Longrightarrow \quad x = p.$$

Proof (i) $\Longrightarrow$ (ii) By the definition of exposed points, there exists a functional $\Lambda \in \mathcal{X}^*$ such that (6.11) and (6.12) hold. Given any element $x \in \mathfrak{B}(\mathcal{X})$, there is a $\zeta \in \mathbb{T}$ such that $\zeta \Lambda x = |\Lambda x|$ (if $\mathcal{X}$ is a real linear space, then $\lambda \in \{\pm 1\}$). Hence, using (6.11), we have

$$|\Lambda x| = \Lambda(\zeta x) = \Re(\Lambda(\zeta x)) \le \Re(\Lambda p) \le |\Lambda p| \qquad (x \in \mathfrak{B}(\mathcal{X})). \tag{6.13}$$

In particular, we see that $\Lambda p \neq 0$, since otherwise $\Lambda \equiv 0$ and we obtain a contradiction with (6.12). Thus, we can consider the linear continuous functional $\Lambda_1 = \Lambda/\Lambda p$. Let us show that Λ_1 satisfies the required properties. First, it follows immediately from (6.13) that $|\Lambda_1(x)| \le 1$ for all $x \in \mathfrak{B}(\mathcal{X})$. Thus, $\|\Lambda_1\| \le 1$. Since $\Lambda_1(p) = 1$, we deduce that

$$\Lambda_1(p) = \|\Lambda_1\| = 1.$$

Second, if $x \in \mathfrak{B}(\mathcal{X})$ is such that $\Lambda_1(x) = 1$, then $\Lambda x = \Lambda p$, which, by (6.12), implies that $x = p$.

(ii) $\Longrightarrow$ (i) Assuming the existence of $\Lambda \in \mathcal{X}^*$ with the properties mentioned in (ii), for each $x \in \mathfrak{B}(\mathcal{X})$, we have

$$\Re(\Lambda x) \le |\Lambda x| \le \|\Lambda\| \times \|x\| \le 1 = \Lambda p = \Re(\Lambda p).$$

Moreover, if $x \in \mathfrak{B}(\mathcal{X})$ is such that $\Re(\Lambda x) = \Re(\Lambda p) = 1$, then the above relations imply that $\Re(\Lambda x) = |\Lambda x| = 1$. Hence, we must have $\Lambda x = 1$, which, by the hypothesis, implies that $x = p$. Therefore, p is an exposed point of the closed unit ball of $\mathcal{X}$. □

A linear functional Λ satisfying the condition (ii) of Lemma 6.11 is called an *exposing functional* for p. We also say that Λ *exposes the point* p. The exposing functional for an exposed point $p \in \mathfrak{B}(\mathcal{X})$ is not necessarily unique. However, if the dual space $\mathcal{X}^*$ is strictly convex, Lemma 6.9 shows that an exposed point of $\mathfrak{B}(\mathcal{X})$ possesses a unique exposing functional.

Exercises

Exercise 6.5.1 Let $\mathcal{X}$ be a normed linear space, and let p be an exposed point of its closed unit ball $\mathfrak{B}(\mathcal{X})$. Show that, for each $x \in \mathfrak{B}(\mathcal{X}) \setminus \mathbb{T}p$ and for any exposing functional Λ, we have $|\Lambda(x)| < 1$.

Exercise 6.5.2 In $\mathbb{R}^2$, let Ω be the union of the square $[-1, 1] \times [-1, 1]$ and the half-disk

$$\{(x, y) \in \mathbb{R}^2 : x^2 + (y-1)^2 \leq 1 \text{ and } y \geq 1\}.$$

Find all the extreme points and exposed points of Ω. In particular, show that $p = (1, 1)$ is an extreme point, but not an exposed point, of Ω.

Exercise 6.5.3 Let f be an element of the closed unit ball of $\mathcal{A}(\mathbb{D})$ with $\|f\|_\infty = 1$. The aim of this exercise is to show that f is an exposed point of the unit ball of $\mathcal{A}(\mathbb{D})$ if and only if $|f| = 1$ on a set of positive measure.

(i) Assume that $E = \{\zeta \in \mathbb{T} : |f(\zeta)| = 1\}$ has positive Lebesgue measure. Show that

$$\begin{array}{rccc} \Lambda : & \mathcal{A}(\mathbb{D}) & \longrightarrow & \mathbb{C} \\ & g & \longmapsto & \dfrac{1}{|E|} \displaystyle\int_E g\bar{f}\, dm \end{array}$$

exposes f.

(ii) Conversely, assume that f is an exposed point of the unit ball of $\mathcal{A}(\mathbb{D})$ and let Λ be an exposing functional for f.

(a) Show that there exists a probability measure μ with support in the set $E = \{\zeta \in \mathbb{T} : |f(\zeta)| = 1\}$ such that

$$\Lambda(g) = \int_{\mathbb{T}} g\bar{f}\, d\mu$$

for all $g \in \mathcal{A}(\mathbb{D})$.
(b) Assume that $|E| = 0$ and find a contradiction.

Hint: Use the fact that E is a *peak set* for the disk algebra.

Exercise 6.5.4 Show that

$$Ext(\mathcal{A}(\mathbb{D})) = \left\{ f \in \mathcal{A}(\mathbb{D}) : \|f\|_\infty \leq 1 \text{ and } \int_{\mathbb{T}} \log(1 - |f|)\, dm = -\infty \right\}.$$

Hint: Use Exercise 6.1.1 and Theorem 6.7 to obtain one inclusion. For the reverse inclusion, let $f \in \mathcal{A}(\mathbb{D})$, $\|f\|_\infty \leq 1$ be such that $\int_{\mathbb{T}} \log(1 - |f|)\, dm > -\infty$, and let E be the set of all points $e^{is} \in \mathbb{T}$ at which $|f(e^{is})| = 1$. Construct a continuous function ϕ on $\mathbb{T}$ such that:

1. $0 \leq \phi(e^{is}) \leq 1 - |f(e^{is})|$;
2. $\int_{\mathbb{T}} \log \phi\, dm > -\infty$;
3. on every closed subarc of $\mathbb{T}$ that is disjoint from E, ϕ has a bounded derivative.

Put $g = [\phi]$. Show that $g \in \mathcal{A}(\mathbb{D})$ and conclude as in Theorem 6.7.

6.6 Strongly exposed points of $\mathfrak{B}(\mathcal{X})$

Let Ω be a convex subset of a normed linear space $\mathcal{X}$. We say that a point $p \in \Omega$ is a *strongly exposed point* of Ω if there is a linear continuous functional $\Lambda \in \mathcal{X}^*$ such that

$$\Re(\Lambda x) < \Re(\Lambda p) \tag{6.14}$$

for each $x \in \Omega \setminus \{p\}$, and if $(x_n)_{n \geq 1}$ is a sequence in Ω such that

$$\Re(\Lambda x_n) \longrightarrow \Re(\Lambda p),$$

as $n \longrightarrow \infty$, then

$$\|x_n - p\| \longrightarrow 0. \tag{6.15}$$

Clearly, a strongly exposed point of Ω is always an exposed point of Ω, but the converse is not true (see Exercise 6.6.1).

Similar to Lemma 6.11, if Ω is the closed unit ball of $\mathcal{X}$, we can give a useful characterization of strongly exposed points of $\mathfrak{B}(\mathcal{X})$.

Lemma 6.12 *Let $\mathcal{X}$ be a normed linear space, and let $p \in \mathcal{X}$ with $\|p\| = 1$. Then the following are equivalent.*

(i) *Point p is a strongly exposed point of $\mathfrak{B}(\mathcal{X})$.*

(ii) *There exists a functional* $\Lambda \in \mathcal{X}^*$ *such that*

$$\Lambda p = \|\Lambda\| = 1,$$

and if $(x_n)_{n\geq 1}$ *is a sequence in* $\mathfrak{B}(\mathcal{X})$ *with*

$$\Lambda x_n \longrightarrow 1 = \Lambda p,$$

then

$$\|x_n - p\| \longrightarrow 0.$$

Proof (i) $\Longrightarrow$ (ii) Let $\Lambda \in \mathcal{X}^*$ be such that (6.14) and (6.15) hold. Then, considering $\Lambda_1 = \Lambda/\Lambda p$, we know from the proof of Lemma 6.11 that Λ_1 exposes the point p. Moreover, if $(x_n)_{n\geq 1}$ is a sequence in $\mathfrak{B}(\mathcal{X})$ such that $\Lambda_1 x_n \longrightarrow 1$, as $n \longrightarrow \infty$, then surely $\Re(\Lambda x_n) \longrightarrow \Re(\Lambda p)$. Thus, by (6.15), we have $\|x_n - p\| \longrightarrow 0$.

(ii) $\Longrightarrow$ (i) Let $\Lambda \in \mathcal{X}^*$ satisfy (ii). Using the proof of Lemma 6.11, we know that

$$\Re(\Lambda x) < \Re(\Lambda p)$$

for each $x \in \mathfrak{B}(\mathcal{X}) \setminus \{p\}$. Moreover, if $(x_n)_{n\geq 1}$ is in $\mathfrak{B}(\mathcal{X})$ such that $\Re(\Lambda x_n) \longrightarrow \Re(\Lambda p) = 1$, then, since

$$\Re(\Lambda x_n) \leq |\Lambda x_n| \leq 1,$$

we must have $\Lambda x_n \longrightarrow 1$, as $n \longrightarrow \infty$. Hence, $\|x_n - p\| \longrightarrow 0$. Therefore, p is a strongly exposed point of $\mathfrak{B}(\mathcal{X})$. □

Parallel to the Kreĭn–Milman theorem, Phelps has shown that, in a separable dual space, every bounded closed convex set is the closed convex hull of its strongly exposed points. As in the Kreĭn–Milman theorem, the theorem of Phelps ensures the existence of many exposed points, but to decide whether a particular point in the boundary of a convex set is extreme, exposed or strongly exposed is a quite different story and could be very difficult. In the following, we discuss in detail the special but important case of the closed unit ball of H^1. For this set, we have seen in Theorem 6.8 that the extreme points are well known and quite easy to characterize. We will see in Theorem 29.20 a rather simple criterion for strongly exposed points of the closed unit ball of H^1. This criterion is expressed in terms of the Helson–Szegő condition. On the other hand, for the exposed points of the closed unit ball of H^1, the situation is not that satisfactory. In fact, we will consider three approaches, which will be based on:

(i) function theoretic methods (Corollary 6.21);
(ii) the theory of Toeplitz operators (Corollary 12.33);
(iii) the theory of $\mathcal{H}(b)$ spaces (Corollary 29.8).

However, it should be pointed out that these three criteria are not easy to check in practice, and it still remains an open problem to find a tractable criterion to characterize the exposed points of the closed unit ball of H^1. The key point for all the criteria mentioned above is the link with another subclass of H^1, the so-called rigid functions, which we introduce in the next section.

Exercise

Exercise 6.6.1 Let $\ell^2 = \ell^2(\mathbb{N})$, let $\Omega = \{(x_k)_{k\geq 1} \in \ell^2 : x_k \geq 0\}$ and let $p = (0, 0, 0, \dots)$.

(i) Show that Ω is a convex subset of ℓ^2.

(ii) Show that p is an exposed point of Ω.
Hint: Consider the linear functional Λ on ℓ^2 defined by

$$\Lambda x = -\sum_{k=1}^{\infty} \frac{x_k}{k} \qquad (x = (x_k)_{k\geq 1} \in \ell^2).$$

(iii) Show that p is not a strongly exposed point of Ω.
Hint: For any linear functional L on ℓ^2, by the Riesz representation theorem, there is a vector x_0 in ℓ^2 such that

$$L(x) = \langle x, x_0 \rangle_{\ell^2} \qquad (x \in \ell^2).$$

Hence, for the canonical basis of ℓ^2, $\mathfrak{e}_n = (\delta_{nk})_{k\geq 1} \in \Omega$, we surely have $L(\mathfrak{e}_n) \longrightarrow Lp = 0$, whereas $\|\mathfrak{e}_n\| \not\rightarrow 0$.

6.7 Equivalence of rigidity and exposed points in H^1

As we saw in Lemma 6.9, in a normed linear space $\mathcal{X}$, where $\mathcal{X}^*$ is strictly convex, an exposed point of $\mathfrak{B}(\mathcal{X})$ has a unique exposing functional. But we know that the dual of H^1 is not strictly convex. Nevertheless, an exposed point of H^1 still has a unique exposing functional.

Lemma 6.13 *Let $f \in H^1$, $\|f\|_1 = 1$, and assume that f is an exposed point of H^1. Then there is a unique exposing functional Λ for f, which is given by the explicit formula*

$$\Lambda(g) = \frac{1}{2\pi} \int_0^{2\pi} g(e^{it}) \frac{\overline{f(e^{it})}}{|f(e^{it})|} \, dt \qquad (g \in H^1).$$

Proof By Lemma 6.11, there is a continuous linear functional $\Lambda \in (H^1)^*$ such that $\|\Lambda\| = \Lambda(f) = 1$, and if $g \in H^1$, with $\|g\|_1 \le 1$, is such that $\Lambda(g) = \Lambda(f)$, then $g = f$. By the Hahn–Banach theorem, we can extend Λ to a linear functional on $L^1(\mathbb{T})$ without increasing its norm. Moreover, since $f \in H^1$, $\|f\| = 1$, Lemma 4.30 implies that $f \neq 0$ a.e. on $\mathbb{T}$. Then it follows from Theorem 6.10 that Λ is of the form

$$\Lambda(g) = \frac{1}{2\pi}\int_0^{2\pi} g(e^{it})\frac{\overline{f(e^{it})}}{|f(e^{it})|}\,dt \qquad (g \in H^1). \qquad \square$$

If $f \in H^1$, $f \not\equiv 0$, then $f \neq 0$ at almost all points of $\mathbb{T}$ and thus $\arg f$ is a well-defined measurable function on $\mathbb{T}$. Without loss of generality, we can even assume that $-\pi < \arg f \le \pi$. Hence, if f and g are two functions in H^1, it makes sense to write $\arg g = \arg f$ (a.e. on $\mathbb{T}$). The identity $\arg g = \arg f$ has several equivalent forms. In fact, one can easily verify that the following assertions are equivalent:

(i) $\arg g = \arg f$;
(ii) $g/|g| = f/|f|$;
(iii) $g\bar{f} = |gf|$.

All identities are assumed to hold almost everywhere on $\mathbb{T}$.

Using Lemma 6.13, we easily see that exposed points of the unit ball of H^1 have an interesting property.

Corollary 6.14 *Let $f \in H^1$, $\|f\|_1 = 1$, be an exposed point of the closed unit ball of H^1. Let $g \in H^1$, $g \not\equiv 0$, be such that* $\arg g = \arg f$ *(a.e. on $\mathbb{T}$). Then there is a positive real constant λ such that $g = \lambda f$.*

Proof Let Λ be the exposing functional for f. The relation $\arg g = \arg f$ is equivalent to $g\bar{f} = |gf|$ almost everywhere on $\mathbb{T}$. Hence, using Lemma 6.13, we have

$$\Lambda(g) = \frac{1}{2\pi}\int_0^{2\pi} g(e^{it})\frac{\overline{f(e^{it})}}{|f(e^{it})|}\,dt = \frac{1}{2\pi}\int_0^{2\pi} |g(e^{it})|\,dt = \|g\|_1.$$

Therefore, we have $\Lambda(g/\|g\|_1) = 1 = \Lambda(f)$ and, since f is an exposed point of the closed unit ball of H^1, we deduce that $g = \|g\|_1 f$. $\square$

Corollary 6.14 motivates the following definition. Let $f \in H^1$, $f \neq 0$. We say that f is *rigid* if, for any function $g \in H^1$, $g \not\equiv 0$, the assumption

$$\arg g = \arg f \qquad \text{(a.e. on } \mathbb{T}) \tag{6.16}$$

implies the existence of a positive real constant λ such that $g = \lambda f$. Hence, up to a positive multiplicative constant, rigid functions of H^1 are uniquely determined by their arguments. Note that, if f is any function in H^1, $f \not\equiv 0$, then f is rigid if and only if λf is rigid for some (all) positive number(s) λ.

Corollary 6.14 can be rephrased by saying that each exposed point of the closed unit ball of H^1 is a rigid function. One of our goals is to show that the converse is essentially true (Theorem 6.15). More precisely, the set of rigid functions of unit norm of H^1 and the set of exposed points of the closed unit ball of H^1 coincide. The class of rigid functions is interesting in its own right, but it also offers a different point of view for studying the exposed points of the closed unit ball of H^1. This point of view is essential in establishing certain results about exposed points (see Section 6.8).

Theorem 6.15 *Let $f \in H^1$, $f \not\equiv 0$. Then the following are equivalent:*

(i) *f is an exposed point of the closed unit ball of H^1;*
(ii) *f is a rigid function and $\|f\|_1 = 1$.*

Proof (i) $\Longrightarrow$ (ii) This is discussed in Corollary 6.14.

(ii) $\Longrightarrow$ (i) Assume now that f is a rigid function of unit norm in H^1, and consider the linear functional Λ defined on $L^1(\mathbb{T})$ by

$$\Lambda(g) = \frac{1}{2\pi}\int_0^{2\pi} g(e^{it})\frac{\overline{f(e^{it})}}{|f(e^{it})|}\,dt \qquad (g \in L^1(\mathbb{T})).$$

We have $\|\Lambda\| = 1$ and $\Lambda(f) = \|f\|_1 = 1$. Thus, according to Lemma 6.11, it remains to check that, if $g \in \mathfrak{B}(\mathcal{X})$ with $\Lambda(g) = 1$, then $g = f$. But note that

$$1 = \Lambda(g) = \frac{1}{2\pi}\int_0^{2\pi} g(e^{it})\frac{\overline{f(e^{it})}}{|f(e^{it})|}\,dt \le \frac{1}{2\pi}\int_0^{2\pi} |g(e^{it})|\,dt = \|g\|_1 \le 1.$$

Thus, we must have $g(e^{it})\overline{f(e^{it})} = |f(e^{it})g(e^{it})|$ almost everywhere on $\mathbb{T}$, which is equivalent to $\arg g = \arg f$. Therefore, since f is assumed to be rigid, there exists a positive real constant λ such that $g = \lambda f$. But, $1 = \|g\|_1 = \lambda\|f\|_1 = \lambda$, and hence $g = f$. This means that f is an exposed point of the closed unit ball of H^1. □

The following result is an easy consequence of Theorems 6.8 and 6.15 and gives a necessary condition for rigidity.

Corollary 6.16 *Let $f \in H^1$, $f \not\equiv 0$. If f is rigid, then f is an outer function.*

Proof Without loss of generality, we may assume that $\|f\|_1 = 1$. Now, by Theorem 6.15, f is an exposed point of the closed unit ball of H^1 and thus, in particular, it is an extreme point of it. It remains to apply Theorem 6.8 to conclude that f is an outer function. □

6.8 Properties of rigid functions

According to Theorem 6.15, given a function f of unit norm in H^1, then f is a rigid function if and only if it is an exposed point of the closed unit ball of H^1. Naively speaking, in this section we will witness that the definition of rigid function is more user-friendly. Most of the results are stated for rigid functions, but obviously they can be rephrased in terms of exposed points of the closed unit ball of H^1.

The first lemma shows that, in the definition of a rigid function, we can assume that the function g, which is supposed to fulfill $\arg g = \arg f$, a.e. on $\mathbb{T}$, is an outer function.

Lemma 6.17 *Let $f \in H^1$, $f \not\equiv 0$. Then the following are equivalent.*

(i) *The function f is rigid.*
(ii) *For any outer function g in H^1, $g \not\equiv 0$, the condition*

$$\arg g = \arg f \qquad (a.e.\ on\ \mathbb{T}),$$

implies that there is a positive real constant λ such that $f = \lambda g$.

Proof (i) $\Longrightarrow$ (ii) This is trivial.

(ii) $\Longrightarrow$ (i) We show that, if f is not rigid, then there exists an outer function $g \in H^1$ such that $\arg g = \arg f$ a.e. on $\mathbb{T}$, but f/g is not a positive constant. Hence, assume that f is not rigid. Then, according to the definition of rigidity, there is a function $\varphi \in H^1$ satisfying $\arg \varphi = \arg f$ a.e. on $\mathbb{T}$, but $\varphi \neq \lambda f$ for any $\lambda > 0$. If φ is outer, we are done. If not, by the canonical factorization theorem (Theorem 4.19), we can write $\varphi = \Theta h$, with Θ inner and nonconstant, and h outer. Consider then

$$g_1 = (1+\Theta)^2 h \quad \text{and} \quad g_2 = -(1-\Theta)^2 h.$$

It is clear that g_1 and g_2 are in H^1 and, according to Corollaries 4.25 and 4.24, the functions g_1 and g_2 are outer. Moreover, using the fact that $|\Theta| = 1$ a.e. on $\mathbb{T}$, we have

$$(1+\Theta)^2 = 4\Theta \cos^2(\arg \Theta/2).$$

Thus,

$$\arg(1+\Theta)^2 = \arg \Theta \qquad (\text{a.e. on } \mathbb{T}). \tag{6.17}$$

By the same token, $\arg(1-\Theta)^2 = \pi + \arg \Theta$. Thus, we have

$$\arg g_1 = \arg g_2 = \arg(\Theta h) = \arg f.$$

If $g_1 \neq \lambda f$ or $g_2 \neq \lambda f$, for any $\lambda > 0$, then they are the outer candidate that we are looking for. Otherwise, there are constants $\lambda_1 > 0$ and $\lambda_2 > 0$ such that $f = \lambda_1 g_1 = \lambda_2 g_2$. The last identity implies $\lambda_1(1+\Theta)^2 = -\lambda_2(1-\Theta)^2$, and thus Θ is constant, which is absurd. □

The next result shows that the rigidity is preserved for the divisors.

Lemma 6.18 *Let $f \in H^1$, $f \not\equiv 0$, and assume that $f = f_1 f_2$ with $f_1, f_2 \in H^1$. If f is rigid, then both factors f_1 and f_2 are also rigid.*

Proof Since $f \not\equiv 0$, we have $f_1 \not\equiv 0$ and $f_2 \not\equiv 0$, which in return implies that $f_1 \neq 0$ and $f_2 \neq 0$ almost everywhere on $\mathbb{T}$ (see Lemma 4.30).

Let $g \in H^1$, $g \not\equiv 0$, be such that $\arg g = \arg f_1$ a.e. on $\mathbb{T}$. Then $\arg(gf_2) = \arg(f_1 f_2) = \arg f$ a.e. on $\mathbb{T}$. Hence, by assumption (the rigidity of f), there exists a real constant $\lambda > 0$ such that $gf_2 = \lambda f = \lambda f_1 f_2$. Hence $g = \lambda f_1$ and f_1 is rigid. The same argument applies also to f_2. □

The converse of Lemma 6.18 is not true, that is, the product of two rigid functions is not necessarily rigid. For example, at the end of this section, we will see that $f(z) = 1 - z$ is rigid, but $f^2(z) = (1-z)^2$ is not.

Corollary 6.16 shows that a rigid function is outer. First, Exercise 6.8.1 provides an example of an outer function that is not rigid. Second, rephrasing Corollary 6.16, one can say that, if f is an H^1 function that is not outer, then f is not rigid. In fact, if f is not outer, using similar techniques to those used in the proof of Lemma 6.17, we can construct explicitly a function $g \in H^1$ with the same argument as f and such that $g \neq \lambda f$, for any $\lambda > 0$. To do so, let $f = \Theta h$ be the inner–outer decomposition of f, with Θ a nonconstant inner function. In particular, $\Theta \not\equiv -1$. Consider now the function $g = (1+\Theta)^2 h$ and check that it satisfies the required conditions. First, since $\Theta \in H^\infty$ and $h \in H^1$, the function g belongs to H^1. Moreover, by Corollary 4.25, $1+\Theta$ is outer and thus Corollary 4.24 implies that g is an outer function. Hence, by (6.17), we obtain

$$\arg g = \arg((1+\Theta)^2 h) = \arg \Theta h = \arg f.$$

To finish, assume that there exists a $\lambda > 0$ such that $g = \lambda f$. Hence, we have $(1+\Theta)^2 = \lambda\Theta$, and remembering that the left-hand side is an outer function and, up to a multiplicative unimodular constant, the canonical factorization is unique, we deduce that Θ has to be a constant. This is absurd.

The idea used in the previous argument also provides a necessary condition for rigidity.

Lemma 6.19 *Let $f \in H^1$, $f \not\equiv 0$. If f is a rigid function, then, for each nonconstant inner function Θ, the function $f/(1+\Theta)^2$ is not in H^1.*

Proof Suppose that there is a nonconstant inner function Θ such that the function $f/(1+\Theta)^2$ belongs to H^1, and then put

$$g = \Theta f/(1+\Theta)^2 \in H^1.$$

Since, by (6.17), $\arg\Theta = \arg(1+\Theta)^2$, we must have $\arg g = \arg f$ a.e. on $\mathbb{T}$. The rigidity of f implies now the existence of a constant $\lambda > 0$ such that $g = \lambda f$. Hence, we obtain $\Theta = \lambda(1+\Theta)^2$, which, as already mentioned, is absurd. □

The converse of Lemma 6.19 is not true. More explicitly, at the end of this section, we construct a function f in the unit ball of H^1, which is not an exposed point and which has the property that $f/(1+\Theta)^2$ does not belong to H^1 for any nonconstant inner function Θ. Nevertheless, a variation of Lemma 6.19 characterizes rigid functions.

Theorem 6.20 *Let f be an outer function in H^1. The following are equivalent.*

(i) *Function f is a rigid function.*
(ii) *For all inner functions Θ_1 and Θ_2, not both constants and $\Theta_1 \neq -\Theta_2$, the function $f/(\Theta_1+\Theta_2)^2$ is not in H^1.*

Proof (i) $\Longrightarrow$ (ii) The proof of this implication has the same flavor as the proof of Lemma 6.19. Assume that f is a rigid function and that there are two inner functions Θ_1 and Θ_2, not both constants, such that the function $f/(\Theta_1+\Theta_2)^2$ is in H^1. Then, using the fact that $|\Theta_1| = |\Theta_2| = 1$ a.e. on $\mathbb{T}$, we have

$$(\Theta_1+\Theta_2)^2 = 4\Theta_1\Theta_2\cos^2(\arg(\Theta_1\overline{\Theta_2})/2) \qquad \text{(a.e on } \mathbb{T}).$$

Thus

$$\arg((\Theta_1+\Theta_2)^2) = \arg(\Theta_1\Theta_2) \qquad \text{(a.e on } \mathbb{T}).$$

Now consider the function

$$g = \frac{\Theta_1\Theta_2}{(\Theta_1+\Theta_2)^2} f.$$

By hypothesis, $g \in H^1$, and moreover it satisfies $\arg g = \arg f$, a.e. on $\mathbb{T}$. The rigidity of f implies the existence of a constant $\lambda > 0$ such that $g = \lambda f$. Hence we obtain $\Theta_1\Theta_2 = \lambda(\Theta_1+\Theta_2)^2$. In particular, the last identity shows that the function $f/(\Theta_1\Theta_2)$ must be in H^1. Since f is outer, appealing to the uniqueness of the canonical factorization, we deduce that $\Theta_1\Theta_2$ is constant, and thus both functions Θ_1 and Θ_2 are constant, which is absurd.

(ii) $\Longrightarrow$ (i) Assume that f is not rigid. By Lemma 6.17 there exists an outer function g in H^1, $g \not\equiv 0$, such that $\arg f = \arg g$ and $g \neq \lambda f$ for all $\lambda > 0$. Denote by f_1 and g_1 the outer functions in H^2 such that $f_1^2 = f$, $g_1^2 = g$ and $\arg f_1 = \arg g_1$. Then, since $g_1 + if_1$ and $g_1 - if_1$ have the same modulus, there exist an outer function $h \in H^2$ and two inner functions Θ_1 and Θ_2 such that

$$g_1 + if_1 = \Theta_1 h \quad \text{and} \quad g_1 - if_1 = \Theta_2 h.$$

Hence,

$$2if_1 = (\Theta_1 - \Theta_2)h \quad \text{and} \quad 2g_1 = (\Theta_1 + \Theta_2)h.$$

Since f_1 is outer, the factor $\Theta_1 - \Theta_2$ also has to be outer, and

$$\frac{f_1}{\Theta_1 - \Theta_2} = \frac{1}{2i}h \in H^2.$$

Therefore,

$$\frac{f}{(\Theta_1 - \Theta_2)^2} = \left(\frac{f_1}{\Theta_1 - \Theta_2}\right)^2 \in H^1.$$

The inner functions Θ_1 and Θ_2 are not both constants, since otherwise there exists a constant γ such that $g_1 = \gamma f_1$. But since $\arg f_1 = \arg g_1$, we necessarily have $\gamma \in \mathbb{R}$. Thus we get that $g = \gamma^2 f$ and $\gamma^2 > 0$, which is absurd. □

Using Theorem 6.15, we can rephrase Theorem 6.20 as follows.

Corollary 6.21 *Let $f \in H^1$, $\|f\|_1 = 1$. The following are equivalent.*

(i) *Function f is an exposed point of the closed unit ball of H^1.*
(ii) *For all inner functions Θ_1 and Θ_2, not both constants and $\Theta_1 \neq -\Theta_2$, the function $f/(\Theta_1 + \Theta_2)^2$ is not in H^1.*

Using Lemma 6.19, it is easy to construct an outer function that is not rigid. Indeed, we can, for instance, take $f(z) = (1 + z)^2$, $z \in \mathbb{D}$. By Corollary 4.25, f is an outer function, and if we take the inner function $\Theta(z) = z$, then $f/(1 + \Theta)^2 = 1$ and Lemma 6.19 implies that f is not rigid. Hence, the set of rigid functions is a proper subclass of the family of extreme points of the closed unit ball of H^1. The drawback of Theorem 6.20, or equivalently Corollary 6.21, is that its condition is difficult to check in practice. To partially overcome this difficulty, we provide two different sufficient conditions for rigidity.

Theorem 6.22 *Let $f \in H^1$, $f \not\equiv 0$. If $1/f \in H^1$, then f is a rigid function.*

Proof Let g be a function in H^1 satisfying $\arg g = \arg f$ a.e. on $\mathbb{T}$. Consider the function $h = g/f$. Since g and $1/f$ are in H^1, the function h is in $H^{1/2}$. Moreover, since $\arg g = \arg f$, we have $\arg h = 0$ a.e. on $\mathbb{T}$. This means that h is a positive real function on $\mathbb{T}$. Thus, by Theorem 4.29, h is a positive constant, which proves that f is rigid. □

The condition $1/f \in H^1$ in Theorem 6.22 is sharp in the sense that the assumption $1/f \in H^p$, for some $p < 1$, is not enough to imply the rigidity (see Exercise 6.8.3).

Theorem 6.23 *Let $f \in H^1$, $f \not\equiv 0$. Suppose that there exists $h \in H^\infty$, $h \not\equiv 0$, such that*

$$\Re(fh) \geq 0 \qquad (a.e.\ on\ \mathbb{T}).$$

Then f is a rigid function.

Proof According to Lemma 6.18, we may assume that $h \equiv 1$. Hence, assume that f is a function in H^1, not identically, zero and such that $\Re(f) \geq 0$ a.e. on $\mathbb{T}$. Now, take any function g in H^1, $g \not\equiv 0$, such that $\arg g = \arg f$ a.e. on $\mathbb{T}$. This is equivalent to $f/g = |f|/|g|$ a.e. on $\mathbb{T}$. In particular, the function f/g is a positive real function on $\mathbb{T}$. Thus,

$$\frac{f}{g} + 1 \geq 1 \qquad (\text{a.e. on } \mathbb{T}). \tag{6.18}$$

We claim that $f + g$ is an outer function in H^1. Since $\Re(f) \geq 0$ and $\arg g = \arg f$ a.e. on $\mathbb{T}$, we have $\Re(g) \geq 0$ a.e. on $\mathbb{T}$, whence

$$\Re(f + g) = \Re(f) + \Re(g) \geq 0 \qquad (\text{a.e. on } \mathbb{T}).$$

Since $f + g \in H^1$, we get from (4.8) that

$$\Re(f + g) \geq 0 \qquad (\text{on } \mathbb{D}).$$

If there exists a point $z_0 \in \mathbb{D}$ such that $\Re(f(z_0) + g(z_0)) = 0$, then, by the maximum principle for harmonic functions, $\Re(f + g) = 0$ on $\mathbb{D}$, and so $f = g = 0$, which is not possible. Hence, there does not exist any point $z_0 \in \mathbb{D}$ such that $\Re(f(z_0) + g(z_0)) = 0$, i.e. $\Re(f + g) > 0$ on $\mathbb{D}$, and Corollary 4.25 implies that $f + g$ is an outer function.

Consider now the function $\varphi = g/(f+g)$. According to (6.18), $\varphi \in L^\infty(\mathbb{T})$, and thus, by Corollary 4.28, we have in fact $\varphi \in H^\infty$. But since f/g is a real function on $\mathbb{T}$, the function φ is also real on $\mathbb{T}$, and (4.12) implies that φ is constant. Thus f/g is a positive constant, which proves that f is rigid. □

As an application of Theorem 6.23, we give an (important) example of exposed points of the closed unit ball of H^1. This example shows that the condition of Theorem 6.22 is not necessary for rigidity.

Example 6.24 *The function $f(z) = 1 - z$ is rigid.*

In fact, since $\Re(f(z)) \geq 0$ for all $z \in \mathbb{T}$, by Theorem 6.23, we deduce that f is a rigid function in H^1. However, note that $1/f$ does not belong to H^1. We can even slightly generalize Example 6.24 to characterize the polynomials that are rigid functions in H^1 (see Exercise 6.8.4).

Corollary 6.25 *Let f be an outer function in H^1. If*

$$\operatorname{dist}_{L^\infty(\mathbb{T})}(\arg f,\ H^\infty + \mathcal{C}(\mathbb{T})) < \frac{\pi}{2},$$

then f is a rigid function.

Proof Note that $\arg f$ is defined a.e. on $\mathbb{T}$ since it is the conjugate harmonic of $\log|f|$, which is integrable. By assumption, there are an $\varepsilon > 0$, $h \in H^\infty$ and $g \in \mathcal{C}(\mathbb{T})$ such that

$$\|\arg f - h - g\|_\infty < \frac{\pi}{2} - \varepsilon.$$

Since trigonometric polynomials are dense in $\mathcal{C}(\mathbb{T})$, in the above estimation, we can assume that g is a trigonometric polynomial, e.g.

$$g = \sum_{n=-N}^{N} a_n \chi_n.$$

Put

$$k = h + \sum_{n=-N}^{-1} \bar{a}_n \chi_{-n} + \sum_{n=0}^{N} a_n \chi_n.$$

Then $k \in H^\infty$ and satisfies $\Re(k) = \Re(h) + \Re(g)$. Hence, since $\arg f$ is real-valued,

$$\|\arg f - \Re(k)\|_\infty = \|\arg f - \Re(h) - \Re(g)\|_\infty < \frac{\pi}{2} - \varepsilon.$$

Then $\exp(-ik)$ belongs to H^∞ and

$$\begin{aligned}\Re(\exp(-ik)f) &= \Re(|f|\exp(i(\arg f - k)))\\ &= |f|\exp(\Im(k))\cos(\arg f - \Re(k)) > 0.\end{aligned}$$

Thus, by Theorem 6.23, f is a rigid function. □

We end this section by showing that the converse of Lemma 6.19 is not true.

Theorem 6.26 *There exists a function G in the unit ball of H^1, which is not an exposed point of the closed unit ball of H^1 and which has the property that $G/(1+\Theta)^2$ does not belong to H^1 for any nonconstant inner function Θ.*

Proof Since

$$\left|1 - \frac{z - e^{i/n^2}}{z-1}\right| = \frac{|1 - e^{i/n^2}|}{|1-z|} = O(1/n^2) \qquad (n \geq 1), \tag{6.19}$$

the product

$$h(z) = \prod_{n\geq 1} \frac{z - e^{i/n^2}}{z-1} \tag{6.20}$$

defines an analytic function on $\mathbb{C} \setminus \{1\}$ with zeros precisely at e^{i/n^2}, $n \geq 1$. In fact, h is also outer. To verify this fact, put

$$h_n(z) = \frac{1 - e^{-i/n^2} z}{1-z} \qquad (n \geq 1).$$

Since this factor is an outer function, we can write

$$h_n(z) = \exp\left\{ \int_{\mathbb{T}} \frac{\zeta + z}{\zeta - z} \log |h_n(\zeta)| \, dm(\zeta) \right\} \qquad (z \in \mathbb{D}).$$

By direct verification,

$$\int_{\mathbb{T}} \log |h_n(\zeta)| \, dm(\zeta) = 0.$$

Hence,

$$\int_{\mathbb{T}} \big| \log |h_n(\zeta)| \big| \, dm(\zeta) = 2 \int_{\mathbb{T}} \log^- |h_n(\zeta)| \, dm(\zeta).$$

Since

$$|h_n(e^{i\theta})| = \left| \frac{\sin(\frac{1}{2}(\theta - 1/n^2))}{\sin(\frac{1}{2}\theta)} \right|,$$

the inequality $|h_n| \leq 1$ happens on the interval $[1/(2n^2),\ \pi + 1/(2n^2)]$. Thus,

$$\begin{aligned}
\int_{\mathbb{T}} \log^- |h_n(\zeta)| \, dm(\zeta) &= -\frac{1}{2\pi} \int_{1/(2n^2)}^{\pi + 1/(2n^2)} \log \left| \frac{\sin(\frac{1}{2}(\theta - 1/n^2))}{\sin(\frac{1}{2}\theta)} \right| d\theta \\
&= -\frac{1}{\pi} \int_{-1/(4n^2)}^{\pi/2 - 1/(4n^2)} \log |\sin \vartheta| \, d\vartheta \\
&\quad + \frac{1}{\pi} \int_{1/(4n^2)}^{\pi/2 + 1/(4n^2)} \log |\sin \vartheta| \, d\vartheta \\
&= -\frac{2}{\pi} \int_{0}^{1/(4n^2)} \log |\tan \vartheta| \, d\vartheta.
\end{aligned}$$

Integration by parts reveals that

$$\int_{\mathbb{T}} \log^- |h_n(\zeta)| \, dm(\zeta) \asymp \frac{\log n}{n^2} \qquad (n \geq 2).$$

Therefore, the sum

$$\varphi = \sum_{n \geq 1} \log |h_n|$$

is convergent in $L^1(\mathbb{T})$. This allows us to write

$$\frac{h(z)}{h(0)} = \prod_{n \geq 1} \frac{1 - e^{-i/n^2} z}{1 - z} = \exp\left\{ \int_{\mathbb{T}} \frac{\zeta + z}{\zeta - z} \varphi(\zeta) \, dm(\zeta) \right\} \qquad (z \in \mathbb{D}),$$

which means that h is an outer function.

Now, define

$$F(z) = \prod_{n \geq 1} \frac{(z - e^{i/n^2})(1 - e^{-i/n^2} z)}{(z - 1)(1 - z)} = e^{-i\pi^2/6} \left(\prod_{n \geq 1} \frac{z - e^{i/n^2}}{z - 1} \right)^2. \tag{6.21}$$

Hence, F is an analytic function on $\mathbb{C} \setminus \{1\}$, outer on $\mathbb{D}$, with zeros of order two at e^{i/n^2}, $n \geq 1$. Moreover, F enjoys one further interesting property. For each $e^{i\theta} \in \mathbb{T}$, we have

$$\frac{(e^{i\theta} - e^{i/n^2})(1 - e^{-i/n^2} e^{i\theta})}{(e^{i\theta} - 1)(1 - e^{i\theta})} = \left| \frac{1 - e^{i(\theta - 1/n^2)}}{1 - e^{i\theta}} \right|^2 \geq 0.$$

Hence,

$$F(e^{i\theta}) \geq 0 \qquad (e^{i\theta} \in \mathbb{T} \setminus \{1\}). \tag{6.22}$$

If the function F, given by (6.21), had been in H^1, we would have obtained a nonexposed outer function in H^1. In fact, (6.22) shows that $\arg F$ is the same as the argument of a constant function, and thus $F/\|F\|_1$ is not rigid. But this is not the case. According to Theorem 4.29, F cannot even be in $H^{1/2}$. To overcome this difficulty, we construct an outer function G such that G and FG are both in H^1.

Consider a small arc centered at e^{i/n^2}, say $I_n = (e^{1/n^2 - \varepsilon_n}, e^{1/n^2 + \varepsilon_n})$, such that the arcs are disjoint and

$$|F(e^{i\theta})| \leq 1 \qquad (e^{i\theta} \in I_n).$$

Then define

$$\phi(e^{i\theta}) = \begin{cases} \min\left\{1, \dfrac{1}{|F(e^{i\theta})|}\right\} & \text{if} \quad e^{i\theta} \in \mathbb{T} \setminus \bigcup_{n \geq 1} I_n, \\ \dfrac{1}{(n^4 \varepsilon_n)} & \text{if} \quad e^{i\theta} \in I_n. \end{cases}$$

First, we have

$$\begin{aligned} \int_{\mathbb{T}} \phi(e^{i\theta})\, d\theta &= \int_{\mathbb{T} \setminus \bigcup_{n \geq 1} I_n} \phi(e^{i\theta})\, d\theta + \sum_{n \geq 1} \int_{I_n} \phi(e^{i\theta})\, d\theta \\ &\leq \int_{\mathbb{T} \setminus \bigcup_{n \geq 1} I_n} d\theta + \sum_{n \geq 1} \frac{2}{n^4} < \infty. \end{aligned} \tag{6.23}$$

Second, since $\varepsilon_n < 1/n^2$,

$$\begin{aligned} \int_{\mathbb{T}} |\log \phi(e^{i\theta})|\, d\theta &= \int_{\mathbb{T} \setminus \bigcup_{n \geq 1} I_n} |\log \phi(e^{i\theta})|\, d\theta + \sum_{n \geq 1} \int_{I_n} |\log \phi(e^{i\theta})|\, d\theta \\ &\leq \int_{\mathbb{T}} \big|\log |F(e^{i\theta})|\big|\, d\theta + \sum_{n \geq 1} 2\varepsilon_n\, |\log(1/(n^4 \varepsilon_n))| \\ &\leq \|\log |F|\,\|_1 + 2 \sum_{n \geq 1} \varepsilon_n \log(1/\varepsilon_n) + 2 \sum_{n \geq 1} 4\varepsilon_n \log n \\ &\leq \|\log |F|\,\|_1 + 2 \sum_{n \geq 1} (2 \log n)/n^2 + 2 \sum_{n \geq 1} (4 \log n)/n^2 \\ &< \infty. \end{aligned}$$

Hence, we can construct the unique outer function G such that $|G| = \phi$, almost everywhere on $\mathbb{T}$, and $G(0) > 0$, i.e.

$$G(z) = \exp\left\{\int_{\mathbb{T}} \frac{\zeta + z}{\zeta - z} \log \phi(\zeta)\, dm(\zeta)\right\} \qquad (z \in \mathbb{D}). \tag{6.24}$$

By (6.23), we have $G \in H^1$. Moreover, the construction of G is such that

$$|F(e^{i\theta})G(e^{i\theta})| \leq \begin{cases} 1 & \text{if} \quad e^{i\theta} \in \mathbb{T} \setminus \bigcup_{n\geq 1} I_n, \\ 1/(n^4\varepsilon_n) & \text{if} \quad e^{i\theta} \in I_n, \end{cases}$$

which, as in (6.23), shows that

$$\int_{\mathbb{T}} |F(e^{i\theta})G(e^{i\theta})|\, d\theta \leq \int_{\mathbb{T}\setminus\bigcup_{n\geq 1} I_n} d\theta + \sum_{n\geq 1} 2/n^4 < \infty.$$

Therefore, the outer function FG is also in H^1.

By (6.22), the normalized outer function $FG/\|FG\|_1$ has the same argument as the normalized outer function $G/\|G\|_1$ at almost all points of $\mathbb{T}$. Therefore, $G/\|G\|_1$ is a function in the unit ball of H^1, which is not an exposed point. It remains to prove that $G/(1+\Theta)^2$ does not belong to H^1 for any nonconstant inner function Θ. We first show that G has the following property: for each δ, $0 < \delta < 2$, we have

$$\inf\{|G(z)| : z \in \mathbb{D},\ |z - 1| \geq \delta\} > 0. \tag{6.25}$$

Choose a continuous function φ on $\mathbb{T} \setminus \{1\}$ such that

$$0 \leq \varphi(e^{i\theta}) \leq \phi(e^{i\theta}),$$

with $\log \varphi \in L^1(\mathbb{T})$. Let H be the harmonic extension of $\log \varphi$ to $\bar{\mathbb{D}} \setminus \{1\}$. For each $z \in \mathbb{D}$, we have

$$\begin{aligned} \log|G(z)| &= \int_{\mathbb{T}} \frac{1 - |z|^2}{|\zeta - z|^2} \log \phi(\zeta)\, dm(\zeta) \\ &\geq \int_{\mathbb{T}} \frac{1 - |z|^2}{|\zeta - z|^2} \log \phi(\zeta)\, dm(\zeta) = H(z). \end{aligned}$$

Hence,

$$\inf\{|G(z)| : z \in \mathbb{D},\ |z - 1| \geq \delta\} \geq \inf\{e^{H(z)} : z \in \mathbb{D},\ |z - 1| \geq \delta\},$$

and the right infimum is strictly positive because $z \longmapsto e^{H(z)}$ is continuous on the compact set $\{z \in \bar{\mathbb{D}} : |z - 1| \geq \delta\}$. We now explain how to get the desired property for G from (6.25). First, let us check that

$$z \longmapsto \frac{G(z)}{(1 - z)^2} \notin H^1.$$

We have

$$
\begin{aligned}
\int_{\mathbb{T}} \frac{|G(e^{i\theta})|}{|1-e^{i\theta}|^2}\, d\theta &\geq \sum_{n\geq 1} \int_{I_n} \frac{|G(e^{i\theta})|}{|1-e^{i\theta}|^2}\, d\theta \\
&= \sum_{n\geq 1} \int_{I_n} \frac{\phi(e^{i\theta})}{|1-e^{i\theta}|^2}\, d\theta \\
&= \sum_{n\geq 1} \frac{1}{n^4\varepsilon_n} \int_{1/n^2-\varepsilon_n}^{1/n^2+\varepsilon_n} \frac{d\theta}{4\sin^2(\theta/2)}\, d\theta \\
&\geq \sum_{n\geq 1} \frac{1}{n^4} \frac{1}{2\sin^2(1/n^2)} \\
&= \frac{1}{2} \sum_{n\geq 1} \left(\frac{n^{-2}}{\sin(n^{-2})} \right)^2 = \infty,
\end{aligned}
$$

and hence $G(z)/(1-z)^2 \notin H^1$. Now, we consider the general case. Suppose that $G/(1-\Theta)^2 \in H^1$ for some nonconstant inner function Θ. The above argument shows that we can, of course, assume that $\Theta(z) \neq z$. Using the fact that $|\Theta| = 1$ a.e. on $\mathbb{T}$, we easily see that $-\Theta/(1-\Theta)^2$ is nonnegative a.e. on $\mathbb{T}$. Moreover, writing

$$
\frac{\Theta}{(1-\Theta)^2} = \frac{G\Theta}{(1-\Theta)^2} \frac{1}{G}
$$

and using (6.25), we apply Corollary 4.23 to get that $\Theta/(1-\Theta)^2$ can be extended analytically beyond every point of $\mathbb{T} \setminus \{1\}$. Therefore, Θ cannot have any zeros accumulating at any point of $T \setminus \{1\}$ (uniqueness principle). For the same reason, the measure of its singular part cannot have any positive mass on $T \setminus \{1\}$. Since, if so, at least at one point of $T \setminus \{1\}$ all derivatives of Θ would go radially to zero. Hence, $\sigma(\Theta) \subset \{1\}$. We claim that the equation $\Theta(\zeta) = 1$ has some solutions on $\mathbb{T} \setminus \{1\}$.

If Θ is a finite Blaschke product, then the result is obvious (remember that $\Theta(z) \neq z$). If not, Θ is a combination of the singular inner function

$$
\Theta_1(z) = \exp\left(-a\frac{1+z}{1-z} \right) \qquad (a > 0)
$$

and an infinite Blaschke product whose zeros converge to 1. Of course, one of these two factors may be missing, but not both. In any case, the argument of Θ is a $\mathcal{C}^\infty$ function on $T \setminus \{1\}$ and increases when we move from 0^+ to $2\pi-$. More importantly, it is an unbounded function, owing to the presence of infinitely many zeros, or the singular factor, or both. In fact, each zero of the Blaschke product produces a change of 2π in the argument when we turn once around $\mathbb{T}$. Hence, infinitely many zeros produces an unbounded argument. For the singular part Θ_1, a direct verification reveals the same behavior. Hence, the equation $\Theta(z) = 1$ has in fact infinitely many solutions on $\mathbb{T} \setminus \{1\}$ in this situation. That implies that $G/(1-\Theta)^2 \notin H^1$, contradicting the assumption above. $\square$

Exercises

Exercise 6.8.1 Let Θ be a nonconstant inner function. Show that the function $(1+\Theta)^2$ is an outer function that is not rigid. In particular, dividing by its H^1 norm, it gives an example of an extreme point of the closed unit ball of H^1 that is not an exposed point.
Hint: Use (6.17) and Theorem 6.8.

Exercise 6.8.2 Give an example of an outer function f that is rigid, but f^2 is not rigid.
Hint: Use Exercise 6.8.1 and Theorem 6.23.

Exercise 6.8.3 Let $f(z) = (1-z)^\alpha$, $\alpha > 1$. Show that $1/f \in H^s$ for all $s < 1/\alpha$, but f is not rigid.
Hint: Use Lemma 6.19 with $\Theta(z) = -z$.

Exercise 6.8.4 Let p be an analytic polynomial. Then p is a rigid function in H^1 if and only if p has no zeros on $\mathbb{D}$ and its zeros on $\mathbb{T}$ (if any) are simple.
Hint: Find a smooth function φ such that the argument of $\varphi - p$ is in $(-\pi/2, \pi/2)$. Then consider $p\exp(\tilde{\varphi} - i\varphi)$ and apply Theorem 6.23. For the other part, if p has a zero $z_0 \in \mathbb{T}$ of multiplicity $m \geq 2$, then the function $p/(1-\bar{z}_0 z)^2$ belongs to H^1. Then consider Lemma 6.19.

Exercise 6.8.5 The aim of this exercise is to give another proof of Theorem 6.23. Let $f \in H^1$, $f \not\equiv 0$, and suppose that there exists $h \in H^\infty$, $h \not\equiv 0$, such that $\Re(fh) \geq 0$ (a.e. on $\mathbb{T}$). Let $g \in H^1$ have the same argument as f a.e. on $\mathbb{T}$. Multiplying f and g by suitable constants, we may assume that $\|hf\|_1 = \|hg\|_1 = 1$.

(i) Show that $(hf + hg)/2$ is an outer function of unit norm in H^1.
(ii) Conclude that $f = g$.

Hint: Note that $\Re(hg) \geq 0$ and thus $\Re((hf+hg)/2) \geq 0$ on $\mathbb{T}$. As in the proof of Theorem 6.23, we have $\Re((hf+hg)/2) > 0$ on $\mathbb{D}$. Hence, $(fh+hg)/2$ is outer. Observe that, since $\arg f = \arg g$, then

$$\frac{|fh+hg|}{2} = \frac{|fh|}{2} + \frac{|gh|}{2}.$$

Thus,

$$\left\| \frac{fh+hg}{2} \right\|_1 = \frac{\|fh\|_1}{2} + \frac{\|hg\|_1}{2} = \frac{1}{2} + \frac{1}{2} = 1.$$

Therefore, $(fh + hg)/2$ is an extreme function of the unit ball of H^1. This implies that $fh = gh$ and thus $f = g$.

6.9 Strongly exposed points of H^1

In this section, we describe the strongly exposed points of the closed unit ball of H^1. First, we show that there are exposed points that are not strongly exposed. Let $h(z) = z + 1$, and put $f = h/\|h\|_1$. Then, by Theorem 6.23, f is an exposed point of the closed unit ball of H^1. Now, we can show that f is not a strongly exposed point. Let

$$f_n(z) = -c_n(z+1)^{1+1/n} \left(\frac{z-1}{z+1} \right)^2,$$

where $c_n > 0$ is chosen such that $\|f_n\|_1 = 1$. Since

$$-\left(\frac{z-1}{z+1} \right)^2 \geq 0 \qquad (z \in \mathbb{T} \setminus \{-1\}),$$

we have

$$\arg f_n = (1 + 1/n) \arg f \qquad (z \in \mathbb{T} \setminus \{1, -1\}).$$

Thus,

$$\int_{\mathbb{T}} f_n \frac{\bar{f}}{|f|} \, dm = \int_{\mathbb{T}} |f_n| \, e^{(i/n) \arg f} \, dm \longrightarrow 1,$$

as $n \longrightarrow \infty$. On the other hand, the functions f_n are tending pointwise to 0 in $\mathbb{D}$, because the constants c_n are decreasing to 0 in order to keep f_n in the closed unit ball of H^1. Therefore, $\|f_n - f\|_1 \not\longrightarrow 0$, as $n \longrightarrow \infty$. This proves that f is not a strongly exposed point of the closed unit ball of H^1.

The following theorem provides a rather simple necessary condition for being a strongly exposed point in H^1.

Theorem 6.27 *Let f be a function in H^1, $\|f\|_1 = 1$. Assume that f is a strongly exposed point of the closed unit ball of H^1. Then*

$$\operatorname{dist}\left(\frac{\bar{f}}{|f|}, H^\infty \right) < 1.$$

Proof Let $\varphi = \bar{f}/|f|$, and assume that $\mathrm{dist}(\varphi, H^\infty) = 1$. Consider $\Lambda' = \Lambda|H_0^1$, where Λ is the (unique) exposing functional for f given by

$$\Lambda(g) = \int_{\mathbb{T}} g\varphi\, dm \qquad (g \in H^1).$$

According to (1.8), we have

$$\|\Lambda'\| = \mathrm{dist}(\Lambda, (H_0^1)^\perp),$$

and then (4.33) implies that

$$\|\Lambda'\| = \mathrm{dist}(\varphi, H^\infty) = 1.$$

Hence, there is a sequence $(f_n)_{n\geq 1}$ in H_0^1, $\|f_n\|_1 = 1$, such that $\Lambda(f_n) \longrightarrow 1$, as $n \longrightarrow \infty$. But the sequence $(f_n)_{n\geq 1}$ cannot converge to f in norm, because norm convergence implies the pointwise convergence at $z = 0$ and $f_n(0) = 0$, whereas $f(0) \neq 0$ (remember that, by Theorem 6.8, f is outer). Thus, f is not strongly exposed, which is a contradiction. Therefore, $\mathrm{dist}(\varphi, H^\infty) < 1$. □

The following result gives a sufficient condition for being a strongly exposed point of the closed unit ball of H^1. The characterization is in terms of the distance to $H^\infty + \mathcal{C}(\mathbb{T})$. We recall that, in Section 5.3, we saw that the space $H^\infty + \mathcal{C}(\mathbb{T})$ is a closed subalgebra of $L^\infty(\mathbb{T})$ (see Theorem 5.10).

Theorem 6.28 *Let f be an exposed point of the closed unit ball of H^1 such that*

$$\mathrm{dist}\left(\frac{\bar{f}}{|f|},\ H^\infty + \mathcal{C}(\mathbb{T})\right) < 1.$$

Then f is a strongly exposed point of the closed unit ball of H^1.

Proof Put $h = \bar{f}/|f|$. According to Lemma 6.13, the (unique) exposing functional for f is given by

$$\Lambda_h(g) = \int_{\mathbb{T}} gh\, dm \qquad (g \in H^1).$$

We must show that, if $(f_n)_{n\geq 1}$ is a sequence in the closed unit ball of H^1 such that $\Lambda_h(f_n) \longrightarrow 1$, then $\|f_n - f\|_1 \longrightarrow 0$, as $n \longrightarrow \infty$.

We claim that $(f_n)_{n\geq 1}$ is sequentially weakly compact, that is, every subsequence of $(f_n)_{n\geq 1}$ has a weakly convergent subsequence. Assuming this for a moment, consider a subsequence $(f_k)_{k\geq 1}$ converging weakly to some $F \in H^1$. Thus, $\|F\|_1 \leq 1$ and

$$\int_{\mathbb{T}} Fh\, dm = \lim_{k\to\infty} \int_{\mathbb{T}} f_k h\, dm = 1.$$

Since f is an exposed point in the unit ball of H^1, that gives $F = f$. Thus, every subsequence of $(f_n)_{n\geq 1}$ has a subsequence that converges weakly to f. Then we conclude that $(f_n)_{n\geq 1}$ converges weakly to f (see Exercise 1.7.9). In particular, f_n tends to f uniformly on compact subsets of $\mathbb{D}$. Furthermore, since

$$|\Lambda_h(f_n)| \leq \|f_n\|_1 \leq 1,$$

we deduce that $\|f_n\|_1 \longrightarrow 1 = \|f\|_1$ as $n \longrightarrow \infty$. It follows from Theorem 4.34 that $\|f_n - f\|_1 \longrightarrow 0$ as $n \longrightarrow \infty$. This proves that f is a strongly exposed point of the closed unit ball of H^1.

Now, assume that the claim does not hold, i.e. $(f_n)_{n\geq 1}$ is not sequentially weakly compact. According to Theorem 1.29, we can find subsets $E_j \subset \mathbb{T}$ such that $|E_j| \longrightarrow 0$, and a subsequence $(f_{\varphi(j)})_{j\geq 1}$ of $(f_j)_{j\geq 1}$ satisfying

$$\left| \int_{E_j} f_{\varphi(j)}\, dm \right| \geq \beta > 0. \tag{6.26}$$

Applying Theorem 4.39, we construct a sequence $(g_j)_j$ of functions in H^∞ and $\varepsilon_j \longrightarrow 0$ such that

(i) $\sup_{E_j} |g_j| \longrightarrow 0,\ j \longrightarrow \infty$,
(ii) $g_j(0) \longrightarrow 1,\ j \longrightarrow \infty$,
(iii) $|g_j| + |1 - g_j| \leq 1 + \varepsilon_j\ (j \geq 1)$.

For $j \geq 1$, we now define by induction a sequence of H^1 functions by setting $H_{0,j} = f_{\varphi(j)}$ and

$$H_{n,j} = \frac{(1 - g_j)H_{n-1,j}}{(1 + \varepsilon_j)\|H_{n-1,j}\|_1} \qquad (n \geq 1).$$

Then, we define

$$G_{n,j} = \frac{g_j H_{n-1,j}}{(1 + \varepsilon_j)\|H_{n-1,j}\|_1} \qquad (n, j \geq 1).$$

We now show that, for every $n \geq 1$, the following properties hold:

(a) $\|G_{n,j}\|_1 + \|H_{n,j}\|_1 \leq 1$;
(b) for j large enough, $\|H_{n,j}\|_1 \geq \beta/2$;
(c) $$\left| \|G_{n,j}\|_1 \int_{\mathbb{T}} h \frac{G_{n,j}}{\|G_{n,j}\|_1}\, dm + \|H_{n,j}\|_1 \int_{\mathbb{T}} h \frac{H_{n,j}}{\|H_{n,j}\|_1}\, dm \right| \longrightarrow 1,$$
as $j \longrightarrow \infty$;
(d) $$\left| \int_{\mathbb{T}} h \frac{H_{n,j}}{\|H_{n,j}\|_1}\, dm \right| \longrightarrow 1, \text{ as } j \longrightarrow +\infty;$$
(e) for $0 \leq \ell < n$, we have $\lim_{j\to+\infty} H_{n,j}^{(\ell)}(0) = 0$.

To begin, by (iii), we have

$$\|G_{n,j}\|_1 + \|H_{n,j}\|_1 \leq \frac{\|1-g_j\|_\infty}{1+\varepsilon_j} + \frac{\|g_j\|_\infty}{1+\varepsilon_j} \leq \frac{1+\varepsilon_j}{1+\varepsilon_j} = 1,$$

which proves (a). We proceed to check the properties (b)–(e) for $n = 1$. We first have

$$\|H_{1,j}\|_1 = \left\| \frac{(1-g_j)f_{\varphi(j)}}{(1+\varepsilon_j)\|f_{\varphi(j)}\|_1} \right\|_1 \geq \int_{E_j} \left| \frac{(1-g_j)f_{\varphi(j)}}{1+\varepsilon_j} \right| dm.$$

By (i), the sequence of functions $(1-g_j)/(1+\varepsilon_j)$ tends to 1 uniformly on E_j. Hence, for sufficiently large j, we have

$$\left| \frac{1-g_j}{1+\varepsilon_j} \right| \geq \frac{1}{2} \qquad (\text{on } E_j),$$

which gives by (6.26)

$$\|H_{1,j}\|_1 \geq \frac{1}{2} \int_{E_j} |f_{\varphi(j)}|\, dm \geq \frac{1}{2} \left| \int_{E_j} f_{\varphi(j)}\, dm \right| \geq \frac{\beta}{2}.$$

That proves (b) for $n = 1$. To prove (c), note that

$$\begin{aligned} \int_{\mathbb{T}} hG_{1,j}\, dm + \int_{\mathbb{T}} hH_{1,j}\, dm &= \int_{\mathbb{T}} h \frac{f_{\varphi(j)}}{(1+\varepsilon_j)\|f_{\varphi(j)}\|_1}\, dm \\ &= \frac{\Lambda_h(f_{\varphi(j)})}{(1+\varepsilon_j)\|f_{\varphi(j)}\|_1}, \end{aligned}$$

and the latter tends to 1, as $j \longrightarrow \infty$, by hypothesis. To prove (d) for $n = 1$, let

$$\mathfrak{h} = \liminf_{j\to\infty} \int_{\mathbb{T}} h \frac{H_{1,j}}{\|H_{1,j}\|_1}\, dm.$$

Since

$$\left| \int_{\mathbb{T}} h \frac{H_{1,j}}{\|H_{1,j}\|_1}\, dm \right| \leq 1,$$

it is sufficient to prove that $|\mathfrak{h}| = 1$. Assume, on the contrary, that $|\mathfrak{h}| < 1$ and take a subsequence $\theta(j)$ such that

$$\int_{\mathbb{T}} h \frac{H_{1,\theta(j)}}{\|H_{1,\theta(j)}\|_1}\, dm \longrightarrow \mathfrak{h} \qquad (j \longrightarrow \infty).$$

Taking a subsequence if necessary, we can also assume that

$$\|G_{1,\theta(j)}\|_1 \longrightarrow \alpha \in \mathbb{R}, \quad \|H_{1,\theta(j)}\|_1 \longrightarrow \gamma \in \mathbb{R} \qquad (j \longrightarrow \infty)$$

and

$$\int_{\mathbb{T}} h \frac{G_{1,\theta(j)}}{\|G_{1,\theta(j)}\|_1}\, dm \longrightarrow \mathfrak{h}_1 \in \mathbb{C} \qquad (j \longrightarrow \infty).$$

Note that $|\mathfrak{h}_1| \leq 1$. By (c), we have

$$1 = |\alpha\mathfrak{h}_1 + \gamma\mathfrak{h}| \leq \alpha|\mathfrak{h}_1| + \gamma|\mathfrak{h}|,$$

and since $\gamma \neq 0$ by (b), we get $1 < \alpha + \gamma$. But, by (a), $\alpha + \gamma \leq 1$, which gives a contradiction. Hence, $|\mathfrak{h}| = 1$, which implies (d). To prove (e) for $n = 1$, we should check that $\lim_{j\to\infty} H_{1,j}(0) = 0$. But

$$H_{1,j}(0) = \frac{(1 - g_j(0))f_{\varphi(j)}(0)}{1 + \varepsilon_j}$$

and $1 - g_j(0) \longrightarrow 0$, $j \longrightarrow \infty$ by (ii). It is thus sufficient to note that $(f_{\varphi(j)}(0))_j$ is bounded, which is a consequence of Cauchy's formula.

Let us now assume that properties (b)–(e) are fulfilled for all ℓ, $1 \leq \ell \leq n$. We prove that they are also satisfied for $n+1$. From (a), we have $\|H_{\ell,j}\|_1 \leq 1$, and since

$$H_{n+1,j} = \frac{(1 - g_j)^{n+1} f_{\varphi(j)}}{(1 + \varepsilon_j)^{n+1} \prod_{\ell=1}^{n} \|H_{\ell,j}\|_1},$$

we get

$$\begin{aligned} \|H_{n+1,j}\|_1 &\geq \frac{1}{(1 + \varepsilon_j)^{n+1}} \|(1 - g_j)^{n+1} f_{\varphi(j)}\|_1 \\ &\geq \frac{1}{(1 + \varepsilon_j)^{n+1}} \int_{E_j} |(1 - g_j)^{n+1} f_{\varphi(j)}| \, dm. \end{aligned}$$

By (i), for j large enough, we have

$$\frac{|1 - g_j|^{n+1}}{(1 + \varepsilon_j)^{n+1}} \geq \frac{1}{2} \qquad \text{(uniformly on } E_j).$$

Hence, by (6.26)

$$\|H_{n+1,j}\|_1 \geq \frac{1}{2} \int_{E_j} |f_{\varphi(j)}| \, dm \geq \frac{\beta}{2},$$

for j large enough. That proves (b). Property (c) follows immediately from the induction hypothesis, because

$$\begin{aligned} &\|G_{n+1,j}\|_1 \int_{\mathbb{T}} h \frac{G_{n+1,j}}{\|G_{n+1,j}\|_1} \, dm + \|H_{n+1,j}\|_1 \int_{\mathbb{T}} h \frac{H_{n+1,j}}{\|H_{n+1,j}\|_1} \, dm \\ &= \int_{\mathbb{T}} h(G_{n+1,j} + H_{n+1,j}) \, dm \\ &= \int_{\mathbb{T}} h \frac{H_{n,j}}{\|H_{n,j}\|_1} \, dm. \end{aligned}$$

Property (d) follows again as above from (a), (b) and (c). It remains to check (e). For $0 \le i < n+1$, we have

$$H_{n+1,j}^{(i)}(0) = \frac{1}{(1+\varepsilon_j)\|H_{n,j}\|_1} \sum_{\ell=0}^{i} \binom{i}{\ell} (1-g_j)^{(\ell)}(0) H_{n,j}^{(i-\ell)}(0). \tag{6.27}$$

If $i \le n-1$, by induction, we have $H_{n,j}^{(i-\ell)}(0) \longrightarrow 0$, for each $0 \le \ell \le i$, as $j \longrightarrow \infty$. Furthermore, $(1-g_j)^{(\ell)}(0)$ is bounded as $j \longrightarrow \infty$, because

$$(1-g_j)^{(\ell)}(0) = \frac{\ell!}{2\pi} \int_0^{2\pi} \frac{(1-g_j)(e^{it})}{e^{i\ell t}}\, dt,$$

which gives

$$|(1-g_j)^{(\ell)}(0)| \le \frac{\ell!}{2\pi}(1+\varepsilon_j).$$

We deduce that $H_{n+1,j}^{(i)}(0) \longrightarrow 0$ as $j \longrightarrow \infty$. For $i = n$ and $\ell > 0$, then all the terms in the sum (6.27) tend to 0 for the same reason. For $i = n$ and $\ell = 0$, then $(1-g_j)(0) \longrightarrow 0$ by (ii) and $H_{n,j}^{(i)}(0)$ is bounded again by Cauchy's formula. So all the terms tend to 0 and we have (e).

Now taking $H_n = H_{n,k_n}/\|H_{n,k_n}\|_1$ with k_n large enough, we obtain a sequence in the unit ball of H^1 that has the properties

$$\left| \int_{\mathbb{T}} h H_n\, dm \right| \longrightarrow 1 \qquad (n \longrightarrow \infty), \tag{6.28}$$

and, for all $m \ge 0$,

$$\int_0^{2\pi} H_n(e^{it}) e^{-imt} \frac{dt}{2\pi} \longrightarrow 0 \qquad (n \longrightarrow \infty). \tag{6.29}$$

Let us check that, for any $g \in \mathcal{C}(\mathbb{T}) + H^\infty$, we have

$$\int_{\mathbb{T}} H_n g\, dm \longrightarrow 0 \qquad (n \longrightarrow \infty). \tag{6.30}$$

Write $g = g_0 + h$, with $g_0 \in \mathcal{C}(\mathbb{T})$ and $h \in H_0^\infty$. Since continuous functions can be uniformly approximated by trigonometric polynomials on $\mathbb{T}$, for any $\varepsilon > 0$, we can find $p_0(e^{it}) = \sum_{k=-N}^{N} a_k e^{ikt}$, such that $\|p_0 - g_0\|_\infty \le \varepsilon$. Writing

$$p(e^{it}) = \sum_{k=-N}^{0} a_k e^{ikt},$$

we have $p - p_0 \in H_0^\infty$, and since $(H^1)^\perp = H_0^\infty$, we have

$$\int_{\mathbb{T}} H_n(p-p_0)\, dm = 0 \quad \text{and} \quad \int_{\mathbb{T}} H_n h\, dm = 0.$$

Thus

$$\begin{aligned}\left|\int_{\mathbb{T}} H_n g\,dm\right| &= \left|\int_{\mathbb{T}} H_n g_0\,dm\right| \\ &= \left|\int_{\mathbb{T}} H_n(g_0 - p_0)\,dm + \int_{\mathbb{T}} H_n p_0\,dm\right| \\ &\le \varepsilon + \left|\int_{\mathbb{T}} H_n p_0\,dm\right| \\ &= \varepsilon + \left|\int_{\mathbb{T}} H_n p\,dm\right|.\end{aligned}$$

By (6.29), for n sufficiently large, we have $|\int_{\mathbb{T}} H_n p\,dm| \le \varepsilon$, which gives

$$\left|\int_{\mathbb{T}} H_n g\,dm\right| \le 2\varepsilon,$$

and proves (6.30).

Now, by assumption, there is $g \in \mathcal{C}(\mathbb{T}) + H^\infty$ with $\|h - g\|_\infty < 1$. Using (6.28) and (6.30), we have

$$\begin{aligned}1 &= \lim_{n\to\infty}\left|\int_{\mathbb{T}} hH_n\,dm\right| \\ &= \lim_{n\to\infty}\left|\int_{\mathbb{T}} H_n(h-g)\,dm\right| \\ &\le \limsup_{n\to\infty} \|H_n\|_1 \|h-g\|_\infty < 1,\end{aligned}$$

which is a contradiction. That proves the claim and so the proof of Theorem 6.28 is complete. □

If an exposed point f satisfies $\operatorname{dist}(\bar{f}/|f|,\ H^\infty + \mathcal{C}(\mathbb{T})) < 1$, then, by Theorem 6.28, it is a strongly exposed point and hence, by Theorem 6.27, it also satisfies $\operatorname{dist}(\bar{f}/|f|,\ H^\infty) < 1$. However, the inverse is not true (see Exercise 11.2.2).

Corollary 6.29 *Let f be a function in H^1, $\|f\|_1 = 1$. The following assertions are equivalent:*

(i) *f is strongly exposed;*
(ii) *f is an exposed point of H^1 and*

$$\operatorname{dist}\left(\frac{\bar{f}}{|f|},\ H^\infty + \mathcal{C}(\mathbb{T})\right) < 1.$$

In Chapter 29, we will use this result to give a characterization of strongly exposed points in terms of Helson–Szegő weights. We know that, if f is a function in H^1 of unit norm such that $1/f$ is also in H^1, then f is an exposed point of the closed unit ball of H^1 (see Theorem 6.22). Strengthening the hypotheses a little bit, we can say more.

Corollary 6.30 *Let f be a function in H^∞ such that $1/f$ also belongs to H^∞. Assume that $\|f\|_1 = 1$. Then f is a strongly exposed point of the closed unit ball of H^1.*

Proof Let δ and M be two positive constants such that $\delta < \|f\|_\infty < M$. Then we have

$$\left\| \frac{\bar{f}}{|f|} - \frac{\delta}{f} \right\|_\infty = \left\| 1 - \frac{\delta}{|f|} \right\|_\infty \leq 1 - \frac{\delta}{M} < 1.$$

Moreover, since $1/f$ belongs to $H^\infty \subset H^1$, by Theorem 6.22, f is an exposed point of the closed unit ball of H^1. Finally, since

$$\operatorname{dist}_{L^\infty(\mathbb{T})}\left(\frac{\bar{f}}{|f|},\ H^\infty + \mathcal{C}(\mathbb{T}) \right) \leq \operatorname{dist}_{L^\infty(\mathbb{T})}\left(\frac{\bar{f}}{|f|},\ H^\infty \right) < 1,$$

Theorem 6.28 ensures that f is a strongly exposed point of the closed unit ball of H^1. □

We will see in Chapter 29 that, if f is a strongly exposed point of the closed unit ball of H^1, then $1/f$ should belong to H^1.

Exercises

Exercise 6.9.1 Let $f(z) = c(z-1)(\log(z-1))^2$, $z \in \mathbb{D}$, where $c > 0$ is chosen such that $\|f\|_1 = 1$. Show that the following assertions hold:

(i) $1/f \in H^1$;
(ii) f is an exposed point of the closed unit ball of H^1;
(iii) f is not a strongly exposed point of the closed unit ball of H^1.

Hint: Consider

$$f_n(z) = -c_n \left(\frac{z+1}{z-1} \right)^2 (f(z))^{1+1/n} \qquad (z \in \mathbb{D}),$$

where $c_n > 0$ is chosen such that $\|f_n\|_1 = 1$. Show that $f_n \longrightarrow 0$, as $n \longrightarrow \infty$, uniformly on compact subsets of $\mathbb{D}$ and

$$\int_{\mathbb{T}} f_n \frac{\bar{f}}{|f|}\, dm \longrightarrow 1,$$

as $n \longrightarrow \infty$.

Exercise 6.9.2 The aim of this exercise is to give, in a special case, a simpler proof of Theorem 6.28.

Let f be an exposed point of the closed unit ball of H^1 and let $\varphi = \bar{f}/|f|$. Assume that $\varphi \in H^\infty + C(\mathbb{T})$. Suppose that $(f_n)_{n\geq 1}$ is a sequence in the unit ball of H^1 such that $\Lambda_\varphi(f_n) \longrightarrow 1$ as $n \longrightarrow \infty$, where

$$\Lambda_\varphi(g) = \int_{\mathbb{T}} g\varphi \, dm, \qquad g \in H^1.$$

(i) Show that there is a subsequence $(f_{n_k})_{k\geq 1}$ and a Borel measure μ on $\mathbb{T}$, $\|\mu\| \leq 1$, such that

$$\int_{\mathbb{T}} g f_{n_k} \, dm \longrightarrow \int_{\mathbb{T}} g \, d\mu, \qquad g \in \mathcal{C}(\mathbb{T}).$$

(ii) Show that there exists $F \in H^1$, $\|F\|_1 \leq 1$, such that $d\mu = F \, dm$.
Hint: Apply the Riesz brothers' theorem (Theorem 4.4).

(iii) Show that

$$\lim_{n\to\infty} \Lambda_\varphi(f_{n_k}) = \Lambda_\varphi(F).$$

(iv) Conclude that $F = f$.

(v) Show that $\|f_{n_k} - f\|_1 \longrightarrow 0$ as $n \longrightarrow \infty$.

(vi) Conclude that f is a strongly exposed point of the unit ball of H^1.

Notes on Chapter 6

Section 6.1

The notion of extreme points, which goes back to Minkowski [362], emerged in the 1950s as an important tool in functional analysis, mainly due to the Kreĭn–Milman theorem [326]. This theorem states that, if $\mathcal{X}$ is a topological vector space on which $\mathcal{X}^*$ separates the points, and if K is a nonempty compact convex subset of $\mathcal{X}$, then K is the closed convex hull of its extreme points. See [441, theorem 3.23]. In particular, the set K has plenty of extreme points. However, the Kreĭn–Milman theorem is an "existence result" and it is of considerable interest to find the extreme points of a given convex set. Let us also mention that the notion of extreme points of a compact convex set plays a crucial role in the classification of the Banach spaces of continuous functions on a compact set. For a more detailed study of extreme points and applications, we refer the reader to [83, 209, 219, 299].

Section 6.3

Theorem 6.8 is due to de Leeuw and Rudin [169]. The extreme points of H^∞ (Theorem 6.7) were identified jointly by Arens, Buck, Carleson, Hoffman and

Royden, during the Princeton Conference on Functions of a Complex Variable, in September 1957.

Section 6.4

The notion of strict convexity is a standard notion in the theory of Banach spaces. We refer to [83, 209] for a more detailed account on this notion.

Section 6.5

The notion of exposed points goes back to Straszewicz [497]. We refer to the work of Klee [317] for more details concerning exposed points. The characterization of exposed points of the unit ball of the disk algebra given in Exercise 6.5.3 is due to Fisher [217]. Amar and Lederer [41], using the maximal ideal space of H^∞, extend Fisher's result to H^∞.

Section 6.6

The notion of exposed points and strongly exposed points became an important tool in functional analysis through a result of Lindenstrauss [337] and Troyanski [527], which says that a bounded closed convex subset of a reflexive space $\mathcal{X}$ is the closed convex hull of its strongly exposed points. See also the paper of Bourgain [114] for a geometric proof of the Lindenstrauss–Troyanski result in the context of weakly compact sets in Banach spaces and also a theorem of Phelps [411]. However, none of these results provide a characterization of exposed, or strongly exposed, points among the set of extreme points. As a general rule, any result that characterizes exposed points (or strongly exposed points) is of particular interest.

Section 6.7

Rigid functions arise in a problem in prediction theory, namely, the problem of characterizing the spectral densities of a completely nondeterministic Gaussian process. That connection, along with the characterization in terms of exposed points (Theorem 6.15) can be found in a paper of Bloomfield, Jewell and Hayashi [101]. See also [169]. Example 6.24 is taken from [101].

Section 6.8

As we know from Lemma 6.19, if f is a rigid function, then

$$f/(1+\Theta)^2 \notin H^1 \qquad \text{for all nonconstant inner functions } \Theta. \qquad (\star)$$

An outer function $f \in H^1$ is called *strongly outer* if f satisfies ($\star$) for all inner functions of the form $\Theta(z) = \lambda z$, where $\lambda \in \mathbb{T}$. In [369, 370], it was conjectured that strongly outer functions are rigid. This conjecture was disproved by Hayashi [270], who gave an example of a nonconstant inner function Θ for which $(1+\Theta)^2$ is strongly outer. A few years later, Sarason [455, 456] conjectured that condition ($\star$) in its full strength is equivalent to rigidity, i.e. that a function $f \in H^1$ is rigid if and only if, for no nonconstant inner function Θ, $f/(1+\Theta)^2 \in H^1$. However, this was again disproved by Inoue [296] by constructing the function G given in Theorem 6.26. But, to verify that the function h, defined by (6.20), is outer, he used another method by appealing to the Zygmund's $L \log L$ theorem. See [355] for a generalization of this example.

As already mentioned, as far as we know, there is no clear-cut function theoretic characterization of exposed points of the unit ball of H^1. The necessary and sufficient condition given in Theorem 6.20 probably comes closest, and is due to Helson [280]. De Leeuw and Rudin [169] found the following sufficient condition: if $f \in H^1$, $\|f\|_1 = 1$ and $1/f \in H^\infty$, then f is an exposed point of the unit ball of H^1. Yabuta [560] strengthened this by proving Theorem 6.22. The proof given in this text is due to Nakazi [370]. Theorem 6.23 is also due to Yabuta [561]. The proof presented in Exercise 6.8.5 is due to Nakazi [370].

Yabuta [559, 561] has shown that a polynomial without zeros in $\mathbb{D}$ and without multiple zeros on $\mathbb{T}$ is a rigid function. The proof of this fact presented in Exercise 6.8.4 is due to Inoue and Nakazi [297].

Section 6.9

As we saw, there exist many exposed points in the unit ball of the function algebras $\mathcal{A}(\mathbb{D})$ and H^∞. In fact, the exposed points are dense in the boundaries of the respective unit balls. For strongly exposed points, however, the situation is quite the opposite. Beneker and Wiegerinck [89] showed that there are no strongly exposed points in the unit ball of any infinite-dimensional function algebra in $\mathcal{C}(K)$. In particular, neither the unit ball of $\mathcal{A}(\mathbb{D})$ nor that of H^∞ have strongly exposed points. Theorem 6.27 is due to Beneker [88]. Theorem 6.28 and Corollary 6.29 are due to Temme and Wiegerinck [520, 553]. In [520], and based on the theory of uniform algebras, an alternative proof of Theorem 6.28 is given.

7

More advanced results in operator theory

This chapter contains some advanced topics on operator theory that are needed in the study of $\mathcal{H}(b)$ spaces. We start by developing a functional calculus for the self-adjoint operators. Then we use this theory to define the square root of a positive operator and explore its properties. For example, we obtain our first intertwining formula (Theorem 7.15), and establish the Douglas criterion, which treats the existence of a contraction connecting two given operators. Möbius transformations of operators and the Julia operator are topics that are needed in studying Hankel operators. Then we study the Wold–Kolmogorov decomposition of an isometry in detail. Partial isometries are also vital in the theory of $\mathcal{H}(b)$ spaces. As an application for partial isometries, we present the polar decomposition of an operator. We give a characterization of contractions on the sequence space ℓ^2. This result is needed in the treatment of Nehari's problem. Then we introduce densely defined operators and treat some of their main properties. We will exploit them in the theory of Toeplitz operators with L^2 symbols. Another fundamental concept is Fredholm operators. The remaining sections contain a brief introduction to this theory, but we just develop the results that are needed in future discussions. For example, the connection between Fredholm operators and the essential spectrum is thoroughly discussed.

7.1 The functional calculus for self-adjoint operators

As we have seen in Section 1.6 (in a more general context), if $A \in \mathcal{L}(\mathcal{H})$ and p is a complex polynomial, then the meaning of $p(A)$ is clear and, moreover, we have

$$\sigma(p(A)) = p(\sigma(A)). \tag{7.1}$$

In this section, for self-adjoint operators, we extend this polynomial functional calculus to continuous functions. The key point is the following lemma.

Lemma 7.1 *Let p be a complex polynomial and let A be a self-adjoint operator. Then*

$$\|p(A)\| = \sup_{t\in\sigma(A)} |p(t)|.$$

Proof It is clear that, since A is self-adjoint, then $p(A)$ is normal and, according to Theorem 2.17, $\|p(A)\| = r(p(A))$. But, using (7.1), we have

$$\begin{aligned} r(p(A)) &= \sup\{|\lambda| : \lambda \in \sigma(p(A))\} \\ &= \sup\{|\lambda| : \lambda \in p(\sigma(A))\} \\ &= \sup\{|p(t)| : t \in \sigma(A)\}, \end{aligned}$$

which gives the result. □

Recall that $\mathcal{C}(\sigma(A))$ denotes the space of continuous functions on $\sigma(A)$, equipped with the supremum norm. If A is a self-adjoint operator, then, according to Lemma 2.11, $\sigma(A)$ is a compact set of $\mathbb{R}$ and thus, by the Weierstrass approximation theorem, any continuous function on $\sigma(A)$ can be uniformly approximated by a sequence of complex polynomials. Therefore, by writing the identity in Lemma 7.1 as

$$\|p(A)\|_{\mathcal{H}} = \|p\|_{L^\infty(\sigma(A))},$$

we easily get the following result. We recall that $\mathcal{P}_+$ denotes the family of complex (analytic) polynomials.

Theorem 7.2 *Let A be a self-adjoint operator on a Hilbert space $\mathcal{H}$, and let Φ_A be the algebra homomorphism defined by*

$$\begin{array}{rccl} \Phi_A : & \mathcal{P}_+ & \longrightarrow & \mathcal{L}(\mathcal{H}) \\ & p & \longmapsto & p(A). \end{array}$$

Then Φ_A extends to an isometric $*$-homomorphism $\tilde{\Phi}_A : \mathcal{C}(\sigma(A)) \longrightarrow \mathcal{L}(\mathcal{H})$.

In the following, for each $f \in \mathcal{C}(\sigma(A))$, we naturally denote $\tilde{\Phi}_A(f)$ by $f(A)$. Based on Theorem 7.2, given $f \in \mathcal{C}(\sigma(A))$, the most important feature of $f(A)$ is the existence of a sequence of polynomials p_n such that

$$\|p_n(A) - f(A)\|_{\mathcal{H}} \longrightarrow 0 \tag{7.2}$$

and $\|p_n - f\|_{L^\infty(\sigma(A))} \longrightarrow 0$, as $n \longrightarrow \infty$. We also have

$$\|f(A)\|_{\mathcal{H}} = \|f\|_{L^\infty(\sigma(A))}.$$

Corollary 7.3 *Let A be a self-adjoint operator on a Hilbert space $\mathcal{H}$, and let $B \in \mathcal{L}(\mathcal{H})$ be such that $AB = BA$. Then*

$$f(A)B = Bf(A)$$

for any continuous function f on $\sigma(A)$.

Proof Since $AB = BA$, by induction, we obtain $A^n B = BA^n$, $n \geq 0$. Thus,

$$p(A)B = Bp(A)$$

for any analytic polynomial p. Now, let $f \in \mathcal{C}(\sigma(A))$. Then there is a sequence of polynomials $(p_n)_{n\geq 1}$ that fulfills (7.2), and thus $f(A)B = Bf(A)$. □

The spectral mapping theorem (Theorem 1.22) extends also to continuous functions.

Theorem 7.4 *Let A be a self-adjoint operator, and let $f \in \mathcal{C}(\sigma(A))$. Then*

$$\sigma\big(f(A)\big) = f\big(\sigma(A)\big).$$

Proof Let $\lambda \notin f(\sigma(A))$. Then $f - \lambda$ is a continuous function that does not vanish on $\sigma(A)$. We can thus consider the function $g = 1/(f - \lambda)$, which is a continuous function on $\sigma(A)$. Since $g(f - \lambda) = (f - \lambda)g = 1$, we have $g(A)(f(A) - \lambda I) = (f(A) - \lambda I)g(A) = I$. Hence, $\lambda \notin \sigma(f(A))$. We have thus proved that $\sigma(f(A)) \subset f(\sigma(A))$.

To establish the other inclusion, let $\lambda \in \sigma(A)$. Consider a sequence of complex polynomials $(p_n)_{n\geq 1}$ that tends to f uniformly on $\sigma(A)$. By Theorem 7.2, we have

$$\|(p_n(\lambda) - f(\lambda))I - (p_n(A) - f(A))\| \leq 2\|p_n - f\|_\infty,$$

where the supremum norm is taken on $\sigma(A)$. Thus, by the uniform convergence, we deduce that

$$\lim_{n\to\infty} \|(p_n(\lambda)I - p_n(A)) - (f(\lambda)I - f(A))\| = 0.$$

For each $n \geq 1$, since $p_n(\lambda) \in p_n(\sigma(A)) = \sigma(p_n(A))$, the operator $p_n(\lambda)I - p_n(A)$ is not invertible. Remember that the set of invertible elements of $\mathcal{L}(\mathcal{H})$ is an open set. Hence, we deduce that $f(\lambda)I - f(A)$ cannot be invertible, i.e. $f(\lambda) \in \sigma(f(A))$. □

Exercises

Exercise 7.1.1 Let A be a self-adjoint operator with $\sigma(A) = \{\lambda_1, \ldots, \lambda_n\}$. Put

$$p(z) = (z - \lambda_1)\cdots(z - \lambda_n).$$

Show that $p(A) = 0$.
Hint: Apply Lemma 7.1.

Exercise 7.1.2 Let A be a self-adjoint and compact operator on a Hilbert space $\mathcal{H}$. Let $f \in \mathcal{C}(\sigma(A))$ with $f(0) = 0$. Show that $f(A)$ is compact on $\mathcal{H}$. Hint: Use (7.2), and note that we can take p_n with $p_n(0) = 0$.

7.2 The square root of a positive operator

An important consequence of Theorem 7.2 is the following result.

Theorem 7.5 *Let $A \in \mathcal{L}(\mathcal{H})$ be a positive operator on a Hilbert space $\mathcal{H}$. Then there is a unique positive operator $B \in \mathcal{L}(\mathcal{H})$ such that $B^2 = A$.*

Proof Since A is positive, we know from Lemma 2.11 that $\sigma(A)$ is a compact subset of $[0, +\infty)$. Hence, the function $f : x \longmapsto \sqrt{x}$ is continuous on $\sigma(A)$ and, according to Theorem 7.2, $f(A)$ defines a bounded operator on $\mathcal{H}$. Moreover, since $\tilde{\Phi}_A$ is a isometric $*$-homomorphism and f is real, we have

$$f(A)^2 = f^2(A) = A$$

and

$$f(A)^* = \bar{f}(A) = f(A).$$

Hence, $B = f(A)$ is a self-adjoint operator that satisfies $B^2 = A$. Moreover, by Theorem 7.4, we have $\sigma(B) = f(\sigma(A)) \subset [0, +\infty)$. Then an application of Corollary 2.15 implies that B is positive and thus satisfies the required properties.

It remains to show that B is unique. Suppose that there is an operator $C \in \mathcal{L}(\mathcal{H})$ such that $C^2 = A$ and $C \geq 0$. Then $AC = C^2C = CC^2 = CA$ and Corollary 7.3 implies that $BC = CB$. Fix $x \in \mathcal{H}$, and let $y = (B - C)x$. Then

$$\langle (B + C)y,\ y\rangle = \langle (B^2 - C^2)x,\ y\rangle = 0.$$

Since B and C are positive operators, we thus have

$$\langle By, y\rangle = \langle Cy, y\rangle = 0.$$

But, by Lemma 2.12, these assumptions imply that $By = Cy = 0$. Finally,

$$\begin{aligned}\|(B - C)x\|^2 &= \langle (B - C)x,\ (B - C)x\rangle \\ &= \langle (B - C)^2x,\ x\rangle = \langle (B - C)y,\ x\rangle = 0.\end{aligned}$$

Therefore, $B = C$. □

The operator B, which was introduced in Theorem 7.5, is called the *positive square root* of A and is denoted by $A^{1/2}$ or $\sqrt{A}$. In the following, we establish several important properties of the square root of positive operators. We start with an intuitively clear convergence property.

Lemma 7.6 *Let $(A_n)_{n\geq 1}$ be a sequence of positive operators that tends to the positive operator A in $\mathcal{L}(\mathcal{H})$. Then the sequence $(\sqrt{A_n})_{n\geq 1}$ tends to $\sqrt{A}$ in $\mathcal{L}(\mathcal{H})$.*

Proof Since $A_n \longrightarrow A$ in $\mathcal{L}(\mathcal{H})$, there exists a constant M such that

$$\|A_n\| \leq M \qquad (n \geq 1).$$

In particular, we have

$$\sigma(A) \subset [0, M] \quad \text{and} \quad \sigma(A_n) \subset [0, M] \qquad (n \geq 1).$$

Given $\varepsilon > 0$, we can find a polynomial p such that

$$\sup_{t\in[0,M]} |p(t) - \sqrt{t}| \leq \frac{\varepsilon}{3}.$$

Thus, by Theorem 7.2, we have

$$\|\sqrt{A_n} - p(A_n)\| = \sup_{t\in\sigma(A_n)} |\sqrt{t} - p(t)| \leq \sup_{t\in[0,M]} |\sqrt{t} - p(t)| \leq \frac{\varepsilon}{3}$$

and, similarly,

$$\|\sqrt{A} - p(A)\| = \sup_{t\in\sigma(A)} |\sqrt{t} - p(t)| \leq \sup_{t\in[0,M]} |\sqrt{t} - p(t)| \leq \frac{\varepsilon}{3}.$$

Then the inequality

$$\|\sqrt{A_n} - \sqrt{A}\| \leq \|\sqrt{A_n} - p(A_n)\| + \|p(A_n) - p(A)\| + \|p(A) - \sqrt{A}\|$$

implies that

$$\|\sqrt{A_n} - \sqrt{A}\| \leq \frac{2\varepsilon}{3} + \|p(A_n) - p(A)\| \qquad (n \geq 1).$$

But $p(A_n) \longrightarrow p(A)$, as $n \longrightarrow \infty$, for any fixed complex polynomial p. Thus, there is $n_0 \in \mathbb{N}$ such that

$$\|p(A_n) - p(A)\| \leq \frac{\varepsilon}{3} \qquad (n \geq n_0),$$

which gives

$$\|\sqrt{A_n} - \sqrt{A}\| \leq \varepsilon \qquad (n \geq n_0).$$

□

Theorem 7.7 *Let $A, B \in \mathcal{L}(\mathcal{H})$ be positive operators with*

$$AB = BA. \tag{7.3}$$

Then the following hold.

(i) $A^{1/2}B^{1/2} = B^{1/2}A^{1/2}$.
(ii) $AB \geq 0$ *and* $(AB)^{1/2} = A^{1/2}B^{1/2}$.
(iii) *If A is invertible, then $A^{-1} \geq 0$. Moreover, $A^{1/2}$ is also invertible and* $(A^{1/2})^{-1} = (A^{-1})^{1/2}$.

(iv) *If* $A \geq B$ *and* A, B *are invertible, then* $B^{-1} \geq A^{-1}$.
(v) *If* $A \geq B$, *then* $A^{1/2} \geq B^{1/2}$.

Proof (i) According to Corollary 7.3, we have $AB^{1/2} = B^{1/2}A$ and then, by the same corollary, $A^{1/2}B^{1/2} = B^{1/2}A^{1/2}$.

(ii) By part (i), $AB = B^{1/2}AB^{1/2}$, and thus $AB \geq 0$. Moreover, by (i), $(A^{1/2}B^{1/2})^2 = AB$, and the uniqueness of the square root shows that $(AB)^{1/2} = A^{1/2}B^{1/2}$.

(iii) We have

$$\langle A^{-1}x, x\rangle = \langle A^{-1}x, AA^{-1}x\rangle = \langle AA^{-1}x, A^{-1}x\rangle \geq 0$$

for each $x \in \mathcal{H}$. Hence, A^{-1} is a positive operator. To show that $A^{1/2}$ is invertible, we apply the spectral mapping theorem (Theorem 1.22). This result gives

$$\sigma(A^{1/2}) = \sqrt{\sigma(A)} \subset (0, +\infty),$$

and thus $0 \notin \sigma(A^{1/2})$. To show that $(A^{-1})^{1/2} = (A^{1/2})^{-1}$, note that $(A^{1/2})^{-1} \geq 0$ and

$$(A^{1/2})^{-1}(A^{1/2})^{-1} = (A^{1/2}A^{1/2})^{-1} = A^{-1}.$$

Then, by the uniqueness of the positive square root, we obtain $(A^{-1})^{1/2} = (A^{1/2})^{-1}$.

(iv) Write

$$B^{-1} - A^{-1} = A^{-1}(A - B)B^{-1}$$

and note that (7.3) implies that A^{-1}, B^{-1} and $A - B$ commute. Hence, the result follows from (ii) and (iii).

(v) First, assume that there is a constant $\delta > 0$ such that $B \geq \delta I$. By (i) and (ii), we know that $A^{1/2}B^{1/2} = B^{1/2}A^{1/2} \geq 0$. Then, for each $x \in \mathcal{H}$, we have

$$\begin{aligned}\|(A^{1/2} + B^{1/2})x\|^2 &= \langle (A^{1/2} + B^{1/2})^2 x,\ x\rangle \\ &= \langle (A + B + 2B^{1/2}A^{1/2})x,\ x\rangle \\ &\geq \langle Bx, x\rangle \geq \delta\|x\|^2.\end{aligned}$$

Thus, we get that $A^{1/2} + B^{1/2}$ is bounded below, and, since it is self-adjoint, it is invertible and $(A^{1/2} + B^{1/2})^{-1} \geq 0$ by (iii). Now, note that

$$(A^{1/2} + B^{1/2})(A^{1/2} - B^{1/2}) = A - B,$$

whence

$$A^{1/2} - B^{1/2} = (A^{1/2} + B^{1/2})^{-1}(A - B).$$

An application of (ii) implies that $A^{1/2} - B^{1/2} \geq 0$, which proves the result in the case $B \geq \delta I$.

For the general case, apply the preceding result to the operators $A + \delta I$ and $B + \delta I$, where $\delta > 0$, to obtain

$$(A + \delta I)^{1/2} \geq (B + \delta I)^{1/2}.$$

Letting $\delta \longrightarrow 0$ and using Lemma 7.6, we obtain the required result. □

An easy consequence of Theorem 7.7 is the following corollary.

Corollary 7.8 *Let $A, B, C \in \mathcal{L}(\mathcal{H})$ be self-adjoint operators. Assume that $A \geq B$, $C \geq 0$ and $C(A - B) = (A - B)C$. Then $AC \geq BC$.*

Proof It is sufficient to write

$$AC - BC = (A - B)C$$

and apply part (ii) of Theorem 7.7. □

We also get the following characterization of invertible positive operators from Theorem 7.7.

Corollary 7.9 *Let A be a positive operator on a Hilbert space $\mathcal{H}$. Then the following are equivalent.*

(i) *A is invertible.*
(ii) *A is bounded below.*
(iii) *There exists a constant $\delta > 0$ such that $A \geq \delta I$.*

Proof The equivalence of (i) and (ii) has already been established (and in fact they hold for self-adjoint operators). The implication (iii) $\Longrightarrow$ (ii) follows from (2.10). It remains to prove (i) $\Longrightarrow$ (iii). If A is invertible, then we know from part (iii) of Theorem 7.7 that $A^{1/2}$ is also invertible, and then, in particular, it is bounded below. Hence, there is a constant $\delta > 0$ such that $\|A^{1/2}x\| \geq \delta\|x\|$, for all $x \in \mathcal{H}$. Now it suffices to note that

$$\langle Ax, x \rangle = \|A^{1/2}x\|^2 \geq \delta^2\|x\|^2,$$

which precisely means $A \geq \delta^2 I$. □

The following result is rather obvious if the operator A is also invertible. However, it has nice applications even if A is not invertible, but can be approached by invertible operators.

Theorem 7.10 *Let $A, A_n \in \mathcal{L}(\mathcal{H})$, $n \geq 1$, be such that:*

(i) *$A_n A_m = A_m A_n$, for all $n, m \geq 1$;*
(ii) *$\lim_{n \to \infty} \|A_n - A\| = 0$;*
(iii) *$A_1 \geq A_2 \geq \cdots \geq A \geq 0$;*
(iv) *for each $n \geq 1$, A_n is invertible.*

Let $y \in \mathcal{H}$. Then $y \in \mathcal{R}(A^{1/2})$ if and only if

$$\sup_{n\geq 1} \|A_n^{-1/2} y\| < \infty.$$

Moreover, if $y = A^{1/2}x$, with $x \in (\ker A)^{\perp}$, then

$$\lim_{n\to\infty} \|A_n^{-1/2} y - x\| = 0.$$

Proof Let us first treat some consequences of assumptions (i)–(iv). Using (i), (iii), (iv) and Theorem 7.7, we have

$$0 \leq A_n^{-1} \leq A_{n+1}^{-1} \qquad (n \geq 1)$$

and thus

$$\|A_n^{-1/2} y\|^2 = \langle A_n^{-1} y, y\rangle \leq \langle A_{n+1}^{-1} y, y\rangle = \|A_{n+1}^{-1/2} y\|^2.$$

Hence, for each $y \in \mathcal{H}$, the sequence $(\|A_n^{-1/2} y\|)_{n\geq 1}$ is increasing. Moreover, it follows from (i) and (ii) that $AA_n = A_n A$ and thus, by Corollary 7.3 and (iv), that

$$A^{1/2} A_n^{-1} = A_n^{-1} A^{1/2} \qquad (n \geq 1). \tag{7.4}$$

The assumptions (iii), (iv) and Theorem 7.7 also imply that

$$A_n^{-1} A \leq I. \tag{7.5}$$

For $m \geq n$, we have

$$\|A_m^{-1/2} y - A_n^{-1/2} y\|^2 = \|A_m^{-1/2} y\|^2 + \|A_n^{-1/2} y\|^2 - 2\,\langle A_m^{-1/2} A_n^{-1/2} y,\, y\rangle.$$

But, since $A_n^{-1/2} \leq A_m^{-1/2}$, we have $A_n^{-1} \leq A_m^{-1/2} A_n^{-1/2}$, and thus

$$\|A_m^{-1/2} y - A_n^{-1/2} y\|^2 \leq \|A_m^{-1/2} y\|^2 - \|A_n^{-1/2} y\|^2 \qquad (m \geq n \geq 1). \tag{7.6}$$

Assume that $y = A^{1/2}x$ and $x \perp \ker(A^{1/2})$. Then, by (7.4) and (7.5), we have

$$\begin{aligned} \|A_n^{-1/2} y\|^2 &= \langle A_n^{-1} y, y\rangle \\ &= \langle A_n^{-1} A^{1/2} x, A^{1/2} x\rangle \\ &= \langle A_n^{-1} A x, x\rangle \\ &\leq \|x\|^2 \qquad (n \geq 1). \end{aligned} \tag{7.7}$$

Therefore, $\sup_{n\geq 1} \|A_n^{-1/2} y\| < \infty$. Since the sequence $(\|A_n^{-1/2} y\|)_{n\geq 1}$ is bounded and increasing, it converges. Thus, the right side of (7.6) tends to zero as $n \longrightarrow \infty$. This proves that $(A_n^{-1/2} y)_{n\geq 1}$ is a Cauchy sequence, whence it converges to a vector $z \in \mathcal{H}$. It remains to prove that $z = x$. We have

$$A^{1/2}x - A^{1/2}z = y - A^{1/2}z = A_n^{1/2}(A_n^{-1/2} y - z) + (A_n^{1/2} - A^{1/2})z.$$

The first term tends to zero because

$$\begin{aligned}\|A_n^{1/2}(A_n^{-1/2}y - z)\| &\le \|A_n^{1/2}\| \times \|A_n^{-1/2}y - z\| \\ &\le \|A_1^{1/2}\| \times \|A_n^{-1/2}y - z\|\end{aligned}$$

and $A_n^{-1/2}y \longrightarrow z$ as $n \longrightarrow \infty$. The second term also tends to zero since $A_n \longrightarrow A$, whence $A_n^{1/2} \longrightarrow A^{1/2}$ (see Lemma 7.6). Finally, we get $A^{1/2}x = A^{1/2}z$, whence $z - x \in \ker(A^{1/2})$. Write $z = x + u$, with $u \in \ker(A^{1/2})$. We have

$$\|z\|^2 = \|x\|^2 + \|u\|^2 \ge \|x\|^2,$$

and, according to (7.7), we also have $\|z\|^2 \le \|x\|^2$, whence $u = 0$ and $z = x$.

It remains to prove that, if

$$\sup_{n\ge 1} \|A_n^{-1/2}y\| < +\infty, \tag{7.8}$$

then y belongs to the range of $A^{1/2}$. But since the sequence $(\|A_n^{-1/2}y\|)_{n\ge 1}$ is increasing, the condition (7.8) implies that this sequence converges. Using (7.6), we get that $(A_n^{-1/2}y)_{n\ge 1}$ converges, say, to $z \in \mathcal{H}$. Hence,

$$\begin{aligned}\|y - A^{1/2}z\| &\le \|y - A_n^{1/2}z\| + \|A_n^{1/2}z - A^{1/2}z\| \\ &= \|A_n^{1/2}(A_n^{-1/2}y - z)\| + \|A_n^{1/2}z - A^{1/2}z\| \\ &\le \|A_n^{1/2}\|\,\|A_n^{-1/2}y - z\| + \|A_n^{1/2}z - A^{1/2}z\| \\ &\le \|A_1^{1/2}\|\,\|A_n^{-1/2}y - z\| + \|A_n^{1/2}z - A^{1/2}z\| \longrightarrow 0,\end{aligned}$$

which shows that $y = A^{1/2}z$. □

Given $A \in \mathcal{L}(\mathcal{H}_1, \mathcal{H})$ and $B \in \mathcal{L}(\mathcal{H}_2, \mathcal{H})$, in certain applications, we need to know if there is a contraction $C \in \mathcal{L}(\mathcal{H}_1, \mathcal{H}_2)$ such that $A = BC$ (see Figure 7.1). If such a contraction exists, then

$$BB^* - AA^* = BB^* - BCC^*B^* = B(I - CC^*)B^* \ge 0.$$

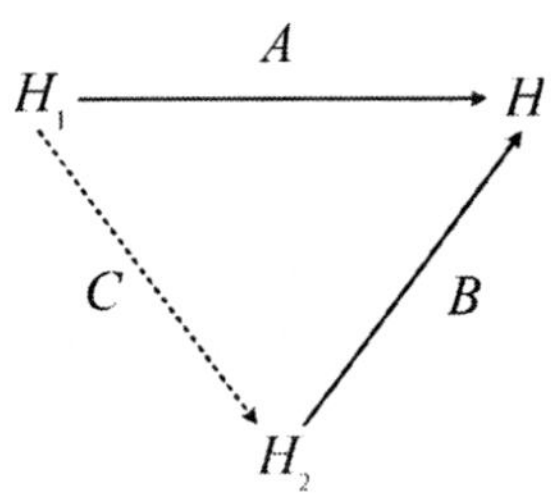

Figure 7.1 The factorization $A = BC$.

Douglas showed that the condition $AA^* \leq BB^*$ is also sufficient for the existence of C. Douglas's result plays a crucial role in the development of $\mathcal{H}(b)$ spaces.

Theorem 7.11 *Let $\mathcal{H}$, $\mathcal{H}_1$ and $\mathcal{H}_2$ be Hilbert spaces, let $A \in \mathcal{L}(\mathcal{H}_1, \mathcal{H})$ and $B \in \mathcal{L}(\mathcal{H}_2, \mathcal{H})$, and let $c > 0$. Then there is a bounded operator $C \in \mathcal{L}(\mathcal{H}_1, \mathcal{H}_2)$, with $\|C\| \leq c$, such that $A = BC$ if and only if $AA^* \leq c^2 BB^*$.*

Proof Suppose that there is a bounded operator $C \in \mathcal{L}(\mathcal{H}_1, \mathcal{H}_2)$, $\|C\| \leq c$, such that $A = BC$. Since $\|C\| \leq c$, the operator $c^2 I - CC^*$ is positive. Then we have

$$c^2 BB^* - AA^* = c^2 BB^* - BCC^*B^* = B(c^2 I - CC^*)B^* \geq 0.$$

The other direction is a bit more delicate. Suppose that $AA^* \leq c^2 BB^*$. Hence, by definition,

$$\|A^*x\|_{\mathcal{H}_1} \leq c\|B^*x\|_{\mathcal{H}_2} \qquad (x \in \mathcal{H}). \tag{7.9}$$

This inequality enables us to define the mapping D from the range of B^* to the range of A^* by

$$\begin{array}{rccc} D: & \mathcal{R}(B^*) & \longrightarrow & \mathcal{R}(A^*) \\ & B^*x & \longmapsto & A^*x. \end{array}$$

If an element in $\mathcal{R}(B^*)$ has two representations, i.e. $z = B^*x = B^*y$, then $B^*(x - y) = 0$. Hence, by (7.9), $A^*(x - y) = 0$. In other words, $Dz = A^*x = A^*y$ is well defined, and, moreover,

$$\|D(B^*x)\|_{\mathcal{H}_1} \leq c\|B^*x\|_{\mathcal{H}_2} \qquad (x \in \mathcal{H}).$$

Therefore, by continuity, D extends to a bounded operator from the closure of $\mathcal{R}(B^*)$ in $\mathcal{H}_2$ into $\mathcal{H}_1$, and the norm of extension is still less than or equal to c. In the last step of extension, we extend D to a bounded operator from $\mathcal{H}_2$ to $\mathcal{H}_1$ by defining

$$D(z) = 0 \qquad (z \in \mathcal{R}(B^*)^\perp).$$

Hence, the norm of D will not increase. According to our primary definition of D, this operator satisfies $DB^* = A^*$ and $\|D\| \leq c$. Hence, $A = BD^*$. Take $C = D^*$. □

In Theorem 7.11, the operators A and B have the same codomain. In some applications, we need a version of Douglas's result in which the operators have the same domain. But such a version is easy to get.

Corollary 7.12 *Let $\mathcal{H}$, $\mathcal{H}_1$ and $\mathcal{H}_2$ be Hilbert spaces, let $A \in \mathcal{L}(\mathcal{H}, \mathcal{H}_1)$ and $B \in \mathcal{L}(\mathcal{H}, \mathcal{H}_2)$, and let $c > 0$. Then there is a bounded operator $C \in \mathcal{L}(\mathcal{H}_2, \mathcal{H}_1)$, with $\|C\| \leq c$, such that $A = CB$ if and only if $A^*A \leq c^2 B^*B$.*

Proof Apply Theorem 7.11 to operators A^* and B^*. $\square$

Given two operators $A \in \mathcal{L}(\mathcal{H}, \mathcal{H}_1)$ and $B \in \mathcal{L}(\mathcal{H}, \mathcal{H}_2)$, consider the operator

$$\begin{bmatrix} A \\ B \end{bmatrix} : \begin{array}{ccc} \mathcal{H} & \longrightarrow & \mathcal{H}_1 \oplus \mathcal{H}_2 \\ x & \longmapsto & (Ax, Bx). \end{array}$$

This operator is a contraction if and only if

$$\|Ax\|^2_{\mathcal{H}_1} + \|Bx\|^2_{\mathcal{H}_2} \le \|x\|^2_{\mathcal{H}} \qquad (x \in \mathcal{H}). \tag{7.10}$$

Hence, surely A and B must be contractions. Using Douglas's result, we can obtain a characterization of this situation.

Corollary 7.13 *Let $A \in \mathcal{L}(\mathcal{H}, \mathcal{H}_1)$ and $B \in \mathcal{L}(\mathcal{H}, \mathcal{H}_2)$. Then $\begin{bmatrix} A \\ B \end{bmatrix}$ is a contraction from $\mathcal{H}$ to $\mathcal{H}_1 \oplus \mathcal{H}_2$ if and only if there is a contraction $C \in \mathcal{L}(\mathcal{H}, \mathcal{H}_1)$ such that*

$$A = C(I - B^*B)^{1/2}.$$

Proof By (7.10), the operator $\begin{bmatrix} A \\ B \end{bmatrix}$ is a contraction if and only if

$$A^*A + B^*B \le I_{\mathcal{H}}.$$

Write the inequality as $A^*A \le I_{\mathcal{H}} - B^*B$, and then apply Corollary 7.12. $\square$

In a similar manner, if we have two operators $A \in \mathcal{L}(\mathcal{H}_1, \mathcal{H})$ and $B \in \mathcal{L}(\mathcal{H}_2, \mathcal{H})$, we can consider the operator

$$\begin{bmatrix} A & B \end{bmatrix} : \begin{array}{ccc} \mathcal{H}_1 \oplus \mathcal{H}_2 & \longrightarrow & \mathcal{H} \\ (x_1, x_2) & \longmapsto & Ax_1 + Bx_2. \end{array}$$

Then this operator is a contraction if and only if

$$\|Ax_1 + Bx_2\|^2_{\mathcal{H}} \le \|x_1\|^2_{\mathcal{H}_1} + \|x_2\|^2_{\mathcal{H}_2} \qquad (x_i \in \mathcal{H}_i). \tag{7.11}$$

Hence, surely A and B must be contractions. The following result is a characterization of this situation.

Corollary 7.14 *Let $A \in \mathcal{L}(\mathcal{H}_1, \mathcal{H})$ and $B \in \mathcal{L}(\mathcal{H}_2, \mathcal{H})$. Then $\begin{bmatrix} A & B \end{bmatrix}$ is a contraction from $\mathcal{H}_1 \oplus \mathcal{H}_2$ to $\mathcal{H}$ if and only if there is a contraction $C \in \mathcal{L}(\mathcal{H}_1, \mathcal{H})$ such that*

$$A = (I - BB^*)^{1/2}C.$$

Proof Either apply Corollary 7.13 to operators A^* and B^*, or, as we did for Corollary 7.13, give a direct proof by using Theorem 7.11. $\square$

Given a contraction $A \in \mathcal{L}(\mathcal{H}_1, \mathcal{H}_2)$, the operator $\mathcal{D}_A = (I - A^*A)^{1/2}$ is called the *defect operator* of the contraction A. Its range $\mathcal{D}_A = \mathrm{Clos}_{\mathcal{H}_1}(\mathcal{R}(\mathcal{D}_A))$ is called the *defect space* of the contraction A and the number

$$\partial_A = \dim(\mathcal{D}_A)$$

is called the *defect index* of the contraction A. These objects are used to measure how far the contraction A is from being an isometry.

The following result, which has many interesting applications in operator theory, says that A intertwines the operators $\mathcal{D}_A$ and $\mathcal{D}_{A^*}$.

Theorem 7.15 (Intertwining formula) *Let $A \in \mathcal{L}(\mathcal{H}_1, \mathcal{H}_2)$ be a Hilbert space contraction. Then*

$$A(I - A^*A)^{1/2} = (I - AA^*)^{1/2}A.$$

Proof We obviously have $A(I - A^*A) = (I - AA^*)A$. Hence, by induction, $A(I - A^*A)^n = (I - AA^*)^n A$ holds for all integer $n \geq 0$. Therefore, for any analytic polynomial p, we have

$$Ap(I - A^*A) = p(I - AA^*)A.$$

By Theorem 7.2, there is a sequence of polynomials $(p_n)_{n\geq 1}$ such that

$$p_n(I - A^*A) \longrightarrow (I - A^*A)^{1/2} \quad \text{and} \quad p_n(I - AA^*) \longrightarrow (I - AA^*)^{1/2}.$$

Thus, the required identity follows immediately. □

Replacing A by A^* in the intertwining formula, we obtain the equivalent identity

$$A^*(I - AA^*)^{1/2} = (I - A^*A)^{1/2}A^*. \tag{7.12}$$

Exercises

Exercise 7.2.1 Let $A \in \mathcal{L}(\mathcal{H}, \mathcal{H}_1)$ and $B \in \mathcal{L}(\mathcal{H}, \mathcal{H}_2)$. Then $\begin{bmatrix} A \\ B \end{bmatrix}$ is a contraction from $\mathcal{H}$ to $\mathcal{H}_1 \oplus \mathcal{H}_2$ if and only if there is a contraction $D \in \mathcal{L}(\mathcal{H}, \mathcal{H}_2)$ such that

$$B = D(I - A^*A)^{1/2}.$$

Hint: Modify the proof of Corollary 7.13.

Exercise 7.2.2 Let $A \in \mathcal{L}(\mathcal{H}_1, \mathcal{H})$ and $B \in \mathcal{L}(\mathcal{H}_2, \mathcal{H})$. Then $\begin{bmatrix} A & B \end{bmatrix}$ is a contraction from $\mathcal{H}_1 \oplus \mathcal{H}_2$ to $\mathcal{H}$ if and only if there is a contraction $C \in \mathcal{L}(\mathcal{H}_2, \mathcal{H})$ such that

$$B = (I - AA^*)^{1/2} C.$$

Hint: Modify the proof of Corollary 7.14.

Exercise 7.2.3 Let $A, B, C \in \mathcal{L}(\mathcal{H})$, and suppose that $A \leq B$. Show that $C^* AC \leq C^* BC$.

7.3 Möbius transformations and the Julia operator

For certain values of the parameters $a, b, c, d \in \mathbb{C}$, the Möbius transformation

$$z \longmapsto \frac{az + b}{cz + d}$$

maps the unit disk $\mathbb{D}$ into itself. If $A,\ B,\ C,\ D$ and X are operators in $\mathcal{L}(\mathcal{H})$ such that $CX + D$ is invertible, then we can form combinations like

$$(AX + B)(CX + D)^{-1} \quad \text{or} \quad (CX + D)^{-1}(AX + B).$$

However, owing to lack of commutativity, the above combinations do not necessarily give the same result. Moreover, as we can write

$$\frac{az + b}{cz + d} = a' + \frac{b' z}{cz + d},$$

many other combinations, e.g.

$$B - AX(I + CX)^{-1} D,$$

exist. Our goal is to find a combination that sends contractions to themselves. A possible approach is outlined below.

Let T be a unitary operator on $\mathcal{H}_1 \oplus \mathcal{H}_2$, i.e. $TT^* = T^*T = I_{\mathcal{H}_1 \oplus \mathcal{H}_2}$. Then we can write T in the matrix form

$$T = \begin{bmatrix} A & B \\ C & D \end{bmatrix} : \begin{array}{ccc} \mathcal{H}_1 \oplus \mathcal{H}_2 & \longrightarrow & \mathcal{H}_1 \oplus \mathcal{H}_2 \\ (x_1, x_2) & \longmapsto & (Ax_1 + Bx_2,\ Cx_1 + Dx_2), \end{array}$$

where $A,\ B,\ C$ and D are bounded operators from $\mathcal{H}_i$ to $\mathcal{H}_j$ (with appropriate choices of i and j in each case). Then we define the operator-valued Möbius transformation $\Psi_T(X)$ by

$$\Psi_T(X) = B - AX(I + CX)^{-1} D$$

on the set

$$\{X \in \mathcal{L}(\mathcal{H}_2, \mathcal{H}_1) : I + CX \text{ is invertible in } \mathcal{L}(\mathcal{H}_2)\},$$

which we refer to as the domain of Ψ_T. Clearly, this set contains all strict contractions, i.e. X with $\|X\| < 1$. Moreover, if $\|C\| < 1$, it contains all the contractions. The magic of $\Psi_T(X)$ is contained in the following lemma.

Lemma 7.16 *Let T be a unitary operator on $\mathcal{H}_1 \oplus \mathcal{H}_2$. Then, for each X in the domain of Ψ_T, we have*

$$I - \Psi_T(X)^*\Psi_T(X) = E^*(I - X^*X)E,$$

where $E = (I + CX)^{-1}D$.

Proof This is just a matter of calculation. The assumption $T^*T = I$, or

$$\begin{bmatrix} A & B \\ C & D \end{bmatrix}^* \begin{bmatrix} A & B \\ C & D \end{bmatrix} = \begin{bmatrix} I & 0 \\ 0 & I \end{bmatrix},$$

is equivalent to

$$\begin{aligned} A^*A + C^*C &= I, \\ A^*B + C^*D &= 0, \\ B^*A + D^*C &= 0, \\ B^*B + D^*D &= I. \end{aligned}$$

Now, we work out $\Delta = I - \Psi_T(X)^*\Psi_T(X)$ and exploit the above four relations. Hence,

$$\begin{aligned} \Delta &= I - \{B - AX(I + CX)^{-1}D\}^* \{B - AX(I + CX)^{-1}D\} \\ &= I - \{B^* - D^*(I + X^*C^*)^{-1}X^*A^*\} \{B - AX(I + CX)^{-1}D\} \\ &= (I - B^*B) + D^*(I + X^*C^*)^{-1}X^*A^*B + B^*AX(I + CX)^{-1}D \\ &\quad - D^*(I + X^*C^*)^{-1}X^*A^*AX(I + CX)^{-1}D \\ &= D^*D - D^*(I + X^*C^*)^{-1}X^*C^*D - D^*CX(I + CX)^{-1}D \\ &\quad - D^*(I + X^*C^*)^{-1}X^*A^*AX(I + CX)^{-1}D \\ &= E^*\{(I + X^*C^*)(I + CX) - X^*C^*(I + CX) \\ &\quad - (I + X^*C^*)CX - X^*A^*AX\}E \\ &= E^*\{I - X^*(A^*A + C^*C)X\}E \\ &= E^*(I - X^*X)E. \end{aligned}$$

This completes the proof. □

In a similar manner, one can show that

$$I - \Psi_T(X)\Psi_T(X)^* = F(I - XX^*)F^*, \tag{7.13}$$

where $F = A(I + XC)^{-1}$. The main result that we were looking for is at our disposal now.

Theorem 7.17 *Let X be in the domain of Ψ_T. If X is a contraction, then so is $\Psi_T(X)$.*

Proof If X is a contraction, then $I - X^*X \geq 0$. Hence, $E^*(I - X^*X)E \geq 0$. Therefore, by Lemma 7.16, $I - \Psi_T(X)^*\Psi_T(X) \geq 0$. This means that $\Psi_T(X)$ is a contraction. □

To use Theorem 7.17 effectively, we should have a unitary operator T at our disposal. One way to construct such an operator is to apply the Julia transformation. Fix a contraction $B \in \mathcal{L}(\mathcal{H}_2, \mathcal{H}_1)$. Then the *Julia operator* $J(B) \in \mathcal{L}(\mathcal{H}_1 \oplus \mathcal{H}_2)$ is defined by

$$J(B) = \begin{bmatrix} (I - BB^*)^{1/2} & B \\ -B^* & (I - B^*B)^{1/2} \end{bmatrix}.$$

The main characteristic of $J(B)$ is mentioned in the following theorem.

Theorem 7.18 *Let B be a contraction in $\mathcal{L}(\mathcal{H}_2, \mathcal{H}_1)$. Then $J(B)$ is a unitary operator on $\mathcal{H}_1 \oplus \mathcal{H}_2$.*

Proof We need to check that $J(B)J(B)^* = J(B)^*J(B) = I$. Hence,

$$\begin{aligned} & J(B)^*J(B) \\ &= \begin{bmatrix} (I - BB^*)^{1/2} & B \\ -B^* & (I - B^*B)^{1/2} \end{bmatrix}^* \begin{bmatrix} (I - BB^*)^{1/2} & B \\ -B^* & (I - B^*B)^{1/2} \end{bmatrix} \\ &= \begin{bmatrix} (I - BB^*)^{1/2} & -B \\ B^* & (I - B^*B)^{1/2} \end{bmatrix} \begin{bmatrix} (I - BB^*)^{1/2} & B \\ -B^* & (I - B^*B)^{1/2} \end{bmatrix}, \end{aligned}$$

and doing the matrix multiplication gives us

$$\begin{bmatrix} I & (I - BB^*)^{1/2}B - B(I - B^*B)^{1/2} \\ B^*(I - BB^*)^{1/2} - (I - B^*B)^{1/2}B^* & I \end{bmatrix}.$$

That the two components in the cross-diagonal are zero follows from Theorem 7.15. Therefore $J(B)^*J(B) = I$. The verification of $J(B)J(B)^* = I$ is similar. □

Knowing that $J(B)$ is a unitary operator on $\mathcal{H}_1 \oplus \mathcal{H}_2$, we can apply Theorem 7.17 to obtain the following valuable result.

Corollary 7.19 *Let B and X be contractions in $\mathcal{L}(\mathcal{H}_2, \mathcal{H}_1)$ such that $I - B^*X$ is invertible in $\mathcal{L}(\mathcal{H}_2)$. Then*

$$\Psi_{J(B)}(X) = B - (I - BB^*)^{1/2}X(I - B^*X)^{-1}(I - B^*B)^{1/2}$$

is also a contraction in $\mathcal{L}(\mathcal{H}_2, \mathcal{H}_1)$.

Corollary 7.19 is a truly beautiful result, and it has several profound applications. We outline one of its applications below, which will be needed in studying Hankel operators. Let us first explain the question. Assume that

$$\begin{bmatrix} A & B \\ C & D \end{bmatrix} : \quad \mathcal{H}_1 \oplus \mathcal{H}_2 \quad \longrightarrow \quad \mathcal{H}_3 \oplus \mathcal{H}_4$$

is a contraction. Hence, the restricted operator

$$\begin{bmatrix} B \\ D \end{bmatrix} : \quad \mathcal{H}_2 \quad \longrightarrow \quad \mathcal{H}_3 \oplus \mathcal{H}_4$$

is also a contraction. If we use the adjoint (twice, in fact), we see that

$$\begin{bmatrix} C & D \end{bmatrix} : \quad \mathcal{H}_1 \oplus \mathcal{H}_2 \quad \longrightarrow \quad \mathcal{H}_4$$

also has to be a contraction. Knowing these two facts, we can now pose our question. Assume that the operators B, C and D are given, and are such that

$$\begin{bmatrix} B \\ D \end{bmatrix} \in \mathcal{L}(\mathcal{H}_2, \mathcal{H}_3 \oplus \mathcal{H}_4) \quad \text{and} \quad \begin{bmatrix} C & D \end{bmatrix} \in \mathcal{L}(\mathcal{H}_1 \oplus \mathcal{H}_2, \mathcal{H}_4)$$

are contractions. Is there an operator $A \in \mathcal{L}(\mathcal{H}_1, \mathcal{H}_3)$ such that

$$\begin{bmatrix} A & B \\ C & D \end{bmatrix} \in \mathcal{L}(\mathcal{H}_1 \oplus \mathcal{H}_2, \mathcal{H}_3 \oplus \mathcal{H}_4)$$

is a contraction? This is an absolutely nontrivial question. We answer it affirmatively in the following theorem.

Theorem 7.20 *Let*

$$\begin{bmatrix} B \\ D \end{bmatrix} \in \mathcal{L}(\mathcal{H}_2, \mathcal{H}_3 \oplus \mathcal{H}_4) \quad \textit{and} \quad \begin{bmatrix} C & D \end{bmatrix} \in \mathcal{L}(\mathcal{H}_1 \oplus \mathcal{H}_2, \mathcal{H}_4)$$

be contractions. Then there is an operator $A \in \mathcal{L}(\mathcal{H}_1, \mathcal{H}_3)$ such that

$$\begin{bmatrix} A & B \\ C & D \end{bmatrix} \in \mathcal{L}(\mathcal{H}_1 \oplus \mathcal{H}_2, \mathcal{H}_3 \oplus \mathcal{H}_4)$$

is a contraction.

Proof According to Corollaries 7.13 and 7.14, there are contractions $E \in \mathcal{L}(\mathcal{H}_2, \mathcal{H}_3)$ and $F \in \mathcal{L}(\mathcal{H}_1, \mathcal{H}_4)$ such that

$$B = E(I - D^*D)^{1/2} \quad \text{and} \quad C = (I - D^*D)^{1/2}F.$$

To exploit Corollary 7.19, consider $\mathbf{X}, \mathbf{B} \in \mathcal{L}(\mathcal{H}_1 \oplus \mathcal{H}_2, \mathcal{H}_3 \oplus \mathcal{H}_4)$ given by

$$\mathbf{X} = \begin{bmatrix} 0 & E \\ F & 0 \end{bmatrix} \quad \text{and} \quad \mathbf{B} = \begin{bmatrix} 0 & 0 \\ 0 & -D \end{bmatrix}.$$

Since E, F and D are contractions, and

$$I_{\mathcal{H}_1 \oplus \mathcal{H}_2} - \mathbf{X}^*\mathbf{X} = \begin{bmatrix} I_{\mathcal{H}_1} - F^*F & 0 \\ 0 & I_{\mathcal{H}_2} - E^*E \end{bmatrix}$$

and

$$I_{\mathcal{H}_1 \oplus \mathcal{H}_2} - \mathbf{B}^*\mathbf{B} = \begin{bmatrix} I_{\mathcal{H}_1} & 0 \\ 0 & I_{\mathcal{H}_2} - D^*D \end{bmatrix},$$

we deduce that $\mathbf{X}$ and $\mathbf{B}$ are also contractions in $\mathcal{L}(\mathcal{H}_1 \oplus \mathcal{H}_2, \mathcal{H}_3 \oplus \mathcal{H}_4)$. Moreover,

$$I_{\mathcal{H}_1 \oplus \mathcal{H}_2} - \mathbf{B}^*\mathbf{X} = I_{\mathcal{H}_1 \oplus \mathcal{H}_2} - \begin{bmatrix} 0 & 0 \\ 0 & -D \end{bmatrix}^* \begin{bmatrix} 0 & E \\ F & 0 \end{bmatrix} = \begin{bmatrix} I_{\mathcal{H}_1} & 0 \\ D^*F & I_{\mathcal{H}_2} \end{bmatrix}.$$

Hence, $I_{\mathcal{H}_1 \oplus \mathcal{H}_2} - \mathbf{B}^*\mathbf{X}$ is invertible in $\mathcal{L}(\mathcal{H}_1 \oplus \mathcal{H}_2, \mathcal{H}_3 \oplus \mathcal{H}_4)$ and

$$(I_{\mathcal{H}_1 \oplus \mathcal{H}_2} - \mathbf{B}^*\mathbf{X})^{-1} = \begin{bmatrix} I_{\mathcal{H}_1} & 0 \\ -D^*F & I_{\mathcal{H}_2} \end{bmatrix}.$$

Thus, by Corollary 7.19, $\Psi_{J(\mathbf{B})}(\mathbf{X})$ is a contraction in $\mathcal{L}(\mathcal{H}_1 \oplus \mathcal{H}_2, \mathcal{H}_3 \oplus \mathcal{H}_4)$. It remains to compute this last creature. To do so, note that

$$(I_{\mathcal{H}_1 \oplus \mathcal{H}_2} - \mathbf{B}\mathbf{B}^*)^{1/2} = \begin{bmatrix} I_{\mathcal{H}_1} & 0 \\ 0 & (I_{\mathcal{H}_2} - DD^*)^{1/2} \end{bmatrix}.$$

Hence, we have

$$\begin{aligned}
&\Psi_{J(\mathbf{B})}(\mathbf{X}) \\
&\quad = \mathbf{B} - (I_{\mathcal{H}_1 \oplus \mathcal{H}_2} - \mathbf{B}\mathbf{B}^*)^{1/2}\mathbf{X}(I_{\mathcal{H}_1 \oplus \mathcal{H}_2} - \mathbf{B}^*\mathbf{X})^{-1}(I_{\mathcal{H}_1 \oplus \mathcal{H}_2} - \mathbf{B}^*\mathbf{B})^{1/2} \\
&\quad = \begin{bmatrix} 0 & 0 \\ 0 & -D \end{bmatrix} - \begin{bmatrix} I_{\mathcal{H}_1} & 0 \\ 0 & (I_{\mathcal{H}_2} - DD^*)^{1/2} \end{bmatrix} \begin{bmatrix} 0 & E \\ F & 0 \end{bmatrix} \\
&\qquad \times \begin{bmatrix} I_{\mathcal{H}_1} & 0 \\ -D^*F & I_{\mathcal{H}_2} \end{bmatrix} \begin{bmatrix} I_{\mathcal{H}_1} & 0 \\ 0 & (I_{\mathcal{H}_2} - D^*D)^{1/2} \end{bmatrix} \\
&\quad = \begin{bmatrix} -ED^*F & B \\ C & D \end{bmatrix}.
\end{aligned}$$

Therefore, if we take $A = -ED^*F$, then we obtain a contraction in $\mathcal{L}(\mathcal{H}_1 \oplus \mathcal{H}_2, \mathcal{H}_3 \oplus \mathcal{H}_4)$. □

We would like to emphasize that the operator A is not necessarily unique. Theorem 7.20 just presents one of the candidates. The general solution is of the form

$$A = (I - EE^*)^{1/2}G(I - FF^*)^{1/2} - ED^*F,$$

where $G \in \mathcal{L}(\mathcal{H}_1, \mathcal{H}_3)$ is a contraction. But we do not need this extended result.

Exercises

Exercise 7.3.1 The formula (7.13) is valid for all X in the domain of Ψ_T. However, in the definition of the domain of Ψ_T, we assume that $I + CX$ is invertible, while in (7.13) we use the fact that $I + XC$ is invertible. Show that this is always the case.
Hint: If Y is the inverse of $I + CX$, work out $Z(I + XC)$ and $(I + XC)Z$, where $Z = I - XYC$.

Exercise 7.3.2 Let $B, C, D \in \mathcal{L}(\mathcal{H}_i, \mathcal{H}_j)$ (with appropriate choices of i and j in each case). Put

$$c = \max\left\{ \left\| \begin{bmatrix} B \\ D \end{bmatrix} \right\|_{\mathcal{L}(\mathcal{H}_2, \mathcal{H}_3 \oplus \mathcal{H}_4)}, \left\| \begin{bmatrix} C & D \end{bmatrix} \right\|_{\mathcal{L}(\mathcal{H}_1 \oplus \mathcal{H}_2, \mathcal{H}_4)} \right\}.$$

Show that

$$\inf_{A \in \mathcal{L}(\mathcal{H}_1, \mathcal{H}_3)} \left\| \begin{bmatrix} A & B \\ C & D \end{bmatrix} \right\|_{\mathcal{L}(\mathcal{H}_1 \oplus \mathcal{H}_2, \mathcal{H}_3 \oplus \mathcal{H}_4)} = c.$$

Hint: The $\geq$ case is elementary. For the $\leq$ case, use Theorem 7.20.

7.4 The Wold–Kolmogorov decomposition

In this section, we discuss the well-known Wold–Kolmogorov decomposition of an isometry. It is rather elementary that the direct sum of a unitary operator and some copies of the unilateral shift is an isometry. The Wold–Kolmogorov decomposition says that the converse is also true.

We say that $A \in \mathcal{L}(\mathcal{H})$ is *completely nonunitary* if, for any subspace $E \subset \mathcal{H}$, $E \neq \{0\}$ and $AE \subset E$, the restriction operator $A|E$ is not unitary. If A is an isometry, a subspace $D \subset \mathcal{H}$ is called *wandering* for A if $A^m D \perp A^n D$, for all $m, n \geq 0$, $m \neq n$.

Theorem 7.21 *Let $A \in \mathcal{L}(\mathcal{H})$ be an isometry, and let $E \subset \mathcal{H}$ be an invariant subspace of $\mathcal{H}$, i.e. $AE \subset E$ (in particular, we can pick $E = \mathcal{H}$). Let D be the orthogonal complement of AE in E, i.e.*

$$D = E \ominus AE.$$

Then D is a wandering subspace for A. Knowing this, put

$$E_0 = \bigoplus_{n \geq 0} A^n D \quad \textit{and} \quad E_\infty = \bigcap_{n \geq 0} A^n E.$$

Then the following hold:

(i) $AE_0 \subset E_0$;
(ii) $AE_\infty \subset E_\infty$;
(iii) *the restriction* $A|E_\infty$ *is unitary;*
(iv) *the restriction* $A|E_0$ *is completely nonunitary; and*
(v) *finally,* $E = E_0 \oplus E_\infty$.

Proof To show that D is a wandering subspace for A, we have to prove that $A^n x \perp A^m y$, for every x, y in D and $m \neq n$, $m, n \geq 0$. Without loss of generality, assume that $n > m$. Then, since A is an isometry, we have

$$\langle A^n x, A^m g\rangle = \langle A^{n-m} x, y\rangle.$$

But $A^{n-m}x \in AE$ and $y \in D = E \ominus AE$, and thus

$$\langle A^n x, A^m y\rangle = 0.$$

Hence, D is wandering. Since D is wandering, E_0 is well defined as the orthogonal direct sum of the subspaces $A^n D$, $n \geq 0$.

Parts (i) and (ii) are trivial.

(iii) We just need to show that $A|E_\infty$ is surjective. So pick any $y \in E_\infty$. By definition, there are $x_n \in E$, $n \geq 1$, such that $y = A^n x_n$. But, since A is an isometry, $Ax_1 = A^n x_n$, $n \geq 2$, implies that $x_1 = A^{n-1}x_n$. This means that $x_1 \in E_\infty$. Therefore, $y = Ax_1 \in AE_\infty$.

We prove (v) and then do (iv). Let us start by showing that $E \ominus E_0 \subset E_\infty$. Take $x \in E$, $x \perp E_0$. In particular, we have $x \perp D$, which means that $x \in AE$. Hence, there exists $x_1 \in E$ such that $x = Ax_1$. But, again by the assumption, we also have $x = Ax_1 \perp AD$. Since A is an isometry, we get $x_1 \perp D$. Thus, $x_1 \in AE$ and $x \in A^2E$. By induction, we deduce that $x \in A^n E$, $n \geq 0$, and the desired inclusion $E \ominus E_0 \subset E_\infty$ follows.

To establish the reverse inclusion $E_\infty \subset E \ominus E_0$, let $x \in E_\infty$. Then, for all $n \geq 1$, there exists $x_n \in E$ such that $x = A^n x_n$. Since $AE \perp D$, for each $y \in D$, we get

$$\langle x, A^n y\rangle = \langle A^{n+1}x_{n+1}, A^n y\rangle = \langle Ax_{n+1}, y\rangle = 0 \qquad (n \geq 0).$$

Thus, $x \perp A^n D$, $n \geq 0$, and hence $E_\infty \subset E \ominus E_0$.

(iv) Let $E' \subset E_0$ be a subspace such that $AE' = E'$. Then $A^n E' = E'$, for each $n \geq 0$, which implies that $E' \subset E_\infty$. Hence, $E' \subset E_0 \cap E_\infty = \{0\}$. □

7.5 Partial isometries and polar decomposition

It happens that an operator fails to be an isometry on the whole space $\mathcal{H}_1$ but, nevertheless, it is an isometry on some nontrivial closed subspace $\mathcal{E}$ of $\mathcal{H}_1$.

For instance, we saw in Lemma 2.16 that a contraction $T \in \mathcal{L}(\mathcal{H}_1, \mathcal{H}_2)$ is an isometry on $\mathcal{E} = \ker(I - T^*T)$. Note that we always have

$$\mathcal{E} = \ker(I - T^*T) \subset \mathcal{R}(T^*) \subset (\ker T)^\perp.$$

A particularly interesting case is when $\mathcal{E} = (\ker T)^\perp$. This motivates the following definition.

The operator $A \in \mathcal{L}(\mathcal{H}_1, \mathcal{H}_2)$ is called a *partial isometry* if

$$\|Ax\|_{\mathcal{H}_2} = \|x\|_{\mathcal{H}_1} \qquad (x \in (\ker A)^\perp).$$

In other words, A is a partial isometry if

$$\|Ax\|_{\mathcal{H}_2} = \|P_{(\ker A)^\perp} x\|_{\mathcal{H}_1} \qquad (x \in \mathcal{H}_1). \tag{7.14}$$

The advantage of the second formulation is that it holds for all $x \in \mathcal{H}_1$. A partial isometry is clearly a contraction. The subspace $(\ker A)^\perp$ is called the *initial space* of A, and its range $\mathcal{R}(A)$, which is a closed subspace of $\mathcal{H}_2$, is called the *final space* of A. Clearly, an isometry is a partial isometry that is one-to-one, and a unitary operator is a partial isometry that is bijective.

We have already seen that $A \in \mathcal{L}(\mathcal{H}_1, \mathcal{H}_2)$ is an isometry if and only if $A^*A = I_{\mathcal{H}_1}$. In the case when A is a partial isometry, what is A^*A? The following result answers this question.

Theorem 7.22 *Let $\mathcal{H}_1$ and $\mathcal{H}_2$ be Hilbert spaces and let $A \in \mathcal{L}(\mathcal{H}_1, \mathcal{H}_2)$. Then the following are equivalent:*

(i) *A is a partial isometry;*
(ii) *A^* is a partial isometry;*
(iii) *AA^* is an orthogonal projection;*
(iv) *A^*A is an orthogonal projection;*
(v) $\ker A^* = \mathcal{R}(I - AA^*)$;
(vi) $\ker A = \mathcal{R}(I - A^*A)$.

Proof (i) $\Longrightarrow$ (iv) Since A is a contraction, for all $x \in \mathcal{H}_1$,

$$\begin{aligned} \langle (I - A^*A)x,\ x\rangle_{\mathcal{H}_1} &= \langle x, x\rangle_{\mathcal{H}_1} - \langle A^*Ax, x\rangle_{\mathcal{H}_1} \\ &= \langle x, x\rangle_{\mathcal{H}_1} - \langle Ax, Ax\rangle_{\mathcal{H}_2} \\ &= \|x\|^2_{\mathcal{H}_1} - \|Ax\|^2_{\mathcal{H}_2} \\ &\geq 0. \end{aligned}$$

Hence, $I - A^*A$ is a positive operator on $\mathcal{H}_1$. Moreover, since A is a partial isometry with the initial space $(\ker A)^\perp$, the last identity gives

$$\langle (I - A^*A)x,\ x\rangle_{\mathcal{H}_1} = 0 \qquad (x \in (\ker A)^\perp). \tag{7.15}$$

Then, by Lemma 2.12, (7.15) implies that

$$(I - A^*A)x = 0 \qquad (x \in (\ker A)^\perp).$$

Thus,

$$A^*Ax = x \qquad (x \in (\ker A)^\perp)$$

and clearly

$$A^*Ax = 0 \qquad (x \in \ker A),$$

which proves that A^*A is the orthogonal projection on $(\ker A)^\perp$, i.e.

$$A^*A = P_{(\ker A)^\perp}. \tag{7.16}$$

(iv) $\Longrightarrow$ (i) We recall that $\ker A^*A = \ker A$. Since A^*A is an orthogonal projection, $A^*Ax = x$ for all $x \in (\ker A)^\perp$. Hence $\langle A^*Ax, x\rangle_{\mathcal{H}_1} = \langle x, x\rangle_{\mathcal{H}_1}$, which is equivalent to

$$\|Ax\|_{\mathcal{H}_2} = \|x\|_{\mathcal{H}_1} \qquad (x \in (\ker A)^\perp).$$

Hence, A is a partial isometry.

(ii) $\Longleftrightarrow$ (iii) Reverse the roles of A and A^* in the preceding argument to get the equivalence of (ii) and (iii).

(iv) $\Longrightarrow$ (iii) If A^*A is an orthogonal projection, then, by (7.16),

$$(AA^*)^2 = A(A^*A)A^* = AP_{(\ker A)^\perp}A^* = AA^*.$$

Therefore, AA^* is also an orthogonal projection.

(iii) $\Longrightarrow$ (iv) The proof of this assertion is similar to the proof of the preceding case.

Note that, so far, we have proved that (i), (ii), (iii) and (iv) are equivalent.

(iii) $\Longrightarrow$ (v) If AA^* is an orthogonal projection, then it is clear that $I - AA^*$ is an orthogonal projection onto $(\mathcal{R}(AA^*))^\perp$. But $(\mathcal{R}(AA^*))^\perp = \ker AA^* = \ker A^*$. Hence we get that $\mathcal{R}(I - AA^*) = \ker A^*$.

(v) $\Longrightarrow$ (ii) Let $x_2 \in \mathcal{H}_2$ be such that $x_2 \perp \ker A^*$. Then, in particular, we have $x_2 \perp (I - AA^*)x_2$. Thus,

$$\begin{aligned}
\|A^*x_2\|_{\mathcal{H}_1}^2 &= \langle A^*x_2, A^*x_2\rangle_{\mathcal{H}_1}\\
&= \langle AA^*x_2, x_2\rangle_{\mathcal{H}_2}\\
&= \langle x_2 - (I - AA^*)x_2,\ x_2\rangle_{\mathcal{H}_2}\\
&= \langle x_2, x_2\rangle_{\mathcal{H}_2}\\
&= \|x_2\|_{\mathcal{H}_2}^2,
\end{aligned}$$

which proves that A^* is a partial isometry.

In a similar manner, one can show that (iv) $\Longrightarrow$ (vi) $\Longrightarrow$ (i). □

Using Theorem 1.30, we see from the proof of Theorem 7.22 that, if A is a partial isometry, then $\mathcal{R}(A)$ and $\mathcal{R}(A^*)$ are closed subspaces and, moreover,

$$A^*A = P_{(\ker A)^\perp} = P_{\mathcal{R}(A^*)} \quad \text{and} \quad AA^* = P_{(\ker A^*)^\perp} = P_{\mathcal{R}(A)}. \tag{7.17}$$

These identities imply the following special but important case.

Corollary 7.23 *Let $\mathcal{H}_1$ and $\mathcal{H}_2$ be Hilbert spaces and let $A \in \mathcal{L}(\mathcal{H}_1, \mathcal{H}_2)$. Then the following are equivalent:*

(i) *A is a surjective partial isometry;*
(ii) *A^* is an isometry;*
(iii) *$AA^* = I_{\mathcal{H}_2}$.*

Under the equivalent conditions of Corollary 7.23, A is called a *co-isometry* from $\mathcal{H}_1$ onto $\mathcal{H}_2$.

For each $A \in \mathcal{L}(\mathcal{H}_1, \mathcal{H}_2)$, clearly $A^*A \geq 0$, and thus $(A^*A)^{1/2}$ is a well-defined positive operator. We denote this operator by

$$|A| = (A^*A)^{1/2}$$

and call it the *absolute value* of A. For each $x \in \mathcal{H}_1$, we have

$$\begin{aligned} \|\,|A|x\,\|_{\mathcal{H}_1}^2 &= \langle (A^*A)^{1/2}x,\, (A^*A)^{1/2}x\rangle_{\mathcal{H}_1} \\ &= \langle A^*Ax, x\rangle_{\mathcal{H}_1} = \langle Ax, Ax\rangle_{\mathcal{H}_2} = \|Ax\|_{\mathcal{H}_2}^2. \end{aligned}$$

Therefore,

$$\|\,|A|x\,\|_{\mathcal{H}_1} = \|Ax\|_{\mathcal{H}_2} \qquad (x \in \mathcal{H}_1). \tag{7.18}$$

In particular, this identity has the following consequences:

(i) $\|\,|A|\,\| = \|A\|$;
(ii) $\ker |A| = \ker A$;
(iii) $\mathrm{Clos}_{\mathcal{H}_1}(\mathcal{R}(|A|)) = (\ker A)^\perp$.

If z is a complex number and we put $r = (\bar{z}z)^{1/2}$, then $r \geq 0$ and there is a complex number ζ of modulus one such that we have the polar decomposition $z = \zeta r$. A similar decomposition exists for operators on a Hilbert space. The proof has the same flavor as the proof of Douglas's factorization theorem (Theorem 7.11).

Theorem 7.24 *For each $A \in \mathcal{L}(\mathcal{H}_1, \mathcal{H}_2)$, there is a partial isometry $U \in \mathcal{L}(\mathcal{H}_1, \mathcal{H}_2)$, with the initial space $(\ker A)^\perp$ and the final space $\mathrm{Clos}_{\mathcal{H}_2}(\mathcal{R}(A))$, such that*

$$A = U|A|.$$

Proof Define

$$\begin{array}{rccc} U: & \mathcal{R}(|A|) & \longrightarrow & \mathcal{H}_2 \\ & |A|x & \longmapsto & Ax. \end{array}$$

If $|A|x_1 = |A|x_2$, then according to (7.18), we must have $Ax_1 = Ax_2$. Thus, U is a well-defined linear mapping. Moreover, using (7.18) once more, we can write

$$\|U(|A|x)\|_{\mathcal{H}_2} = \|Ax\|_{\mathcal{H}_2} = \|\,|A|x\,\|_{\mathcal{H}_1}.$$

This means that U is an isometry from $\mathcal{R}(|A|)$ onto $\mathcal{R}(A)$. Thus, by continuity, we can extend U to an isometry from $\mathrm{Clos}_{\mathcal{H}_1}(\mathcal{R}(|A|)) = (\ker A)^{\perp}$ onto $\mathrm{Clos}_{\mathcal{H}_2}(\mathcal{R}(A))$. Finally, we extend U to a bounded operator from $\mathcal{H}_1$ into $\mathcal{H}_2$ by defining

$$Ux = 0 \qquad (x \in \ker A).$$

According to the construction, we have $U|A|x = Ax$, for each $x \in \mathcal{H}_1$. Moreover, U is an isometry on $(\ker A)^{\perp}$ with $\ker U = \ker A$. In other words, U is a partial isometry, with initial space $(\ker A)^{\perp}$ and final space $\mathrm{Clos}_{\mathcal{H}_2}(\mathcal{R}(A))$. □

Theorem 7.24 clearly states the initial and final spaces of the partial isometry U. Hence, by (7.17), we have

$$U^*U = P_{(\ker U)^{\perp}} = P_{(\ker A)^{\perp}} \quad \text{and} \quad UU^* = P_{\mathcal{R}(U)} = P_{\overline{\mathcal{R}(A)}}, \tag{7.19}$$

where $\overline{\mathcal{R}(A)} = \mathrm{Clos}_{\mathcal{H}_2}(\mathcal{R}(A))$.

In the case of invertible operators, we can say a little bit more about their polar decompositions.

Corollary 7.25 *Let A be an invertible operator in $\mathcal{L}(\mathcal{H}_1, \mathcal{H}_2)$. Then there is a unitary operator $U \in \mathcal{L}(\mathcal{H}_1, \mathcal{H}_2)$ such that*

$$A = U|A|.$$

Proof According to Theorem 7.24, we know that there is a partial isometry $U \in \mathcal{L}(\mathcal{H}_1, \mathcal{H}_2)$, with the initial space $(\ker A)^{\perp}$ and the final space $\overline{\mathcal{R}(A)}$, such that $A = U|A|$. Since A is invertible, then we have $\ker A = \{0\}$ and $\overline{\mathcal{R}(A)} = \mathcal{R}(A) = \mathcal{H}_2$. Hence, U is an isometry from $\mathcal{H}_1$ onto $\mathcal{H}_2$, which means that U is a unitary operator. □

Exercises

Exercise 7.5.1 Let $\mathcal{M}$ and $\mathcal{N}$ be closed subspaces of the Hilbert spaces $\mathcal{H}_1$ and $\mathcal{H}_2$, respectively. Suppose that $\dim \mathcal{M} = \dim \mathcal{N}$. Show that there is a partial isometry $A \in \mathcal{L}(\mathcal{H}_1, \mathcal{H}_2)$ with the initial space $\mathcal{M}$ and final space $\mathcal{N}$.

Hint: Let $(x_\iota)_{\iota\in I}$ and $(y_\iota)_{\iota\in I}$ be orthonormal bases for $\mathcal{M}$ and $\mathcal{N}$, respectively. Each $x \in \mathcal{H}_1$ has the unique representation $x = x' + \sum_\iota \alpha_\iota x_\iota$, where $x' \perp \mathcal{M}$. Define $Ax = \sum_\iota \alpha_\iota y_\iota$.

Exercise 7.5.2 Show that, if $A \in \mathcal{L}(\mathcal{H}_1, \mathcal{H}_2)$ is a partial isometry, then $\mathcal{R}(A)$ is a closed subspace of $\mathcal{H}_2$. More generally, show that the image of any closed subspace of $\mathcal{H}_1$ is a closed subspace of $\mathcal{H}_2$.

Exercise 7.5.3 Let $A \in \mathcal{L}(\mathcal{H})$ be a partial isometry. Show that

$$(\ker A)^\perp = \mathcal{R}(A^*).$$

Hint: Use Exercise 7.5.2.

Exercise 7.5.4 Let $A \in \mathcal{L}(\mathcal{H}_1, \mathcal{H})$ and $B \in \mathcal{L}(\mathcal{H}_2, \mathcal{H})$. Show that the following are equivalent:

(i) there is a partial isometry $C \in \mathcal{L}(\mathcal{H}_1, \mathcal{H}_2)$, with $\ker(A)^\perp$ as initial space and $\ker(B)^\perp$ as final space, such that $A = BC$;

(ii) $AA^* = BB^*$.

Hint: Look at the proof of Douglas's factorization theorem (Theorem 7.11).

Exercise 7.5.5 Let $A \in \mathcal{L}(\mathcal{H})$. Show that there is a partial isometry $V \in \mathcal{L}(\mathcal{H})$ such that:

(i) $A = |A^*|V$;

(ii) $V^*V = P_{(\ker A)^\perp}$;

(iii) $VV^* = P_{\overline{R(A)}}$.

Hint: Apply the polar decomposition theorem to A^* (see also (7.19)).

Exercise 7.5.6 Let $A \in \mathcal{L}(\mathcal{H})$. Show that there is a sequence of real polynomials $(p_n)_{n\geq 1}$, with $p_n(0) = 0$, such that

$$p_n(|A|^2) \longrightarrow |A| \qquad (\text{in } \mathcal{L}(\mathcal{H})).$$

Hint: Use Theorem 7.2.

Exercise 7.5.7 Let $A \in \mathcal{L}(\mathcal{H}_1, \mathcal{H}_2)$ be a Hilbert space contraction. Show that

$$A(I - |A|^2)^{1/2} = (I - |A^*|^2)^{1/2} A.$$

Hint: See Theorem 7.15.

7.6 Characterization of contractions on $\ell^2(\mathbb{Z})$

If we equip $\ell^2(\mathbb{Z})$ with the standard orthonormal basis $(\mathfrak{e}_n)_{n\in\mathbb{Z}}$, then each bounded operator on $\ell^2(\mathbb{Z})$ is represented by a doubly infinite matrix $[a_{ij}]_{i,j=-\infty}^{\infty}$. Similarly, equipping ℓ^2 with the basis $(\mathfrak{e}_n)_{n\geq 0}$, we see that each operator on ℓ^2 is represented by a singly infinite matrix. Hence, after making the required change of indices, any submatrix $[a_{ij}]_{i=i_0,j=j_0}^{\infty}$, where $i_0, j_0 \in \mathbb{Z}$, can be interpreted as the matrix of an operator on ℓ^2. In particular, if A is a contraction on $\ell^2(\mathbb{Z})$, then any submatrix $[a_{ij}]_{i=i_0,j=j_0}^{\infty}$ will give a contraction on ℓ^2. It is interesting that the converse is also true.

Theorem 7.26 *Let $[a_{ij}]_{i,j=-\infty}^{\infty}$ be a doubly infinite matrix such that, for each fixed $i_0, j_0 \in \mathbb{Z}$, the submatrix $[a_{ij}]_{i=i_0,j=j_0}^{\infty}$ is a contraction on ℓ^2. Then $[a_{ij}]_{i,j=-\infty}^{\infty}$ is a contraction on $\ell^2(\mathbb{Z})$.*

Proof If A is a contraction on a Hilbert space $\mathcal{H}$, and $[\alpha_{mn}]_{m,n\geq 1}$ is the matrix representation of A with respect to an orthonormal basis in $\mathcal{H}$, then the nth column of $[\alpha_{mn}]_{m,n\geq 1}$ contains precisely the components of the image of the nth element in the basis. Hence, each column belongs to ℓ^2 and its norm does not exceed 1. Applying the same argument to A^* reveals that each row also belongs to ℓ^2 with norm at most 1. Therefore, based on the hypothesis of the theorem, we have

$$\sum_{j=j_0}^{\infty} |a_{ij}|^2 \leq 1 \qquad (i, j_0 \in \mathbb{Z}).$$

Let $j_0 \longrightarrow -\infty$ to deduce that

$$\sum_{j=-\infty}^{\infty} |a_{ij}|^2 \leq 1 \qquad (i \in \mathbb{Z}). \tag{7.20}$$

If $x = (x_n)_{n\in\mathbb{Z}} \in \ell^2(\mathbb{Z})$ is given, then (7.20) ensures that, for each fixed $i \in \mathbb{Z}$, the quantity

$$y_i = \sum_{j=-\infty}^{\infty} a_{ij}x_j$$

is well defined and finite. If we put $y = (y_n)_{n\in\mathbb{Z}}$, it is easy to see that $y = [a_{ij}]x$. Hence, to finish the proof, we need to show that $y \in \ell^2(\mathbb{Z})$ and, moreover, its norm does not exceed $\|x\|$.

Fix $x = (x_n)_{n\in\mathbb{Z}} \in \ell^2(\mathbb{Z})$ with $\|x\|_{\ell^2(\mathbb{Z})} \leq 1$. Then, for each $i_0, j_0 \in \mathbb{Z}$, the action of contraction $[a_{ij}]_{i=i_0,j=j_0}^{\infty}$ on the vector $(x_{j_0}, x_{j_0+1}, x_{j_0+2}, \dots)$ gives

$$\sum_{i=i_0}^{\infty} \left| \sum_{j=j_0}^{\infty} a_{ij}x_j \right|^2 \leq 1.$$

We emphasize that the above inequality holds uniformly with respect to i_0 and j_0. In the first step, as we saw, (7.20) allows us to let $j_0 \longrightarrow -\infty$ and obtain

$$\sum_{i=i_0}^{\infty} |y_i|^2 \leq 1.$$

In the second step, let $i_0 \longrightarrow -\infty$, to get

$$\|y\|_{\ell^2(\mathbb{Z})}^2 = \sum_{i=-\infty}^{\infty} |y_i|^2 \leq 1. \qquad \square$$

7.7 Densely defined operators

Up to now, we have only considered bounded operators on a Hilbert space $\mathcal{H}$. In the first place, such an operator is defined at all points of $\mathcal{H}$. Second, it fulfills the boundedness property

$$\|Ax\| \leq C\|x\| \qquad (x \in \mathcal{H}). \tag{7.21}$$

As a consequence of this property, the graph of A, i.e.

$$\mathcal{G}(A) = \{(x, Ax) : x \in \mathcal{H}\},$$

is a closed subspace of $\mathcal{H} \oplus \mathcal{H}$. More specifically, the latter property means that, if $(x_n, y_n) \in \mathcal{G}(A)$, i.e. $y_n = Ax_n$, and $(x_n, y_n) \longrightarrow (x, y)$, then $(x, y) \in \mathcal{G}(A)$, i.e. $y = Ax$.

In the discussion of $\mathcal{H}(b)$ spaces, we also need to consider mappings that are linear but are not defined at all points of $\mathcal{H}$. Hence, we have a linear manifold $\mathcal{D}_{(A)} \subset \mathcal{H}$ at our disposal as the *domain* of A, and the mapping $A : \mathcal{D}_{(A)} \longrightarrow \mathcal{H}$, which is linear. However, we do not assume that A satisfies an inequality like (7.21). That is why such mappings are sometimes called *unbounded operators*. But one must be careful, since it is probable that A has a bounded extension to $\mathcal{H}$. However, this terminology does not create a serious problem.

In order to compare unbounded operators, we define the partial ordering $\prec$ on this family in the following way. Given two unbounded operators A and B, we write $A \prec B$ if $\mathcal{D}_{(A)} \subset \mathcal{D}_{(B)}$ and, moreover, $B = A$ on $\mathcal{D}_{(A)}$. In other words, $A \prec B$ means that B is an extension of A to larger domain.

An operator A is called *closed* if $\mathcal{G}_{(A)}$ is a closed subspace of $\mathcal{H} \oplus \mathcal{H}$. More explicitly, this means that, if $(x_n)_{n\geq 1} \subset \mathcal{D}_{(A)}$, and both limits $x = \lim_{n\to\infty} x_n$ and $y = \lim_{n\to\infty} Ax_n$ exist, then we have $x \in \mathcal{D}_{(A)}$ and $y = Ax$. Note that for a bounded operator, the existence of $x = \lim_{n\to\infty} x_n$ automatically implies the existence of $y = \lim_{n\to\infty} Ax_n$ (and the equality $y = Ax$ holds). But this is not necessarily true for closed operators. That is why in the definition above we assumed that both limits exist. To get an interesting theory for unbounded operators, we usually assume that the domain $\mathcal{D}_{(A)}$ is a dense linear manifold in $\mathcal{H}$ and that A is a closed operator.

In the definition of the adjoint of a bounded operator, we considered the bounded functional $x \longmapsto \langle Ax, y \rangle$ on $\mathcal{H}$ and then we appealed to Riesz's representation theorem to obtain the unique element A^*y for which $\langle Ax, y \rangle = \langle x, A^*y \rangle$. For an unbounded operator, the functional $x \longmapsto \langle Ax, y \rangle$ is defined on $\mathcal{D}_{(A)}$ and, moreover, it is not necessarily a continuous mapping. Hence, we cannot apply Riesz's theorem. However, we might still have pairs y and y^* for which the identity

$$\langle Ax, y \rangle = \langle x, y^* \rangle \qquad (x \in \mathcal{D}_{(A)}) \tag{7.22}$$

holds. For example, the above identity is valid with $y = y^* = 0$. Two simple observations are in order. First, if we assume that $\mathcal{D}_{(A)}$ is dense in $\mathcal{H}$, then, for a given y, if y^* exists, it is unique. Indeed, if

$$\langle Ax, y \rangle = \langle x, y_1^* \rangle = \langle x, y_2^* \rangle \qquad (x \in \mathcal{D}_{(A)}),$$

then

$$\langle x,\ y_1^* - y_2^* \rangle = 0 \qquad (x \in \mathcal{D}_{(A)}).$$

Since $\mathcal{D}_{(A)}$ is a dense manifold in $\mathcal{H}$, we must have $y_1^* = y_2^*$. Second, as in the bounded case, it is easy to verify that the correspondence between y and y^* is linear. Hence, we define the linear mapping $A^*y = y^*$ on the domain $\mathcal{D}_{(A^*)}$ where (7.22) holds, i.e. $\mathcal{D}_{(A^*)}$ is the set of points $y \in \mathcal{H}$ for which there is a (unique) point A^*y such that

$$\langle Ax, y \rangle = \langle x, A^*y \rangle. \tag{7.23}$$

The above identity holds for all $x \in \mathcal{D}_{(A^*)}$ and all $y \in \mathcal{D}_{(A^*)}$. Note that $\mathcal{D}_{(A^*)}$ is not necessarily dense in $\mathcal{H}$. In the following, we study some elementary properties of the adjoint A^*.

The above discussion reveals the importance of operators defined on a dense subset of $\mathcal{H}$. In the following, we refer to them as *densely defined operators*. The ambient space is always a fixed Hilbert space $\mathcal{H}$.

Lemma 7.27 *Let A and B be densely defined operators with $A \prec B$. Then $B^* \prec A^*$.*

Proof By definition, we have

$$\langle Bx, y \rangle = \langle x, B^*y \rangle \qquad (x \in \mathcal{D}_{(B)},\ y \in \mathcal{D}_{(B^*)}).$$

Since $B = A$ on $\mathcal{D}_{(A)} \subset \mathcal{D}_{(B)}$, we can write

$$\langle Ax, y \rangle = \langle x, B^*y \rangle \qquad (x \in \mathcal{D}_{(A)},\ y \in \mathcal{D}_{(B^*)}).$$

This identity shows that $y \in \mathcal{D}_{(A^*)}$ and, moreover, $A^*x = B^*x$. The conclusion precisely means that $B^* \prec A^*$. □

Lemma 7.28 *Let A be a densely defined operator. Then A^* is closed.*

Proof Assume that $(y_n)_{n\geq 1} \subset \mathcal{D}_{(A^*)}$, and that both limits $y = \lim_{n\to\infty} y_n$ and $z = \lim_{n\to\infty} A^*y_n$ exist. By definition, we have

$$\langle Ax, y_n\rangle = \langle x, A^*y_n\rangle \qquad (x \in \mathcal{D}_{(A)},\ n \geq 1).$$

Let $n \longrightarrow \infty$ to get

$$\langle Ax, y\rangle = \langle x, z\rangle \qquad (x \in \mathcal{D}_{(A)}).$$

This identity reveals that $y \in \mathcal{D}_{(A^*)}$ and $A^*y = z$. Hence, A^* is closed. □

Note that, in the above lemma, we did not assume that A is closed. Nevertheless, A^* is always closed, whether this is the case or not for A.

An operator A is called *closable* if the closure of $\mathcal{G}(A)$ in $\mathcal{H} \oplus \mathcal{H}$ is the graph of some linear operator A_0. If this holds, we call A_0 (whose uniqueness is rather trivial) the *closure* of A and denote it by $\bar{A}$. Hence, the closure of A is such that

$$\mathrm{Clos}_{\mathcal{H}\oplus\mathcal{H}}\, \mathcal{G}(A) = \mathcal{G}(\bar{A}). \tag{7.24}$$

The definition above also shows that $\bar{A}$, if it exists, is a closed operator satisfying $A \prec \bar{A}$. Hence, a closable operator has a closed extension. In fact, this is a characterization of closable operators. More precisely, A is closable if and only if there is a closed operator B such that $A \prec B$. One way was demonstrated above, and the other way is a consequence of the following topological fact. A detailed examination of the definition of closable operators shows that A is closable if and only if whenever $x_n, t_n \in \mathcal{D}_{(A)}$, $x_n \longrightarrow x$, $t_n \longrightarrow x$, with $x \in \mathcal{H}$ (not necessarily in $\mathcal{D}_{(A)}$), and both $(Ax_n)_{n\geq 1}$ and $(At_n)_{n\geq 1}$ are convergent, then $\lim_{n\to\infty} Ax_n = \lim_{n\to\infty} At_n$. Surely these conditions are fulfilled whenever $A \prec B$ for some closed operator B. It also reveals that

$$D_{(\bar{A})} = \{x \in \mathcal{H} : \exists\, (x_n)_{n\geq 1} \subset \mathcal{D}_{(A)},\ x_n \longrightarrow x \text{ and } (Ax_n)_{n\geq 1} \text{ is convergent}\}.$$

Lemma 7.29 *Let A be a densely defined and closable operator. Then we have $(\bar{A})^* = A^*$.*

Proof Since $A \prec \bar{A}$, Lemma 7.27 implies that $(\bar{A})^* \prec A^*$. Going back to (7.23), the defining property of A^*, we have

$$\langle Ax, y\rangle = \langle x, A^*y\rangle \qquad (x \in \mathcal{D}_{(A)},\ y \in \mathcal{D}_{(A^*)}). \tag{7.25}$$

If (x_n, Ax_n), $n \geq 1$, is any convergent sequence in $\mathcal{G}(A)$, say $x_n \longrightarrow x$ and $Ax_n \longrightarrow z$, then $(x, z) \in \mathrm{Clos}_{\mathcal{H}\oplus\mathcal{H}}\ \mathcal{G}(A)$, which, by (7.24), implies that

$x \in D_{(\bar{A})}$ and $\bar{A}x = z$. Therefore, if we put $x = x_n$ in (7.25) and then let $n \longrightarrow \infty$, we obtain

$$\langle \bar{A}x, y \rangle = \langle x, A^* y \rangle \qquad (x \in D_{(\bar{A})},\ y \in \mathcal{D}_{(A^*)}).$$

This identity shows that $\mathcal{D}_{(A^*)} \subset D_{((\bar{A})^*)}$ and $(\bar{A})^* y = A^* y$ for all $y \in \mathcal{D}_{A^*}$. In other words, $A^* \prec (\bar{A})^*$. □

To define A^*, we had to assume that $\mathcal{D}_{(A)}$ is dense in $\mathcal{H}$. By the same token, $A^{**} = (A^*)^*$ can be defined provided that $\mathcal{D}_{(A)}$ and $\mathcal{D}_{(A^*)}$ are dense submanifolds of $\mathcal{H}$. If we write the identity (7.23) as

$$\langle A^* y, x \rangle = \langle y, Ax \rangle \qquad (y \in \mathcal{D}_{A^*},\ x \in \mathcal{D}_A),$$

it says that $x \in \mathcal{D}_{(A^{**})}$ and $A^{**}x = Ax$. This precisely means that

$$A \prec A^{**}. \tag{7.26}$$

By Lemma 7.28, A^{**} is closed. Hence, A is closable. But more can be said on this issue.

Theorem 7.30 *Let A be a densely defined operator. Then A is closable if and only if $\mathcal{D}_{(A^*)}$ is dense in $\mathcal{H}$. In this case, we have*

$$\bar{A} = A^{**}.$$

Proof If $\mathcal{D}_{(A^*)}$ is dense in $\mathcal{H}$, then, as we discussed above, (7.26) holds and thus A is closable. Moreover, we have $\bar{A} \prec A^{**}$ (see also Exercise 7.7.2).

To deal with the other part, let us find a description for $\mathcal{G}(A)^\perp$. By definition, $(u, v) \perp \mathcal{G}(A)^\perp$ if and only if

$$\langle u, x \rangle + \langle v, Ax \rangle = 0 \qquad (x \in \mathcal{D}_{(A)}).$$

Write this identity as

$$\langle Ax, v \rangle = \langle x, -u \rangle \qquad (x \in \mathcal{D}_{(A)}),$$

to conclude that

$$\mathcal{G}(A)^\perp = \{(-A^* v, v) : v \in \mathcal{D}_{(A^*)}\}. \tag{7.27}$$

Now, suppose that A is closable. Let $x \perp \mathcal{D}_{(A^*)}$. Hence, for each $v \in \mathcal{D}_{(A^*)}$, we have $(0, x) \perp (-A^* v, v)$. According to (7.24) and (7.27), this means that

$$(0, x) \in \mathcal{G}(A)^{\perp\perp} = \mathrm{Clos}_{\mathcal{H}\oplus\mathcal{H}}\, \mathcal{G}(A) = \mathcal{G}(\bar{A}).$$

Hence, $x = \bar{A}(0) = 0$. In other words, $\mathcal{D}_{(A^*)}$ is dense in $\mathcal{H}$.

The characterization (7.27) has another consequence. According to the definition of adjoint, we have

$$\langle A^* x, y \rangle = \langle x, A^{**} y \rangle \qquad (x \in \mathcal{D}_{(A^*)},\ y \in \mathcal{D}_{(A^{**})}).$$

Write this identity in $\mathcal{H} \oplus \mathcal{H}$ as

$$\langle(-A^*x, x), (y, A^{**}y)\rangle = 0 \qquad (x \in \mathcal{D}_{(A^*)},\ y \in \mathcal{D}_{(A^{**})}).$$

Hence, again by (7.24) and (7.27), we have

$$(y, A^{**}y) \in \mathcal{G}(A)^{\perp\perp} = \mathrm{Clos}_{\mathcal{H}\oplus\mathcal{H}}\, \mathcal{G}(A) = \mathcal{G}(\bar{A}).$$

Therefore, $\mathcal{D}_{(A^{**})} \subset D_{(\bar{A})}$ and $\bar{A}(y) = A^{**}y$ for all $y \in \mathcal{D}_{(A^{**})}$. This precisely means that $A^{**} \prec \bar{A}$. □

Exercises

Exercise 7.7.1 Let A be a densely defined injective operator. Assume furthermore that $\mathcal{R}(A)$ is dense in $\mathcal{H}$. Define A^{-1} on $\mathcal{D}_{(A^{-1})} = \mathcal{R}(A)$ as usual. Show that

$$(A^{-1})^* = (A^*)^{-1}.$$

Exercise 7.7.2 Let A be a densely defined closable operator. Show that $\bar{A}$ is the smallest closed extension of A.

7.8 Fredholm operators

If $A \in \mathcal{L}(\mathcal{H}_1, \mathcal{H}_2)$ is an invertible operator, then clearly $\ker A = \{0\}$ and $\ker A^* = (\mathcal{R}(A))^{\perp} = \{0\}$. Hence, $\dim\ker A = \dim\ker A^* = 0$. To obtain a larger class of operators, which still have some interesting properties, we relax these requirements in the following way. An operator $A \in \mathcal{L}(\mathcal{H}_1, \mathcal{H}_2)$ is called *Fredholm* if

(i) $\mathcal{R}(A)$ is a closed subspace of $\mathcal{H}_2$,
(ii) $\dim\ker A < \infty$, and
(iii) $\dim\ker A^* < \infty$.

The difference

$$\operatorname{ind} A = \dim\ker A - \dim\ker A^* \tag{7.28}$$

is called the *index* of A. Clearly, by (ii) and (iii), the index is an integer. In the finite-dimensional case, all operators are Fredholm.

Lemma 7.31 *If $\mathcal{H}_1$ and $\mathcal{H}_2$ are finite-dimensional Hilbert spaces and if $A : \mathcal{H}_1 \longrightarrow \mathcal{H}_2$ is any operator, then A is Fredholm and* $\operatorname{ind} A = \dim\mathcal{H}_1 - \dim\mathcal{H}_2$.

Proof Clearly, A is a Fredholm operator. Moreover, from the rank-nullity theorem, we know that

$$\dim \mathcal{H}_1 = \dim \mathcal{R}(A) + \dim \ker A.$$

Moreover, by Theorem 1.30,

$$\dim \mathcal{H}_2 = \dim \mathcal{R}(A) + \dim \mathcal{R}(A)^{\perp} = \dim \mathcal{R}(A) + \dim \ker A^*.$$

Thus,

$$\operatorname{ind} A = \dim \ker A - \dim \ker A^* = \dim \mathcal{H}_1 - \dim \mathcal{H}_2. \qquad \square$$

The following result is an immediate consequence of Theorem 1.33.

Theorem 7.32 *Let $A \in \mathcal{L}(\mathcal{H}_1, \mathcal{H}_2)$. Then*

$$A \text{ is Fredholm} \iff A^* \text{ is Fredholm.}$$

Moreover, in this situation, we have

$$\operatorname{ind} A^* = -\operatorname{ind} A.$$

An operator $A \in \mathcal{L}(\mathcal{H}_1, \mathcal{H}_2)$ is called *left semi-Fredholm* if it satisfies (i) and (ii) and *right semi-Fredholm* if it satisfies (i) and (iii). Finally, A is called *semi-Fredholm* if it is either left or right semi-Fredholm. Clearly, A is left (respectively right) semi-Fredholm if and only if A^* is right (respectively left) semi-Fredholm. Moreover, A is Fredholm if it is both left and right semi-Fredholm. Note also that, if A is left or right semi-Fredholm, then one could still define its index by formula (7.28), but, in this situation, the index is an element of $\mathbb{Z} \cup \{\pm\infty\}$. More precisely, if A is left semi-Fredholm, then $\operatorname{ind} A \in \mathbb{Z} \cup \{-\infty\}$; and if it is right semi-Fredholm, then $\operatorname{ind} A \in \mathbb{Z} \cup \{+\infty\}$. In the following, we shall focus on Fredholm operators, but most of the results have an analog for semi-Fredholm operators.

An invertible operator is surely a Fredholm operator of index 0. However, it is easy to get Fredholm operators (of index 0) that are not invertible. Instead of constructing a concrete example, let us explain a general construction. Consider a compact operator K that has 1 in its spectrum. Then according to the Fredholm alternative (see Theorem 2.6), the operator $I - K$ is a Fredholm operator of index 0. But, since $1 \in \sigma_p(K)$, the operator $I - K$ is not invertible. The following result is a slight generalization of the preceding example.

Lemma 7.33 *Assume that $A \in \mathcal{L}(\mathcal{H})$ has the decomposition $A = V + K$, where $V \in \mathcal{L}(\mathcal{H})$ is invertible and $K \in \mathcal{L}(\mathcal{H})$ is compact. Then A is Fredholm and*

$$\operatorname{ind} A = 0.$$

Proof We know that, for any compact operator S, the operator $I+S$ is a Fredholm operator with $\operatorname{ind}(I+S)=0$. Now $V^{-1}A = I+V^{-1}K$ and $V^{-1}K$ is a compact operator. Thus, $V^{-1}A$ is a Fredholm operator and $\operatorname{ind}(V^{-1}A)=0$. It is easy to verify that a Fredholm operator composed with an invertible operator, from either right or left, remains Fredholm with the same index. Hence, $A = VV^{-1}A$ is also a Fredholm operator with $\operatorname{ind} A = \operatorname{ind}(V^{-1}A)=0$. □

The following result makes an interesting and useful connection between the essential spectrum and the property of being Fredholm. Naively speaking, since the family of Fredholm operators forms a generalization of invertible operators, the following result says that the essential spectrum contains the points λ for which $A-\lambda I$ is far from being invertible.

Theorem 7.34 *Let $\mathcal{H}$ be a Hilbert space and let $A \in \mathcal{L}(\mathcal{H})$. Then*

$$\sigma_{ess}(A) = \{\lambda \in \mathbb{C} : A-\lambda I \textit{ is not Fredholm}\}.$$

Proof Let us take $\lambda \notin \sigma_{ess}(A)$ and prove that $A-\lambda I$ is Fredholm. According to the definition of essential spectrum, there is a bounded operator $B \in \mathcal{L}(\mathcal{H})$ such that

$$B(A-\lambda I) = I+K_1 \quad \text{and} \quad (A-\lambda I)B = I+K_2, \tag{7.29}$$

for some compact operators K_1, K_2 on $\mathcal{H}$. But, by Lemma 7.33, $\ker(A-\lambda I) \subset \ker B(A-\lambda I) = \ker(I+K_1) < \infty$. Similarly, using the second identity in (7.29), we obtain $\dim\ker(A-\lambda I)^* < \infty$. Lemma 7.33 also implies that $\mathcal{R}(B(A-\lambda I)) = \mathcal{R}(I+K_1)$ is closed. Hence, the operator $B(A-\lambda I)$ is bounded below on the closed subspace $\mathcal{M}^\perp$, where $\mathcal{M} = \ker(B(A-\lambda I))$. In other words, there is a constant $c>0$ such that

$$\|B(A-\lambda I)x\| \geq c\|x\| \qquad (x \perp \mathcal{M}).$$

This relation immediately implies that

$$\|(A-\lambda I)x\| \geq \frac{c}{\|B\|}\,\|x\| \qquad (x \perp \mathcal{M}).$$

Thus, the subspace $(A-\lambda I)(\mathcal{M}^\perp)$ is closed. But we have

$$\mathcal{R}(A-\lambda I) = (A-\lambda I)\mathcal{H} = (A-\lambda I)(\mathcal{M}) + (A-\lambda I)(\mathcal{M}^\perp),$$

and, since $\mathcal{M}$ is finite-dimensional, $(A-\lambda I)(\mathcal{M})$ is also finite-dimensional. Therefore, remembering that the sum of a closed subspace with a finite-dimensional subspace gives a closed subspace, we conclude that $\mathcal{R}(A-\lambda I)$ is closed. In short, we have shown that the operator $A-\lambda I$ is Fredholm.

Reciprocally, assume that the operator $A - \lambda I$ is Fredholm, and we proceed to show that $\lambda \notin \sigma_{ess}(A)$. First, set $\mathcal{N} = \ker(A - \lambda I)^\perp$ and define

$$\begin{array}{rccl} A_1: & \mathcal{N} & \longrightarrow & R(A - \lambda I) \\ & x & \longmapsto & (A - \lambda I)x. \end{array}$$

It is easy to check that A_1 is one-to-one and onto, and then, by the open mapping theorem (Theorem 1.14), A_1 is invertible. Now let P be the projection of $\mathcal{H}$ onto $\mathcal{R}(A - \lambda I)$ and define $B : \mathcal{H} \longrightarrow \mathcal{H}$ by $B = iA_1^{-1}P$, where i the canonical injection of $\mathcal{N}$ into $\mathcal{H}$. Let us check that

$$B(A - \lambda I) = I - P_{\mathcal{N}^\perp},$$

where $P_{\mathcal{N}^\perp}$ is the orthogonal projection of $\mathcal{H}$ onto $\mathcal{N}^\perp = \ker(A - \lambda I)$. Indeed, let $x \in \mathcal{H}$ and write $x = P_{\mathcal{N}^\perp}x + (I - P_{\mathcal{N}^\perp})x$. Using the fact that $P_{\mathcal{N}^\perp}x \in \ker(A - \lambda I)$ and $(I - P_{\mathcal{N}^\perp})x \in \mathcal{N}$, we have

$$\begin{aligned} B(A - \lambda I)x &= B(A - \lambda I)P_{\mathcal{N}^\perp}x + B(A - \lambda I)(I - P_{\mathcal{N}^\perp})x \\ &= B(A - \lambda I)(I - P_{\mathcal{N}^\perp})x \\ &= A_1^{-1}P(A - \lambda I)(I - P_{\mathcal{N}^\perp})x \\ &= A_1^{-1}(A - \lambda I)(I - P_{\mathcal{N}^\perp})x \\ &= (I - P_{\mathcal{N}^\perp})x. \end{aligned}$$

Thus, $B(A - \lambda I) = I - P_{\mathcal{N}^\perp}$. Similarly, we can check that $(A - \lambda I)B = I - P_{\mathcal{M}^\perp}$, where $P_{\mathcal{M}^\perp}$ is the orthogonal projection of $\mathcal{H}$ onto $\ker(A - \lambda I)^*$. Since $\dim\ker(A - \lambda I) < \infty$ and $\dim\ker(A - \lambda I)^* < \infty$, we get that $P_{\mathcal{N}^\perp}$ and $P_{\mathcal{M}^\perp}$ are finite-rank operators. Therefore, $\pi(P_{\mathcal{N}^\perp}) = \pi(P_{\mathcal{M}^\perp}) = 0$ and we obtain

$$\pi(A - \lambda I)\pi(B) = \pi(B)\pi(A - \lambda I) = I_{\mathcal{L}(\mathcal{H})/\mathcal{K}(\mathcal{H})},$$

i.e. $\pi(A - \lambda I)$ is invertible. This means that $\lambda \notin \sigma_{ess}(A)$. □

Another way to interpret Theorem 7.34 is the following corollary. In fact, it is equivalent to that theorem.

Corollary 7.35 *Let $\mathcal{H}$ be a Hilbert space and let $A \in \mathcal{L}(\mathcal{H})$. Then A is a Fredholm operator if and only if $\pi(A)$ is invertible in $\mathcal{L}(\mathcal{H})/\mathcal{K}(\mathcal{H})$.*

Proof By Theorem 7.34, A is a Fredholm operator if and only if 0 is not in $\sigma_{ess}(A)$. Therefore, A is a Fredholm operator if and only if $\pi(A)$ is invertible in $\mathcal{L}(\mathcal{H})/\mathcal{K}(\mathcal{H})$. □

Under the condition of Corollary 7.35, there are a bounded operator $B \in \mathcal{L}(\mathcal{H})$ and two compact operators K_1 and K_2 such that

$$BA = I + K_1 \quad \text{and} \quad AB = I + K_2.$$

This means that $\pi(A)\pi(B) = \pi(B)\pi(A) = \pi(I)$. But, if we closely examine the proof of Theorem 7.34, we see that B could be chosen so that $K_1 = P_{\ker A}$ and $K_2 = P_{\ker A^*}$. Therefore, in particular, K_1 and K_2 are finite-rank operators. The characterization given in Corollary 7.35 immediately implies the following result, which generalizes a part of Lemma 7.33.

Corollary 7.36 *Let $\mathcal{H}$ be a Hilbert space, let $A \in \mathcal{L}(\mathcal{H})$ be Fredholm, and let $K \in \mathcal{L}(\mathcal{H})$ be compact. Then $A + K$ is Fredholm.*

Proof Just note that $\pi(A + K) = \pi(A)$. □

Exercises

Exercise 7.8.1 Let U be the unilateral forward shift on ℓ^2, that is

$$U(x_0, x_1, x_2, \dots) = (0, x_0, x_1, x_2, x_3, \dots).$$

Show that U is a Fredholm operator and compute its index.

Exercise 7.8.2 Let $A \in \mathcal{L}(\mathcal{H})$ be a normal operator.

(i) Show that $\ker A = \ker A^*$.
(ii) Deduce that A is semi-Fredholm if and only A is Fredholm.

Hint: Use (2.16).

Exercise 7.8.3 Let $A \in \mathcal{L}(\mathcal{H})$ be a normal Fredholm operator. Show that $\operatorname{ind} A = 0$.
Hint: Use Exercise 7.8.2.

Exercise 7.8.4 Let A be a normal operator on a Hilbert space $\mathcal{H}$. Show that A is a Fredholm operator if and only if 0 is not a limit point of $\sigma(A)$ and $\dim \ker A < \infty$.

Exercise 7.8.5 Let $\mathcal{H}$ be a Hilbert space and let $A \in \mathcal{L}(\mathcal{H})$. Show that

$$\sigma_{ess}^{\ell}(A) = \{\lambda \in \mathbb{C} : A - \lambda I \text{ is left semi-Fredholm}\}.$$

Hint: See Theorem 7.34.

Exercise 7.8.6 Let $A \in \mathcal{L}(\mathcal{H})$. Show that A is left semi-Fredholm if and only if there is no sequence $(x_n)_{n\geq 1}$ of unit vectors in $\mathcal{H}$ such that $x_n \xrightarrow{w} 0$ (weakly) and $\|Ax_n\| \longrightarrow 0$, as $n \longrightarrow \infty$.

Exercise 7.8.7 Let $A \in \mathcal{L}(\mathcal{H})$. Show that $A\mathcal{M}$ is closed, for every closed subspace $\mathcal{M}$ of $\mathcal{H}$, if and only if either the rank of A is finite or A is left semi-Fredholm.

Exercise 7.8.8 Let A be a bounded operator on the Banach space $\mathcal{X}$, and let K be a compact operator on $\mathcal{X}$. Assume that λ is not in the spectrum of A, but it is in the spectrum of $A+K$. Show that λ is an eigenvalue of $A+K$.
Hint: Since λ is not in the spectrum of A, the operator $A - \lambda I$ is invertible. Then, by Lemma 7.33, $A + K - \lambda I$ is a Fredholm operator of index 0. This means that this operator has closed range and we have

$$\dim\, \ker(A + K - \lambda I) = \dim\, \ker(A + K - \lambda I)^*.$$

If λ is not an eigenvalue of $A+K$, the above identity ensures that $A+K-\lambda I$ and its adjoint are both one-to-one. Thus, $A + K - \lambda I$ would be invertible, which is absurd by hypothesis. Hence, λ is an eigenvalue of $A + K$.

7.9 Essential spectrum of block-diagonal operators

Theorem 7.34 has a useful consequence concerning the essential spectrum of a block-diagonal operator.

Theorem 7.37 *Let $\mathcal{H} = \mathcal{M} \oplus \mathcal{N}$ and $\mathcal{H}' = \mathcal{M}' \oplus \mathcal{N}'$, and let $A \in \mathcal{L}(\mathcal{H}, \mathcal{H}')$ admit the matrix representation*

$$A = \begin{bmatrix} A_1 & 0 \\ 0 & A_2 \end{bmatrix} : \begin{matrix} \mathcal{M} \\ \oplus \\ \mathcal{N} \end{matrix} \longrightarrow \begin{matrix} \mathcal{M}' \\ \oplus \\ \mathcal{N}' \end{matrix}.$$

Then $\sigma_{ess}(A) = \sigma_{ess}(A_1) \cup \sigma_{ess}(A_2)$.

Proof Let $\lambda \in \mathbb{C}$. Then according to Theorem 7.34, $\lambda \notin \sigma_{ess}(A)$ if and only if $A-\lambda I$ is a Fredholm operator. The latter means that the following properties are satisfied:

(a) $\dim(\ker(A - \lambda I)) < \infty$;
(b) $\dim(\ker(A^* - \bar{\lambda} I)) < \infty$; and
(c) $A - \lambda I$ has a closed range.

According to (2.6), condition (a) is equivalent to

$$\dim(\ker(A_1 - \lambda I)) < \infty \quad \text{and} \quad \dim(\ker(A_2 - \lambda I)) < \infty,$$

and, considering (1.40), condition (b) is equivalent to

$$\dim(\ker(A_1^* - \bar{\lambda} I)) < \infty \quad \text{and} \quad \dim(\ker(A_2^* - \bar{\lambda} I)) < \infty.$$

Moreover, by Lemma 1.38, condition (c) is equivalent to saying that both operators $A_1 - \lambda I$ and $A_2 - \lambda I$ have closed ranges. In other words, $A - \lambda I$ is a Fredholm operator if and only if both $A_1 - \lambda I$ and $A_2 - \lambda I$ are Fredholm operators. Therefore, again by Theorem 7.34, this means that

$$\mathbb{C} \setminus \sigma_{ess}(A) = \mathbb{C} \setminus (\sigma_{ess}(A_1) \cup \sigma_{ess}(A_2)),$$

which gives the desired assertion. □

The following result has a similar flavor. However, note that the matrix representation is not necessarily diagonal.

Theorem 7.38 *Let $\mathcal{H} = \mathcal{M} \oplus \mathcal{N}$ and $\mathcal{H}' = \mathcal{M}' \oplus \mathcal{N}'$, and let $A \in \mathcal{L}(\mathcal{H}, \mathcal{H}')$ admit the matrix representation*

$$A = \begin{bmatrix} A_1 & A_3 \\ 0 & A_2 \end{bmatrix} : \begin{matrix} \mathcal{M} \\ \oplus \\ \mathcal{N} \end{matrix} \longrightarrow \begin{matrix} \mathcal{M}' \\ \oplus \\ \mathcal{N}' \end{matrix}.$$

If A_1 is invertible, and $\mathcal{N}$ and $\mathcal{N}'$ are finite-dimensional, then A is a Fredholm operator and

$$\operatorname{ind} A = \dim \mathcal{N} - \dim \mathcal{N}'.$$

Proof It is easy to see that $\mathcal{R}(A) = A_1\mathcal{M} + (A_3 + A_2)\mathcal{N}$. But A_1 is invertible, whence $A_1\mathcal{M} = \mathcal{M}'$ is closed. Moreover, since $\mathcal{N}$ is finite-dimensional, the subspace $(A_3 + A_2)\mathcal{N}$ is also finite-dimensional. Then we deduce that $\mathcal{R}(A)$ is closed. Moreover, according to Lemma 7.31, the operator A_2 is Fredholm and

$$\operatorname{ind} A_2 = \dim \mathcal{N} - \dim \mathcal{N}'. \tag{7.30}$$

We now show that

$$\ker A^* = 0 \oplus \ker A_2^* \tag{7.31}$$

and

$$\dim \ker A = \dim \ker A_2. \tag{7.32}$$

For the first equality, let $f' \in \mathcal{M}'$ and $g' \in \mathcal{N}'$. Then, using (1.40), we have

$$A^*(f' \oplus g') = A_1^* f' \oplus (A_3^* f' + A_2^* g').$$

So $f' \oplus g' \in \ker A^*$ if and only if

$$A_1^* f' = 0 \quad \text{and} \quad A_2^* g' = -A_3^* f'.$$

But A_1 is invertible, which implies that A_1^* is invertible and thus $f' = 0$, and hence $A_2^* g' = 0$. From here it is clear that $\ker A^* = 0 \oplus \ker A_2^*$. To establish the second equality, consider the mapping

$$\begin{array}{rccl} \varphi : & \ker A_2 & \longrightarrow & \ker A \\ & g & \longmapsto & -A_1^{-1}A_3 g \oplus g. \end{array}$$

In the first place, if $g \in \ker A_2$, then

$$A(-A_1^{-1}A_3 g \oplus g) = (-A_3 g + A_3 g) \oplus A_2 g = 0,$$

i.e. $-A_1^{-1}A_3 g \oplus g \in \ker A$, and thus the mapping φ is well defined. Second, let $h_1 \in \mathcal{M}$ and $h_2 \in \mathcal{N}$ be such that $h_1 \oplus h_2 \in \ker A$. Then

$$0 = A(h_1 \oplus h_2) = (A_1 h_1 + A_3 h_2) \oplus A_2 h_2,$$

which gives

$$A_2 h_2 = 0 \quad \text{and} \quad A_1 h_1 + A_3 h_2 = 0.$$

Hence, $h_2 \in \ker A_2$ and $h_1 = -A_1^{-1}A_2 h_2$. Therefore, by taking $g = h_2$, we get

$$h_1 \oplus h_2 = -A_1^{-1}A_3 g \oplus g.$$

This means that φ is surjective. It is clear that φ is also injective. In short, φ is a bijection between $\ker A_2$ and $\ker A$, and this fact implies (7.32).

The relations (7.31) and (7.32) immediately imply that

$$\dim \ker A < \infty \quad \text{and} \quad \dim \ker A^* < \infty,$$

and thus A is a Fredholm operator. Moreover, using (7.30), we obtain

$$\begin{aligned} \operatorname{ind} A &= \dim \ker A - \dim \ker A^* \\ &= \dim \ker A_2 - \dim \ker A_2^* \\ &= \operatorname{ind} A_2 \\ &= \dim \mathcal{N} - \dim \mathcal{N}'. \end{aligned}$$

This completes the proof. □

The sum of two Fredholm operators is not necessarily Fredholm. For example, on an infinite-dimensional Hilbert space, the operator 0 is not Fredholm, and thus the sum of A and $-A$, where A is any Fredholm operator, e.g. $A = I$, is not Fredholm. However, this family is closed under multiplication. This is established below.

Theorem 7.39 *If $A, B \in \mathcal{L}(\mathcal{H})$ are Fredholm operators, then AB and BA are also Fredholm operators and*

$$\operatorname{ind} AB = \operatorname{ind} BA = \operatorname{ind} A + \operatorname{ind} B.$$

Proof Clearly, reversing the role of A and B, it is sufficient to prove the result for BA. Using Theorem 7.34, we have $0 \notin \sigma_{ess}(A)$ and $0 \notin \sigma_{ess}(B)$. Hence, there are operators $A', B' \in \mathcal{L}(\mathcal{H})$ such that

$$A'A = I + K_1, \qquad AA' = I + K_1',$$

$$B'B = I + K_2, \qquad BB' = I + K_2',$$

for some compact operators K_1, K_1', K_2, K_2' on $\mathcal{H}$. On the one hand, we have

$$A'B'(BA) = A'(I + K_2)A = A'A + A'K_2A = I + K_1 + A'K_2A,$$

and $K_1 + A'K_2A$ is a compact operator. On the other hand, we have

$$BA(A'B') = B(I + K_1')B' = BU + BK_1'B' = I + K_2' + BK_1'B',$$

and $K_2' + BK_1'U$ is a compact operator. Hence, $0 \notin \sigma_{ess}(BA)$ and using Theorem 7.34 once more, we deduce that BA is a Fredholm operator.

It remains to prove the formula for the index, which is slightly more sophisticated to verify. We will use the following three orthogonal decompositions of $\mathcal{H}$:

$$\mathcal{H} = \mathcal{M} \oplus \mathcal{N}, \qquad \mathcal{H} = \mathcal{M}' \oplus \mathcal{N}', \qquad \mathcal{H} = \mathcal{M}'' \oplus \mathcal{N}'',$$

where

$$\mathcal{M}' = \mathcal{R}(A) \cap (\ker B)^{\perp}, \qquad \mathcal{N}' = \mathcal{H} \ominus \mathcal{M}',$$

$$\mathcal{M} = A^{-1}(\mathcal{M}') \cap (\ker A)^{\perp}, \qquad \mathcal{N} = \mathcal{H} \ominus \mathcal{M}$$

and

$$\mathcal{M}'' = B\mathcal{M}', \qquad \mathcal{N}'' = \mathcal{H} \ominus \mathcal{M}''.$$

Note that $A\mathcal{M} = \mathcal{M}'$. Therefore, with respect to these decompositions, the operators A, B and BA have the matrix representations

$$A = \begin{bmatrix} A_1 & A_3 \\ 0 & A_2 \end{bmatrix} : \begin{matrix} \mathcal{M} \\ \oplus \\ \mathcal{N} \end{matrix} \longrightarrow \begin{matrix} \mathcal{M}' \\ \oplus \\ \mathcal{N}' \end{matrix},$$

$$B = \begin{bmatrix} B_1 & B_3 \\ 0 & B_2 \end{bmatrix} : \begin{matrix} \mathcal{M}' \\ \oplus \\ \mathcal{N}' \end{matrix} \longrightarrow \begin{matrix} \mathcal{M}'' \\ \oplus \\ \mathcal{N}'' \end{matrix}$$

and

$$BA = \begin{bmatrix} B_1A_1 & C \\ 0 & B_2A_2 \end{bmatrix} : \begin{matrix} \mathcal{M} \\ \oplus \\ \mathcal{N} \end{matrix} \longrightarrow \begin{matrix} \mathcal{M}'' \\ \oplus \\ \mathcal{N}'' \end{matrix}.$$

In order to exploit Theorem 7.38, we proceed to show that $\mathcal{N}$, $\mathcal{N}'$ and $\mathcal{N}''$ are finite-dimensional and that A_1 and B_1 are invertible.

A_1 and B_1 are invertible. We have

$$\ker A_1 = \ker A \cap \mathcal{M} = \ker A \cap (A^{-1}(\mathcal{M}') \cap (\ker A)^\perp) = \{0\}$$

and $A_1\mathcal{M} = A\mathcal{M} = \mathcal{M}'$. Thus A_1 is invertible. Similarly, since

$$\ker B_1 = \ker B \cap \mathcal{M}' = \ker B \cap (\mathcal{R}(A) \cap (\ker B)^\perp) = \{0\}$$

and $B_1\mathcal{M}' = B\mathcal{M}' = \mathcal{M}''$, we deduce that B_1 is also invertible.

$\mathcal{N}'$ is finite-dimensional. According to (1.28), (1.29) and Theorem 1.30, we have

$$\mathcal{N}' = \mathcal{M}'^\perp = \mathcal{R}(A)^\perp + \ker B = \ker A^* + \ker B.$$

Since both $\ker A^*$ and $\ker B$ are finite-dimensional (because A and B are Fredholm), we get $\dim \mathcal{N}' < \infty$.

$\mathcal{N}$ is finite-dimensional. Consider the operator

$$\begin{array}{rccc} \mathbf{A}: & (\ker A)^\perp & \longrightarrow & \mathcal{R}(A) \\ & x & \longmapsto & Ax. \end{array}$$

It is easy to see that $\mathbf{A}$ is invertible and we have $\mathbf{A}\mathcal{M} = \mathcal{M}'$. Therefore,

$$\begin{aligned} \dim[(\ker A)^\perp \cap \mathcal{M}^\perp] &= \dim[\mathcal{R}(A) \cap \mathcal{M}'^\perp] \\ &\le \dim \mathcal{M}'^\perp = \dim \mathcal{N}' < \infty. \end{aligned}$$

Moreover, since A is a Fredholm operator, $\dim \ker A < \infty$ and then

$$\dim[\ker A \cap \mathcal{M}^\perp] < \infty,$$

which implies that

$$\dim \mathcal{N} = \dim \mathcal{M}^\perp < \infty.$$

$\mathcal{N}''$ is finite-dimensional. We use a similar argument by considering the operator

$$\begin{array}{rccc} \mathbf{B}: & (\ker B)^\perp & \longrightarrow & \mathcal{R}(B) \\ & x & \longmapsto & Bx. \end{array}$$

It is easy to see that $\mathbf{B}$ is invertible and we have $\mathbf{B}\mathcal{M}' = B\mathcal{M}' = \mathcal{M}''$. Therefore,

$$\begin{aligned} \dim[\mathcal{R}(B) \cap \mathcal{M}''^\perp] &= \dim[(\ker B)^\perp \cap \mathcal{M}'^\perp] \\ &\le \dim \mathcal{M}'^\perp = \dim \mathcal{N}' < \infty. \end{aligned}$$

Moreover, since B is a Fredholm operator, $\dim \mathcal{R}(B)^\perp = \dim \ker B^* < \infty$, which implies that

$$\dim[\mathcal{R}(B)^\perp \cap \mathcal{M}''^\perp] < \infty,$$

whence

$$\dim \mathcal{N}'' = \dim \mathcal{M}''^\perp < \infty.$$

We can now apply Theorem 7.38, which gives

$$\begin{aligned} \operatorname{ind} A &= \dim \mathcal{N} - \dim \mathcal{N}', \\ \operatorname{ind} B &= \dim \mathcal{N}' - \dim \mathcal{N}'', \\ \operatorname{ind} BA &= \dim \mathcal{N} - \dim \mathcal{N}'', \end{aligned}$$

whence

$$\operatorname{ind}(BA) = \operatorname{ind} A + \operatorname{ind} B.$$ □

We immediately obtain the following corollary.

Corollary 7.40 *Let $A \in \mathcal{L}(\mathcal{H})$ be a Fredholm operatorand let V be an invertible operator in $\mathcal{L}(\mathcal{H})$. Then VAV^{-1} is a Fredholm operator and*

$$\operatorname{ind}(VAV^{-1}) = \operatorname{ind} A.$$

Proof Clearly, an invertible operator is Fredholm. Hence, by Theorem 7.39,

$$\begin{aligned} \operatorname{ind}(VAV^{-1}) &= \operatorname{ind} V + \operatorname{ind} A + \operatorname{ind} V^{-1} \\ &= \operatorname{ind} A + (\operatorname{ind} V + \operatorname{ind} V^{-1}) \\ &= \operatorname{ind} A + \operatorname{ind}(VV^{-1}) \\ &= \operatorname{ind} A + \operatorname{ind} I \\ &= \operatorname{ind} A. \end{aligned}$$

This completes the proof. □

We can now add a bit more to the contents of Corollary 7.36.

Corollary 7.41 *Let $A \in \mathcal{L}(\mathcal{H})$ be a Fredholm operator and let $K \in \mathcal{L}(\mathcal{H})$ be a compact operator. Then $A + K$ is a Fredholm operator and*

$$\operatorname{ind}(A + K) = \operatorname{ind} A.$$

Proof By Corollary 7.35, there is an operator $B \in \mathcal{L}(\mathcal{H})$ such that

$$\pi(A)\pi(B) = \pi(B)\pi(A) = \pi(I).$$

Hence, B is also Fredholm. By the same result, since

$$\pi(AB) = \pi((A + K)B) = \pi(I),$$

both AB and $(A+K)B$ are also Fredholm. Moreover, by Lemma 7.33,

$$\text{ind}(AB) = \text{ind}((A+K)B) = 0.$$

But, by Theorem 7.34,

$$\text{ind}(AB) = \text{ind}\, A + \text{ind}\, B$$

and

$$\text{ind}((A+K)B) = \text{ind}(A+K) + \text{ind}\, B,$$

which gives $\text{ind}(A+K) = \text{ind}\, A$. □

Using Theorem 7.34, it is easy to see that the set of Fredholm operators on $\mathcal{H}$ is an open set in the space $\mathcal{L}(\mathcal{H})$. We shed more light on this fact by showing that the index is in fact unchanged under small perturbations.

Theorem 7.42 *Let $A \in \mathcal{L}(\mathcal{H})$ be a Fredholm operator. Then there exists an $\varepsilon = \varepsilon(A) > 0$ such that, if $B \in \mathcal{L}(\mathcal{H})$, with $\|B\| \le \varepsilon$, then $A+B$ is a Fredholm operator and, moreover,*

$$\text{ind}(A+B) = \text{ind}\, A.$$

Proof Consider the decompositions

$$\mathcal{H} = (\ker A)^{\perp} \oplus \ker A \quad \text{and} \quad \mathcal{H} = \mathcal{R}(A) \oplus \ker A^*.$$

Then, with respect to these decompositions, the operator A has the matrix representation

$$A = \begin{bmatrix} A_1 & 0 \\ 0 & 0 \end{bmatrix} : \begin{matrix} (\ker A)^{\perp} \\ \oplus \\ \ker A \end{matrix} \longrightarrow \begin{matrix} \mathcal{R}(A) \\ \oplus \\ \ker A^* \end{matrix},$$

where the operator $A_1 : (\ker A)^{\perp} \longrightarrow \mathcal{R}(A)$ is invertible. But the collection of invertible elements of $\mathcal{L}((\ker A)^{\perp}, \mathcal{R}(A))$ is an open subset, and thus there exists an $\varepsilon > 0$ such that, if $\|B_1\| < \varepsilon$, then $A_1 + B_1$ is invertible.

Now, let $B \in \mathcal{L}(\mathcal{H})$ such that $\|B\| < \varepsilon$ and write its matrix decomposition relative to the preceding decomposition of $\mathcal{H}$:

$$B = \begin{bmatrix} B_1 & B_2 \\ B_3 & B_4 \end{bmatrix} : \begin{matrix} (\ker A)^{\perp} \\ \oplus \\ \ker A \end{matrix} \longrightarrow \begin{matrix} \mathcal{R}(A) \\ \oplus \\ \ker A^* \end{matrix}.$$

Thus,

$$A + B = \begin{bmatrix} A_1 + B_1 & B_2 \\ B_3 & B_4 \end{bmatrix} = \begin{bmatrix} A_1 + B_1 & B_2 \\ 0 & B_4 \end{bmatrix} + \begin{bmatrix} 0 & 0 \\ B_3 & 0 \end{bmatrix}.$$

Since $\|B_1\| = \|P_{\mathcal{R}(A)} B i_{(\ker A)^\perp}\| \le \|B\| \le \varepsilon$, we conclude that $A_1 + B_1$ is invertible. Moreover, since $\ker A$ and $\ker A^*$ are finite-dimensional, Theorem 7.38 implies that the first matrix

$$\begin{bmatrix} A_1 + B_1 & B_2 \\ 0 & B_4 \end{bmatrix}$$

is a Fredholm operator with

$$\operatorname{ind} \begin{bmatrix} A_1 + B_1 & B_2 \\ 0 & B_4 \end{bmatrix} = \dim\ker A - \dim\ker A^* = \operatorname{ind} A.$$

Finally, the second matrix

$$\begin{bmatrix} 0 & 0 \\ B_3 & 0 \end{bmatrix}$$

gives an operator of finite rank. Thus, an application of Corollary 7.41 shows that $A + B$ is a Fredholm operator and

$$\operatorname{ind}(A+B) = \operatorname{ind} \begin{bmatrix} A_1 + B_1 & B_2 \\ 0 & B_4 \end{bmatrix} = \operatorname{ind} A. \qquad \square$$

Corollary 7.43 *The set of Fredholm operators is an open subset of $\mathcal{L}(\mathcal{H})$.*

Exercise

Exercise 7.9.1 Show that the index is a constant function on each connected component of the set of Fredholm operators.
Hint: Use Theorem 7.42.

7.10 The dilation theory

Among the family of contractions, the isometric and unitary operators have more interesting properties. In fact, arbitrary contractions are related to isometries via dilations. In the main result of this section, we show that any Hilbert space contraction has an isometric dilation. This result is an important tool in operator theory and will be used in the next section to obtain an abstract commutant lifting theorem.

We recall that, if $\mathcal{K}$ is a Hilbert space, $\mathcal{H}$ is a (closed) subspace of $\mathcal{K}$, $S \in \mathcal{L}(\mathcal{K})$ and $T \in \mathcal{L}(\mathcal{H})$, then S is called a *dilation* of T if

$$T^n = P_{\mathcal{H}} S^n i_{\mathcal{H}} \qquad (n \ge 0).$$

If, in addition, S is an isometry (unitary operator), then S is called an isometric (unitary) dilation of T. An isometric (unitary) dilation S of T is said to be *minimal* if no restriction of S to an invariant subspace is an isometric (unitary) dilation of T.

Lemma 7.44 *Let $S \in \mathcal{L}(\mathcal{K})$ be an isometric (unitary) dilation of $T \in \mathcal{L}(\mathcal{H})$. Then S is a minimal isometric dilation of T if and only if*

$$\bigvee_{n=0}^{\infty} S^n \mathcal{H} = \mathcal{K}.$$

Similarly, S is a minimal unitary dilation of T if and only if

$$\bigvee_{n=-\infty}^{\infty} S^n \mathcal{H} = \mathcal{K}.$$

Proof Assume, first, that S is a minimal isometric dilation, and let

$$\mathcal{K}' = \bigvee_{n=0}^{\infty} S^n \mathcal{H}.$$

It is clear that $S\mathcal{K}' \subset \mathcal{K}'$. Moreover, for any $n \geq 0$ and $h \in \mathcal{H}$, we have

$$P_{\mathcal{H}} S_{\mathcal{K}'}^n h = P_{\mathcal{H}} S^n h = T^n h,$$

where $S_{\mathcal{K}'} = S_{|\mathcal{K}'} : \mathcal{K}' \longrightarrow \mathcal{K}'$. Hence, $S_{\mathcal{K}'}$ is a dilation of T. Therefore, by the definition of minimality, we must have $\mathcal{K}' = \mathcal{K}$.

Conversely, assume that S is an isometric dilation of T and that $\mathcal{K} = \bigvee_{n=0}^{\infty} S^n \mathcal{H}$. Let $\mathcal{H} \subset \mathcal{K}' \subset \mathcal{K}$ be such that $S\mathcal{K}' \subset \mathcal{K}'$ and $S_{\mathcal{K}'} = S_{|\mathcal{K}'} : \mathcal{K}' \longrightarrow \mathcal{K}'$ is an isometric dilation of T. Since $\mathcal{H} \subset \mathcal{K}'$, we have

$$S^n \mathcal{H} \subset S^n \mathcal{K}' \subset \mathcal{K}' \qquad (n \geq 0).$$

Hence,

$$\mathcal{K} = \bigvee_{n=0}^{\infty} S^n \mathcal{H} \subset \mathcal{K}',$$

which implies that $\mathcal{K} = \mathcal{K}'$. Thus S is a minimal isometric dilation of T. The proof for unitary dilations is similar. □

Let T be a contraction on $\mathcal{H}$. We recall the following notation:

$$\begin{aligned} D_T &= (I - T^*T)^{1/2}, \\ D_{T^*} &= (I - TT^*)^{1/2}, \\ \mathcal{D}_T &= \mathrm{Clos}(D_T \mathcal{H}), \\ \mathcal{D}_{T^*} &= \mathrm{Clos}(D_{T^*} \mathcal{H}). \end{aligned}$$

The operator D_T is called the *defect operator* of T and $\mathcal{D}_T$ is called the *defect space* (see Section 7.2). Since

$$\|x\|^2 - \|Tx\|^2 = \langle x, x\rangle - \langle T^*Tx, x\rangle \qquad (x \in \mathcal{H}),$$

we deduce that

$$\|x\|^2 = \|Tx\|^2 + \|D_T x\|^2 \qquad (x \in \mathcal{H}). \tag{7.33}$$

This identity will be exploited repeatedly.

Theorem 7.45 *Every contraction $T \in \mathcal{L}(\mathcal{H})$ has a minimal isometric dilation. This dilation is unique in the following sense. If $S \in \mathcal{L}(\mathcal{K})$ and $S' \in \mathcal{L}(\mathcal{K}')$ are two minimal isometric dilations for T, then there exists a unitary operator $U : \mathcal{K} \longrightarrow \mathcal{K}'$ such that*

$$Ux = x \qquad (x \in \mathcal{H})$$

and $S'U = US$.

Proof (Existence part) Let us form the Hilbert space

$$\mathcal{K}_+ = \mathcal{H} \oplus \bigoplus_{n=0}^{\infty} \mathcal{D}_n, \tag{7.34}$$

where $D_n = \mathcal{D}_T$, and whose elements are the vectors $\mathbf{h} = (h, h_0, h_1, \dots)$ with $h \in \mathcal{H}$, $h_n \in \mathcal{D}_n$, and

$$\|\mathbf{h}\|^2 = \|h\|^2 + \sum_{n=0}^{\infty} \|h_n\|^2 < \infty.$$

We embed $\mathcal{H}$ into $\mathcal{K}_+$ as a subspace by identifying the element $h \in \mathcal{H}$ with the element $(h, 0, 0, \dots) \in \mathcal{K}_+$. Next, we define the operator $U_+ \in \mathcal{L}(\mathcal{K}_+)$ by

$$U_+\mathbf{h} = (Th, D_T h, h_0, h_1, \dots), \qquad \mathbf{h} = (h, h_0, h_1, \dots) \in \mathcal{K}_+. \tag{7.35}$$

By (7.33), we easily see that U_+ is an isometry. Let $\mathcal{N} = \{0\} \oplus \bigoplus_{n=0}^{\infty} \mathcal{D}_n$. It is clear that $U_+\mathcal{N} \subset \mathcal{N}$ and

$$\mathcal{K}_+ = \mathcal{H} \oplus \mathcal{N}.$$

By Theorem 1.41, we deduce that U_+ is a dilation of T. It remains to show that U_+ is minimal. Let $h \in \mathcal{H}$. Then

$$U_+h = (Th, D_T h, 0, 0, \dots).$$

Hence

$$\mathcal{H} \vee U_+\mathcal{H} \subset \mathcal{H} \oplus \mathcal{D}_T \oplus \{0\} \oplus \{0\} \oplus \cdots.$$

Conversely, let $\mathbf{h} = (x, D_T h, 0, 0, \dots) \in \mathcal{H} \oplus \mathcal{D}_T \oplus \{0\} \oplus \{0\} \oplus \cdots$, where $x, h \in \mathcal{H}$. Then

$$\begin{aligned}\mathbf{h} &= (x - Th, 0, 0, \dots) + (Th, D_T h, 0, 0, \dots) \\ &= (x - Th, 0, 0, \dots) + U_+ h,\end{aligned}$$

which proves that $\mathbf{h} \in \mathcal{H} \vee U_+\mathcal{H}$. Thus,

$$\mathcal{H} \vee U_+\mathcal{H} = \mathcal{H} \oplus \mathcal{D}_T \oplus \{0\} \oplus \{0\} \oplus \cdots .$$

It now follows easily from the definition of U_+ that

$$\bigvee_{j=0}^{n} U_+^j \mathcal{H} = \mathcal{H} \oplus \underbrace{\mathcal{D}_T \oplus \cdots \oplus \mathcal{D}_T}_{n \text{ times}} \oplus \{0\} \oplus \{0\} \oplus \cdots .$$

and the minimality of U_+ follows from Lemma 7.44.

(Uniqueness part) Let $S \in \mathcal{L}(\mathcal{K})$ and $S' \in \mathcal{L}(\mathcal{K}')$ be two minimal isometric dilations of T. By Lemma 7.44, we have

$$\mathcal{K} = \bigvee_{n=0}^{\infty} S^n \mathcal{H} \quad \text{and} \quad \mathcal{K}' = \bigvee_{n=0}^{\infty} S'^n \mathcal{H}.$$

If $(x_j)_{j \ge 0}$ is a finitely supported sequence of vectors in $\mathcal{H}$, we have

$$\left\| \sum_{j=0}^{\infty} S^j x_j \right\|^2 = \sum_{j,k=0}^{\infty} \langle S^j x_j, S^k x_k \rangle.$$

Since S is an isometry, we have

$$\langle S^j x_j, S^k x_k \rangle = \langle S^{j-k} x_j, x_k \rangle \qquad (j \ge k)$$

and

$$\langle S^j x_j, S^k x_k \rangle = \langle x_j, S^{k-j} x_k \rangle \qquad (j < k).$$

Hence,

$$\begin{aligned}\left\| \sum_{j=0}^{\infty} S^j x_j \right\|^2 &= \sum_{j \ge k} \langle S^{j-k} x_j, x_k \rangle + \sum_{0 \le j < k} \langle x_j, S^{k-j} x_k \rangle \\ &= \sum_{j \ge k} \langle P_{\mathcal{H}} S^{j-k} x_j, x_k \rangle + \sum_{0 \le j < k} \langle x_j, P_{\mathcal{H}} S^{k-j} x_k \rangle \\ &= \sum_{j \ge k} \langle T^{j-k} x_j, x_k \rangle + \sum_{0 \le j < k} \langle x_j, T^{k-j} x_k \rangle,\end{aligned}$$

where we have used the fact that S is a dilation of T. A similar computation for S' shows that

$$\left\| \sum_{j=0}^{\infty} S^j x_j \right\|^2 = \left\| \sum_{j=0}^{\infty} S'^j x_j \right\|^2.$$

This fact immediately implies the existence of an isometry U of $\mathcal{K}$ onto $\mathcal{K}'$ satisfying

$$U\Big(\sum_{j=0}^{\infty} S^j x_j\Big) = \sum_{j=0}^{\infty} S'^j x_j$$

for every finitely supported sequence $(x_j)_{j\geq 0}$ in $\mathcal{H}$. This mapping clearly satisfies $Ux = x$, $x \in \mathcal{H}$ and $S'U = US$.

This completes the proof of Theorem 7.45. □

The following result is the counterpart of Theorem 7.45 for unitary dilations.

Theorem 7.46 *Every contraction $T \in \mathcal{L}(\mathcal{H})$ has a minimal unitary dilation. This dilation is unique in the following sense. If $S \in \mathcal{L}(\mathcal{K})$ and $S' \in \mathcal{L}(\mathcal{K}')$ are two minimal unitary dilations for T, then there exists a unitary operator $U : \mathcal{K} \longrightarrow \mathcal{K}'$ such that*

$$Ux = x \qquad (x \in \mathcal{H})$$

and $S'U = US$.

Proof The uniqueness is proved using the same calculation as in the proof of Theorem 7.45 except that one considers sums of the form

$$\sum_{j=-\infty}^{\infty} S^j x_j \qquad (x_j \in \mathcal{H}).$$

In order to prove the existence of a minimal unitary dilation, we define the space $\mathcal{K}$ and the operator $U \in \mathcal{L}(\mathcal{K})$ by

$$\mathcal{K} = \Big(\bigoplus_{j=-\infty}^{0} \mathcal{E}_j\Big) \oplus \mathcal{H} \oplus \Big(\bigoplus_{j=0}^{\infty} \mathcal{D}_j\Big),$$

where $\mathcal{E}_j = \mathcal{D}_{T^*}$ and $\mathcal{D}_j = \mathcal{D}_T$, and

$$\begin{aligned} &U(\dots, e_{-2}, e_{-1}, e_0, x, d_0, d_1, d_2, \dots) \\ &\quad = (\dots, e'_{-2}, e'_{-1}, e'_0, x', d'_0, d'_1, d'_2, \dots), \end{aligned}$$

where

$$\begin{aligned} x' &= Tx + D_{T^*} e_0, \\ d'_0 &= -T^* e_0 + D_T x, \\ d'_j &= d_{j-1} \qquad (j \geq 1), \\ e'_j &= e_{j-1} \qquad (j \leq 0). \end{aligned}$$

We see that $\mathcal{K}_+$ constructed in (7.34) can be identified with $\{0\} \oplus \mathcal{K}_+ \subset \mathcal{K}$ and the minimal isometric dilation for T satisfies

$$U_+ = U_{|\mathcal{K}_+}.$$

If we identify $\mathcal{H}$ with $\{0\} \oplus \mathcal{H} \oplus \{0\} \subset \mathcal{K}$, then we have

$$T^n = P_{\mathcal{H}} U^n_{+|\mathcal{H}} = P_{\mathcal{H}} U^n{}_{|\mathcal{H}} \qquad (n \geq 0).$$

Hence, U is a dilation of T. Let us show that U is isometric. This is equivalent to

$$\|Tx + D_{T^*} e_0\|^2 + \| - T^* e_0 + D_T x\|^2 = \|e_0\|^2 + \|x\|^2 \qquad (e_0 \in \mathcal{D}_{T^*},\ x \in \mathcal{H}).$$

The left-hand side of this identity can be rewritten as

$$\|Tx\|^2 + \|D_{T^*} e_0\|^2 + \|T^* e_0\|^2 + \|D_T x\|^2 + 2\,\Re(\langle T_x, D_{T^*} e_0\rangle - \langle T^* e_0, D_T x\rangle),$$

which using (7.33) is equal to

$$\|x\|^2 + \|e_0\|^2 + 2\,\Re(\langle D_{T^*} Tx - D_T^* T^* x, e_0\rangle).$$

Therefore, using Theorem 7.15, we conclude that U is an isometry.

Let us check that U is onto. Let $\mathbf{h}' = (\dots, e'_{-2}, e'_{-1}, e'_0, x', d'_0, d'_1, d'_2, \dots) \in \mathcal{K}$. We want to find an $\mathbf{h} = (\dots, e_{-2}, e_{-1}, e_0, x, d_0, d_1, d_2, \dots) \in \mathcal{K}$ such that $U\mathbf{h} = \mathbf{h}'$. This last equation is equivalent to the system

$$\begin{aligned} e_{j-1} &= e'_j \qquad (j \leq 0),\\ Tx + D_{T^*} e_0 &= x',\\ -T^* e_0 + D_T x &= d'_0,\\ d_{j-1} &= d'_j \qquad (j \geq 1). \end{aligned} \tag{7.36}$$

On the one hand, this implies that

$$\begin{aligned} e_j &= e'_{j+1} \qquad (j \leq -1),\\ d_j &= d'_{j+1} \qquad (j \geq 0). \end{aligned} \tag{7.37}$$

On the other, if we apply the operator T^* to the second equation of (7.36) and the operator D_T to the third one, we obtain

$$\begin{aligned} T^*Tx + T^* D_{T^*} e_0 &= T^* x',\\ -D_T T^* e_0 + D_T^2 x &= D_T d'_0. \end{aligned}$$

Since $T^* D_{T^*} = D_T T^*$, adding the two last equations, we obtain

$$x = D_T d'_0 + T^* x'. \tag{7.38}$$

Similarly, if we apply the operator D_{T^*} to the second equation of (7.36) and the operator T to the third one, we obtain

$$\begin{aligned} D_{T^*} Tx + D_{T^*}^2 e_0 &= D_{T^*} x',\\ -TT^* e_0 + T D_T x &= T d'_0. \end{aligned}$$

Using now $D_{T^*}T = TD_T$ and subtracting the two equations, we get

$$e_0 = D_{T^*}x' - Td'_0. \tag{7.39}$$

It is now easy to check that the vector $\mathbf{h}$ defined by the sets of relations (7.37), (7.38) and (7.39) satisfies the required properties in (7.36). Thus, $U\mathbf{h} = \mathbf{h}'$.

It remains to check the minimality of U. Arguing as in the proof of Theorem 7.45, we first note that

$$U^{-1}\mathcal{H} \vee \mathcal{H} \vee U\mathcal{H} = \cdots \{0\} \oplus \{0\} \oplus \mathcal{D}_{T^*} \oplus \mathcal{H} \oplus \mathcal{D}_T \oplus \{0\} \oplus \{0\} \cdots .$$

Hence, by definition of U, we have

$$\begin{aligned}\bigvee_{j=-n}^{n} U^j\mathcal{H} = \cdots \{0\} \oplus \{0\} \oplus \underbrace{\mathcal{D}_{T^*} \oplus \cdots \oplus \mathcal{D}_{T^*}}_{n \text{ times}} \oplus \mathcal{H} \\ \oplus \underbrace{\mathcal{D}_T \oplus \cdots \oplus \mathcal{D}_T}_{n \text{ times}} \oplus \{0\} \oplus \{0\} \oplus \cdots ,\end{aligned}$$

and the minimality of U follows from Lemma 7.44.

This completes the proof of Theorem 7.46. □

We now study in further detail the space of the minimal isometric dilation of a contraction $T \in \mathcal{L}(\mathcal{H})$. Let us $\mathcal{H}_n$ be the subspace of $\mathcal{K}_+$ defined as $\mathcal{H}_0 = \mathcal{H}$ and

$$\mathcal{H}_n = \mathcal{H} \oplus \underbrace{\mathcal{D}_T \oplus \cdots \oplus \mathcal{D}_T}_{n \text{ times}} \oplus \{0\} \oplus \{0\} \oplus \cdots \qquad (n \geq 1).$$

Note that each $\mathcal{H}_n$ is invariant under U_+^*. Indeed, for $n = 0$, we have $\mathcal{H} \in \text{Lat}(U_+^*)$ if and only if $\mathcal{H}^\perp \in \text{Lat}(U_+)$. But

$$\mathcal{H}^\perp = \{0\} \oplus \bigoplus_{n=0}^{\infty} \mathcal{D}_T,$$

and it is clear that $U_+\mathcal{H}^\perp \subset \mathcal{H}^\perp$. Now, since $\mathcal{H}_n = \bigvee_{j=0}^n U_+^j\mathcal{H}$, we have $U_+^*\mathcal{H}_n \subset \bigvee_{j=0}^n U_+^*U_+^j\mathcal{H}$ and, using $U_+^*U_+ = I$, we get

$$U_+^*\mathcal{H}_n \subset \bigvee_{j=0}^{n-1} U_+^j\mathcal{H} \subset \mathcal{H}_n. \tag{7.40}$$

If we set

$$T_n = P_{\mathcal{H}_n}U_{+|\mathcal{H}_n}, \tag{7.41}$$

then T_{n+1} is a dilation of T_n for each $n \geq 0$.

The contraction T_n can be viewed differently. For an arbitrary contraction $S \in \mathcal{L}(\mathcal{H})$, we can construct a dilation $S^\sharp$ of S on $\mathcal{H} \oplus \mathcal{D}_S$ defined by

$$S^\sharp(x, y) = (Sx, D_Sx). \tag{7.42}$$

Clearly, $S^\sharp$ is a partial isometry and $\mathcal{D}_{S^\sharp} = \ker S^\sharp = \{0\} \oplus \mathcal{D}_S$. If we repeat this procedure, we can construct a partial isometry $S^{\sharp\sharp} = (S^\sharp)^\sharp$, which dilates $S^\sharp$ on $\mathcal{H} \oplus \mathcal{D}_S \oplus \mathcal{D}_S$ and is defined by

$$S^{\sharp\sharp}(x, y, z) = (Sx, D_S x, y).$$

Moreover, we have $\ker S^{\sharp\sharp} = \{0\} \oplus \{0\} \oplus \mathcal{D}_S$. It is now clear that the contractions T_n defined above satisfy the relations

$$T_{n+1} = T_n^\sharp \qquad (n \geq 0).$$

Lemma 7.47 *Let $T \in \mathcal{L}(\mathcal{H})$, and let $U_+ \in \mathcal{L}(\mathcal{K}_+)$ be the minimal isometric dilation of T. Then we have*

$$U_+\mathbf{h} = \lim_{n\to\infty} T_n P_{\mathcal{H}_n}\mathbf{h} \qquad (\mathbf{h} \in \mathcal{K}_+)$$

and

$$U_+^*\mathbf{h} = \lim_{n\to\infty} T_n^*\mathbf{h} \qquad (\mathbf{h} \in \mathcal{K}_+),$$

where T_n is defined by (7.41).

Proof First, note that

$$\lim_{n\to\infty} P_{\mathcal{H}_n}\mathbf{h} = \mathbf{h} \qquad (\mathbf{h} \in \mathcal{K}_+).$$

Indeed, if $\mathbf{h} = (h, h_0, h_1, \dots) \in \mathcal{K}_+$, then

$$P_{\mathcal{H}_n}\mathbf{h} = (h, h_0, h_1, \dots, h_n, 0, 0, \dots)$$

and

$$\mathbf{h} - P_{\mathcal{H}_n}\mathbf{h} = (0, 0, \dots, 0, h_{n+1}, h_{n+2}, \dots).$$

Hence,

$$\|\mathbf{h} - P_{\mathcal{H}_n}\mathbf{h}\|^2 = \sum_{k\geq n+1} \|h_k\|^2,$$

and the last term tends to 0 as n tends to ∞, because the series is convergent.

To obtain the first equation, it is sufficient to note that, since $U_+\mathcal{H}_n^\perp \subset \mathcal{H}_n^\perp$ (see (7.40)), we have

$$T_n P_{\mathcal{H}_n}\mathbf{h} = P_{\mathcal{H}_n} U_+ P_{\mathcal{H}_n}\mathbf{h} = P_{\mathcal{H}_n} U_+\mathbf{h}.$$

For the second equation, note that according to (7.40) and (7.41), we have

$$T_n^* = P_{\mathcal{H}_n} U^*_{+|\mathcal{H}_n} = U^*_{+|\mathcal{H}_n}.$$

Take any $\mathbf{h} \in \mathcal{H}_\ell$, $\ell \geq 0$. Then, for each $n \geq \ell$, we have $\mathcal{H}_\ell \subset \mathcal{H}_n$ and

$$T_n^*\mathbf{h} = U_+^*\mathbf{h}.$$

Thus,

$$U_+^* \mathbf{h} = \lim_{n \to \infty} T_n^* \mathbf{h} \qquad (\mathbf{h} \in \mathcal{H}_\ell,\ \ell \geq 0).$$

It remains to use that $\bigvee_{\ell=0}^{\infty} \mathcal{H}_\ell = \mathcal{K}_+$ to conclude that the last equality is true for any $\mathbf{h} \in \mathcal{K}_+$. □

An important consequence of the existence of the unitary dilation is the following inequality known as von Neumann's inequality.

Theorem 7.48 *Let $T \in \mathcal{L}(\mathcal{H})$ be a contraction and let p be an analytic polynomial. Then*

$$\|p(T)\| \leq \sup_{|z| \leq 1} |p(z)|.$$

Proof Consider the (minimal) unitary dilation U of T (which exists by Theorem 7.46). Since U is a dilation of T, we have $p(T) = P_{\mathcal{H}} p(U)_{|\mathcal{H}}$, which gives

$$\|p(T)\| \leq \|p(U)\|.$$

To conclude, it suffices to use Theorem 2.17. □

7.11 The abstract commutant lifting theorem

We now exploit the results of the preceding section to prove a general and very useful lifting theorem.

Lemma 7.49 *Let $T \in \mathcal{L}(\mathcal{H})$ and $T' \in \mathcal{L}(\mathcal{H}')$ be two contractions, and let $X \in \mathcal{L}(\mathcal{H}, \mathcal{H}')$ satisfy the intertwining relation*

$$T'X = XT.$$

Then there exists an operator $Y \in \mathcal{L}(\mathcal{H} \oplus \mathcal{D}_T, \mathcal{H}' \oplus \mathcal{D}_{T'})$ such that:

(i) $Y(\{0\} \oplus \mathcal{D}_T) \subset \{0\} \oplus \mathcal{D}_{T'}$;
(ii) $P_{\mathcal{H}'} Y_{|\mathcal{H}} = X$;
(iii) $\|Y\| = \|X\|$;
(iv) $T'^\sharp Y = Y T^\sharp$, *where $T^\sharp$ and $T'^\sharp$ are the dilations of T and T' as described by* (7.42).

Proof We may assume, without loss of generality, that $\|X\| = 1$. In order to satisfy (i) and (ii), Y must have the form

$$Y(x, y) = (Xx, Z(x, y)) \qquad (x, y) \in \mathcal{H} \oplus \mathcal{D}_T,$$

where $Z \in \mathcal{L}(\mathcal{H} \oplus \mathcal{D}_T, \mathcal{D}_{T'})$. The condition $\|Y\| \leq 1$ is easily seen to be equivalent to

$$\|Z(x,y)\|^2_{\mathcal{D}_{T'}} \leq \|x\|^2_{\mathcal{H}} - \|Xx\|^2_{\mathcal{H}'} + \|y\|^2_{\mathcal{D}_T} = \|(D_X x, y)\|^2_{\mathcal{H}\oplus\mathcal{D}_T}.$$

Thus, there must exist a contraction $C \in \mathcal{L}(\mathcal{H} \oplus \mathcal{D}_T, \mathcal{D}_{T'})$, such that

$$Z = C(D_X \oplus I).$$

Moreover, for $(x,y) \in \mathcal{H} \oplus \mathcal{D}_T$, we have

$$T'^{\sharp} Y(x,y) = T'^{\sharp}(Xx, Z(x,y)) = (T'Xx, D_{T'}Xx)$$

and

$$\begin{aligned} YT^{\sharp}(x,y) &= Y(Tx, D_T x) = (XTx, Z(Tx, D_T x)) \\ &= (XTx, C(D_X Tx, D_T x)). \end{aligned}$$

Since $T'Xx = XTx$, the condition (iv) is translated into

$$C(D_X Tx, D_T x) = D_{T'}Xx \qquad (x \in \mathcal{H}).$$

Let us prove that

$$\|D_{T'}Xx\|_{\mathcal{D}_{T'}} \leq \|(D_X Tx, D_T x)\|_{\mathcal{H}\oplus\mathcal{D}_T}. \tag{7.43}$$

Using (7.33) three times, this is equivalent to

$$\begin{aligned} \|Xx\|^2_{\mathcal{H}'} - \|T'Xx\|^2_{\mathcal{H}'} &\leq \|Tx\|^2_{\mathcal{H}} - \|XTx\|^2_{\mathcal{H}'} + \|x\|^2_{\mathcal{H}} - \|Tx\|^2_{\mathcal{H}} \\ &= \|x\|^2_{\mathcal{H}} - \|XTx\|^2_{\mathcal{H}} \\ &= \|x\|^2_{\mathcal{H}} - \|T'Xx\|^2_{\mathcal{H}'}. \end{aligned}$$

In other words, (7.43) is equivalent to

$$\|Xx\|_{\mathcal{H}'} \leq \|x\|_{\mathcal{H}},$$

which follows from the fact that we assumed that $\|X\| = 1$. Hence, we can define C on the linear manifold

$$\mathcal{E} = \text{Lin}\{(D_X Tx, D_T x) : x \in \mathcal{H}\}$$

by

$$C(D_X Tx, D_T x) = D_{T'}Xx \qquad (x \in \mathcal{H}).$$

The inequality (7.43) enables us to extend C to a contraction on the closure of $\mathcal{E}$ into $\mathcal{H} \oplus \mathcal{D}_T$, and we extend it into a contraction on $\mathcal{H} \oplus \mathcal{D}_T$ by defining it to be zero on $\mathcal{E}^{\perp}$.

It remains to prove that $\|Y\| = 1$. We already know that $\|Y\| \leq 1$. Moreover, there exists $x_n \in \mathcal{H}$, $\|x_n\|_{\mathcal{H}} = 1$, such that $\|Xx_n\|_{\mathcal{H}'} \to \|X\| = 1$ as $n \to \infty$. Hence

$$\begin{aligned}\|Y(x_n, 0)\|^2_{\mathcal{H}'\oplus\mathcal{D}_{T'}} &= \|(Xx_n, Z(x_n, 0))\|^2_{\mathcal{H}'\oplus\mathcal{D}_{T'}} \\ &= \|Xx_n\|^2_{\mathcal{H}'} + \|Z(x_n, 0)\|^2_{\mathcal{D}_{T'}} \\ &\geq \|Xx_n\|^2_{\mathcal{H}'}.\end{aligned}$$

Since $\|(x_n, 0)\|_{\mathcal{H}\oplus\mathcal{D}_T} = 1$, we have

$$\|Y(x_n, 0)\|_{\mathcal{H}'\oplus\mathcal{D}_{T'}} \longrightarrow 1$$

and thus $\|Y\| = 1$. □

We can now prove the following general lifting theorem.

Theorem 7.50 *Let $T \in \mathcal{L}(\mathcal{H})$ and $T' \in \mathcal{L}(\mathcal{H}')$ be two contractions, and let $U_+ \in \mathcal{L}(\mathcal{K}_+)$ and $U'_+ \in \mathcal{L}(\mathcal{K}'_+)$ be the minimal isometric dilations of T and T', respectively. Then, for every $X \in \mathcal{L}(\mathcal{H}, \mathcal{H}')$ satisfying $T'X = XT$, there exists $Y \in \mathcal{L}(\mathcal{K}_+, \mathcal{K}'_+)$ such that:*

(i) $U'_+Y = YU_+$;
(ii) $\|Y\| = \|X\|$;
(iii) $X = P_{\mathcal{H}'}Y_{|\mathcal{H}}$;
(iv) $Y(\mathcal{K}_+ \ominus \mathcal{H}) \subset \mathcal{K}'_+ \ominus \mathcal{H}'$.

Proof Denote the compression of U_+ to $\mathcal{H}_n = \bigvee_{j=0}^{n} U^j\mathcal{H}$ by T_n. The compression T'_n is defined similarly. Put $Y_0 = X$. Observe that by Lemma 7.49 we can find bounded operators $Y_n \in \mathcal{L}(\mathcal{H}_n, \mathcal{H}'_n)$, $n \geq 0$, such that

$$\begin{aligned}P_{\mathcal{H}'_n}Y_{n+1|\mathcal{H}_n} &= Y_n, \\ Y_{n+1}(\mathcal{H}_{n+1} \ominus H_n) &\subset \mathcal{H}'_{n+1} \ominus \mathcal{H}'_n, \\ \|Y_{n+1}\| &= \|X\|, \\ T'_nY_n &= Y_nT_n.\end{aligned}$$

The above equations imply that

$$Y_n^* = Y^*_{n+1|\mathcal{H}'_n}, \tag{7.44}$$

$$Y^*_{n+1}\mathcal{H}'_n \subset \mathcal{H}_n, \tag{7.45}$$

$$\|Y^*_{n+1}\| = \|X\|, \tag{7.46}$$

$$Y_n^*{T'_n}^* = T_n^*Y_n^*. \tag{7.47}$$

Let $h \in \bigcup_{\ell\geq 0}\mathcal{H}'_\ell$. Thus, there exists $\ell \geq 0$ such that $h \in \mathcal{H}'_\ell$. Observe that, for any $n \geq \ell$, $\mathcal{H}'_\ell \subset \mathcal{H}'_n$, and by (7.44), we have

$$Y_n^*h = Y_\ell^*h.$$

In particular, the sequence $(Y_n^*h)_{n\geq\ell}$ is convergent and we can define

$$Y^*h = \lim_{n\to\infty} Y_n^*h \qquad \left(h \in \bigcup_{\ell\geq 0} \mathcal{H}'_\ell\right).$$

Since

$$\|Y_n^*h\| \leq \|Y_n^*\|\,\|h\| = \|X\|\,\|h\|,$$

we get

$$\|Y^*h\| \leq \|X\|\,\|h\| \qquad \left(h \in \bigcup_{\ell\geq 0} \mathcal{H}'_\ell\right).$$

Thus, Y^* defines a bounded operator from $\bigcup_{\ell\geq 0}\mathcal{H}'_\ell$ into $\mathcal{K}_+$. Since $\bigcup_{\ell\geq 0}\mathcal{H}'_\ell$ is dense in $\mathcal{K}'_+$, Y^* extends to a bounded operator from $\mathcal{K}'_+$ into $\mathcal{K}_+$ and we have

$$\|Y^*\| \leq \|X\|.$$

Moreover, for any $h \in \mathcal{H}'_0$ and any $n \geq 0$, we have

$$Y_n^*h = Y_0^*h = X^*h.$$

Letting n tend to ∞ gives $Y^*h = X^*h$. Thus,

$$Y^*_{|\mathcal{H}'} = X^*. \tag{7.48}$$

Hence,

$$\|X\| = \|X^*\| \leq \|Y^*\| \leq \|X\|,$$

which gives $\|Y\| = \|X\|$. The equality (7.48) can be translated into

$$X = P_{\mathcal{H}'}Y_{|\mathcal{H}}.$$

Moreover, (7.48) implies that $Y^*\mathcal{H}' \subset \mathcal{H}$, which is equivalent to

$$Y(\mathcal{K}_+ \ominus \mathcal{H}) \subset \mathcal{K}'_+ \ominus \mathcal{H}'.$$

Finally, letting $n \to \infty$ in (7.47) and using Lemma 7.47, we obtain

$$Y^*{U'_+}^* = U_+^*Y^*,$$

which can be written as

$$U'_+Y = YU_+. \qquad \square$$

Exercises

Exercise 7.11.1 Let $U \in \mathcal{L}(\mathcal{K})$ be a minimal unitary dilation of $T \in \mathcal{L}(\mathcal{H})$ and set $U_+ = U_{|\mathcal{K}_+}$, where $\mathcal{K}_+ = \bigvee_{n\geq 0} U^n\mathcal{H}$. Show that U is the minimal isometric dilation of U_+.

Exercise 7.11.2 Let $V \in \mathcal{L}(\mathcal{H})$ be an isometry. Show that the minimal isometric dilation of V^* is a unitary operator.

Exercise 7.11.3 Let $T \in \mathcal{L}(\mathcal{H})$ and $T' \in \mathcal{L}(\mathcal{H}')$ be two contractions, and let $U_+ \in \mathcal{L}(\mathcal{K}_+)$ and $U'_+ \in \mathcal{L}(\mathcal{K}'_+)$ be the minimal isometric dilations of T and T', respectively. Let $Y \in \mathcal{L}(\mathcal{K}_+, \mathcal{K}'_+)$ be such that:

(i) $U'_+ Y = YU_+$;
(ii) $\|Y\| = \|X\|$;
(iii) $Y(\mathcal{K}_+ \ominus \mathcal{H}) \subset \mathcal{K}'_+ \ominus \mathcal{H}'$.

Denote by $X := P_{\mathcal{H}'} Y_{|\mathcal{H}}$. Show that $T'X = XT$.

Notes on Chapter 7

The notions developed in this chapter are rather standard and are presented in many classic textbooks. We refer the reader to [11, 55, 187, 421] for further information.

Section 7.1

In this section, we developed the functional calculus for self-adjoint operators, essentially to pave the way for the existence of the square root for positive operators. Nevertheless, the reader should know that there exists a complete spectral theory for normal operators. See [11, 177, 187, 250, 314, 421, 431, 441, 564] for a complete account of the spectral theorem for bounded normal operators. The spectral theorem (Theorem 7.4) for self-adjoint operators on finite-dimensional vector spaces is also nicely described in [250].

Section 7.2

We have deduced the existence of the square root of a positive operator (Theorem 7.5) from the functional calculus for self-adjoint operators. One can also construct the square root directly. See Exercise 2.4.8 and [55, 421] for such an approach. In [347], Löwner completely characterized operator-monotone functions, that is, real-valued continuous functions f satisfying

$$A \geq B \geq 0 \quad \Longrightarrow \quad f(A) \geq f(B).$$

In particular, he showed that $f(t) = t^{\alpha}$ is operator-monotone if $\alpha \in [0, 1]$, while it is not operator-monotone if $\alpha > 1$. In [275], Heinz discovered an alternative proof. Part (v) of Theorem 7.7 is known as the Löwner–Heinz inequality.

Theorem 7.10 is borrowed from [451]. Theorem 7.11 is due to R. Douglas and appeared in [173]. Theorem 7.15 is due to P. Halmos.

Section 7.3

The main part of this section is taken from [565]. Theorem 7.18 is due to Julia [301] and Halmos [248]. This theorem was substantially strengthened by Sz.-Nagy [505], who showed that any contraction B on a Hilbert space $\mathcal{H}$ admits a unitary dilation, that is, there exists a unitary operator U on a larger space $\mathcal{K}$ satisfying

$$P_{\mathcal{H}} U^n{}_{|\mathcal{H}} = B^n \qquad (n \geq 1)$$

(see Section 7.10). Note that $J(B)$ is not a dilation of B, even if $P_{\mathcal{H}} J(B)_{|\mathcal{H}} = B$. Sz.-Nagy's theorem is the foundation of the theory of unitary dilations and the model theory for Hilbert space contractions; see [388, 508]. Theorem 7.20 is due to Parrott [402]. Möbius operators and Julia operators have become standard tools in the theory of linear systems. In particular, when $X = -z$, then $\Psi_T(z) = B + zA(I - zC)^{-1}D$ represents the *transfer function* of the linear system

$$\begin{cases} x'(t) = Cx(t) + Du(t) \\ y(t) = Ax(t) + Bu(t) \\ x(0) = x_0 \end{cases},$$

where the functions $t \longmapsto u(t)$, $t \longmapsto x(t)$ and $t \longmapsto y(t)$ are respectively the input, the state and the output functions associated with the system (see e.g. [19, 388]).

Section 7.4

Theorem 7.21 on the decomposition induced by an isometry was stated first in probabilistic language by Wold [558] and Kolmogorov [318]. For more information on the stated form, see [251].

Section 7.5

Partial isometries form an attractive and important class of operators. They play a vital role in operator theory. They enter, for instance, in the theory of the polar decomposition of arbitrary operators, and they form the cornerstone of the dimension theory of von Neumann algebras. Partial isometries appeared in [366]. Theorem 7.22 is partially contained in [366]. The polar decomposition (Theorem 7.24) is due to von Neumann [537]. For linear transformations on $\mathbb{R}^n$, it has a simple geometric meaning. Any linear transformation A on $\mathbb{R}^n$ can be written as $A = OS$, where O is orthogonal and S is self-adjoint, and, by the spectral theorem, S can be seen as a multiplication operator.

Section 7.7

The theory of unbounded operators was stimulated by attempts in the late 1920s to put quantum mechanics on a rigorous mathematical foundation. The systematic development of the theory is due to von Neumann [535] and Stone [496]. The technique of using graphs to analyze unbounded operators was introduced by von Neumann [537].

Section 7.8

Fredholm only studied compact integral operators. It was Noether [397] who introduced the notion of index and implicitly the class that is now called Fredholm operators. The connection between this class and the Calkin algebra was made by Atkinson [50]. In particular, he obtained Theorem 7.34. Finally, Gohberg and Kreĭn [241] systematized and extended the theory of Fredholm operators to its present form. Further results can be found in [431] and [334]. Left semi-Fredholm and right semi-Fredholm operators were first considered by Yood [563], who called them operators with property A and B, respectively.

Note that a slightly different definition of Fredholm operators can be found in some textbooks: more precisely, it is said that an operator $A \in \mathcal{L}(\mathcal{H}_1, \mathcal{H}_2)$ is Fredholm if and only if $\dim(\ker A) < \infty$ and A has a closed range of finite codimension ($\dim(\mathcal{H}_2/\mathcal{R}(A)) < \infty$). However, considering Theorems 1.11 and 1.28, this is equivalent to our definition. Moreover, with this definition, one can avoid the condition that "$\mathcal{R}(A)$ be closed", since as mentioned in [164, theorem 4.3.4] or [1, lemma 4.38] it is implied by the facts that $\dim(\ker A) < \infty$ and $\dim(\mathcal{H}_2/\mathcal{R}(A)) < \infty$.

Section 7.9

The main result of this section is taken from [159, chap. XI]. Corollary 7.41 is due to Yood [563], with a different proof. Theorem 7.42 is due to Dieudonné

[171]. In [50], Atkinson re-proved the results of Yood and Dieudonné, and also obtained Theorem 7.39. The same year, Gohberg [238] independently obtained the same results. We refer to [172] for a historical survey of work done in the first half of the twentieth century on index theory for linear operators in Banach spaces.

Section 7.10

Theorems 7.45 and 7.46 are due to Sz.-Nagy. These two results were the starting point of an important branch in operator theory. All the material in this section is covered in the books by Sz.-Nagy and Foiaş [508] and Bercovici [90].

The inequality proved in Theorem 7.48 is known as von Neumann's inequality [540]. Another proof has already been presented in Exercise 2.5.1 without the theory of dilation.

Section 7.11

The abstract commutant lifting theorem (Theorem 7.50) was proved by Sz.-Nagy and Foiaş [507]. The earliest form of the commutant lifting theorem appeared in Sarason [447] and was motivated by the study of interpolation problems for bounded analytic functions in the unit disk. The general form of the theorem has also been used in a variety of interpolation problems. There are also interesting applications in control problems and systems theory.

8

The shift operator

The shift operator is the most studied operator in mathematics. It also plays a profound role in the development of $\mathcal{H}(b)$ spaces. In this chapter, we mainly study some variations of this operator on the Lebesgue spaces $L^2(\mu)$ and the Hardy space $H^2(\mu)$. We introduce the bilateral forward shift operator Z on $L^2(\mathbb{T})$ and its generalization, the bilateral forward shift operator Z_μ on $L^2(\mu)$, the unilateral forward shift operator S on H^2 and its generalization, the unilateral forward shift operator S_μ on $H^2(\mu)$, and their adjoints, which are respectively addressed as the bilateral backward shift and the unilateral backward shift. In each case, we discuss spectrum and essential spectrum, reducing invariant subspaces, simply invariant subspaces, the commutant and cyclic vectors. In fact, a comprehensive study of the invariant (simply or reducing) subspaces of the shift operators in the form presented below is new and stems from several research papers. We also study the possibility of $L^p(\mu) = H^p(\mu)$.

8.1 The bilateral forward shift operator Z_μ

Let μ be a finite and positive Borel measure on $\mathbb{T}$, and consider the Hilbert space $L^2(\mu)$. Then the *bilateral forward shift operator* Z_μ on $L^2(\mu)$ is defined by

$$\begin{array}{rccc} Z_\mu : & L^2(\mu) & \longrightarrow & L^2(\mu) \\ & f & \longmapsto & \chi_1 f. \end{array}$$

More explicitly, we can write

$$(Z_\mu f)(\zeta) = \zeta f(\zeta) \qquad (\zeta \in \mathbb{T}),$$

i.e. Z_μ is the operator of multiplication by the independent variable ζ and, using the notation of Section 2.8, we have

$$Z_\mu = M_{\chi_1} \in \mathcal{L}(L^2(\mu)). \tag{8.1}$$

On some occasions, we also write M_z for this operator. If μ is absolutely continuous with respect to the normalized Lebesgue measure, i.e. $d\mu = w\,dm$, $w \in L^1(\mathbb{T})$, then we write Z_w for Z_μ. In the particular, but important, case when μ is the normalized Lebesgue measure m on $\mathbb{T}$, for simplicity of notation we just write Z.

Remember that, in Section 2.8, we proved that $M_\varphi^* = M_{\bar{\varphi}}$, where $M_\varphi \in \mathcal{L}(L^2(\mu))$ is the operator of multiplication by $\varphi \in L^\infty(\mu)$. Moreover, in Theorem 2.23, we showed that

$$\sigma(M_\varphi) = \sigma_a(M_\varphi) = \mathcal{R}_e^\mu(\varphi), \tag{8.2}$$

where $\mathcal{R}_e^\mu(\varphi) = \{\lambda \in \mathbb{C} : \forall\, r > 0,\ \mu(\varphi^{-1}(D(\lambda, r))) > 0\}$ is the essential range of φ with respect to μ. Thus, according to (8.1), we have

$$Z_\mu^* = M_{\chi_1}^* = M_{\overline{\chi_1}} = M_{\chi_{-1}},$$

and this fact implies that

$$Z_\mu Z_\mu^* = Z_\mu^* Z_\mu = I_{L^2(\mu)}. \tag{8.3}$$

Comparing the Fourier coefficient of f, Zf and Z^*f reveals why Z and Z^*, respectively, are called the bilateral forward and backward shift operator (see Exercise 8.1.1).

The identity (8.3) means that Z_μ is a unitary operator. Moreover, remembering the definition of the support of a measure, we see that

$$\mathcal{R}_e^\mu(\chi_1) = \{\lambda \in \mathbb{C} : \forall\, r > 0,\ \mu(D(\lambda, r)) > 0\} = \operatorname{supp} \mu. \tag{8.4}$$

Thus, from (8.2) and (8.4), we immediately obtain the following description of the spectrum of Z_μ.

Theorem 8.1 *Let μ be a finite and positive Borel measure on $\mathbb{T}$. Then we have*

$$\sigma(Z_\mu) = \sigma_a(Z_\mu) = \operatorname{supp} \mu.$$

As a special case, Theorem 8.1 implies that

$$\sigma(Z) = \sigma_a(Z) = \mathbb{T}. \tag{8.5}$$

We can also easily determine the eigenvalues and eigenvectors of Z_μ.

Theorem 8.2 *Let μ be a finite and positive Borel measure on $\mathbb{T}$. Then*

$$\sigma_p(Z_\mu) = \{\lambda \in \mathbb{T} : \mu(\{\lambda\}) > 0\}.$$

Moreover, if $\lambda \in \sigma_p(Z_\mu)$, then $\ker(Z_\mu - \lambda I)$ is the one-dimensional space spanned by the characteristic function $\chi_{\{\lambda\}}$.

Proof By Theorem 8.1, $\sigma_p(Z_\mu) \subset \sigma(Z_\mu) = \operatorname{supp}\mu \subset \mathbb{T}$. Let $\lambda \in \sigma_p(Z_\mu)$. This means that there exists a function $f \in L^2(\mu)$, $f \neq 0$, such that

$$(z - \lambda)f(z) = 0$$

for μ-almost all $z \in \mathbb{T}$. This relation has two immediate consequences. It implies the equality $f = c\,\chi_{\{\lambda\}}$ in $L^2(\mu)$, for some constant $c \in \mathbb{C} \setminus \{0\}$, and moreover $\mu(\{\lambda\}) > 0$. Therefore, we have proved that

$$\sigma_p(Z_\mu) \subset \{\lambda \in \mathbb{T} : \mu(\{\lambda\}) > 0\}$$

and

$$\ker(Z_\mu - \lambda I) \subset \mathbb{C}\chi_{\{\lambda\}}.$$

Conversely, let $\lambda \in \mathbb{T}$ be such that $\mu(\{\lambda\}) > 0$. Then $\chi_{\{\lambda\}} \neq 0$ in $L^2(\mu)$ and we easily see that

$$(Z_\mu - \lambda I)\chi_{\{\lambda\}} = 0,$$

which proves that $\lambda \in \sigma_p(Z_\mu)$ and $\mathbb{C}\chi_{\{\lambda\}} \subset \ker(Z_\mu - \lambda I)$. □

We recall that, if A is a normal operator on a Hilbert space $\mathcal{H}$, then we have

$$\sigma_p(A^*) = \overline{\sigma_p(A)} \quad \text{and} \quad \ker(A - \lambda I) = \ker(A^* - \bar{\lambda} I).$$

Applying these facts to the operator Z_μ and appealing to Theorem 8.2, we obtain the following result.

Corollary 8.3 *Let μ be a finite and positive Borel measure on $\mathbb{T}$. Then*

$$\sigma_p(Z_\mu^*) = \{\lambda \in \mathbb{T} : \mu(\{\bar{\lambda}\}) > 0\}$$

and, for each $\lambda \in \sigma_p(Z_\mu^*)$, we have

$$\ker(Z_\mu^* - \lambda I) = \mathbb{C}\chi_{\{\bar{\lambda}\}}.$$

As a special case, Theorem 8.2 and Corollary 8.3 together imply that

$$\sigma_p(Z) = \sigma_p(Z^*) = \emptyset. \tag{8.6}$$

We now give an explicit description of the essential spectrum of the operator Z_μ. See Section 2.3 for the definition of the essential spectrum of a Hilbert space operator. It is trivial that $\sigma_{ess}(Z_\mu) \subset \sigma(Z_\mu) = \operatorname{supp}(\mu)$.

Theorem 8.4 *Let μ be a finite and positive Borel measure on $\mathbb{T}$, and let $\lambda \in \operatorname{supp}(\mu)$. Then the following are equivalent:*

(i) *$\lambda \in \sigma_{ess}(Z_\mu)$;*
(ii) *either $\mu(\{\lambda\}) = 0$, or $\mu(\{\lambda\}) > 0$ but there exists a sequence $(\lambda_n)_{n\geq 1}$ such that $\lambda_n \in \operatorname{supp}(\mu)$, $\lambda_n \neq \lambda$, and $\lambda_n \longrightarrow \lambda$, as $n \longrightarrow \infty$.*

In other words, by putting aside all points of $\operatorname{supp}(\mu)$ *at which there is an isolated Dirac measure, we obtain the essential spectrum of* Z_μ.

Proof To establish (i) $\Longleftrightarrow$ (ii), we consider three cases.

Case I. Isolated points of $\operatorname{supp}(\mu)$ at which there is a Dirac measure do not belong to the essential spectrum of Z_μ.

Assume that $\mu(\{\lambda\}) > 0$, but there does not exist a sequence $(\lambda_n)_{n\geq 1}$ such that $\lambda_n \in \operatorname{supp}(\mu)$, $\lambda_n \neq \lambda$ and $\lambda_n \longrightarrow \lambda$, as $n \longrightarrow \infty$. Hence, there exists $\varepsilon > 0$ such that

$$D'(\lambda, \varepsilon) \cap \operatorname{supp}(\mu) = \emptyset,$$

where $D'(\lambda,\varepsilon) = \{z \in \mathbb{C} : 0 < |z-\lambda| < \varepsilon\} = D(\lambda,\varepsilon) \setminus \{\lambda\}$. Then we have

$$(Z_\mu - \lambda I)L^2(\mu) = \{f \in L^2(\mu) : f(\lambda) = 0\}. \tag{8.7}$$

That the range of the operator $Z_\mu - \lambda I$ is contained in $\{f \in L^2(\mu) : f(\lambda) = 0\}$ is in fact trivial. Reciprocally, let $f \in L^2(\mu)$ with $f(\lambda) = 0$, and define

$$g(z) = \begin{cases} \dfrac{f(z)}{z-\lambda} & \text{if} \quad z \in \mathbb{T} \setminus \{\lambda\}, \\ 0 & \text{if} \quad z = \lambda. \end{cases}$$

To check that $g \in L^2(\mu)$, write

$$\begin{aligned} \int_{\mathbb{T}} |g(z)|^2\, d\mu(z) &= \int_{\mathbb{T}\cap D(\lambda,\varepsilon)} |g(z)|^2\, d\mu(z) + \int_{\mathbb{T}\setminus D(\lambda,\varepsilon)} |g(z)|^2\, d\mu(z) \\ &= \int_{\mathbb{T}\setminus D(\lambda,\varepsilon)} \left|\frac{f(z)}{z-\lambda}\right|^2 d\mu(z) \\ &\leq \frac{1}{\varepsilon^2}\int_{\mathbb{T}} |f(z)|^2\, d\mu(z) < \infty. \end{aligned}$$

The $g \in L^2(\mu)$, and this function is defined such that

$$((Z_\mu - \lambda I)g)(z) = (z-\lambda)g(z) = f(z) \qquad (z \in \mathbb{T}).$$

Therefore, we have $f \in (Z_\mu - \lambda I)L^2(\mu)$.

We exploit the identity (8.7) to show that $Z_\mu - \lambda I$ has a closed range. In fact, since $\mu(\{\lambda\}) > 0$, the linear functional $f \longmapsto f(\lambda)$ is well defined and continuous on $L^2(\mu)$. Hence, its kernel, which is the set $\{f \in L^2(\mu) : f(\lambda) = 0\}$, is closed.

According to Theorem 8.2 and Corollary 8.3, the assumption $\mu(\{\lambda\}) > 0$ also implies that

$$\dim(\ker(Z_\mu - \lambda I)) = \dim(\ker(Z_\mu^* - \bar{\lambda} I)) = 1.$$

Therefore, we deduce that the operator $Z_\mu - \lambda I$ is Fredholm, which, by Theorem 7.34, means that $\lambda \notin \sigma_{ess}(Z_\mu)$.

Case II. Any point of $\text{supp}(\mu)$ at which there is no Dirac measure belongs to the essential spectrum of Z_μ.

Assume that $\lambda \in \text{supp}(\mu)$, but $\mu(\{\lambda\}) = 0$. By Theorem 8.2, we have $\lambda \notin \sigma_p(Z_\mu)$. In other words, $\ker(Z_\mu - \lambda I) = \{0\}$, and since Z_μ is normal, we also have $\ker(Z_\mu^* - \bar{\lambda} I) = \{0\}$. Hence, the operator $Z_\mu - \lambda I$ is one-to-one and has a dense range. Since $\lambda \in \text{supp}(\mu) = \sigma(Z_\mu)$, this operator cannot have a closed range. In particular, $Z_\mu - \lambda I$ is not Fredholm. Using Theorem 7.34, we thus conclude that $\lambda \in \sigma_{ess}(Z_\mu)$.

Case III. Non-isolated points of $\text{supp}(\mu)$ at which there is a Dirac measure belong to the essential spectrum of Z_μ.

Let $\lambda \in \text{supp}(\mu)$, with $\mu(\{\lambda\}) > 0$, be such that there exists a sequence $(\lambda_n)_{n \geq 1}$ with $\lambda_n \in \text{supp}(\mu)$, $\lambda_n \neq \lambda$ and $\lambda_n \longrightarrow \lambda$, as $n \longrightarrow \infty$. Let

$$\begin{array}{rccl} A: & (\ker(Z_\mu - \lambda I))^\perp & \longrightarrow & \text{Clos}_{L^2(\mu)}(\mathcal{R}(Z_\mu - \lambda I)) \\ & f & \longmapsto & (Z_\mu - \lambda I) f. \end{array}$$

If the operator $Z_\mu - \lambda I$ had a closed range, then A would be invertible by an application of the Banach isomorphism theorem. But, since λ is not isolated, we are able to construct a sequence of unit functions $(f_n)_{n \geq 1} \in (\ker(Z_\mu - \lambda I))^\perp$ such that $A f_n \longrightarrow 0$. Hence, A is not invertible, and as a consequence, the operator $Z_\mu - \lambda I$ does not have a closed range. Therefore, $Z_\mu - \lambda I$ is not Fredholm, which, by Theorem 7.34, implies $\lambda \in \sigma_{ess}(Z_\mu)$.

The construction of $(f_n)_{n \geq 1}$ is as follows. For each fixed integer n, we have three possibilities:

(a) $\ker(Z_\mu - \lambda_n I) \neq \{0\}$;
(b) $\ker(Z_\mu - \lambda_n I) = \{0\}$ and $\mathcal{R}(Z_\mu - \lambda_n I)$ is not closed;
(c) $\ker(Z_\mu - \lambda_n I) = \{0\}$ and $\mathcal{R}(Z_\mu - \lambda_n I)$ is closed.

In fact, (c) cannot happen, since otherwise the normal operator $Z_\mu - \lambda_n I$ would be invertible, which contradicts the assumption $\lambda_n \in \text{supp}(\mu) = \sigma(Z_\mu)$.

If (a) holds, consider

$$f_n = \frac{\chi_{\{\lambda_n\}}}{\sqrt{\lambda_n}}.$$

The function f_n satisfies $\|f_n\|_{L^2(\mu)} = 1$ and

$$\|(Z_\mu - \lambda_n I) f_n\|_{L^2(\mu)} = 0.$$

Assuming (b), since the operator $Z_\mu - \lambda_n I$ is not bounded below, for each $\varepsilon > 0$, there also exists a unit vector $f_n \in L^2(\mu)$ such that

$$\|(Z_\mu - \lambda_n I) f_n\|_{L^2(\mu)} < \varepsilon.$$

Therefore, in both possible cases (a) and (b), there exists a vector $f_n \in L^2(\mu)$, $\|f_n\|_{L^2(\mu)} = 1$, such that

$$\|(Z_\mu - \lambda_n I)f_n\|_{L^2(\mu)} \leq |\lambda - \lambda_n|^2. \tag{8.8}$$

We will shortly see why $\varepsilon = |\lambda - \lambda_n|^2$, which is actually strictly positive by assumption, was chosen.

Decompose the unit function f_n as $f_n = g_n + h_n$, where $g_n \in \ker(Z_\mu - \lambda I)$ and $h_n \in (\ker(Z_\mu - \lambda I))^\perp$. Hence,

$$(Z_\mu - \lambda_n I)g_n = (Z_\mu - \lambda I)g_n + (\lambda - \lambda_n)g_n = (\lambda - \lambda_n)g_n.$$

Using the fact that Z_μ is normal, we have

$$(Z_\mu - \lambda I)(Z_\mu - \lambda I)^* g_n = (Z_\mu - \lambda I)^*(Z_\mu - \lambda I)g_n = 0,$$

whence $(Z_\mu - \lambda I)^* g_n \in \ker(Z_\mu - \lambda I)$. Since $h_n \perp \ker(Z_\mu - \lambda I)$, we obtain

$$\langle g_n, (Z_\mu - \lambda I)h_n\rangle = \langle (Z_\mu - \lambda I)^* g_n, h_n\rangle = 0.$$

Thus,

$$\begin{aligned}\|(Z_\mu - \lambda_n I)f_n\|^2_{L^2(\mu)} &= \|(Z_\mu - \lambda_n I)g_n + (Z_\mu - \lambda_n I)h_n\|^2_{L^2(\mu)}\\ &= \|(\lambda - \lambda_n)g_n + (Z_\mu - \lambda_n I)h_n\|^2_{L^2(\mu)}\\ &= |\lambda - \lambda_n|^2\|g_n\|^2_{L^2(\mu)} + \|(Z_\mu - \lambda_n I)h_n\|^2_{L^2(\mu)}.\end{aligned}$$

Therefore, according to (8.8),

$$|\lambda - \lambda_n|^4 \geq |\lambda - \lambda_n|^2\|g_n\|^2_{L^2(\mu)},$$

which implies that

$$\lim_{n\to\infty} \|g_n\|_{L^2(\mu)} = 0.$$

This fact in return gives

$$\lim_{n\to\infty} \|(Z_\mu - \lambda_n I)h_n\|_{L^2(\mu)} = 0.$$

But since

$$\|g_n\|^2_{L^2(\mu)} + \|h_n\|^2_{L^2(\mu)} = \|f_n\|^2_{L^2(\mu)} = 1,$$

we also have

$$\lim_{n\to\infty} \|h_n\|_{L^2(\mu)} = 1.$$

Hence, we may replace the original sequence $(f_n)_{n\geq 1}$ with $(h_n/\|h_n\|_{L^2(\mu)})_{n\geq 1}$, which means that we can assume $f_n \perp \ker(Z_\mu - \lambda I)$ and still

$$\lim_{n\to\infty} \|(Z_\mu - \lambda_n I)f_n\|_{L^2(\mu)} = 0.$$

Clearly, this property implies that

$$\lim_{n\to\infty} \|Af_n\|_{L^2(\mu)} = \lim_{n\to\infty} \|(Z_\mu - \lambda I)f_n\|_{L^2(\mu)} = 0.$$

This completes the proof of Theorem 8.4. □

Theorem 8.4 says that a point λ in the support of the measure μ belongs to the essential spectrum of Z_μ if and only if either λ is not an eigenvalue or it is an eigenvalue but is not isolated in the spectrum of Z_μ. It follows immediately from Theorem 8.4 that

$$\sigma_{ess}(Z) = \sigma(Z) = \mathbb{T}.$$

In the following, we will need a special consequence of Theorem 8.4.

Corollary 8.5 *Let μ be a finite and positive Borel measure on $\mathbb{T}$, and let I_0 be an open subarc of $\mathbb{T}$. Assume that, on I_0, μ consists only of a finite number of Dirac measures, i.e.*

$$\mu_{|I_0} = \sum_{i=1}^{n} a_i \delta_{\zeta_i},$$

where $\zeta_i \in I_0$ and $a_i > 0$, $1 \le i \le n$. Then $\sigma_{ess}(Z_\mu) \cap I_0 = \emptyset$.

Proof By Theorem 8.1 and our main assumption, we have

$$\sigma_{ess}(Z_\mu) \cap I_0 \subset \sigma(Z_\mu) \cap I_0 = \{\zeta_1, \dots, \zeta_n\}.$$

Hence, we end up showing that none of the points $\zeta_1, \dots, \zeta_n$ could be in $\sigma_{ess}(Z_\mu)$. But, since I_0 is open, there exists an $\varepsilon > 0$ such that $D(\zeta_i, \varepsilon) \cap \operatorname{supp}(\mu) = \{\zeta_i\}$. Therefore, according to Theorem 8.4, $\zeta_i \notin \sigma_{ess}(Z_\mu)$. □

Exercises

Exercise 8.1.1 The space $\ell^2(\mathbb{Z})$ was defined in Exercise 1.1.5. To distinguish the zeroth position in a sequence, we write

$$\mathfrak{a} = (\dots, a_{-2}, a_{-1}, \boldsymbol{a_0}, a_1, a_2, \dots).$$

Define the operator $W : \ell^2(\mathbb{Z}) \longrightarrow \ell^2(\mathbb{Z})$ by

$$W\mathfrak{a} = (\dots, a_{-3}, a_{-2}, \boldsymbol{a_{-1}}, a_0, a_1, \dots).$$

Show that

$$W^*\mathfrak{a} = (\dots, a_{-1}, a_0, \boldsymbol{a_1}, a_2, a_3, \dots)$$

and that W is unitarily equivalent to Z.

Hint: Consider the mapping

$$\begin{array}{rccc} A: & \ell^2(\mathbb{Z}) & \longrightarrow & L^2(\mathbb{T}) \\ & (a_n)_{n\in\mathbb{Z}} & \longmapsto & \displaystyle\sum_{n=-\infty}^{\infty} a_n\zeta^n. \end{array}$$

Remark: This exercise shows why Z and W (respectively, Z^* and W^*) are called bilateral forward (respectively, backward) shift operators.

Exercise 8.1.2 Show that $\sigma(Z_\mu^*) = \sigma_a(Z_\mu^*) = \overline{\text{supp}(\mu)}$, where the bar stands for complex conjugate. In particular, we have $\sigma(Z^*) = \sigma_a(Z^*) = \mathbb{T}$. Hint: Use Theorem 1.30 , (2.17) and Theorem 8.1.

8.2 The unilateral forward shift operator S

In Section 2.7, we introduced the unilateral forward shift operator on $\ell^2 = \ell^2(\mathbb{N})$. In this section, we study its analog on the Hardy space H^2. The *unilateral forward shift* operator S is defined by

$$\begin{array}{rccc} S: & H^2 & \longrightarrow & H^2 \\ & f & \longmapsto & \chi_1 f, \end{array}$$

where we recall that $\chi_1(\zeta) = \zeta$, $\zeta \in \mathbb{T}$. Its adjoint S^* is called the *unilateral backward shift* operator. Since $|\chi_1| = 1$ on $\mathbb{T}$, it is trivial that

$$\|Sf\|_2 = \|f\|_2 \qquad (f \in H^2). \tag{8.9}$$

In other words, S is an isometry on H^2. Figure 8.1 reveals the relation between forward shifts Z and S. Note that, contrary to the bilateral forward shift operator Z, the operator S is not unitary. For instance, the constant function χ_0 is

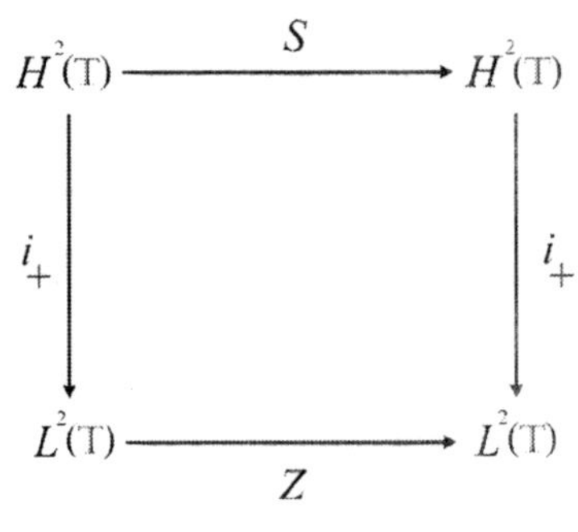

Figure 8.1 The relation between S and Z.

not in the range of S. Now, let $f(\zeta) = \sum_{n=0}^{\infty} \hat{f}(n)\zeta^n \in H^2$. Then the formula $Sf = \chi_1 f$ is rewritten as

$$(Sf)(\zeta) = \zeta f(\zeta) = \sum_{n=1}^{\infty} \hat{f}(n-1)\zeta^n. \tag{8.10}$$

Considering the way S acts on the Fourier coefficients of a function in $H^2(\mathbb{T})$, or equivalently on the Taylor coefficient of a function in $H^2(\mathbb{D})$, we realize why this operator is called the unilateral forward shift operator. By induction, we also obtain the useful formula

$$\hat{f}(n) = \langle f, \chi_n \rangle_2 = \langle f, S^n \chi_0 \rangle_2 \qquad (f \in H^2,\ n \geq 0). \tag{8.11}$$

As already mentioned, a closely related shift operator on ℓ^2 was defined in Section 2.7. It is easy to establish the connection between these operators. The mapping

$$\begin{array}{rccc} A: & \ell^2 & \longrightarrow & H^2 \\ & (a_0, a_1, \cdots) & \longmapsto & \displaystyle\sum_{n=0}^{\infty} a_n\, \zeta^n \end{array}$$

is a unitary operator between ℓ^2 and H^2. Then the definitions of S and U reveal that (see Figure 8.2)

$$S = AUA^*.$$

The operators P_+ and i_+ introduced in Section 4.3 can be used to clarify the relation between the shift operators $S \in \mathcal{L}(H^2)$ and $Z \in \mathcal{L}(L^2(\mathbb{T}))$. Both operators are multiplication by the independent variable z, but on different function spaces. Indeed, we have

$$S = P_+ Z i_+ = P_+ M_z i_+ \tag{8.12}$$

(see Figure 8.3). By (4.27), we thus have

$$S^* = P_+ Z^* i_+ = P_+ M_{\bar{z}} i_+. \tag{8.13}$$

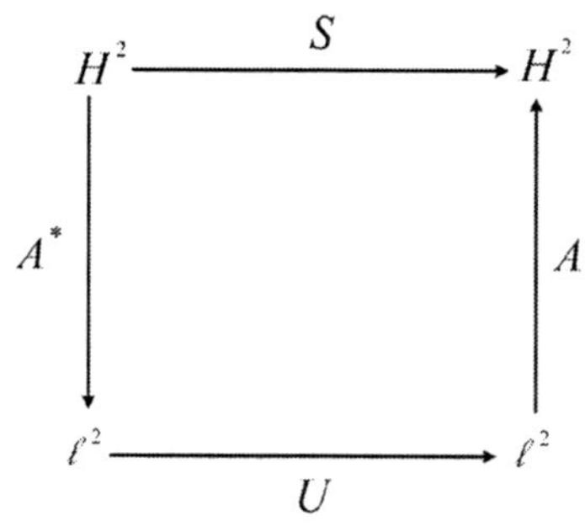

Figure 8.2 The relation between S and U.

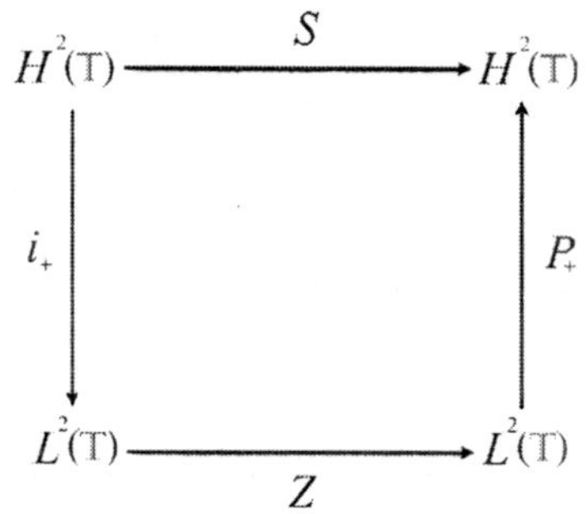

Figure 8.3 The relation between S, Z, P_+ and i_+.

Knowing the adjoint of U on ℓ^2 and the preceding unitary operator A, we can obtain the adjoint of S on H^2. However, we proceed to compute S^* directly. To start, note that $\chi_0 \perp \chi_1 H^2$. Hence, for each $f, g \in H^2$, we have

$$\begin{aligned}\langle Sg, f\rangle_{H^2} &= \langle \chi_1 g, f\rangle_{H^2}\\ &= \langle \chi_1 g, f - f(0)\rangle_{H^2}\\ &= \langle \chi_1 g, f - f(0)\rangle_{L^2}\\ &= \langle g, (f - f(0))\chi_{-1}\rangle_{L^2}.\end{aligned}$$

But the function $z \longmapsto (f(z) - f(0))/z$ is in $H^2(\mathbb{D})$, and its trace on the unit circle is precisely $(f - f(0))\chi_{-1}$. In fact, that is why we replaced f by $f - f(0)$ in the above calculation. Hence, we can write

$$\langle Sg, f\rangle_{H^2} = \langle g, (f - f(0))\chi_{-1}\rangle_{H^2} \qquad (f, g \in H^2).$$

This identity reveals that

$$S^* f = (f - f(0))\chi_{-1} \qquad (f \in H^2),$$

or, more explicitly,

$$(S^* f)(z) = \frac{f(z) - f(0)}{z} = \sum_{n=0}^{\infty} \hat{f}(n+1) z^n \qquad (z \in \mathbb{D}), \tag{8.14}$$

where $f(z) = \sum_{n=0}^{\infty} \hat{f}(n) z^n$. By induction, we easily obtain

$$(S^{*k} f)(z) = \sum_{n=0}^{\infty} \hat{f}(n+k) z^n \qquad (z \in \mathbb{D},\ k \geq 0). \tag{8.15}$$

In light of (4.17), this identity implies that

$$\|S^{*k} f\|_2^2 = \sum_{n=k}^{\infty} |\hat{f}(n)|^2 \tag{8.16}$$

for each $f \in H^2$ and $k \geq 0$. In particular, $\|S^* f\|_2^2 = \|f\|_2^2 - |f(0)|^2$. We have already noticed that S is an isometry, i.e.

$$S^* S = I. \tag{8.17}$$

However, it is not a unitary operator, that is, $SS^* \neq I$. More precisely, for each $f \in H^2$, we have $(SS^*)f = f - f(0)$, which can be written in the more compact form

$$SS^* = I - \chi_0 \otimes \chi_0. \tag{8.18}$$

We also have

$$\langle (SS^*)f, \chi_0 \rangle_{H^2} = 0 \qquad (f \in H^2).$$

This is another indirect manifestation of $SS^* \neq I$.

Since S and U are unitarily equivalent, Theorem 2.19 is written as follows for the operator S. Compare the following result with (8.5) and (8.6).

Lemma 8.6 *We have*

$$\sigma_p(S) = \emptyset,$$
$$\sigma_p(S^*) = \mathbb{D},$$
$$\sigma(S) = \sigma(S^*) = \bar{\mathbb{D}}$$

and, moreover, for each $\lambda \in \mathbb{D}$,

$$\ker(S^* - \bar{\lambda} I) = \mathbb{C} k_\lambda.$$

Proof We just need to verify $\ker(S^* - \bar{\lambda} I) = \mathbb{C} k_\lambda$. It easily follows from (8.14) that $S^* k_\lambda = \bar{\lambda} k_\lambda$, which implies that $\mathbb{C} k_\lambda \subset \ker(S^* - \bar{\lambda} I)$. For the converse inclusion, let $f \in \ker(S^* - \bar{\lambda} I)$. In other words, f satisfies the relation $S^* f = \bar{\lambda} f$. By induction, we have

$$S^{*n} f = \bar{\lambda}^n f \qquad (n \geq 0).$$

Therefore, by (8.11),

$$\hat{f}(n) = \langle f, \chi_n \rangle_2 = \langle f, S^n \chi_0 \rangle_2 = \langle S^{*n} f, \chi_0 \rangle_2 = \bar{\lambda}^n \langle f, \chi_0 \rangle_2 = \bar{\lambda}^n \hat{f}(0)$$

for each $n \geq 0$. Hence, from the Taylor expansion of f on $\mathbb{D}$, we obtain

$$f(z) = \sum_{n=0}^{\infty} \hat{f}(n) z^n = \hat{f}(0) \sum_{n=0}^{\infty} \bar{\lambda}^n z^n = \frac{\hat{f}(0)}{1 - \bar{\lambda} z} = \hat{f}(0) k_\lambda(z). \qquad \square$$

For the following lemma, let $\mathcal{P}_k$ denote the linear space of all analytic polynomials of degree at most $k - 1$. Note that the dimension of this space is k.

Lemma 8.7 *Let* $k \geq 1$. *Then the following hold:*

(i) $\ker S^{*k} = \mathcal{P}_k$;

(ii) $S^{*k}H^2 = H^2$;
(iii) S^{*k} *is a Fredholm operator of index* k;
(iv) S^k *is a Fredholm operator of index* $-k$.

Proof (i) By the representation (8.15),

$$(S^{*k}f)(z) = \sum_{n=0}^{\infty} \hat{f}(n+k)z^n = 0$$

if and only if

$$\hat{f}(n+k) = 0 \qquad (n \geq 0).$$

Hence, an element of the kernel has the form

$$f(z) = \sum_{n=0}^{k-1} \hat{f}(n)z^n,$$

i.e. an arbitrary polynomial of degree at most $k-1$.

(ii) It is sufficient to note that, given a function f in H^2, since S is an isometry (see (8.17)), we have $f = S^{*k}S^k f$, and thus $f \in S^{*k}H^2$.

(iii) By (i) and (ii), we already know that S^{*k} has a closed range and dim ker $S^{*k} = k < \infty$. Moreover, since S is an isometry, we have $\dim\ker S^k = 0$. Hence, S^{*k} is a Fredholm operator with

$$\operatorname{ind}(S^{*k}) = \dim\ker S^{*k} - \dim\ker S^k = k.$$

(iv) This follows immediately from (iii) and Theorem 7.32. □

According to Lemma 8.7 and Theorem 7.34, we have $0 \notin \sigma_{ess}(S)$. In fact, at this point we can determine precisely the essential spectrum of the unilateral shift operators. For the sake of completeness, we repeat the contents of Lemma 8.6.

Theorem 8.8 *We have*

$$\sigma_p(S) = \emptyset,$$
$$\sigma_a(S) = \sigma_c(S) = \sigma_{ess}(S) = \mathbb{T}$$

and

$$\sigma(S) = \bar{\mathbb{D}}.$$

Proof By Corollary 2.9 and Lemma 8.6, $\sigma_p(S) = \emptyset$ and thus

$$\mathbb{T} = \partial\sigma(S) \subset \sigma_a(S) = \sigma_c(S) \subset \sigma_{ess}(S) \subset \sigma(S) = \bar{\mathbb{D}}.$$

To show that $\sigma_{ess}(S) \cap \mathbb{D} = \emptyset$, fix $\lambda \in \mathbb{D}$. It is enough to verify that $S - \lambda I$ is a Fredholm operator. Using Lemma 8.6 once more, we know that

$$\dim(\ker(S-\lambda I)) = 0 \quad \text{and} \quad \dim(\ker(S-\lambda I)^*) = 1.$$

Thus, it remains to verify that $S - \lambda I$ has a closed range. Using the triangle inequality, we easily derive

$$(1 - |\lambda|)\, \|f\|_2 \leq \|(S - \lambda I)f\|_2 \qquad (f \in H^2).$$

Hence, the operator $S - \lambda I$ is bounded below. Thus, in particular, its range is closed. According to Theorem 7.34, we conclude that $\lambda \notin \sigma_{ess}(S)$. This means that

$$\sigma_{ess}(S) \subset \mathbb{T},$$

which finishes the proof. □

Theorem 8.9 *We have*

$$\sigma_p(S^*) = \mathbb{D},$$
$$\sigma_c(S^*) = \sigma_{ess}(S^*) = \mathbb{T}$$

and

$$\sigma_a(S^*) = \sigma(S) = \bar{\mathbb{D}}.$$

Proof The equality $\sigma_{ess}(S^*) = \mathbb{T}$ is an immediate consequence of (2.4) and $\sigma_{ess}(S) = \mathbb{T}$, which was established in Theorem 8.8.

By Theorem 2.8 and Lemma 8.6,

$$\mathbb{D} = \sigma_p(S^*) \subset \sigma_a(S^*) \subset \sigma(S^*) = \bar{\mathbb{D}}$$

and

$$\mathbb{T} = \partial\sigma(S^*) \subset \sigma_a(S^*) \subset \sigma(S^*) = \bar{\mathbb{D}}.$$

Hence, $\sigma_a(S^*) = \bar{\mathbb{D}}$.

By the same set of theorems,

$$\sigma_c(S^*) \subset \sigma_{ess}(S^*) = \mathbb{T}$$

and

$$\bar{\mathbb{D}} = \sigma_a(S^*) = \sigma_p(S^*) \cup \sigma_c(S^*) = \mathbb{D} \cup \sigma_c(S^*).$$

Hence, $\sigma_c(S^*) = \mathbb{T}$. This completes the proof. □

For each $w, z \in \mathbb{D}$,

$$\begin{aligned} ((I - \bar{w}S)k_w)(z) &= k_w(z) - \bar{w}(Sk_w)(z) \\ &= \frac{1}{1 - \bar{w}z} - \bar{w}z\frac{1}{1 - \bar{w}z} = 1 = k_0(z). \end{aligned}$$

Hence, $(I - \bar{w}S)k_w = k_0$. But, according to Lemma 8.6, for each $w \in \mathbb{D}$, the operator $I - \bar{w}S$ is invertible. Thus,

$$k_w = (I - \bar{w}S)^{-1}k_0$$

and, for each $f \in H^2$, we have

$$f(w) = \langle f, k_w \rangle_2 = \langle f,\ (I - \bar{w}S)^{-1} k_0 \rangle_2 = \langle (I - wS^*)^{-1} f,\ k_0 \rangle_2. \quad (8.19)$$

With this background, we define

$$Q_w = (I - wS^*)^{-1} S^* \in \mathcal{L}(H^2), \quad (8.20)$$

where the parameter w runs over $\mathbb{D}$. It is also easy to see that

$$Q_w = \sum_{n=1}^{\infty} w^{n-1} S^{*n}. \quad (8.21)$$

The series is convergent in $\mathcal{L}(H^2)$.

This family of operators will enter our discussion on several occasions. Here we study some of their elementary properties. The first result shows why Q_w is called the *difference quotient operator*. Note that $Q_0 = S^*$.

Theorem 8.10 *Let $w \in \mathbb{D}$. Then, for each $f \in H^2$,*

$$(Q_w f)(z) = \frac{f(z) - f(w)}{z - w} \qquad (z \in \mathbb{D}).$$

Proof Let

$$f(z) = \sum_{m=0}^{\infty} a_m z^m \qquad (z \in \mathbb{D}).$$

Then, for all $n \geq 1$,

$$(S^{*n} f)(z) = \sum_{m=0}^{\infty} a_{m+n} z^m \qquad (z \in \mathbb{D}).$$

Using (8.21), we obtain

$$\begin{aligned}
(Q_w f)(z) &= \sum_{n=1}^{\infty} w^{n-1} (S^{*n} f)(z) \\
&= \sum_{n=1}^{\infty} w^{n-1} \left(\sum_{m=0}^{\infty} a_{m+n} z^m \right) \\
&= \sum_{k=1}^{\infty} a_k \left(\sum_{m+n=k} z^m w^{n-1} \right) \\
&= \sum_{k=1}^{\infty} a_k \frac{z^k - w^k}{z - w} \\
&= \frac{\sum_{k=1}^{\infty} a_k z^k - \sum_{k=1}^{\infty} a_k w^k}{z - w} \\
&= \frac{f(z) - f(w)}{z - w}.
\end{aligned}$$

This completes the proof. □

Corollary 8.11 *Let $w \in \mathbb{D}$ and let $f, g \in H^2$ be such that the product fg also belongs to H^2. Then*

$$Q_w(fg) = fQ_w g + g(w)Q_w f.$$

In particular, for $w = 0$,

$$S^*(fg) = fS^*g + g(0)S^*f.$$

Proof By Theorem 8.10, for each $z \in \mathbb{D}$,

$$Q_w(fg)(z) = \frac{f(z)g(z) - f(w)g(w)}{z - w}.$$

Thus, rearranging the terms, we see that

$$Q_w(fg)(z) = f(z)\frac{g(z) - g(w)}{z - w} + g(w)\frac{f(z) - f(w)}{z - w}.$$

Again, by Theorem 8.10, this is precisely the first identity that we want to prove. Since $Q_0 = S^*$, the second identity is a special case of the first one. □

Exercises

Exercise 8.2.1 Show that

$$Zi_+ = i_+S.$$

Exercise 8.2.2 Show that

$$P_+(\chi_1 f) = \hat{f}(-1) + \chi_1 P_+ f \qquad (f \in L^2(\mathbb{T})).$$

Verify that this identity can be rewritten as

$$P_{H^2}Z = \chi_0 \otimes \chi_{-1} + ZP_{H^2}$$

in $\mathcal{L}(L^2(\mathbb{T}))$.
Hint: Use the main definition (4.21).

8.3 Commutants of Z and S

For an operator A on a Hilbert space $\mathcal{H}$, the set

$$A' = \{B \in \mathcal{L}(\mathcal{H}) : AB = BA\}$$

is called the *commutant* of A. It is easy to verify that A' is a closed subalgebra of $\mathcal{L}(\mathcal{H})$. If p is any analytic polynomial, then we clearly have

$$p(A)B = Bp(A) \qquad (B \in A'). \tag{8.22}$$

If A is invertible, we have $(A^{-1})' = A'$. In this situation, (8.22) holds for each trigonometric polynomial. In this section, we describe the commutants of S and Z. For this purpose, it will be useful to note that, for any analytic polynomial p, we have

$$p(S)f = pf \qquad (f \in H^2). \tag{8.23}$$

Similarly, for any trigonometric polynomial p, we have

$$p(Z)f = pf \qquad (f \in L^2(\mathbb{T})). \tag{8.24}$$

Lemma 8.12 *Let $\varphi \in L^2(\mathbb{T})$. Assume that there exists a constant $c > 0$ such that*

$$\|p\varphi\|_2 \le c\|p\|_2 \tag{8.25}$$

for all trigonometric polynomials p. Then $\varphi \in L^\infty(\mathbb{T})$ and, moreover, we have $\|\varphi\|_\infty \le c$.

Proof The first step is to show that (8.25) actually holds for the larger class $L^2(\mathbb{T})$, which contains trigonometric polynomials. To verify this, fix any $f \in L^2(\mathbb{T})$. Then there exists a sequence of trigonometric polynomials $(p_n)_{n\ge1}$ such that $p_n \longrightarrow f$ in $L^2(\mathbb{T})$, as $n \longrightarrow \infty$. According to (8.25), we easily see that the sequence $(p_n\varphi)_{n\ge1}$ is a Cauchy sequence in $L^2(\mathbb{T})$ and thus it converges to a function g in $L^2(\mathbb{T})$. Hence, we can find a subsequence, say $(p_{n_j})_{j\ge1}$, such that $p_{n_j} \longrightarrow f$ and $p_{n_j}\varphi \longrightarrow g$ almost everywhere on $\mathbb{T}$. Therefore, $f\varphi = g$ almost everywhere on $\mathbb{T}$ and $f\varphi \in L^2(\mathbb{T})$. Moreover, since

$$\|p_n\varphi\|_2 \le c\|p_n\|_2,$$

letting $n \longrightarrow \infty$, we obtain

$$\|f\varphi\|_2 \le c\|f\|_2 \qquad (f \in L^2(\mathbb{T})).$$

It suffices to apply Theorems 2.21 and 2.20 to get the result. □

For the following result, for each $\varphi \in L^\infty(\mathbb{T})$, the symbol M_φ denotes the operator of multiplication by φ on $L^2(\mathbb{T}) = L^2(\mathbb{T}, m)$ (see Section 2.8).

Theorem 8.13 *For the bilateral forward shift operator $Z \in \mathcal{L}(L^2(\mathbb{T}))$, we have*

$$Z' = \{M_\varphi : \varphi \in L^\infty(\mathbb{T})\}.$$

Proof Since $Z = M_{\chi_1} \in \mathcal{L}(L^2(\mathbb{T}))$ and

$$M_{\chi_1} M_\varphi = M_\varphi M_{\chi_1} = M_{\chi_1 \varphi},$$

we clearly have $\{M_\varphi : \varphi \in L^\infty(\mathbb{T})\} \subset Z'$. To prove the inverse, let $B \in Z'$, and define $\varphi = B\chi_0$. By definition, $\varphi \in L^2(\mathbb{T})$. We show that $\varphi \in L^\infty(\mathbb{T})$ and $B = M_\varphi$.

Since Z is invertible on $L^2(\mathbb{T})$, in fact $Z^{-1} = Z^*$, (8.22) holds for each trigonometric polynomial. Hence, using (8.24), we have

$$Bp = Bp(Z)\chi_0 = p(Z)B\chi_0 = p(Z)\varphi = p\varphi,$$

for any trigonometric polynomial p. We can extract all the required conclusions from the identity $Bp = p\varphi$. In the first place, this identity implies that

$$\|p\varphi\|_2 \le \|B\|\,\|p\|_2 \tag{8.26}$$

for all trigonometric polynomials p, and Lemma 8.12 assures that $\varphi \in L^\infty(\mathbb{T})$. Now, knowing that $\varphi \in L^\infty(\mathbb{T})$, we can rewrite the identity $Bp = p\varphi$ as $Bp = M_\varphi p$. Since the set of trigonometric polynomials is dense in $L^2(\mathbb{T})$, we conclude that $B = M_\varphi$. □

The second result has the same flavor. Note that M_φ in Theorem 8.13 represents an element of $\mathcal{L}(L^2(\mathbb{T}))$, while M_φ in Theorem 8.14 below refers to an element of $\mathcal{L}(H^2)$. Later on, after introducing Toeplitz operators, we will see that we can write T_φ instead of M_φ in the following result and thus avoid such possible confusion.

Theorem 8.14 *For the unilateral forward shift operator $S \in \mathcal{L}(H^2)$, we have*

$$S' = \{M_\varphi : \varphi \in H^\infty\}.$$

Proof Since $S = M_{\chi_1} \in \mathcal{L}(H^2)$ and

$$M_{\chi_1} M_\varphi = M_\varphi M_{\chi_1} = M_{\chi_1 \varphi},$$

we clearly have $\{M_\varphi : \varphi \in H^\infty\} \subset S'$. To prove the inverse, let $B \in S'$, and define $\varphi = B\chi_0$. By definition, $\varphi \in H^2$. We show that $\varphi \in H^\infty$ and $B = M_\varphi$.

The identity (8.22) holds for each analytic polynomial. Hence, we have

$$Bp = Bp(S)\chi_0 = p(S)B\chi_0 = p(S)\varphi = p\varphi.$$

This identity implies that

$$\|p\varphi\|_2 \le \|B\|\,\|p\|_2$$

for all analytic polynomials p. Since $\|\chi_n f\|_2 = \|f\|_2$, $n \in \mathbb{Z}$, we see that the preceding inequality holds in fact for all trigonometric polynomials. Hence, using Lemma 8.12, $\varphi \in L^\infty(\mathbb{T}) \cap H^2 = H^\infty$. Second, knowing that $\varphi \in H^\infty$, we can rewrite the identity $Bp = p\varphi$ as $Bp = M_\varphi p$. Since the set of analytic polynomials is dense in H^2, we conclude that $B = M_\varphi$. □

Lemma 8.15 *Let $\mathcal{E}$ be a reducing invariant subspace of $Z \in \mathcal{L}(L^2(\mathbb{T}))$. Let $T : \mathcal{E} \longrightarrow L^2(\mathbb{T})$ be a linear and bounded operator such that*

$$TZ_{|\mathcal{E}} = ZT.$$

Then there exists $\varphi \in L^\infty(\mathbb{T})$ such that

$$Tf = \varphi f \qquad (f \in \mathcal{E}).$$

Proof Define $\tilde{T} : L^2(\mathbb{T}) \longrightarrow L^2(\mathbb{T})$ by

$$\tilde{T}(f \oplus g) = Tf,$$

for any $f \oplus g \in \mathcal{E} \oplus \mathcal{E}^\perp = L^2(\mathbb{T})$. Then $\tilde{T}$ is a linear and bounded operator on $L^2(\mathbb{T})$. Moreover, since $\mathcal{E}$ is a reducing subspace of Z, we have by definition $Z\mathcal{E} \subset \mathcal{E}$ and $Z\mathcal{E}^\perp \subset \mathcal{E}^\perp$. Hence, for any $f \oplus g \in \mathcal{E} \oplus \mathcal{E}^\perp$, we have

$$Z(f \oplus g) = Zf \oplus Zg,$$

along with $Zf \in \mathcal{E}$ and $Zg \in \mathcal{E}^\perp$. Then

$$\tilde{T}Z(f \oplus g) = \tilde{T}(Zf \oplus Zg) = TZf = ZTf = Z\tilde{T}(f \oplus g).$$

We thus get $\tilde{T}Z = Z\tilde{T}$, that is, $\tilde{T} \in \{Z\}'$. Theorem 8.13 now implies the existence of a function $\varphi \in L^\infty(\mathbb{T})$ such that $\tilde{T} = M_\varphi$. For any $f \in \mathcal{E}$, we finally obtain

$$\varphi f = \tilde{T}f = \tilde{T}(f \oplus 0) = Tf.$$ □

Exercises

Exercise 8.3.1 Show that

$$(Z^*)' = \{M_\varphi : \varphi \in L^\infty(\mathbb{T})\} \quad \text{and} \quad (S^*)' = \{M_\varphi^* : \varphi \in H^\infty\},$$

where in the second set M_φ denotes the multiplication operator by φ viewed as an operator of $\mathcal{L}(H^2)$.
Hint: Use Theorems 8.13 and 8.14.
Remark: Using the notation of Toeplitz operators (see Chapter 12), we can write

$$(S^*)' = \{T_{\bar{\varphi}} : \varphi \in H^\infty\}.$$

Exercise 8.3.2 Let $A \in \mathcal{L}(L^2(\mathbb{T}))$. Show that the following are equivalent.

(i) $Z^*AZ = A$.
(ii) $AZ = ZA$.
(iii) The matrix of A with respect to the standard basis $(\chi_n)_{n\in\mathbb{Z}}$ is a doubly infinite Toeplitz matrix.
(iv) There is a function $\varphi \in L^\infty(\mathbb{T})$ such that $A = M_\varphi$.

Hint: (i) $\Longleftrightarrow$ (ii) follows from $Z^* = Z^{-1}$. (i) $\Longleftrightarrow$ (iv) was discussed in Theorem 8.13. For (iv) $\Longrightarrow$ (iii), see (2.30). Finally, for (iii) $\Longrightarrow$ (i), the assumption (iii) means that

$$\langle A\chi_n, \chi_m\rangle_{L^2(\mathbb{T})} = \langle A\chi_{n+1}, \chi_{m+1}\rangle_{L^2(\mathbb{T})} \qquad (m, n \in \mathbb{Z}).$$

Since $\chi_{k+1} = Z\chi_k$, we can rewrite the preceding relation as

$$\langle A\chi_n, \chi_m\rangle_{L^2(\mathbb{T})} = \langle Z^*AZ\chi_n, \chi_m\rangle_{L^2(\mathbb{T})} \qquad (m, n \in \mathbb{Z}).$$

Exercise 8.3.3 Let $A \in \mathcal{L}(H^2(\mathbb{T}))$. Show that the matrix of A with respect to the standard basis $(\chi_n)_{n\geq 0}$ is a singly infinite Toeplitz matrix if and only if

$$S^*AS = A.$$

Hint: The matrix of A is a singly infinite Toeplitz matrix if and only if

$$\langle A\chi_n, \chi_m\rangle_{H^2(\mathbb{T})} = \langle A\chi_{n+1}, \chi_{m+1}\rangle_{H^2(\mathbb{T})} \qquad (m, n \geq 0).$$

Since $\chi_{k+1} = S\chi_k$, we can rewrite the preceding relation as

$$\langle A\chi_n, \chi_m\rangle_{H^2(\mathbb{T})} = \langle S^*AS\chi_n, \chi_m\rangle_{H^2(\mathbb{T})} \qquad (m, n \geq 0).$$

Remark: Note that the two identities $S^*AS = A$ and $AS = SA$ are not the same. Since $S^*S = I$, the relations $AS = SA$ implies that $S^*AS = A$, but the inverse implication does not generally hold. An example is $A = S^*$.

Exercise 8.3.4 Let $A = M_\varphi \in \mathcal{L}(H^2(\mathbb{T}))$, where $\varphi \in H^\infty(\mathbb{T})$. Show that the matrix of A with respect to the standard basis $(\chi_n)_{n\geq 0}$ is a doubly infinite Toeplitz matrix. Give an example $A \in \mathcal{L}(H^2(\mathbb{T}))$ whose matrix with respect to the standard basis is a doubly infinite Toeplitz matrix, but $A \notin \{M_\varphi : \varphi \in H^\infty(\mathbb{T})\}$.

Hint: Use Exercise 8.3.3 and Theorem 8.14.

Remark: Similar to Exercise 8.3.2, this is a partial result for operators on H^2 that have a Toeplitz form. For the general result, we need some tools that will be developed in Chapter 12 (see Exercise 12.1.2).

8.4 Cyclic vectors of S

We now characterize the cyclic vectors of S. See Section 1.10 for the definition of cyclic vectors in a general context. In our situation, instead of the notation $\langle f \rangle$, where $f \in H^2$, we use $\mathcal{P}[f]$ for the smallest closed invariant subspace of H^2 that contains f. In other words,

$$\mathcal{P}[f] = \text{Span}\{S^n f : n \geq 0\} = \text{Span}\{\chi_n f : n \geq 0\},$$

where we recall that $\chi_n(\zeta) = \zeta^n$, $\zeta \in \mathbb{T}$, $n \geq 0$. Then f is a cyclic vector of the unilateral forward shift operator S if and only if $\mathcal{P}[f] = H^2$. In fact, we even consider the general case of $f \in H^p$, and then let $\mathcal{P}[f]$ denote the closed subspace of H^p generated by $\chi_n f$, $n \geq 0$. Since an analytic polynomial is a finite linear combination of the χ_n, $n \geq 0$, we can say that

$$\mathcal{P}[f] = \text{Clos}_{H^p}\{\mathfrak{p} f : \mathfrak{p} \in \mathcal{P}_+\},$$

where $\mathcal{P}_+$ denotes the family of all analytic polynomials. Describing the subspace $\mathcal{P}[f]$ is a problem of polynomial approximation. In fact, using the representation of linear functionals on H^p, we can obtain a complete description of $\mathcal{P}[f]$.

Theorem 8.16 *Let $f \in H^p$, $0 < p < \infty$, and suppose that $f = \Theta h$ is the inner–outer decomposition of f. Then*

$$\mathcal{P}[f] = \Theta H^p.$$

Proof In the first step, we show that $\mathcal{P}[f] = \Theta \mathcal{P}[h]$ and then we complete the proof by verifying that $\mathcal{P}[h] = H^p$.

Let $g \in \mathcal{P}[f]$. Thus, by definition, there are analytic polynomials $\mathfrak{p}_n$, $n \geq 1$, such that

$$\lim_{n \to \infty} \|\mathfrak{p}_n f - g\|_p = 0. \tag{8.27}$$

In particular, $(\mathfrak{p}_n f)_{n \geq 1}$ is a Cauchy sequence in H^p. But, since the inner functions are unimodular on $\mathbb{T}$, we have

$$\|\mathfrak{p}_n f - \mathfrak{p}_m f\|_p = \|\Theta(\mathfrak{p}_n h - \mathfrak{p}_m h)\|_p = \|\mathfrak{p}_n h - \mathfrak{p}_m h\|_p,$$

which shows that $(\mathfrak{p}_n h)_{n \geq 1}$ is also a Cauchy sequence in H^p. Since H^p is complete, there is a $g_1 \in H^p$ such that

$$\lim_{n \to \infty} \|\mathfrak{p}_n h - g_1\|_p = 0. \tag{8.28}$$

By definition, this means that $g_1 \in \mathcal{P}[h]$. Moreover, (8.27) and (8.28) imply that $g = \Theta g_1$. This identity ensures that $\mathcal{P}[f] \subset \Theta \mathcal{P}[h]$.

On the other hand, let $g \in \Theta\mathcal{P}[h]$. Hence, $g = \Theta g_1$, for some $g_1 \in \mathcal{P}[h]$. Therefore, there are polynomials $\mathfrak{p}_n$, $n \geq 1$, such that (8.28) holds. Thus,

$$\|g - \mathfrak{p}_n f\|_p = \|\Theta(g_1 - \mathfrak{p}_n h)\|_p = \|g_1 - \mathfrak{p}_n h\|_p \longrightarrow 0 \qquad (n \longrightarrow \infty),$$

which yields $g \in \mathcal{P}[f]$. In other words, we also have $\Theta\mathcal{P}[h] \subset \mathcal{P}[f]$.

It remains to verify that $\mathcal{P}[h] = H^p$. To show that the inclusion $\mathcal{P}[h] \subset H^p$ is not proper, we use a standard functional analysis method when $1 \leq p < \infty$. The proof for the case $0 < p < 1$ is based on Zygmund's trick.

Suppose that $1 \leq p < \infty$, and let Λ be a linear functional on H^p such that

$$\Lambda(g) = 0 \tag{8.29}$$

for all $g \in \mathcal{P}[h]$. We show that $\Lambda \equiv 0$, and, since $\mathcal{P}[h]$ is a closed subspace of H^p, this is enough to conclude that $\mathcal{P}[h] = H^p$. Using the Hahn–Banach theorem and Riesz's representation theorem for linear functionals on $L^p(\mathbb{T})$, there is a function $\varphi \in L^q(\mathbb{T})$, where $1/p + 1/q = 1$, such that

$$\Lambda(g) = \frac{1}{2\pi}\int_0^{2\pi} g(e^{it})\varphi(e^{it})\,dt \qquad (g \in H^p).$$

Now, the assumption (8.29) is written as

$$\int_0^{2\pi} e^{int} h(e^{it})\varphi(e^{it})\,dt = 0 \qquad (n \geq 0). \tag{8.30}$$

Since $h \in H^p \subset L^p(\mathbb{T})$ and $\varphi \in L^q(\mathbb{T})$, then $h\varphi \in L^1(\mathbb{T})$. Therefore, (8.30) means that $h\varphi \in H^1_0$. Hence,

$$h(e^{it})\varphi(e^{it}) = e^{it}\psi(e^{it})$$

for some $\psi \in H^1$. We thus can write

$$\varphi(e^{it}) = e^{it}\,\frac{\psi(e^{it})}{h(e^{it})}. \tag{8.31}$$

Since h is an outer function, $\psi \in H^1$ and $|\psi/h| = |\varphi| \in L^q(\mathbb{T})$, by Corollary 4.28, we have $\psi/h \in H^q$, and thus $\varphi \in H^q_0$. This fact ensures that $g\varphi \in H^1_0$ for any $g \in H^p$. This property implies in particular that

$$\int_0^{2\pi} g(e^{it})\varphi(e^{it})\,dt = 0 \qquad (g \in H^p).$$

But this precisely means that $\Lambda(g) = 0$ for any $g \in H^p$. This finishes the proof for the case $1 \leq p < \infty$.

To establish the case $0 < p < 1$, we prove the following statement by induction: *for each $m \geq 1$, if $p \geq 1/2^{m-1}$ and $h \in H^p$, h outer, then $\mathcal{P}[h] = H^p$.* In fact, our proof up to here shows that the statement is true for $m = 1$. Now suppose that it holds for m and we proceed to show that it also holds for $m + 1$. To do so, we only need to verify the claim for $1/2^m \leq p < 1/2^{m-1}$.

To show that $\mathcal{P}[h] = H^p$, $1/2^m \le p < 1/2^{m-1}$, fix $g \in H^p$ and let $\varepsilon > 0$. Let $\mathfrak{p}$, $\mathfrak{q}$ and $\mathfrak{r}$ be analytic polynomials. These polynomials are going to be determined soon, but, for the time being, they are arbitrary. We have

$$\|g - \mathfrak{p}h\|_p \le \|g - \mathfrak{r}\|_p + \|\mathfrak{r} - \mathfrak{q}h^{1/2}\|_p + \|\mathfrak{q}h^{1/2} - \mathfrak{p}h\|_p.$$

Since $h^{1/2}$ is an outer function in H^{2p}, we surely have

$$\|\mathfrak{r} - \mathfrak{q}h^{1/2}\|_p \le \|\mathfrak{r} - \mathfrak{q}h^{1/2}\|_{2p}.$$

Moreover, by the Cauchy–Schwarz inequality,

$$\|\mathfrak{q}h^{1/2} - \mathfrak{p}h\|_p \le \|h\|_p^{1/2} \|\mathfrak{q} - \mathfrak{p}h^{1/2}\|_{2p}.$$

Therefore, we obtain

$$\|g - \mathfrak{p}h\|_p \le \|g - \mathfrak{r}\|_p + \|\mathfrak{r} - \mathfrak{q}h^{1/2}\|_{2p} + \|h\|_p^{1/2} \|\mathfrak{q} - \mathfrak{p}h^{1/2}\|_{2p}.$$

Now, we are ready to choose $\mathfrak{p}$, $\mathfrak{q}$ and $\mathfrak{r}$ properly. First, since analytic polynomials are dense in H^p, there is a polynomial $\mathfrak{r}$ such that

$$\|g - \mathfrak{r}\|_p < \varepsilon.$$

Second, $2p \ge 1/2^{m-1}$, and thus by the induction hypothesis, there is a polynomial $\mathfrak{q}$ such that

$$\|\mathfrak{r} - \mathfrak{q}h^{1/2}\|_{2p} < \varepsilon.$$

Third, again by the induction hypothesis, there is a polynomial $\mathfrak{p}$ such that

$$\|\mathfrak{q} - \mathfrak{p}h^{1/2}\|_{2p} < \varepsilon.$$

Therefore, we conclude that

$$\|g - \mathfrak{p}h\|_p \le (2 + \|h\|_p^{1/2})\varepsilon. \qquad \square$$

Based on the above theorem, we are able to completely characterize the cyclic vectors for the unilateral forward shift operator S on H^2. Despite many efforts, it should be noted that, for the time being, a complete characterization of cyclic vectors for the unilateral forward shift operator on many other function spaces in the unit disk, e.g. Dirichlet space, is not available. Thus, it is remarkable that such a result exists for the shift operator S on H^2.

Corollary 8.17 *Let $f \in H^p$, $0 < p < \infty$. Then*

$$\mathcal{P}[f] = \mathrm{Clos}_{H^p}\{\mathfrak{p}f : \mathfrak{p} \in \mathcal{P}_+\} = H^p$$

if and only if f is outer.

Considering the case $p = 2$, Corollary 8.17 says that a function $f \in H^2$ is a cyclic vector for S if and only if f is outer. We know that

$$\text{Span}\{k_w : w \in \mathbb{D}\} = H^2,$$

where k_w is the Cauchy kernel. The following result shows that we can replace analytic polynomials by Cauchy kernels in the above corollary.

Corollary 8.18 *Let h be an outer function in H^2. Then*

$$\text{Span}\{hk_w : w \in \mathbb{D}\} = H^2.$$

Proof Let $f \in H^2$, and assume that

$$\langle f, k_w h\rangle_2 = 0 \qquad (w \in \mathbb{D}).$$

By Theorem 5.5(ii), the linear manifold created by $\{k_w,\ w \in \mathbb{D}\}$ is uniformly dense in $\mathcal{A}$. Hence, we also have

$$\langle f, \mathfrak{p}h\rangle_2 = 0 \qquad (\mathfrak{p} \in \mathcal{P}_+).$$

But, according to Theorem 8.16, $\mathcal{P}[h] = H^2$, and we conclude that $f = 0$. □

Exercises

Exercise 8.4.1 Let h be a function in H^2 and suppose that

$$\text{Span}\{hk_w : w \in \mathbb{D}\} = H^2.$$

Then show that h is outer. In particular, the converse of Corollary 8.18 is true. Hint: Write $h = \Theta h_1$, where Θ is inner and h_1 is outer, and then use Corollary 8.18.

Exercise 8.4.2 Let h be an outer function in H^p, $1 \le p < \infty$. Show that

$$\text{Span}\{hk_w : w \in \mathbb{D}\} = H^p.$$

Hint: Let $\Lambda \in (H^p)^*$ such that $\Lambda(hk_w) = 0$, $w \in \mathbb{D}$. Show that $\Lambda \equiv 0$. See the proofs of Theorem 8.16 and Corollary 8.18.

8.5 When do we have $H^p(\mu) = L^p(\mu)$?

We recall that $H^p(\mu)$ was defined as the closure of all analytic polynomials $\mathcal{P}_+$ in $L^p(\mu)$, and, if $\mathcal{P}_{0+}$ denotes the space of all analytic polynomials $\mathfrak{p}$ such that $\mathfrak{p}(0) = 0$, then $H^p_0(\mu)$ is the closure of $\mathcal{P}_{0+}$ in $L^p(\mu)$. In this section, we discuss the problem of characterizing measures μ such that $H^p(\mu) = L^p(\mu)$. In Lemma 5.12, we proved that $H^p(\mu) = L^p(\mu)$ if and only if $\chi_0 \in H_0^p(\mu)$. Therefore, we should study the distance

$$\operatorname{dist}(\chi_0, H_0^p(\mu)) = \inf_{f \in H_0^p(\mu)} \left(\int_{\mathbb{T}} |1 - f|^p \, d\mu \right)^{1/p},$$

and we shall determine when this is zero. Note that, since $\mathcal{P}_{0+}$ is dense in $H_0^p(\mu)$, we also have

$$\operatorname{dist}(\chi_0, H_0^p(\mu)) = \inf_{\mathfrak{p} \in \mathcal{P}_{0+}} \left(\int_{\mathbb{T}} |1 - \mathfrak{p}|^p \, d\mu \right)^{1/p}.$$

Our goal is to give an explicit formula for this distance in terms of the absolutely continuous part of μ. First, let us treat the case of singular measures.

Lemma 8.19 *Let σ be a finite positive Borel measure on $\mathbb{T}$. Suppose that σ is singular with respect to the Lebesgue measure. Then*

$$\operatorname{dist}(\chi_0, H_0^p(\sigma)) = \inf_{\mathfrak{p} \in \mathcal{P}_{0+}} \int_{\mathbb{T}} |1 - \mathfrak{p}|^p \, d\sigma = 0$$

and thus $H^p(\sigma) = L^p(\sigma)$.

Proof We use a duality argument. Assume that $\Lambda \in (L^p(\sigma))^*$ is such that $\Lambda_{|H^p(\sigma)} \equiv 0$. Hence, according to the Riesz representation theorem, there is a function $\varphi \in L^q(\sigma)$, where $1/p + 1/q = 1$, such that

$$\Lambda(g) = \int_{\mathbb{T}} g\varphi \, d\sigma \qquad (g \in L^p(\sigma)).$$

Since $\Lambda_{|H^p(\sigma)} \equiv 0$, we have

$$\int_{\mathbb{T}} \zeta^n \varphi(\zeta) \, d\sigma(\zeta) = 0 \qquad (n \geq 0).$$

Now, on the one hand, this assumption along with the theorem of F. and M. Riesz (Theorem 4.4) imply that the measure $\varphi \, d\sigma$ is absolutely continuous. On the other hand, remembering that $L^q(\sigma) \subset L^1(\sigma)$, we see that $\varphi \, d\sigma$ is a singular measure whose support is contained in the support of σ. Therefore, we must have $\varphi \, d\sigma \equiv 0$, i.e. $\Lambda = 0$.

The above duality argument shows that $H^p(\sigma)$ is dense in $L^p(\sigma)$. Since $H^p(\sigma)$ is closed, we conclude that $H^p(\sigma) = L^p(\sigma)$. For the distance, note that

$$\operatorname{dist}(\chi_0, H_0^p(\sigma)) = \operatorname{dist}(\chi_{-1}, H^p(\sigma)) = \operatorname{dist}(\chi_{-1}, L^p(\sigma)) = 0. \qquad \square$$

If $\mu = \mu_a + \mu_s$ is the Lebesgue decomposition of a positive Borel measure μ, then by Lemmas 8.19 and 5.12 and (5.17), we have

$$H^p(\mu) = H^p(\mu_a) \oplus L^p(\mu_s).$$

To obtain the formula for $\mathrm{dist}(\chi_0, H_0^p(\mu))$ in the general case, we proceed into two steps. The first step shows that the singular part of μ plays no role. The second step treats the case of absolutely continuous measure and uses Theorem 8.16.

Lemma 8.20 *Let μ be a finite positive Borel measure on $\mathbb{T}$ with the Lebesgue decomposition $d\mu = w\,dm + d\sigma$, $w \in L^1(\mathbb{T})$, $w \geq 0$ and $d\sigma \perp dm$. Then, for each $1 \leq p < \infty$,*

$$\mathrm{dist}(\chi_0, H_0^p(\mu)) = \inf_{\mathfrak{p} \in \mathcal{P}_{0+}} \int_{\mathbb{T}} |1 - \mathfrak{p}|^p \, d\mu = \inf_{\mathfrak{p} \in \mathcal{P}_{0+}} \int_{\mathbb{T}} |1 - \mathfrak{p}|^p w \, dm.$$

Proof Since μ is positive, for each polynomial $\mathfrak{p}$, we have

$$\int_{\mathbb{T}} |1 - \mathfrak{p}|^p \, d\mu \geq \int_{\mathbb{T}} |1 - \mathfrak{p}|^p w \, dm$$

and thus

$$\inf_{\mathfrak{p} \in \mathcal{P}_{0+}} \int_{\mathbb{T}} |1 - \mathfrak{p}|^p \, d\mu \geq \inf_{\mathfrak{p} \in \mathcal{P}_{0+}} \int_{\mathbb{T}} |1 - \mathfrak{p}|^p w \, dm.$$

To establish the inverse inequality, put

$$C = \inf_{\mathfrak{p} \in \mathcal{P}_{0+}} \int_{\mathbb{T}} |1 - \mathfrak{p}|^p w \, dm.$$

It is sufficient to show that, for any fixed $\varepsilon > 0$, there exists a polynomial $q \in \mathcal{P}_{0+}$ such that

$$\int_{\mathbb{T}} |1 - \mathfrak{q}|^p \, d\mu < C + 5\varepsilon. \tag{8.32}$$

On the one hand, by the definition of C, there is $\mathfrak{q}_1 \in \mathcal{P}_{0+}$ such that

$$\int_{\mathbb{T}} |1 - \mathfrak{q}_1|^p w \, dm < C + \varepsilon. \tag{8.33}$$

On the other hand, by Lemma 8.19, there is $\mathfrak{q}_2 \in \mathcal{P}_{0+}$ such that

$$\int_{\mathbb{T}} |1 - \mathfrak{q}_2|^p \, d\sigma < \varepsilon. \tag{8.34}$$

We show that a polynomial combination of $\mathfrak{q}_1$ and $\mathfrak{q}_2$ does the job.

Take a closed set E in the support of σ such that

$$\int_{\mathbb{T} \setminus E} (1 + |\mathfrak{q}_1| + |\mathfrak{q}_2 - \mathfrak{q}_1|)^p \, d\sigma < \varepsilon. \tag{8.35}$$

The reason for such a choice is clarified below.

Since σ is a singular measure, we certainly have $|E| = 0$. Let f be a Fatou function for E, i.e. a function in the disk algebra such that $f = 1$ on E, while

$|f| < 1$ on $\mathbb{D} \cup (\mathbb{T} \setminus E)$. Based on these properties of f, we can get an upper estimate for

$$I = \int_{\mathbb{T}} |1 - (1 - f^n)\mathfrak{q}_1 - f^n \mathfrak{q}_2|^p \, d\sigma.$$

Write

$$I = I_E + I_{\mathbb{T}\setminus E} = \int_E + \int_{\mathbb{T}\setminus E}.$$

Since $f \equiv 1$ on E, by (8.34), we have

$$I_E = \int_E |1 - \mathfrak{q}_2|^p \, d\sigma < \varepsilon,$$

and, by (8.35), on its complement $\mathbb{T} \setminus E$, the estimate

$$I_{\mathbb{T}\setminus E} \le \int_{\mathbb{T}\setminus E} (1 + |\mathfrak{q}_1| + |\mathfrak{q}_2 - \mathfrak{q}_1|)^p \, d\sigma < \varepsilon$$

holds. Therefore, we obtain

$$I = \int_{\mathbb{T}} |1 - (1 - f^n)\mathfrak{q}_1 - f^n \mathfrak{q}_2|^p \, d\sigma < 2\varepsilon \tag{8.36}$$

for all $n \ge 1$.

We need to find a similar estimate when $d\sigma$ is replaced by $w\,dm$. In this case, without loss of generality, we can replace $\int_{\mathbb{T}}$ by $\int_{\mathbb{T}\setminus E}$, and since $|f| < 1$ on $\mathbb{T} \setminus E$, by the dominated convergence theorem,

$$\lim_{n\to\infty} \int_{\mathbb{T}} |1 - (1 - f^n)\mathfrak{q}_1 - f^n \mathfrak{q}_2|^p w \, dm = \int_{\mathbb{T}} |1 - \mathfrak{q}_1|^p w \, dm.$$

Thus, by (8.33), there is a sufficiently large n such that

$$\int_{\mathbb{T}} |1 - (1 - f^n)\mathfrak{q}_1 - f^n \mathfrak{q}_2|^p w \, dm < C + 2\varepsilon. \tag{8.37}$$

Therefore, (8.36) and (8.37) together imply that

$$\int_{\mathbb{T}} |1 - (1 - f^n)\mathfrak{q}_1 - f^n \mathfrak{q}_2|^p \, d\mu < C + 4\varepsilon.$$

We are almost done. The only obstacle is that the combination $(1 - f^n)\mathfrak{q}_1 - f^n \mathfrak{q}_2$ is not a polynomial. But f^n is an element of the disk algebra $\mathcal{A}$ and thus it can be uniformly approximated by analytic polynomials. Hence, there is an analytic polynomial $\mathfrak{q}_3$ such that

$$\int_{\mathbb{T}} |1 - (1 - \mathfrak{q}_3)\mathfrak{q}_1 - \mathfrak{q}_3 \mathfrak{q}_2|^p \, d\mu < C + 5\varepsilon.$$

Remember that $\mathfrak{q}_1, \mathfrak{q}_2 \in \mathcal{P}_{0+}$. Thus $(1 - \mathfrak{q}_3)\mathfrak{q}_1 + \mathfrak{q}_3\mathfrak{q}_2$ belongs to $\mathcal{P}_{0+}$, and this is the desired polynomial that we were looking for to establish (8.32). □

Knowing that the singular part plays no role in determining the value of $\mathrm{dist}(1, H_0^p(\mu))$, we now proceed to study the case of absolutely continuous measures.

Lemma 8.21 *Let $w \in L^1(\mathbb{T})$, $w \geq 0$, and let $1 \leq p < \infty$. Then*

$$\mathrm{dist}(\chi_0, H_0^p(w)) = \inf_{\mathfrak{p} \in \mathcal{P}_{0+}} \int_{\mathbb{T}} |1 - \mathfrak{p}|^p w \, dm = \exp\left(\int_{\mathbb{T}} \log w \, dm \right).$$

Proof Since $w \in L^1(\mathbb{T})$, we certainly have $\log^+ w \in L^1(\mathbb{T})$. But we have no control over $\log^- w$. Hence, we consider two cases. In the first case, we assume that $\log^- w \in L^1(\mathbb{T})$. In other words, we treat the situation $\log w \in L^1(\mathbb{T})$. At the end of the proof, we will consider the case $\log^- w \notin L^1(\mathbb{T})$.

For simplicity, we consider the normalized function $w_1 = Cw$, where

$$C = \exp\left(- \int_{\mathbb{T}} \log w \, dm \right).$$

The normalization is done such that

$$\int_{\mathbb{T}} \log w_1 \, dm = 0. \tag{8.38}$$

Thus, to establish the theorem, we need to show that

$$\inf_{\mathfrak{p} \in \mathcal{P}_{0+}} \int_{\mathbb{T}} |1 - \mathfrak{p}|^p w_1 \, dm = 1.$$

Consider the outer function

$$h(z) = \exp\left(\int_{\mathbb{T}} \frac{\zeta + z}{\zeta - z} \log w_1(\zeta) \, dm(\zeta) \right).$$

The condition (8.38) implies that $h(0) = 1$, and since

$$|h| = w_1 \qquad \text{(a.e. on } \mathbb{T}),$$

we have $h \in H^1$. Hence, for each $\mathfrak{p} \in \mathcal{P}_{0+}$, the function $g = (1 - \mathfrak{p})h^{1/p}$ is in H^p, with $g(0) = 1$. Therefore, by (4.16),

$$\int_{\mathbb{T}} |1 - \mathfrak{p}|^p w_1 \, dm \geq 1 \qquad (\mathfrak{p} \in \mathcal{P}_{0+}).$$

Now, we show that, for an appropriate sequence $(\mathfrak{p}_n)_{n \geq 1} \subset \mathcal{P}_{0+}$, the last integral actually tends to 1 and thus its infimum when $\mathfrak{p}$ runs over $\mathcal{P}_{0+}$ is precisely 1. Since $h^{1/p}$ is an outer function in H^p, by Theorem 8.16, there is a sequence of analytic polynomials $\mathfrak{q}_n$ such that

$$\lim_{n \to \infty} \|\mathfrak{q}_n h^{1/p} - 1\|_p = 0.$$

Moreover, $h^{1/p}(0) = 1$, and thus we necessarily have $\mathfrak{q}_n(0) \longrightarrow 1$. Hence, we can consider the normalized polynomials $\mathfrak{q}_n/\mathfrak{q}_n(0)$, which can be written as

$$\frac{\mathfrak{q}_n}{\mathfrak{q}_n(0)} = 1 - \mathfrak{p}_n,$$

where $\mathfrak{p}_n \in \mathcal{P}_+^0$. With the above new notation, we also have

$$\lim_{n\to\infty} \|(1-\mathfrak{p}_n)h^{1/p} - 1\|_p = 0.$$

Since $|h| = w_1$ a.e. on $\mathbb{T}$, this fact implies that

$$\lim_{n\to\infty} \int_{\mathbb{T}} |1-\mathfrak{p}_n|^p w_1 \, dm = 1.$$

To finish the proof, it remains to consider the case $\log^- w \notin L^1(\mathbb{T})$, i.e.

$$\int_{\mathbb{T}} \log w \, dm = -\infty,$$

and we must show that

$$\inf_{\mathfrak{p}\in\mathcal{P}_{0+}} \int_{\mathbb{T}} |1-\mathfrak{p}|^p w \, dm = 0. \tag{8.39}$$

Let

$$w_n = \max\{w, 1/n\}.$$

Then $w \le w_n$, $\log w_n \in L^1(\mathbb{T})$ and, moreover, by the monotone convergence theorem,

$$\lim_{n\to\infty} \int_{\mathbb{T}} \log w_n \, dm = -\infty.$$

By the first part of the proof, we have

$$\begin{aligned} \inf_{\mathfrak{p}\in\mathcal{P}_{0+}} \int_{\mathbb{T}} |1-\mathfrak{p}|^p w \, dm &\le \inf_{\mathfrak{p}\in\mathcal{P}_{0+}} \int_{\mathbb{T}} |1-\mathfrak{p}|^p w_n \, dm \\ &= \exp\left(\int_{\mathbb{T}} \log w_n \, dm\right). \end{aligned}$$

Now, let $n \longrightarrow \infty$ to obtain (8.39). □

Now, we have gathered enough tools to establish the promised formula for the distance between 1 and $H_0^p(\mu)$ in the space $H^p(\mu)$. Putting together Lemmas 8.20 and 8.21, we obtain the following general result.

Theorem 8.22 *Let* $1 \le p < \infty$, *and let* μ *be a finite and positive Borel measure on* $\mathbb{T}$ *with Lebesgue decomposition*

$$d\mu = d\mu_a + d\mu_s = w \, dm + d\mu_s,$$

where $w \in L^1(\mathbb{T})$, $w \geq 0$ and $d\mu_s \perp dm$. Then

$$\operatorname{dist}(\chi_0, H_0^p(\mu)) = \operatorname{dist}(\chi_0, H_0^p(\mu_a)) = \exp\left(\int_{\mathbb{T}} \log w \, dm\right).$$

In Lemma 8.19, we saw that $H^p(\sigma) = L^p(\sigma)$, whenever σ is a singular measure. However, in light of Lemma 5.12, the formula provided in Theorem 8.22 enables us to completely characterize measures with such a property.

Corollary 8.23 *Let $1 \leq p < \infty$, and let μ be a finite and positive Borel measure on $\mathbb{T}$ with Lebesgue decomposition*

$$d\mu = d\mu_a + d\mu_s = w\, dm + d\mu_s,$$

where $w \in L^1(\mathbb{T})$, $w \geq 0$ and $d\mu_s \perp dm$. Then the following are equivalent:

(i) $H^p(\mu) = L^p(\mu)$;
(ii) $\chi_0 \in H_0^p(\mu)$;
(iii) $\displaystyle\int_{\mathbb{T}} \log w \, dm = -\infty$.

8.6 The unilateral forward shift operator S_μ

In Section 8.2, we introduced the unilateral forward shift operator S on the Hardy space H^2. Now we define a generalization of this operator on $H^2(\mu)$, where μ is a finite and positive Borel measure on $\mathbb{T}$, and study its main spectral properties.

The *unilateral forward shift operator* S_μ is defined by

$$\begin{array}{rccc} S_\mu: & H^2(\mu) & \longrightarrow & H^2(\mu) \\ & f & \longmapsto & \chi_1 f. \end{array}$$

In other words, we have $S_\mu = Z_\mu | H^2(\mu)$, i.e.

$$S_\mu = P_{H^2(\mu)} Z_\mu i_{H^2(\mu)},$$

where $P_{H^2(\mu)}$ is the orthogonal projection of $L^2(\mu)$ onto $H^2(\mu)$ and $i_{H^2(\mu)}$ is the inclusion of $H^2(\mu)$ into $L^2(\mu)$. Its adjoint S_μ^* is called the *unilateral backward shift operator*. Note that, according to (1.36), we have

$$S_\mu^* = P_{H^2(\mu)} Z_\mu^* i_{H^2(\mu)}.$$

If μ is absolutely continuous with respect to the Lebesgue measure, i.e. $d\mu(\zeta) = w(\zeta)\, dm(\zeta)$, with $w \in L^1(\mathbb{T})$, then we write S_w instead of S_μ. In the particular but important case when μ is the normalized Lebesgue measure m on $\mathbb{T}$, we have already used the simpler notation S instead of S_m.

Since $|\chi_1| = 1$, it is trivial to see that

$$\|S_\mu f\|_{H^2(\mu)} = \|f\|_{H^2(\mu)} \qquad (f \in H^2(\mu)).$$

This means that S_μ is an isometry on $H^2(\mu)$. In particular, this property implies that $\sigma_p(S_\mu) \subset \mathbb{T}$. In fact, one can easily explicitly describe the point spectrum of the operator S_μ.

Lemma 8.24 *Let μ be a positive and finite Borel measure on $\mathbb{T}$. Then*

$$\sigma_p(S_\mu) = \{\zeta \in \mathbb{T} : \mu(\zeta) > 0\}.$$

Moreover, if $\zeta \in \sigma_p(S_\mu)$, then $\chi_{\{\zeta\}} \in H^2(\mu)$ and

$$\ker(S_\mu - \zeta I) = \mathbb{C}\chi_{\{\zeta\}}.$$

Proof Since $\sigma_p(S_\mu) \subset \sigma_p(Z_\mu)$, Theorem 8.2 implies that each $\zeta \in \sigma_p(S_\mu)$ must fulfill $\mu(\{\zeta\}) > 0$. Conversely, let $\zeta \in \mathbb{T}$ be such that $\mu(\{\zeta\}) > 0$. If $\mu = \mu_a + \mu_s$ is the Lebesgue decomposition of the measure μ, with $\mu_a \ll m$ and $\mu_s \perp m$, then we necessarily have $\mu_s(\{\zeta\}) > 0$. Hence, $\chi_{\{\zeta\}} \in L^2(\mu_s) = H^2(\mu_s) \subset H^2(\mu)$, with $\chi_{\{\zeta\}} \neq 0$. Moreover, we have

$$(S_\mu - \zeta I)\chi_{\{\zeta\}} = (z - \zeta)\chi_{\{\zeta\}} = 0.$$

Thus, $\zeta \in \sigma_p(S_\mu)$ and

$$\mathbb{C}\chi_{\{\zeta\}} \subset \ker(S_\mu - \zeta I).$$

But since

$$\ker(S_\mu - \zeta I) \subset \ker(Z_\mu - \zeta I) = \mathbb{C}\chi_{\{\zeta\}},$$

we conclude that $\ker(S_\mu - \zeta I) = \mathbb{C}\chi_{\{\zeta\}}$. □

Let $d\mu = d\mu_a + d\mu_s = w\,dm + d\mu_s$ be the Lebesgue decomposition of μ with respect to m, with $w \geq 0$, $w \in L^1(\mathbb{T})$ and $\mu_s \perp m$. Studying the spectral properties of the operator S_μ splits into two cases depending on whether $\log w$ is integrable or not on the unit circle. In the case when $\log w \notin L^1(\mathbb{T})$, then, according to Corollary 8.23, we have $H^2(\mu) = L^2(\mu)$, and thus the operators S_μ and Z_μ coincide. In particular, S_μ is a unitary operator, and its spectral properties have already been studied in Section 8.1. In the opposite case, that is, when $\log w \in L^1(\mathbb{T})$, the spectral properties of S_μ are dramatically different. The main reason is that in this case the operator S_{μ_a} is completely nonunitary. The key point in the study of the spectral properties of the operator S_μ is the following result.

Theorem 8.25 *Let $d\mu = w\,dm$, with $w \geq 0$, $w \in L^1(\mathbb{T})$ and $\log w \in L^1(\mathbb{T})$. Let $h = [w^{1/2}]$, i.e. h is the outer function constructed with $w^{1/2}$. Then S_μ is unitarily equivalent to S and*

$$H^2(\mu) = \frac{1}{h} H^2(\mathbb{T}).$$

Proof Since $w \in L^1(\mathbb{T})$ and $\log w \in L^1(\mathbb{T})$, the function $h = [w^{1/2}]$ is an outer function in H^2, which satisfies $|h|^2 = w$ a.e. on $\mathbb{T}$. Consider the mapping

$$\begin{array}{rccc} V: & L^2(\mu) & \longrightarrow & L^2(\mathbb{T}) \\ & f & \longmapsto & hf. \end{array}$$

It is easy to see that, for each $f \in L^2(\mu)$,

$$\begin{aligned} \|Vf\|_2^2 &= \|hf\|_2^2 \\ &= \int_{\mathbb{T}} |f(\zeta)|^2 |h(\zeta)|^2 \, dm(\zeta) \\ &= \int_{\mathbb{T}} |f(\zeta)|^2 w(\zeta) \, dm(\zeta) \\ &= \int_{\mathbb{T}} |f(\zeta)|^2 \, d\mu(\zeta) \\ &= \|f\|_{L^2(\mu)}^2. \end{aligned}$$

Hence, V is an isometry from $L^2(\mu)$ into $L^2(\mathbb{T})$. Moreover, for each $g \in L^2(\mathbb{T})$, define $f = g/h$. The function f is measurable on $\mathbb{T}$ and we have

$$\int_{\mathbb{T}} |f(\zeta)|^2 \, d\mu(\zeta) = \int_{\mathbb{T}} |f(\zeta)|^2 |h(\zeta)|^2 \, dm(\zeta) = \int_{\mathbb{T}} |g(\zeta)|^2 \, dm(\zeta) < \infty.$$

Thus, the function f belongs to $L^2(\mu)$ and is defined such that $Vf = hf = g$. Therefore, V is onto and is a unitary operator from $L^2(\mu)$ into $L^2(\mathbb{T})$.

Using Theorem 8.16 and the fact that V is an isometry, we have

$$\begin{aligned} VH^2(\mu) &= V\mathrm{Clos}_{L^2(\mu)}(\mathcal{P}_+) = \mathrm{Clos}_{L^2(\mathbb{T})}(V\mathcal{P}_+) \\ &= \mathrm{Clos}_{L^2(\mathbb{T})}(h\mathcal{P}_+) = H^2(\mathbb{T}). \end{aligned}$$

Hence, the mapping

$$\begin{array}{rccc} U: & H^2(\mu) & \longrightarrow & H^2(\mathbb{T}) \\ & f & \longmapsto & hf \end{array} \tag{8.40}$$

is well defined and is in fact a unitary operator. It remains to note that, for each $f \in H^2(\mu)$, we have

$$(SUf)(\zeta) = \zeta h(\zeta) f(\zeta) = (US_\mu f)(\zeta) \qquad (\zeta \in \mathbb{T}),$$

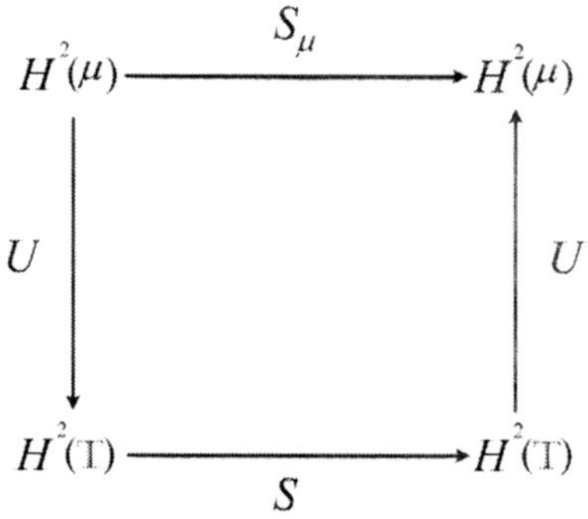

Figure 8.4 The relation between S_μ and S.

which gives $SU = US_\mu$, or

$$S_\mu = U^* SU$$

(see Figure 8.4). □

Theorem 8.25 enables us to give an explicit and useful description of the space $H^2(\mu)$ in the case when $\log w$ is integrable on $\mathbb{T}$. Let $d\mu = d\mu_a + d\mu_s = w\,dm + d\mu_s$ be the Lebesgue decomposition of the measure μ with respect to m, with $w \geq 0$, $w \in L^1(\mathbb{T})$ and $\mu_s \perp m$. We recall that

$$H^2(\mu) = H^2(\mu_a) \oplus H^2(\mu_s) = H^2(\mu_a) \oplus L^2(\mu_s).$$

Assuming furthermore that $\log w \in L^1(\mathbb{T})$, according to Theorem 8.25, we get

$$H^2(\mu) = h^{-1}H^2 \oplus L^2(\mu_s). \tag{8.41}$$

Since $H^2(\mu_a)$ and $L^2(\mu_s)$ are S_μ-invariant, we see that the matrix of the operator S_μ with respect to the decomposition $H^2(\mu) = H^2(\mu_a) \oplus L^2(\mu_s)$ is given by

$$S_\mu = \begin{bmatrix} U^*SU & 0 \\ 0 & Z_{\mu_s} \end{bmatrix} : \begin{matrix} H^2(\mu_a) \\ \oplus \\ L^2(\mu_s) \end{matrix} \longrightarrow \begin{matrix} H^2(\mu_a) \\ \oplus \\ L^2(\mu_s) \end{matrix}, \tag{8.42}$$

where U is given by (8.40). Using this matrix representation, we can apply the results of Sections Section 8.1 and Section 8.2 and obtain the spectral properties of the operator S_μ and its conjugate.

Theorem 8.26 *Let $d\mu = d\mu_a + d\mu_s = w\,dm + d\mu_s$ be the Lebesgue decomposition of a positive and finite Borel measure μ on $\mathbb{T}$, with $w \in L^1(\mathbb{T})$ and $d\mu_s \perp dm$. Then the following hold.*

(i) *If* $\log w \notin L^1(\mathbb{T})$, *then*

$$\sigma(S_\mu) = \mathrm{supp}(\mu),$$

and, for $\lambda \in \mathrm{supp}(\mu)$, *we have* $\lambda \in \sigma_{ess}(S_\mu)$ *if and only if either* $\mu(\{\lambda\}) = 0$, *or* $\mu(\{\lambda\}) > 0$ *but there exists a sequence* $(\lambda_n)_{n\geq 1}$ *such that* $\lambda_n \in \mathrm{supp}(\mu)$, $\lambda_n \neq \lambda$, *and* $\lambda_n \longrightarrow \lambda$, *as* $n \longrightarrow \infty$.

(ii) *If* $\log w \in L^1(\mathbb{T})$, *then*

$$\sigma(S_\mu) = \bar{\mathbb{D}} \quad \textit{and} \quad \sigma_{ess}(S_\mu) = \mathbb{T}.$$

Proof The case $\log w \notin L^1(\mathbb{T})$ follows from Theorems 8.1 and 8.4. Hence, let us now assume that $\log w \in L^1(\mathbb{T})$. Then, according to (8.42), (2.8) and Theorem 7.37, we have

$$\sigma(S_\mu) = \sigma(U^*SU) \cup \sigma(Z_{\mu_s}) \tag{8.43}$$

and

$$\sigma_{ess}(S_\mu) = \sigma_{ess}(U^*SU) \cup \sigma_{ess}(Z_{\mu_s}). \tag{8.44}$$

But, by Theorems 8.1 and 8.4,

$$\sigma(Z_{\mu_s}) \subset \mathbb{T} \quad \text{and} \quad \sigma_{ess}(Z_{\mu_s}) \subset \mathbb{T}. \tag{8.45}$$

Moreover, according to Lemma 8.6 and Theorem 8.8, we have

$$\sigma(U^*SU) = \sigma(S) = \bar{\mathbb{D}} \tag{8.46}$$

and

$$\sigma_{ess}(U^*SU) = \sigma_{ess}(S) = \mathbb{T}. \tag{8.47}$$

Hence, combining (8.43), (8.44), (8.45), (8.46) and (8.47) gives the result. □

The above result reveals the great difference concerning the spectrum of the unilateral shift operator S_μ depending on whether the Radon–Nikodym derivative of the absolute part of μ is log-integrable or not. In one case, the spectrum of S_μ is the whole closed unit disk. In the other case, it is a subset of the unit circle, possibly very small. For instance, if μ is a finite sum of Dirac measures, i.e.

$$\mu = \sum_{i=1}^{n} a_i \delta_{\{\zeta_i\}},$$

$a_i > 0$, $\zeta_i \in \mathbb{T}$, then $\sigma(S_\mu) = \sigma_{ess}(S_\mu) = \sigma_p(S_\mu) = \{\zeta_i : 1 \leq i \leq n\}$. We can also easily describe the spectrum and the essential spectrum of the operator S_μ^*.

Theorem 8.27 *Let* $d\mu = d\mu_a + d\mu_s = w\,dm + d\mu_s$ *be the Lebesgue decomposition of a positive and finite Borel measure* μ *on* $\mathbb{T}$, *with* $w \in L^1(\mathbb{T})$ *and* $d\mu_s \perp dm$. *Then the following hold.*

(i) *If* $\log w \notin L^1(\mathbb{T})$, *then*

$$\sigma_p(S_\mu^*) = \{\zeta \in \mathbb{T} : \mu(\{\bar{\zeta}\}) > 0\}.$$

Moreover, if $\zeta \in \mathbb{T}$, *with* $\mu(\{\bar{\zeta}\}) > 0$, *then*

$$\ker(S_\mu^* - \zeta I) = \mathbb{C}\chi_{\{\bar{\zeta}\}}.$$

(ii) *If* $\log w \in L^1(\mathbb{T})$, *then*

$$\sigma_p(S_\mu^*) = \mathbb{D} \cup \{\zeta \in \mathbb{T} : \mu(\{\bar{\zeta}\}) > 0\}.$$

Moreover, if $\lambda \in \mathbb{D}$, *then*

$$\ker(S_\mu^* - \bar{\lambda} I) = \mathbb{C}k_\lambda/h,$$

where $h = [w^{1/2}]$, *and if* $\zeta \in \mathbb{T}$, *with* $\mu(\{\bar{\zeta}\}) > 0$, *then*

$$\ker(S_\mu^* - \zeta I) = \mathbb{C}\chi_{\{\bar{\zeta}\}}.$$

Proof In the case when $\log w \notin L^1(\mathbb{T})$, as already mentioned, $H^2(\mu) = L^2(\mu)$ and the operator S_μ coincides with Z_μ. Hence, the result follows immediately from Corollary 8.3. So let us now assume that $\log w \in L^1(\mathbb{T})$. Then, according to (8.42) and (2.7), we have

$$\sigma_p(S_\mu^*) = \sigma_p(U^* S^* U) \cup \sigma_p(Z_{\mu_s}^*). \tag{8.48}$$

Using Corollary 8.3 once more, we have

$$\sigma_p(Z_{\mu_s}^*) = \{\zeta \in \mathbb{T} : \mu_s(\{\bar{\zeta}\}) > 0\} = \{\zeta \in \mathbb{T} : \mu(\{\bar{\zeta}\}) > 0\}, \tag{8.49}$$

and for such points we have

$$\ker(Z_{\mu_s}^* - \zeta I) = \mathbb{C}\chi_{\{\bar{\zeta}\}}. \tag{8.50}$$

On the other hand, by Lemma 8.6, we have

$$\sigma_p(U^* S^* U) = \sigma_p(S^*) = \mathbb{D}. \tag{8.51}$$

Moreover, if $\lambda \in \mathbb{D}$, then

$$\begin{aligned} \ker(U^* S^* U - \bar{\lambda} I) &= U^* \ker(S^* - \bar{\lambda} I) = U^* \mathbb{C} k_\lambda \\ &= U^{-1} \mathbb{C} k_\lambda = \mathbb{C} k_\lambda / h. \end{aligned} \tag{8.52}$$

Combining (8.48), (8.49) and (8.51) gives

$$\sigma_p(S_\mu^*) = \mathbb{D} \cup \{\zeta \in \mathbb{T} : \mu(\{\bar{\zeta}\}) > 0\}.$$

Moreover, according to (2.6), (8.50) and (8.52), for each $\lambda \in \mathbb{D}$, we have

$$\ker(S_\mu^* - \bar{\lambda} I) = \ker(U^* S^* U - \bar{\lambda} I) \oplus \ker(Z_{\mu_s} - \bar{\lambda} I) = \mathbb{C} h^{-1} k_\lambda \oplus 0,$$

and if $\zeta \in \mathbb{T}$, with $\mu(\{\bar{\zeta}\}) > 0$, then

$$\ker(S_\mu^* - \zeta I) = \ker(U^* S^* U - \zeta I) \oplus \ker(Z_{\mu_s} - \zeta I) = 0 \oplus \mathbb{C}\chi_{\{\bar{\zeta}\}}. \qquad \square$$

We finish this section with a result that characterizes the situation when eigenvectors form an orthogonal basis.

Corollary 8.28 *Let μ be a finite and positive Borel measure on $\mathbb{T}$. Then the following are equivalent:*

(i) *$H^2(\mu)$ has an orthogonal basis consisting of eigenvectors of S_μ;*
(ii) *μ is purely atomic.*

Moreover, in this case, where μ is the discrete measure

$$\mu = \sum_{n=1}^{\infty} a_n \delta_{\{\zeta_n\}},$$

with $a_n = \mu(\{\zeta_n\}) > 0$, $n \geq 1$, the sequence $(\zeta_n)_{n\geq 1}$ is precisely the set of eigenvalues of S_μ.

Proof (i) $\Longrightarrow$ (ii) Assume that there exists a sequence $(f_n)_{n\geq 1}$ of functions in $H^2(\mu)$, $f_n \neq 0$, and a sequence of complex numbers $(\zeta_n)_{n\geq 1}$ such that

$$S_\mu f_n = \zeta_n f_n \qquad (n \geq 1),$$

and $(f_n)_{n\geq 1}$ is an orthogonal basis of $H^2(\mu)$. According to Lemma 8.24, we must have $\zeta_n \in \mathbb{T}$, $\mu(\{\zeta_n\}) > 0$ and $f_n = c_n \chi_{\{\zeta_n\}}$, where $c_n \in \mathbb{C} \setminus \{0\}$. Without loss of generality, to get a normalized eigenvector, we can take $c_n = \|\chi_{\zeta_n}\|_{H^2(\mu)}^{-1} = (\mu(\{\zeta_n\}))^{-1/2}$. Hence, $(f_n)_{n\geq 1}$ is an orthonormal basis of $H^2(\mu)$. Let p be any analytic polynomial. Then, since $(f_n)_{n\geq 1}$ is an orthonormal basis of $H^2(\mu)$, we have

$$\|p\|_{H^2(\mu)}^2 = \sum_{n=1}^{\infty} |\langle p, f_n\rangle_{H^2(\mu)}|^2$$

and

$$\langle p, f_n\rangle_{H^2(\mu)} = \frac{1}{\mu(\zeta_n)^{1/2}} \int_{\mathbb{T}} p\chi_{\{\zeta_n\}}\, d\mu = p(\zeta_n)\mu(\zeta_n)^{1/2} \qquad (n \geq 1).$$

Therefore,

$$\|p\|_{H^2(\mu)}^2 = \sum_{n=1}^{\infty} \mu(\zeta_n)|p(\zeta_n)|^2.$$

This identity shows that

$$\|p\|_{H^2(\mu)} = \|p\|_{H^2(\nu)} \qquad (p \in \mathcal{P}_+),$$

where ν is the atomic measure

$$\nu = \sum_{n=1}^{\infty} \mu(\zeta_n)\delta_{\{\zeta_n\}}.$$

Now, let p be a trigonometric polynomial. Then there exists $N \in \mathbb{N}$ such that $z^N p \in \mathcal{P}_+$, whence

$$\|p\|_{L^2(\mu)} = \|z^N p\|_{H^2(\mu)} = \|z^N p\|_{H^2(\nu)} = \|p\|_{L^2(\nu)}.$$

Since the trigonometric polynomials are dense in $L^2(\mu)$ and $L^2(\nu)$, we must have $\mu = \nu$.

(ii) $\Longrightarrow$ (i) Assume that μ is purely atomic and write

$$\mu = \sum_{n=1}^{\infty} a_n \delta_{\{\zeta_n\}},$$

with

$$a_n = \mu(\{\zeta_n\}) = \|\chi_{\{\zeta_n\}}\|^2_{L^2(\mu)} > 0 \qquad (n \geq 1).$$

Then, for all $f \in H^2(\mu)$, we have

$$\|f\|^2_{H^2(\mu)} = \sum_{n=1}^{\infty} a_n |f(\zeta_n)|^2. \tag{8.53}$$

But, by Lemma 8.24, $\chi_{\{\zeta_n\}} \in H^2(\mu)$. Hence, we can compute

$$|\langle f, \chi_{\{\zeta_n\}}\rangle_{H^2(\mu)}|^2 = \left| \int_{\mathbb{T}} f(\zeta)\chi_{\{\zeta_n\}}(\zeta)\, d\mu(\zeta) \right|^2 = a_n^2 |f(\zeta_n)|^2.$$

Plugging this relation into (8.53) gives

$$\|f\|^2_{H^2(\mu)} = \sum_{n=1}^{\infty} |\langle f, f_n\rangle_{H^2(\mu)}|^2 \qquad (f \in H^2(\mu)),$$

where $f_n = a_n^{-1/2} \chi_{\{\zeta_n\}}$. This identity implies that $(\chi_{\{\zeta_n\}})_{n\geq 1}$ is an orthogonal basis of $H^2(\mu)$. To conclude, remember that $\mu(\{\zeta_n\}) > 0$ and then, by Lemma 8.24, $\chi_{\{\zeta_n\}}$ is an eigenvector of S_μ associated with the eigenvalue ζ_n. □

Exercises

Exercise 8.6.1 Let $d\mu = w\, dm + d\mu_s$ be the Lebesgue decomposition of the measure μ, with $w \geq 0$, $w \in L^1(\mathbb{T})$ and $d\mu_s \perp dm$. Show that S_μ is a unitary operator if and only if $\log w \notin L^1(\mathbb{T})$.
Hint: Use Corollary 8.23.

Exercise 8.6.2 Let $d\mu = w\, dm + d\mu_s$ be the Lebesgue decomposition of the measure μ, with $w \geq 0$, $w \in L^1(\mathbb{T})$ and $d\mu_s \perp dm$. Assume that $\log w \in L^1(\mathbb{T})$. Show that $\sigma_a(S_\mu) = \mathbb{T}$.

Hint: Use $\partial\sigma(S_\mu) \subset \sigma_a(S_\mu)$ and Theorem 8.26 to obtain that $\mathbb{T} \subset \sigma_a(S_\mu)$. For the reverse inclusion, show that $\sigma_a(S_\mu) \subset \sigma_a(Z_\mu)$ and use Theorem 8.1.

Exercise 8.6.3 Let $d\mu = w\,dm + d\mu_s$ be the Lebesgue decomposition of the measure μ, with $w \geq 0$, $w \in L^1(\mathbb{T})$ and $d\mu_s \perp dm$. Assume that $\log w \in L^1(\mathbb{T})$. Show that $\sigma_a(S_\mu^*) = \bar{\mathbb{D}}$.
Hint: Use $\sigma_p(S_\mu^*) \subset \sigma_a(S_\mu^*)$ and Theorem 8.27 to get that $\mathbb{D} \subset \sigma_a(S_\mu^*)$. Remember that $\sigma_a(S_\mu^*)$ is closed.

Exercise 8.6.4 Let μ be a positive and finite measure having compact support in the complex plane $\mathbb{C}$, and let Z_μ be the multiplication operator

$$(Z_\mu f)(z) = z f(z)$$

defined on $L^2(\mu)$. Show that:

(i) Z_μ is a bounded operator on $L^2(\mu)$;
(ii) $\mathcal{P}_+ \subset L^2(\mu)$;
(iii) $Z_\mu H^2(\mu) \subset H^2(\mu)$.

Remark 1: As usual, we define $H^2(\mu) = \mathrm{Clos}_{L^2(\mu)}(\mathcal{P}_+)$.
Remark 2: The restriction of Z_μ to $H^2(\mu)$ is denoted by S_μ.

Exercise 8.6.5 Let μ be a positive and finite measure having compact support in the complex plane $\mathbb{C}$. We use the notation from the previous exercise. A point λ in $\mathbb{C}$ is called a *bounded point evaluation* for $H^2(\mu)$ if there exists a constant $M > 0$ such that

$$|p(\lambda)| \leq M \|p\|_{H^2(\mu)}$$

for all analytic polynomials p. We denote the set of bounded point evaluations for $H^2(\mu)$ by $bpe(H^2(\mu))$. We also highlight that, in this exercise, the bar in notation like $\bar{E}$ stands for the complex conjugate.

(i) Let $\lambda \in bpe(H^2(\mu))$ and let $f \in H^2(\mu)$. Take any sequence $(p_n)_{n\geq 1}$ of analytic polynomials such that $\|p_n - f\|_{H^2(\mu)} \longrightarrow 0$, as $n \longrightarrow \infty$. Show that the sequence $(p_n(\lambda))_{n\geq 1}$ is convergent. If $(q_n)_{n\geq 1}$ is another sequence of analytic polynomials such that $\|q_n - f\|_{H^2(\mu)} \longrightarrow 0$, as $n \longrightarrow \infty$, then show that

$$\lim_{n\to\infty} p_n(\lambda) = \lim_{n\to\infty} q_n(\lambda).$$

Knowing this, we define $f(\lambda) = \lim_{n\to\infty} p_n(\lambda)$.

(ii) Show that, for each bounded point evaluation λ, there exists $k_\lambda^\mu \in H^2(\mu)$ such that

$$f(\lambda) = \langle f, k_\lambda^\mu \rangle_{H^2(\mu)} \qquad (f \in H^2(\mu)).$$

(iii) Let $\lambda \in bpe(H^2(\mu))$. Show that $\bar{\lambda} \in \sigma_p(S_\mu^*)$ and

$$\mathbb{C}k_\lambda^\mu \subset \ker(S_\mu^* - \bar{\lambda}I).$$

(iv) Conversely, let $\lambda \in \overline{\sigma_p(S_\mu^*)}$ and let $g_\lambda \in H^2(\mu)$, $g_\lambda \neq 0$, be such that $S_\mu^* g_\lambda = \bar{\lambda} g_\lambda$. Show that, for any analytic polynomial p, we have

$$\langle p, g_\lambda \rangle_{H^2(\mu)} = p(\lambda)\langle 1, g_\lambda \rangle_{H^2(\mu)}.$$

Deduce that $\lambda \in bpe(H^2(\mu))$ and

$$g_\lambda = \overline{\langle 1, g_\lambda \rangle_{H^2(\mu)}} k_\lambda^\mu.$$

(v) Deduce from (iii) and (iv) that

$$\sigma_p(S_\mu^*) = \overline{bpe(H^2(\mu))}$$

and, if $\lambda \in bpe(H^2(\mu))$, then

$$\ker(S_\mu^* - \bar{\lambda}I) = \mathbb{C}k_\lambda^\mu.$$

(vi) Show that, if $\lambda \in \operatorname{supp}(\mu)$ is such that $\mu(\{\lambda\}) > 0$, then $\bar{\lambda} \in \sigma_p(S_\mu^*)$.

Exercise 8.6.6 Let μ be a positive and finite Borel measure on $\mathbb{T}$. Determine $bpe(H^2(\mu))$.
Remark: $bpe(H^2(\mu))$ was defined and discussed in Exercise 8.6.5.
Hint: Use Exercise 8.6.5 and Theorem 8.27.

8.7 Reducing invariant subspaces of Z_μ

The aim of this section and Section 8.9 is to give a complete description of the reducing subspaces of the bilateral and unilateral forward shift operators Z_μ and S_μ. See Section 1.10 for the general definition of reducing subspaces for a Hilbert space operator.

We first begin with the bilateral forward shift operator Z_μ on $L^2(\mu)$. In this situation, since Z_μ is a unitary operator, it is easy to see that a closed subspace $\mathcal{E}$ is reducing for Z_μ if and only if $Z_\mu\mathcal{E} = \mathcal{E}$, that is, $z\mathcal{E} = \mathcal{E}$. The following celebrated theorem of Wiener classifies all the reducing subspaces of the bilateral forward shift operator on $L^2(\mu)$.

Theorem 8.29 *Let $\mathcal{E}$ be a closed subspace of $L^2(\mu)$. Then $z\mathcal{E} = \mathcal{E}$ if and only if there exists a measurable set $E \subset \mathbb{T}$ such that*

$$\mathcal{E} = \chi_E L^2(\mu) = \{f \in L^2(\mu) : f = 0,\ \mu\text{-a.e. on } \mathbb{T} \setminus E\}.$$

Proof The "if" part is obvious. For the "only if" part, we first make an elementary observation about $\mathcal{E}$. If $\phi \in \mathcal{E}$, then, by the assumption $z\mathcal{E} = \mathcal{E}$, we have

$$p\phi \in \mathcal{E}, \tag{8.54}$$

where p is any trigonometric polynomial. Moreover, since trigonometric polynomials are dense in $L^2(\mu)$, if ϕ is a bounded function in $\mathcal{E}$, we deduce that

$$\phi L^2(\mu) \subset \mathcal{E}. \tag{8.55}$$

If $\mathcal{E} = L^2(\mu)$, it is enough to take $E = \mathbb{T}$. Hence, suppose that $\mathcal{E}$ is a proper subspace of $L^2(\mu)$. In this case, we must have $1 \notin \mathcal{E}$, since otherwise, by (8.55), we get a contradiction. Let ϕ be the orthogonal projection of 1 onto $\mathcal{E}$. Therefore, we have $1-\phi \perp \mathcal{E}$. According to (8.54), we thus obtain $1-\phi \perp p\psi$ for all trigonometric polynomials p and all $\psi \in \mathcal{E}$. This fact is rewritten as

$$\int_{\mathbb{T}} (1 - \overline{\phi(e^{it})})\psi(e^{it})p(e^{it})\, d\mu(e^{it}) = 0.$$

Thus, we must have

$$(1 - \bar{\phi})\psi = 0 \tag{8.56}$$

μ-almost everywhere on $\mathbb{T}$. In particular, with $\psi = \phi$, we obtain $\phi = |\phi|^2$, which holds if and only if there is a measurable set E such that $\phi = \chi_E$. Then, in the first place, by (8.55), we have

$$\chi_E L^2(\mu) \subset \mathcal{E}.$$

Second, by (8.56), we also have

$$(1 - \chi_E)\mathcal{E} = 0.$$

Hence, in fact, $\mathcal{E} = \chi_E L^2(\mu)$. □

It is clear that the μ-measurable set E, given in Theorem 8.29, is unique up to a set of μ-measure zero. Wiener's theorem reveals that the operator Z_μ has a rich structure of reducing subspaces. The situation for the operator S_μ is more complicated. This will be discussed in Section 8.9.

8.8 Simply invariant subspaces of Z_μ

In this section and Section 8.10, we respectively discuss the description of simply invariant subspaces of the operators Z_μ and S_μ. See Section 1.10 for the general definition of simply invariant subspaces for a Hilbert space operator.

In our situation, using the fact that Z_μ is a unitary operator, we easily see that a closed subspace $\mathcal{E}$ of $L^2(\mu)$ is simply invariant for Z_μ if and only $Z_\mu\mathcal{E} \subsetneqq \mathcal{E}$, that is

$$z\mathcal{E} \subsetneqq \mathcal{E}.$$

We first concentrate on the case when μ is the normalized Lebesgue measure on $\mathbb{T}$.

Theorem 8.30 *Let $\mathcal{E}$ be a closed subspace of $L^2(\mathbb{T})$. Then $z\mathcal{E} \subsetneqq \mathcal{E}$ if and only if there is a unimodular function $\Theta \in L^\infty(\mathbb{T})$, i.e. fulfilling the property*

$$|\Theta| = 1 \qquad (a.e.\ on\ \mathbb{T}),$$

such that

$$\mathcal{E} = \Theta H^2(\mathbb{T}).$$

Proof The proof has certain similarities to the proof of Theorem 8.29. Suppose that $\mathcal{E}$ is a closed subspace of $L^2(\mathbb{T})$ such that $z\mathcal{E} \subsetneqq \mathcal{E}$. If $\phi \in \mathcal{E}$, then we have

$$p\phi \in \mathcal{E}, \tag{8.57}$$

where p is any analytic polynomial. (We emphasize "analytic", not "trigonometric" – this is the main difference between the two proofs.) Moreover, since analytic polynomials are dense in H^2, if ϕ is bounded, we also have

$$\phi H^2 \subset \mathcal{E}. \tag{8.58}$$

Since $\mathcal{E}$ is closed and Z is a unitary operator, $z\mathcal{E}$ is a closed subspace of $\mathcal{E}$. Pick $\Theta \in \mathcal{E}$, $\Theta \neq 0$, such that $\Theta \perp z\mathcal{E}$. The assumption $z\mathcal{E} \subsetneqq \mathcal{E}$ guarantees that such a function exists. Therefore, we have

$$\int_0^{2\pi} e^{int}\phi(e^{it})\overline{\Theta(e^{it})}\,dt = 0 \tag{8.59}$$

for all $n \geq 1$ and all $\phi \in \mathcal{E}$. First, we take $\phi = \Theta$ to get

$$\int_0^{2\pi} e^{int}|\Theta(e^{it})|^2\,dt = 0 \qquad (n \geq 1).$$

Taking the conjugate of both sides shows that we in fact have

$$\int_0^{2\pi} e^{int}|\Theta(e^{it})|^2\,dt = 0 \qquad (n \in \mathbb{Z},\ n \neq 0).$$

Thus, by the uniqueness theorem for Fourier coefficients, we conclude that $|\Theta|$ is a constant function. Without loss of generality, we can assume that $|\Theta| = 1$. According to (8.58), we see that $\Theta H^2 \subset \mathcal{E}$. But (8.59) can be interpreted as $\bar{\Theta}\mathcal{E} \subset H^2$. Therefore, we have $\mathcal{E} = \Theta H^2$.

Reciprocally, let $\Theta \in L^\infty(\mathbb{T})$, satisfying $|\Theta| = 1$ a.e. on $\mathbb{T}$, and let $\mathcal{E} = \Theta H^2$. Since multiplication by Θ is an isometry on $L^2(\mathbb{T})$, we get immediately that $\mathcal{E}$ is a closed subspace of $L^2(\mathbb{T})$ that is invariant under Z. If $\mathcal{E}$ is a reducing subspace for Z_μ, then the function $\Theta\chi_{-1}$ belongs to $\mathcal{E} = \Theta H^2(\mathbb{T})$. Since $\Theta \neq 0$ a.e. on $\mathbb{T}$, we get that $\chi_{-1} \in H^2(\mathbb{T})$, which is absurd. Hence, $\mathcal{E}$ is a simply invariant subspace of Z. □

Owing to the nature of our applications, we also need to characterize the closed subspaces of $L^2(\mathbb{T})$ that are invariant under the bilateral backward shift operator Z^*. However, based on Theorems 8.29 and 8.30, this task is straightforward.

Corollary 8.31 *Let $\mathcal{E}$ be a closed subspace of $L^2(\mathbb{T})$. Then $\bar{z}\mathcal{E} \subset \mathcal{E}$ if and only if either there is a unimodular measurable function Θ such that*

$$\mathcal{E} = \Theta\overline{H^2(\mathbb{T})}$$

or there is a measurable set E such that

$$\mathcal{E} = \chi_E L^2(\mathbb{T}).$$

Proof It is enough to observe that the set $\bar{\mathcal{E}}$ fulfills the property $Z\bar{\mathcal{E}} \subset \bar{\mathcal{E}}$. Now, either $Z\bar{\mathcal{E}} = \bar{\mathcal{E}}$ or $Z\bar{\mathcal{E}} \subsetneq \bar{\mathcal{E}}$, and thus we respectively apply Theorems 8.29 and 8.30 to the preceding two cases to obtain the reducing and simply invariant subspaces of $L^2(\mathbb{T})$ under Z^*. □

The following result easily follows from Theorem 8.30. However, owing to its importance and its central role in the study of our main objects, i.e. $\mathcal{H}(b)$ spaces, we state it as a theorem.

Theorem 8.32 *Let $\mathcal{E}$ be a nonzero closed subspace of H^2. Then $z\mathcal{E} \subset \mathcal{E}$ if and only if there is an inner function Θ such that*

$$\mathcal{E} = \Theta H^2.$$

Proof The "if" part is obvious. For the "only if" part, let us interpret H^2 as a closed subspace of $L^2(\mathbb{T})$ and thus apply the preceding results for the invariant subspaces of $L^2(\mathbb{T})$. We cannot have $z\mathcal{E} = \mathcal{E}$, since otherwise we would have $z^n\phi \in \mathcal{E}$ for all $n \in \mathbb{Z}$ and all $\phi \in \mathcal{E}$. Picking a nonzero $\phi \in \mathcal{E}$ and n large enough such that $z^{-n}\phi$ has a nonzero negative spectrum, we obtain a contradiction. Remember that $\mathcal{E} \subset H^2$.

Therefore, we must have $z\mathcal{E} \subsetneqq \mathcal{E}$. Applying Theorem 8.30, we obtain a unimodular function $\Theta \in \mathcal{E}$ such that $\mathcal{E} = \Theta H^2$. But, $\Theta \in H^2$ and is unimodular. Hence, it is an inner function. □

The subspaces of the form ΘH^2, where Θ is a nonconstant inner function, are called *Beurling subspaces* of $H^2(\mathbb{D})$. Therefore, Beurling subspaces of H^2 are precisely the simply invariant subspaces of H^2 under the unilateral forward shift S. Using the preceding result and Lemma 1.39, we immediately obtain the following result.

Corollary 8.33 *Let $\mathcal{E}$ be a proper closed subspace of H^2, and satisfying*

$$S^*\mathcal{E} \subset \mathcal{E}.$$

Then there is a nonconstant inner function Θ, unique up to a multiplicative constant of modulus one, such that

$$\mathcal{E} = (\Theta H^2)^\perp.$$

In the above results, we have described explicitly the simply invariant subspaces of the operators Z_μ and S_μ when μ is the normalized Lebesgue measure on $\mathbb{T}$. For the rest of this section and Section 8.10, we study the general case. So let μ be a finite and positive Borel measure on $\mathbb{T}$ and let $\mu = \mu_a + \mu_s$ be its Lebesgue decomposition, where $\mu_a \ll m$ and $\mu_s \perp m$. Remember that

$$L^2(\mu) = L^2(\mu_a) \oplus L^2(\mu_s) \quad \text{and} \quad H^2(\mu) = H^2(\mu_a) \oplus L^2(\mu_s),$$

and all spaces $L^2(\mu_a)$, $L^2(\mu_s)$ and $H^2(\mu_a)$ are invariant under multiplication by the independent variable z, i.e.

$$zL^2(\mu_a) \subset L^2(\mu_a), \quad zL^2(\mu_s) \subset L^2(\mu_s), \quad zH^2(\mu_a) \subset H^2(\mu_a).$$

The following result, which gives a description of simply invariant subspaces of Z_μ, is a generalization of Theorem 8.30. It is based on the Wold–Kolmogorov decomposition of Z_μ.

Theorem 8.34 *Let $d\mu = d\mu_a + d\mu_s = w\,dm + d\mu_s$ be the Lebesgue decomposition of a finite positive Borel measure μ on $\mathbb{T}$, where $w \geq 0$, $w \in L^1(\mathbb{T})$ and $d\mu_s \perp d\mu_a$. Let $\mathcal{E}$ be a closed subspace of $L^2(\mu)$ such that $z\mathcal{E} \subsetneqq \mathcal{E}$. Then there exists a measurable subset σ of $\mathbb{T}$ and a measurable function Θ such that the following assertions hold:*

(i) $\mathcal{E} = \Theta H^2 \oplus \chi_\sigma L^2(\mu_s)$;
(ii) $m(\sigma) = 0$;
(iii) $|\Theta|^2 w \equiv 1$, *m-a.e. on* $\mathbb{T}$;
(iv) $\Theta H^2 \subset L^2(\mu_a)$;
(v) $\chi_\sigma L^2(\mu_s) \subset L^2(\mu_s)$.

Conversely, if σ and Θ satisfy the above properties, then $\Theta H^2 \oplus \chi_\sigma L^2(\mu_s)$ is a closed simply invariant subspace for the operator Z_μ.

Proof First, let $\mathcal{E}$ be a closed subspace of $L^2(\mu)$ such that $z\mathcal{E} \subsetneqq \mathcal{E}$, and set $\mathcal{D} = \mathcal{E} \ominus z\mathcal{E}$. According to Theorem 7.21, $\mathcal{D}$ is a wandering subspace of Z_μ and

$$\mathcal{E} = \mathcal{E}_0 \oplus \mathcal{E}_\infty,$$

where

$$\mathcal{E}_0 = \bigoplus_{n \geq 0} Z_\mu^n \mathcal{D} \quad \text{and} \quad \mathcal{E}_\infty = \bigcap_{n \geq 0} Z_\mu^n \mathcal{D}.$$

Moreover, the restriction $Z_\mu | \mathcal{E}_\infty$ is unitary and the restriction $Z_\mu | \mathcal{E}_0$ is completely nonunitary. Since, by hypothesis, $\mathcal{D} \neq \{0\}$, we can pick a function $\Theta \in \mathcal{D}$, $\Theta \neq 0$, such that $\|\Theta\|_{L^2(\mu)} = 1$. By definition, we have $\Theta \in \mathcal{E}$, $\Theta \perp z\mathcal{E}$ and $z^n\Theta \in z\mathcal{E}$, $n \geq 1$. Hence,

$$\int_{\mathbb{T}} z^n |\Theta|^2 \, d\mu = \int_{\mathbb{T}} (z^n \Theta) \bar{\Theta} \, d\mu = 0 \qquad (n \geq 1).$$

By conjugation, the above identity remains true for $n \leq -1$ and thus the measure $|\Theta|^2 \, d\mu$ satisfies

$$(\widehat{|\Theta|^2 \, d\mu})(n) = 0 \qquad (n \in \mathbb{Z} \setminus \{0\}).$$

Since

$$1 = \|\Theta\|_{L^2(\mu)}^2 = \int_{\mathbb{T}} |\Theta|^2 \, d\mu = \int_{\mathbb{T}} dm,$$

we thus get

$$(\widehat{|\Theta|^2 \, d\mu})(n) = \widehat{dm}(n) \qquad (n \in \mathbb{Z}).$$

Hence, the uniqueness theorem for the Fourier coefficients implies that

$$|\Theta|^2 \, d\mu = dm.$$

Using the Lebesgue decomposition of μ, we can write

$$dm = |\Theta|^2 \, d\mu = |\Theta|^2 \, d\mu_a + |\Theta|^2 \, d\mu_s = |\Theta|^2 w \, dm + |\Theta|^2 \, d\mu_s.$$

Hence, we must have $|\Theta| = 0$, μ_s-a.e., and $|\Theta|^2 \, d\mu_a = |\Theta|^2 w \, dm = dm$. These identities have several consequences. First, we have $\Theta \in L^2(\mu_a)$, whence,

$$\mathcal{D} \subset L^2(\mu_a),$$

since Θ is an arbitrary element of $\mathcal{D}$ of unit norm. Second, $|\Theta|^2 w = 1$, m-a.e. on $\mathbb{T}$. Since $Z_\mu | \mathcal{E}_\infty$ is unitary, we have $z\mathcal{E}_\infty = \mathcal{E}_\infty$, and by Theorem 8.29, this gives

$$\mathcal{E}_\infty = \chi_\sigma L^2(\mu),$$

for some measurable set $\sigma \subset \mathbb{T}$. As $\Theta \in \mathcal{D} \subset \mathcal{E}_0 \perp \mathcal{E}_\infty$, we have $\Theta \perp \chi_\sigma L^2(\mu)$. Thus, $\Theta = 0$, μ-a.e., on σ. But $\Theta \neq 0$ m-a.e. (because $|\Theta|^2 w = 1$ m-a.e.), whence $m(\sigma) = 0$. In particular, $\chi_\sigma L^2(\mu_a) = \{0\}$ and then we obtain that

$$\mathcal{E}_\infty = \chi_\sigma L^2(\mu) = \chi_\sigma L^2(\mu_s)$$

and

$$\chi_\sigma L^2(\mu_s) \subset L^2(\mu_s).$$

It remains to prove that $\mathcal{E}_0 = \Theta H^2$ and $\Theta H^2 \subset L^2(\mu_a)$. We have already proved that $\mathcal{D} \subset L^2(\mu_a)$ and, since $L^2(\mu_a)$ is invariant under Z_μ, we thus obtain

$$\mathcal{E}_0 \subset L^2(\mu_a).$$

We claim that

$$\mathcal{E}_0 = \text{Span}_{L^2(\mu)}\{z^n \Theta : n \geq 0\}.$$

The inclusion $\text{Span}_{L^2(\mu)}\{z^n \Theta : n \geq 0\} \subset \mathcal{E}_0$ is obvious. Now, let $f \in \mathcal{E}_0$, $f \perp z^n \Theta$, $n \geq 0$, and let us prove that $f \equiv 0$. We have

$$\int_{\mathbb{T}} f \bar{\Theta} z^n \, d\mu = 0 \qquad (n \leq 0). \tag{8.60}$$

Moreover, $\Theta \perp z\mathcal{E}$ and, as $f \in \mathcal{E}$, we have $z^n f \perp \Theta$, $n \geq 1$. Hence,

$$\int_{\mathbb{T}} f \bar{\Theta} z^n \, d\mu = 0 \qquad (n \geq 1). \tag{8.61}$$

The two relations (8.60) and (8.61) imply that $f \bar{\Theta} \, d\mu = 0$. As $\Theta \neq 0$, m-a.e., and $f \in L^2(\mu_a)$, we obtain that $f = 0$, μ-a.e, which proves the claims.

Remember that $\mathcal{D}$ is a wandering subspace and thus $z^n \Theta \perp z^k \Theta$ for $n, k \geq 0$, $n \neq k$. Therefore, by Parseval's identity,

$$\mathcal{E}_0 = \left\{ \sum_{n=0}^{\infty} a_n z^n \Theta : \sum_{n=0}^{\infty} |a_n|^2 < \infty \right\}.$$

Now, note that, for any function $f \in L^2(\mathbb{T})$, we have

$$\int_{\mathbb{T}} |f|^2 \, dm = \int_{\mathbb{T}} |f|^2 |\Theta|^2 w \, dm = \int_{\mathbb{T}} |f\Theta|^2 \, d\mu_a,$$

which proves that the operator $f \longmapsto \Theta f$ is an isometry from $L^2(\mathbb{T})$ onto $L^2(\mu_a) = L^2(w \, dm)$. Therefore,

$$\mathcal{E}_0 = \Theta \left\{ \sum_{n=0}^{\infty} a_n z^n : \sum_{n=0}^{\infty} |a_n|^2 < \infty \right\} = \Theta H^2.$$

That concludes the first part of the theorem.

Reciprocally, let σ be a measurable subset of $\mathbb{T}$, and let Θ be a measurable function on $\mathbb{T}$ satisfying the aforementioned properties. It is clear that $\mathcal{E}$ is a closed invariant subspace. The only thing to check is that $z\mathcal{E} \neq \mathcal{E}$. Assume on the contrary that $z\mathcal{E} = \mathcal{E}$. Then, in particular, we must have $\Theta \in z\mathcal{E}$, which gives $\Theta \in z\Theta H^2 \oplus \chi_\sigma L^2(\mu_s)$. Since $\Theta \in L^2(\mu_a)$, we get $\Theta \in z\Theta H^2$. In other words, there exists $h \in H^2$ such that $\Theta = z\Theta h$. Now remembering that $\Theta \neq 0$ a.e. on $\mathbb{T}$, we deduce that $1 = zh$, which is absurd. □

Note that if $d\mu = w\,dm + d\mu_s$ and

$$m(\{\zeta \in \mathbb{T} : w(\zeta) = 0\}) > 0,$$

then every invariant subspace of Z_μ is necessarily a reducing subspace. This is a trivial consequence of the required property $|\Theta|^2 w \equiv 1$. Therefore, in the case when $m(\{\zeta \in \mathbb{T} : w(\zeta) = 0\}) > 0$, the operator Z_μ has plenty of reducing subspaces, but it has no simply invariant subspace. In particular, this is the case if μ is singular. On the other hand, in the case when $w > 0$ a.e. on $\mathbb{T}$, then Z_μ has plenty of simply invariant subspaces. Indeed, in that case, we define a measurable function Θ on $\mathbb{T}$ such that $|\Theta|^2 = 1/w$ a.e. on $\mathbb{T}$. Hence, $\Theta H^2 \subset L^2(\mu_a)$, and, for any measurable set $\sigma \subset \mathbb{T}$, $\Theta H^2 \oplus \chi_\sigma L^2(\mu_s)$ is a simply invariant subspace of Z_μ.

The following result shows that we may decompose the invariant subspaces of Z_μ.

Corollary 8.35 *Let $d\mu = d\mu_a + d\mu_s = w\,dm + d\mu_s$ be the Lebesgue decomposition of a finite and positive Borel measure μ on $\mathbb{T}$. Let $\mathcal{E}$ be a closed subspace of $L^2(\mu)$ such that $z\mathcal{E} \subset \mathcal{E}$. Then there exist $\mathcal{E}_a \subset L^2(\mu_a)$ and $\mathcal{E}_s \subset L^2(\mu_s)$ satisfying the following properties:*

(i) $\mathcal{E} = \mathcal{E}_a \oplus \mathcal{E}_s$;

(ii) $z\mathcal{E}_a \subset \mathcal{E}_a$;

(iii) $z\mathcal{E}_s \subset \mathcal{E}_s$.

Furthermore, the above decomposition is unique, and, if $\mathcal{E} \subset H^2(\mu)$, then $\mathcal{E}_a \subset H^2(\mu_a)$.

Proof The first part follows immediately from Theorems 8.29 and 8.34. The uniqueness of the decomposition follows from $L^2(\mu_a) \perp L^2(\mu_s)$, whereas the inclusion $\mathcal{E}_a \subset H^2(\mu_a)$ comes from the decomposition $H^2(\mu) = H^2(\mu_a) \oplus L^2(\mu_s)$, with $H^2(\mu_a) \subset L^2(\mu_a)$. □

Exercises

Exercise 8.8.1 Show that every simply invariant subspace of the unilateral forward shift S is cyclic.
Hint: Use Theorem 8.32.

Exercise 8.8.2 Let $\mathcal{E}$ be a closed subspace of $\overline{H^2(\mathbb{T})}$ such that $\bar{z}\mathcal{E} \subset \mathcal{E}$. Show that there is an inner function Θ such that

$$\mathcal{E} = \overline{\Theta H^2(\mathbb{T})}.$$

Hint: Apply Theorem 8.32.

Exercise 8.8.3 Let $\Theta_1, \Theta_2 \in L^\infty(\mathbb{T})$ be such that

$$|\Theta_1| = |\Theta_2| = 1 \qquad \text{(a.e. on } \mathbb{T}\text{)}.$$

Suppose that $\Theta_1 H^2 = \Theta_2 H^2$. Show that there is a constant $\gamma \in \mathbb{T}$ such that $\Theta_1 = \gamma\Theta_2$. Deduce that the unimodular function corresponding to a simply invariant subspace (Theorems 8.30 and 8.32) is essentially unique.
Hint: If $f, \bar{f} \in H^2$, then f is constant (see (4.12)).

Exercise 8.8.4 Let $\mu = \mu_s$ be a singular measure and let $\mathcal{E} \subset L^2(\mu_s)$ be such that $z\mathcal{E} \subset \mathcal{E}$. Show that there exists a measurable set $\sigma \subset \mathbb{T}$ such that $\mathcal{E} = \chi_\sigma L^2(\mu_s)$.
Hint: Show that $z\mathcal{E} = \mathcal{E}$ and use Theorem 8.29.

Exercise 8.8.5 Let $d\mu = d\mu_a = w\,dm$ be a finite Borel measure on $\mathbb{T}$, with $w \geq 0$ and $w \in L^1(\mathbb{T})$. Let $\mathcal{E} \subset L^2(\mu_a)$ be such that $z\mathcal{E} \subset \mathcal{E}$. Show that either there exists a measurable set $\sigma \subset \mathbb{T}$ such that

$$\mathcal{E} = \chi_\sigma L^2(\mu_a)$$

or there exists a measurable function Θ on $\mathbb{T}$ such that $|\Theta|^2 w = 1$, a.e. on $\mathbb{T}$, and

$$\mathcal{E} = \Theta H^2.$$

Exercise 8.8.6 Let μ be a finite and positive Borel measure on $\mathbb{T}$, and let $f \in L^2(\mu)$. Assume that there exists a measurable set $E \subset \mathbb{T}$ such that $m(E) > 0$ and $f|E = 0$. Show that

$$\mathcal{E}_f := \text{Span}_{L^2(\mu)}\{z^n f : n \geq 0\}$$

is a reducing subspace for Z_μ.
Hint: Argue by absurdity and apply Theorem 8.34.

8.9 Reducing invariant subspaces of S_μ

It happens that, for some measures μ, the operator S_μ has a lot of reducing subspaces, and, for other measures μ, it does not have any nontrivial reducing subspace. The study of reducing subspaces of the operator S_μ depends on the Lebesgue decomposition of the measure μ,

$$d\mu = w\, dm + d\mu_s,$$

where $w \geq 0$, $w \in L^1(\mathbb{T})$ and $d\mu_s \perp dm$. If $\log w \notin L^1(\mathbb{T})$, then we know from Corollary 8.23 that $H^2(\mu) = L^2(\mu)$ and the operators S_μ and Z_μ coincide. In that case, Theorem 8.29 describes all the reducing subspaces of S_μ. In the opposite case, that is, when $\log w \in L^1(\mathbb{T})$, we decompose the study of reducing subspaces into two steps. The first step concerns the particular situation of the unilateral forward shift operator S on H^2, corresponding to $w \equiv 1$ and $\mu_s = 0$. In the second step, we use the matrix representation (8.42) and Lemma 1.40 to deduce the general case. The following result shows that the unilateral forward shift operator S on H^2 has no nontrivial reducing subspace.

Theorem 8.36 *Let $\mathcal{E}$ be a closed subspace of H^2 such that*

$$S\mathcal{E} \subset \mathcal{E} \quad \textit{and} \quad S^*\mathcal{E} \subset \mathcal{E}.$$

Then either $\mathcal{E} = \{0\}$ or $\mathcal{E} = H^2$.

Proof Suppose that $\mathcal{E} \neq \{0\}$. Pick any $f \in \mathcal{E}$, $f \not\equiv 0$. Let $\hat{f}(k)$ be the first nonzero Fourier coefficient of f. Replacing f by $f/\hat{f}(k)$, we may assume that $\hat{f}(k) = 1$. Then, since $\mathcal{E}$ is invariant under S^*, we have $g = S^{*k} f \in \mathcal{E}$. Note that $\hat{g}(0) = 1$. Hence, by (8.18),

$$\chi_0 = g - SS^* g \in \mathcal{E}.$$

Since $\mathcal{E}$ is invariant under S, we deduce that $\chi_n = S^n \chi_0 \in \mathcal{E}$ for each $n \geq 0$. Therefore, $\mathcal{E} = H^2$. □

Now we can complete the picture of reducing subspaces of the unilateral forward shift operator S_μ.

Theorem 8.37 *Let $d\mu = d\mu_a + d\mu_s = w\, dm + d\mu_s$ be the Lebesgue decomposition of a finite and positive Borel measure μ on $\mathbb{T}$, and assume that $\log w \in L^1(\mathbb{T})$. Let $\mathcal{E}$ be a closed subspace of $H^2(\mu)$. Then $\mathcal{E}$ is a reducing subspace of S_μ if and only if there exists a measurable set $\sigma \subset \mathbb{T}$ such that either*

$$\mathcal{E} = H^2(\mu_a) \oplus \chi_\sigma L^2(\mu_s)$$

or

$$\mathcal{E} = \{0\} \oplus \chi_\sigma L^2(\mu_s).$$

Proof Remember that

$$H^2(\mu) = H^2(\mu_a) \oplus L^2(\mu_s). \tag{8.62}$$

Moreover, the condition $\log w \in L^1(\mathbb{T})$ implies the existence of a unitary operator U from $H^2(\mu_a)$ onto H^2 such that the operator S_μ has the matrix representation

$$S_\mu = \begin{pmatrix} U^*SU & 0 \\ 0 & Z_{\mu_s} \end{pmatrix}$$

with respect to the decomposition (8.62). Let $\mathcal{E}$ be a reducing subspace of S_μ. Then, according to Lemma 1.40 and Corollary 8.35, there exist a closed subspace $\mathcal{E}_1 \subset H^2(\mu_a)$ and a closed subspace $\mathcal{E}_2 \subset L^2(\mu_s)$ such that

$$\mathcal{E} = \mathcal{E}_1 \oplus \mathcal{E}_2$$

and $\mathcal{E}_1$ is reducing for U^*SU and $\mathcal{E}_2$ is reducing for Z_{μ_s}. Then, on the one hand, according to Theorem 8.29, there is a measurable set $\sigma \subset \mathbb{T}$ such that $\mathcal{E}_2 = \chi_\sigma L^2(\mu_s)$. On the other hand, if $\mathcal{E}_1$ is reducing for U^*SU, then $U\mathcal{E}_1$ is reducing for S. But, according to Theorem 8.36, we then have $U\mathcal{E}_1 = \{0\}$ or $U\mathcal{E}_1 = H^2$. Hence $\mathcal{E}_1 = \{0\}$ or $\mathcal{E}_1 = H^2(\mu_a)$. Therefore, we have either

$$\mathcal{E} = H^2(\mu_a) \oplus \chi_\sigma L^2(\mu_s)$$

or

$$\mathcal{E} = \{0\} \oplus \chi_\sigma L^2(\mu_s).$$

The converse follows easily from Lemma 1.40 and the fact that $\{0\}$ and $H^2(\mu_a)$ are reducing subspaces of U^*SU and $\chi_\sigma L^2(\mu_s)$ is a reducing subspace of Z_{μ_s}. □

We see from Theorems 8.29 and 8.37 that S_μ has nontrivial reducing subspaces if and only if either $\log w \notin L^1(\mathbb{T})$ or $\log w \in L^1(\mathbb{T})$ but $\mu_s \neq 0$. In other words, if μ is absolutely continuous with respect to the Lebesgue measure m and is such that $\log(d\mu/dm) \in L^1(\mathbb{T})$, then S_μ has no nontrivial reducing subspace. In the opposite case, S_μ has plenty of reducing subspaces.

8.10 Simply invariant subspaces of S_μ

We now study the simply invariant subspaces for the operator S_μ. As already done for the reducing subspaces, we distinguish two cases depending on whether $\log w$ is integrable or not on the unit circle. If $\log w \notin L^1(\mathbb{T})$, then $H^2(\mu) = L^2(\mu)$ and the operators S_μ and Z_μ coincide. Thus, in that case, the simply invariant subspaces for S_μ are described by Theorem 8.34. In the opposite case, that is, when $\log w \in L^1(\mathbb{T})$, we use the matrix representation (8.42) to establish the following result.

Theorem 8.38 *Let $d\mu = d\mu_a + d\mu_s = w\,dm + d\mu_s$ be the Lebesgue decomposition of a finite positive Borel measure μ on $\mathbb{T}$, where $w \geq 0$, $w \in L^1(\mathbb{T})$ and $d\mu_s \perp d\mu_a$. Assume furthermore that $\log w \in L^1(\mathbb{T})$. Then $\mathcal{E}$ is a simply invariant subspace of S_μ if and only if there exist a nonconstant inner function Θ and a measurable set $\sigma \subset \mathbb{T}$ such that*

$$\mathcal{E} = \Theta H^2(\mu_a) \oplus \chi_\sigma L^2(\mu_s).$$

Proof Remembering that $H^2(\mu) = H^2(\mu_a) \oplus L^2(\mu_s)$ and if $h = [w^{1/2}]$, then the operator

$$\begin{array}{rccc} U: & H^2(\mu_a) & \longrightarrow & H^2 \\ & f & \longmapsto & hf \end{array}$$

is a unitary operator. Moreover, with respect to the preceding decomposition, the operator S_μ has the matrix representation

$$S_\mu = \begin{bmatrix} U^*SU & 0 \\ 0 & Z_{\mu_s} \end{bmatrix}.$$

Let $\mathcal{E}$ be a simply invariant subspace for S_μ. Then, according to Lemma 1.40 and Corollary 8.35, we can write

$$\mathcal{E} = \mathcal{E}_1 \oplus \mathcal{E}_2,$$

where $U^*SU\mathcal{E}_1 \subset \mathcal{E}_1 \subset H^2(\mu_a)$, $Z_{\mu_s}\mathcal{E}_2 \subset \mathcal{E}_2 \subset L^2(\mu_s)$ and either $\mathcal{E}_1$ is simply invariant under U^*SU or $\mathcal{E}_2$ is simply invariant under Z_{μ_s}. Since μ_s is singular, according to Theorem 8.34, Z_{μ_s} has no simply invariant subspace. See also the comment after the proof of Theorem 8.34. Hence, $\mathcal{E}_2$ is necessarily a reducing subspace for Z_{μ_s}, and, according to Theorem 8.29, there exists a measurable set $\sigma \subset \mathbb{T}$ such that

$$\mathcal{E}_2 = \chi_\sigma L^2(\mu_s).$$

Since $\mathcal{E}_2$ is reducing, then necessarily $\mathcal{E}_1$ is simply invariant under U^*SU, that is, $U\mathcal{E}_1$ is simply invariant under S. Then, by Theorem 8.32, there exists a nonconstant inner function Θ such that $U\mathcal{E}_1 = \Theta H^2$. Hence,

$$\mathcal{E}_1 = U^{-1}(\Theta H^2) = h^{-1}\Theta H^2 = \Theta h^{-1} H^2 = \Theta H^2(\mu_a).$$

Therefore, we get

$$\mathcal{E} = \Theta H^2(\mu_a) \oplus \chi_\sigma L^2(\mu_s).$$

Reciprocally, let Θ be a nonconstant inner function, let σ be a measurable subset of $\mathbb{T}$, and let $\mathcal{E} = \Theta H^2(\mu_a) \oplus \chi_\sigma L^2(\mu_s)$. It is clear that $\mathcal{E}$ is a closed invariant subspace of S_μ. Assume that $\mathcal{E}$ is reducing for S_μ. Then, using

(8.42) and Lemma 1.40 once more, we deduce that $\Theta H^2(\mu_a)$ is reducing for $S_{\mu_a} = U^*SU$. Hence, $U(\Theta H^2(\mu_a))$ is reducing for S. But, according to Theorem 8.36, we must have either $U(\Theta H^2(\mu_a)) = H^2$ or $U(\Theta H^2(\mu_a)) = \{0\}$. We finally get that either $\Theta H^2 = H^2$ or $\Theta H^2 = \{0\}$. The second case is impossible, whereas the first case can occur only if Θ is constant, which contradicts the hypothesis on Θ. Thus $\mathcal{E}$ is not a reducing subspace, whence it is a simply invariant subspace for S_μ. □

8.11 Cyclic vectors of Z_μ and S^*

In this section, our goal is to characterize the cyclic vectors for the bilateral forward shift operator Z_μ and then define the important concept of a model subspace. See Section 1.10 for the definition of cyclic vectors in a general context. In our situation, a vector $f \in L^2(\mu)$ is cyclic (or 1-cyclic) if

$$\text{Span}_{L^2(\mu)}\{z^n f : n \geq 0\} = L^2(\mu)$$

and 2-cyclic if

$$\text{Span}_{L^2(\mu)}\{z^n f : n \in \mathbb{Z}\} = L^2(\mu).$$

The characterization of 2-cyclic vectors is quite easy to establish.

Theorem 8.39 *Let μ be a finite and positive Borel measure on $\mathbb{T}$, and let $f \in L^2(\mu)$. Then f is a 2-cyclic vector for Z_μ if and only if $f \neq 0$ μ-a.e. on $\mathbb{T}$.*

Proof Put

$$\mathcal{E} = \text{Span}_{L^2(\mu)}\{z^n f : n \in \mathbb{Z}\}.$$

It is obvious that $\mathcal{E}$ is a doubly invariant subspace for Z_μ. Hence, according to Theorem 8.29, there exists a measurable set $\sigma \subset \mathbb{T}$ such that $\mathcal{E} = \chi_\sigma L^2(\mu)$.

Assume that $f \neq 0$, μ-a.e. on $\mathbb{T}$. Since $f \in \mathcal{E}$, we have $f = 0$, μ-a.e. on $\mathbb{T} \setminus \sigma$, and thus $\mu(\mathbb{T} \setminus \sigma) = 0$. Hence, $\mathcal{E} = L^2(\mu)$.

Reciprocally, assume that $\mathcal{E} = L^2(\mu)$ and denote

$$E = \{\zeta \in \mathbb{T} : f(\zeta) = 0\}.$$

Let $\varepsilon > 0$. Then there exists a trigonometric polynomial p such that

$$\|\chi_E - pf\|^2_{L^2(\mu)} \leq \varepsilon.$$

In particular,

$$\int_E |\chi_E - pf|^2 \, d\mu \leq \|\chi_E - pf\|^2_{L^2(\mu)} \leq \varepsilon.$$

But, since $f = 0$, μ-a.e. on E and $\chi_E = 1$ on E, we get $\mu(E) \leq \varepsilon$. Hence, $\mu(E) = 0$, which means that $f \neq 0$ μ-a.e. on $\mathbb{T}$. □

For the characterization of 1-cyclic vectors, we recall that, if $d\mu = d\mu_a + d\mu_s = w\,dm + d\mu_s$ is the Lebesgue decomposition of the measure μ, then $L^2(\mu)$ decomposes as

$$L^2(\mu) = L^2(\mu_a) \oplus L^2(\mu_s),$$

and each function $f \in L^2(\mu)$ has a unique orthogonal decomposition

$$f = f_a + f_s \qquad (f_a \in L^2(\mu_a),\ f_s \in L^2(\mu_s)).$$

Furthermore, since

$$\begin{aligned}&\text{Span}_{L^2(\mu)}\{z^n f : n \geq 0\}\\ &\quad\subset \text{Span}_{L^2(\mu_a)}\{z^n f_a : n \geq 0\} \oplus \text{Span}_{L^2(\mu_s)}\{z^n f_s : n \geq 0\},\end{aligned}$$

if f is 1-cyclic for Z_μ, then f_a is 1-cyclic for Z_{μ_a} and also f_s is 1-cyclic for Z_{μ_s}.

Theorem 8.40 *A function f in $L^2(\mu)$ is 1-cyclic for Z_μ if and only if the following two conditions hold:*

(i) $f \neq 0$, *μ-a.e. on $\mathbb{T}$;*
(ii) $\log|f_a w^{1/2}| \notin L^1(\mathbb{T})$.

Proof We assume that the two conditions (i) and (ii) are satisfied and we show that the closed subspace

$$\mathcal{E} = \text{Span}_{L^2(\mu)}\{z^n f : n \geq 0\}$$

is the whole space $L^2(\mu)$. Clearly, $\mathcal{E}$ is Z_μ-invariant. We first prove that $\mathcal{E}$ is doubly invariant. Argue by absurdity and assume that $\mathcal{E}$ is a simply invariant subspace for the operator Z_μ. Hence, according to Theorem 8.34, there exist a measurable subset σ of $\mathbb{T}$ and a measurable function Θ such that $m(\sigma) = 0$ and

$$\begin{gathered}\mathcal{E} = \Theta H^2 \oplus \chi_\sigma L^2(\mu_s),\\ \Theta H^2 \subset L^2(\mu_a),\\ \chi_\sigma L^2(\mu_s) \subset L^2(\mu_s),\\ |\Theta|^2 w \equiv 1 \qquad (m\text{-a.e. on } \mathbb{T}).\end{gathered}$$

In particular, $f_a \in \Theta H^2$. In other words, there exists a function $g \in H^2$ such that $f_a = \Theta g$. But

$$|f_a w^{1/2}| = (|\Theta|^2 w)^{1/2}|g| = |g| \qquad (m\text{-a.e. on } \mathbb{T}).$$

Thus, we get from (ii) that $\log|g| \notin L^1(\mathbb{T})$, which, by Lemma 4.30, implies that $g \equiv 0$, a.e. on $\mathbb{T}$. Then $f_a \equiv 0$, a.e. on $\mathbb{T}$, and we have $f = f_s$. Hence,

$$\mathcal{E} = \text{Span}\{z^n f_s : n \geq 0\} \subset L^2(\mu_s)$$

and $\Theta H^2 = \{0\}$. This is a contradiction because $\Theta \neq 0$, a.e. on $\mathbb{T}$. Therefore, $\mathcal{E}$ is a doubly invariant subspace for Z_μ. According to Theorem 8.29, there exists a measurable set $\sigma \subset \mathbb{T}$ such that $\mathcal{E} = \chi_\sigma L^2(\mu)$. In particular, $f \in \chi_\sigma L^2(\mu)$. Hence, $f = 0$, μ-a.e. on $\mathbb{T} \setminus \sigma$. But, the hypothesis (i) implies that $\mu(\mathbb{T} \setminus \sigma) = 0$. In other words,

$$\mathcal{E} = \chi_\sigma L^2(\mu) = L^2(\mu).$$

Conversely, let f be a function in $L^2(\mu)$ and assume that f is 1-cyclic for Z_μ, that is, $\mathcal{E} = L^2(\mu)$. Then, since

$$\mathcal{E} \subset \text{Span}\{z^n f : n \in \mathbb{Z}\},$$

we get from Theorem 8.39 that $f \neq 0$, μ-a.e. on $\mathbb{T}$. If the measure μ is singular, then the proof is complete. If $\mu_a \neq 0$, we prove that $\log|f_a w^{1/2}| \notin L^1(\mathbb{T})$. Arguing by absurdity, assume on the contrary that $\log|f_a w^{1/2}| \in L^1(\mathbb{T})$. Then there exists an outer function g in H^2 such that $|g| = |f_a w^{1/2}|$, a.e. on $\mathbb{T}$. Define $\Theta = f_a/g$. Then Θ is a measurable function that satisfies

$$|\Theta|^2 w = \frac{|f_a|^2}{|g|^2} w = \frac{|f_a|^2}{|f_a|^2 w} w = 1 \qquad \text{(a.e. on } \mathbb{T}). \tag{8.63}$$

Moreover, since $f_a = \Theta g \in \Theta H^2$, we get

$$\text{Span}_{L^2(\mu_a)}\{z^n f_a : n \geq 0\} \subset \Theta H^2.$$

Furthermore, for each function $h \in H^2$, by Theorem 8.16, there exists a sequence $(p_n)_n$ of analytic polynomials such that $\|p_n g - h\|_2 \longrightarrow 0$, as $n \longrightarrow \infty$. But, using (8.63), we have

$$\begin{aligned}
\|\Theta h - \Theta p_n g\|_{L^2(\mu_a)}^2 &= \int_{\mathbb{T}} |\Theta h - \Theta p_n g|^2 w \, dm \\
&= \int_{\mathbb{T}} |h - p_n g|^2 |\Theta|^2 w \, dm \\
&= \int_{\mathbb{T}} |h - p_n g|^2 \, dm \\
&= \|h - p_n g\|_2^2,
\end{aligned}$$

whence, $\|\Theta h - \Theta p_n g\|_{L^2(\mu)} \longrightarrow 0$, as $n \longrightarrow \infty$. This implies that

$$\Theta H^2 \subset \text{Span}_{L^2(\mu_a)}\{z^n f_a : n \geq 0\},$$

and finally the two subspaces coincide. But the function f_a should be cyclic for Z_{μ_a}, whence $\Theta H^2 = L^2(\mu_a) = L^2(w\, dm)$. Now, let σ be a measurable subset of $\mathbb{T}$ such that

$$\mu_a(\sigma) = \mu_a(\mathbb{T} \setminus \sigma) = \frac{\mu(\mathbb{T})}{2} > 0.$$

Since $\chi_\sigma \in L^2(\mu_a)$, then there exists a function $h \in H^2$ such that $\chi_\sigma = \Theta h$. Using (8.63), we have

$$\|h\|_2^2 = \int_{\mathbb{T}} |\Theta|^2 |h|^2 w \, dm = \int_{\mathbb{T}} |\chi_\sigma|^2 \, d\mu_a = \mu_a(\sigma) > 0,$$

and, in particular, $h \not\equiv 0$. Moreover, $h = 0$, m-a.e. on $\mathbb{T} \setminus \sigma$. Thus, Lemma 4.30 implies $m(\mathbb{T} \setminus \sigma) = 0$ and then $\mu_a(\mathbb{T} \setminus \sigma) = 0$, which contradicts the definition of σ. □

We immediately get the following application.

Corollary 8.41 *Let $f \in L^2(\mathbb{T})$. Then f is 1-cyclic for Z if and only if $f \neq 0$ a.e. on $\mathbb{T}$ and*

$$\int_{\mathbb{T}} \log |f| \, dm = -\infty.$$

Using the concept of pseudocontinuation, the cyclic vectors of the backward shift operator S^* were completely characterized by Douglas, Shapiro and Shields. We explain this concept for the unit disk. However, it can be defined for any two disjoint Jordan domains with a common smooth boundary on the Riemann sphere.

Let f_1 and f_2, respectively, be meromorphic functions on $\mathbb{D}$ and $\mathbb{D}_e = \{z : 1 < |z| \leq \infty\}$. Note that $\infty \in \mathbb{D}_e$. In particular, this means that we assume that f_2 behaves fairly well at ∞. More precisely, it is either analytic or has a finite pole at ∞. In other words, f_2 does not have an essential singularity at ∞. Then f_1 and f_2 are pseudocontinuations of one another across a set $E \subset \mathbb{T}$, with $|E| > 0$, if both have nontangential boundary values almost everywhere on E, and moreover their boundary values are equal almost everywhere on E. We recall that f_2 is said to have a nontangential limit equal to ℓ at $\zeta \in \mathbb{T}$ if, for each $C \geq 1$,

$$f_2(z) \longrightarrow L \qquad \text{as } z \longrightarrow \zeta, \; z \in S_C^*(\zeta),$$

where $S_C^*(\zeta) = \{z \in \mathbb{D}_e : |z - \zeta| < C(|z| - 1)\}$. This is equivalent to saying that the meromorphic function

$$z \longmapsto \overline{f_2(1/\bar{z})}$$

in $\mathbb{D}$ has a nontangential limit equal to L at ζ (in the sense defined in Section 3.1).

Given f_1, there is no guarantee that it has a pseudocontinuation f_2. However, based on the Lusin–Privalov uniqueness theorem (see Theorem 4.32), whenever f_2 exists, it is certainly unique. Moreover, if f_1 has an analytic continuation and a pseudocontinuation, both across a set $E \subset \mathbb{T}$, with positive measure, then these extensions must coincide.

It is easy to show that each inner function on $\mathbb{D}$ has a pseudocontinuation across $\mathbb{T}$. In fact, if $f_1 = \Theta$ is an inner function in the open unit disk, then define

$$f_2(z) = \frac{1}{\overline{\Theta(1/\bar{z})}} \qquad (z \in \mathbb{D}_e). \tag{8.64}$$

The function f_2 is the unique pseudocontinuation of Θ across $\mathbb{T}$. Abusing notation, we will also write Θ for f_2.

We recall that a meromorphic function f, defined on a domain Ω of the Riemann sphere, is of bounded characteristic (or bounded type) if it is of the form $f = \varphi/\psi$, where φ and ψ are bounded and analytic on Ω. Certainly, we must have $\psi \not\equiv 0$. However, ψ may have some isolated zeros in Ω. If φ and ψ have no common zeros, the zeros of ψ are in fact the poles of f.

Theorem 8.42 *Let $f \in H^2$. Then the following are equivalent.*

(i) *Function f is noncyclic for S^*.*
(ii) *There exist $g, h \in \bigcup_{p>0} H^p$ such that*

$$f = \frac{\bar{h}}{\bar{g}} \qquad (a.e. \text{ on } \mathbb{T}).$$

(iii) *Function f has a bounded-type meromorphic pseudocontinuation $\tilde{f}$ across $\mathbb{T}$ to $\mathbb{D}_e$.*

Proof (i) $\Longrightarrow$ (ii) Suppose that f is noncyclic for S^*. Hence, by definition, the closed subspace

$$\text{Span}\{S^{*n} f : n \ge 0\}$$

is not equal to H^2. In particular, there is a nonzero function $g \in H^2$ such that

$$g \perp S^{*n} f \qquad (n \ge 0).$$

Hence,

$$0 = \langle g, S^{*n} f\rangle_2 = \langle S^n g, f\rangle_2 = \langle \chi_n g, f\rangle_2 \qquad (n \ge 0).$$

Therefore, we have

$$\int_{\mathbb{T}} \overline{f(\zeta)} g(\zeta) \chi_n(\zeta)\, dm(\zeta) = 0 \qquad (n \ge 0).$$

This identity suggests that we consider $h = \bar{f} g$. Clearly, $h \in L^1(\mathbb{T})$ and the above identity means that $h \in H^1_0(\mathbb{T})$. Hence, we can write $f = \bar{h}/\bar{g}$, with $h \in H^1_0$ and $g \in H^2$.

(ii) $\Longrightarrow$ (iii) Assume that there are two functions $g, h \in \bigcup_{p>0} H^p$ such that $f = \bar{h}/\bar{g}$, a.e. on $\mathbb{T}$. By Lemma 5.1, we can assume that h and g belong to H^∞. Now define

$$\tilde{f}(z) = \frac{\overline{h(1/\bar{z})}}{\overline{g(1/\bar{z})}} \qquad (|z| > 1).$$

It is clear that $\tilde{f}$ is meromorphic and of bounded type on $\mathbb{D}_e$. Moreover, it has nontangential limits almost everywhere on $\mathbb{T}$ and

$$\tilde{f}(\zeta) = \frac{\overline{h(\zeta)}}{\overline{g(\zeta)}} = f(\zeta) \qquad \text{(a.e. on } \mathbb{T}).$$

(iii) $\Longrightarrow$ (i) Assume that f has a bounded-type meromorphic pseudocontinuation $\tilde{f}$ across $\mathbb{T}$ to the domain $\mathbb{D}_e$. Hence, $\tilde{f} = h/g$, where g and h are bounded and analytic on $\mathbb{D}_e$. Define

$$g_1(z) = \overline{g(1/\bar{z})} \quad \text{and} \quad h_1(z) = \overline{h(1/\bar{z})} \qquad (z \in \mathbb{D}).$$

Then $g_1, h_1 \in H^\infty(\mathbb{D})$. Moreover, by the pseudocontinuity assumption, we have

$$f(e^{it}) = \tilde{f}(e^{it}) = \frac{h(e^{it})}{g(e^{it})} = \frac{\overline{h_1(e^{it})}}{\overline{g_1(e^{it})}} \qquad \text{(a.e. on } \mathbb{T}). \tag{8.65}$$

Since, if needed, we can multiply h_1 and g_1 by e^{it} in the representation (8.65), we can assume without loss of generality that $h_1 \in H_0^\infty$, that is, $h_1(0) = 0$, and thus we can say that

$$\langle \bar{h}_1, \chi_n \rangle_2 = 0 \qquad (n \geq 0).$$

Hence, for any $n \geq 0$, we have

$$\langle S^{*n} f, g_1 \rangle_2 = \langle f, S^n g_1 \rangle_2 = \langle f, \chi_n g_1 \rangle_2 = \langle f \bar{g}_1, \chi_n \rangle_2 = 0.$$

Thus, the function g_1, which is a nonzero element of $H^2(\mathbb{D})$, is orthogonal to f and all its backward shifts $S^{*n} f$, $n \geq 0$. Hence, f is not cyclic for S^*. □

Based on Theorem 5.8, a rational function $r \in H^2$ is analytic on $\bar{\mathbb{D}}$, that is, analytic on a domain containing the closed unit disk. Furthermore, if we write

$$r(z) = \frac{\prod_{i=1}^n (z - w_i)}{\prod_{i=1}^m (z - z_i)},$$

then we have

$$r(z) = \frac{z^n \prod_{i=1}^n (1 - w_i/z)}{z^m \prod_{i=1}^n (1 - z_i/z)} = \frac{\overline{p(z)}}{\overline{q(z)}} \qquad (z \in \mathbb{T}),$$

with $p(z) = z^m \prod_{i=1}^n (1 - \overline{w}_i z)$ and $q(z) = z^n \prod_{i=1}^n (1 - \bar{z}_i z)$. Since p and q belong to H^∞, Theorem 8.42 ensures that r is noncyclic for S^*. Hence, each rational function in H^2 is analytic on $\bar{\mathbb{D}}$ and noncyclic for S^*. In fact, the converse is also true.

Corollary 8.43 *Let f be a function analytic on a domain Ω that contains $\bar{\mathbb{D}}$, and assume that f is noncyclic for S^*. Then f is a rational function.*

Proof Let $\tilde{f}$ be the pseudocontinuation of f promised in Theorem 8.42. By Privalov's uniqueness theorem, we have $\tilde{f}(z) = f(z)$, $z \in \Omega \setminus \bar{\mathbb{D}}$. Hence, the function

$$g(z) = \begin{cases} f(z) & \text{if} \quad z \in \Omega, \\ \tilde{f}(z) & \text{if} \quad z \in \mathbb{D}_e \end{cases}$$

is well defined and meromorphic on the Riemann sphere. But it is well known that the only meromorphic functions on the Riemann sphere are the rational functions. □

Let Θ be an inner function on $\mathbb{D}$. Then the *model subspace* of H^2 generated by Θ is defined by

$$K_\Theta = (\Theta H^2)^\perp.$$

Owing to the Hilbert space structure of H^2, we can give another useful description of the space K_Θ. Considering Corollary 8.33, the following representation will shed more light on the nature of these S^*-invariant subspaces.

Theorem 8.44 *Let Θ be an inner function. Then*

$$K_\Theta = H^2 \cap (\Theta \overline{H_0^2}) = (\Theta H^2)^\perp.$$

Proof Let f be a function in H^2. Then, by definition, f belongs to $(\Theta H^2)^\perp$ if and only if

$$\langle f, \Theta h \rangle_{H^2} = 0$$

for every $h \in H^2$. Using the fact that $|\Theta| = 1$ almost everywhere on $\mathbb{T}$, this is equivalent to

$$\langle \bar{\Theta} f, h \rangle_{L^2} = 0$$

for every $h \in H^2$. Note that $\bar{\Theta} f$ remains in $L^2(\mathbb{T})$. But the last relation holds if and only if $\bar{\Theta} f \in \overline{H_0^2}$. In other words, $f \in \Theta \overline{H_0^2}$. □

The description of K_Θ given in Theorem 8.44 can be exploited to generalize the definition in the Banach setting. Hence, if Θ is an inner function on $\mathbb{D}$, and $0 < p \leq \infty$, then the *model subspace* of H^p generated by Θ is defined by

$$K_\Theta^p = H^p \cap (\Theta \overline{H_0^p}).$$

If $p = 2$, we surely have $K_\Theta^2 = K_\Theta$.

If $f \in H^2$ is a noncyclic function for S^*, then, by definition, the closed subspace

$$\mathcal{E} = \text{Span}\{S^{*n} f : n \geq 0\}$$

is not equal to H^2. Clearly, $\mathcal{E}$ is invariant under S^*. Hence, by Corollary 8.33, there is an inner function Θ such that $\mathcal{E} = K_\Theta$. Conversely, for any given Θ and for all $f \in K_\Theta$, we have

$$\text{Span}\{S^{*n} f : n \geq 0\} \subset K_\Theta \neq H^2.$$

Therefore, the set

$$K = \bigcup_{\Theta \text{ inner}} K_\Theta$$

is precisely the aggregate of all noncyclic elements of H^2 for the operator S^*. Moreover, Theorem 8.42 says that

$$K = \{f \in H^2 : f \text{ has a bounded-type meromorphic pseudocontinuation across } \mathbb{T} \text{ to } \mathbb{D}_e\}.$$

Based on the alternative representation given in Theorem 8.44, we can give a better picture of the pseudocontinuation $\tilde{f}$ that was promised in Theorem 8.42. If f is noncyclic, then $f \in K_\Theta$ for some inner function Θ. Thus, by Theorem 8.44, $g = \Theta \bar{f} \in H_0^2$. Put

$$\tilde{f}(z) = \frac{\overline{g(1/\bar{z})}}{\overline{\Theta(1/\bar{z})}} \qquad (z \in \mathbb{D}_e).$$

Clearly, $\tilde{f}$ is a meromorphic function on $\mathbb{D}_e$ and

$$\tilde{f} = \frac{\bar{g}}{\bar{\Theta}} = f \qquad \text{(a.e. on } \mathbb{T}).$$

Moreover, using Lemma 5.1, we see that $\tilde{f}$ is also of bounded type on $\mathbb{D}_e$.

Exercises

Exercise 8.11.1 Show that $f(e^{i\theta}) = |1 - e^{i\theta}|^\alpha$, $\alpha > -1/2$, is not 1-cyclic for Z.
Hint: Apply Corollary 8.41.

Exercise 8.11.2 Show that $f(e^{i\theta}) = \exp(-1/|1 - e^{i\theta}|)$ is 1-cyclic for Z.
Hint: Apply Corollary 8.41.

Exercise 8.11.3 Show that $f(z) = -\log(1-z)$ and $g(z) = (1-z)^{1/2}$ are cyclic for S^*.
Hint: They both have an essential singularity at 1.

Exercise 8.11.4 Show that $f(z) = e^z$ is cyclic for S^*.
Hint: Use Corollary 8.43.

Exercise 8.11.5 Show that $f(z) = e^{1/(z-2)}$ is cyclic for S^*.
Hint: Its analytic continuation across $\mathbb{T}$ is not meromorphic.

Exercise 8.11.6 Show that

$$f(z) = \sum_{n=1}^{\infty} \frac{1}{n^3} \times \frac{1}{z-(1+1/n)}$$

is cyclic for S^*.
Hint: What is the growth restriction on the poles of a meromorphic function of bounded type?

Exercise 8.11.7 Let f, g, h be functions in H^2. Assume that f and g are noncyclic vectors for S^* and h is a cyclic vector for S^*. Show that:

(a) $f+g$ is noncyclic;
(b) $f+h$ is cyclic;
(c) if $fg \in H^2$, then fg is noncyclic;
(d) if $f/g \in H^2$, then f/g is noncyclic;
(e) if $fh \in H^2$, then fh is cyclic;
(f) if $f/h \in H^2$, then f/h is cyclic.

Exercise 8.11.8 Let $f \in H^2$. Show that f is the sum of two cyclic vectors for S^*.
Hint: If f is cyclic, write $f = f/2 + f/2$. If f is noncyclic, take any cyclic vector g, write $f = (f-g)+g$ and use Exercise 8.11.7.

Exercise 8.11.9 Show that the set of noncyclic vectors is a linear manifold in H^2. Show that this manifold is dense in H^2.
Hint: For the first assertion, use Exercise 8.11.7. For the second assertion, use the analytic polynomials.

Exercise 8.11.10 Show that the set of cyclic vectors is dense in H^2.
Hint: Take f be a fixed cyclic vector and observe that the set $\{f+p : p \in \mathcal{P}_+\}$ consists only of cyclic vectors and is dense in H^2.

Exercise 8.11.11 Let $1 \leq p < \infty$, and let q be its conjugate exponent. Show that

$$K_\Theta^p = (\Theta H^q)^\perp,$$

where $\perp$ stands for the annihilator.

Exercise 8.11.12 Let $f(z) = \sum_{n=0}^{\infty} a_n z^n$ be an analytic function such that $|a_n|^{1/n} \longrightarrow 0$ as $n \longrightarrow \infty$. Assume that f is not a polynomial. Show that f is a cyclic vector for S^*.
Hint: Apply Corollary 8.43.

Notes on Chapter 8

Section 8.1

The importance of the bilateral forward shift operator $Z_\mu : L^2(\mu) \longrightarrow L^2(\mu)$ can be viewed through the following spectral theorem. Let A be a normal operator on a Hilbert space $\mathcal{H}$ which admits a vector $x_0 \in \mathcal{H}$ such that $\mathcal{H}$ is the smallest reducing subspace for A that contains x_0. Then A is unitarily equivalent to Z_μ for some positive and finite Borel measure μ on $\sigma(A)$. See for instance [159, p. 269] for a proof of this classic result. The spectral study of the operator Z_μ is well understood and the results presented in this section are classic. Here we restricted our study to the case where μ is a positive and finite Borel measure on $\mathbb{T}$, but most of the results can be generalized to an arbitrary positive and finite Borel measure on a compact subset of $\mathbb{C}$.

Section 8.2

The unilateral forward shift S is one of the most important operators in functional analysis. It can be viewed as a concrete representation of linear isometries A on an abstract Hilbert space $\mathcal{H}$ that have a cyclic vector and satisfy

$$\bigcap_{n \geq 0} A^n \mathcal{H} = \{0\}.$$

The cyclicity requirement can be dropped if one considers vector-valued versions of H^2; see [437]. This analytic representation appeared probably for the first time in the famous paper of Beurling [97]. Nowadays, shift operators on Hilbert spaces of analytic functions play an important role in the study of bounded linear operators on Hilbert spaces since they often serve as *models* for various classes of linear operators. For a general treatment of the backward shift operator on H^2 and its connection to problems in operator theory, we refer to the book [386]. See also [142] for a study in the Banach space setting.

Lemma 8.6 can be found in [253] and [97]. The spectral properties of $S: H^2 \longrightarrow H^2$ that appear in Lemma 8.6 have been generalized to various classes of Banach spaces of analytic functions in [18].

Section 8.3

Theorems 8.13 and 8.14 are taken from [253].

Section 8.4

Theorem 8.16 is due to Beurling, Srinivasan and Wang. Beurling [97] studied the case $p = 2$, and then Srinivasan and Wang [493] generalized this result for $0 < p < \infty$. The implication of Corollary 8.17, saying that an outer function is cyclic for the shift operator, was obtained by Smirnov in [488]. In [490] he gave another proof of this fact, essentially the same as that rediscovered by Beurling in [97].

Section 8.5

For a finite Borel measure μ, the distance $\mathrm{dist}(\chi_0, H_0^2(\mu))$ is called the *Szegő infimum* and appeared in [511] for absolutely continuous measures on the unit circle $\mathbb{T}$. The general case and its reduction to the absolutely continuous measure (Lemma 8.20) are due to Kolmogorov [318] and Kreĭn [324]. Theorem 8.22 was proved by Szegő [511] for absolutely continuous measures and by Kolmogorov [318] for the general case.

For general measures μ having compact support in $\mathbb{C}$, the problem of completeness of the polynomials in $L^p(\mu)$ was a long outstanding and difficult problem. It was eventually solved by Thomson in the remarkable paper [521].

Section 8.6

A Hilbert space operator is said to be *subnormal* if it has a normal extension – in other words, if there is a normal operator acting on a space containing the given one as a subspace and coinciding with the given operator in that subspace. This notion was introduced by Halmos in [248]. By the spectral theorem, one easily sees that all cyclic subnormal operators are unitarily equivalent to S_μ on $H^2(\mu)$ for some compactly supported measure μ in the complex plane. Thus, the study of subnormal operators quickly leads to questions about the unilateral shift operator S_μ. We refer to the book of Conway [160] for a comprehensive account of subnormal operators.

The notion of bounded point evaluation, appearing in Exercise 8.6.5, is a central one in operator theory. More precisely, it appears naturally in the invariant subspace problem for subnormal operators. This problem asks whether every subnormal operator has a nontrivial invariant subspace. Because it has an obvious positive answer for noncyclic operators, this problem is equivalent to the question whether, for every compactly supported measure μ in the complex plane, the space $H^2(\mu)$ has a nontrivial invariant subspace (under multiplication by z). In the case $H^2(\mu) = L^2(\mu)$, the answer is clear. If $H^2(\mu) \neq L^2(\mu)$ and if $H^2(\mu)$ has a bounded point evaluation at λ, then, as we have seen in Exercise 8.6.5, λ is an eigenvalue of S_μ^* associated with an eigenvector k_λ^μ. In particular, $\mathbb{C}k_\lambda^\mu$ is invariant under S_μ^* and thus $(\mathbb{C}k_\lambda^\mu)^\perp$ is a nontrivial invariant subspace. The question of the existence of bounded point evaluation has a long story and has attracted much attention. See for instance the work of Brennan [116, 117]. Finally, it was completely solved by Thomson [521]. A point in the complex plane is called an *analytic bounded point evaluation* of $H^2(\mu)$ if it belongs to an open set of bounded point evaluation on which any function in $H^2(\mu)$ is analytic. Thomson showed that, if $H^2(\mu) \neq L^2(\mu)$, then $H^2(\mu)$ not only has bounded point evaluation, but also has an abundance of analytic bounded point evaluation. However, we should say that, given a compactly supported measure μ in the complex plane, it is not easy to determine if a given point is a bounded (or analytic bounded) point evaluation for $H^2(\mu)$. Even though a nice characterization of bounded point evaluation for an arbitrary measure μ seems to be out of reach, for some natural classes of measures, it is possible to study the structure of bounded point evaluation. For instance, in the case when μ is supported on the unit circle, using Theorem 8.27, one can obtain a simple and explicit characterization of bounded point evaluation (see Exercise 8.6.6). We mention the paper of Miller, Smith and Yang [361] for further references and interesting results concerning the question of existence of bounded point evaluation. The result proved in Exercise 8.3.3 can be found in Wintner [555].

The description of $H^2(\mu)$ given in (8.41) is due to Smirnov [490].

Section 8.7

The study of invariant subspaces of an operator is the main subject of spectral theory. Theorem 8.29 is contained in Wiener's result on translation-invariant subspaces [554].

Section 8.8

Theorem 8.30 is due to Helson and Srinivasan [279] and Theorem 8.32 is the classic result of Beurling on invariant subspaces of the shift operator. It was originally proved in [97]. It can be generalized in the Banach space setting for Hardy spaces H^p, $0 < p < \infty$ (see [230] and [320, p. 79]). Theorem 8.34 is due to Helson [279]. The geometric approach to the invariant subspace problem, making use of the Wold–Kolmogorov decomposition, was first proposed by Helson and Lowdenslager [281].

Section 8.9

There exists a criterion due to Sarason [448] for the existence of nonreducing invariant subspaces for S_μ. This criterion also describes the noncompleteness of the analytic polynomials in the space $L^\infty(\mu)$ equipped with the weak-star topology.

Section 8.11

The problem of characterizing the cyclic vectors of S^* on H^2 was officially posed by Sarason in 1965 at a conference in Lexington, Kentucky. The problem was solved by Douglas, Shapiro and Shields in a famous paper [180], which served as the motivation for much of the work on the backward shift. The concept of pseudocontinuation was developed earlier by Shapiro in [476]. However, before the paper of Douglas, Shapiro and Shields, there were some earlier results in the literature [475] concerning this question of cyclicity for the backward shift operator. In particular, the result proved in Exercise 8.11.12 was obtained in [475]. Several other exercises at the end this section are taken from [180].

As we noticed at the end of this section, if $\mathcal{M}$ is a nontrivial S^*-invariant subspace, then every $f \in \mathcal{M}$ has a bounded-type pseudomeromorphic continuation across $\mathbb{D}$ to $\mathbb{D}_e$. This result has been generalized to other spaces of holomorphic functions. In particular, in [18], a similar description has been obtained for the backward shift invariant subspaces on the Bergman spaces L^p_a, $1 < p < \infty$. See [142, chap. 5] for a discussion on that subject.

9

Analytic reproducing kernel Hilbert spaces

Our main object in this work, i.e. an $\mathcal{H}(b)$ space, is an analytic reproducing kernel Hilbert space. Hence, this chapter is devoted to the study of reproducing kernel spaces from a general point of view. We start by introducing the notion of reproducing kernel in a Hilbert space $\mathcal{H}$ of analytic functions defined on a domain $\Omega \subset \mathbb{C}$. Then we study the multipliers of $\mathcal{H}$, individually and as a collection, which leads to the creation of the commutative Banach algebra $\mathfrak{Mult}(\mathcal{H})$. Among other things, we observe that $\mathfrak{Mult}(\mathcal{H}) \subset \mathcal{H} \cap H^\infty(\Omega)$, and that equality holds for the classic Hardy space $\mathcal{H} = H^2$. We also treat the notion of weak kernel, which will be needed in more advanced topics. At the end, we discuss the analytic reproducing kernel Hilbert spaces $\mathcal{H}$ that are invariant under multiplication by the independent variable. In particular, we study the commutant and the invariant subspaces of the abstract forward shift operator $S_{\mathcal{H}}$. The characterization of equality between the algebra of multipliers of $\mathcal{H}$ and the algebra of analytic and bounded functions on Ω is also expressed using the operator $S_{\mathcal{H}}$.

9.1 The reproducing kernel

In our discussions, we encounter Hilbert spaces whose elements are functions on a set Ω. We say that a Hilbert space $\mathcal{H}$ of functions on Ω is a *reproducing kernel Hilbert space* if it satisfies the following two properties:

(i) for each $z \in \Omega$, the mapping

$$\begin{array}{rccc} \Lambda_z : & \mathcal{H} & \longrightarrow & \mathbb{C} \\ & f & \longmapsto & f(z) \end{array}$$

is a continuous linear functional on $\mathcal{H}$;

(ii) for each $z \in \Omega$, there is an $f_z \in \mathcal{H}$ such that $f_z(z) \neq 0$.

For our applications, the set Ω is a domain, i.e. an open and connected set, in the complex plane, and the functions in $\mathcal{H}$ are analytic on Ω. In this case, $\mathcal{H}$ is said to be an *analytic reproducing kernel Hilbert space*. In the following, $\mathrm{Hol}(\Omega)$ denotes the family of all analytic functions on Ω, and $H^\infty(\Omega)$ represents the subclass consisting of all bounded functions in $\mathrm{Hol}(\Omega)$.

According to the Riesz representation theorem (Theorem 1.24) and the assumption (i) above, for each $z \in \Omega$, there is a unique element $k_z^{\mathcal{H}} \in \mathcal{H}$ such that

$$f(z) = \langle f, k_z^{\mathcal{H}} \rangle_{\mathcal{H}} \qquad (f \in \mathcal{H}). \tag{9.1}$$

The function $k_z^{\mathcal{H}}$ is called the *reproducing kernel* of $\mathcal{H}$ at the point z.

If we consider elements of $\ell^2(\mathbb{N})$ as functions defined on the set $\mathbb{N}$, then $\ell^2(\mathbb{N})$ is a reproducing kernel Hilbert space whose reproducing kernel at point $n \in \mathbb{N}$ is given by $\mathfrak{e}_n = (\delta_{n,k})_{k\geq 1}$. In Section 4.2, we introduced the Hardy space H^2. It turns out that H^2 is an analytic reproducing kernel Hilbert space whose reproducing kernel at point $z \in \mathbb{D}$ is precisely the Cauchy kernel $k_z(w) = (1 - \bar{z}w)^{-1}$; see (4.19).

The defining identity (9.1) yields several elementary properties of $k_z^{\mathcal{H}}$. First of all, by taking $f = k_z^{\mathcal{H}}$, we obtain

$$\|k_z^{\mathcal{H}}\|_{\mathcal{H}}^2 = k_z^{\mathcal{H}}(z) \qquad (z \in \Omega). \tag{9.2}$$

Note that, by Riesz's theorem, the norm of the evaluation functional (9.1) is equal to $\|k_z^{\mathcal{H}}\|$. Considering the assumption (ii) above, we must have

$$k_z^{\mathcal{H}} \neq 0 \qquad (\text{in } \mathcal{H}), \tag{9.3}$$

since otherwise, by (9.1), we would have $f(z) = 0$, for all $f \in \mathcal{H}$, which is absurd. Then (9.2) implies that

$$k_z^{\mathcal{H}}(z) \neq 0 \qquad (z \in \Omega). \tag{9.4}$$

Therefore, a candidate for f_z is $k_z^{\mathcal{H}}$.

By (9.1), we have

$$k_w^{\mathcal{H}}(z) = \langle k_w^{\mathcal{H}}, k_z^{\mathcal{H}} \rangle \qquad \text{and} \qquad k_z^{\mathcal{H}}(w) = \langle k_z^{\mathcal{H}}, k_w^{\mathcal{H}} \rangle \tag{9.5}$$

for each $z, w \in \Omega$. Hence,

$$k_z^{\mathcal{H}}(w) = \overline{k_w^{\mathcal{H}}(z)} \qquad (z, w \in \Omega). \tag{9.6}$$

The representation (9.1) also shows that the linear manifold of all finite linear combinations of $k_z^{\mathcal{H}}$, $z \in \Omega$, is dense in $\mathcal{H}$. In other words, we have

$$\mathrm{Lin}\{k_z^{\mathcal{H}} : z \in \Omega\}^\perp = \{0\}. \tag{9.7}$$

If $\mathcal{E}$ is a closed subset of $\mathcal{H}$, and the sequence $(f_n)_{n\geq 1} \subset \mathcal{E}$ is pointwise convergent, say to f, we cannot claim that the limiting function f belongs to $\mathcal{E}$. In fact, we cannot even claim that f belongs to $\mathcal{H}$. For example, let us consider the Hardy space H^2 and the sequence $f_n(z) = (1 - \bar{\lambda}_n z)^{-1}$, $z \in \mathbb{D}$, where λ_n is any sequence in $\mathbb{D}$ tending to 1. Then, for each fixed $z \in \mathbb{D}$, $f_n(z) \longrightarrow f(z) = (1 - z)^{-1}$, but the function f does not even belong to H^2. However, if the sequence $(f_n)_{n \geq 1}$ is norm-bounded, the desired conclusion holds.

Lemma 9.1 *Let $\mathcal{H}$ be a reproducing kernel Hilbert space on Ω, and let $\mathcal{E}$ be a closed subspace of $\mathcal{H}$. Let $(f_n)_{n \geq 1}$ be a sequence of functions in $\mathcal{E}$. Assume that*

$$\sup_{n \geq 1} \|f_n\|_{\mathcal{H}} < \infty$$

and that the function $f : \Omega \longrightarrow \mathbb{C}$ is such that

$$\lim_{n \to \infty} f_n(z) = f(z)$$

for every $z \in \Omega$. Then $f \in \mathcal{E}$ and

$$\|f\|_{\mathcal{H}} \leq \liminf_{n \to \infty} \|f_n\|_{\mathcal{H}}.$$

Proof Since $(f_n)_{n \geq 1}$ is a bounded sequence in $\mathcal{H}$, there is a subsequence $(f_{n_j})_{j \geq 1}$ of $(f_n)_{n \geq 1}$ that weakly converges to a function $g \in \mathcal{H}$. But, since $\mathcal{E}$ is closed in norm, it is also weakly closed, and thus g in fact belongs to $\mathcal{E}$. Now, weak convergence implies, in particular, that

$$\lim_{j \to \infty} f_{n_j}(z) = \langle f_{n_j}, k_z^{\mathcal{H}} \rangle_{\mathcal{H}} \longrightarrow \langle g, k_z^{\mathcal{H}} \rangle_{\mathcal{H}} = g(z)$$

for every $z \in \Omega$. Hence, by uniqueness of the limit, we deduce that $f = g$, and thus $f \in \mathcal{E}$. Since f_{n_j} weakly converges to f, by (1.30), we deduce that $\|f\|_{\mathcal{H}} \leq \liminf_{n \to \infty} \|f_n\|_{\mathcal{H}}$. □

If $\mathcal{H}$ is a given set, can we put two *essentially* different reproducing kernel Hilbert space structures on it? Suppose that there are two such structures that we put on the ambient space $\mathcal{H}$, and call the corresponding Hilbert spaces by $\mathcal{H}_1$ and $\mathcal{H}_2$. Now, consider the bijective mapping

$$\begin{array}{ccc} \mathcal{H}_1 & \longrightarrow & \mathcal{H}_2 \\ f & \longmapsto & f. \end{array}$$

The closed graph theorem (Corollary 1.18) immediately implies that this mapping is continuous. Hence, there are constants c_1 and c_2 such that

$$c_1 \|f\|_{\mathcal{H}_1} \leq \|f\|_{\mathcal{H}_2} \leq c_2 \|f\|_{\mathcal{H}_1} \tag{9.8}$$

for all $f \in \mathcal{H}$. In other words, the two structures are necessarily equivalent.

Exercises

Exercise 9.1.1 Let $\mathcal{H}$ be a reproducing kernel Hilbert space on Ω, and let $(e_i)_{i\in I}$ be any orthonormal basis for $\mathcal{H}$. Show that

$$k_z^{\mathcal{H}} = \sum_{i\in I} \overline{e_i(z)}\, e_i \qquad (z \in \Omega),$$

where the series is convergent in $\mathcal{H}$.
Hint: Fix $z \in \Omega$. Observe that

$$\sum_{i\in I} |e_i(z)|^2 = \sum_{i\in I} |\langle e_i, k_z^{\mathcal{H}}\rangle_{\mathcal{H}}|^2 = \|k_z^{\mathcal{H}}\|_{\mathcal{H}}^2 < \infty.$$

Hence, $\sum_{i\in I} \overline{e_i(z)}\, e_i$ converges in $\mathcal{H}$. Now, evaluate the identity

$$f = \sum_{i\in I} \langle f, e_i\rangle_{\mathcal{H}} e_i$$

at z.

Exercise 9.1.2 Let $\mathcal{H}$ be a reproducing kernel Hilbert space on Ω, and let $\mathcal{E}$ be a closed subspace of $\mathcal{H}$. Show that $\mathcal{E}$ is a reproducing kernel Hilbert space on Ω and find the relation between $k_z^{\mathcal{E}}$ and $k_z^{\mathcal{H}}$, $z \in \Omega$.
Hint: Use the orthogonal projection from $\mathcal{H}$ onto $\mathcal{E}$.

Exercise 9.1.3 Let $\mathcal{H}_1$ and $\mathcal{H}_2$ be two reproducing kernel Hilbert spaces on Ω, and let $\mathcal{H}_2$ be boundedly contained in $\mathcal{H}_1$, that is, $\mathcal{H}_2 \subset \mathcal{H}_1$ and there exists a positive constant $c > 0$ such that

$$\|f\|_{\mathcal{H}_1} \le c\|f\|_{\mathcal{H}_2} \qquad (f \in \mathcal{H}_2).$$

Find a formula between $k_z^{\mathcal{H}_2}$ and $k_z^{\mathcal{H}_1}$, $z \in \Omega$.
Hint: Use the embedding

$$\begin{array}{rccc} i: & \mathcal{H}_2 & \longrightarrow & \mathcal{H}_1 \\ & f & \longmapsto & f. \end{array}$$

Remark: Note that Exercise 9.1.2 can be considered as a special case of this exercise.

Exercise 9.1.4 Let Θ be an inner function. Show that $\mathcal{H} = \Theta H^2$ is an analytic reproducing kernel Hilbert space on $\mathbb{D}$ whose reproducing kernel is given by

$$k_\lambda^{\mathcal{H}}(z) = \frac{\overline{\Theta(\lambda)}\,\Theta(z)}{1 - \bar{\lambda} z} \qquad (z, \lambda \in \mathbb{D}).$$

Exercise 9.1.5 Let Θ be an inner function. Show that $\mathcal{H} = (\Theta H^2)^\perp = H^2 \ominus \Theta H^2$ is an analytic reproducing kernel Hilbert space on $\mathbb{D}$ whose reproducing kernel is given by

$$k_\lambda^{\mathcal{H}}(z) = \frac{1 - \overline{\Theta(\lambda)}\Theta(z)}{1 - \bar{\lambda} z} \qquad (z, \lambda \in \mathbb{D}).$$

Exercise 9.1.6 Let $\mathcal{H}$ be an analytic reproducing kernel Hilbert space on a domain Ω in $\mathbb{C}$. Given $T \in \mathcal{L}(\mathcal{H})$, its *Berezin transform* B_T is the function B_T defined on Ω by

$$B_T(z) = \frac{\langle T k_z^{\mathcal{H}}, k_z^{\mathcal{H}} \rangle_{\mathcal{H}}}{\|k_z^{\mathcal{H}}\|_{\mathcal{H}}^2} \qquad (z \in \Omega).$$

Show that, if $T, S \in \mathcal{L}(\mathcal{H})$ such that

$$B_T(z) = B_S(z) \qquad (z \in \Omega),$$

then $T = S$.
Hint: Without loss of generality, we can assume that $S = 0$, and that $0 \in \Omega$. Define

$$\begin{array}{rccl} f: & \bar{\Omega} \times \Omega & \longrightarrow & \mathbb{C} \\ & (z, w) & \longmapsto & \langle T k_{\bar{z}}^{\mathcal{H}}, k_w^{\mathcal{H}} \rangle_{\mathcal{H}}. \end{array}$$

Show that f is analytic on $\bar{\Omega} \times \Omega$. By assumption

$$f(\bar{z}, z) = 0 \qquad (z \in \Omega).$$

Then, by considering the Taylor expansion of $f(t\bar{z}, tz)$, where t is real and small, deduce that $f \equiv 0$.

Exercise 9.1.7 Let $A^2(\mathbb{D})$ be the space of analytic functions $f : \mathbb{D} \longrightarrow \mathbb{C}$ satisfying

$$\|f\|_2^2 = \frac{1}{\pi} \iint_{\mathbb{D}} |f(x + iy)|^2 \, dx \, dy < \infty.$$

We easily see that $A^2(\mathbb{D})$ equipped with the inner product

$$\langle f, g \rangle = \frac{1}{\pi} \iint_{\mathbb{D}} f(x + iy) \, \overline{g(x + iy)} \, dx \, dy$$

is a Hilbert space. The space $A^2(\mathbb{D})$ is called the *Bergman space*. Show that $A^2(\mathbb{D})$ is a reproducing kernel Hilbert space whose kernel is given by

$$k_\lambda^{A^2}(z) = \frac{1}{(1 - \bar{\lambda} z)^2}.$$

Exercise 9.1.8 Let $\beta = (\beta_n)_{n\geq 0}$, with $\beta_n > 0$, and $\liminf_{n\to\infty} |\beta_n|^{1/n} \geq 1$. Also, let H^2_β be the space of analytic functions $f : \mathbb{D} \longrightarrow \mathbb{C}$, $f(z) = \sum_{n=0}^{\infty} a_n z^n$, $z \in \mathbb{D}$, satisfying

$$\|f\|_\beta^2 = \sum_{n=0}^{\infty} \beta_n^2 |a_n|^2 < \infty.$$

Show that H^2_β is a reproducing kernel Hilbert space on $\mathbb{D}$, whose reproducing kernel is given by

$$k_\lambda^{H^2_\beta}(z) = \sum_{n=0}^{\infty} \frac{\bar{\lambda}^n}{\beta_n^2} z^n.$$

Hint: Use Exercise 9.1.1.

Exercise 9.1.9 Let $\omega \in L^1(\mathbb{T})$, $w \geq 0$ and $\log \omega \in L^1(\mathbb{T})$. Put $d\mu = w\, dm$ and recall that $H^2(\mu) = \overline{\operatorname{Span}\{z^n : n \geq 0\}}$, where the closure is taken with respect to the norm in $L^2(\mu)$. Show that $H^2(\mu)$ is a reproducing kernel Hilbert space on $\mathbb{D}$ whose reproducing kernel is given by

$$k_\lambda^{H^2(\mu)} = \frac{k_\lambda}{\overline{f(\lambda)} f}$$

or, more explicitly,

$$k_\lambda^{H^2(\mu)}(z) = \frac{1}{1-\bar{\lambda} z} \frac{1}{\overline{f(\lambda)} f(z)},$$

where $f = [w^{1/2}]$ is the outer function with modulus $|w^{1/2}|$ a.e. on $\mathbb{T}$.
Hint: Use Theorem 8.25. Remember that, for each $f_1, f_2 \in H^2$, we have

$$\left\langle \frac{f_1}{f}, \frac{f_2}{f} \right\rangle_{H^2(\mu)} = \langle f_1, f_2 \rangle_{H^2}.$$

9.2 Multipliers

A function φ on Ω is called a *multiplier* for $\mathcal{H}$ if $\varphi f \in \mathcal{H}$ for all $f \in \mathcal{H}$. For each multiplier φ, we define the mapping

$$\begin{array}{rccc} M_\varphi : & \mathcal{H} & \longrightarrow & \mathcal{H} \\ & f & \longmapsto & \varphi f. \end{array}$$

Clearly, this mapping is linear. Moreover, by the closed graph theorem (Corollary 1.18), it is continuous. In other words, the multiplication operator M_φ is an element of $\mathcal{L}(\mathcal{H})$. We now show that each reproducing kernel $k_z^{\mathcal{H}}$ is an eigenvector of M_φ^* with the corresponding eigenvalue $\overline{\varphi(z)}$. Despite its simple proof, this result has many profound applications.

Theorem 9.2 *Let $\mathcal{H}$ be a reproducing kernel Hilbert space on Ω, and let φ be a multiplier of $\mathcal{H}$. Then*

$$M_\varphi^* k_z^{\mathcal{H}} = \overline{\varphi(z)} k_z^{\mathcal{H}} \qquad (z \in \Omega).$$

Proof For each $f \in \mathcal{H}$, we have

$$\langle f, M_\varphi^* k_z^{\mathcal{H}} \rangle = \langle M_\varphi f, k_z^{\mathcal{H}} \rangle = \langle \varphi f, k_z^{\mathcal{H}} \rangle.$$

But, by (9.1),

$$\varphi(z) f(z) = \langle \varphi f, k_z^{\mathcal{H}} \rangle \quad \text{and} \quad f(z) = \langle f, k_z^{\mathcal{H}} \rangle.$$

Hence,

$$\langle f, M_\varphi^* k_z^{\mathcal{H}} \rangle = \varphi(z) f(z) = \varphi(z) \langle f, k_z^{\mathcal{H}} \rangle = \langle f, \overline{\varphi(z)} k_z^{\mathcal{H}} \rangle,$$

which implies that $M_\varphi^* k_z^{\mathcal{H}} = \overline{\varphi(z)} k_z^{\mathcal{H}}$. □

By Theorem 9.2, each $k_z^{\mathcal{H}}$ is an eigenvector of the adjoint of every multiplication operator on $\mathcal{H}$. It is rather surprising that no other operator in $\mathcal{L}(\mathcal{H})$ has this property.

Theorem 9.3 *Let $\mathcal{H}$ be a reproducing kernel Hilbert space on Ω, and let $A \in \mathcal{L}(\mathcal{H})$ be such that, for each $z \in \Omega$, $k_z^{\mathcal{H}}$ is an eigenvector of A^*. Then there is a unique multiplier φ of $\mathcal{H}$ such that*

$$A = M_\varphi.$$

Proof Let φ be the function defined on Ω by the identity

$$A^* k_z^{\mathcal{H}} = \overline{\varphi(z)} k_z^{\mathcal{H}} \qquad (z \in \Omega).$$

According to our assumption, φ is at least a well-defined function on Ω. But more is true. For each $f \in \mathcal{H}$ and $z \in \Omega$, we have

$$\begin{aligned}(Af)(z) &= \langle Af, k_z^{\mathcal{H}} \rangle = \langle f, A^* k_z^{\mathcal{H}} \rangle = \langle f, \overline{\varphi(z)} k_z^{\mathcal{H}} \rangle \\ &= \varphi(z) \langle f, k_z^{\mathcal{H}} \rangle = \varphi(z) f(z).\end{aligned}$$

Therefore, for each $f \in \mathcal{H}$,

$$\varphi f = Af \in \mathcal{H}.$$

This identity means that φ is a multiplier of $\mathcal{H}$ and that $A = M_\varphi$. □

Theorems 9.2 and 9.3 together immediately imply the following characterization of multipliers.

Corollary 9.4 *Let $\mathcal{H}$ be a reproducing kernel Hilbert space on Ω, and let φ be a function on Ω. Then φ is a multiplier of $\mathcal{H}$ if and only if the mapping*

$$k_z^{\mathcal{H}} \longmapsto \overline{\varphi(z)} k_z^{\mathcal{H}},$$

where z runs over Ω, extends to a continuous linear operator in $\mathcal{L}(\mathcal{H})$.

9.3 The Banach algebra $\mathfrak{Mult}(\mathcal{H})$

The space of all multipliers of $\mathcal{H}$ is denoted by $\mathfrak{Mult}(\mathcal{H})$. We saw that, for each multiplier φ, there is a corresponding multiplication operator $M_\varphi \in \mathcal{L}(\mathcal{H})$. In fact, considering the elementary properties

$$M_{\alpha\varphi+\beta\psi} = \alpha M_\varphi + \beta M_\psi$$

and

$$M_\varphi = 0 \iff \varphi = 0,$$

we see that the linear mapping

$$\begin{array}{rcl} \mathfrak{Mult}(\mathcal{H}) & \longrightarrow & \mathcal{L}(\mathcal{H}) \\ \varphi & \longmapsto & M_\varphi \end{array}$$

is bijective. Naively speaking, this means that the above mapping puts a copy of $\mathfrak{Mult}(\mathcal{H})$ inside $\mathcal{L}(\mathcal{H})$. Moreover, we have

$$M_\varphi M_\psi = M_\psi M_\varphi = M_{\varphi\psi}. \tag{9.9}$$

We exploit this bijection to put a topological structure on $\mathfrak{Mult}(\mathcal{H})$. For each $\varphi \in \mathfrak{Mult}(\mathcal{H})$, define the norm

$$\|\varphi\|_{\mathfrak{Mult}(\mathcal{H})} = \|M_\varphi\|_{\mathcal{L}(\mathcal{H})}. \tag{9.10}$$

Lemma 9.5 *The space $(\mathfrak{Mult}(\mathcal{H}), \|\cdot\|_{\mathfrak{Mult}(\mathcal{H})})$, equipped with the pointwise multiplication, is a commutative Banach algebra.*

Proof All the required properties are easy to establish. However, the only fact that we verify is that, if $(\varphi_n)_{n\geq 1}$ is a Cauchy sequence in $\mathfrak{Mult}(\mathcal{H})$, then it is convergent to a multiplier in the norm of $\mathfrak{Mult}(\mathcal{H})$. Hence, assume that $(\varphi_n)_{n\geq 1}$ is a Cauchy sequence in $\mathfrak{Mult}(\mathcal{H})$. By definition, this means that $(M_{\varphi_n})_{n\geq 1}$ is a Cauchy sequence in $\mathcal{L}(\mathcal{H})$. Since $\mathcal{L}(\mathcal{H})$ is complete, there is an operator $A \in \mathcal{L}(\mathcal{H})$ such that

$$\|M_{\varphi_n} - A\|_{\mathcal{L}(\mathcal{H})} \longrightarrow 0 \tag{9.11}$$

as $n \longrightarrow \infty$. Therefore, for each $z \in \Omega$, we have

$$\lim_{n\to\infty} M_{\varphi_n}^* k_z^{\mathcal{H}} = A^* k_z^{\mathcal{H}}.$$

But, according to Theorem 9.2, $M_{\varphi_n}^* k_z^{\mathcal{H}} = \overline{\varphi_n(z)} k_z^{\mathcal{H}}$. Therefore, in the first place, $\varphi(z) = \lim_{n\to\infty} \varphi_n(z)$ exists, and second it fulfills the identity

$$A^* k_z^{\mathcal{H}} = \overline{\varphi(z)} k_z^{\mathcal{H}} \qquad (z \in \mathbb{D}).$$

Theorems 9.2 and 9.3 now imply that $A = M_\varphi$. The relation (9.11) is now rewritten as

$$\|\varphi_n - \varphi\|_{\mathfrak{Mult}(\mathcal{H})} \longrightarrow 0$$

as $n \longrightarrow \infty$. □

According to Theorem 9.2, $\overline{\varphi(z)}$ is an eigenvalue of M_φ^*. We use this property to deduce that the elements of $\mathfrak{Mult}(\mathcal{H})$ are bounded functions on Ω.

Lemma 9.6 *Let $\mathcal{H}$ be a reproducing kernel Hilbert space on Ω, and let $\varphi \in \mathfrak{Mult}(\mathcal{H})$. Then φ is bounded on Ω and, moreover,*

$$\sup_{z\in\Omega} |\varphi(z)| \leq \|\varphi\|_{\mathfrak{Mult}(\mathcal{H})}.$$

Proof Fix $z \in \Omega$. By Theorem 9.2,

$$|\varphi(z)| \, \|k_z^{\mathcal{H}}\|_{\mathcal{H}} = \|\overline{\varphi(z)} k_z^{\mathcal{H}}\|_{\mathcal{H}} = \|M_\varphi^* k_z^{\mathcal{H}}\|_{\mathcal{H}} \leq \|M_\varphi^*\|_{\mathcal{L}(\mathcal{H})} \|k_z^{\mathcal{H}}\|_{\mathcal{H}}.$$

By (9.3), $\|k_z^{\mathcal{H}}\|_{\mathcal{H}} \neq 0$, and thus we can divide both sides by $\|k_z^{\mathcal{H}}\|_{\mathcal{H}}$. To conclude, it remains to remember that, by Theorem 1.30, we have $\|M_\varphi^*\|_{\mathcal{L}(\mathcal{H})} = \|M_\varphi\|_{\mathcal{L}(\mathcal{H})}$. □

If the elements of $\mathcal{H}$ are analytic functions, we can say a bit more about $\mathfrak{Mult}(\mathcal{H})$.

Corollary 9.7 *Let $\mathcal{H}$ be an analytic reproducing kernel Hilbert space on a domain Ω. Then $\mathfrak{Mult}(\mathcal{H}) \subset H^\infty(\Omega)$ and, moreover,*

$$\|\varphi\|_{H^\infty(\Omega)} \leq \|\varphi\|_{\mathfrak{Mult}(\mathcal{H})}.$$

Proof Let $\varphi \in \mathfrak{Mult}(\mathcal{H})$. Fix $z_0 \in \Omega$. Since $M_\varphi k_{z_0}^{\mathcal{H}} = \varphi k_{z_0}^{\mathcal{H}}$, we have

$$\varphi(z) = \frac{(M_\varphi k_{z_0}^{\mathcal{H}})(z)}{k_{z_0}^{\mathcal{H}}(z)} \qquad (z \in \Omega).$$

By (9.4), $k_{z_0}^{\mathcal{H}}(z_0) \neq 0$. Therefore, since $k_{z_0}^{\mathcal{H}}$ and $M_\varphi k_{z_0}^{\mathcal{H}}$ are in $\mathrm{Hol}(\Omega)$, the above identity shows that φ is analytic in a neighborhood of z_0. Hence, φ is analytic on Ω and Lemma 9.6 ensures that $\varphi \in H^\infty(\Omega)$. □

The reproducing kernel Hilbert spaces that we face in practice usually contain the analytic polynomials. In particular, they have the constant function 1 as an element. In this case, we immediately deduce that $\mathfrak{Mult}(\mathcal{H}) \subset \mathcal{H}$. In fact, by Corollary 9.7, in this situation we can even say that

$$\mathfrak{Mult}(\mathcal{H}) \subset \mathcal{H} \cap H^\infty(\Omega). \tag{9.12}$$

A basic example of this situation is the Hardy space H^2 of the unit disk $\mathbb{D}$. In this case, we actually have

$$\mathfrak{Mult}(H^2) = H^\infty. \tag{9.13}$$

In fact, since $H^\infty \subset H^2$, (9.12) implies that $\mathfrak{Mult}(H^2) \subset H^\infty$. The inclusion $H^\infty \subset \mathfrak{Mult}(H^2)$ is trivial.

According to Theorem 1.30, the spectrum of an operator is closely related to the spectrum of its adjoint. Hence, Theorem 9.2 facilitates the detection of a part of the spectrum of a multiplication operator. We recall that, for a function φ, $\mathcal{R}(\varphi)$ is the range of φ, i.e.

$$\mathcal{R}(\varphi) = \{\varphi(z) : z \in \Omega\}$$

and $\mathrm{Clos}_{\mathbb{C}}\,\mathcal{R}(\varphi)$ represents the closure of $\mathcal{R}(\varphi)$ in the complex plane $\mathbb{C}$.

Theorem 9.8 *Let $\mathcal{H} \subset \mathrm{Hol}(\Omega)$ be an analytic reproducing kernel Hilbert space, and let $\varphi \in \mathfrak{Mult}(\mathcal{H})$. Then*

$$\mathrm{Clos}_{\mathbb{C}}\,\mathcal{R}(\varphi) \subset \sigma(M_\varphi).$$

If, furthermore, $\mathfrak{Mult}(\mathcal{H}) = H^\infty(\Omega)$, then

$$\sigma(M_\varphi) = \mathrm{Clos}_{\mathbb{C}}\,\mathcal{R}(\varphi).$$

Proof By Theorem 9.2, $\overline{\varphi(z)} \in \sigma(M_\varphi^*)$, which, by Theorem 1.30, gives $\varphi(z) \in \sigma(M_\varphi)$. Therefore, $\mathrm{Clos}_{\mathbb{C}}\,\mathcal{R}(\varphi) \subset \sigma(M_\varphi)$. If, furthermore, $\mathfrak{Mult}(\mathcal{H}) = H^\infty(\Omega)$ holds, then, for each $\lambda \notin \mathrm{Clos}_{\mathbb{C}}\,\mathcal{R}(\varphi)$, we can consider the bounded function $\psi = 1/(\lambda - \varphi) \in H^\infty(\Omega)$ as a multiplier on $\mathcal{H}$. Clearly, $(\lambda - \varphi)\psi = 1$, which implies that $(\lambda I - M_\varphi)M_\psi = I$. Hence, the operator $\lambda I - M_\varphi$ is invertible and $\lambda \notin \sigma(M_\varphi)$. □

Exercises

Exercise 9.3.1 Let Θ be an inner function. Determine $\mathfrak{Mult}(\Theta H^2)$.
Remark: Compare this result with Theorem 14.40.

Exercise 9.3.2 Let $\mathcal{H}$ be a reproducing kernel Hilbert space on Ω, and let $\varphi \in \mathfrak{Mult}(\mathcal{H})$. Show that

$$\sup_{z \in \Omega} |\varphi(z)| \leq r(M_\varphi), \tag{9.14}$$

where $r(M_\varphi)$ denotes the spectral radius of the operator M_φ (see Section 1.6).
Hint: By Lemma 9.6,

$$\sup_{z \in \Omega} |\varphi^n(z)| \leq \|M_{\varphi^n}\| = \|M_\varphi^n\| \qquad (n \geq 1).$$

Hence,

$$\sup_{z \in \Omega} |\varphi(z)| \leq \|M_\varphi^n\|^{1/n} \qquad (n \geq 1).$$

Let $n \longrightarrow \infty$.

9.4 The weak kernel

Let Ω be a set and let $K : \Omega \times \Omega \longrightarrow \mathbb{C}$ be a function. We say that K is a *kernel* if the following three conditions hold:

(i) $K(z, w) = \overline{K(w, z)}$, for all $z, w \in \Omega$;
(ii) K is positive semidefinite, i.e.

$$\sum_{i,j=1}^{n} \bar{a}_i a_j K(\lambda_i, \lambda_j) \geq 0$$

for all finite sets $\{\lambda_1, \dots, \lambda_n\}$ of points in Ω and all complex numbers $a_1, a_2, \dots, a_n$;
(iii) $K(z, z) \neq 0$, $z \in \Omega$.

A *weak kernel* on Ω is a function K that satisfies conditions (i) and (ii). We can exploit the reproducing kernel of a Hilbert space to produce a kernel in the sense described above.

Lemma 9.9 *Let $\mathcal{H}$ be a reproducing kernel Hilbert space on Ω with the reproducing kernel $k_z^{\mathcal{H}}$. Put*

$$K(z, w) = k_w^{\mathcal{H}}(z) \qquad (z, w \in \Omega).$$

Then K is a kernel on Ω.

Proof First, by (9.6), we surely have $K(z, w) = \overline{K(w, z)}$, for all $z, w \in \Omega$. Second, pick any finite set $\{\lambda_1, \dots, \lambda_n\}$ of points in Ω and any set of complex numbers $a_1, a_2, \dots, a_n$. Then, by (9.5),

$$\begin{aligned}
\sum_{i,j=1}^{n} \bar{a}_i a_j K(\lambda_i, \lambda_j) &= \sum_{i,j=1}^{n} \bar{a}_i a_j k_{\lambda_j}^{\mathcal{H}}(\lambda_i) \\
&= \sum_{i,j=1}^{n} \bar{a}_i a_j \langle k_{\lambda_j}^{\mathcal{H}}, k_{\lambda_i}^{\mathcal{H}} \rangle_{\mathcal{H}} \\
&= \left\langle \sum_{i=1}^{n} a_j k_{\lambda_j}^{\mathcal{H}}, \sum_{i=1}^{n} a_i k_{\lambda_i}^{\mathcal{H}} \right\rangle_{\mathcal{H}} \\
&= \left\| \sum_{i=1}^{n} a_i k_{\lambda_i}^{\mathcal{H}} \right\|_{\mathcal{H}} \geq 0.
\end{aligned}$$

Finally, (9.4) shows that $K(z, z) \neq 0$, for each $z \in \Omega$. □

The kernel introduced in Lemma 9.9 is not the only possible kernel on a reproducing kernel Hilbert space $\mathcal{H}$. The following result reveals the relation between the multipliers of $\mathcal{H}$ and such kernels.

Theorem 9.10 *Let $\mathcal{H}$ be a reproducing kernel Hilbert space on Ω. Fix $\rho > 0$, and let φ be a complex function on Ω. Then the following are equivalent.*

(i) *Function* $\varphi \in \mathfrak{Mult}(\mathcal{H})$ *and* $\|\varphi\|_{\mathfrak{Mult}(\mathcal{H})} \le \rho$.

(ii) *The function*

$$K(z,w) = (\rho^2 - \varphi(z)\overline{\varphi(w)})k_w^{\mathcal{H}}(z) \qquad (z,w \in \Omega)$$

is a weak kernel on Ω.

Proof Fix any finite set $\{\lambda_1, \dots, \lambda_n\}$ of points in Ω and any set of complex numbers $a_1, a_2, \dots, a_n$. Then

$$\begin{aligned}
\sum_{i,j=1}^{n} \bar{a}_i a_j K(\lambda_i, \lambda_j) &= \sum_{i,j=1}^{n} \bar{a}_i a_j (\rho^2 - \varphi(\lambda_i)\overline{\varphi(\lambda_j)}) k_{\lambda_j}^{\mathcal{H}}(\lambda_i) \\
&= \sum_{i,j=1}^{n} \bar{a}_i a_j (\rho^2 - \varphi(\lambda_i)\overline{\varphi(\lambda_j)}) \langle k_{\lambda_j}^{\mathcal{H}}, k_{\lambda_i}^{\mathcal{H}} \rangle_{\mathcal{H}} \\
&= \rho^2 \sum_{i,j=1}^{n} \bar{a}_i a_j \langle k_{\lambda_j}^{\mathcal{H}}, k_{\lambda_i}^{\mathcal{H}} \rangle_{\mathcal{H}} - \sum_{i,j=1}^{n} \bar{a}_i a_j \varphi(\lambda_i)\overline{\varphi(\lambda_j)} \langle k_{\lambda_j}^{\mathcal{H}}, k_{\lambda_i}^{\mathcal{H}} \rangle_{\mathcal{H}} \\
&= \rho^2 \Big\| \sum_{i=1}^{n} a_i k_{\lambda_i}^{\mathcal{H}} \Big\|_{\mathcal{H}}^2 - \Big\| \sum_{i=1}^{n} a_i \overline{\varphi(\lambda_i)} k_{\lambda_i}^{\mathcal{H}} \Big\|_{\mathcal{H}}^2.
\end{aligned}$$

Hence,

$$\sum_{i,j=1}^{n} \bar{a}_i a_j K(\lambda_i, \lambda_j) = \rho^2 \Big\| \sum_{i=1}^{n} a_i k_{\lambda_i}^{\mathcal{H}} \Big\|_{\mathcal{H}}^2 - \Big\| \sum_{i=1}^{n} a_i \overline{\varphi(\lambda_i)} k_{\lambda_i}^{\mathcal{H}} \Big\|_{\mathcal{H}}^2. \tag{9.15}$$

(i) $\Longrightarrow$ (ii) By (9.6), we see that $K(z,w) = \overline{K(w,z)}$, for all $z, w \in \Omega$. By Theorem 9.2, we can write (9.15) as

$$\sum_{i,j=1}^{n} \bar{a}_i a_j K(\lambda_i, \lambda_j) = \rho^2 \|f\|_{\mathcal{H}}^2 - \|M_\varphi^* f\|_{\mathcal{H}}^2,$$

where

$$f = \sum_{i=1}^{n} a_i k_{\lambda_i}^{\mathcal{H}}.$$

Hence, by assumption,

$$\sum_{i,j=1}^{n} \bar{a}_i a_j K(\lambda_i, \lambda_j) \ge (\rho^2 - \|M_\varphi^*\|_{\mathcal{L}(\mathcal{H})}) \|f\|_{\mathcal{H}}^2 \ge 0.$$

(ii) $\Longrightarrow$ (i) By assumption and (9.15), we have

$$\Big\| \sum_{i=1}^{n} a_i \overline{\varphi(\lambda_i)} k_{\lambda_i} \Big\|_{\mathcal{H}} \le \rho \Big\| \sum_{i=1}^{n} a_i k_{\lambda_i}^{\mathcal{H}} \Big\|_{\mathcal{H}}$$

for all possible choices of a_i and λ_i. This means that the mapping

$$k_\lambda \longmapsto \overline{\varphi(\lambda)} k_\lambda^{\mathcal{H}},$$

where λ runs over Ω, extends to a continuous linear operator in $\mathcal{L}(\mathcal{H})$ and, moreover, the norm of this operator is at most ρ. Hence, by Corollary 9.4, φ is a multiplier of $\mathcal{H}$ whose norm is at most ρ. □

In a reproducing kernel Hilbert space $\mathcal{H}$, we can construct a family of weak kernels parameterized by the unit ball of $\mathcal{H}$.

Theorem 9.11 *Let $\mathcal{H}$ be a reproducing kernel Hilbert space on Ω and $f \in \mathcal{H}$. Denote by $K(z,w) = k_w^{\mathcal{H}}(z)$ and $\widetilde{K}(z,w) = K(z,w) - \overline{f(w)} f(z)$. Then the following are equivalent:*

(i) *we have $\|f\|_{\mathcal{H}} \le 1$;*
(ii) *the function $\widetilde{K}$ is a weak kernel on Ω.*

Proof Note that the function $\widetilde{K}$ is positive semidefinite if and only if

$$\sum_{i,j=1}^{n} \bar{a_i}\, a_j \overline{f(\lambda_j)} f(\lambda_i) \le \sum_{i,j=1}^{n} \bar{a_i}\, a_j K(\lambda_i, \lambda_j),$$

for all finite sets $\{\lambda_1, \dots, \lambda_n\}$ of points in Ω and all complex numbers $a_1, a_2, \dots, a_n$. But, on the one hand, we have

$$\begin{aligned} \sum_{i,j=1}^{n} \bar{a_i}\, a_j \overline{f(\lambda_j)} f(\lambda_i) &= \left| \sum_{i=1}^{n} \bar{a_i} f(\lambda_i) \right|^2 \\ &= \left| \sum_{i=1}^{n} \bar{a_i} \langle f, k_{\lambda_i}^{\mathcal{H}} \rangle_{\mathcal{H}} \right|^2 \\ &= \left| \left\langle f, \sum_{i=1}^{n} a_i k_{\lambda_i}^{\mathcal{H}} \right\rangle_{\mathcal{H}} \right|^2 \end{aligned}$$

and, on the other, we have

$$\begin{aligned} \sum_{i,j=1}^{n} \bar{a_i}\, a_j K(\lambda_i, \lambda_j) &= \sum_{i,j=1}^{n} \bar{a_i}\, a_j k_{\lambda_j}^{\mathcal{H}}(\lambda_i) \\ &= \sum_{i,j=1}^{n} \bar{a_i}\, a_j \langle k_{\lambda_j}^{\mathcal{H}}, k_{\lambda_i}^{\mathcal{H}} \rangle_{\mathcal{H}} \\ &= \left\| \sum_{i=1}^{n} a_i k_{\lambda_i}^{\mathcal{H}} \right\|_{\mathcal{H}}^2. \end{aligned}$$

Hence, the function $\widetilde{K}$ is positive semidefinite if and only if

$$\left| \left\langle f, \sum_{i=1}^{n} a_i k_{\lambda_i}^{\mathcal{H}} \right\rangle_{\mathcal{H}} \right| \le \left\| \sum_{i=1}^{n} a_i k_{\lambda_i}^{\mathcal{H}} \right\|_{\mathcal{H}},$$

which is equivalent to $\|f\|_{\mathcal{H}} \le 1$. □

Lemma 9.12 *Let $K_1, K_2 : \Omega \times \Omega \longrightarrow \mathbb{C}$ be two (weak) kernels. Then $K_1 K_2$ is also a (weak) kernel.*

Proof Assume that K_1 and K_2 are two weak kernels. Then, the matrices $(K_\ell(\lambda_i, \lambda_j))_{1 \le i,j \le n}$ are positive semidefinite for $\ell = 1, 2$ and for all finite sets $\{\lambda_1, \dots, \lambda_n\}$ of points in Ω. By the Schur product theorem (see Section 2.4), we thus obtain that the matrix $(K_1(\lambda_i, \lambda_j) K_2(\lambda_i, \lambda_j))_{1 \le i,j \le n}$ is positive semidefinite. That proves that $K_1 K_2$ is a weak kernel. Now it is clear that, if K_1 and K_2 are two kernels, then $(K_1 K_2)(z, z) = K_1(z, z) K_2(z, z) \neq 0$ for any $z \in \Omega$, and we conclude that $K_1 K_2$ is also a kernel. □

Lemma 9.13 *Let $b : \mathbb{D} \longrightarrow \mathbb{D}$. Then the function*

$$K(z, w) = \frac{1}{1 - \overline{b(w)} b(z)}$$

is a kernel on $\mathbb{D} \times \mathbb{D}$.

Proof Note that $K(z, w) = k_{b(w)}(b(z))$, where $k_\lambda = k_\lambda^{H^2}$ is the kernel of the classic Hardy space H^2 of the unit disk. Hence,

$$\sum_{i,j=1}^{n} \bar{a_i}\, a_j K(\lambda_i, \lambda_j) = \left\| \sum_{i=1}^{n} a_i k_{b(\lambda_i)} \right\|_2^2 \ge 0,$$

for all finite sets $\{\lambda_1, \dots, \lambda_n\}$ of points in $\mathbb{D}$ and all complex numbers $a_1, a_2, \dots, a_n$. □

Exercises

Exercise 9.4.1 Let $\mathcal{H}$ be an analytic reproducing kernel Hilbert space on Ω with the kernel k_z, and let $\varphi \in \mathfrak{Mult}(\mathcal{H})$, with φ nonconstant, such that $\|\varphi\|_{\mathfrak{Mult}(\mathcal{H})} \le \rho$. Show that the function

$$K(z, w) = (\rho^2 - \varphi(z)\overline{\varphi(w)}) k_w(z)$$

is a kernel on Ω.
Hint: Use the maximum principle, Corollary 9.7 and Theorem 9.10.

Exercise 9.4.2 Let $b \in H^\infty$, $\|b\|_\infty \le 1$, b nonconstant. Show that

$$K(z, w) = \frac{1 - \overline{b(w)} b(z)}{1 - \bar{w} z}$$

is a kernel on $\mathbb{D}$.
Hint: Note that $b \in \mathfrak{Mult}(H^2) = H^\infty$ and use Exercise 9.4.1.

9.5 The abstract forward shift operator $S_{\mathcal{H}}$

In the rest of this chapter, we will consider analytic reproducing kernel Hilbert spaces $\mathcal{H}$ on the open unit disk $\mathbb{D}$. We make two additional assumptions:

(H1) The function

$$\begin{array}{rccc} \chi_1: & \mathbb{D} & \longrightarrow & \mathbb{C} \\ & z & \longmapsto & z \end{array}$$

is a multiplier of $\mathcal{H}$. We will denote the corresponding multiplication operator by $S_{\mathcal{H}}$, i.e.

$$\begin{array}{rccc} S_{\mathcal{H}}: & \mathcal{H} & \longrightarrow & \mathcal{H} \\ & f & \longmapsto & \chi_1 f. \end{array}$$

In other words, $S_{\mathcal{H}} = M_{\chi_1} = M_z$.

(H2) For each $\lambda \in \mathbb{D}$, we assume that $\ker(S_{\mathcal{H}}^* - \bar{\lambda} I) = \mathbb{C} k_\lambda^{\mathcal{H}}$.

According to Lemma 8.6, we see that the Hardy space H^2 satisfies the assumptions (H1) and (H2). Note that, according to Theorem 9.2, the inclusion

$$\mathbb{C} k_\lambda^{\mathcal{H}} \subset \ker(S_{\mathcal{H}}^* - \bar{\lambda} I)$$

is valid in any analytic reproducing kernel Hilbert space on $\mathbb{D}$ that satisfies (H1). Therefore, the assumption (H2) is equivalent to saying that $k_\lambda^{\mathcal{H}}$ is a simple eigenfunction of the operator $S_{\mathcal{H}}^*$. We also highlight that, if an analytic reproducing kernel Hilbert space $\mathcal{H}$ satisfies (H1), then, since the set of multipliers is an algebra, $\mathfrak{Mult}(\mathcal{H})$ contains the analytic polynomials. In fact, if p is an analytic polynomial, we easily see that

$$p(S_{\mathcal{H}}) = M_p.$$

In practice, verifying the assumption (H2) can be difficult. Hence, we introduce a third assumption that implies (H2).

(H3) For each $\lambda \in \mathbb{D}$ and $f \in \mathcal{H}$, the function $Q_\lambda f$ defined by

$$(Q_\lambda f)(z) = \frac{f(z) - f(\lambda)}{z - \lambda}$$

belongs to $\mathcal{H}$.

Using the closed graph theorem (Corollary 1.18), we easily see that the linear mapping

$$\begin{array}{rccc} Q_{\lambda,\mathcal{H}}: & \mathcal{H} & \longrightarrow & \mathcal{H} \\ & f & \longmapsto & Q_\lambda f \end{array}$$

is continuous. For simplicity, we write $X_{\mathcal{H}} = Q_{0,\mathcal{H}}$. More explicitly, $X_{\mathcal{H}}$ is the backward shift, which acts on $\mathcal{H}$ as

$$(X_{\mathcal{H}}f)(z) = \frac{f(z) - f(0)}{z}.$$

If $\mathcal{H} = H^2$, then we know that $S_{\mathcal{H}}^* = X_{\mathcal{H}}$. However, one should keep in mind that, in an arbitrary reproducing kernel Hilbert space, this is far from being true.

Lemma 9.14 *Let $\mathcal{H}$ be an analytic reproducing kernel Hilbert space on $\mathbb{D}$ satisfying (H1) and (H3). Let $f \in \mathcal{H}$, and let $\lambda \in \mathbb{D}$. Then $f \perp k_\lambda^{\mathcal{H}}$ if and only if there is an element $g \in \mathcal{H}$ such that $f(z) = (z - \lambda)g(z)$, $z \in \mathbb{D}$.*

Proof If $f(z) = (z - \lambda)g(z)$, $z \in \mathbb{D}$, for some $g \in \mathcal{H}$, then $f(\lambda) = 0$, which, in light of (9.1), means that $f \perp k_\lambda^{\mathcal{H}}$. For the other direction, fix $f \in \mathcal{H}$ with $f \perp k_\lambda^{\mathcal{H}}$. This assumption means that $f(\lambda) = 0$, and thus the function g, defined by

$$g(z) = \frac{f(z)}{z - \lambda} \qquad (z \in \mathbb{D}),$$

is precisely the function $g = Q_\lambda f$. By (H3), g belongs to $\mathcal{H}$. Therefore, $f(z) = (z - \lambda)g(z)$, with $g \in \mathcal{H}$. □

The above lemma plays an essential role in the proof of the following result. Note that the relation $f(z) = (z - \lambda)g(z)$, $z \in \mathbb{D}$, can be equivalently written as $f = (S_{\mathcal{H}} - \lambda I)g$. Hence, Lemma 9.14 can be stated as

$$\mathcal{R}(S_{\mathcal{H}} - \lambda I) = (\mathbb{C}k_\lambda^{\mathcal{H}})^\perp. \tag{9.16}$$

Theorem 9.15 *Let $\mathcal{H}$ be an analytic reproducing kernel Hilbert space on $\mathbb{D}$ satisfying (H1) and (H3). Then $\mathcal{H}$ satisfies (H2), that is,*

$$\ker(S_{\mathcal{H}}^* - \bar{\lambda} I) = \mathbb{C}k_\lambda^{\mathcal{H}}$$

for every $\lambda \in \mathbb{D}$.

Proof By (9.16),

$$\ker(S_{\mathcal{H}}^* - \bar{\lambda} I) = (\mathcal{R}(S_{\mathcal{H}} - \lambda I))^\perp = \mathbb{C}k_\lambda^{\mathcal{H}}.$$ □

Exercises

Exercise 9.5.1 Let Θ be a singular inner function. Show that $\mathcal{H} = \Theta H^2$ is an analytic reproducing kernel Hilbert space that satisfies (H1) and (H2) but not (H3). What happens if Θ is not singular and has a zero at some point $\lambda \in \mathbb{D}$?

Hint: Use Theorem 8.36. See also Exercise 9.1.4 for the reproducing kernels of $\mathcal{H}$. Note that

$$S_{\mathcal{H}}^* f = (\Theta S^* \bar{\Theta}) f \qquad (f \in \Theta H^2).$$

Exercise 9.5.2 Let Θ be an inner function. Show that $\mathcal{H} = (\Theta H^2)^\perp = H^2 \ominus \Theta H^2$ is a reproducing kernel Hilbert space that does not satisfy (H1). Hint: Use Theorem 8.36.

Exercise 9.5.3 Let $\beta = (\beta_n)_{n \geq 0}$, with $\beta_n > 0$ and $\liminf_{n\to\infty} |\beta_n|^{1/n} \geq 1$, and let H^2_β be the analytic reproducing kernel Hilbert space introduced in Exercise 9.1.8. Show that H^2_β satisfies (H1) if and only if

$$\sup_{n \geq 0} \frac{\beta_{n+1}}{\beta_n} < \infty.$$

9.6 The commutant of $S_{\mathcal{H}}$

We recall that the commutant of $A \in \mathcal{L}(\mathcal{H})$ is the set of all operators $B \in \mathcal{L}(\mathcal{H})$ such that $AB = BA$. The following result gives a characterization of the commutant of $S_{\mathcal{H}}$.

Theorem 9.16 *Let $\mathcal{H}$ be an analytic reproducing kernel Hilbert space on $\mathbb{D}$ satisfying (H1) and (H2). Let $A \in \mathcal{L}(\mathcal{H})$. Then the following are equivalent:*

(i) $AS_{\mathcal{H}} = S_{\mathcal{H}}A$;
(ii) *there is a multiplier $\varphi \in \mathfrak{Mult}(\mathcal{H})$ such that $A = M_\varphi$.*

Proof (i) $\Longrightarrow$ (ii) By Theorem 9.2 and the fact that $S_{\mathcal{H}} = M_z$, we have $S_{\mathcal{H}}^* k_\lambda^{\mathcal{H}} = \bar{\lambda} k_\lambda^{\mathcal{H}}$, for each $\lambda \in \mathbb{D}$. Therefore, using the commutant relation $AS_{\mathcal{H}} = S_{\mathcal{H}}A$, we get

$$S_{\mathcal{H}}^* A^* k_\lambda^{\mathcal{H}} = A^* S_{\mathcal{H}}^* k_\lambda^{\mathcal{H}} = \bar{\lambda} A^* k_\lambda^{\mathcal{H}}.$$

In other words, we have $A^* k_\lambda^{\mathcal{H}} \in \ker(S_{\mathcal{H}}^* - \bar{\lambda} I)$. Hence, by (H2), we must have

$$A^* k_\lambda^{\mathcal{H}} = \alpha k_\lambda^{\mathcal{H}},$$

where α is an appropriate constant. But this means that, for every $\lambda \in \mathbb{D}$, the reproducing kernel $k_\lambda^{\mathcal{H}}$ is an eigenvector of A^*. Knowing this property, Theorem 9.3 now ensures that there is $\varphi \in \mathfrak{Mult}(\mathcal{H})$ such that $A = M_\varphi$.

(ii) $\Longrightarrow$ (i) Since $S_{\mathcal{H}} = M_z$, in light of (9.9), this is trivial. □

Corollary 9.17 *Let $\mathcal{H}$ be an analytic reproducing kernel Hilbert space on $\mathbb{D}$ satisfying (H1) and (H2). Relative to the norm topology, the set*

$$\mathcal{F} = \{M_\varphi : \varphi \in \mathfrak{Mult}(\mathcal{H})\}$$

is a closed subspace of $\mathcal{L}(\mathcal{H})$. Moreover, if $\mathfrak{Mult}(\mathcal{H}) = H^\infty(\mathbb{D})$, then there is a universal constant $c > 0$ such that

$$\|\varphi\|_\infty \leq \|M_\varphi\| \leq c\|\varphi\|_\infty, \tag{9.17}$$

for every $\varphi \in H^\infty(\mathbb{D})$.

Proof That $\mathcal{F}$ is a linear manifold is rather obvious. Let $(\varphi_n)_{n\geq 1}$ be a sequence in $\mathfrak{Mult}(\mathcal{H})$ such that M_{φ_n} converges to A in the norm topology of $\mathcal{L}(\mathcal{H})$. For each $n \geq 1$, we trivially have

$$M_{\varphi_n} S_{\mathcal{H}} = S_{\mathcal{H}} M_{\varphi_n}.$$

Let $n \longrightarrow \infty$ to get $AS_{\mathcal{H}} = S_{\mathcal{H}}A$. But now Theorem 9.16 implies that there exists a multiplier $\varphi \in \mathfrak{Mult}(\mathcal{H})$ such that $A = M_\varphi$. Hence, $\mathcal{F}$ is a closed subspace of $\mathcal{L}(\mathcal{H})$.

To prove the second assertion of the theorem, consider the well-defined mapping

$$\begin{array}{rccc} T: & H^\infty(\mathbb{D}) & \longrightarrow & \mathcal{F} \\ & \varphi & \longmapsto & M_\varphi. \end{array}$$

To prove that the linear map T is continuous, we use the closed graph theorem (Corollary 1.18). Hence, let $(\varphi_n)_{n\geq 1}$ be a sequence in $H^\infty(\mathbb{D})$ such that $(\varphi_n)_{n\geq 1}$ converges to φ in $H^\infty(\mathbb{D})$ and $(M_{\varphi_n})_{n\geq 1}$ converges to M_ψ in $\mathcal{L}(\mathcal{H})$. The assumption $M_{\varphi_n} \longrightarrow M_\psi$ implies that

$$\lim_{n\to\infty} \|\varphi_n f - \psi f\|_{\mathcal{H}} = 0 \qquad (f \in \mathcal{H}).$$

Since functional evaluations are continuous, we have, on the one hand,

$$\lim_{n\to\infty} (\varphi_n f)(z) = \psi(z) f(z)$$

and, on the other,

$$\lim_{n\to\infty} \varphi_n(z) = \varphi(z).$$

Thus, we deduce that $\varphi = \psi$. In other words, $M_\varphi = M_\psi$ and thus the closed graph theorem (Corollary 1.18) ensures that T is continuous. The continuity means that there is a constant $c > 0$ such that

$$\|M_\varphi\| \leq c\|\varphi\|_\infty \qquad (\varphi \in H^\infty(\mathbb{D})). \tag{9.18}$$

The other inequality in (9.17) follows from Lemma 9.6. □

9.7 When do we have $\mathfrak{Mult}(\mathcal{H}) = H^\infty$?

In Corollary 9.7, we showed that $\mathfrak{Mult}(\mathcal{H}) \subset H^\infty$. One may naturally wonder when equality holds. In the following, we establish a characterization of this situation. We recall that an operator $A \in \mathcal{L}(\mathcal{H})$ is said to be "polynomially bounded" if there is a constant $c > 0$ such that

$$\|p(A)\|_{\mathcal{L}(\mathcal{H})} \le c\|p\|_\infty$$

for all analytic polynomials p. The quantity $\|p\|_\infty$ refers to the maximum of $|p|$ on the closed unit disk.

Theorem 9.18 *Let $\mathcal{H}$ be an analytic reproducing kernel Hilbert space on $\mathbb{D}$ satisfying (H1) and (H2). The following assertions are equivalent:*

(i) *$S_\mathcal{H}$ is polynomially bounded;*
(ii) $\mathfrak{Mult}(\mathcal{H}) = H^\infty(\mathbb{D})$.

Moreover, if the constant c is such that

$$\|p(S_\mathcal{H})\|_{\mathcal{L}(\mathcal{H})} = \|M_p\|_{\mathcal{L}(\mathcal{H})} = \|p\|_{\mathfrak{Mult}(\mathcal{H})} \le c\|p\|_\infty,$$

where p is any analytic polynomial, then for each $\varphi \in H^\infty(\mathbb{D})$ we also have

$$\|M_\varphi\|_{\mathcal{L}(\mathcal{H})} = \|\varphi\|_{\mathfrak{Mult}(\mathcal{H})} \le c\|\varphi\|_\infty.$$

Proof (i) $\Longrightarrow$ (ii) Assume that there exists a positive constant c such that

$$\|p(S_\mathcal{H})\|_{\mathcal{L}(\mathcal{H})} \le c\|p\|_\infty$$

for every analytic polynomial p. Therefore, for every function f in $\mathcal{H}$, we have

$$\|pf\|_\mathcal{H} = \|p(S_\mathcal{H})f\|_\mathcal{H} \le c\|f\|_\mathcal{H}\|p\|_\infty. \tag{9.19}$$

Fix $f \in \mathcal{H}$. Let φ be any function in $H^\infty(\mathbb{D})$ and let $(p_n)_{n\ge1}$ be the sequence of Fejér means of the power series of φ, that is,

$$p_n = \frac{1}{n+1}(s_0 + s_1 + \cdots + s_n), \tag{9.20}$$

with $s_j = \sum_{k=0}^{j} \hat{\varphi}(k)\chi_k$. Then, recall that the sequence $(p_n)_{n\ge1}$ is uniformly bounded and converges to φ in the weak-star topology of $L^\infty(\mathbb{T})$. More precisely, we have $\|p_n\|_\infty \le \|\varphi\|_\infty$ and

$$\lim_{n\to\infty} \frac{1}{2\pi}\int_0^{2\pi} p_n(e^{i\theta})h(e^{i\theta})\,d\theta = \frac{1}{2\pi}\int_0^{2\pi} \varphi(e^{i\theta})h(e^{i\theta})\,d\theta, \tag{9.21}$$

for every $h \in L^1(\mathbb{T})$. Hence, by (9.19),

$$\|p_n f\|_\mathcal{H} \le c\|f\|_\mathcal{H}\|\varphi\|_\infty \qquad (n \ge 1).$$

Moreover, applying (9.21) to $h(e^{i\theta}) = (1 - ze^{-i\theta})^{-1}$, $z \in \mathbb{D}$, gives

$$\lim_{n\to\infty} p_n(z) = \varphi(z).$$

Now, Lemma 9.1 implies that $\varphi f \in \mathcal{H}$ and

$$\|\varphi f\|_{\mathcal{H}} \le c\|f\|_{\mathcal{H}}\|\varphi\|_\infty.$$

This means that $\varphi \in \mathfrak{Mult}(\mathcal{H})$ and thus $H^\infty(\mathbb{D}) \subset \mathfrak{Mult}(\mathcal{H})$. The inclusion $\mathfrak{Mult}(\mathcal{H}) \subset H^\infty(\mathbb{D})$ was shown in Corollary 9.7.

(ii) $\Longrightarrow$ (i) Conversely, assume that $\mathfrak{Mult}(\mathcal{H}) = H^\infty(\mathbb{D})$. Then, by Corollary 9.17, there exists a constant $c > 0$ such that

$$\|\varphi\|_\infty \le \|M_\varphi\|_{\mathcal{L}(\mathcal{H})} \le c\|\varphi\|_\infty \qquad (\varphi \in H^\infty(\mathbb{D})).$$

In particular, for every polynomial p, we have

$$\|M_p\|_{\mathcal{L}(\mathcal{H})} \le c\|p\|_\infty,$$

and since $p(S_\mathcal{H}) = M_p$, we get the result. □

Exercises

Exercise 9.7.1 Let Θ be an inner function and let $\mathcal{H} = \Theta H^2$. Show that

$$\|p(S_\mathcal{H})\| = \|p\|_\infty$$

for any analytic polynomial p.

Exercise 9.7.2 Let $\mathcal{H}$ be an analytic reproducing kernel Hilbert space on $\mathbb{D}$ satisfying (H1) and (H2). Suppose furthermore that $S_\mathcal{H}$ is a contraction. Show that $\mathfrak{Mult}(\mathcal{H}) = H^\infty(\mathbb{D})$. Moreover, show that

$$\|\varphi\|_{\mathfrak{Mult}(\mathcal{H})} = \|\varphi\|_\infty$$

for each $\varphi \in H^\infty(\mathbb{D})$.
Hint: Use the von Neumann inequality (see Exercise 2.5.1) and Corollary 9.7 to show that

$$\|p(S_\mathcal{H})\| = \|p\|_\infty$$

for any analytic polynomial p. Then apply Theorem 9.18.

9.8 Invariant subspaces of $S_{\mathcal{H}}$

The proof of the following result has the same flavor as the proof of Theorem 9.18. It says that the $S_{\mathcal{H}}$-closed invariant subspaces are also invariant under all multiplication operators.

Theorem 9.19 *Let $\mathcal{H}$ be an analytic reproducing kernel Hilbert space on $\mathbb{D}$ satisfying (H1) and (H2). Suppose that $\mathfrak{Mult}(\mathcal{H}) = H^\infty(\mathbb{D})$. Let $\mathcal{E}$ be a closed subspace of $\mathcal{H}$ that is invariant under $S_{\mathcal{H}}$. Then, for every $\varphi \in H^\infty(\mathbb{D})$, the subspace $\mathcal{E}$ is also invariant under M_φ.*

Proof Let p be a polynomial. Then $\mathcal{E}$ is clearly invariant under $p(S_{\mathcal{H}}) = M_p$. Now let $\varphi \in H^\infty$ and let $(p_n)_{n\geq 1}$ be the sequence of Fejér means of the power series of φ. Since $\mathfrak{Mult}(\mathcal{H}) = H^\infty(\mathbb{D})$, according to (9.17), we have

$$\|M_{p_n} f\|_{\mathcal{H}} \leq \|M_{p_n}\|_{\mathcal{L}(\mathcal{H})} \|f\|_{\mathcal{H}} \leq c\|p_n\|_\infty \|f\|_{\mathcal{H}} \leq c\|\varphi\|_\infty \|f\|_{\mathcal{H}} \qquad (f \in \mathcal{E}).$$

Therefore, the sequence $(M_{p_n} f)_{n\geq 1}$ is bounded in the norm of $\mathcal{H}$. Moreover, the weak convergence of $(p_n)_{n\geq 1}$ to φ gives

$$\lim_{n\to\infty} (M_{p_n} f)(z) = (M_\varphi f)(z) \qquad (z \in \mathbb{D}).$$

Then we apply Lemma 9.1 to deduce that $M_\varphi f \in \mathcal{E}$. □

Among the lattice of closed invariant subspaces of a Hilbert space operator $A \in \mathcal{L}(\mathcal{H})$, one often distinguishes those which are also invariant under all operators that commute with A. In this context, we say that a closed subspace $\mathcal{E}$ of $\mathcal{H}$ is *hyper-invariant* under A if $B\mathcal{E} \subset \mathcal{E}$ for every bounded operator B on $\mathcal{H}$ such that $AB = BA$. Clearly, a hyper-invariant subspace is always an invariant subspace. It is amazing that the converse is true for the operator $S_{\mathcal{H}}$.

Corollary 9.20 *Let $\mathcal{H}$ be an analytic reproducing kernel Hilbert space on $\mathbb{D}$ satisfying (H1) and (H2). Suppose also that $\mathfrak{Mult}(\mathcal{H}) = H^\infty(\mathbb{D})$. Let $\mathcal{E}$ be a closed subspace of $\mathcal{H}$ that is invariant under $S_{\mathcal{H}}$. Then $\mathcal{E}$ is a hyper-invariant subspace for $S_{\mathcal{H}}$.*

Proof Let $A \in \mathcal{L}(\mathcal{H})$ such that $AS_{\mathcal{H}} = S_{\mathcal{H}}A$. By Theorem 9.16, there is a multiplier φ such that $A = M_\varphi$. Since $\mathfrak{Mult}(\mathcal{H}) = H^\infty(\mathbb{D})$, Theorem 9.19 implies that $\mathcal{E}$ is invariant under A. □

Notes on Chapter 9

The theory of reproducing kernels is a significant and powerful tool in many fields of mathematical analysis. In this chapter, we focused on analytic reproducing kernel Hilbert spaces and developed only a few notions that will be useful for us in the study of $\mathcal{H}(b)$ spaces. For a complete account of this beautiful theory, we refer to the seminal paper of Aronszajn [45]. See [20] or [6, chap. 2] for a good illustration of the variety of fields where the reproducing kernels arise.

Section 9.1

The early signs of the theory of reproducing kernels appeared at the beginning of the twentieth century with the pioneering work of Zaremba [567], Bochner [104] and Szegő [510]. In particular, Zaremba was the first to highlight the existence of a reproducing kernel for a space of functions. The family he considered was the space of harmonic functions on a complex domain that are elements of L^2. Then the theory of reproducing kernel functions was decisively influenced by the work of Bergman [93], who introduced and intensively studied what is now called the *Bergman kernel*. The abstract theory of reproducing kernels goes back to a famous paper [45] of Aronszajn published in 1950. He investigated in depth the general theory of reproducing kernel Hilbert spaces. A more systematic approach, in spaces of harmonic and holomorphic functions, was developed by Bergman – see [92]. The identity proved in Exercise 9.1.1 is due to him. The result of Exercise 9.1.9 is from Szegő [509].

Section 9.2

Theorems 9.2 and 9.3 are due to Shields and Wallen [483] and Halmos [253]. They are also quoted in the thesis of G. D. Taylor [519].

Section 9.3

Lemma 9.6 is also valid for functional Banach spaces (that is, Banach spaces of functions defined on Ω such that the point evaluation functionals are bounded). In the context of Hilbert spaces, it was proved in 1960 by Shields, Wallen and Shapiro, but was not published. It then appeared in [483] and [253].

Section 9.4

The theory of kernels as positive definite functions was initiated by Mercer [360]. As we saw in Lemma 9.9, every reproducing kernel Hilbert space on Ω gives rise to a kernel on Ω. Conversely, one can start with a positive definite kernel and construct a reproducing kernel Hilbert space that has the given kernel as its reproducing kernel. This one-to-one correspondence between reproducing kernel Hilbert spaces and positive definite kernels was discovered by Moore [364]. The positive definite kernels occur naturally in an enormous variety of contexts, e.g. Fourier analysis, probability theory, operator theory, complex function theory, moment problems, integral equations, boundary value problems, orthogonal polynomials, dissipative linear systems, harmonic analysis, information theory and statistical learning theory.

Section 9.6

Theorem 9.16 is due to Shields and Wallen [483] and Halmos [253]. See also [519].

Sections 9.7 and 9.8

The idea of Theorems 9.18 and 9.19 comes from [451], where the results are proved in the special case where $\mathcal{H} = \mathcal{H}(b)$ is a de Branges–Rovnyak space.

10

Bases in Banach spaces

In the framework of Hilbert spaces, an orthonormal basis quite often appears naturally. Nevertheless, even though we know that a countable orthonormal basis always exists in each separable Hilbert space, explicitly finding one could be a difficult problem. In this situation, we wonder what would happen if we deviate slightly from orthonormality. This course leads to the notion of Riesz basis, which makes sense even in the context of Banach spaces. In this chapter, we introduce minimal sequences, asymptotically orthonormal sequences, Bessel sequences, Schauder basis, unconditional basis and Riesz basis and, moreover, we discuss several useful interconnections between these properties. We also explore an interesting connection with an abstract interpolation problem, the so-called Feichtinger conjecture.

10.1 Minimal sequences

Let $\mathfrak{X} = (x_n)_{n\geq 1}$ be a sequence in a Banach space $\mathcal{X}$. Then $\mathfrak{X}$ is said to be

(i) *complete* in $\mathcal{X}$ if

$$\text{Span}\{x_n : n \geq 1\} = \mathcal{X};$$

(ii) *finitely linearly independent* if any finite subsequence of $\mathfrak{X}$ is linearly independent, that is, for any finitely supported sequence $(\alpha_n)_{n\geq 1} \subset \mathbb{C}$, we have

$$\sum_{n\geq 1} \alpha_n x_n = 0 \quad \Longrightarrow \quad \alpha_n = 0 \qquad (n \geq 1);$$

(iii) *w-topologically linearly independent* if, for any sequence $(\alpha_n)_{n\geq 1} \subset \mathbb{C}$, we have

$$\lim_{N\to\infty} \left\| \sum_{n\geq 1}^{N} \alpha_n x_n \right\|_{\mathcal{X}} = 0 \quad \Longrightarrow \quad \alpha_n = 0 \qquad (n \geq 1);$$

(iv) *minimal* if, for any $n \geq 1$,

$$x_n \notin \text{Span}\{x_k : k \geq 1,\ k \neq n\};$$

(v) *uniformly minimal* if $\delta(\mathfrak{X}) > 0$, where

$$\delta(\mathfrak{X}) = \inf_{n\geq 1} \text{dist}\left(\frac{x_n}{\|x_n\|},\ \text{Span}\{x_k : k \geq 1,\ k \neq n\}\right)$$

is called the *constant of uniform minimality* of the sequence $\mathfrak{X}$.

Finally, if $\mathfrak{X}^* = (x_n^*)_{n\geq 1}$ is a sequence in the dual of $\mathcal{X}$, we say that $\mathfrak{X}^*$ is a *biorthogonal* associated with $\mathfrak{X}$ if

$$x_k^*(x_n) = \delta_{n,k} = \begin{cases} 1, & \text{if } n = k, \\ 0, & \text{otherwise.} \end{cases}$$

There exists a link between minimality, uniform minimality and the existence of a biorthogonal. More precisely, we can easily obtain the following characterization of minimal and uniformly minimal sequences of $\mathcal{X}$.

Lemma 10.1 *Let $\mathcal{X}$ be a Banach space, and let $\mathfrak{X} = (x_n)_{n\geq 1}$ be a sequence in $\mathcal{X}$. Then the following hold.*

(i) *The sequence $\mathfrak{X}$ is minimal if and only if $\mathfrak{X}$ has a biorthogonal sequence $\mathfrak{X}^*$.*

(ii) *The biorthogonal sequence $\mathfrak{X}^*$ is unique if and only if $\mathfrak{X}$ is complete in $\mathcal{X}$.*

(iii) *For each biorthogonal sequence $\mathfrak{X}^* = (x_n^*)_{n\geq 1}$, we have*

$$\delta(\mathfrak{X}) \geq \frac{1}{\sup_{n\geq 1} \|x_n\|\,\|x_n^*\|}.$$

(iv) *There is a biorthogonal sequence $\mathfrak{X}^* = (x_n^*)_{n\geq 1}$ such that*

$$\delta(\mathfrak{X}) = \frac{1}{\sup_{n\geq 1} \|x_n\|\,\|x_n^*\|}.$$

(v) *The sequence $\mathfrak{X}$ is uniformly minimal if and only if $\mathfrak{X}$ has a biorthogonal sequence $\mathfrak{X}^* = (x_n^*)_{n\geq 1}$ such that*

$$\sup_{n\geq 1} \|x_n\|\,\|x_n^*\| < \infty.$$

Proof (i) and (ii) These follow from Theorem 1.7.

(iii) Theorem 1.7 again implies that, for each $n \geq 1$, we have

$$\min\{\|f\| : f \in \mathcal{X}^*,\ f(x_n) = 1,\ f(x_k) = 0,\ k \neq n\} = \frac{1}{\text{dist}(x_n, \text{Span}\{x_k : k \neq n\})}.$$

Hence, for an arbitrary biorthogonal sequence $(x_n^*)_{n\geq 1}$, we have

$$\|x_n^*\| \geq \frac{1}{\text{dist}(x_n, \text{Span}\{x_k : k \neq n\})} \qquad (n \geq 1).$$

Thus,

$$\text{dist}\left(\frac{x_n}{\|x_n\|}, \text{Span}\{x_k : k \neq n\}\right) \geq \frac{1}{\|x_n\|\,\|x_n^*\|} \qquad (n \geq 1).$$

Taking the infimum with respect to n gives

$$\delta(\mathfrak{X}) \geq \frac{1}{\sup_{n\geq 1} \|x_n\|\,\|x_n^*\|}.$$

(iv) In part (iii), there is a choice of x_n^* for which the minimum is attained, i.e.

$$\|x_n^*\| = \frac{1}{\text{dist}(x_n, \text{Span}\{x_k : k \neq n\})} \qquad (n \geq 1).$$

Thus,

$$\text{dist}\left(\frac{x_n}{\|x_n\|}, \text{Span}\{x_k : k \neq n\}\right) = \frac{1}{\|x_n\|\,\|x_n^*\|} \qquad (n \geq 1).$$

Taking the infimum with respect to n gives

$$\delta(\mathfrak{X}) = \frac{1}{\sup_{n\geq 1} \|x_n\|\,\|x_n^*\|}.$$

(v) This follows from part (iii). □

If $\mathfrak{X} = (x_n)_{n\geq 1}$ is a minimal and complete sequence in $\mathcal{X}$ and if $(x_n^*)_{n\geq 1}$ is its unique biorthogonal sequence, then Lemma 10.1 implies that

$$\delta(\mathfrak{X}) = \frac{1}{\sup_{n\geq 1} \|x_n\|\,\|x_n^*\|}. \tag{10.1}$$

It is clear that uniform minimality implies minimality, that any minimal sequence is w-topologically linearly independent, and finally that any w-topologically linearly independent sequence is finitely independent. For finite sequences, all these properties coincide. Nevertheless, for infinite sequences, the reversed implications are not valid in general. See Exercises 10.1.1, 10.1.2 and 10.1.3.

Corollary 10.2 *Let $\mathcal{X}_1$ and $\mathcal{X}_2$ be two Banach spaces, let $A : \mathcal{X}_1 \longrightarrow \mathcal{X}_2$ be a bounded operator, and let $\mathfrak{X} = (x_n)_{n\geq 1}$ be a sequence in $\mathcal{X}_1$ for which there exists a constant $\epsilon > 0$ satisfying*

$$\|Ax_n\|_{\mathcal{X}_2} \geq \epsilon \|x_n\|_{\mathcal{X}_1} \qquad (n \geq 1). \tag{10.2}$$

If $(Ax_n)_{n\geq 1}$ is uniformly minimal in $\mathcal{X}_2$, then $(x_n)_{n\geq 1}$ is uniformly minimal in $\mathcal{X}_1$.

Proof Since $(Ax_n)_{n\geq 1}$ is uniformly minimal, by Lemma 10.1, there exists a sequence $(y_n^*)_{n\geq 1}$ in $\mathcal{X}_2^*$ such that

$$y_m^*(Ax_n) = \delta_{nm} \qquad (m, n \geq 1)$$

and

$$\sup_{n\geq 1} \|Ax_n\| \, \|y_n^*\| < \infty.$$

Hence, $x_m^* = y_m^* A \in \mathcal{X}_1^*$ and, writing the previous identity as

$$x_m^*(x_n) = \delta_{nm} \qquad (m, n \geq 1),$$

we see that $\mathfrak{X} = (x_n)_{n\geq 1}$ is a minimal sequence and $(x_n^*)_{n\geq 1}$ is a biorthogonal sequence associated with $\mathfrak{X}$. Moreover, using (10.2), we have

$$\|x_n^*\| \, \|x_n\| \leq \epsilon^{-1} \|A\| \times \|y_n^*\| \, \|Ax_n\| \qquad (n \geq 1),$$

which implies that $\sup_{n\geq 1} \|x_n^*\| \, \|x_n\| < \infty$. Thus, by Lemma 10.1, $(x_n)_{n\geq 1}$ is uniformly minimal. □

Exercises

Exercise 10.1.1 Let $(\mathfrak{e}_n)_{n\geq 1}$ be the canonical basis of ℓ^2 and define

$$x_n = \mathfrak{e}_1 + \frac{\mathfrak{e}_n}{n} \qquad (n \geq 2).$$

Show that $(x_n)_{n\geq 2}$ is complete and minimal but not uniformly minimal.
Hint: Show that its biorthogonal sequence is given by $(n\mathfrak{e}_n)_{n\geq 2}$ and apply Lemma 10.1.

Exercise 10.1.2 Let $\mathcal{X} = \mathcal{C}[0,1]$ be the space of continuous functions on $[0,1]$, equipped with the uniform norm

$$\|f\| = \sup_{t\in[0,1]} |f(t)|.$$

Let $(\lambda_n)_{n\geq 1}$ be a strictly increasing sequence of positive numbers satisfying

$$\sum_{n\geq 1} \frac{1}{\lambda_n} = \infty,$$

and define a sequence $(x_n)_{n\geq 0}$ in $\mathcal{X}$ by

$$x_0(t) = 1 \quad \text{and} \quad x_n(t) = t^{\lambda_n} \qquad (t \in [0,1],\ n \geq 1).$$

Show that the sequence $(x_n)_{n\geq 0}$ is complete, nonminimal and w-topologically linearly independent in $\mathcal{X}$.

Hint: To show that $(x_n)_{n\geq 0}$ is complete but nonminimal, apply the Müntz–Szasz theorem [442, p. 293].

Exercise 10.1.3 Let $\mathcal{H}$ be a separable Hilbert space, let $(x_n)_{n\geq 1}$ be an orthonormal basis of $\mathcal{H}$, and let x be a vector of $\mathcal{H}$ such that

$$\langle x, x_n \rangle \neq 0$$

for all $n \geq 1$. For example, we can take

$$x = \sum_{n=1}^{\infty} \frac{\mathfrak{e}_n}{n} \in \ell^2.$$

Define a sequence $(y_n)_{n\geq 1}$ by

$$y_1 = x \quad \text{and} \quad y_n = x_{n-1} \qquad (n \geq 2).$$

Then show that $(y_n)_{n\geq 1}$ is complete, finitely linearly independent, but not w-topologically linearly independent.
Hint: Note that

$$x = \sum_{n\geq 1} \langle x, x_n \rangle x_n \,.$$

10.2 Schauder basis

A sequence $(x_n)_{n\geq 1}$ in a Banach space $\mathcal{X}$ is called a *Schauder basis* (or simply a *basis*) of $\mathcal{X}$ if, for any vector $x \in \mathcal{X}$, there exists a unique sequence $(c_n)_{n\geq 1} = (c_n(x))_{n\geq 1}$ of complex numbers such that

$$x = \sum_{n=1}^{\infty} c_n x_n, \tag{10.3}$$

where the series is convergent in the norm topology. For a Schauder basis $(x_n)_{n\geq 1}$, the sequence of linear functionals $(f_n)_{n\geq 1}$ defined by

$$f_n(x) = c_n \qquad (n \geq 1),$$

where x is represented by (10.3), is called the *sequence of functional coordinates* associated with the basis $(x_n)_{n\geq 1}$. Therefore, if $(x_n)_{n\geq 1}$ is a basis of a Banach space $\mathcal{X}$ and if $(f_n)_{n\geq 1}$ is its sequence of functional coordinates, then, for any $x \in \mathcal{X}$, we can write

$$x = \sum_{n=1}^{\infty} f_n(x) x_n.$$

We show that the functional coordinates are continuous on $\mathcal{X}$ and thus they are elements of the dual of $\mathcal{X}$. The key point in the proof of this result is the following observation.

Lemma 10.3 *Let $(x_n)_{n\geq 1}$ be a sequence of nonzero vectors in the Banach space $\mathcal{X}$. Let $\mathcal{A}_1$ be the linear space of complex sequences defined by*

$$\mathcal{A}_1 = \left\{ (c_n)_{n\geq 1} \subset \mathbb{C} : \sum_{n=1}^{\infty} c_n x_n \text{ is norm convergent} \right\}$$

equipped with the norm

$$\|(c_n)_{n\geq 1}\| = \sup_{N\geq 1} \left\| \sum_{n=1}^{N} c_n x_n \right\|. \tag{10.4}$$

Then $\mathcal{A}_1$ is a Banach space.

Proof First note that, if $(c_n)_{n\geq 1} \in \mathcal{A}_1$, then

$$\sup_{N\geq 1} \left\| \sum_{n=1}^{N} c_n x_n \right\| < \infty$$

because, by definition, the sequence $\sum_{n=1}^{N} c_n x_n$ is convergent as $N \longrightarrow \infty$. Moreover, since $x_n \neq 0$, $n \geq 1$, it is not difficult to check that (10.4) defines a norm on the linear space $\mathcal{A}_1$. It remains to prove that $\mathcal{A}_1$ is complete.

Put $\mathbf{c}_k = (c_{kn})_{n\geq 1}$ and assume that $(\mathbf{c}_k)_{k\geq 1}$ is a Cauchy sequence in $\mathcal{A}_1$. Hence, for each $\varepsilon > 0$, there exists a positive integer $K(\varepsilon)$ such that

$$\|\mathbf{c}_k - \mathbf{c}_{k'}\| = \sup_{N\geq 1} \left\| \sum_{n=1}^{N} (c_{kn} - c_{k'n}) x_n \right\| < \varepsilon \qquad (k, k' > K(\varepsilon)).$$

Hence, fixing any $m \geq 1$, for each $k, k' > K(\varepsilon)$, we have

$$\begin{aligned} \|(c_{km} - c_{k'm}) x_m\| &\leq \left\| \sum_{n=1}^{m} (c_{kn} - c_{k'n}) x_n \right\| + \left\| \sum_{n=1}^{m-1} (c_{kn} - c_{k'n}) x_n \right\| \\ &< 2\varepsilon. \end{aligned}$$

Since $x_m \neq 0$, we get

$$|c_{km} - c_{k'm}| < \frac{2\varepsilon}{\|x_m\|} \qquad (k, k' > K(\varepsilon)).$$

Thus, the sequence of complex numbers $(c_{km})_{k\geq 1}$ is a Cauchy sequence in $\mathbb{C}$, and thus it converges to a complex number, say c_m, i.e.

$$c_m = \lim_{k\to\infty} c_{km} \qquad (m \geq 1).$$

Put $\mathbf{c} = (c_m)_{m\geq 1}$. In the inequality

$$\left\| \sum_{n=1}^{N} (c_{kn} - c_{k'n}) x_n \right\| < \varepsilon \qquad (k, k' > K(\varepsilon),\ N \geq 1)$$

letting $k' \longrightarrow \infty$ gives

$$\left\| \sum_{n=1}^{N} (c_{kn} - c_n) x_n \right\| \leq \varepsilon \qquad (k > K(\varepsilon),\ N \geq 1). \tag{10.5}$$

In particular, for any $n_2 \geq n_1 \geq 1$, we have

$$\left\| \sum_{n=n_1}^{n_2} c_n x_n \right\| \leq 2\varepsilon + \left\| \sum_{n=n_1}^{n_2} c_{kn} x_n \right\| \qquad (k > K(\varepsilon)).$$

Since the series $\sum_{n=1}^{\infty} c_{kn} x_n$ is convergent and $\mathcal{X}$ is complete, it follows that the series $\sum_{n=1}^{\infty} c_n x_n$ is also convergent, that is, $\mathbf{c} \in \mathcal{A}_1$. Moreover, (10.5) implies that

$$\|\mathbf{c}_k - \mathbf{c}\| = \sup_{N\geq 1} \left\| \sum_{n=1}^{N} (c_{kn} - c_n) x_n \right\| \leq \varepsilon \qquad (k > K(\varepsilon)),$$

which proves that the Cauchy sequence $(\mathbf{c}_k)_{k\geq 1}$ converges to c in the space $\mathcal{A}_1$. □

Lemma 10.4 *Let $\mathcal{X}$ be a Banach space, let $(x_n)_{n\geq 1}$ be a Schauder basis of $\mathcal{X}$, and let $(f_n)_{n\geq 1}$ be its sequence of functional coordinates. Then the following assertions hold.*

(i) *If $\mathcal{A}_1$ is the Banach space introduced in Lemma 10.3, the map*

$$(c_n)_{n\geq 1} \longmapsto \sum_{n=1}^{\infty} c_n x_n \tag{10.6}$$

defines an isomorphism from $\mathcal{A}_1$ onto $\mathcal{X}$.

(ii) *For $x \in \mathcal{X}$, put*

$$|||x||| = \sup_{N\geq 1} \left\| \sum_{n=1}^{N} f_n(x) x_n \right\|. \tag{10.7}$$

Then $|||\cdot|||$ defines a norm on $\mathcal{X}$, which is equivalent to the norm $\|\cdot\|$.

Proof (i) Since $(x_n)_{n\geq 1}$ is a Schauder basis, we necessarily have $x_n \neq 0$, $n \geq 1$ (using the uniqueness in the decomposition (10.3)). Hence, Lemma 10.3 implies that $\mathcal{A}_1$ is a Banach space. That $(x_n)_{n\geq 1}$ is a Schauder basis of $\mathcal{X}$ implicitly means that the mapping (10.6) is bijective. It is also clearly linear and of norm at most 1. Thus, the Banach isomorphism theorem implies that this map is an isomorphism from $\mathcal{A}_1$ onto $\mathcal{X}$.

(ii) According to (i), there exists a constant $C \geq 1$ such that

$$\left\| \sum_{n=1}^{\infty} c_n x_n \right\| \leq \sup_{N \geq 1} \left\| \sum_{n=1}^{N} c_n x_n \right\| \leq C \left\| \sum_{n=1}^{\infty} c_n x_n \right\|.$$

To conclude, just observe that, for any $x = \sum_{n=1}^{\infty} c_n x_n \in \mathcal{X}$, we have, by definition, $f_n(x) = c_n$, $n \geq 1$. □

Theorem 10.5 *Let $(x_n)_{n\geq 1}$ be a Schauder basis of a Banach space $\mathcal{X}$, and let $(f_n)_{n\geq 1}$ be its corresponding sequence of functional coordinates. Then, for each $n \geq 1$, we have that $f_n \in \mathcal{X}^*$. Moreover, there exists a constant $M \geq 1$ such that*

$$1 \leq \|x_n\|_{\mathcal{X}} \times \|f_n\|_{\mathcal{X}^*} \leq M \qquad (n \geq 1).$$

Proof Let $|||\cdot|||$ be the norm on $\mathcal{X}$ defined by (10.7). Hence, by Lemma 10.4(ii), there exists a constant $C \geq 1$ such that

$$|||x||| \leq C\|x\| \qquad (x \in \mathcal{X}).$$

Fix any $m \geq 1$. Since $x_m \neq 0$, we have

$$\begin{aligned} |f_m(x)| &= \frac{\|f_m(x)x_m\|}{\|x_m\|} \\ &\leq \frac{\|\sum_{n=1}^{m} f_n(x)x_n\| + \|\sum_{n=1}^{m-1} f_n(x)x_n\|}{\|x_m\|} \\ &\leq \frac{2\,|||x|||}{\|x_m\|} \leq \frac{2C}{\|x_m\|}\|x\| \qquad (x \in \mathcal{X}), \end{aligned}$$

which proves that $f_m \in \mathcal{X}^*$ and $\|f_m\|\,\|x_m\| \leq 2C$. To conclude, note that, by definition of f_m, we have $1 = f_m(x_m) \leq \|f_m\|\,\|x_m\|$. □

Corollary 10.6 *Let $(x_n)_{n\geq 1}$ be a Schauder basis of a Banach space $\mathcal{X}$, and let $(f_n)_{n\geq 1}$ be its corresponding sequence of functional coordinates. Then $(x_n)_{n\geq 1}$ is uniformly minimal and its biorthogonal is the sequence $(f_n)_{n\geq 1}$.*

Proof According to Theorem 10.5, $f_n \in \mathcal{X}^*$, $n \geq 1$, and

$$\sup_{n\geq 1} \|x_n\|\,\|f_n\| < \infty.$$

Moreover, by definition of f_n, it is clear that $f_n(x_m) = \delta_{n,m}$, $m, n \geq 1$. Hence, the sequence $(f_n)_{n\geq 1}$ is the biorthogonal associated with $(x_n)_{n\geq 1}$ (and it is unique because $(x_n)_{n\geq 1}$ is complete in $\mathcal{X}$). Since $\sup_{n\geq 1} \|x_n\|\,\|f_n\| < \infty$, Lemma 10.1(v) implies that $(x_n)_{n\geq 1}$ is uniformly minimal. □

According to the above discussion, if $\mathfrak{X} = (x_n)_{n\geq 1}$ is a Schauder basis of a Banach space $\mathcal{X}$, then, for any $x \in \mathcal{X}$, we have

$$x = \sum_{n\geq 1} x_n^*(x) x_n, \tag{10.8}$$

where $\mathfrak{X}^* = (x_n^*)_{n\geq 1}$ is the biorthogonal sequence associated with $\mathfrak{X}$. The coefficients $x_n^*(x)$ are called the *generalized Fourier coefficients* of x and the series (10.8) is called the *generalized Fourier series* of x relative to the sequences $\mathfrak{X}$ and $\mathfrak{X}^*$.

We know from Theorem 10.5 and Corollary 10.6 that, if $(x_n)_{n\geq 1}$ is a Schauder basis of a Banach space $\mathcal{X}$, then it is uniformly minimal, but the converse is not true in general (see Exercise 10.2.1). However, we will see that the two properties coincide for the sequences of normalized reproducing kernels of the Hardy space H^2.

The following result, which is an easy consequence of the Banach–Steinhaus theorem, gives a characterization of the basis property in terms of the partial sums of the generalized Fourier series.

Theorem 10.7 *Let $(x_n)_{n\geq 1}$ be a minimal sequence in a Banach space $\mathcal{X}$. Denote by $(x_n^*)_{n\geq 1}$ a biorthogonal sequence associated with $(x_n)_{n\geq 1}$ and define the sequence of bounded operators $(S_n)_{n\geq 1}$ on $\mathcal{X}$ by*

$$S_n(x) = \sum_{k=1}^{n} x_k^*(x)x_k \qquad (x \in \mathcal{X},\ n \geq 1). \tag{10.9}$$

Then the following are equivalent:

(i) *$(x_n)_{n\geq 1}$ is a Schauder basis of $\mathcal{X}$;*
(ii) *for each $x \in \mathcal{X}$, $\lim_{n\to\infty} S_n(x) = x$;*
(iii) *$(x_n)_{n\geq 1}$ is complete in $\mathcal{X}$ and, for each $x \in \mathcal{X}$, $\sup_{n\geq 1} \|S_n(x)\|_{\mathcal{X}} < \infty$;*
(iv) *$(x_n)_{n\geq 1}$ is complete in $\mathcal{X}$ and $\sup_{n\geq 1} \|S_n\|_{\mathcal{L}(\mathcal{X})} < \infty$.*

Proof (i) $\Longrightarrow$ (ii) This follows from Corollary 10.6.

(ii) $\Longrightarrow$ (i) This follows from the definition of a Schauder basis. Note that the uniqueness of coordinates follows from the minimality of $(x_n)_{n\geq 1}$.

(ii) $\Longrightarrow$ (iii) For each $x \in \mathcal{X}$, the sequence $(S_n(x))_{n\geq 1}$ is convergent and thus bounded. The completeness is trivial, since the assumption $S_n(x) \longrightarrow x$ implies that $\operatorname{Span}\{x_n : n \geq 1\} = \mathcal{X}$.

(iii) $\Longrightarrow$ (iv) This an immediate application of the Banach–Steinhaus theorem.

(iv) $\Longrightarrow$ (ii) For any finite linear combination $p = \sum_{k=1}^{m} c_k x_k$ and for all $n \geq m$, we have

$$S_n(p) = \sum_{k=1}^{n} x_k^*(p)x_k = \sum_{k=1}^{m} c_k x_k = p,$$

whence $\lim_{n\to\infty} S_n(p) = p$. Since $(x_n)_{n\geq 1}$ is complete in $\mathcal{X}$ and since $\sup_{n\geq 1} \|S_n\| < \infty$, a standard argument now shows that the relation $\lim_{n\to\infty} S_n(x) = x$ holds in fact for all $x \in \mathcal{X}$. □

The operator S_n that was defined in (10.9) may be rewritten as

$$S_n = \sum_{k=1}^{n} x_k \otimes x_k^* \qquad (n \geq 1) \tag{10.10}$$

and, under the conditions of Theorem 10.7, S_n strongly tends to the identity operator I.

The following technical result will be useful for us in the study of the basis of reproducing kernels of $\mathcal{H}(b)$ spaces.

Lemma 10.8 *Let $\mathcal{X}_1$ and $\mathcal{X}_2$ be two Banach spaces, and let $A \in \mathcal{L}(\mathcal{X}_1, \mathcal{X}_2)$. If $(x_n)_{n\geq 1}$ and $(Ax_n)_{n\geq 1}$ are Schauder bases in their closed linear spans, then*

$$\dim(\mathcal{X}_1/\text{Span}\{x_n : n \geq 1\}) \geq \dim\ker A.$$

Proof Put

$$k = \dim(\mathcal{X}_1/\text{Span}\{x_n : n \geq 1\}).$$

If $k = \infty$, the result is trivial. Hence, assume that k is finite. Then we know from Theorem 1.10 that there exists a closed subspace $\mathcal{E}_1$ of $\mathcal{X}_1$ such that

$$\mathcal{X}_1 = \text{Span}\{x_n : n \geq 1\} \oplus \mathcal{E}_1$$

and $\dim \mathcal{E}_1 = k$. Argue by absurdity and assume that $\dim\ker(A) > k$. Then there exists a set $\{a_1, \dots, a_{k+1}\}$ of linearly independent vectors in $\ker(A)$. Write

$$a_i = a_i^{(1)} + a_i^{(2)} \qquad (1 \leq i \leq k+1),$$

with $a_i^{(1)} \in \text{Span}\{x_n : n \geq 1\}$ and $a_i^{(2)} \in \mathcal{E}_1$.

If, for any i, it happens that $a_i^{(2)} = 0$, then we get $a_i \in \text{Span}\{x_n : n \geq 1\} \cap \ker(A)$. This is absurd because A is one-to-one on $\text{Span}\{x_n : n \geq 1\}$. This fact is an immediate consequence of the assumption that $(x_n)_{n\geq 1}$ and $(Ax_n)_{n\geq 1}$ are Schauder bases in their closed linear spans. Therefore, $a_i^{(2)} \neq 0$, for all $1 \leq i \leq k+1$.

We claim that $\{a_1^{(2)}, \dots, a_{k+1}^{(2)}\}$ are linearly independent. Indeed, if there are scalars $\lambda_1, \dots, \lambda_{k+1}$ such that

$$\sum_{i=1}^{k+1} \lambda_i a_i^{(2)} = 0,$$

then $\sum_{i=1}^{k+1} \lambda_i a_i \in \ker(A) \cap \text{Span}\{x_n : n \geq 1\}$. Hence, owing to the injectivity of A on $\text{Span}\{x_n : n \geq 1\}$, we deduce that

$$\sum_{i=1}^{k+1} \lambda_i a_i = 0.$$

But, since $\{a_1, \dots, a_{k+1}\}$ are linearly independent, then $\lambda_i = 0$ for $1 \le i \le k+1$. We have thus found a linearly independent system of $k+1$ vectors in $\mathcal{E}_1$ that is of dimension k. This is absurd, and hence $\dim \ker(A) \le k$. □

Exercises

Exercise 10.2.1 Let $\mathcal{X} = \mathcal{C}(\mathbb{T})$ be the Banach space of continuous functions on $\mathbb{T}$, equipped with the uniform norm $\|f\| = \sup_{\zeta \in \mathbb{T}} |f(\zeta)|$. Let

$$\chi_n(\zeta) = \zeta^n \qquad (\zeta \in \mathbb{T},\ n \in \mathbb{Z}).$$

Show that $(\chi_n)_{n\in\mathbb{Z}}$ is complete, uniformly minimal, but not a Schauder basis of $\mathcal{X}$.
Hint: For the completeness, use the Stone–Weierstrass theorem. For the uniform minimality, use the biorthogonal sequence $\chi_n^*(f) = \hat{f}(n)$, $n \in \mathbb{Z}$. Finally, for declining the basis property, use the du Bois–Reymond theorem (see [442, p. 97]).

Exercise 10.2.2 Let $(x_n)_{n\ge1}$ be a sequence of nonzero vectors in the Banach space $\mathcal{X}$, and consider the Banach space $\mathcal{A}_1$ equipped with the norm (10.4). Let $\mathbf{c}_k = (c_{kn})_{n\ge1} \in \mathcal{A}_1$ be such that, for each fixed $n \ge 1$,

$$\lim_{k\to\infty} c_{kn} = 0.$$

Can we deduce that

$$\lim_{k\to\infty} \mathbf{c}_k = 0 \qquad (\text{in } \mathcal{A}_1)?$$

Hint: Let $(x_n)_{n\ge1}$ be an orthonormal sequence in a Hilbert space and consider

$$c_{kn} = \begin{cases} 1/\sqrt{n} & \text{if} \quad k \le n, \\ 0 & \text{if} \quad k > n. \end{cases}$$

Exercise 10.2.3 Let $\mathcal{X}$ be a Banach space. We say that two Schauder bases $(x_n)_{n\ge1}$ and $(y_n)_{n\ge1}$ are "equivalent" if

$$\sum_{n=1}^{\infty} c_n x_n \text{ is convergent} \iff \sum_{n=1}^{\infty} c_n y_n \text{ is convergent.}$$

Show that two Schauder bases $(x_n)_{n\ge1}$ and $(y_n)_{n\ge1}$ are equivalent if and only if there exists a bounded invertible operator $T : \mathcal{X} \longrightarrow \mathcal{X}$ such that

$$T x_n = y_n \qquad (n \ge 1).$$

Hint: Use Theorem 10.5 and the closed graph theorem (Corollary 1.18).

Exercise 10.2.4 Let $(x_n)_{n\geq 1}$ be a Schauder basis of a Banach space $\mathcal{X}$, and let $(x_n^*)_{n\geq 1}$ be its biorthogonal sequence. Show that

$$\Lambda = \sum_{n=1}^{\infty} \Lambda(x_n)x_n^*$$

for every $\Lambda \in \text{Span}\{x_n^* : n \geq 1\} \subset \mathcal{X}^*$. In particular, show that $(x_n^*)_{n\geq 1}$ is a Schauder basis for its closed linear span.
Hint: Consider the bounded operator S_n defined in (10.9) and verify that

$$S_n^*\Lambda = \sum_{k=1}^{n} \Lambda(x_k)x_k^*.$$

To find S_n^*, use the representation (10.10). Then show that $S_n^*\Lambda \longrightarrow \Lambda$ for every $\Lambda \in \text{Span}\{x_n^* : n \geq 1\}$.

Exercise 10.2.5 Let $\mathcal{X}$ be a reflexive Banach space, and let $(x_n)_{n\geq 1}$ be a Schauder basis of $\mathcal{X}$. Show that $(x_n^*)_{n\geq 1}$ is a Schauder basis of $\mathcal{X}^*$.
Hint: Show that $(x_n^*)_{n\geq 1}$ is complete in $\mathcal{X}^*$ and use Exercise 10.2.4.

Exercise 10.2.6 Let $(x_n)_{n\geq 1}$ be a Schauder basis for a Banach space $\mathcal{X}$ and suppose that $(y_n)_{n\geq 1}$ is a sequence of elements of $\mathcal{X}$ such that, for any $m \geq 1$,

$$\left\| \sum_{n=1}^{m} c_n(x_n - y_n) \right\| \leq \lambda \left\| \sum_{n=1}^{m} c_n x_n \right\|$$

for some constant λ, $0 \leq \lambda < 1$, and all choices of scalars $c_1, c_2, \ldots, c_n$. Show that $(y_n)_{n\geq 1}$ is a Schauder basis for $\mathcal{X}$ equivalent to $(x_n)_{n\geq 1}$.
Hint: Let $x = \sum_{n=1}^{\infty} c_n x_n$ and define

$$Sx = \sum_{n=1}^{\infty} c_n(x_n - y_n).$$

Observe that S is a well-defined linear and bounded operator on $\mathcal{X}$ and $\|S\| < 1$. Then use Exercise 10.2.3 with $T = I - S$.

Exercise 10.2.7 Let $(x_n)_{n\geq 1}$ be a Schauder basis for a Banach space $\mathcal{X}$, let $(x_n^*)_{n\geq 1}$ be its biorthogonal sequence, and let $(y_n)_{n\geq 1}$ be a sequence of vectors in $\mathcal{X}$ satisfying

$$\sum_{n=1}^{\infty} \|x_n - y_n\| \, \|x_n^*\| < 1.$$

Show that $(y_n)_{n\geq 1}$ is a Schauder basis of $\mathcal{X}$ equivalent to $(x_n)_{n\geq 1}$.
Hint: Use Exercise 10.2.6.

10.3 The multipliers of a sequence

In this section, we introduce the notion of multipliers associated with a minimal and complete sequence in a Banach space $\mathcal{X}$. This notion enables us to give a new characterization of the basis property. Let $\mathfrak{X} = (x_n)_{n\geq 1}$ be a complete and minimal sequence in a Banach space $\mathcal{X}$. Using minimality, we easily see that the linear mapping

$$\begin{aligned} \text{Lin}\{x_n : n \geq 1\} &\longrightarrow \text{Lin}\{x_n : n \geq 1\} \\ \sum_n a_n x_n &\longmapsto \sum_n a_n \mu_n x_n \end{aligned}$$

is well defined for any complex sequence $\mu = (\mu_n)_{n\geq 1}$. The sequence μ is called a *multiplier* of $(x_n)_{n\geq 1}$ if this linear map extends to a bounded operator, say M_μ, on $\mathcal{X}$. In other words, $\mu = (\mu_n)_{n\geq 1} \subset \mathbb{C}$ is a multiplier of $(x_n)_{n\geq 1}$ if there exists an operator $M_\mu \in \mathcal{L}(\mathcal{X})$ such that $M_\mu(x_n) = \mu_n x_n$, $n \geq 1$. Note that the completeness of $(x_n)_{n\geq 1}$ ensures the uniqueness of the operator M_μ (if it exists). The set of multipliers of a sequence $\mathfrak{X} = (x_n)_{n\geq 1}$ is denoted by $\mathfrak{Mult}(\mathfrak{X})$. If $\mu = (\mu_n)_{n\geq 1} \in \mathfrak{Mult}(\mathfrak{X})$, we put

$$\|\mu\|_{\mathfrak{Mult}(\mathfrak{X})} = \|M_\mu\|.$$

Theorem 10.9 *Let $\mathfrak{X} = (x_n)_{n\geq 1}$ be a complete and minimal sequence of a Banach space $\mathcal{X}$. Then $(\mathfrak{Mult}(\mathfrak{X}), \|\cdot\|_{\mathfrak{Mult}(\mathfrak{X})})$ is a Banach space and*

$$\mathfrak{Mult}(\mathfrak{X}) \hookrightarrow \ell^\infty,$$

that is, $\mathfrak{Mult}(\mathfrak{X})$ is contractively contained in ℓ^∞, the Banach space of bounded complex sequences.

Proof Let us first show that $\mathfrak{Mult}(\mathfrak{X})$ is contractively contained in ℓ^∞. If $(\mu_n)_{n\geq 1} \in \mathfrak{Mult}(\mathfrak{X})$, then we have

$$|\mu_n| \times \|x_n\| = \|M_\mu(x_n)\| \leq \|M_\mu\|_{\mathcal{L}(\mathcal{X})} \|x_n\| \qquad (n \geq 1).$$

Since $(x_n)_{n\geq 1}$ is minimal, we have $x_n \neq 0$, and thus we get

$$\sup_{n\geq 1} |\mu_n| \leq \|M_\mu\|_{\mathcal{L}(\mathcal{X})}. \tag{10.11}$$

Hence,

$$\|\mu\|_{\ell^\infty} = \sup_{n\geq 1} |\mu_n| \leq \|M_\mu\| = \|\mu\|_{\mathfrak{Mult}(\mathfrak{X})}, \tag{10.12}$$

for all $\mu = (\mu_n)_{n\geq 1} \in \mathfrak{Mult}(\mathfrak{X})$, which, technically speaking, means that $\mathfrak{Mult}(\mathfrak{X})$ embeds contractively into ℓ^∞.

We can easily check that $\mathfrak{Mult}(\mathfrak{X})$ is a linear space and $\|\cdot\|_{\mathfrak{Mult}(\mathfrak{X})}$ defines a norm on it. The only fact that needs to be checked is that the norm is complete.

Hence, let $(\mu^{(m)})_{m\geq 1}$ be a Cauchy sequence in $\mathfrak{Mult}(\mathfrak{X})$. Note that, for each fixed m, $\mu^{(m)}$ itself is a sequence in $\mathbb{C}$, and we write

$$\mu^{(m)} = (\mu_n^{(m)})_{n\geq 1}.$$

Being Cauchy in $\mathfrak{Mult}(\mathfrak{X})$ means that, for each $\varepsilon > 0$, there exists an integer $M \geq 1$ such that

$$\|\mu^{(k)} - \mu^{(\ell)}\|_{\mathfrak{Mult}(\mathfrak{X})} = \|M_{\mu^{(k)}} - M_{\mu^{(\ell)}}\|_{\mathcal{L}(\mathcal{X})} < \varepsilon \qquad (k, \ell > M).$$

Thus, the sequence $(M_{\mu^{(m)}})_{m\geq 1}$ is a Cauchy sequence in $\mathcal{L}(\mathcal{X})$ that is complete. Hence $(M_{\mu^{(m)}})_{m\geq 1}$ is convergent, say, to $T \in \mathcal{L}(\mathcal{X})$. But, by (10.11), for each fixed $n \geq 1$, we have

$$|\mu_n^{(k)} - \mu_n^{(\ell)}| \leq \|M_{\mu^{(k)}} - M_{\mu^{(\ell)}}\|_{\mathcal{L}(\mathcal{X})} < \varepsilon \quad (k, \ell > M),$$

which shows that $(\mu_n^{(m)})_{m\geq 1}$ is a Cauchy sequence in $\mathbb{C}$. Thus, it converges, say, to the complex number μ_n. Knowing this, we see that

$$T(x_n) = \lim_{m\to\infty} M_{\mu^{(m)}}(x_n) = \lim_{m\to\infty} \mu_n^{(m)} x_n = \mu_n x_n \qquad (n \geq 1).$$

Since $T \in \mathcal{L}(\mathcal{X})$, this ensures that $\mu = (\mu_n)_{n\geq 1} \in \mathfrak{Mult}(\mathfrak{X})$ and $M_\mu = T$. Now, we can write

$$\|\mu^{(m)} - \mu\|_{\mathfrak{Mult}(\mathfrak{X})} = \|M_{\mu^{(m)}} - M_\mu\|_{\mathcal{L}(\mathcal{X})} = \|M_{\mu^{(m)}} - T\|_{\mathcal{L}(\mathcal{X})},$$

and thus $(\mu^{(m)})_{m\geq 1}$ converges in $\mathfrak{Mult}(\mathfrak{X})$ to μ. Therefore, we conclude that $(\mathfrak{Mult}(\mathfrak{X}), \|\cdot\|_{\mathfrak{Mult}(\mathfrak{X})})$ is a Banach space. □

As already mentioned, the space of multipliers is used for a characterization of the basis property. For that purpose, we introduce the space

$$BV = \left\{ (\mu_n)_{n\geq 1} \subset \mathbb{C} : \sum_{n\geq 1} |\mu_n - \mu_{n+1}| < \infty \right\}.$$

An element of BV is said to be of *bounded variation*. The convergence of the series $\sum_{n\geq 1} |\mu_n - \mu_{n+1}|$ ensures that the sequence $\mu_N = \mu_1 - \sum_{n=1}^{N-1}(\mu_n - \mu_{n+1})$ is also convergent. Hence, for a sequence $\mu = (\mu_n)_{n\geq 1} \in BV$, we put

$$\|\mu\|_{BV} = \sum_{n\geq 1} |\mu_n - \mu_{n+1}| + \lim_{n\to\infty} |\mu_n|.$$

Then, it is well known (and easy to check) that $(BV, \|\cdot\|_{BV})$ is a Banach space, and we have the following characterization.

Theorem 10.10 *Let $\mathfrak{X} = (x_n)_{n\geq 1}$ be a minimal and complete sequence in the Banach space $\mathcal{X}$. Then $\mathfrak{X}$ is a Schauder basis for $\mathcal{X}$ if and only if $BV \subset \mathfrak{Mult}(\mathfrak{X})$.*

Proof Denote by $(x_n^*)_{n\geq 1}$ the biorthogonal sequence associated with $(x_n)_{n\geq 1}$ and by $(S_n)_{n\geq 1}$ the sequence of operators defined by (10.9). Let us first assume that $(x_n)_{n\geq 1}$ is a Schauder basis of $\mathcal{X}$ and then we prove that any sequence of bounded variation $\mu = (\mu_n)_{n\geq 1}$ is a multiplier of the sequence $(x_n)_{n\geq 1}$. Using Abel's method, we can write

$$\begin{aligned}\sum_{k=1}^{n} \mu_k x_k^*(x) x_k &= \sum_{k=1}^{n} \mu_k \left(\sum_{j=1}^{k} x_j^*(x) x_j - \sum_{j=1}^{k-1} x_j^*(x) x_j \right) \\ &= \sum_{k=1}^{n} \mu_k (S_k(x) - S_{k-1}(x)) \\ &= \sum_{k=1}^{n-1} (\mu_k - \mu_{k+1}) S_k(x) + \mu_n S_n(x)\end{aligned}$$

for each $x \in \mathcal{X}$. According to Theorem 10.7, there exists a constant $C > 0$ such that $\|S_n\| \leq C$, $n \geq 1$, which gives

$$\left\| \sum_{k=1}^{n} x_k^*(x) \mu_k x_k \right\| \leq C\|x\| \left(\sum_{k=1}^{n} |\mu_k - \mu_{k+1}| + |\mu_{n+1}| \right) \leq C\|x\| \, \|\mu\|_{BV}.$$

Hence, the linear mapping

$$\sum_k a_k x_k \longmapsto \sum_k \mu_k a_k x_k,$$

defined on $\mathrm{Lin}\{x_n : n \geq 1\}$, is continuous, with norm less than or equal to $C\|\mu\|_{BV}$. Since the sequence $(x_n)_{n\geq 1}$ is complete, we can extend this linear map into a continuous linear map T on $\mathcal{X}$, with norm $\|T\| \leq C\|\mu\|_{BV}$. Therefore, $\mu \in \mathfrak{Mult}(\mathfrak{X})$, with $M_\mu = T$.

Reciprocally, assume that $BV \subset \mathfrak{Mult}(\mathfrak{X})$ and consider the canonical injection

$$\begin{array}{ccc} BV & \longrightarrow & \mathfrak{Mult}(\mathfrak{X}) \\ \mu & \longmapsto & \mu. \end{array}$$

Then, the closed graph theorem (Corollary 1.18) ensures that this linear map is continuous. Hence, there exists a constant $C > 0$ such that, for any sequence $\mu \in BV$, we have

$$\|\mu\|_{\mathfrak{Mult}(\mathfrak{X})} \leq C\|\mu\|_{BV}.$$

We recall that $\mathfrak{e}_k$ is the sequence whose components are all equal to 0 except in the kth place, which is equal to 1. By hypothesis, $\mathfrak{e}_k \in \mathfrak{Mult}(\mathfrak{X})$ and we have $M_{\mathfrak{e}_k}(x_j) = \delta_{k,j} x_j = x_k^*(x_j)\, x_k$, $j \geq 1$. Hence, by linearity and continuity, we have

$$M_{\mathfrak{e}_k}(x) = x_k^*(x) x_k \qquad (x \in \mathcal{X}).$$

Put

$$\mu^{(m)} = \sum_{k=1}^{m} \mathfrak{e}_k$$

and note that

$$\|\mu^{(m)}\|_{BV} = 1.$$

Then

$$\begin{aligned}\|S_m(x)\| = \left\| \sum_{k=1}^{m} x_k^*(x) x_k \right\| &= \left\| \sum_{k=1}^{m} M_{\mathfrak{e}_k}(x) \right\| \\ &= \|M_{\mu^{(m)}}(x)\| \\ &\leq \|\mu^{(m)}\|_{\mathfrak{Mult}(\mathcal{X})} \|x\| \\ &\leq C \|\mu^{(m)}\|_{BV} \|x\| = C\|x\|,\end{aligned}$$

which proves, in light of Theorem 10.7, that $(x_n)_{n\geq 1}$ is a basis of $\mathcal{X}$, and completes the proof of Theorem 10.10. □

10.4 Symmetric, nonsymmetric and unconditional basis

A sequence $(x_n)_{n\in\mathbb{Z}}$ of a Banach space $\mathcal{X}$ is called a *symmetric basis* of $\mathcal{X}$ if, for any vector $x \in \mathcal{X}$, there exists a unique sequence $(c_n)_{n\in\mathbb{Z}} = (c_n(x))_{n\in\mathbb{Z}}$ of complex numbers such that

$$x = \lim_{N\to\infty} \sum_{n=-N}^{N} c_n x_n, \tag{10.13}$$

where the limit is taken with respect to the norm topology. Using techniques similar to those developed in Section 10.2, we can exploit the uniqueness of the coefficients in (10.13) to deduce that a symmetric basis is minimal.

Parallel to this definition, we say that $(x_n)_{n\in\mathbb{Z}}$ is a *nonsymmetric basis* of $\mathcal{X}$ if, for any vector $x \in \mathcal{X}$, there exists a unique sequence $(c_n)_{n\in\mathbb{Z}} = (c_n(x))_{n\in\mathbb{Z}}$ of complex numbers such that

$$x = \lim_{M,N\to\infty} \sum_{n=-M}^{N} c_n x_n. \tag{10.14}$$

In fact, the notion of nonsymmetric basis is not new and it is precisely a Schauder basis that is indexed by $\mathbb{Z}$. The terminology is also a bit misleading, since a nonsymmetric basis is surely a symmetric basis too. But the inverse is not true. The family of symmetric bases form a much larger collection.

Let $(x_n)_{n\in\mathbb{Z}}$ be a minimal sequence in a Banach space $\mathcal{X}$. Let $(x_n^*)_{n\in\mathbb{Z}}$ denote a biorthogonal sequence associated with $(x_n)_{n\in\mathbb{Z}}$. Adopting the same circle of ideas that led to the creation of S_n in (10.9), we proceed and define

$$S_{m,n}(x) = \sum_{k=m}^{n} x_k^*(x)x_k \qquad (m,n \in \mathbb{Z},\ x \in \mathcal{X})$$

or, equivalently

$$S_{m,n} = \sum_{k=m}^{n} x_k \otimes x_k^* \qquad (m,n \in \mathbb{Z},\ x \in \mathcal{X}).$$

Note that, for a symmetric basis, which is automatically minimal, the assumption (10.13) implies that $x_n^*(x) = c_n(x)$, $n \in \mathbb{Z}$.

Theorem 10.11 *Let $(x_n)_{n\in\mathbb{Z}}$ be a minimal sequence in a Banach space $\mathcal{X}$. Then the following statements are equivalent:*

(i) $(x_n)_{n\in\mathbb{Z}}$ *is a nonsymmetric basis;*

(ii) $(x_n)_{n\in\mathbb{Z}}$ *is complete and* $\sup_{m,n\geq 1} \|S_{-m,n}\|_{\mathcal{L}(\mathcal{X})} < \infty$;

(iii) $\lim_{m,n\to\infty} S_{-m,n}(x) = x \qquad (x \in \mathcal{X}).$

Proof Since a nonsymmetric basis is a Schauder basis indexed by $\mathbb{Z}$, this is precisely Theorem 10.7 in disguised form. □

Theorem 10.12 *Let $(x_n)_{n\in\mathbb{Z}}$ be a minimal sequence in a Banach space $\mathcal{X}$. Then the following statements are equivalent:*

(i) $(x_n)_{n\in\mathbb{Z}}$ *is a symmetric basis;*

(ii) $(x_n)_{n\in\mathbb{Z}}$ *is complete and* $\sup_{n\geq 1} \|S_{-n,n}\|_{\mathcal{L}(\mathcal{X})} < \infty$;

(iii) $\lim_{n\to\infty} S_{-n,n}(x) = x \qquad (x \in \mathcal{X}).$

Proof Similar to the proof of Theorem 10.7 (see also Exercises 10.4.1 and 10.4.2). □

Theorem 10.13 *Let $(x_n)_{n\in\mathbb{Z}}$ be a symmetric basis in a Banach space $\mathcal{X}$. Then the following statements hold:*

(i) $x_n^*(x) = 0\ (n \in \mathbb{Z}) \implies x = 0$;

(ii) *for any subset E of $\mathbb{Z}$,*

$$\text{Span}\{x_n : n \in E\} = \{x \in \mathcal{X} : x_k^*(x) = 0,\ k \notin E\}.$$

Proof (i) This implication follows immediately from Theorem 10.12(iii).

(ii) Since, for each $n \in E$, we have $x_n \in \{x \in \mathcal{X} : x_k^*(x) = 0,\ k \notin E\}$, the inclusion

$$\mathrm{Span}\{x_n : n \in E\} \subset \{x \in \mathcal{X} : x_k^*(x) = 0,\ k \notin E\}$$

is trivial. The reverse inclusion is a consequence of Theorem 10.12(iii). □

Let $\mathcal{X}$ be a topological vector space, equipped with a topology τ. A sequence $(x_n)_{n\geq 1}$ in $\mathcal{X}$ is called an *unconditional basis* in $(\mathcal{X}, \tau)$ if, for each vector $x \in \mathcal{X}$, there exists a unique sequence of complex numbers $(c_n)_{n\geq 1} = (c_n(x))_{n\geq 1}$ such that

$$x = \sum_n c_n x_n, \tag{10.15}$$

where the series is unconditionally convergent in $(\mathcal{X}, \tau)$. More specifically, this means that, for each open neighborhood of 0, there exists $\sigma_0 \in \mathfrak{F}$, the family of all finite subsets of $\mathbb{N}$, such that, for all $\sigma \in \mathfrak{F}$ satisfying $\sigma_0 \subset \sigma$, we have

$$x - \sum_{n\in\sigma} c_n x_n \in V.$$

The "unconditional convergence" refers to the freedom that we have in choosing σ_0 and σ. In the series that we have seen up to now, we take the limit of $\sum_{n=1}^{N}$ as $N \longrightarrow \infty$. In other words, in the traditional case, we just consider the finite subsets $\sigma = \{1, 2, \dots, N\}$, $N \geq 1$.

It is rather elementary to restate this definition in a more familiar way if $\mathcal{X}$ is a Banach space. In this case, a sequence $(x_n)_{n\geq 1}$ is an unconditional basis of $\mathcal{X}$ if and only if, for each $x \in \mathcal{X}$, there exists a unique sequence of complex numbers $(c_n)_{n\geq 1} = (c_n(x))_{n\geq 1}$ such that, for each $\varepsilon > 0$, there is $\sigma_0 \in \mathfrak{F}$ such that, for all $\sigma \in \mathfrak{F}$ satisfying $\sigma_0 \subset \sigma$, we have

$$\left\| x - \sum_{n\in\sigma} c_n x_n \right\| < \varepsilon. \tag{10.16}$$

Therefore, if $(x_n)_{n\geq 1}$ is an unconditional basis of $\mathcal{X}$, then it is also surely a Schauder basis of $\mathcal{X}$. (The inverse is not true; see Theorem 10.14.) Hence, using (10.8), we see that, if $(x_n)_{n\geq 1}$ is a minimal and complete sequence in $\mathcal{X}$ and if $(x_n^*)_{n\geq 1}$ denotes its biorthogonal, then $(x_n)_{n\geq 1}$ is an unconditional basis of $\mathcal{X}$ if and only if, for each $x \in \mathcal{X}$ and $\varepsilon > 0$, there is $\sigma_0 \in \mathfrak{F}$ such that, for all $\sigma \in \mathfrak{F}$ satisfying $\sigma_0 \subset \sigma$, we have

$$\left\| x - \sum_{n\in\sigma} x_n^*(x) x_n \right\| < \varepsilon. \tag{10.17}$$

In other words, if the sequence $(c_n(x))_{n\geq 1}$ that appears in (10.16) exists, then necessarily we must have $c_n(x) = x_n^*(x)$, $n \geq 1$.

We now give a characterization of unconditional basis that is an analog of Theorems 10.7 and 10.10. To state this characterization, we will introduce some notation. For $\sigma \in \mathfrak{F}$, if $(x_n)_{n\geq 1}$ is a minimal and complete sequence in a Banach space $\mathcal{X}$ and $(x_n^*)_{n\geq 1}$ denotes its biorthogonal, then S_σ is the bounded operator on $\mathcal{X}$ defined by

$$S_\sigma(x) = \sum_{n\in\sigma} x_n^*(x) x_n \qquad (x \in \mathcal{X}) \tag{10.18}$$

or, equivalently,

$$S_\sigma(x) = \sum_{n\in\sigma} x_n \otimes x_n^*. \tag{10.19}$$

This is a generalization of the notion S_n, which was introduced in (10.9). In fact, this is a known operator in disguised form. To clarify the situation, let $\mu^\sigma = (\mu_n^\sigma)_{n\geq 1}$ be the sequence defined by

$$\mu_n^\sigma = \begin{cases} 1 & \text{if} \quad n \in \sigma, \\ 0 & \text{if} \quad n \notin \sigma. \end{cases}$$

Using this notation and the multiplication operator defined in Section 10.3, we can write

$$\begin{aligned} S_\sigma(x_p) &= \sum_{n\in\sigma} x_n^*(x_p) x_n \\ &= \sum_{n\in\sigma} \delta_{n,p} x_n \\ &= \mu_p^\sigma x_p \\ &= M_{\mu^\sigma}(x_p) \qquad (p \geq 1). \end{aligned}$$

Hence, we have

$$S_\sigma = M_{\mu^\sigma} \qquad (\sigma \in \mathfrak{F}). \tag{10.20}$$

Theorem 10.14 *Let $\mathcal{X}$ be a Banach space, let $\mathfrak{X} = (x_n)_{n\geq 1}$ be a minimal and complete sequence in $\mathcal{X}$, and let $(x_n^*)_{n\geq 1}$ be the biorthogonal sequence associated with $\mathfrak{X}$. Then the following are equivalent.*

(i) *The sequence $(x_n)_{n\geq 1}$ is an unconditional basis of $\mathcal{X}$.*
(ii) *For each $x \in \mathcal{X}$, we have $\sup_{\sigma\in\mathfrak{F}} \|S_\sigma(x)\|_{\mathcal{X}} < \infty$.*
(iii) *We have $\sup_{\sigma\in\mathfrak{F}} \|S_\sigma\|_{\mathcal{L}(\mathcal{X})} < \infty$.*
(iv) *The sequence $(x_n^*)_{n\geq 1}$ is an unconditional basis of $\mathcal{X}^*$, endowed with the weak-star topology.*
(v) $\mathfrak{Mult}(\mathfrak{X}) = \ell^\infty$.

Proof Our strategy is to show that (i) $\Longrightarrow$ (ii) $\Longrightarrow$ (iii) $\Longrightarrow$ (i), then (i) $\Longrightarrow$ (iv) $\Longrightarrow$ (ii), and finally (iii) $\Longrightarrow$ (v) $\Longrightarrow$ (iii).

(i) $\Longrightarrow$ (ii) Fix $x \in \mathcal{X}$. According to (10.17), there exists $\sigma_0 \in \mathfrak{F}$ such that, for all $\sigma \in \mathfrak{F}$ satisfying $\sigma_0 \subset \sigma$, we have

$$\left\| x - \sum_{n\in\sigma} x_n^*(x)x_n \right\| < 1.$$

For each $\sigma \in \mathfrak{F}$, write

$$\begin{aligned} \|S_\sigma(x)\| &= \left\| \sum_{n\in\sigma} x_n^*(x)x_n \right\| \\ &= \left\| \sum_{n\in\sigma\cup\sigma_0} x_n^*(x)x_n - \sum_{n\in\sigma_0\setminus(\sigma\cap\sigma_0)} x_n^*(x)x_n \right\|. \end{aligned}$$

Hence,

$$\begin{aligned} \|S_\sigma(x)\| &\le \left\| \sum_{n\in\sigma\cup\sigma_0} x_n^*(x)x_n \right\| + \sum_{n\in\sigma_0\setminus(\sigma\cap\sigma_0)} |x_n^*(x)| \times \|x_n\| \\ &< 1 + \|x\| + \sum_{n\in\sigma_0} |x_n^*(x)| \times \|x_n\|. \end{aligned}$$

(ii) $\Longrightarrow$ (iii) This follows immediately from the Banach–Steinhaus theorem.

(iii) $\Longrightarrow$ (i) Put $C = \sup_{\sigma\in\mathfrak{F}} \|S_\sigma\| < \infty$ and fix $x \in \mathcal{X}$. Then, for each $\varepsilon > 0$, there exists $\widetilde{x} \in \text{Lin}\{x_n : n \ge 1\}$ such that

$$\|x - \widetilde{x}\| \le \min\left(\frac{\varepsilon}{2C}, \frac{\varepsilon}{2}\right).$$

We can write

$$\widetilde{x} = \sum a_n x_n,$$

where the sum runs over some $\widetilde{\sigma} \in \mathfrak{F}$. Then it is easy to see that, for any $\sigma \in \mathfrak{F}$, $\widetilde{\sigma} \subset \sigma$, we have $S_\sigma(\widetilde{x}) = \widetilde{x}$. Hence, using the triangle inequality, we get

$$\begin{aligned} \|x - S_\sigma(x)\| &\le \|x - \widetilde{x}\| + \|\widetilde{x} - S_\sigma(x)\| \\ &= \|x - \widetilde{x}\| + \|S_\sigma(\widetilde{x}) - S_\sigma(x)\| \\ &\le \frac{\varepsilon}{2} + C\frac{\varepsilon}{2C} = \varepsilon \qquad (\sigma \in \mathfrak{F},\ \widetilde{\sigma} \subset \sigma). \end{aligned}$$

(i) $\Longrightarrow$ (iv) By the definition of weak-star topology, the sequence $(x_n^*)_{n\ge1}$ is an unconditional basis of $\mathcal{X}^*$, equipped with the weak-star topology, if and only if, for each $x^* \in \mathcal{X}^*$, $x \in \mathcal{X}$ and $\varepsilon > 0$, there exists $\sigma_0 \in \mathfrak{F}$ such that, for all $\sigma \in \mathfrak{F}$, with $\sigma_0 \subset \sigma$, we have

$$\left| x^*(x) - \sum_{n\in\sigma} x^*(x_n)x_n^*(x) \right| < \varepsilon, \tag{10.21}$$

and observe that

$$x^*(x) - \sum_{n\in\sigma} x^*(x_n)x_n^*(x) = x^*(x - S_\sigma(x)).$$

But, by hypothesis, for each $x^* \in \mathcal{X}^*$, $x \in \mathcal{X}$ and $\varepsilon > 0$, there exists $\sigma_0 \in \mathfrak{F}$ such that, for all $\sigma \in \mathfrak{F}$, with $\sigma_0 \subset \sigma$, we have

$$\|x - S_\sigma(x)\| < \frac{\varepsilon}{1 + \|x^*\|}.$$

Hence,

$$|x^*(x - S_\sigma(x))| < \|x^*\| \frac{\varepsilon}{1 + \|x^*\|} \le \varepsilon,$$

which proves (10.21). Hence, the sequence $(x_n^*)_{n\ge 1}$ is an unconditional basis of $\mathcal{X}^*$.

(iv) $\Longrightarrow$ (ii) Fix $x \in \mathcal{X}$, and let $x^* \in \mathcal{X}^*$. Then there exists $\sigma_0 \in \mathfrak{F}$ such that

$$|x^*(x - S_\sigma(x))| \le 1$$

for all $\sigma \in \mathfrak{F}$, with $\sigma_0 \subset \sigma$. For any such $\sigma \in \mathfrak{F}$, we then decompose S_σ as in the proof of (i) $\Longrightarrow$ (ii) to obtain

$$\begin{aligned} |x^*(x - S_\sigma(x))| &= \left| x^*\left(x - \sum_{n\in\sigma} x_n^*(x)x_n \right) \right| \\ &\le \left| x^*\left(x - \sum_{n\in\sigma\cup\sigma_0} x_n^*(x)x_n \right) \right| \\ &\quad + \left| x^*\left(\sum_{n\in\sigma_0\setminus(\sigma\cap\sigma_0)} x_n^*(x)x_n \right) \right| \\ &< 1 + \|x^*\| \sum_{n\in\sigma_0} |x_n^*(x)| \times \|x_n\|. \end{aligned}$$

Therefore,

$$\sup_{\sigma\in\mathfrak{F}} |x^*(x - S_\sigma(x))| \le C \tag{10.22}$$

for some positive constant $C = C(x, x^*)$. This means that the set $\{x - S_\sigma(x) : \sigma \in \mathfrak{F}\}$ is weakly bounded and thus, by Theorem 1.27, it is norm-bounded.

(iii) $\Longrightarrow$ (v) This part is quite long. On the one hand, Theorem 10.9 ensures the inclusion $\mathfrak{Mult}(\mathfrak{X}) \subset \ell^\infty$. On the other, by (10.20), our hypothesis says that

$$C = \sup_{\sigma\in\mathfrak{F}} \|\mu^\sigma\|_{\mathfrak{Mult}(\mathfrak{X})} < \infty.$$

Let $\mathcal{E}$ be the space of finitely supported complex sequences, i.e. the linear span of μ^σ, $\sigma \in \mathfrak{F}$, in ℓ^∞. For each $\mu \in \mathcal{E}$, there exist $a_1, \dots, a_n \in \mathbb{C}$ such that

$$\mu = \sum_{k=1}^{n} a_k \mu^{\sigma_k}, \quad \sigma_k \in \mathfrak{F}, \quad \sigma_k \cap \sigma_j = \emptyset, \ k \neq j. \tag{10.23}$$

Since $\mathfrak{Mult}(\mathfrak{X})$ is a linear space and $\mu^\sigma \in \mathfrak{Mult}(\mathfrak{X})$, $\sigma \in \mathfrak{F}$, we clearly have $\mathcal{E} \subset \mathfrak{Mult}(\mathfrak{X})$. We show furthermore that there exists a constant $C > 0$ such that

$$\|\mu\|_{\mathfrak{Mult}(\mathfrak{X})} \leq C\|\mu\|_\infty \qquad (\mu \in \mathcal{E}).$$

To prove this, we start with a special case. In the decomposition (10.23), assume that all the coefficients a_k are real, $1 \leq k \leq n$. Without loss of generality, we may assume a_k are enumerated such that $a_1 \leq a_2 \leq \cdots \leq a_n$. Then we can write

$$\mu = a_1 \mu^{\sigma'_1} + \sum_{k=2}^{n} (a_k - a_{k-1}) \mu^{\sigma'_k},$$

where

$$\sigma'_k = \bigcup_{i=k}^{n} \sigma_i \qquad (1 \leq k \leq n).$$

Hence,

$$\begin{aligned}\|\mu\|_{\mathfrak{Mult}(\mathfrak{X})} &\leq |a_1| \, \|\mu^{\sigma'_1}\|_{\mathfrak{Mult}(\mathfrak{X})} + \sum_{k=2}^{n} (a_k - a_{k-1}) \|\mu^{\sigma'_k}\|_{\mathfrak{Mult}(\mathfrak{X})} \\ &\leq C(|a_1| + a_n - a_1) \leq 3C\|\mu\|_\infty.\end{aligned}$$

If the coefficients a_k are complex, separating real and imaginary parts, we easily obtain

$$\|\mu\|_{\mathfrak{Mult}(\mathfrak{X})} \leq 6C\|\mu\|_\infty \qquad (\mu \in \mathcal{E}). \tag{10.24}$$

Now, fix any $\mu \in \ell^\infty$. Consider the linear map M_μ defined on $\mathrm{Lin}\{x_n : n \geq 1\}$ by $M_\mu x_n = \mu_n x_n$, $n \geq 1$. To show that $\mu \in \mathfrak{Mult}(\mathfrak{X})$, we must show that M_μ can be extended to a bounded operator on $\mathcal{X}$. To do so, note that, for any $x = \sum_{k=1}^{N} a_k x_k \in \mathrm{Lin}\{x_n : n \geq 1\}$, we have

$$M_\mu x = \sum_{k=1}^{N} a_k \mu_k x_k = M_{\mu^{(N)}} x,$$

where $\mu^{(N)} = (\mu_1, \mu_2, \dots, \mu_N, 0, 0, \dots)$. Since $\mu^{(N)} \in \mathcal{E}$, and we showed that $\mu^{(N)} \in \mathfrak{Mult}(\mathfrak{X})$ with (10.24), then we get

$$\|M_\mu x\| = \|M_{\mu^{(N)}} x\| \leq \|\mu^{(N)}\|_{\mathfrak{Mult}(\mathfrak{X})} \|x\| \leq 6C\|\mu^{(N)}\|_\infty \|x\|.$$

But $\|\mu^{(N)}\|_\infty \le \|\mu\|_\infty$ $(N \ge 1)$, whence we obtain

$$\|M_\mu x\| \le 6C\|\mu\|_\infty \|x\| \qquad (x \in \mathrm{Lin}\{x_n : n \ge 1\}).$$

Therefore, since the sequence $(x_n)_{n\ge1}$ is complete, we can extend M_μ to a bounded operator on $\mathcal{X}$ such that

$$\|\mu\|_{\mathfrak{Mult}(\mathfrak{X})} \le 6C\|\mu\|_\infty.$$

(v) $\Longrightarrow$ (iii) Let J denote the canonical injection from ℓ^∞ into $\mathfrak{Mult}(\mathfrak{X})$. Using Theorem 10.9 and the closed graph theorem (Corollary 1.18), we easily see that the linear mapping J is continuous. Hence, there exists a constant $c > 0$ such that

$$\|\mu\|_{\mathfrak{Mult}(\mathfrak{X})} \le c\|\mu\|_\infty \qquad (\mu \in \ell^\infty). \tag{10.25}$$

In particular, for each $\sigma \in \mathfrak{F}$, we have

$$\sup_{\sigma\in\mathfrak{F}} \|\mu^\sigma\|_{\mathfrak{Mult}(\mathfrak{X})} \le c.$$

and, in light of (10.20), we deduce (iii).

This completes the proof of Theorem 10.14. □

In the framework of Hilbert spaces, we will see that, in the assertion (iv) of Theorem 10.14, we can replace the weak-star topology by the norm topology. Before proving this result, let us recall that, as usual, we identify the Hilbert space $\mathcal{H}$ with its dual. In particular, given a minimal sequence $\mathfrak{X} = (x_n)_{n\ge1}$ in $\mathcal{H}$, a biorthogonal sequence associated with $\mathfrak{X}$ is a sequence $(x_n^*)_{n\ge1}$ of elements of $\mathcal{H}$ such that

$$\langle x_n, x_k^* \rangle = \delta_{n,k}.$$

Note that there is a unique biorthogonal sequence that lies in Span$\{x_n : n \ge 1\}$. Indeed, for the existence, take any biorthogonal sequence and project onto Span$\{x_n : n \ge 1\}$. As is easily seen, the projected sequence still remains a biorthogonal sequence. Moreover, it is unique by the Hahn–Banach theorem.

Corollary 10.15 *Let $\mathcal{H}$ be a Hilbert space, let $\mathfrak{X} = (x_n)_{n\ge1}$ be a minimal and complete sequence in $\mathcal{H}$, and let $(x_n^*)_{n\ge1}$ be the biorthogonal sequence associated with $\mathfrak{X}$. Then the following are equivalent.*

(i) *The sequence $(x_n)_{n\ge1}$ is an unconditional basis of $\mathcal{H}$.*
(ii) *The sequence $(x_n^*)_{n\ge1}$ is an unconditional basis of $\mathcal{H}$.*

Proof Put

$$V_\sigma = \sum_{n\in\sigma} x_n^* \otimes x_n, \qquad (\sigma \in \mathfrak{F}).$$

By (10.19), we have $V_\sigma = S_\sigma^*$. According to Theorem 10.14,

$$\text{(i)} \quad \Longleftrightarrow \quad \sup_{\sigma \in \mathfrak{F}} \|S_\sigma\| < \infty$$

and

$$\text{(ii)} \quad \Longleftrightarrow \quad \left(\sup_{\sigma \in \mathfrak{F}} \|V_\sigma\| < \infty \text{ and } (x_n^*)_{n\geq 1} \text{ is complete in } \mathcal{H} \right).$$

Hence, we just need to verify that, if $(x_n)_{n\geq 1}$ is an unconditional basis of $\mathcal{H}$, then the sequence $(x_n^*)_{n\geq 1}$ is complete in $\mathcal{H}$. So, let $x \in \mathcal{H}$ be such that $\langle x, x_p^* \rangle = 0$ for all $p \geq 1$. Then, for each $\sigma \in \mathfrak{F}$, $S_\sigma(x) = 0$, and since $(x_n)_{n\geq 1}$ is an unconditional basis, we must have $x = 0$. Thus, $(x_n^*)_{n\geq 1}$ is complete. □

Exercises

Exercise 10.4.1 Let $\mathcal{X}$ be a Banach space, and let $x_1, x_2 \in \mathcal{X}$ be independent. Suppose that the scalar sequences $(a_n)_{n\geq 1}$ and $(b_n)_{n\geq 1}$ are such that the sequence $(a_n x_1 + b_n x_2)_{n\geq 1}$ is Cauchy in $\mathcal{X}$. Show that $(a_n)_{n\geq 1}$ and $(b_n)_{n\geq 1}$ are both convergent.
Hint: Consider the two-dimensional space generated by x_1 and x_2, and remember that, in a finite-dimensional space, all norms are equivalent.

Exercise 10.4.2 Let $\mathcal{X}$ and $\mathcal{Y}$ be Banach spaces, and let $A \in \mathcal{L}(\mathcal{X}, \mathcal{Y})$ be such that $\mathcal{R}(A) \subset \text{Span}\{y_1, y_2\}$, where $y_1, y_2 \in \mathcal{Y}$ are independent. Decompose A as

$$Ax = (A_1 x)y_1 + (A_2 x)y_2 \qquad (x \in X).$$

Show that $A_1, A_2 \in \mathcal{X}^*$.
Hint: Apply the same trick as in Exercise 10.4.1.

10.5 Riesz basis

We say that a sequence $\mathfrak{X} = (x_n)_{n\geq 1}$ in a Hilbert space $\mathcal{H}$ is a *Riesz basis* of $\mathcal{H}$ if there exists an isomorphism V from $\mathcal{H}$ onto itself such that $(Vx_n)_{n\geq 1}$ is an orthonormal basis of $\mathcal{H}$. In light of the Riesz–Fischer theorem, we easily see that a sequence $(x_n)_{n\geq 1}$ is a Riesz basis of $\mathcal{H}$ if and only if there exists an isomorphism

$$\begin{array}{rccc} U_{\mathfrak{X}}: & \mathcal{X} & \longrightarrow & \ell^2 \\ & x_n & \longmapsto & \mathfrak{e}_n, \end{array}$$

where $(\mathfrak{e}_n)_{n\geq 1}$ is the canonical orthonormal basis of ℓ^2. The operator $U_{\mathfrak{X}}$ is called the *orthogonalizer* of the Riesz basis $\mathfrak{X}$. Note that, since the sequence $\mathfrak{X}$ is complete, the operator $U_{\mathfrak{X}}$ is unique.

In some cases, we need to consider a sequence that is not complete and yet has a Riesz basis-type property. More precisely, the sequence $\mathfrak{X} = (x_n)_{n\geq 1}$ is called a *Riesz sequence* if it is a Riesz basis of its closed linear span. Hence, a sequence $(x_n)_{n\geq 1}$ is a Riesz sequence of $\mathcal{H}$ if and only if there exists an isomorphism $U_{\mathfrak{X}}$ from $\text{Span}\{x_n : n \geq 1\}$ onto ℓ^2 such that $U_{\mathfrak{X}} x_n = \mathfrak{e}_n$, $n \geq 1$.

It is easy to check that an orthonormal basis is an unconditional basis. Hence, if $(x_n)_{n\geq 1}$ is a Riesz basis of $\mathcal{H}$ and if $U_{\mathfrak{X}}$ denotes its orthogonalizer, then

$$U_{\mathfrak{X}} x = \sum_{n\geq 1} \langle U_{\mathfrak{X}} x, \mathfrak{e}_n \rangle_{\ell^2} \mathfrak{e}_n = \sum_{n\geq 1} \langle U_{\mathfrak{X}} x, U_{\mathfrak{X}} x_n \rangle_{\ell^2} U_{\mathfrak{X}} x_n \qquad (x \in \mathcal{H}), \tag{10.26}$$

where the series is unconditionally convergent in ℓ^2. Using the continuity of $U_{\mathfrak{X}}^{-1}$, we thus obtain

$$\begin{aligned} x = U_{\mathfrak{X}}^{-1} U_{\mathfrak{X}} x &= U_{\mathfrak{X}}^{-1} \left(\sum_{n\geq 1} \langle U_{\mathfrak{X}} x, U_{\mathfrak{X}} x_n \rangle_{\ell^2} U_{\mathfrak{X}} x_n \right) \\ &= \sum_{n\geq 1} \langle x, U_{\mathfrak{X}}^* U_{\mathfrak{X}} x_n \rangle_{\mathcal{H}} x_n, \end{aligned}$$

where the series is unconditionally convergent in $\mathcal{H}$. Therefore, a Riesz basis is in particular an unconditional basis. In Section 10.9, we will see that these two notions are in fact equivalent in the framework of Hilbert spaces.

Theorem 10.16 *Let $\mathcal{H}$ be a Hilbert space, let $\mathfrak{X} = (x_n)_{n\geq 1}$ be a Riesz basis of $\mathcal{H}$, and let $U_{\mathfrak{X}}$ be the orthogonalizer of $\mathfrak{X}$. Then the following hold.*

(i) *The sequence $(x_n)_{n\geq 1}$ is uniformly minimal and its biorthogonal sequence $(x_n^*)_{n\geq 1}$ is given by*

$$x_n^* = U_{\mathfrak{X}}^* U_{\mathfrak{X}} x_n \qquad (n \geq 1).$$

(ii) *The sequence $\mathfrak{X}^* = (x_n^*)_{n\geq 1}$ is also a Riesz basis of $\mathcal{H}$.*

Proof (i) We have

$$\langle U_{\mathfrak{X}}^* U_{\mathfrak{X}} x_m, x_n \rangle_{\mathcal{H}} = \langle U_{\mathfrak{X}} x_m, U_{\mathfrak{X}} x_n \rangle_{\ell^2} = \langle \mathfrak{e}_m, \mathfrak{e}_n \rangle_{\ell^2} = \delta_{m,n} \qquad (m, n \geq 1).$$

Hence, $(x_n)_{n\geq 1}$ is minimal and $(U_{\mathfrak{X}}{}^* U_{\mathfrak{X}} x_n)_{n\geq 1}$ is its biorthogonal. Note that the biorthogonal is unique, because $(x_n)_{n\geq 1}$ is complete in $\mathcal{H}$. Moreover, we have

$$\begin{aligned}\|x_n\| \times \|x_n^*\| &= \|x_n\| \times \|U_{\mathfrak{X}}^* U_{\mathfrak{X}} x_n\| \\ &= \|U_{\mathfrak{X}}^{-1} U_{\mathfrak{X}} x_n\| \times \|U_{\mathfrak{X}}^* U_{\mathfrak{X}} x_n\| \\ &= \|U_{\mathfrak{X}}^{-1} \mathfrak{e}_n\| \times \|U_{\mathfrak{X}}^* \mathfrak{e}_n\| \\ &\leq \|U_{\mathfrak{X}}^{-1}\| \times \|U_{\mathfrak{X}}^*\| \qquad (n \geq 1),\end{aligned}$$

which, by Lemma 10.1, ensures that $(x_n)_{n\geq1}$ is uniformly minimal.

(ii) The operator $U_{\mathfrak{X}}^{*\,-1}$ is an isomorphism from $\mathcal{H}$ onto ℓ^2 that sends the sequence $(x_n^*)_{n\geq1}$ onto the canonical orthonormal basis $(\mathfrak{e}_n)_{n\geq1}$ of ℓ^2. Hence, by definition, $(x_n^*)_{n\geq1}$ is a Riesz basis of $\mathcal{H}$. □

The following result shows that a small perturbation of an orthonormal basis gives a Riesz basis.

Theorem 10.17 *Let $(e_n)_{n\geq1}$ be an orthonormal basis of a Hilbert space $\mathcal{H}$, and let $(x_n)_{n\geq1}$ be a sequence in $\mathcal{H}$ such that*

$$\sum_{n=1}^{\infty} \|x_n - e_n\|^2 < 1.$$

Then $(x_n)_{n\geq1}$ is a Riesz basis of $\mathcal{H}$.

Proof Let U be the linear map defined on the orthonormal basis by $Ue_n = x_n$, $n \geq 1$. Then, for any $x = \sum_{n=1}^{N} a_n e_n \in \mathcal{H}$, we have

$$\begin{aligned}\|(I-U)x\| &= \left\| \sum_{n=1}^{N} a_n (x_n - e_n) \right\| \\ &\leq \sum_{n=1}^{N} |a_n| \times \|x_n - e_n\| \\ &\leq \left(\sum_{n=1}^{N} |a_n|^2 \right)^{1/2} \left(\sum_{n=1}^{N} \|x_n - e_n\|^2 \right)^{1/2} \\ &\leq \left(\sum_{n=1}^{\infty} \|x_n - e_n\|^2 \right)^{1/2} \|x\|.\end{aligned}$$

Hence, $I - U$ extends to a bounded operator on $\mathcal{H}$ of norm strictly smaller than 1. Therefore, $U = I - (I - U)$ is an isomorphism from $\mathcal{H}$ onto itself, which implies that the sequence $(x_n)_{n\geq1}$ is a Riesz basis of $\mathcal{H}$. □

In Corollary 10.34, we slightly improve the conclusion of Theorem 10.17.

Exercises

Exercise 10.5.1 Let $\mathfrak{X} = (x_n)_{n\geq1}$ be a Riesz basis of a Hilbert space $\mathcal{H}$. Show that there exists a constant $\delta > 0$ such that, for any sequence $\mathfrak{X}' = (x_n')_{n\geq1}$ satisfying

$$\sum_{n\geq 1} \|x_n - x'_n\|^2 \leq \delta,$$

the sequence $\mathfrak{X}'$ is also a Riesz basis of $\mathcal{H}$.
Hint: Consider the orthogonalizer $U_{\mathfrak{X}}$ and that, with $x = \sum_{n=1}^N a_n x_n$,

$$\left(\sum_{n=1}^{N} |a_n|^2 \right)^{1/2} = \|U_{\mathfrak{X}} x\| \leq \|U_{\mathfrak{X}}\| \, \|x\|.$$

Then use a similar argument as in the proof of Theorem 10.17.

Exercise 10.5.2 Two sequences of vectors $(x_n)_{n\geq 1}$ and $(y_n)_{n\geq 1}$ in a Banach space $\mathcal{X}$ are said to be *quadratically close* if

$$\sum_{n=1}^{\infty} \|x_n - y_n\|^2 < \infty.$$

Let $\mathcal{H}$ be a separable Hilbert space and let $(e_n)_{n\geq 1}$ be an orthonormal basis for $\mathcal{H}$. Let $(x_n)_{n\geq 1}$ be a w-topologically linearly independent sequence in $\mathcal{H}$ that is quadratically close to $(e_n)_{n\geq 1}$. Show that $(x_n)_{n\geq 1}$ is a Riesz basis of $\mathcal{H}$.
Hint: Define

$$Tx = \sum_{n=1}^{N} a_n (e_n - x_n) = x - \sum_{n=1}^{N} a_n x_n,$$

for $x = \sum_{n=1}^N a_n e_n \in \mathcal{H}$. Show that T extends to a linear bounded operator on $\mathcal{H}$ such that

$$\sum_{n=1}^{\infty} \|T e_n\|^2 < \infty,$$

and thus deduce that T is compact. Show that $\ker(I - T) = \{0\}$ and apply the Fredholm alternative (Theorem 2.6) to conclude that $I - T$ is invertible.
Remark: See Section 10.1 for the definition of a w-topologically linearly independent sequence.

10.6 The mappings $J_{\mathfrak{X}}$, $V_{\mathfrak{X}}$ and $\Gamma_{\mathfrak{X}}$

In order to make the connection between the Riesz basis property and an abstract interpolation problem, we define the operator $J_{\mathfrak{X}}$, where $\mathfrak{X} = (x_n)_{n\geq 1}$ is a sequence in a Hilbert space $\mathcal{H}$, as the linear mapping from $\mathcal{H}$ into the linear space of complex sequences, by

$$J_{\mathfrak{X}}(x) = (\langle x, x_n \rangle)_{n\geq 1} \qquad (x \in \mathcal{H}). \tag{10.27}$$

The following properties are immediate to check.

(i) The sequence $\mathfrak{X}$ is complete in $\mathcal{H}$ if and only if the operator $J_{\mathfrak{X}}$ is one-to-one.
(ii) The sequence $\mathfrak{X}$ is minimal in $\mathcal{H}$ if and only if the range of the operator $J_{\mathfrak{X}}$ contains all the sequences $(\delta_{m,n})_{m\geq 1}$, $n \geq 1$.

Our first result gives a simple sufficient condition for the boundedness of $J_{\mathfrak{X}}$. We recall that $\mathcal{R}(J_{\mathfrak{X}}) = J_{\mathfrak{X}}\mathcal{H}$.

Lemma 10.18 *Let $\mathfrak{X} = (x_n)_{n\geq 1}$ be a sequence in a Hilbert space $\mathcal{H}$ such that $\mathcal{R}(J_{\mathfrak{X}}) \subset \ell^2$. Then $J_{\mathfrak{X}}$ is bounded as an operator from $\mathcal{H}$ onto ℓ^2.*

Proof Considering the map

$$\begin{array}{rccl} J_{\mathfrak{X}}: & \mathcal{H} & \longrightarrow & \ell^2 \\ & x & \longmapsto & (\langle x, x_n\rangle)_{n\geq 1}, \end{array}$$

we use the closed graph theorem (Corollary 1.18) to show that $J_{\mathfrak{X}}$ is continuous. So, let $(g_n)_{n\geq 1}$ be a sequence in $\mathcal{H}$ such that $g_n \longrightarrow g$ in $\mathcal{H}$ and $J_{\mathfrak{X}} g_n \longrightarrow a = (a_k)_{k\geq 1}$ in ℓ^2, as $n \longrightarrow \infty$. Then, for each $m \geq 1$, we have

$$|\langle g_n, x_m\rangle - a_m|^2 \leq \sum_{k\geq 1} |\langle g_n, x_k\rangle - a_k|^2 = \|J_{\mathfrak{X}} g_n - a\|^2.$$

Hence, on the one hand, we have

$$\lim_{n\to\infty} \langle g_n, x_m\rangle = a_m \qquad (m \geq 1)$$

and, on the other, we have

$$\lim_{n\to\infty} \langle g_n, x_m\rangle = \langle g, x_m\rangle \qquad (m \geq 1).$$

Thus $a_m = \langle g, x_m\rangle$, whence $a = J_{\mathfrak{X}} g$. Consequently, the closed graph theorem (Corollary 1.18) implies that the operator $J_{\mathfrak{X}}$ is continuous. □

Let $\mathcal{H}_1$ and $\mathcal{H}_2$ be two Hilbert spaces, and let $A : \mathcal{H}_1 \longrightarrow \mathcal{H}_2$ be a linear map. If A is continuous on a dense subset $\mathcal{E}_1$ of $\mathcal{H}_1$, then it is easy to see that A can be extended into a continuous operator on $\mathcal{H}_1$. However, there is no reason for the extension to coincide with A on the whole space $\mathcal{H}_1$. This property is true for $A = J_{\mathfrak{X}}$.

Lemma 10.19 *Let $\mathfrak{X} = (x_n)_{n\geq 1}$ be a sequence in a Hilbert space $\mathcal{H}$. Assume that there exist a dense subset $\mathcal{E}$ in $\mathcal{H}$ and a constant $c > 0$ such that*

$$\|J_{\mathfrak{X}} x\|_{\ell^2} \leq c\|x\|, \tag{10.28}$$

for all $x \in \mathcal{E}$. Then $\mathcal{R}(J_{\mathfrak{X}}) \subset \ell^2$ and

$$\|J_{\mathfrak{X}} x\|_{\ell^2} \leq c\|x\| \qquad (x \in \mathcal{H}).$$

Proof Let $x \in \mathcal{H}$. Then there exists a sequence $(u_m)_{m\geq 1} \subset \mathcal{E}$ such that $u_m \longrightarrow x$ in $\mathcal{H}$, as $m \longrightarrow \infty$. By (10.28), we see that $(J_{\mathfrak{X}}u_m)_{m\geq 1}$ is a Cauchy sequence in ℓ^2, whence it converges to an element $a \in \ell^2$. By (10.27),

$$\langle x, x_n \rangle = \lim_{m\to\infty} \langle u_m, x_n \rangle = \lim_{m\to\infty} \langle J_{\mathfrak{X}}u_m, \mathfrak{e}_n \rangle = \langle a, \mathfrak{e}_n \rangle \qquad (n \geq 1).$$

Hence, $J_{\mathfrak{X}}x = a$. In particular, $J_{\mathfrak{X}}x \in \ell^2$. Moreover, letting $m \longrightarrow \infty$ in

$$\|J_{\mathfrak{X}}u_m\|_{\ell^2} \leq c\|u_m\|$$

gives $\|J_{\mathfrak{X}}x\|_{\ell^2} \leq c\|x\|$. □

The last result shows that $J_{\mathfrak{X}}$ is a familiar object whenever $\mathfrak{X} = (x_n)_{n\geq 1}$ is a Riesz basis.

Lemma 10.20 *Let $\mathcal{H}$ be a Hilbert space, let $\mathfrak{X} = (x_n)_{n\geq 1}$ be a Riesz basis of $\mathcal{H}$, and let $U_{\mathfrak{X}}$ be the orthogonalizer of $\mathfrak{X}$. Then we have*

$$J_{\mathfrak{X}} = U_{\mathfrak{X}^*} \quad \textit{and} \quad J_{\mathfrak{X}^*} = U_{\mathfrak{X}}.$$

In particular, the mappings $J_{\mathfrak{X}^*}$ and $J_{\mathfrak{X}}$ are isomorphisms from $\mathcal{H}$ onto ℓ^2.

Proof By definition, for each $x \in \mathcal{H}$, we have

$$\begin{aligned} J_{\mathfrak{X}^*}x &= (\langle x, x_n^* \rangle)_{n\geq 1} \\ &= (\langle x, U_{\mathfrak{X}}{}^*U_{\mathfrak{X}}x_n \rangle)_{n\geq 1} \\ &= (\langle U_{\mathfrak{X}}x, U_{\mathfrak{X}}x_n \rangle)_{n\geq 1}. \end{aligned}$$

The identity (10.26) now reveals that $J_{\mathfrak{X}^*} = U_{\mathfrak{X}}$. In particular, $J_{\mathfrak{X}^*}$ is an isomorphism from $\mathcal{H}$ onto ℓ^2. Moreover, by Theorem 10.16(ii), we can reverse the role of $\mathfrak{X}$ and $\mathfrak{X}^*$, which shows that $J_{\mathfrak{X}}$ is also an isomorphism from $\mathcal{H}$ onto ℓ^2 and $J_{\mathfrak{X}} = U_{\mathfrak{X}^*}$. □

Besides the operator $J_{\mathfrak{X}}$, there are two other important operators that appear naturally in the theory of the Riesz basis. In the following, the space of complex finitely supported sequences, equipped with the norm of ℓ^2, is denoted by $\ell^2_{\mathfrak{F}}$. Fix $\mathfrak{X} = (x_n)_{n\geq 1} \in \mathcal{H}$.

The first important operator is the linear mapping

$$\begin{aligned} V_{\mathfrak{X}}: \ \ell^2_{\mathfrak{F}} &\longrightarrow \mathcal{H} \\ a &\longmapsto \sum_{n\geq 1} a_n x_n, \end{aligned}$$

where $a = (a_n)_{n\geq 1} \in \ell^2_{\mathfrak{F}}$. Since a is finitely supported, the sum is well defined. For each $b \in \mathcal{H}$, we have

$$\langle V_{\mathfrak{X}}a, b \rangle_{\mathcal{H}} = \left\langle \sum_{n\geq 1} a_n x_n, b \right\rangle_{\mathcal{H}} = \sum_{n\geq 1} a_n \overline{\langle b, x_n \rangle_{\mathcal{H}}}.$$

If $J_{\mathfrak{X}}b \in \ell^2$, this identity can be rewritten as

$$\langle V_{\mathfrak{X}}a, b\rangle_{\mathcal{H}} = \langle a, J_{\mathfrak{X}}b\rangle_{\ell^2}. \tag{10.29}$$

Furthermore, we have

$$J_{\mathfrak{X}^*}V_{\mathfrak{X}}a = a \qquad (a \in \ell^2_{\mathfrak{F}}) \tag{10.30}$$

and

$$V_{\mathfrak{X}}J_{\mathfrak{X}^*}x = x \qquad (x \in \mathrm{Lin}\{x_n : n \geq 1\}). \tag{10.31}$$

The second important operator, called the *Gram matrix* of the sequence $\mathfrak{X}$, is the singly infinite matrix

$$\Gamma_{\mathfrak{X}} = (\langle x_i, x_j\rangle_{\mathcal{H}})_{i,j\geq 1}.$$

It is clear that $\Gamma_{\mathfrak{X}}$ is a self-adjoint matrix and, for each $a \in \ell^2_{\mathfrak{F}}$,

$$\begin{aligned} \|V_{\mathfrak{X}}a\|^2_{\mathcal{H}} &= \left\langle \sum_{n\geq 1} a_n x_n, \sum_{k\geq 1} a_k x_k \right\rangle_{\mathcal{H}} \\ &= \sum_{n,k\geq 1} a_n \overline{a_k} \langle x_n, x_k\rangle_{\mathcal{H}} \\ &= \langle \Gamma_{\mathfrak{X}}a, a\rangle_{\ell^2}. \end{aligned} \tag{10.32}$$

Since a linear mapping A on a Hilbert space $\mathcal{K}$ is continuous if and only if

$$\sup_{x\in\mathcal{K},\, \|x\|\leq 1} |\langle Ax, x\rangle_{\mathcal{K}}| < \infty,$$

we easily deduce from (10.32) that $V_{\mathfrak{X}}$ extends to a bounded operator from ℓ^2 into $\mathcal{H}$ if and only if $\Gamma_{\mathfrak{X}}$ defines a bounded operator on ℓ^2. Furthermore, using (10.29), we see that $V_{\mathfrak{X}}$ extends to a bounded operator from ℓ^2 into $\mathcal{H}$ if and only if $J_{\mathfrak{X}}$ is bounded from $\mathcal{H}$ into ℓ^2. Therefore, we conclude that the mappings $J_{\mathfrak{X}} : \mathcal{H} \longrightarrow \ell^2$, $V_{\mathfrak{X}} : \ell^2 \longrightarrow \mathcal{H}$ and $\Gamma_{\mathfrak{X}} : \ell^2 \longrightarrow \ell^2$ are simultaneously bounded. In this situation, by (10.32) and (10.29), we have

$$V_{\mathfrak{X}} = J^*_{\mathfrak{X}} \quad \text{and} \quad \Gamma_{\mathfrak{X}} = V^*_{\mathfrak{X}}V_{\mathfrak{X}}. \tag{10.33}$$

Exercises

Exercise 10.6.1 Let $\mathfrak{X} = (x_n)_{n\geq 1}$ be a sequence of a Hilbert space $\mathcal{H}$. Assume that $\ell^2 \subset \mathcal{R}(J_{\mathfrak{X}})$.

(i) Show that there exists a constant $M > 0$ such that, for each $(c_n)_{n\geq 1} \in \ell^2$, the interpolation problem

$$\langle x, x_n\rangle = c_n \qquad (n \geq 1)$$

has at least one solution $x \in \mathcal{H}$ satisfying

$$\|x\|^2 \le M \sum_{n=1}^{\infty} |c_n|^2.$$

Hint: Apply the open mapping theorem (Theorem 1.14).

(ii) Deduce that $\mathfrak{X}$ is uniformly minimal.

Exercise 10.6.2 Let $\mathcal{H} = L^2(0,1)$ and let $\mathfrak{X} = (x_j)_{j\ge 0}$ be the family of vectors in $\mathcal{H}$ defined by

$$x_j(t) = t^j \qquad (j \ge 0,\ t \in [0,1]).$$

The associated Gram matrix $\Gamma_{\mathfrak{X}}$ is called the *Hilbert matrix*.

(i) Let $c_0, c_1, \dots, c_N$ be $(N+1)$ nonnegative real numbers and let $f(z) = \sum_{n=0}^{N} c_n z^n$.

(a) Using Cauchy's theorem, show that

$$\int_{-1}^{1} |f(x)|^2\, dx \le \frac{1}{2} \int_{-\pi}^{\pi} |f(e^{it})|^2\, dt. \tag{10.34}$$

(b) Deduce that, if $c_n \ge 0$ $(n = 0, 1, \dots)$, then

$$\sum_{n,m=0}^{\infty} \frac{c_n c_m}{n+m+1} \le \pi \sum_{n=0}^{\infty} |c_n|^2. \tag{10.35}$$

(ii) Conclude that the Hilbert matrix defines a bounded linear operator on ℓ^2 of norm not exceeding π.

(iii) Using the Carleson theorem (see Theorem 5.15), give another proof of inequality (10.34).

Remark: Inequality (10.35) is called *Hilbert's inequality*.

Exercise 10.6.3 Let $\mathfrak{X} = (x_n)_{n\ge 1}$ and $\mathfrak{X}^* = (x_n^*)_{n\ge 1}$ be complete biorthogonal sequences belonging to a Hilbert space $\mathcal{H}$. Show that $\mathcal{R}(J_{\mathfrak{X}}) \subset \ell^2$ if and only if $\ell^2 \subset \mathcal{R}(J_{\mathfrak{X}^*})$.

Hint: Use Lemma 10.18 and the remark at the end of the section.

Exercise 10.6.4 Let $\mathfrak{X} = (x_n)_{n\ge 1}$ and $\mathfrak{Y} = (y_n)_{n\ge 1}$ be two complete sequences of a Hilbert space $\mathcal{H}$. Show that there exists a bounded invertible operator $A \in \mathcal{L}(\mathcal{H})$ such that $Ax_n = y_n$, $n \ge 1$, if and only if $\mathcal{R}(J_{\mathfrak{X}}) = \mathcal{R}(J_{\mathfrak{Y}})$.

10.7 Characterization of the Riesz basis

The following result gives a useful characterization of the Riesz basis property in terms of the Gram matrix and the interpolation operator $J_{\mathfrak{X}}$.

Theorem 10.21 (Bari's theorem) *Let $\mathfrak{X} = (x_n)_{n\geq 1}$ be a minimal and complete sequence in a Hilbert space $\mathcal{H}$, and let $\mathfrak{X}^* = (x_n^*)_{n\geq 1}$ be its biorthogonal. Then the following are equivalent.*

(i) *The sequence $\mathfrak{X}$ is a Riesz basis of $\mathcal{H}$.*

(ii) *There exist two constants $c > 0$ and $C > 0$ such that, for any finitely supported complex sequence $(a_n)_{n\geq 1}$, we have*

$$c \sum_{n\geq 1} |a_n|^2 \leq \left\| \sum_{n\geq 1} a_n x_n \right\|^2 \leq C \sum_{n\geq 1} |a_n|^2. \tag{10.36}$$

(iii) *There exist two constants $C > 0$ and $C' > 0$ such that, for any finitely supported complex sequence $(a_n)_{n\geq 1}$, we have*

$$\left\| \sum_{n\geq 1} a_n x_n \right\|^2 \leq C \sum_{n\geq 1} |a_n|^2 \tag{10.37}$$

and

$$\left\| \sum_{n\geq 1} a_n x_n^* \right\|^2 \leq C' \sum_{n\geq 1} |a_n|^2. \tag{10.38}$$

(iv) *$\mathfrak{X}^*$ is complete and there exist two constants $c > 0$ and $c' > 0$ such that, for any finitely supported complex sequence $(a_n)_{n\geq 1}$, we have*

$$c \sum_{n\geq 1} |a_n|^2 \leq \left\| \sum_{n\geq 1} a_n x_n \right\|^2 \tag{10.39}$$

and

$$c' \sum_{n\geq 1} |a_n|^2 \leq \left\| \sum_{n\geq 1} a_n x_n^* \right\|^2. \tag{10.40}$$

(v) *$\mathfrak{X}^*$ is complete in $\mathcal{H}$ and $\mathcal{R}(J_{\mathfrak{X}^*}) = \ell^2$.*

(vi) *$\mathcal{R}(J_{\mathfrak{X}}) = \ell^2$.*

(vii) *$\mathcal{R}(J_{\mathfrak{X}}) \subset \ell^2$ and $\mathcal{R}(J_{\mathfrak{X}^*}) \subset \ell^2$.*

(viii) *The Gram matrices $\Gamma_{\mathfrak{X}}$ and $\Gamma_{\mathfrak{X}^*}$ are bounded operators from ℓ^2 into itself.*

(ix) *The Gram matrix $\Gamma_{\mathfrak{X}}$ is a bounded and invertible operator from ℓ^2 onto itself.*

Proof The proof is quite long. We will proceed as follows: (i) $\Longleftrightarrow$ (ii), (i) $\Longleftrightarrow$ (v), (i) $\Longleftrightarrow$ (vi), (i) $\Longrightarrow$ (iii) $\Longrightarrow$ (vii) $\Longrightarrow$ (viii) $\Longrightarrow$ (ix) $\Longrightarrow$ (i), and finally (i) $\Longrightarrow$ (iv) $\Longrightarrow$ (vii).

(i) $\Longrightarrow$ (ii) By hypothesis, there exists an isomorphism $U_{\mathfrak{X}}$ from $\mathcal{H}$ onto ℓ^2 such that $U_{\mathfrak{X}} x_n = \mathfrak{e}_n$, $n \geq 1$. Hence, for any sequence $(a_n)_{n\geq 1} \in \ell^2_{\mathfrak{F}}$, we have

$$\Big\| \sum_{n\geq 1} a_n x_n \Big\|^2 = \Big\| U_{\mathfrak{X}}^{-1}\Big(\sum_{n\geq 1} a_n \mathfrak{e}_n \Big) \Big\|^2 \asymp \Big\| \sum_{n\geq 1} a_n \mathfrak{e}_n \Big\|^2 = \sum_{n\geq 1} |a_n|^2,$$

which gives assertion (ii).

(ii) $\Longrightarrow$ (i) By hypothesis, there exist two constants $c, C > 0$ such that

$$c\|a\|^2 \leq \|V_{\mathfrak{X}} a\|^2 \leq C\|a\|^2 \qquad (a \in \ell^2_{\mathfrak{F}}).$$

Hence, $V_{\mathfrak{X}}$ is an isomorphism from $\ell^2_{\mathfrak{F}}$ onto $\mathrm{Lin}\{x_n : n \geq 1\}$. By a standard argument, we can now extend $V_{\mathfrak{X}}$ into an isomorphism from ℓ^2 onto $\mathcal{H}$. Since $V_{\mathfrak{X}}^{-1} x_n = \mathfrak{e}_n$, $n \geq 1$, we can conclude that $(x_n)_{n\geq 1}$ is a Riesz basis of $\mathcal{H}$.

(i) $\Longrightarrow$ (v) This is contained in Theorem 10.16 and Lemma 10.20.

(v) $\Longrightarrow$ (i) The completeness of the sequence $(x_n^*)_{n\geq 1}$ in $\mathcal{H}$ assures that the linear map $J_{\mathfrak{X}^*}$ is one-to-one. Hence, we get that $J_{\mathfrak{X}^*}$ is a bijection from $\mathcal{H}$ onto ℓ^2. Moreover, by Lemma 10.18, the operator $J_{\mathfrak{X}^*}$ is continuous from $\mathcal{H}$ onto ℓ^2. Hence, by the Banach isomorphism theorem, it is actually an isomorphism. Since $J_{\mathfrak{X}^*} x_n = \mathfrak{e}_n$, $n \geq 1$, we conclude that $(x_n)_{n\geq 1}$ is a Riesz basis of $\mathcal{H}$.

(i) $\Longleftrightarrow$ (vi) According to Theorem 10.16, the sequence $(x_n)_{n\geq 1}$ is a Riesz basis of $\mathcal{H}$ if and only if the sequence $(x_n^*)_{n\geq 1}$ is also a Riesz basis of $\mathcal{H}$. To conclude, it remains to apply (i) $\Longleftrightarrow$ (v) (which we have just proved) to the sequence $(x_n^*)_{n\geq 1}$.

(i) $\Longrightarrow$ (iii) Since $\mathfrak{X}$ is a Riesz basis, then, by Lemma 10.16, $\mathfrak{X}^*$ is also a Riesz basis. Hence, since (i) $\Longleftrightarrow$ (ii), the inequality (10.36) is satisfied for the sequence $\mathfrak{X}$ and also for the sequence $\mathfrak{X}^*$. In particular, we have (10.37) and (10.38).

(iii) $\Longrightarrow$ (vii) The inequality (10.37) means that the operator $V_{\mathfrak{X}}$ is bounded on $\ell^2_{\mathfrak{F}}$, whereas the inequality (10.37) means that $V_{\mathfrak{X}^*}$ is bounded on $\ell^2_{\mathfrak{F}}$. Hence, these two operators extend into bounded operators from ℓ^2 into $\mathcal{H}$. Hence, $J_{\mathfrak{X}}$ and $J_{\mathfrak{X}^*}$ are bounded operators from $\mathcal{H}$ into ℓ^2. In particular, we have $J_{\mathfrak{X}}\mathcal{H} \subset \ell^2$ and $J_{\mathfrak{X}^*}\mathcal{H} \subset \ell^2$.

(vii) $\Longrightarrow$ (viii) According to Lemma 10.19, the operators $J_{\mathfrak{X}}$ and $J_{\mathfrak{X}^*}$ are continuous from $\mathcal{H}$ into ℓ^2. This is equivalent to saying that $\Gamma_{\mathfrak{X}}$ and $\Gamma_{\mathfrak{X}^*}$ extend to bounded operators on ℓ^2.

(viii) $\Longrightarrow$ (ix) The operator $J_{\mathfrak{X}^*}$ is continuous from $\mathcal{H}$ into ℓ^2 and the operator $V_{\mathfrak{X}}$ is continuous from ℓ^2 into $\mathcal{H}$. Using (10.30) and (10.31), we can conclude that the operator $V_{\mathfrak{X}}$ is an isomorphism from ℓ^2 onto $\mathcal{H}$ and $V_{\mathfrak{X}}^{-1} = J_{\mathfrak{X}^*}$. To conclude that $\Gamma_{\mathfrak{X}}$ is an isomorphism from ℓ^2 onto itself, it remains to use (10.33).

(ix) $\Longrightarrow$ (i) The equality (10.32) shows that $\Gamma_{\mathfrak{X}}$ is a positive operator. Thus, by Corollary 7.9, there exist two positive constants $c, C > 0$ such that

$$c^2\|a\|^2 \le \langle \Gamma_{\mathfrak{X}} a, a\rangle \le C^2\|a\|^2 \qquad (a \in \ell^2).$$

Using (10.32) once more, we get

$$c\|a\| \le \|V_{\mathfrak{X}} a\| \le C\|a\| \quad (a \in \ell^2),$$

which means that $V_{\mathfrak{X}}$ is an isomorphism onto its range. But we have

$$\operatorname{Lin}\{x_n : n \ge 1\} = V_{\mathfrak{X}}\ell^2_{\mathfrak{F}} \subset V_{\mathfrak{X}}\ell^2 \subset \mathcal{H},$$

which implies, by completeness of $(x_n)_{n\ge1}$, that $V_{\mathfrak{X}}\ell^2 = \mathcal{H}$. Therefore $V_{\mathfrak{X}}$ is an isomorphism from ℓ^2 onto $\mathcal{H}$. To conclude that $(x_n)_{n\ge1}$ is a Riesz basis of $\mathcal{H}$, it remains to note that $V_{\mathfrak{X}}\mathfrak{e}_n = x_n$, $n \ge 1$.

(i) $\Longrightarrow$ (iv) This is similar to the proof of (i) $\Longrightarrow$ (iii). It suffices to note that, since $\mathfrak{X}$ is a Riesz basis, then $\mathfrak{X}^*$ is also a Riesz basis, and apply (i) $\Longrightarrow$ (ii).

(iv) $\Longrightarrow$ (vii) The inequality (10.39) means that

$$c\|a\|_{\ell^2} \le \|V_{\mathfrak{X}} a\| \qquad (a \in \ell^2_{\mathfrak{F}}).$$

Using (10.31), we obtain

$$c\|J_{\mathfrak{X}^*}x\| \le \|V_{\mathfrak{X}}J_{\mathfrak{X}^*}x\| = \|x\| \qquad (x \in \operatorname{Lin}\{x_n : n \ge 1\}),$$

which means that $J_{\mathfrak{X}^*}$ is bounded on the dense subset $\operatorname{Lin}\{x_n : n \ge 1\}$. Now Lemma 10.19 implies that $J_{\mathfrak{X}^*}\mathcal{H} \subset \ell^2$. Reversing the role of $\mathfrak{X}$ and $\mathfrak{X}^*$ gives that $J_{\mathfrak{X}}\mathcal{H} \subset \ell^2$.

This completes the proof of Theorem 10.21. □

If $\mathfrak{X} = (x_n)_{n\ge1}$ is a Riesz basis of a Hilbert space $\mathcal{H}$, then we see from the proof of Theorem 10.21 that we have

$$\Gamma_{\mathfrak{X}} = V_{\mathfrak{X}}^*V_{\mathfrak{X}}, \quad U_{\mathfrak{X}} = V_{\mathfrak{X}}^{-1} = J_{\mathfrak{X}^*} \quad \text{and} \quad V_{\mathfrak{X}} = J_{\mathfrak{X}}^*. \tag{10.41}$$

The following consequence of Theorem 10.21 gives two useful necessary conditions for the Riesz basis property.

Corollary 10.22 *Let $\mathfrak{X} = (x_n)_{n\ge1}$ be a normalized sequence of vectors in a Hilbert space $\mathcal{H}$, and let $\Gamma_{\mathfrak{X}} = (\Gamma_{n,p})_{n,p\ge1}$ be its Gram matrix.*

(i) *If $\mathfrak{X}$ satisfies* (10.39)*, then*

$$\sup_{n\neq p} |\Gamma_{n,p}| < 1. \tag{10.42}$$

(ii) *If $\mathfrak{X}$ satisfies* (10.37)*, then*

$$\sup_{n\geq 1} \sum_{p=1}^{\infty} |\Gamma_{n,p}|^2 < +\infty. \tag{10.43}$$

In particular, if $\mathfrak{X}$ is a Riesz sequence, then $\Gamma_{\mathfrak{X}}$ satisfies (10.42) *and* (10.43).

Proof (i) The condition (10.39) says that there exists a constant $\delta > 0$ such that

$$\delta \|a\|_{\ell^2}^2 \leq \|V_{\mathfrak{X}} a\|^2 \qquad (a \in \ell^2_{\mathfrak{F}}).$$

Using (10.32), this is equivalent to

$$\delta \|a\|_{\ell^2}^2 \leq \langle \Gamma_{\mathfrak{X}} a, a\rangle \qquad (a \in \ell^2_{\mathfrak{F}}).$$

Now fix two positive integers n and p, $n \neq p$, and consider the matrix

$$\Gamma_{\mathfrak{X}}^{(n,p)} = \begin{bmatrix} 1 & \Gamma_{p,n} \\ \Gamma_{n,p} & 1 \end{bmatrix}_{2\times 2}.$$

Then it is clear that $\Gamma_{\mathfrak{X}}^{(n,p)}$ is the compression of $\Gamma_{\mathfrak{X}}$ to the two-dimensional subspace $\mathcal{E}_{n,p}$ generated by $\mathfrak{e}_n$ and $\mathfrak{e}_p$, that is,

$$\Gamma_{\mathfrak{X}}^{(n,p)} = P_{\mathcal{E}_{n,p}} \Gamma_{\mathfrak{X}} i_{\mathcal{E}_{n,p}}.$$

Thus, for each $a \in \mathcal{E}_{n,p}$, we have

$$\langle \Gamma_{\mathfrak{X}}^{(n,p)} a, a\rangle = \langle P_{\mathcal{E}_{n,p}} \Gamma_{\mathfrak{X}} a, a\rangle = \langle \Gamma_{\mathfrak{X}} a, a\rangle \geq \delta \|a\|_{\ell^2}^2.$$

Now, by direct verification, it is easy to see that the lowest eigenvalue of $\Gamma_{\mathfrak{X}}^{(n,p)}$ is $1 - |\Gamma_{n,p}|$. Hence, $1 - |\Gamma_{n,p}| \geq \delta$, which gives

$$\sup_{n\neq p} |\Gamma_{n,p}| \leq 1 - \delta < 1.$$

(ii) The condition (10.37) says that $V_{\mathfrak{X}}$ extends to a bounded operator on ℓ^2. This is equivalent to saying that $\Gamma_{\mathfrak{X}}$ defines a bounded operator on ℓ^2. Thus, we have

$$\sum_{p=1}^{\infty} |\Gamma_{n,p}|^2 = \|\Gamma_{\mathfrak{X}} \mathfrak{e}_n\|_{\ell^2}^2 \leq \|\Gamma_{\mathfrak{X}}\|^2 \|\mathfrak{e}_n\|_{\ell^2}^2 = \|\Gamma_{\mathfrak{X}}\|^2,$$

and hence

$$\sup_{n\geq 1} \sum_{p=1}^{\infty} |\Gamma_{n,p}|^2 \leq \|\Gamma_{\mathfrak{X}}\|^2 < \infty.$$

To conclude the proof, it remains to note that, according to Theorem 10.21, if $\mathfrak{X}$ is a Riesz sequence, then it satisfies both conditions (10.37) and (10.39). □

The next result says that we can add a finite number of terms to a Riesz sequence and still keep the Riesz property, provided that we keep the minimality condition.

Corollary 10.23 *Let $\mathcal{H}$ be a Hilbert space, and let $\mathfrak{X} = (x_n)_{n\geq 1}$ be a minimal sequence in $\mathcal{H}$. Assume that there is $n_0 \geq 1$ such that $(x_n)_{n\geq n_0}$ is a Riesz sequence. Then $(x_n)_{n\geq 1}$ is also a Riesz sequence.*

Proof Let $\delta_1 = \inf_{n\leq n_0-1} \|x_n\|$ and $\delta_2 = \sup_{n\leq n_0-1} \|x_n\|$. Clearly, $\delta_1 > 0$ and $\delta_2 < \infty$. Let also $c, C > 0$ be such that

$$c \sum_{n\geq n_0} |a_n|^2 \leq \left\| \sum_{n\geq n_0} a_n x_n \right\|^2 \leq C \sum_{n\geq n_0} |a_n|^2, \tag{10.44}$$

for any sequence $(a_n)_{n\geq n_0} \in \ell^2_{\mathfrak{F}}$. Since, by minimality, $x_{n_0-1} \notin \operatorname{Span}\{x_n : n \geq n_0\}$, we have the direct decomposition

$$\mathcal{E}_{n_0-1} = \operatorname{Span}\{x_n : n \geq n_0 - 1\} = \mathbb{C}\, x_{n_0-1} \oplus \operatorname{Span}\{x_n : n \geq n_0\}.$$

We put on $\mathcal{E}_{n_0-1}$ the norm

$$|||\lambda x_{n_0-1} + u||| = \sqrt{|\lambda|^2 \|x_{n_0-1}\|^2 + \|u\|^2}$$
$$(\lambda \in \mathbb{C},\ u \in \operatorname{Span}\{x_n : n \geq n_0\}).$$

It is well known and easy to check that $(\mathcal{E}_{n_0-1}, |||\cdot|||)$ is a Hilbert space. Now consider the linear mapping

$$\begin{array}{rcccc} j: & (\mathcal{E}_{n_0-1}, |||\cdot|||) & \longrightarrow & (\mathcal{E}_{n_0-1}, \|\cdot\|) \\ & x & \longmapsto & x. \end{array}$$

Then, since

$$\begin{aligned} \|\lambda x_{n_0-1} + u\| &\leq |\lambda|\, \|x_{n_0-1}\| + \|u\| \\ &\leq \sqrt{2}(|\lambda|^2 \|x_{n_0-1}\|^2 + \|u\|^2)^{1/2} \\ &= \sqrt{2}\, |||\lambda x_{n_0-1} + u|||, \end{aligned}$$

the mapping j is continuous. Since it is a bijection, the Banach isomorphism theorem ensures that j is an isomorphism. Hence, there is a constant $K > 0$ such that

$$K\, |||x||| \leq \|x\| \leq \sqrt{2}\, |||x||| \qquad (x \in \mathcal{E}_{n_0-1}). \tag{10.45}$$

Then, using (10.44) and (10.45), for each sequence $(a_n)_{n\geq 1} \in \ell^2_{\mathfrak{F}}$, we have

$$\begin{aligned}\Big\| \sum_{n\geq n_0-1} a_n x_n \Big\|^2 &= \Big\| a_{n_0-1} x_{n_0-1} + \sum_{n\geq n_0} a_n x_n \Big\|^2 \\ &\geq K^2 \left(|a_{n_0-1}|^2 \|x_{n_0-1}\|^2 + \Big\| \sum_{n\geq n_0} a_n x_n \Big\|^2 \right) \\ &\geq K^2 \left(\delta_1^2 |a_{n_0-1}|^2 + c \sum_{n\geq n_0} |a_n|^2 \right) \\ &\geq K^2 \min(\delta_1^2, c) \sum_{n\geq n_0-1} |a_n|^2.\end{aligned}$$

Similarly, we have

$$\begin{aligned}\Big\| \sum_{n\geq n_0-1} a_n x_n \Big\|^2 &\leq \sqrt{2} \left(|a_{n_0-1}|^2 \|x_{n_0-1}\|^2 + \Big\| \sum_{n\geq n_0} a_n x_n \Big\|^2 \right) \\ &\leq \sqrt{2} \left(\delta_2^2 |a_{n_0-1}|^2 + C \sum_{n\geq n_0} |a_n|^2 \right) \\ &\leq \sqrt{2} \max(\delta_2^2, C) \sum_{n\geq n_0-1} |a_n|^2.\end{aligned}$$

Therefore, by Theorem 10.21, $(x_n)_{n\geq n_0-1}$ is a Riesz sequence. The rest is by induction. □

Exercise

Exercise 10.7.1 Show that the sequence of powers $\{1, t, t^2, \dots\}$ is not a Riesz basis of $L^2(0,1)$.
Hint: Show that the associated Gram matrix Γ (which is the Hilbert matrix) is not invertible. For that purpose, show that $\|\Gamma e_n\|_2 \longrightarrow 0$ as $n \longrightarrow \infty$, where $(e_n)_{n\geq 1}$ is the canonical orthonormal basis.

10.8 Bessel sequences and the Feichtinger conjecture

Let $\mathfrak{X} = (x_n)_{n\geq 1}$ be a sequence in a Hilbert space $\mathcal{H}$. We say that $\mathfrak{X}$ is a *Bessel sequence* if there exists a constant $C > 0$ such that

$$\sum_{n=1}^{\infty} |\langle x, x_n\rangle|^2 \leq C\|x\|^2 \tag{10.46}$$

for all $x \in \mathcal{H}$. If $\mathfrak{X}$ is a sequence of unit vectors, then we say it is *separated* if there exists a constant $\gamma < 1$ such that

$$|\langle x_m, x_n \rangle| \le \gamma \qquad (m, n \ge 1,\ m \ne n). \tag{10.47}$$

It is clear from the definition that $\mathfrak{X}$ is a Bessel sequence if and only if the operator $J_{\mathfrak{X}}$ is bounded as a mapping from $\mathcal{H}$ into ℓ^2. Moreover, the norm of $J_{\mathfrak{X}}$ is precisely the best constant C that appears in (10.46). Applying the inequality (10.46) to x_k, for a fixed $k \ge 1$, shows that $\|J_{\mathfrak{X}}\| \ge 1$. According to Lemma 10.19, Theorem 10.21 and Corollary 10.22, we see that, if $\mathfrak{X}$ is a Riesz sequence of unit vectors, then $\mathfrak{X}$ is a separated Bessel sequence. The converse is not true in general (see Exercise 10.8.3). However, we will see in Section 15.2 that it is true for sequences of normalized reproducing kernels of the Hardy space H^2.

In Theorem 10.25, we will see that any Bessel sequence of unit vectors in $\mathcal{H}$ can be partitioned into finitely many separated Bessel sequences. This result is based on a simple combinatorial lemma.

In the following lemma, instead of saying that two integers satisfy (any two) properties P_1 and P_2, but not both, we say that they are either *friends* or *enemies*, but not both. Hence, the result looks more friendly!

Lemma 10.24 *Fix a nonnegative integer k. Assume that any two distinct integers are either friends or enemies, and not both. Moreover, assume that an integer has at most k enemies. Then there is a partition of $\mathbb{N}$ into $k+1$ subsets, $I_1, I_2, \dots, I_{k+1}$, such that the elements of each subset are friends.*

Proof We inductively construct the subsets $I_1, I_2, \dots, I_{k+1}$. At the first step, set them all to be empty. Then put $1 \in I_1$. Assume that we have distributed $\{1, 2, \dots, n\}$ into $I_1, I_2, \dots, I_{k+1}$ such that the elements of each subset are friends. If the integer $n+1$ has an enemy in each subset, then it has at least $k+1$ enemies, which is a contradiction. Therefore, there exists an index $i \in \{1, \dots, k+1\}$ such that the elements of I_i are all friends with $n+1$. (The choice of i is not necessarily unique; we may choose the smallest i.) Then we add $n+1$ to I_i. Inductively, we construct a partition of $\mathbb{N}$ as required. In some situations, we might have some empty partitions. If we would like to have precisely $k+1$ subsets, we can further divide the nonempty partitions. □

Note that we did not assume that "being a friend" is an equivalence relation on $\mathbb{N}$. In that case, the proof is simpler and the partition into cells of friends is unique. The above lemma considers a more general situation and the partitioning is not necessarily unique.

Theorem 10.25 *Let $\mathcal{H}$ be a Hilbert space and let $(x_n)_{n\ge1}$ be a Bessel sequence of unit vectors in $\mathcal{H}$. Then $(x_n)_{n\ge1}$ can be partitioned into finitely many separated Bessel sequences.*

Proof We say that two integers m and n are enemies if

$$|\langle x_m, x_n \rangle|^2 \geq 1/2.$$

Since $(x_n)_{n\geq 1}$ is a Bessel sequence, there exists a constant $C > 0$ such that

$$\sum_{n=1}^{\infty} |\langle x, x_n \rangle|^2 \leq C\|x\|^2.$$

Fix any $m \in \mathbb{N}$. Then for the cardinality of n we get

$$\begin{aligned}\#\{n : n \text{ is an enemy of } m\} &\leq 2 \sum_{n \text{ is an enemy of } m} |\langle x_m, x_n \rangle|^2 \\ &\leq 2 \sum_{n=1}^{\infty} |\langle x_m, x_n \rangle|^2 \\ &\leq 2C\|x_m\| \leq 2C.\end{aligned}$$

It follows that m has at most $[2C] + 1$ enemies, where $[\cdot]$ denotes the integer part. The conclusion follows now from Lemma 10.24. □

Parallel to Bari's theorem (Theorem 10.21), we have the following result for Bessel sequences.

Theorem 10.26 *Let $\mathfrak{X} = (x_n)_{n\geq 1}$ be a sequence in a Hilbert space $\mathcal{H}$. Then the following are equivalent.*

(i) *Sequence $\mathfrak{X}$ is a Bessel sequence in $\mathcal{H}$.*
(ii) $\mathcal{R}(J_{\mathfrak{X}}) \subset \ell^2$.
(iii) *$\Gamma_{\mathfrak{X}}$ is a bounded operator on ℓ^2.*
(iv) *There exists a constant $C > 0$ such that*

$$\Big\| \sum_{n\geq 1} a_n x_n \Big\|_{\mathcal{H}}^2 \leq C \sum_{n\geq 1} |a_n|^2,$$

for all finitely supported complex sequences $(a_n)_{n\geq 1}$.

Proof By Lemma 10.19, the assertion (ii) is equivalent to saying that $J_{\mathfrak{X}}$ is bounded as an operator from $\mathcal{H}$ into ℓ^2. The assertion (iv) means that $V_{\mathfrak{X}}$ extends to a bounded operator from ℓ^2 into $\mathcal{H}$. Now the equivalence between the assertions follows from the comments given at the end of Section 10.6. □

Corollary 10.27 *Let $\mathfrak{X} = (x_n)_{n\geq 1}$ be a Bessel sequence in a Hilbert space $\mathcal{H}$. Then $\mathfrak{X}$ is a Riesz sequence of $\mathcal{H}$ if and only if, for each $a \in \ell^2$, the interpolation problem*

$$J_{\mathfrak{X}} x = a \tag{10.48}$$

has a solution $x \in \mathcal{H}$. Moreover, the solution is unique if and only if the sequence $\mathfrak{X}$ is complete in $\mathcal{H}$.

Proof The equivalence between the uniqueness of the solution to the interpolation problem and the completeness of the sequence $\mathfrak{X}$ is an easy consequence of the Hahn–Banach theorem. The existence of the solution is equivalent to $\ell^2 \subset \mathcal{R}(J_{\mathfrak{X}}) = J_{\mathfrak{X}}\mathcal{H}$. Moreover, the inclusion $\ell^2 \subset J_{\mathfrak{X}}\mathcal{H}$ implies in particular that the sequence $\mathfrak{X}$ is minimal (consider the solutions for sequences $(\delta_{n,m})_{n\geq 1}$, $m \geq 1$). Now the result follows easily from Theorems 10.21 and 10.26. □

Note that the separation can be viewed as a kind of *baby version* of the Riesz sequence condition. In fact, if we consider the *Gram matrix*

$$\begin{pmatrix} 1 & \langle x_n, x_m\rangle \\ \langle x_m, x_n\rangle & 1 \end{pmatrix}$$

of the two unit vectors x_n and x_m, its smallest eigenvalue is $1 - |\langle x_n, x_m\rangle|$. Thus, the condition (10.47) can be viewed as saying that all these two-term sequences are Riesz pairs with a common lower bound in (10.36).

The *Feichtinger conjecture* (FC) states that every Bessel sequence of unit vectors in a Hilbert space can be partitioned into finitely many Riesz sequences. It follows from Theorem 10.25 that, if $(x_n)_{n\geq 1}$ is a Bessel sequence of unit vectors in a Hilbert space $\mathcal{H}$, then $(x_n)_{n\geq 1}$ can be partitioned into finitely many separated Bessel sequences. That allows us to formulate the equivalent *weaker Feichtinger conjecture* (WFC), which states that every separated Bessel sequence of unit vectors in a Hilbert space can be partitioned into finitely many Riesz sequences. For further reference, we gather this result in the following corollary.

Corollary 10.28 *The Feichtinger conjecture (FC) and the weak Feichtinger conjecture (WFC) are equivalent.*

Proof It is clear that FC implies WFC. As we discussed above, the converse follows immediately from Theorem 10.25. □

It is natural to ask at this point if a separated Bessel sequence is always a Riesz sequence. According to Corollary 10.28, this would give a simple solution to the Feichtinger conjecture. But this is false in general (see Exercise 10.8.3). However, we will see that this is true for normalized sequences of reproducing kernels in H^2 (see Corollary 15.8).

Exercises

Exercise 10.8.1 Let $\mathcal{H}$ be a Hilbert space, let $(x_n)_{n\geq 1}$ be a sequence of nonzero vectors in $\mathcal{H}$, and let $A : \mathcal{H} \longrightarrow \mathcal{H}'$ be an invertible operator. Put

$$x'_n = Ax_n, \quad \hat{x}_n = \frac{x_n}{\|x_n\|} \quad \text{and} \quad \widehat{x'_n} = \frac{x'_n}{\|x'_n\|} \qquad (n \geq 1).$$

Show that:

(i) $(\hat{x}_n)_{n\geq 1}$ is a Bessel sequence if and only if $(\widehat{x'_n})_{n\geq 1}$ is a Bessel sequence;
(ii) $(\hat{x}_n)_{n\geq 1}$ is a Riesz sequence if and only if $(\widehat{x'_n})_{n\geq 1}$ is a Riesz sequence;
(iii) $(\hat{x}_n)_{n\geq 1}$ is separated if and only if $(\widehat{x'_n})_{n\geq 1}$ is separated.

Hint: Let $J_{\mathfrak{X}}$ and $J_{\mathfrak{X}'}$ be the operators associated respectively with $(\hat{x}_n)_{n\geq 1}$ and $(\widehat{x'_n})_{n\geq 1}$. Let D be the diagonal operator on ℓ^2 with entries $\|x_n\|/\|Ax_n\|$. Check that $J_{\mathfrak{X}'} = AJ_{\mathfrak{X}}D$.

Exercise 10.8.2 Let $\mathcal{H}$ be a reproducing kernel Hilbert space on a set Ω, and, for each $\lambda \in \Omega$, denote the corresponding normalized reproducing kernel by $\hat{k}_\lambda$. We say that a positive and finite measure μ on Ω is a *Carleson measure* for $\mathcal{H}$ if $\mathcal{H} \subset L^2(\mu)$. Given a sequence $\Lambda = (\lambda_n)_{n\geq 1}$ in Ω, show that $(\hat{k}_{\lambda_n})_{n\geq 1}$ is a Bessel sequence if and only if the measure

$$\mu_\Lambda = \sum_{n\geq 1} \|k_{\lambda_n}\|^{-2} \delta_{\lambda_n}$$

is a Carleson measure for $\mathcal{H}$ (where δ_λ is the Dirac measure at point λ).

Exercise 10.8.3 Let

$$\varphi(z) = 1 + \frac{\zeta + \bar{\zeta}}{2} = 1 + \cos\theta \qquad (\zeta = e^{i\theta} \in \mathbb{T})$$

and let M_φ be the operator of multiplication by φ on $L^2(\mathbb{T})$.

(i) Show that M_φ is a bounded, noninvertible and positive operator.
(ii) Verify that its matrix with respect to the canonical basis $(\chi_n)_{n\in\mathbb{Z}}$ of $L^2(\mathbb{T})$ is $\Gamma = (\Gamma_{n,m})_{n,m\in\mathbb{Z}}$, where

$$\Gamma_{n,m} = \begin{cases} 1 & \text{if } n - m = 0, \\ 1/2 & \text{if } |n - m| = 1, \\ 0 & \text{otherwise.} \end{cases}$$

Deduce that Γ is a bounded nonsurjective operator on ℓ^2.
(iii) Define

$$f_n(\zeta) = \varphi^{1/2}(\zeta)\zeta^n \qquad (n \in \mathbb{Z}).$$

Show that

$$\langle f_n, f_m \rangle_2 = \Gamma_{n,m}.$$

(iv) Deduce that $(f_n)_{n\in\mathbb{Z}}$ is a separated Bessel sequence that is not a Riesz sequence.

Exercise 10.8.4 Show that the sequence of powers $\{1, t, t^2, \ldots\}$ forms a Bessel sequence in $L^2(0,1)$.
Hint: Use Exercise 10.6.2 and Theorem 10.26.

10.9 Equivalence of Riesz and unconditional bases

In Section 10.5, we saw that, if $\mathfrak{X} = (x_n)_{n\geq 1}$ is a Riesz basis of a Hilbert space $\mathcal{H}$, then $\mathfrak{X}$ is an unconditional basis of $\mathcal{H}$; see the discussion before Theorem 10.16. This section is devoted to the converse of this result. We first begin with a technical lemma.

Lemma 10.29 *Let $\mathcal{H}$ be a Hilbert space, and let $(x_k)_{k=1}^n$ be a finite sequence of vectors in $\mathcal{H}$. Then we have*

$$\frac{1}{2^n}\sum_{\varepsilon_1,\ldots,\varepsilon_n\in\{-1,1\}}\left\|\sum_{k=1}^n \varepsilon_k x_k\right\|^2 = \sum_{k=1}^n \|x_k\|^2.$$

Proof We argue by induction. For $n = 1$, the identity is rather trivial. In this case, we have

$$\begin{aligned}\sum_{\varepsilon_1\in\{-1,1\}}\left\|\sum_{k=1}^1 \varepsilon_k x_k\right\|^2 &= \sum_{\varepsilon_1\in\{-1,1\}}\left\|\varepsilon_1 x_1\right\|^2\\ &= \|x_1\|^2 + \|-x_1\|^2 = 2\|x_1\|^2.\end{aligned}$$

To apply the induction hypothesis, we use the parallelogram identity to get

$$\begin{aligned}&\sum_{\varepsilon_1,\ldots,\varepsilon_{n+1}\in\{-1,1\}}\left\|\sum_{k=1}^{n+1} \varepsilon_k x_k\right\|^2\\ &= \sum_{\varepsilon_1,\ldots,\varepsilon_{n+1}\in\{-1,1\}}\left\|\varepsilon_{n+1}x_{n+1} + \sum_{k=1}^n \varepsilon_k x_k\right\|^2\\ &= \sum_{\varepsilon_1,\ldots,\varepsilon_n\in\{-1,1\}}\left(\left\|x_{n+1} + \sum_{k=1}^n \varepsilon_k x_k\right\|^2\right.\\ &\qquad\left.+\left\|-x_{n+1} + \sum_{k=1}^n \varepsilon_k x_k\right\|^2\right)\\ &= \sum_{\varepsilon_1,\ldots,\varepsilon_n\in\{-1,1\}}\left(2\|x_{n+1}\|^2 + 2\left\|\sum_{k=1}^n \varepsilon_k x_k\right\|^2\right).\end{aligned}$$

Now, note that the last sum has 2^n terms. Hence, by induction on the number of vectors, the identity is valid for all integers n. □

Theorem 10.30 *Let $\mathfrak{X} = (x_n)_{n\geq 1}$ be a sequence in a Hilbert space $\mathcal{H}$ such that*

$$0 < \inf_{n\geq 1} \|x_n\| \leq \sup_{n\geq 1} \|x_n\| < \infty.$$

Then the following are equivalent.

(i) *The sequence $\mathfrak{X}$ is a Riesz basis of $\mathcal{H}$.*

(ii) *The sequence $\mathfrak{X}$ is an unconditional basis of $\mathcal{H}$.*

Proof The implication (i) $\Longrightarrow$ (ii) is already discussed. Hence, we proceed to establish (ii) $\Longrightarrow$ (i).

If $\mathfrak{X}$ is an unconditional basis of $\mathcal{H}$, then, by Theorem 10.14, $\mathfrak{Mult}(\mathfrak{X}) = \ell^\infty$. Moreover, according to (10.25), there exists a constant $c > 0$ such that

$$\|\mu\|_{\mathfrak{Mult}(\mathfrak{X})} \leq c\|\mu\|_\infty$$

for all $\mu \in \ell^\infty$. In particular, if M_μ denotes the bounded operator on $\mathcal{H}$ associated with $\mu \in \mathfrak{Mult}(\mathfrak{X})$, we have

$$\begin{aligned}
\left\|\sum_{k=1}^n a_k \mu_k x_k\right\|_{\mathcal{H}} &= \left\|M_\mu\left(\sum_{k=1}^n a_k x_k\right)\right\|_{\mathcal{H}} \\
&\leq \|M_\mu\|_{\mathcal{L}(\mathcal{H})}\left\|\sum_{k=1}^n a_k x_k\right\|_{\mathcal{H}} \\
&= \|\mu\|_{\mathfrak{Mult}(\mathfrak{X})}\left\|\sum_{k=1}^n a_k x_k\right\|_{\mathcal{H}} \\
&\leq c\|\mu\|_\infty\left\|\sum_{k=1}^n a_k x_k\right\|_{\mathcal{H}},
\end{aligned}$$

for each finite sequence $(a_k)_{k=1}^n \subset \mathbb{C}$. Hence, for $\mu = (\varepsilon_k)_{k\geq 1}$, where $\varepsilon_k = \pm 1$, we get

$$\left\|\sum_{k=1}^n \varepsilon_k a_k x_k\right\|_{\mathcal{H}} \leq c\left\|\sum_{k=1}^n a_k x_k\right\|_{\mathcal{H}}.$$

Therefore, in light of Lemma 10.29, we obtain

$$\begin{aligned}\sum_{k=1}^{n} |a_k|^2 \|x_k\|_{\mathcal{H}}^2 &= \sum_{k=1}^{n} \|a_k x_k\|_{\mathcal{H}}^2 \\ &= \frac{1}{2^n} \sum_{\varepsilon_1,\dots,\varepsilon_n \in \{-1,1\}} \left\| \sum_{k=1}^{n} \varepsilon_k a_k x_k \right\|_{\mathcal{H}}^2 \\ &\leq \frac{1}{2^n} \sum_{\varepsilon_1,\dots,\varepsilon_n \in \{-1,1\}} c^2 \left\| \sum_{k=1}^{n} a_k x_k \right\|_{\mathcal{H}}^2 \\ &= c^2 \left\| \sum_{k=1}^{n} a_k x_k \right\|_{\mathcal{H}}^2 .\end{aligned}$$

Put $c' = \inf_{n\geq 1} \|x_n\|$. We immediately deduce that

$$\sum_{k=1}^{n} |a_k|^2 \leq \frac{c^2}{c'^2} \left\| \sum_{k=1}^{n} a_k x_k \right\|_{\mathcal{H}}^2 . \tag{10.49}$$

Now, according to Corollary 10.15, the sequence $(x_n^*)_{n\geq 1}$ is also an unconditional basis of $\mathcal{H}$ and, using Theorem 10.5 and Corollary 10.6, we have

$$0 < \inf_{n\geq 1} \|x_n^*\| \leq \sup_{n\geq 1} \|x_n^*\| < \infty.$$

Thus, reversing the roles of $\mathfrak{X}$ and $\mathfrak{X}^*$ in the above argument shows that

$$\sum_{k=1}^{n} |a_k|^2 \leq C \left\| \sum_{k=1}^{n} a_k x_k^* \right\|_{\mathcal{H}}^2 \tag{10.50}$$

for some constant $C > 0$. Therefore, considering (10.49) and (10.50), Theorem 10.21 ensures that $\mathfrak{X}$ is a Riesz basis for $\mathcal{H}$. □

In Theorem 10.30, the hypothesis

$$0 < \inf_{n\geq 1} \|x_n\| \leq \sup_{n\geq 1} \|x_n\| < \infty$$

is essential. Indeed, the property of a Riesz basis implies, by definition, that the norms of vectors are bounded from below and above, whereas the property of an unconditional basis puts no control on the norms. For instance, if $(x_n)_{n\geq 1}$ is a Riesz basis of a Hilbert space $\mathcal{H}$ and if $y_n = n x_n / \|x_n\|$, $n \geq 1$, then the sequence $(y_n)_{n\geq 1}$ is an unconditional basis of $\mathcal{H}$ but not a Riesz basis of $\mathcal{H}$ (because $\|y_n\| \longrightarrow \infty$, as $n \longrightarrow \infty$).

10.10 Asymptotically orthonormal sequences

In Section 10.7, we saw that, if $\mathfrak{X} = (x_n)_{n\geq 1}$ is a sequence in a Hilbert space $\mathcal{H}$, then $\mathfrak{X}$ is a Riesz sequence if and only if there are positive constants c and C such that

$$c \sum_{n \ge 1} |a_n|^2 \le \left\| \sum_{n \ge 1} a_n x_n \right\|_{\mathcal{H}}^2 \le C \sum_{n \ge 1} |a_n|^2$$

for all finitely supported complex sequences $(a_n)_{n \ge 1}$. In other words, a Riesz sequence is a sequence that "asymptotically" satisfies Parseval's identity. Motivated by this point of view, we introduce a slightly stronger property, which gives birth to a family between Riesz sequences and orthogonal sequences.

The sequence $\mathfrak{X}$ is called an *asymptotically orthonormal sequence* (AOS) if there exists an $N_0 \in \mathbb{N}$, such that, for all $N \ge N_0$, there are constants $c_N, C_N > 0$ satisfying

$$c_N \sum_{n \ge N} |a_n|^2 \le \left\| \sum_{n \ge N} a_n x_n \right\|_{\mathcal{H}}^2 \le C_N \sum_{n \ge N} |a_n|^2, \tag{10.51}$$

for all finitely supported complex sequences $(a_n)_{n \ge 1}$, and

$$\lim_{N \to \infty} c_N = \lim_{N \to \infty} C_N = 1.$$

Naively speaking, this means that, as we ignore more and more of the starting terms of $(x_n)_{n \ge 1}$, it becomes ever more similar to an orthogonal sequence. By the same token, $\mathfrak{X}$ is called an *asymptotically orthonormal basic sequence* (AOB) if it is an AOS with $N_0 = 1$. Obviously, this is equivalent to $(x_n)_{n \ge 1}$ being an AOS as well as a Riesz sequence.

If $(x_n)_{n \ge 1}$ is an AOB, then it is a Riesz sequence, and thus it is minimal. Conversely, if $(x_n)_{n \ge 1}$ is an AOS, then $(x_n)_{n \ge N_0}$ is a Riesz sequence for some N_0. Now, by Corollary 10.23, the minimality ensures that $(x_n)_{n \ge 1}$ is still a Riesz sequence. Therefore, we conclude that $(x_n)_{n \ge 1}$ is an AOB if and only if it is minimal and an AOS.

As in the case of Riesz sequences, several equivalent characterizations are available for AOB sequences in terms of the Gram matrix and orthogonalizer. Before giving this result, we start with a technical lemma.

Lemma 10.31 *Let $\mathfrak{X} = (x_n)_{n \ge 1}$ be a Riesz sequence of a Hilbert space $\mathcal{H}_1$, and let $K : \mathrm{Span}\{x_n : n \ge 1\} \longrightarrow \mathcal{H}_2$ be a compact operator. Put*

$$K_N = K|\mathrm{Span}\{x_n : n \ge N\}.$$

Then

$$\lim_{N \to \infty} \|K_N\| = 0.$$

Proof Since $\mathfrak{X}$ is a Riesz sequence, for each $x \in \mathrm{Span}\{x_n : n \ge N\}$, $N \ge 1$, there exists a unique sequence $(a_n)_{n \ge N}$ in ℓ^2 such that

$$x = \sum_{n \ge N} a_n x_n,$$

where the series is unconditionally convergent and

$$c_1 \sum_{n \ge N} |a_n|^2 \le \|x\|^2 \le c_2 \sum_{n \ge N} |a_n|^2$$

for some absolute constants $c_1, c_2 > 0$. Let $\varepsilon > 0$. Since K is compact, there is a finite-rank operator $R : \text{Span}\{x_n : n \ge 1\} \longrightarrow \mathcal{H}_2$ such that $\|K - R\| \le \varepsilon$. Hence, for each $x \in \text{Span}\{x_n : n \ge 1\}$,

$$\|Kx\| \le \varepsilon \|x\| + \|Rx\|.$$

But $Rx = \sum_{n \ge N} a_n R x_n$ and

$$\begin{aligned} \|Rx\| &\le \sum_{n \ge N} |a_n| \, \|Rx_n\| \\ &\le \bigg(\sum_{n \ge N} |a_n|^2 \bigg)^{1/2} \bigg(\sum_{n \ge N} \|Rx_n\|^2 \bigg)^{1/2} \\ &\le \frac{1}{\sqrt{c_1}} \|x\| \bigg(\sum_{n \ge N} \|Rx_n\|^2 \bigg)^{1/2}. \end{aligned}$$

Hence, we get

$$\|K_N\| \le \varepsilon + \frac{1}{\sqrt{c_1}} \bigg(\sum_{n \ge N} \|Rx_n\|^2 \bigg)^{1/2}.$$

Since $(x_n)_{n \ge 1}$ is a Riesz sequence, for each $x \in \mathcal{H}$, the sequence $(\langle x_n, x \rangle)_{n \ge 1}$ belongs to ℓ^2. Since R is a finite-rank operator, there are $y_k, z_k \in \mathcal{H}$, with $1 \le k \le m$, such that

$$R = \sum_{k=1}^{m} z_k \otimes y_k.$$

Hence, for each $x \in H$,

$$\|Rx\|^2 = \sum_{k=1}^{m} \sum_{\ell=1}^{m} \langle z_k, z_\ell \rangle \, \langle x, y_k \rangle \, \overline{\langle x, y_\ell \rangle}.$$

This representation shows that the sequence $(\|Rx_n\|)_{n \ge 1}$ also belongs to ℓ^2. Thus, there is an $N \ge 1$ such that

$$\sum_{n \ge N} \|Rx_n\|^2 \le c_1 \varepsilon^2,$$

which implies that

$$\|K_N\| \le 2\varepsilon,$$

and this concludes the proof of the lemma. □

If $(x_n)_{n\geq 1}$ is an orthonormal sequence in $\mathcal{H}$, then there exists a unitary operator $U : \operatorname{Span}\{x_n : n \geq 1\} \longrightarrow \ell^2$ such that $U(x_n) = \mathfrak{e}_n$, $n \geq 1$. This fact is generalized below for AOB sequences.

Theorem 10.32 *Let $\mathfrak{X} = (x_n)_{n\geq 1}$ be a sequence in $\mathcal{H}$. Then the following are equivalent.*

(i) *The sequence $(x_n)_{n\geq 1}$ is an AOB.*
(ii) *There are operators $U, K : \operatorname{Span}\{x_n : n \geq 1\} \longrightarrow \ell^2$, with U unitary, K compact and $U + K$ invertible, such that $(U + K)(x_n) = \mathfrak{e}_n$, $n \geq 1$.*
(iii) *The Gram matrix $\Gamma_{\mathfrak{X}}$ defines a bounded invertible operator on ℓ^2 of the form $I + K$, with K compact.*

Proof (i) $\Longrightarrow$ (iii) Since $(x_n)_{n\geq 1}$ is an AOB, it is in particular a Riesz sequence and then, by Theorem 10.21, the Gram matrix $\Gamma_{\mathfrak{X}}$ is a bounded invertible operator on ℓ^2. Denote by P_N the orthogonal projection of ℓ^2 onto $\operatorname{Span}\{\mathfrak{e}_n : n \geq N\}$. Since $P_N(I - \Gamma_{\mathfrak{X}})P_N$ is self-adjoint, we know from Corollary 2.14 that

$$\|P_N(I - \Gamma_{\mathfrak{X}})P_N\|_{\mathcal{L}(\ell^2)} = \sup_{a\in\ell^2,\, \|a\|_2\leq 1} |\langle P_N(I - \Gamma_{\mathfrak{X}})P_N a, a\rangle_2|,$$

and in the above supremum we may even restrict ourselves to the finitely supported sequences. But, using (10.32), we have

$$\begin{aligned}\langle P_N(I - \Gamma_{\mathfrak{X}})P_N a, a\rangle_2 &= \langle (I - \Gamma_{\mathfrak{X}})P_N a, P_N a\rangle_2\\ &= \|P_N a\|_2^2 - \langle \Gamma_{\mathfrak{X}} P_N a, P_N a\rangle_2\\ &= \sum_{n\geq N} |a_n|^2 - \Big\|\sum_{n\geq N} a_n x_n\Big\|_{\mathcal{H}}^2.\end{aligned}$$

Hence, using (10.51), we get that

$$\langle P_N(I - \Gamma_{\mathfrak{X}})P_N a, a\rangle_2 \leq (1 - c_N)\sum_{n\geq N} |a_n|^2.$$

Thus $\|P_N(I - \Gamma_{\mathfrak{X}})P_N\|_{\mathcal{L}(\ell^2)} \leq 1 - c_N$, which implies that

$$\lim_{N\to\infty} \|P_N(I - \Gamma_{\mathfrak{X}})P_N\|_{\mathcal{L}(\ell^2)} = 0. \tag{10.52}$$

To conclude, write

$$I - \Gamma_{\mathfrak{X}} = P_N(I - \Gamma_{\mathfrak{X}})P_N + T_N. \tag{10.53}$$

A simple computation shows that

$$T_N = P_N(I - \Gamma_{\mathfrak{X}})(I - P_N) + (I - P_N)(I - \Gamma_{\mathfrak{X}}),$$

and this representation assures that T_N is a finite-rank operator. By (10.52) and (10.53), we have $T_N \longrightarrow I - \Gamma_{\mathfrak{X}}$, and thus $I - \Gamma_{\mathfrak{X}}$ is a compact operator.

(iii) $\Longrightarrow$ (ii) We defined

$$\begin{array}{rccl} V_{\mathfrak{X}} : & \ell^2_{\mathfrak{F}} & \longrightarrow & \operatorname{Span}\{x_n : n \geq 1\} \\ & a = (a_n)_{n\geq 1} & \longmapsto & \sum_{n\geq 1} a_n x_n. \end{array}$$

Then, since $\Gamma_{\mathfrak{X}}$ is invertible, according to Theorem 10.21, we know that $V_{\mathfrak{X}}$ extends to an isomorphism from ℓ^2 onto $\operatorname{Span}\{x_n : n \geq 1\}$ and $U_{\mathfrak{X}} = V_{\mathfrak{X}}^{-1}$ is the orthogonalizer of $\mathfrak{X}$, that is, $U_{\mathfrak{X}}(x_n) = \mathfrak{e}_n$, $n \geq 1$. It remains to prove that $U_{\mathfrak{X}}$ is the sum of a unitary operator and a compact operator. Consider the polar decomposition of $V_{\mathfrak{X}}$, which, according to Corollary 7.25, is

$$V_{\mathfrak{X}} = A|V_{\mathfrak{X}}|,$$

with A being a unitary operator from ℓ^2 onto $\operatorname{Span}\{x_n : n \geq 1\}$ and $|V_{\mathfrak{X}}| = (V_{\mathfrak{X}}^* V_{\mathfrak{X}})^{1/2}$, which is a positive operator on ℓ^2. Then

$$\Gamma_{\mathfrak{X}} = V_{\mathfrak{X}}^* V_{\mathfrak{X}} = |V_{\mathfrak{X}}|^2,$$

and, by hypothesis, $\Gamma_{\mathfrak{X}} = I + K$, with K a compact operator on ℓ^2. Hence,

$$|V_{\mathfrak{X}}|^2 - I = K.$$

Since $|V_{\mathfrak{X}}| \geq 0$, $\sigma(|V_{\mathfrak{X}}|) \subset [0, +\infty)$ and the operator $I + |V_{\mathfrak{X}}|$ is invertible. Thus,

$$|V_{\mathfrak{X}}| - I = (I + |V_{\mathfrak{X}}|)^{-1} K = K_1,$$

with K_1 a compact operator. We then get

$$V_{\mathfrak{X}} = A|V_{\mathfrak{X}}| = A(I + K_1) = A + AK_1 = A + K_2,$$

with K_2 compact. Finally, since $U_{\mathfrak{X}} = V_{\mathfrak{X}}^{-1}$, we have $U_{\mathfrak{X}}(A + K_2) = I$, that is, $U_{\mathfrak{X}} A + U_{\mathfrak{X}} K_2 = I$. Composing on the right by A^*, we get

$$U_{\mathfrak{X}} = A^* - U_{\mathfrak{X}} K_2 A^* = A^* + K_3,$$

with A^* unitary and K_3 compact.

(ii) $\Longrightarrow$ (i) The existence of U and K ensures that $\mathfrak{X}$ is a Riesz sequence. Moreover, arguing as in the proof of (iii) $\Longrightarrow$ (ii), we easily see that $V_{\mathfrak{X}} = U_{\mathfrak{X}}^{-1} = U_1 + K_1$, with U_1 unitary and K_1 compact. Now, for $N \geq 1$, denote by

$$\varepsilon_N = \|K|\operatorname{Span}\{x_n : n \geq N\}\| \quad \text{and} \quad \widetilde{\varepsilon}_N = \|K_1|\operatorname{Span}\{e_n : n \geq N\}\|.$$

It follows from Lemma 10.31 that $\varepsilon_N, \widetilde{\varepsilon}_N \longrightarrow 0$, as $N \longrightarrow \infty$. Moreover, for each $a = \sum_{n\geq N} a_n e_n \in \ell^2_{\mathfrak{F}}$, we have

$$\begin{aligned}\Big(\sum_{n\geq N}|a_n|^2\Big)^{1/2} &= \Big\|U_{\mathfrak{X}}\Big(\sum_{n\geq N}a_nx_n\Big)\Big\|_{\mathcal{H}}\\ &= \Big\|U\Big(\sum_{n\geq N}a_nx_n\Big)+K\Big(\sum_{n\geq N}a_nx_n\Big)\Big\|_{\mathcal{H}}\\ &\leq (1+\varepsilon_N)\Big\|\sum_{n\geq N}a_nx_n\Big\|_{\mathcal{H}}.\end{aligned}$$

Similarly, we have

$$\begin{aligned}\Big\|\sum_{n\geq N}a_nx_n\Big\|_{\mathcal{H}} &= \Big\|V_{\mathfrak{X}}\Big(\sum_{n\geq N}a_n\mathfrak{e}_n\Big)\Big\|_{\mathcal{H}}\\ &= \Big\|U_1\Big(\sum_{n\geq N}a_n\mathfrak{e}_n\Big)+K_1\Big(\sum_{n\geq N}a_n\mathfrak{e}_n\Big)\Big\|_{\mathcal{H}}\\ &\leq (1+\widetilde{\varepsilon}_N)\Big\|\sum_{n\geq N}a_n\mathfrak{e}_n\Big\|_{\ell^2}\\ &= (1+\widetilde{\varepsilon}_N)\Big(\sum_{n\geq N}|a_n|^2\Big)^{1/2}.\end{aligned}$$

Therefore, we get

$$(1+\varepsilon_N)\Big(\sum_{n\geq N}|a_n|^2\Big)^{1/2} \leq \Big\|\sum_{n\geq N}a_nx_n\Big\| \leq (1+\widetilde{\varepsilon}_N)\Big(\sum_{n\geq N}|a_n|^2\Big)^{1/2},$$

and since $\varepsilon_N, \widetilde{\varepsilon}_N \longrightarrow 0$, as $N \longrightarrow \infty$, we conclude that $\mathfrak{X}$ is an AOB.

This completes the proof of Theorem 10.32. □

Corollary 10.33 *Let $\mathcal{H}$ be a Hilbert space, let $\mathfrak{X} = (x_n)_{n\geq 1}$ be a complete AOB in $\mathcal{H}$, with $\|x_n\| = 1$, $n \geq 1$, and let $(x_n^*)_{n\geq 1}$ be the corresponding biorthogonal sequence. Then*

$$\lim_{n\to\infty}\|x_n^*\| = 1.$$

Proof We have

$$1 = \langle x_n, x_n^*\rangle \leq \|x_n^*\| \qquad (n \geq 1). \tag{10.54}$$

According to Theorem 10.32, there are operators $U, K : \mathcal{H} \longrightarrow \ell^2$, with U unitary, K compact, $U+K$ invertible and $(U+K)(x_n) = \mathfrak{e}_n$, $n \geq 1$. In fact, using the notation introduced before Theorem 10.16, we have $U + K = U_{\mathfrak{X}}$, and thus

$$x_n^* = (U+K)^*(U+K)x_n = (U^*+K^*)\mathfrak{e}_n \qquad (n \geq 1).$$

Hence,

$$\|x_n^*\| \le \|U^* \mathfrak{e}_n\| + \|K^* \mathfrak{e}_n\| = 1 + \|K^* \mathfrak{e}_n\|. \tag{10.55}$$

Since K is a compact operator, $\|K^* \mathfrak{e}_n\| \longrightarrow 0$, as $n \longrightarrow \infty$, and thus, by (10.54) and (10.55), we obtain

$$\lim_{n \to \infty} \|x_n^*\| = 1. \qquad \square$$

One can obtain complete AOBs by slightly perturbing orthonormal bases. This fact is made precise in the following result, which is a more accurate version of Theorem 10.17.

Corollary 10.34 *Let $\mathcal{H}$ be a Hilbert space, let $(x_n)_{n \ge 1}$ be an orthonormal basis in $\mathcal{H}$, and let $(x'_n)_{n \ge 1}$ be a sequence in $\mathcal{H}$, such that*

$$\sum_{n \ge 1} \|x_n - x'_n\|^2 < 1.$$

Then $(x'_n)_{n \ge 1}$ is a complete AOB in $\mathcal{H}$.

Proof According to Theorems 10.17 and 10.32 and the Riesz–Fischer theorem, the only thing to prove is that U is of the form unitary plus compact, where $U : \mathcal{H} \longrightarrow \mathcal{H}$ is defined by $U(x_n) = x'_n$. But the condition in the statement implies in particular that

$$\sum_{n \ge 1} \|(I - U)x_n\|^2 < \infty,$$

which is known to be equivalent to saying that $I - U$ is a Hilbert–Schmidt operator. Thus, in particular, $I - U$ is a compact operator and we are done. $\square$

We end this section with the following result, which guarantees the existence of an AOB subsequence in a weakly convergent normalized sequence.

Corollary 10.35 *Let $(x_n)_{n \ge 1}$ be a normalized sequence in $\mathcal{H}$ tending weakly to 0. Then it has a subsequence $(x_{n_k})_{k \ge 1}$ that is an AOB.*

Proof Choose recursively the sequence $(x_{n_k})_{k \ge 1}$ by requiring that

$$|\langle x_n, x_{n_k} \rangle|^2 \le \frac{1}{2^k} \qquad (n \ge n_k).$$

Hence,

$$\sum_{\ell \ge 1,\, k \ge 1,\, \ell \ne k} |\langle x_{n_k}, x_{n_\ell} \rangle|^2 \le 2.$$

Therefore, if Γ' is the Gram matrix associated with $(x_{n_k})_{k \ge 1}$, then $\Gamma' - I$ has a finite Hilbert–Schmidt norm. Applying Theorem 10.32 to Γ' implies that $(x_{k_n})_{n \ge 1}$ is an AOB. $\square$

Since a Riesz sequence is the image through a left-invertible operator of an orthonormal sequence, it follows that any Riesz sequence tends weakly to 0. Therefore, according to Corollary 10.35, each Riesz sequence contains an AOB subsequence.

Notes on Chapter 10

The publication of Banach's book [67] in 1932 is regarded as the beginning of a systematic study of Banach spaces. Research activity in this area has expanded dramatically since then. Interesting new directions have developed and interplays between Banach space theory and other branches of mathematics have proved to be very fruitful. One of these directions concerns the study of bases, which play a prominent role in the analysis of Banach spaces. The notions of Schauder basis, unconditional basis and Riesz basis, which are central in this chapter, are now classic subjects and there are many excellent textbooks devoted to their study. We refer to [566] and [352] for very comprehensive introductions to this subject, to [486] for an encyclopedic account and to [137] for a more applied point of view.

Section 10.1

The three types of linear independence of sequences (finitely linear independence, w-topologically linear independence and minimality) considered in this section were studied a long time ago in infinite-dimensional Hilbert spaces (see e.g. the book of Kaczmarz and Steinhaus [308]). Lemma 10.1 appeared in the work of Markushevich [351]. This property can also be found in the book of Kaczmarz and Steinhaus [308] in the case where the Banach space is L^p, $p \geq 1$. Note that Banach [67] had used the term *fundamental* for complete sequences. Exercise 10.1.3 is due to Nikolskii [396].

Section 10.2

The notion of Schauder basis considered in this section was first introduced by Schauder [467]. But the first systematic investigation of Schauder bases in Banach spaces was made by Banach [67]. Note that Schauder's original definition of a basis had required the continuity of the functional coordinates. Banach [67] showed that this property comes as a byproduct (Theorem 10.5). The idea of the proof of Theorem 10.5 (through Lemmas 10.3 and 10.4), due to Banach, has become a useful tool in various extensions of this theorem. For instance, this theorem has been extended with a similar method to bases in Fréchet spaces by Newns [378], and in complete metric linear spaces (without the hypothesis of local convexity) by Nikolskii [396], Bessaga and Pełczyński [95] and Arsove [48].

It is clear that a Banach space that admits a Schauder basis must be separable. In 1932, Banach raised the converse problem: Does every separable Banach space have a Schauder basis? This problem remained for a long time as one of the outstanding unsolved problems of functional analysis. Finally, the question was settled in 1973 by Enflo [206], who constructed a separable Banach space with no Schauder basis.

Theorem 10.7, which gives useful characterizations of bases, is essentially due to Banach [67]. Lemma 10.8 appeared in [293]. The example given in Exercise 10.2.1 of a complete and minimal system in $\mathcal{X}$ that is not a basis of $\mathcal{X}$ is well known; e.g. it is given in a paper of Markushevich [351]. Equivalent bases introduced in Exercise 10.2.3 were studied by Arsove [47], who called them *similar bases*. The characterization obtained in Exercise 10.2.3 is proved by Arsove for locally convex Fréchet spaces and, independently, by Bessaga and Pełczyński [94]. The results of Exercises 10.2.4 and 10.2.5 are essentially due to Banach [67].

We know that, if $\mathcal{X}$ is a reflexive Banach space and if $(x_n)_{n\geq 1}$ is a basis of $\mathcal{X}$, then its biorthogonal $(x_n^*)_{n\geq 1}$ is a basis of $\mathcal{X}^*$. Clearly, this result cannot be true if the dual $\mathcal{X}^*$ is no longer separable. But, even if $\mathcal{X}^*$ is separable and $\mathcal{X}$ has a basis, then $\mathcal{X}^*$ need not have a basis. The proof is hard and makes use of the existence of a separable Banach space without a basis (see Lindenstrauss and Tzafriri [338]). A deep result of Johnson, Rosenthal and Zippin [300] states that, if $\mathcal{X}^*$ has a basis, then so does $\mathcal{X}$. This answers a question posed by Karlim [313].

The original form of the result proved in Exercise 10.2.6, with $\mathcal{X} = L^2(a,b)$ and $(x_n)_{n\geq 1}$ an orthonormal basis of $\mathcal{X}$, is due to Paley and Wiener [401]. The generalization given in Exercise 10.2.6 is due to Boas [102]. The result of Paley and Wiener has opened the door on a fruitful direction of research called the theory of nonharmonic Fourier series. The result proved in Exercise 10.2.7 is due to Kreĭn and Liusternik [325]. For further results concerning the stability of bases in Banach spaces, we refer to the excellent book of R. M. Young [566, chap. 1].

Section 10.3

Multipliers were used long ago in analysis (e.g. see the work of Orlicz [398]). In the case when $\mathfrak{X} = (x_n)_{n\geq 1}$ is a Schauder basis of $\mathcal{X}$, Theorem 10.9 is due to Yamazaki [562], and in the general case, essentially, to McGivney and Ruckle [359]. Theorem 10.10 was proved by Yamazaki [562]. In [309], Kadec got a slightly different characterization showing that $\mathfrak{X} = (x_n)_{n\geq 1}$ is a

Schauder basis if and only if $\mathfrak{Mult}(\mathcal{X})$ contains every nonincreasing sequence tending to zero. See also Nikolskii [381] or Singer [486, chap. I, sec. 5].

Section 10.4

In the literature, various families of bases in Banach spaces have been introduced and studied, e.g. monotone and strictly monotone bases, normal bases, k-shrinking bases, unconditional bases, etc. We refer to the book of Singer [486, chap. II] for further details.

That a space has a Schauder basis or not is usually difficult to answer, and, upon responding affirmatively to this question, the possibility of the existence of an unconditional basis is even more difficult. For instance, it was not until 1974 that Bočkarev [105] showed that the disk algebra has a basis that is the Franklin system, i.e. the Gram–Schmidt orthogonalization of the Faber–Schauder system in the Hilbert space $L^2(0,1)$. Soon thereafter, Pełczyński [404] showed that the disk algebra does not have an unconditional basis. But each proof is deep and difficult. Again, the Franklin system was shown by Wojtaszczyk [556] to be an unconditional basis for the Hardy space $H^1(\mathbb{D})$. Maurey earlier [357] had shown that $H^1(\mathbb{D})$ possesses an unconditional basis without explicitly citing one.

The notion of Schauder basis for a general topological linear space was considered by Arsove and Edwards [49]. An analog of the equivalence of (i) and (iv) in Theorem 10.14 for the property of the Schauder basis is proved by Singer [485].

Theorem 10.14 is due to Lorch [341] and to Grinblyum [245]. This result was generalized to bases of subspaces; see [242, chap. VI]. Corollary 10.15 is essentially due to Banach [67], where an analogous result is established for the Schauder basis in the setting of reflexive Banach spaces. See Exercise 10.2.5 and also [566, p. 28] for a presentation of this result.

Section 10.5

As already mentioned, the study of the stability of an (orthonormal) basis in Hilbert spaces led to the theory of nonharmonic Fourier series and was initiated by Paley and Wiener in their celebrated treatise [401]. In particular, they proved the criterion obtained in Exercise 10.2.6 and applied it to trigonometric systems to show that, if $(\lambda_n)_{n\in\mathbb{Z}}$ is a sequence of real numbers satisfying

$$\sup_{n\in\mathbb{Z}} |\lambda_n - n| \leq L < \frac{1}{\pi^2},$$

then $(e^{i\lambda_n t})_{n\in\mathbb{Z}}$ is a Riesz basis of $L^2(-\pi,\pi)$. Then Duffin and Eachus [183] showed that the Paley–Wiener criterion is satisfied whenever $L < (\log 2)/\pi$.

Finally, Kadec [310] showed that the constant $(\log 2)/\pi$ could be replaced by $1/4$. A result of Ingham [295] tells us that the constant $1/4$ is sharp.

The result proved in Exercise 10.5.2 is due to Bari [82]. He also introduced the terminology of Riesz basis [81].

Section 10.6

The Gram matrices and operators introduced in this section have been applied in geometric problems of Hilbert spaces as well as in the study of the accuracy of biorthogonal expansions. See the book of Akhiezer and Glazman [11]. In particular, one can show that, for a finite family $\mathfrak{X} = (x_n : n \in \mathcal{N})$ of vectors in a Hilbert space $\mathcal{H}$, we have

$$\operatorname{dist}(x_n, \operatorname{Span}\{x_i : i \in \mathcal{N}, i \neq n\}) = \frac{\det \Gamma_{\mathfrak{X}}}{\det \Gamma_{\mathfrak{X} \setminus \{x_n\}}}$$

for every $n \in \mathcal{N}$. Moreover, the constant of uniform minimality of the family $\mathfrak{X}$ can be expressed in terms of the Gram matrix.

The sequence $\mathfrak{X} = (x_n)_{n \geq 1}$ of a Hilbert space $\mathcal{H}$ for which $\ell^2 \subset \mathcal{R}(J_{\mathfrak{X}})$ are sometimes called *Riesz–Fischer sequences*. The first detailed investigation of Riesz–Fischer sequences was made by Bari [82].

The suggested proof of Hilbert's inequality (10.35) based on Cauchy's theorem is due to Fejér and F. Riesz [213]. For additional proofs of Hilbert's inequality, together with applications, generalizations and historical remarks, see Hardy, Littlewood and Pólya [258, chap. IX].

The result obtained in Exercise 10.6.4 is essentially due to Gurevich [246]. See also R. M. Young [566, p. 167, theorem 7].

Section 10.7

Theorem 10.21 is due to Bari [82]. The equivalence of assertions (i) and (ix) was also proved independently by Boas [103]. There exists an analog of Theorem 10.21 for bases of subspaces obtained by Nikolskii and Pavlov [382, 392].

Section 10.8

Bessel sequences (as well as Riesz–Fischer sequences) were introduced and extensively studied by Bari [82]. The original equivalence of (i) and (iv) in Theorem 10.26 with $H = L^2(0,1)$ is due to Boas [103]. Then it was rediscovered by Bari, who also gave the other equivalent conditions. Lemma 10.24 and Theorem 10.25 are due to Chalendar, Fricain and Timotin [132].

The Feichtinger conjecture in harmonic analysis was stated in 2003 and appeared in print for the first time in [127]. It has become a topic of great interest and strong activity since it has been shown to be equivalent to the celebrated Kadison–Singer problem [128]. There are many variations of the Feichtinger conjecture, all equivalent, but in this chapter we have just presented the version involving Bessel sequences and Riesz sequences. In Chapter 15, we will show that the sequences of normalized reproducing kernels of H^2 satisfy the Feichtinger conjecture.

Section 10.9

Lemma 10.29 is due to Orlicz [399]. Theorem 10.30 is proved by Köthe and Toeplitz [321]. It was later rediscovered by Lorch [341] and Gelfand [235].

Section 10.10

A particular case of AOB was introduced by Bari [82]. It corresponds to the case where the Gram matrix is of the form $I + K$ with K being a Hilbert–Schmidt operator. Later on, this particular case of AOB was extensively studied by Kreĭn [327], who called them *Bari bases*. They correspond also to w-linearly independent sequences $\mathfrak{X} = (x_n)_{n\geq 1}$ that are quadratically close to an orthonormal basis $(e_n)_{n\geq 1}$, in the sense that

$$\sum_{n\geq 1} \|x_n - e_n\|^2 < \infty.$$

See also Gohberg and Kreĭn [242, chap. VI, sec. 3]. Theorem 10.32 and Corollaries 10.34 and 10.35 appeared in Chalendar, Fricain and Timotin [133].

11

Hankel operators

Finding the distance between a point and a subset in a metric space is an essential problem in metric space theory. In the metric space $L^\infty(\mathbb{T})$, finding the distance between a function $f \in L^\infty(\mathbb{T})$ and the subspace H^∞ is known as Nehari's problem and it leads naturally to Hankel operators. In this chapter, the main focus is on Hankel operators and Nehari's problem. We first define the Hankel operator H_φ and obtain its matrix representation and explicitly calculate its norm as a distant formula. This result has several interesting consequences. For example, we establish Hilbert's inequality, a version of Nehari's problem, the Fejér–Carathéodory problem, and the Nevanlinna–Pick problem. At the end, we treat finite-rank and compact Hankel operators.

11.1 A matrix representation for H_φ

Let $\varphi \in L^\infty(\mathbb{T})$. Then the *Hankel operator* associated with φ, or with symbol φ, is defined by

$$\begin{array}{rccc} H_\varphi : & H^2(\mathbb{T}) & \longrightarrow & H^2(\mathbb{T})^\perp \\ & f & \longmapsto & P_-(\varphi f). \end{array}$$

In other words,

$$H_\varphi = P_- \circ M_\varphi \circ i_+, \tag{11.1}$$

where $M_\varphi : L^2(\mathbb{T}) \longrightarrow L^2(\mathbb{T})$ denotes the multiplication operator introduced in Section 7.2 and i_+ is the inclusion map of $H^2(\mathbb{T})$ into $L^2(\mathbb{T})$. Hence, H_φ is a bounded operator and, by Theorem 2.20,

$$\begin{aligned} &\|H_\varphi\|_{\mathcal{L}(H^2(\mathbb{T}),H^2(\mathbb{T})^\perp)} \\ &\quad \le \|P_-\|_{\mathcal{L}(L^2(\mathbb{T}),H^2(\mathbb{T})^\perp)} \, \|M_\varphi\|_{\mathcal{L}(L^2(\mathbb{T}))} \, \|i_+\|_{\mathcal{L}(H^2(\mathbb{T}),L^2(\mathbb{T}))} \\ &\quad = \|\varphi\|_{L^\infty(\mathbb{T})}. \end{aligned}$$

Thus, in the first step, we have

$$\|H_\varphi\| \leq \|\varphi\|_\infty. \tag{11.2}$$

But, we will say more about $\|H_\varphi\|$.

Lemma 11.1 *Let $\varphi \in L^\infty(\mathbb{T})$. Then*

$$H_\varphi^* = P_+ \circ M_{\bar{\varphi}} \circ i_-.$$

In other words,

$$H_\varphi^* f = P_+(\bar{\varphi} f) \qquad (f \in H^{2\perp}).$$

Proof By (11.1)

$$H_\varphi^* = i_+^* \circ M_\varphi^* \circ P_-^*.$$

We exploit three relations that were developed before. By (4.27), $i_+^* = P_+$. By (2.22), $M_\varphi^* = M_{\bar{\varphi}}$. Finally, by (4.30), $P_-^* = i_-$. Therefore, we deduce that

$$H_\varphi^* = P_+ \circ M_{\bar{\varphi}} \circ i_-. \qquad \square$$

To obtain a matrix representation for H_φ, equip $H^2(\mathbb{T})$ with the standard basis $(\chi_n)_{n\geq 0}$, and $H^2(\mathbb{T})^\perp$ with the basis $(\chi_{-m})_{m\geq 1}$. Note that, in comparison with the standard basis in $L^2(\mathbb{T})$, the ordering in the basis of $H^2(\mathbb{T})^\perp$ is reversed. Then, by Lemma 4.9, we have

$$\begin{aligned}
\langle H_\varphi \chi_n, \chi_{-m} \rangle_{H^2(\mathbb{T})^\perp} &= \langle P_-(\varphi\chi_n), \chi_{-m} \rangle_{H^2(\mathbb{T})^\perp} \\
&= \langle \varphi\chi_n, \chi_{-m} \rangle_{L^2(\mathbb{T})} \\
&= \langle \varphi, \chi_{-m-n} \rangle_{L^2(\mathbb{T})} \\
&= \hat{\varphi}(-m-n).
\end{aligned} \tag{11.3}$$

Hence, in the matrix representation of H_φ with respect to the two above-mentioned bases, the component in the position mn is $\hat{\varphi}(-m-n)$, where $m \geq 1$ and $n \geq 0$. In other words, H_φ is represented by the singly infinite Hankel matrix

$$\begin{bmatrix}
\hat{\varphi}(-1) & \hat{\varphi}(-2) & \hat{\varphi}(-3) & \hat{\varphi}(-4) & \hat{\varphi}(-5) & \cdots \\
\hat{\varphi}(-2) & \hat{\varphi}(-3) & \hat{\varphi}(-4) & \hat{\varphi}(-5) & \hat{\varphi}(-6) & \cdots \\
\hat{\varphi}(-3) & \hat{\varphi}(-4) & \hat{\varphi}(-5) & \hat{\varphi}(-6) & \hat{\varphi}(-7) & \cdots \\
\hat{\varphi}(-4) & \hat{\varphi}(-5) & \hat{\varphi}(-6) & \hat{\varphi}(-7) & \hat{\varphi}(-8) & \cdots \\
\hat{\varphi}(-5) & \hat{\varphi}(-6) & \hat{\varphi}(-7) & \hat{\varphi}(-8) & \hat{\varphi}(-9) & \cdots \\
\vdots & \vdots & \vdots & \vdots & \vdots & \ddots
\end{bmatrix}. \tag{11.4}$$

This should not be a surprise. As we saw in Section 2.8, the matrix of the multiplication operator M_φ with respect to the standard orthonormal basis of $L^2(\mathbb{T})$ on its domain, and the reverse standard orthonormal basis of $L^2(\mathbb{T})$ on its codomain, is a doubly infinite Hankel matrix. The matrix representation (11.4) means that for each

$$f = \sum_{n=0}^{\infty} \hat{f}(n)\chi_n \in H^2(\mathbb{T})$$

we have

$$H_\varphi f = \sum_{m=1}^{\infty} \left(\sum_{n=0}^{\infty} \hat{\varphi}(-m-n)\hat{f}(n) \right) \chi_{-m} \in H^2(\mathbb{T})^\perp. \tag{11.5}$$

In light of (11.6) and (11.4), there is a dangerous temptation to decompose φ into two functions, say $\varphi_+ + \varphi_-$, where $\varphi_+ = \sum_{n\geq 0} \hat{\varphi}(n)\chi_n$ and $\varphi_- = \sum_{n<0} \hat{\varphi}(n)\chi_n$, and thus deduce that $H_\varphi = H_{\varphi_-}$. The main obstacle is that φ_+ and φ_- might not remain bounded. An example is given in Exercise 4.3.1. We will return to this issue in Section 11.4.

The matrix representation enables us to study the effect of multiplication of a Hankel operator by the shift operator.

Lemma 11.2 *Let $\varphi \in L^\infty(\mathbb{T})$, and let $k \geq 1$. Then we have*

$$H_\varphi S^k = H_{\chi_k \varphi}.$$

Proof As for the matrix representation for the Hankel operator, we equip $H^2(\mathbb{T})$ with the standard basis $(\chi_n)_{n\geq 0}$, and $H^2(\mathbb{T})^\perp$ with the basis $(\chi_{-m})_{m\geq 1}$. Hence, by (11.3), the component in the position mn for the operator $H_\varphi S^k$ is given by

$$\langle H_\varphi S^k \chi_n, \chi_{-m} \rangle_{H^2(\mathbb{T})^\perp} = \langle H_\varphi \chi_{n+k}, \chi_{-m} \rangle_{H^2(\mathbb{T})^\perp} = \hat{\varphi}(-m-n-k).$$

But we have

$$\hat{\varphi}(-m-n-k) = \widehat{\chi_k \varphi}(-m-n),$$

and therefore, again by (11.3), we can write

$$\langle H_\varphi S^k \chi_n, \chi_{-m} \rangle_{H^2(\mathbb{T})^\perp} = \langle H_{\chi_k \varphi} \chi_n, \chi_{-m} \rangle_{H^2(\mathbb{T})^\perp}.$$

This identity ensures that $H_\varphi S^k = H_{\chi_k \varphi}$. □

Exercises

Exercise 11.1.1 Let $\varphi \in L^\infty(\mathbb{T})$. Show that there is an operator $A \in \mathcal{L}(H^2(\mathbb{T}))$ whose matrix representation with respect to the standard basis $(\chi_n)_{n\geq 0}$ is

$$\begin{bmatrix} \hat{\varphi}(0) & \hat{\varphi}(1) & \hat{\varphi}(2) & \hat{\varphi}(3) & \hat{\varphi}(4) & \cdots \\ \hat{\varphi}(1) & \hat{\varphi}(2) & \hat{\varphi}(3) & \hat{\varphi}(4) & \hat{\varphi}(5) & \cdots \\ \hat{\varphi}(2) & \hat{\varphi}(3) & \hat{\varphi}(4) & \hat{\varphi}(5) & \hat{\varphi}(6) & \cdots \\ \hat{\varphi}(3) & \hat{\varphi}(4) & \hat{\varphi}(5) & \hat{\varphi}(6) & \hat{\varphi}(7) & \cdots \\ \hat{\varphi}(4) & \hat{\varphi}(5) & \hat{\varphi}(6) & \hat{\varphi}(7) & \hat{\varphi}(8) & \cdots \\ \vdots & \vdots & \vdots & \vdots & \vdots & \ddots \end{bmatrix}.$$

Hint: Take $A = P_+ \circ M_\varphi \circ R \circ i_+$, where $M_\varphi : L^2(\mathbb{T}) \longrightarrow L^2(\mathbb{T})$ denotes the multiplication operator introduced in Section 7.2, i_+ is the inclusion map of $H^2(\mathbb{T})$ into $L^2(\mathbb{T})$, and R is the flip operator

$$\begin{array}{rccc} R: & L^2(\mathbb{T}) & \longrightarrow & L^2(\mathbb{T}) \\ & \chi_n & \longmapsto & \chi_{-n}. \end{array}$$

Note that the relation $A = P_+ \circ M_\varphi \circ R \circ i_+$ also shows that

$$\|A\| \le \|\varphi\|_\infty.$$

Exercise 11.1.2 Let $\Gamma = (\gamma_{i,j})_{i,j\ge 0}$ be a matrix. If $x = (x_k)_{k\ge 0}$ is a sequence of complex numbers, we recall that S and S^* denote the operators defined by

$$Sx = (0, x_0, x_1, \dots) \quad \text{and} \quad S^*x = (x_1, x_2, x_3, \dots).$$

Show that Γ is a Hankel matrix (i.e. there exists a sequence of complex numbers $a = (a_n)_{n\ge 0}$ such that $\gamma_{i,j} = a_{i+j}$, $i, j \ge 0$) if and only if

$$\Gamma S \mathfrak{e}_k = S^* \Gamma \mathfrak{e}_k \qquad (k \ge 0).$$

Remark: We recall that $\mathfrak{e}_k$ is the infinite vector whose components are all zero except for the kth place, where it is one.

Exercise 11.1.3 Let A be a linear operator acting from H^2 into H^2_-. Show that its matrix (with respect to the standard bases $(z^k)_{k\ge 0}$ and $(z^{-k})_{k\ge 1}$) is a single infinite Hankel matrix if and only if

$$ASp = P_- Z A p$$

for all analytic polynomials p.

11.2 The norm of H_φ

The inequality

$$\|H_\varphi\| \le \|\varphi\|_\infty$$

was not difficult to establish (see (11.2)). An important feature of Hankel operators is that the corresponding symbol is not unique. In fact, if $f \in H^2$ and $\psi \in H^\infty$, then $f\psi \in H^2$ and thus, by Lemma 4.10, $H_\psi f = 0$. In other words,

$$H_\psi = 0 \qquad (\psi \in H^\infty).$$

Since H_φ is linear with respect to φ, i.e.

$$H_{\alpha_1\varphi_1+\alpha_2\varphi_2} = \alpha_1 H_{\varphi_1} + \alpha_2 H_{\varphi_2},$$

we immediately deduce that

$$H_\varphi = H_{\varphi-\psi} \qquad (\psi \in H^\infty(\mathbb{T}),\ \varphi \in L^\infty(\mathbb{T})). \tag{11.6}$$

This relation sheds more light on the norm of this Hankel operator and leads us to

$$\|H_\varphi\| = \|H_{\varphi-\psi}\| \le \|\varphi - \psi\|_\infty \qquad (\psi \in H^\infty).$$

Therefore, by taking the infimum with respect to ψ, we obtain

$$\|H_\varphi\| \le \operatorname{dist}(\varphi, H^\infty). \tag{11.7}$$

But, in fact, equality holds in (11.7).

Theorem 11.3 *Let the sequence of complex numbers $(\alpha_n)_{n\ge1}$ be such that the singly infinite Hankel matrix*

$$A = \begin{bmatrix} \alpha_1 & \alpha_2 & \alpha_3 & \cdots \\ \alpha_2 & \alpha_3 & \alpha_4 & \cdots \\ \alpha_3 & \alpha_4 & \alpha_5 & \cdots \\ \vdots & \vdots & \vdots & \ddots \end{bmatrix}$$

defines a bounded linear operator on ℓ^2. Then the following hold.

(i) *There exists a function $\varphi \in L^\infty(\mathbb{T})$ such that*

$$\hat{\varphi}(-n) = \alpha_n \qquad (n \ge 1).$$

(ii) *The choice of φ is not unique. However, for any such φ, we have*

$$\|A\|_{\mathcal{L}(\ell^2)} = \operatorname{dist}_{L^\infty(\mathbb{T})}(\varphi, H^\infty).$$

(iii) *There is a choice of $\varphi \in H^\infty(\mathbb{T})$ such that*

$$\|A\|_{\mathcal{L}(\ell^2)} = \|\varphi\|_\infty.$$

Proof Consider the orthonormal basis $(\mathfrak{e}_k)_{k\ge0}$ on ℓ^2, and then define two unitary operators

$$\begin{array}{cccc} U: & \ell^2 & \longrightarrow & H^2 \\ & \displaystyle\sum_{k=0}^{\infty} a_k \mathfrak{e}_k & \longmapsto & \displaystyle\sum_{k=0}^{\infty} a_k z^k \end{array}$$

and

$$\begin{array}{rccc} V: & \ell^2 & \longrightarrow & H^{2\perp} \\ & \displaystyle\sum_{k=0}^{\infty} a_k \mathfrak{e}_k & \longmapsto & \displaystyle\sum_{k=0}^{\infty} a_k/z^{k+1}. \end{array}$$

Suppose that there is a $\varphi \in L^\infty(\mathbb{T})$ such that (i) holds. Clearly, since only the negative part of the spectrum of φ is relevant, the choice of φ is not unique. However, for any such φ, the matrix representations of H_φ and A show that

$$A = V^* H_\varphi U.$$

Hence, from (11.7), we deduce that

$$\|A\| = \|H_\varphi\| \le \mathrm{dist}(\varphi, H^\infty). \tag{11.8}$$

If φ_1 and φ_2 are two such symbols, then $H_{\varphi_1} = H_{\varphi_2}$, and thus $\varphi_1 - \varphi_2 \in H^\infty(\mathbb{T})$. This fact implies that

$$\mathrm{dist}(\varphi_1, H^\infty) = \mathrm{dist}(\varphi_2, H^\infty). \tag{11.9}$$

Now, we proceed to show that a symbol φ actually exists. Without loss of generality, assume that $\|A\| = 1$. Add one column to the left of A as follows:

$$\begin{bmatrix} \alpha_0 & \alpha_1 & \alpha_2 & \alpha_3 & \alpha_4 & \cdots \\ \alpha_1 & \alpha_2 & \alpha_3 & \alpha_4 & \alpha_5 & \cdots \\ \alpha_2 & \alpha_3 & \alpha_4 & \alpha_5 & \alpha_6 & \cdots \\ \alpha_3 & \alpha_4 & \alpha_5 & \alpha_6 & \alpha_7 & \cdots \\ \alpha_4 & \alpha_5 & \alpha_6 & \alpha_7 & \alpha_8 & \cdots \\ \vdots & \vdots & \vdots & \vdots & \vdots & \ddots \end{bmatrix}. \tag{11.10}$$

The element α_0 is unknown, but it will be properly chosen soon. If we write this matrix as

$$\begin{bmatrix} \alpha_0 & B \\ C & D \end{bmatrix},$$

where

$$B = \begin{bmatrix} \alpha_1 & \alpha_2 & \alpha_3 & \alpha_4 & \cdots \end{bmatrix}, \qquad C = \begin{bmatrix} \alpha_1 \\ \alpha_2 \\ \alpha_3 \\ \alpha_4 \\ \vdots \end{bmatrix}$$

and

$$D = \begin{bmatrix} \alpha_2 & \alpha_3 & \alpha_4 & \alpha_5 & \cdots \\ \alpha_3 & \alpha_4 & \alpha_5 & \alpha_6 & \cdots \\ \alpha_4 & \alpha_5 & \alpha_6 & \alpha_7 & \cdots \\ \alpha_5 & \alpha_6 & \alpha_7 & \alpha_8 & \cdots \\ \vdots & \vdots & \vdots & \vdots & \ddots \end{bmatrix},$$

then the conditions of Theorem 7.20 are fulfilled, and hence there is an $\alpha_0 \in \mathbb{C}$ such that (11.10) is a contraction on ℓ^2. We can now carry out the same process on the matrix (11.10) and obtain $\alpha_{-1} \in \mathbb{C}$ such that

$$\begin{bmatrix} \alpha_{-1} & \alpha_0 & \alpha_1 & \alpha_2 & \alpha_3 & \alpha_4 & \cdots \\ \alpha_0 & \alpha_1 & \alpha_2 & \alpha_3 & \alpha_4 & \alpha_5 & \cdots \\ \alpha_1 & \alpha_2 & \alpha_3 & \alpha_4 & \alpha_5 & \alpha_6 & \cdots \\ \alpha_2 & \alpha_3 & \alpha_4 & \alpha_5 & \alpha_6 & \alpha_7 & \cdots \\ \alpha_3 & \alpha_4 & \alpha_5 & \alpha_6 & \alpha_7 & \alpha_8 & \cdots \\ \vdots & \vdots & \vdots & \vdots & \vdots & \vdots & \ddots \end{bmatrix}$$

is a contraction on ℓ^2. By induction, we obtain the sequence $(\alpha_n)_{n\in\mathbb{Z}}$ such that, for each $n_0 \in \mathbb{Z}$, the singly infinite Hankel matrix formed with the sequence $(\alpha_n)_{n\geq n_0}$ is a contraction on ℓ^2.

Now, form the doubly infinite Hankel matrix with the sequence $(\alpha_n)_{n\in\mathbb{Z}}$ with α_0 at its 00 position. The above discussion shows that the conditions of Theorem 7.26 are fulfilled and hence this matrix represents a contraction on $\ell^2(\mathbb{Z})$. Therefore, according to Theorem 2.25, there is a function $\varphi \in L^\infty(\mathbb{T})$ such that

$$\hat{\varphi}(n) = \alpha_{-n} \qquad (n \in \mathbb{Z}),$$

and the doubly infinite Hankel matrix corresponds to the multiplication operator M_φ on $L^2(\mathbb{T})$, equipped with the standard basis $(\chi_n)_{n\in\mathbb{Z}}$. Moreover,

$$\|\varphi\|_\infty = \|M_\varphi\| \leq 1 = \|A\|. \tag{11.11}$$

The symbol φ, which was obtained above, is a special one. Since

$$\operatorname{dist}(\varphi, H^\infty) \leq \|\varphi\|_\infty \qquad (\varphi \in L^\infty(\mathbb{T})),$$

by (11.8) and (11.11), for this particular symbol, we have

$$\|A\| = \operatorname{dist}(\varphi, H^\infty) = \|\varphi\|_\infty.$$

The relation (11.9) shows that the first identity holds for all symbols φ. This completes the proof of Theorem 11.3. □

Corollary 11.4 *Let $\varphi \in L^\infty(\mathbb{T})$. Then*

$$\|H_\varphi\|_{\mathcal{L}(H^2(\mathbb{T}),H^2(\mathbb{T})^\perp)} = \operatorname{dist}_{L^\infty(\mathbb{T})}(\varphi, H^\infty).$$

Moreover, there is a function $\psi \in H^\infty(\mathbb{T})$ *such that*

$$\|H_\varphi\| = \|\varphi - \psi\|_\infty.$$

Proof By (11.4), the matrix of H_φ is a singly infinite Hankel matrix, which can be interpreted as a bounded operator on ℓ^2. Hence, by Theorem 11.3, there is $\eta \in L^\infty(\mathbb{T})$ such that $H_\varphi = H_\eta$ and

$$\|H_\eta\|_{\mathcal{L}(H^2(\mathbb{T}),H^2(\mathbb{T})^\perp)} = \mathrm{dist}_{L^\infty(\mathbb{T})}(\eta, H^\infty) = \|\eta\|_\infty.$$

Take $\psi = \varphi - \eta$. The relation $H_\varphi = H_\eta$ ensures that $\psi \in H^\infty(\mathbb{T})$. □

Exercises

Exercise 11.2.1 Let A be the operator introduced in Exercise 11.1.1. Show that

$$\|A\|_{\mathcal{L}(H^2(\mathbb{T}))} = \mathrm{dist}_{L^\infty(\mathbb{T})}(\varphi, \overline{H_0^\infty}),$$

where φ is any symbol such that $A = P_+ \circ M_\varphi \circ R \circ i_+$.
Hint: Use Theorem 11.3.

Exercise 11.2.2 Show that

$$\mathrm{dist}(\bar{z}, H^\infty) = 1.$$

Hint: By Theorem 11.3, $\mathrm{dist}(\bar{z}, H^\infty) = \|H_{\bar{z}}\|$.
Remark: Note that, since $\bar{z} \in \mathcal{C}(\mathbb{T})$, we have $\mathrm{dist}(\bar{z}, H^\infty + \mathcal{C}(\mathbb{T})) = 0$.

Exercise 11.2.3 Let $\psi \in L^\infty(\mathbb{T})$. Show that the following assertions are equivalent:

(i) H_ψ has a nontrivial kernel;
(ii) $\mathcal{R}(H_\psi)$ is not dense in H^2_-;
(iii) $\psi = \bar{\Theta}\varphi$, for some inner function Θ and some function $\varphi \in H^\infty$.

Hint: To do so, take the following steps.
(i) $\Longrightarrow$ (iii) Show that $f \in \ker H_\psi \Longrightarrow zf \in \ker H_\psi$. Deduce that there exists an inner function Θ such that $\ker H_\psi = \Theta H^2$. Show that $\Theta\psi \in H^\infty$.
(iii) $\Longrightarrow$ (ii) Show that $\bar{z}\bar{\Theta} \perp \mathcal{R}(H_\psi)$.
(ii) $\Longrightarrow$ (i) Let $g \in H^2_-$, $g \neq 0$, $g \perp \mathcal{R}(H_\psi)$. Write $g = \bar{z}\bar{h}$, where $h \in H^2$. Show that $h \in \ker H_\psi$.

11.3 Hilbert's inequality

The inequality (11.2) can be rewritten as

$$|\langle H_\varphi f, g\rangle_{H^2(\mathbb{T})^\perp}| \le \|\varphi\|_\infty \|f\|_2 \|g\|_2,$$

for all $f \in H^2$ and $g \in H^2(\mathbb{T})^\perp$. By Lemma 4.9,

$$\langle H_\varphi f, g\rangle_{H^2(\mathbb{T})^\perp} = \langle \varphi f, g\rangle_{L^2(\mathbb{T})}.$$

Hence, using Parseval's identity and (11.5), the last inequality is equivalent to

$$\begin{aligned}\Big|\sum_{m=1,\,n=0}^{\infty} \hat{\varphi}(-m-n)\hat{f}(n)\overline{\hat{g}(-m)}\Big| \\ \le \|\varphi\|_\infty \bigg(\sum_{n=0}^{\infty} |\hat{f}(n)|^2\bigg)^{1/2} \bigg(\sum_{m=1}^{\infty} |\hat{g}(-m)|^2\bigg)^{1/2}.\end{aligned}$$

Since, by the Riesz–Fischer theorem, there is a correspondence between H^2, $H^2(\mathbb{T})^\perp$ and ℓ^2, this inequality is equivalent to

$$\Big|\sum_{i,j=1}^{\infty} \hat{\varphi}(-i-j+1)x_i\overline{y_j}\Big| \le \|\varphi\|_\infty \bigg(\sum_{i=1}^{\infty} |x_i|^2\bigg)^{1/2} \bigg(\sum_{j=1}^{\infty} |y_j|^2\bigg)^{1/2} \tag{11.12}$$

for all $(x_n)_{n\ge1}$ and $(y_n)_{n\ge1}$ in ℓ^2.

Now, let us consider a very special case. In Exercise 4.3.1, we saw that the function

$$\varphi(e^{it}) = -i(\pi - t) \qquad (0 < t < 2\pi),$$

has the Fourier coefficients

$$\hat{\varphi}(n) = \begin{cases} -\dfrac{1}{n} & \text{if} \quad n \ne 0, \\ 0 & \text{if} \quad n = 0. \end{cases}$$

Hence, the matrix corresponding to the Hankel operator H_φ is

$$\Gamma = \begin{bmatrix} 1 & \frac{1}{2} & \frac{1}{3} & \frac{1}{4} & \cdots \\ \frac{1}{2} & \frac{1}{3} & \frac{1}{4} & \frac{1}{5} & \cdots \\ \frac{1}{3} & \frac{1}{4} & \frac{1}{5} & \frac{1}{6} & \cdots \\ \frac{1}{4} & \frac{1}{5} & \frac{1}{6} & \frac{1}{7} & \cdots \\ \vdots & \vdots & \vdots & \vdots & \ddots \end{bmatrix}$$

This is the so-called *Hilbert–Hankel matrix*. Since $\|\varphi\|_\infty = \pi$, the inequality (11.12) becomes

$$\left| \sum_{i,j=1}^{\infty} \frac{x_i \overline{y_j}}{i+j-1} \right| \leq \pi \left(\sum_{i=1}^{\infty} |x_i|^2 \right)^{1/2} \left(\sum_{j=1}^{\infty} |y_j|^2 \right)^{1/2}$$

for all $(x_n)_{n\geq 1}$ and $(y_n)_{n\geq 1}$ in ℓ^2. This is equivalent to saying that the matrix Γ defines a bounded operator of ℓ^2 and its norm is at most π, a fact that is not easy to verify directly.

Exercises

Exercise 11.3.1 Let

$$Z_n = \begin{bmatrix} 1 & \frac{1}{2} & \frac{1}{3} & \frac{1}{4} & \cdots & \frac{1}{n} \\ \frac{1}{2} & \frac{1}{3} & \frac{1}{4} & \cdots & \cdots & 0 \\ \frac{1}{3} & \frac{1}{4} & \cdots & \cdots & \cdots & 0 \\ \frac{1}{4} & \cdots & \cdots & \cdots & \cdots & 0 \\ \vdots & \vdots & \vdots & \vdots & \ddots & \vdots \\ \frac{1}{n} & 0 & 0 & 0 & \cdots & 0 \end{bmatrix}.$$

This is a finite Hankel matrix with all components below the main cross-diagonal being zero. Take

$$u_n = \sum_{k=1}^{n} \frac{1}{\sqrt{k(n+1-k)}}.$$

(i) Show that $u_n \longrightarrow \pi$, as $n \longrightarrow \infty$.
Hint: Use Riemann's sum for the integral

$$\int_0^1 \frac{dx}{\sqrt{x(1-x)}}.$$

(ii) Let $v = (1, 1/\sqrt{2}, 1/\sqrt{3}, \ldots, 1/\sqrt{n})$. Show that

$$\|Z_n\| \geq \frac{\langle Z_n v, v\rangle_{\ell^2}}{\|v\|_{\ell^2}^2} = \frac{\sum_{k=1}^{n} u_k/k}{\sum_{k=1}^{n} 1/k}.$$

Then deduce from (i) that

$$\limsup_{n\to\infty} \|Z_n\| \geq \pi.$$

(iii) Let Γ be the Hilbert–Hankel matrix. Show that $\|\Gamma\| = \pi$.
Hint: Use (ii) to show that $\|\Gamma\| \geq \pi$.

Exercise 11.3.2 Let $a = (a_n)_{n\geq 0}$ be a sequence of real numbers, and let Γ_a be the Hankel matrix defined by

$$\Gamma_a = (a_{i+j})_{i,j\geq 0}.$$

The purpose of this exercise is to show that the following two assertions are equivalent.

(i) There exists a positive Borel measure μ on $\mathbb{R}$ such that

$$\int_{\mathbb{R}} |t|^n \, d\mu(t) < \infty \quad \text{and} \quad a_n = \int_{\mathbb{R}} t^n \, d\mu(t) \qquad (n \geq 0).$$

(ii) The Hankel matrix Γ_a is a positive matrix, i.e. we have

$$\sum_{i,j=0}^{n} a_{i+j} x_i \overline{x_j} \geq 0$$

for every $n \geq 0$ and any choice of $x_0, \dots, x_n \in \mathbb{C}$.

Hint: To do so, take the following steps.
(i) $\Longrightarrow$ (ii) Given $x_0, x_1, \dots, x_n \in \mathbb{C}$, consider the polynomial p defined by

$$p(t) = \sum_{k=0}^{n} x_k t^k,$$

and compute

$$\int_{\mathbb{R}} |p(t)|^2 \, d\mu(t).$$

(ii) $\Longrightarrow$ (i) To show this implication requires seven substeps.
Step I: For any polynomial $p(t) = \sum_{k=0}^{n} x_k t^k$, put

$$\varphi(p) = \sum_{j=0}^{n} a_j x_j.$$

In the following, we identify the polynomial p and the finitely compactly supported sequence $(x_0, x_1, \dots, x_n, 0, \dots)$. In particular, $\Gamma_a p$ denotes the image of vector $(x_0, x_1, \dots, x_n, 0, \dots)$ under the matrix Γ_a.

(a) Verify that $\varphi(p) = \langle \Gamma_a p, e_0 \rangle_{\ell^2}$, where $e_0 = (1, 0, \dots)$.
(b) Show that, for any polynomials p and q, we have

$$\Gamma_a(pq) = p(S^*)\Gamma_a q.$$

Hint: Use that $\Gamma_a S = S^* \Gamma_a$ (see Exercise 11.1.2).
(c) Show that, for any polynomial q, we have

$$\varphi(|q|^2) = \langle \Gamma_a q, q \rangle_{\ell^2}.$$

Hint: Use I(b) with $p(t) = \overline{q(t)}$, $t \in \mathbb{R}$.
(d) Deduce that, if p is a polynomial such that $p(t) \geq 0$, $t \in \mathbb{R}$, then $\varphi(p) \geq 0$.
Hint: Use the Fejér–Riesz theorem.

Step II: Let

$$\mathcal{X} = \{f \in C(\mathbb{R}) : \exists\, C > 0,\ \exists\, N \geq 1,\ |f(x)| \leq C(1 + x^2)^N\}$$

and $K = \{f \in \mathcal{X} : f \geq 0\}$. Using Exercise 1.4.7, show that there exists a linear form φ_* on $\mathcal{X}$ such that $\varphi_*(p) = \varphi(p)$ for any real polynomial p and $\varphi_*(f) \geq 0$ for any $f \in K$.

Step III: Let

$$\mathcal{X}_N = \{f \in \mathcal{C}(\mathbb{R}) : |f(x)| = o((1 + x^2)^N),\ |x| \to \infty\},$$

equipped with the norm

$$\|f\|_N = \sup_{x \in \mathbb{R}} \frac{|f(x)|}{(1 + x^2)^N}.$$

Check that $\mathcal{X}_N$ is a Banach space and, if $f \in \mathcal{X}_N$, then

$$|\varphi_*(f)| \leq \|f\|_N \varphi_*((1 + x^2)^N).$$

Step IV: Deduce that φ_* is continuous on $\mathcal{X}_N$.

Step V: Deduce that there exists a positive measure μ_N on $\mathbb{R}$ such that

$$\int_{\mathbb{R}} (1 + x^2)^N \, d\mu_N(x) < +\infty \quad \text{and} \quad \varphi_*(f) = \int_{\mathbb{R}} f \, d\mu_N$$

for any $f \in \mathcal{X}_N$.

Hint: Use the Riesz representation theorem for functionals.

Step VI: Show that $\mu_{N+1} = \mu_N$.

Hint: Use the fact that continuous compactly supported functions are dense in $\mathcal{X}_{N+1}$.

Step VII: Then define $\mu = \mu_N$. Check that (i) is satisfied with the measure μ.

Exercise 11.3.3 Let μ be a positive Borel measure on $[0, 1)$ and let

$$a_n = \int_0^1 t^n \, d\mu(t), \qquad n = 0, 1, \ldots.$$

Show that the following are equivalent:

(i) μ is a Carleson measure;
(ii) $a_n = O(1/n)$, $n \longrightarrow \infty$;
(iii) Γ_a is a bounded operator on ℓ^2.

Hint: Here are the steps to take.

(i) $\Longrightarrow$ (ii) Use the fact that $\mu((t, 1)) \leq K(1 - t)$, $0 \leq t \leq 1$, and, by integration by parts, show that

$$\int_0^1 t^n \, d\mu(t) = \int_0^1 n t^{n-1} \mu((t, 1)) \, dt.$$

(ii) $\Longrightarrow$ (iii) Let $x = (x_n)_{n\geq 0}$ and $y = (y_n)_{n\geq 0}$ be two sequences in ℓ^2. Then we have

$$\langle \Gamma_a x, y\rangle_{\ell^2} = \sum_{i,j\geq 0} a_{i+j} x_i \overline{y_j},$$

and use Hilbert's inequality.
(iii) $\Longrightarrow$ (i) Use $x_r = (1, r, r^2, r^3, \ldots)$ for $0 \leq r < 1$ and then compute $\langle \Gamma_a x_r, x_r\rangle_{\ell^2}$.

Exercise 11.3.4 Let $f(z) = \sum_{n=0}^{\infty} a_n z^n$ be a function in H^1. The purpose of this exercise is to prove *Hardy's inequality*, which says that

$$\sum_{n=0}^{\infty} \frac{|a_n|}{n+1} \leq \pi \|f\|_1. \tag{11.13}$$

(i) Let $\psi(t) = ie^{-it}(\pi - t)$, $0 \leq t < 2\pi$. Show that the Fourier coefficients of ψ are given by

$$\hat{\psi}(n) = \frac{1}{n+1} \qquad (n \geq 0).$$

Show also that $\|\psi\|_\infty = \pi$.
(ii) Factorize f as $f = gh$ with $g, h \in H^2$, $\|g\|_2^2 = \|h\|_2^2 = \|f\|_1$, and write

$$g(z) = \sum_{n=0}^{\infty} b_n z^n \quad \text{and} \quad h(z) = \sum_{n=0}^{\infty} c_n z^n \qquad (z \in \mathbb{D}).$$

(iii) Show that

$$\sum_{k=0}^{N} \frac{|a_n|}{n+1} \leq \sum_{k,n=0}^{\infty} \frac{|b_k|\,|c_n|}{n+k+1}.$$

(iv) Conclude by applying Hilbert's inequality.

11.4 The Nehari problem

The Riesz projection P_+ can also be defined via some approximation theory considerations. In this regard, we look at H^2 as a closed subspace of $L^2(\mathbb{T})$, and, for a given $f \in L^2(\mathbb{T})$, we ask to find $g \in H^2$ such that the quantity $\|f - g\|_2$ is minimal. As we saw implicitly in Section 4.3, Parseval's identity ensures that the *unique* solution to this approximation problem is $g = P_+ f$. In fact, for any $g \in H^2$, we have

$$\|f - g\|_2^2 = \|f - P_+ f\|_2^2 + \|P_+ f - g\|_2^2,$$

and thus the unique solution is obtained by taking $g = P_+ f$. Moreover, we have

$$\inf_{g \in H^2} \|f - g\|_2 = \|f - P_+ f\|_2.$$

The above approximation question can be posed in a more general setting. We can consider a Banach space X and a closed subspace M, and then, given $f \in X$, ask for the existence (and uniqueness) of $g \in M$ such that $\|f - g\|_X$ is minimal. In the first place, in this generality, the question is difficult to treat, and a solution may even fail to exist, or there are cases in which infinitely many solutions exist.

In this section, we are dealing with a case that has a well-behaved theory. The *Nehari problem* is the above approximation problem with $X = L^\infty(\mathbb{T})$ and $M = H^\infty(\mathbb{T})$. More explicitly, for a given $\varphi \in L^\infty(\mathbb{T})$, we seek a function $\psi \in H^\infty$ such that the norm $\|\varphi - \psi\|_\infty$ is minimal. Our goal is to show that the solution exists and, moreover, that it is unique. As the first step, note that Exercise 4.3.1 at least shows that Nehari's solution is not necessarily the same as Riesz's solution. The function $\varphi(e^{i\theta}) = -i(\pi - \theta)$ belongs to $L^\infty(\mathbb{T})$, while $(P_+\varphi)(e^{i\theta}) = \log(1 - e^{i\theta})$ is clearly unbounded.

From a theoretical point of view, the answer to Nehari's problem is implicitly contained in Corollary 11.4. This result gives a formula for the distance between φ and H^∞, i.e.

$$\operatorname{dist}_{L^\infty(\mathbb{T})}(\varphi, H^\infty) = \|H_\varphi\|_{\mathcal{L}(H^2(\mathbb{T}), H^2(\mathbb{T})^\perp)}.$$

Moreover, in its proof, we provided a procedure to construct a function $\psi \in H^\infty(\mathbb{T})$ such that

$$\operatorname{dist}_{L^\infty(\mathbb{T})}(\varphi, H^\infty) = \|\varphi - \psi\|_\infty.$$

To obtain ψ, we must inductively obtain the coefficients $\alpha_0, \alpha_1, \alpha_2, \ldots$, a task that may not be preferable, or even feasible, in applications. Hence, we approach Nehari's problem from another point of view.

Theorem 11.5 *Let $\varphi \in L^\infty(\mathbb{T})$. Assume that the Hankel operator H_φ has a maximizing vector, i.e. a nonzero function $f \in H^2(\mathbb{T})$ that fulfills*

$$\|H_\varphi f\|_2 = \|H_\varphi\| \times \|f\|_2.$$

Put

$$\eta = \frac{H_\varphi f}{f} \qquad (a.e.\ on\ \mathbb{T}).$$

Then the following hold:

(i) *$|\eta|$ is a constant function on $\mathbb{T}$;*
(ii) *we have $H_\varphi = H_\eta$ and*

$$\operatorname{dist}_{L^\infty(\mathbb{T})}(\varphi, H^\infty) = \operatorname{dist}_{L^\infty(\mathbb{T})}(\eta, H^\infty) = \|\eta\|_\infty;$$

(iii) *if* $\omega \in L^\infty(\mathbb{T})$ *is such that* $H_\varphi = H_\omega$ *and* $\operatorname{dist}_{L^\infty(\mathbb{T})}(\omega, H^\infty) = \|\omega\|_\infty$, *then* $\omega = \eta$.

Proof The existence of a symbol of minimal norm is guaranteed (even without the extra assumption of having a maximizing vector) in Corollary 11.4. To establish the uniqueness of this symbol, if a maximizing vector exists, let $\omega \in L^\infty(\mathbb{T})$ be such that $H_\varphi = H_\omega$ and

$$\|H_\varphi\| = \|H_\omega\| = \operatorname{dist}_{L^\infty(\mathbb{T})}(\omega, H^\infty) = \|\omega\|_\infty.$$

Then, for any function $g \in H^2(\mathbb{T})$, we have

$$\begin{aligned}\|H_\varphi g\|_2 &= \|H_\omega g\|_2\\ &= \|P_-(\omega g)\|_2\\ &\le \|\omega g\|_2\\ &\le \|\omega\|_\infty \times \|g\|_2\\ &= \|H_\varphi\| \times \|g\|_2.\end{aligned}$$

Hence, if we replace g by f in the above relations, and appeal to the maximizing property of f, we see that f must satisfy

$$\|P_-(\omega f)\|_2 = \|\omega f\|_2 = \|\omega\|_\infty \times \|f\|_2.$$

Hence, by Corollary 4.13, we have

$$H_\varphi f = P_-(\omega f) = \omega f.$$

Since $f \in H^2$, $f \not\equiv 0$, we surely have $f \neq 0$ almost everywhere on $\mathbb{T}$. Thus, we can divide by f to obtain $\omega = \eta$. The identity $\|\omega f\|_2 = \|\omega\|_\infty \times \|f\|_2$ also reveals that $|\omega|$ is constant on $\mathbb{T}$. □

Two comments are in order. First, there are symbols φ such that H_φ does not have any maximizing vector. In fact, by other methods, we can construct a function $\varphi \in L^\infty(\mathbb{T})$ for which the Nehari problem has several solutions. Hence, in light of Theorem 11.5, H_φ cannot have any maximizing vectors. Second, if f_1 and f_2 are maximizing vectors of H_φ, the above formula for ψ shows that

$$\frac{H_\varphi f_1}{f_1} = \frac{H_\varphi f_2}{f_2} \qquad \text{(a.e. on } \mathbb{T}).$$

In some elementary cases, we can directly guess the maximizing vector and thus solve Nehari's problem. For example, if

$$\varphi(z) = \frac{1 - \bar{\alpha} z}{\alpha - z} \qquad (\alpha \in \mathbb{D},\ z \in \mathbb{T}),$$

then

$$\varphi(z) = \bar{\alpha} + \frac{1 - |\alpha|^2}{\alpha - z},$$

and thus, for each $f \in H^2(\mathbb{T})$, we can write

$$\begin{aligned}\varphi(z)f(z) &= \bar{\alpha}f(z) - (1-|\alpha|^2)\frac{f(z)-f(\alpha)}{z-\alpha} - (1-|\alpha|^2)\frac{f(\alpha)}{z-\alpha}\\ &= \bar{\alpha}f(z) - (1-|\alpha|^2)Q_\alpha f(z) - (1-|\alpha|^2)\frac{f(\alpha)}{z-\alpha},\end{aligned}$$

where Q_α is the operator introduced in Section 8.2. By Theorem 8.10, $Q_\alpha f \in H^2(\mathbb{T})$. Moreover, the Fourier series representation

$$g(z) = \frac{1}{z-\alpha} = \sum_{n=1}^{\infty} \frac{\alpha^{n-1}}{z^n}$$

shows that

$$g \in H^2(\mathbb{T})^\perp$$

with

$$\|g\|_2 = \frac{1}{(1-|\alpha|^2)^{1/2}}. \tag{11.14}$$

Hence, we conclude that

$$H_\varphi f = -(1-|\alpha|^2)f(\alpha)g \qquad (f \in H^2(\mathbb{T})).$$

In light of (4.18), the above identity can be rewritten as

$$H_\varphi = -(1-|\alpha|^2)g \otimes k_\alpha.$$

Now, on the one hand, by (4.20), (11.14) and Theorem 1.37, we have

$$\|H_\varphi\| = (1-|\alpha|^2)\|g\|_2\|k_\alpha\|_2 = 1,$$

and, on the other, if we take $f = k_\alpha$, we get

$$H_\varphi k_\alpha = -g.$$

Hence, $\|H_\varphi k_\alpha\|_2 = \|g\|_2 = \|k_\alpha\|_2$. In other words, k_α is a maximizing vector H_φ. Hence, by Theorem 11.5, the best approximation $\psi \in H^\infty(\mathbb{T})$ to φ is given by

$$\begin{aligned}\psi &= \varphi - \frac{H_\varphi k_\alpha}{k_\alpha}\\ &= \varphi + \frac{g}{k_\alpha}\\ &= \varphi - \varphi = 0.\end{aligned}$$

This means that

$$\operatorname{dist}_{L^\infty(\mathbb{T})}(\varphi, H^\infty) = \|\varphi\|_\infty = 1,$$

and, for every $\eta \in H^\infty(\mathbb{T})$, $\eta \neq 0$, we have

$$\|\varphi - \eta\|_\infty < 1.$$

The above technique can be generalized to rational functions φ with no poles on $\mathbb{T}$.

Exercise

Exercise 11.4.1 Let

$$g(z) = \frac{1}{z-\alpha} \qquad (z \in \mathbb{T}).$$

Show that

$$\mathrm{dist}_{L^\infty(\mathbb{T})}(g, H^\infty) = \frac{1}{1-|\alpha|^2},$$

and the best H^∞ approximation to g is the constant function

$$\psi = \frac{\bar{\alpha}}{1-|\alpha|^2}.$$

Hint: Look at the example treated at the end of this section.

11.5 More approximation problems

Nehari's problem is closely related to several other approximation problems. We mention some simplified versions of two such problems below.

In the *Fejér–Carathéodory problem*, an analytic polynomial

$$p(z) = \alpha_0 + \alpha_1 z + \cdots + \alpha_n z^n$$

is given, and the goal is to determine the coefficients α_k, $k \geq n+1$, such that

$$\sum_{k=0}^{\infty} \alpha_k z^k \in H^\infty$$

and, moreover, the norm

$$\left\| \sum_{k=0}^{\infty} \alpha_k z^k \right\|_\infty$$

is minimized. Writing

$$\sum_{k=0}^{\infty} \alpha_k z^k = p(z) + z^{n+1} f(z),$$

we immediately see that the coefficients α_k, $k \geq n+1$, are precisely the Taylor coefficient of any function $f \in H^\infty$, and thus we should minimize

$$\|p + \chi_{n+1} f\|_\infty$$

when f runs through H^∞. Since χ_{n+1} is unimodular, we have

$$\|p + \chi_{n+1} f\|_{H^\infty(\mathbb{T})} = \|p\chi_{-n-1} + f\|_{L^\infty(\mathbb{T})}$$

and hence

$$\inf_{f \in H^\infty} \|p + \chi_{n+1} f\|_\infty = \operatorname{dist}_{L^\infty(\mathbb{T})}(p\chi_{-n-1}, H^\infty).$$

Therefore, by Corollary 11.4,

$$\inf_{f \in H^\infty} \|p + \chi_{n+1} f\|_\infty = \|H_{p\chi_{-n-1}}\|_{\mathcal{L}(H^2(\mathbb{T}), H^2(\mathbb{T})^\perp)}.$$

But $p\chi_{-n-1}$ is a trigonometric polynomial, i.e.

$$(p\chi_{-n-1})(z) = \alpha_n z^{-1} + \alpha_{n-1} z^{-2} + \cdots + a_0 z^{-n-1},$$

and thus $H_{p\chi_{-n-1}}$ is of finite rank. Hence, we can say that

$$\inf_{f \in H^\infty} \|p + \chi_{n+1} f\|_\infty$$

$$= \left\| \begin{matrix} \alpha_n & \alpha_{n-1} & \alpha_{n-2} & \cdots & \alpha_2 & \alpha_1 & \alpha_0 \\ \alpha_{n-1} & \alpha_{n-2} & \alpha_{n-3} & \cdots & \alpha_1 & \alpha_0 & 0 \\ \alpha_{n-2} & \alpha_{n-3} & \alpha_{n-4} & \cdots & \alpha_0 & 0 & 0 \\ \vdots & \vdots & \vdots & \ddots & \vdots & \vdots & \vdots \\ \alpha_2 & \alpha_1 & \alpha_0 & \ddots & 0 & 0 & 0 \\ \alpha_1 & \alpha_0 & 0 & \ddots & 0 & 0 & 0 \\ \alpha_0 & 0 & 0 & \ddots & 0 & 0 & 0 \end{matrix} \right\|_{\mathcal{L}(\mathbb{C}^{n+1})}$$

Now, we discuss the second approximation problem. Suppose that two finite sets of points $z_1, \dots, z_n \in \mathbb{D}$ and $w_1, \dots, w_n \in \mathbb{C}$ are given. Surely, there are infinitely many bounded analytic functions f on $\mathbb{D}$ such that

$$f(z_k) = w_k \qquad (1 \le k \le n). \tag{11.15}$$

For example, we can use Lagrange interpolating polynomials to construct a polynomial p of degree at most $n - 1$ that does the job. The *Nevanlinna–Pick problem* asks for an interpolating function f to be found for which $\|f\|_\infty$ is minimal.

Now, on the one hand, if f is any interpolating function, and p is the interpolating polynomial that was described above, then $f - p \in H^\infty$ and, moreover,

$$(f - p)(z_k) = 0 \qquad (1 \le k \le n).$$

Hence, $f - p = Bh$, where B is the finite Blaschke product formed with the points $z_1, \dots, z_n$ and $h \in H^\infty$. On the other hand, for any arbitrary $h \in H^\infty$, if we define $f = p + Bh$, then surely f is an interpolating function. Therefore, all solutions of (11.15) are parameterized as

$$f = p + Bh,$$

where p is the Lagrange interpolating polynomial, B is the finite Blaschke product formed with the points $z_1, \dots, z_n$ and h is any function in H^∞. Therefore, in the Nevanlinna–Pick problem, we seek

$$\inf_{h \in H^\infty} \|p + Bh\|_\infty.$$

But, since B is unimodular on $\mathbb{T}$, we have

$$\|p + Bh\|_{H^\infty} = \|p\bar{B} + h\|_{L^\infty},$$

and thus

$$\inf_{h \in H^\infty} \|p + Bh\|_\infty = \mathrm{dist}_{L^\infty(\mathbb{T})}(p\bar{B}, H^\infty).$$

Therefore, this problem reduces to Nehari's problem. It is clear that, in the above discussion, we do not need to stick to the Lagrange interpolating polynomial p. From the beginning, p can be chosen to be any solution of the interpolation problem and the rest is the same.

Exercises

Exercise 11.5.1 The purpose of this exercise is to solve the Fejér–Carathéodory problem for the polynomial

$$p(z) = 1 + 2z.$$

(i) Show that the Hankel matrix

$$\begin{bmatrix} 2 & 1 \\ 1 & 0 \end{bmatrix}$$

has the eigenvalues $1 + \sqrt{2}$ and $1 - \sqrt{2}$.

(ii) Deduce that

$$\left\| \begin{matrix} 2 & 1 \\ 1 & 0 \end{matrix} \right\|_{\mathcal{L}(\mathbb{C}^2)} = 1 + \sqrt{2}.$$

Hint: The matrix is symmetric.

(iii) Show that

$$\inf_{f \in H^\infty} \|p + z^2 f\|_\infty = 1 + \sqrt{2}.$$

Remark: Note that $\|p\|_\infty = 3$.

(iv) Verify that

$$(1 + \sqrt{2}) \frac{z - (1 - \sqrt{2})}{1 - (1 - \sqrt{2})z} = 1 + 2z + O(z^2),$$

and thus conclude that a solution to this Fejér–Carathéodory problem is a multiple of a Blaschke factor.

Exercise 11.5.2 The purpose of this exercise is to solve the Nevanlinna–Pick problem for the interpolating problem

$$f(0) = 1 \qquad \text{and} \qquad f(1/2) = 0.$$

(i) Show that the polynomial $p(z) = -2z + 1$ is a solution of the interpolation problem.

(ii) The finite Blaschke product with zeros at 0 and $1/2$ is

$$B(z) = \frac{z(1-2z)}{2-z}.$$

(iii) Show that

$$p(z)\overline{B(z)} = \frac{2}{z} - 1 \qquad (z \in \mathbb{T}).$$

(iv) Use the preceding identity to deduce that

$$\operatorname{dist}_{L^\infty(\mathbb{T})}(p\bar{B}, H^\infty) = 2.$$

(v) Show that

$$f(z) = p(z) + B(z) = 2\frac{1-2z}{2-z}$$

is a solution to this Nevanlinna–Pick problem.

11.6 Finite-rank Hankel operators

In studying the Fejér–Carathéodory problem in Section 11.5, we are faced with a finite-rank Hankel matrix. In the example after Theorem 11.5, we also saw that, if

$$\varphi(z) = \frac{1-\bar{\alpha}z}{\alpha - z} \qquad (\alpha \in \mathbb{D}),$$

then H_φ is of rank one. We note that

$$\hat{\varphi}(-n) = -(1-|\alpha|^2)\alpha^{n-1} \qquad (n \ge 1)$$

and that

$$\sum_{n=1}^{\infty} \frac{\hat{\varphi}(-n)}{z^n} = \frac{1-|\alpha|^2}{\alpha - z}$$

is a rational function with just one pole inside $\mathbb{D}$. These observations can be generalized for arbitrary finite-rank Hankel operators.

Theorem 11.6 *Let $\varphi \in L^\infty(\mathbb{T})$. Then H_φ has a finite rank if and only if*

$$\eta(z) = \sum_{n=1}^{\infty} \frac{\hat{\varphi}(-n)}{z^n}$$

is a rational function. In this case, the following hold:

(i) $H_\varphi = H_\eta$;
(ii) *all poles of η are in $\mathbb{D}$;*
(iii) *the rank of H_φ is precisely the number of poles of η (counting multiplicities) inside $\mathbb{D}$.*

Proof Suppose that H_φ is of rank r. Then $H_\varphi(\chi_0)$, $H_\varphi(\chi_1)$, ..., $H_\varphi(\chi_r)$ are dependent. Hence, there are constants $\alpha_0, \alpha_1, \dots, \alpha_r$, not all equal to zero, such that

$$\alpha_0 H_\varphi(\chi_0) + \alpha_1 H_\varphi(\chi_1) + \cdots + \alpha_r H_\varphi(\chi_r) = 0.$$

In terms of the matrix representation for H_φ, this is equivalent to saying that the first $r+1$ columns of the matrix of H_φ are dependent, and the previous identity is rewritten as

$$\alpha_0\hat{\varphi}(-m) + \alpha_1\hat{\varphi}(-m-1) + \cdots + \alpha_r\hat{\varphi}(-m-r) = 0 \qquad (m \geq 1). \tag{11.16}$$

This identity implies that, if we work out the multiplication in

$$p(z) = (\alpha_0 + \alpha_1 z + \cdots + \alpha_r z^r)\eta(z),$$

we see that p is a polynomial of degree at most $r-1$. Hence, η is a rational function with at most r poles. The above argument is reversible, and thus H_φ has a finite rank if and only if η is a rational function.

Suppose that η is a rational function. Since, by the Riemann–Lebesque lemma,

$$\lim_{n\to\infty} \hat{\varphi}(n) = 0,$$

the partial fraction expansion of η shows that all the poles of η are inside $\mathbb{D}$. Assume that it has r poles in $\mathbb{D}$. Then we can write

$$\eta(z) = \frac{p(z)}{\alpha_0 + \alpha_1 z + \cdots + \alpha_r z^r},$$

where p is a polynomial of degree at most $r-1$. Hence, if we again work out the multiplication $(\alpha_0 + \alpha_1 z + \cdots + \alpha_r z^r)\,\eta(z)$, we see that (11.16) holds. Therefore, owing to the special form of a Hankel matrix, any consecutive $r+1$ columns are dependent. Hence, H_φ is of rank at most r.

Finally, since $\eta \in L^\infty(\mathbb{T})$ and

$$\hat{\eta}(-n) = \hat{\varphi}(-n) \qquad (n \geq 1),$$

we conclude that $H_\eta = H_\varphi$. □

Since the poles of η, which was obtained above in the proof of Theorem 11.6, are all in $\mathbb{D}$, this function is in fact a rational function in $\overline{H_0^\infty}$. If we denote the rational functions in a Banach space $\mathcal{X}$ by $R\mathcal{X}$, we immediately get the following variation of Theorem 11.6.

Corollary 11.7 *Let $\varphi \in L^\infty(\mathbb{T})$. Then H_φ has a finite rank if and only if*

$$\varphi \in H^\infty + \overline{RH_0^\infty}.$$

As a special case of Theorem 11.6, or Corollary 11.7, if φ is a trigonometric polynomial, e.g.

$$\varphi = \sum_{n=-N}^{M} a_n \chi_n,$$

with $a_{-N} \neq 0$, then H_φ is a Hankel operator of rank N.

Exercises

Exercise 11.6.1 Let $\varphi \in L^\infty(\mathbb{T})$. Show that H_φ has a finite rank if and only if

$$\varphi \in H^\infty + RL^\infty.$$

Hint: Verify that $RL^\infty = RH^\infty + \overline{RH_0^\infty}$. Then use Corollary 11.7.

Exercise 11.6.2 Let $\varphi = p/q$, where p and q are polynomials with no common divisors. Assume that q has no zeros on $\mathbb{T}$, and N zeros in $\mathbb{D}$ (it might have some other zeros outside $\bar{\mathbb{D}}$). Show that H_φ has rank N.
Hint: Use Theorem 11.6.

11.7 Compact Hankel operators

In Section 5.3, the algebra $\mathcal{C}(\mathbb{T}) + H^\infty$ was discussed in detail. This algebra is extensively used below. Pick any $\varphi \in \mathcal{C}(\mathbb{T}) + H^\infty$, and write $\varphi = \omega + \psi$, where $\omega \in \mathcal{C}(\mathbb{T})$ and $\psi \in H^\infty$. Fix $\varepsilon > 0$. Then, using Fejér's polynomials or the classic Weierstrass approximation theorem, there is a trigonometric polynomial p such that $\|\omega - p\|_\infty < \varepsilon$. Hence, by (11.7), we have

$$\|H_\varphi - H_p\| = \|H_{\varphi - p}\| = \|H_{\omega - p}\| \leq \|\omega - p\|_\infty < \varepsilon.$$

But, according to Corollary 11.7, H_p is a finite-rank Hankel operator. Therefore, we conclude that each H_φ, with $\varphi \in \mathcal{C}(\mathbb{T}) + H^\infty$, is a compact Hankel operator. Interestingly, the converse is also true. For this part, Sarason's result is needed.

Theorem 11.8 *Let $\varphi \in L^\infty(\mathbb{T})$. Then H_φ is compact if and only if $\varphi \in \mathcal{C}(\mathbb{T}) + H^\infty$.*

Proof We saw that, if $\varphi \in \mathcal{C}(\mathbb{T}) + H^\infty$, then surely H_φ is a compact Hankel operator. Conversely, assume that H_φ is compact. Hence, according to Theorem 2.7, we have

$$\lim_{k\to\infty} \|H_\varphi S^k\|_{\mathcal{L}(H^2,(H^2)^\perp)} = 0.$$

But, by Lemma 11.2,

$$H_\varphi S^k = H_{\chi_k \varphi},$$

and, by Corollary 11.4, there is $\psi_k \in H^\infty$ such that

$$\|H_{\chi_k\varphi}\| = \|\chi_k\varphi - \psi_k\|_\infty = \|\varphi - \chi_{-k}\psi_k\|_\infty.$$

Therefore,

$$\lim_{k\to\infty} \|\varphi - \chi_{-k}\psi_k\|_\infty = 0.$$

Since $\chi_{-k}\psi_k \in \mathcal{C}(\mathbb{T}) + H^\infty$ and, by Theorem 5.10, $\mathcal{C}(\mathbb{T}) + H^\infty$ is a closed subalgebra of $L^\infty(\mathbb{T})$, we deduce that $\varphi \in \mathcal{C}(\mathbb{T}) + H^\infty$. □

An interesting aspect of Theorem 11.8 merits further clarification. When we say that H_φ is compact, based on the original definition of compact operators, we can just say that there is a sequence of finite-rank operators $A_n \in \mathcal{L}(H^2, (H^2)^\perp)$ such that

$$\|H_\varphi - A_n\| \longrightarrow 0$$

as $n \longrightarrow \infty$. However, the above theorem says that the symbol φ belongs to $\mathcal{C}(\mathbb{T}) + H^\infty$. Now, based on the discussion before the theorem, we can be more specific about the choice of A_n and claim that there exists a sequence of finite-rank Hankel operators H_{p_n}, where p_n is a trigonometric polynomial, such that

$$\|H_\varphi - H_{p_n}\| \longrightarrow 0$$

as $n \longrightarrow \infty$.

Exercises

Exercise 11.7.1 Let $\varphi \in \mathcal{C}(\mathbb{T})$. Show that there exists a unique function $f \in H^\infty$ such that

$$\operatorname{dist}(\varphi, H^\infty) = \|\varphi - f\|_\infty.$$

Hint: A compact operator attains its norm.

Exercise 11.7.2 Let K be a compact operator from H^2 into H^2_-. Show that

$$\lim_{n\to\infty} \|KS^n\| = 0.$$

Hint: First assume that K is a rank-one operator, $K = \eta \otimes \zeta$, $\eta \in H^2_-$ and $\zeta \in H^2$. Then $\|KS^n\| = \|S^{*n}\zeta\|_2 \|\eta\|_2$.

Exercise 11.7.3 We recall that, if $T \in \mathcal{L}(\mathcal{H}_1, \mathcal{H}_2)$, then $\|T\|_e$ denotes its essential norm, that is,

$$\|T\|_e = \inf_{K \in \mathcal{K}(\mathcal{H}_1, \mathcal{H}_2)} \|T - K\|,$$

where $\mathcal{K}(\mathcal{H}_1, \mathcal{H}_2)$ denotes the set of compact operators from $\mathcal{H}_1$ into $\mathcal{H}_2$. Let $\varphi \in L^\infty(\mathbb{T})$.

(i) Show that $\operatorname{dist}(\varphi, H^\infty + \mathcal{C}) \geq \|H_\varphi\|_e$.
Hint: Use Corollary 11.4 and Theorem 11.8.

(ii) Let K be a compact operator from H^2 into H^2_-.

(a) Show that

$$\|H_\varphi - K\| \geq \|H_\varphi S^n\| - \|KS^n\|.$$

(b) Deduce that

$$\|H_\varphi - K\| \geq \operatorname{dist}(\varphi, \overline{z^n} H^\infty) - \|KS^n\|.$$

Hint: Use Lemma 11.2.

(c) Show finally that

$$\|H_\varphi\|_e \geq \operatorname{dist}(\varphi, H^\infty + \mathcal{C}).$$

Hint: Use Exercise 11.7.2.

(iii) Conclude that

$$\|H_\varphi\|_e = \operatorname{dist}(\varphi, H^\infty + \mathcal{C}).$$

Exercise 11.7.4 Let $a = (a_n)_{n\geq 0}$ be a sequence of complex numbers and let $\Gamma_a = (a_{i+j})_{i,j\geq 0}$ be the associated Hankel matrix. Assume that $a_n = o(1/n)$, $n \longrightarrow \infty$. Show that Γ_a is a compact operator on ℓ^2.
Hint: Decompose $\Gamma_a = \Gamma_n^{(1)} + \Gamma_n^{(2)}$, where $\Gamma_n^{(1)} = (a_{i+j}^{(1)})_{i,j\geq 0}$ and

$$a_j^{(1)} = a_j \quad (j \leq n), \qquad a_j^{(1)} = 0 \quad (j > n).$$

Note that $\Gamma_n^{(1)}$ has finite rank and $\|\Gamma_n^{(2)}\| \longrightarrow 0$, as $n \longrightarrow \infty$.

Exercise 11.7.5 Let $a = (a_n)_{n\geq 0}$ be a sequence of complex numbers such that

$$\sum_{n=0}^{\infty} |a_n| < \infty.$$

Show that the associated Hankel matrix $\Gamma_a = (a_{i+j})_{i,j\geq 0}$ is a compact operator on ℓ^2.
Hint: Let $\varphi(z) = \sum_{n=0}^{\infty} a_n z^n$, $z \in \mathbb{T}$. Note that $\varphi \in \mathcal{C}(\mathbb{T})$.

Notes on Chapter 11

As already mentioned in Chapter 2, Hankel [256] was the first to study (finite) matrices whose entries depend only on the sum of the coordinates. These matrices play an important role in many classic problems of analysis and in particular in the so-called *Hamburger moment problem*. Since the work of Nehari [373] and Hartman [259], it has become clear that Hankel operators are an important tool in function theory on the unit circle. Together with Toeplitz operators, they form two of the most important classes of operators on Hardy spaces, which illustrate as well the deep interplay between operator theory and function theory. For the past four decades, the theory of Hankel operators has been growing rapidly. A lot of applications in various domains of mathematics have been found: interpolation problem [2–4], rational approximation [405, 406], stationary processes [409], perturbation theory [407] and the Sz.-Nagy–Foiaş functional model [386]. In the 1980s, the theory of Hankel operators was stimulated by the rapid development of H^∞ control theory and systems theory; see [220, 227, 237]. It has become clear that it is especially important to develop the theory of Hankel operators with matrix-valued (and even operator-valued) symbols. In this chapter, we have presented a few results that will be useful later. We have just touched the surface of the beautiful theory of Hankel operators. For a detailed account, we refer to the book [408], which is a standard reference for Hankel operators.

Section 11.2

Theorem 11.3 is due to Nehari [373]. The original proof given by Nehari involves a reduction to finite-rank Hankel operators and a fair amount of complex function theory computation. The proof presented here (based on Parrott's result, Theorem 7.20) is due to Adamjan, Arov and Kreĭn [3, 5], who were among the founding fathers of the modern theory of Hankel operators. The advantage of the Adamjan–Arov–Kreĭn proof is that it applies with no changes to Hankel operators on $\ell^2(\mathcal{E})$ – the Hilbert space of square summable sequences with values in a fixed Hilbert space $\mathcal{E}$. These operators are defined as the bounded operators on $\ell^2(\mathcal{E})$ that satisfy $TS = S^*T$, where S (respectively S^*) is the forward (respectively backward) unilateral shift operator on $\ell^2(\mathcal{E})$.

Theorem 11.3 reduces the problem of whether a sequence $(\alpha_n)_{n\geq 0}$ determines a bounded Hankel operator Γ_α on ℓ^2 to the question of the existence of an extension of $(\alpha_n)_{n\geq 0}$ to the sequence of Fourier coefficients of a bounded function. However, after the work of C. Fefferman on the space *BMO* (functions of bounded mean oscillation), it has become possible to determine whether Γ_α is bounded in terms of the sequence α itself. Indeed, by Fefferman's theorem (see the book of Garnett [233, chap. VI, sec. 4]), a function φ on the unit circle belongs to *BMO* if and only if it admits a representation

$$\varphi = \varphi_1 + P_+\varphi_2 \qquad \text{with } \varphi_1, \varphi_2 \in L^\infty.$$

Then, using Nehari's theorem and Fefferman's theorem, it is easy to see that the operator Γ_α is bounded on ℓ^2 if and only if the function

$$\varphi(z) = \sum_{n=0}^{\infty} \alpha_n z^n \tag{11.17}$$

belongs to $BMOA = BMO \cap H^1$. Clearly Γ_α is a bounded operator if the function φ defined by (11.17) is bounded. However, the operator Γ_α can be bounded even with an unbounded φ. The Hilbert matrix corresponding to the sequence $\alpha_n = 1/(n+1)$, $n \geq 0$, as introduced in Section 11.3, gives such an example. The function

$$\varphi(z) = \sum_{n=0}^{\infty} \frac{1}{n+1}\, z^n$$

is clearly unbounded on $\mathbb{D}$, but, as we have seen, Γ_α is bounded on ℓ^2. Nevertheless, Theorem 11.3 can be reformulated in the following form: given $\psi \in L^2(\mathbb{T})$, the Hankel operator H_ψ is bounded on H^2 if and only if there is $\varphi \in L^\infty(\mathbb{T})$ such that $H_\psi = H_\varphi$. In others words, the operator H_φ, with $\varphi \in L^\infty(\mathbb{T})$, exhausts the class of bounded Hankel operators.

Section 11.3

The main ingredient of the proof given in Exercise 11.3.1 to compute the norm of the Hilbert–Hankel matrix Γ has appeared in [514]. There are many other known proofs of $\|\Gamma\| = \pi$ in the literature, see e.g. [258, chap. 9, appdx III] and [450, chap. 9]. The characterization of positive definite Hankel matrices in terms of moments of positive measures on $\mathbb{R}$ is due to Hamburger [255]. The relationship between Hankel operators and moment problems was explored in the work of Hankel and Stieltjes as long ago as the 1870s. The result obtained in Exercise 11.3.3 is due to Widom [551]. In the same paper, he also obtained a characterization for the compactness of positive Hankel matrices $\Gamma_a = (a_{i+j})_{i,j\geq 0}$ in terms of the measures μ solving the Hamburger moment problem. Hardy's inequality (Exercise 11.3.4) seems to have appeared first in a paper of Hardy and Littlewood [257].

Section 11.4

Nehari's problem plays a significant role in applications, particularly in control theory. The first results in Nehari's problem was obtained by Havinson [269], who showed that, for a continuous function φ on $\mathbb{T}$, there exists a unique best approximation by analytic functions (see Exercise 11.7.1) and that uniqueness fails in general; see also [233, sec. IV.1]. Theorem 11.5 is due to Adamjan, Arov and Kreĭn [3].

Section 11.6

Theorem 11.6 is due to Kronecker [330]. There is another proof of Theorem 11.6 suggested by Axler and based on the observation that $\ker H_\varphi$ is a closed S-invariant subspace of H^2, whence of the form $\ker H_\varphi = \Theta H^2$, where Θ is an inner function. For this second proof, see Sarason [450] or Peller [408].

Section 11.7

Theorem 11.8 is due to Hartman. There exist analogs of the Nehari and Hartman theorems in the vector-valued setting; see [400] and [111]. A weak analog of the Riesz factorization theorem for Hardy spaces on the ball and sphere in $\mathbb{C}^n$ has been obtained by Coifman, Rochberg and Weiss [158]. This deep result yields boundedness and compactness theorems for Hankel operators in this context. As already mentioned, the result obtained in Exercise 11.7.1 is due to Havinson [269]; see also [435]. The explicit formula for the essential norm of a Hankel operator given in Exercise 11.7.3 was discovered by Adamjan, Arov and Kreĭn [2]. In Section 11.4, we studied the question of the existence and uniqueness of a best H^∞ approximate in the L^∞ norm. The same question can be asked about approximation by $H^\infty + \mathcal{C}$ functions, which is involved in the computation of the essential norm of Hankel operators. This was explicitly posed by Adamjan, Arov and Kreĭn in the problems book [267, pp. 254–258]. However, the situation here is quite different. It was shown by Axler, Berg, Jewell and Shields [52] that, for $\varphi \in L^\infty \setminus H^\infty + \mathcal{C}$, there are infinitely many best approximates in $H^\infty + \mathcal{C}$. See also [348] for a proof of this result using other techniques.

12

Toeplitz operators

Toeplitz operators are cousins of Hankel operators. Despite their similarities and the easy transformation from one type to another, they have serious differences. In the original definition of an $\mathcal{H}(b)$ space, we use two Toeplitz operators. Hence, without doubt, they are extremely important in this context. Here, we first introduce the Toeplitz operator T_φ and calculate its norm. Then we characterize self-adjoint, positive, orthogonal and compact Toeplitz operators. We obtain the kernel of T_φ and also discuss several results concerning its spectrum. Then we apply them to further study rigid functions in H^1. At the end, we generalize this concept by defining the Riesz projection on $L^2(\mu)$ and Toeplitz operators on $H^2(\mu)$, and then exploit them to obtain a characterization of invertibility for Toeplitz operators.

12.1 The operator $T_\varphi \in \mathcal{L}(H^2)$

Let $\varphi \in L^\infty(\mathbb{T})$. Then the *Toeplitz operator* associated with φ, or with symbol φ, is defined by

$$\begin{array}{rccc} T_\varphi : & H^2(\mathbb{T}) & \longrightarrow & H^2(\mathbb{T}) \\ & f & \longmapsto & P_+(\varphi f). \end{array}$$

In other words,

$$T_\varphi = P_+ \circ M_\varphi \circ i_+,$$

where P_+ is the Riesz projection, $M_\varphi : L^2(\mathbb{T}) \longrightarrow L^2(\mathbb{T})$ denotes the multiplication operator introduced in Section 7.2 and i_+ is the inclusion map of $H^2(\mathbb{T})$ into $L^2(\mathbb{T})$; see Figure 12.1.

Hence, T_φ is a bounded operator and, by Theorem 2.20,

$$\begin{aligned} \|T_\varphi\|_{\mathcal{L}(H^2(\mathbb{T}))} &\leq \|P_+\|_{\mathcal{L}(L^2(\mathbb{T}),H^2(\mathbb{T}))} \|M_\varphi\|_{\mathcal{L}(L^2(\mathbb{T}))} \|i_+\|_{\mathcal{L}(H^2(\mathbb{T}),L^2(\mathbb{T}))} \\ &= \|\varphi\|_{L^\infty(\mathbb{T})}. \end{aligned}$$

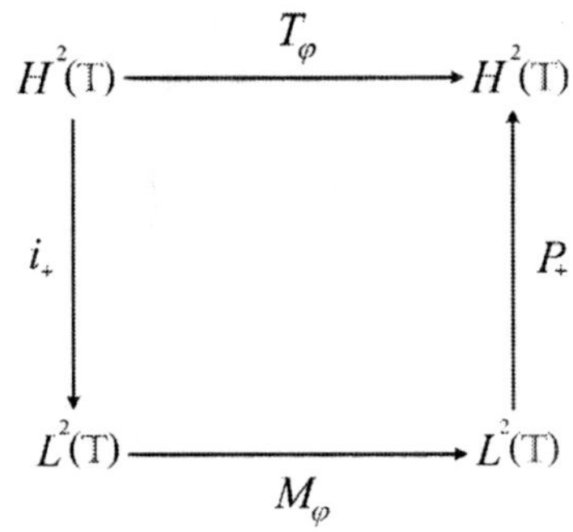

Figure 12.1 The Toeplitz operator T_φ.

However, in fact, we have $\|T_\varphi\| = \|\varphi\|_\infty$. To prove this equality, we need an approximation result for functions in $L^2(\mathbb{T})$.

Lemma 12.1 *Let $\varphi \in L^\infty(\mathbb{T})$, and let $f \in L^2(\mathbb{T})$. Then*

$$\lim_{m\to+\infty} \|\chi_{-m}T_\varphi(P_+(\chi_m f)) - \varphi f\|_2 = 0.$$

In particular, we have

$$\lim_{m\to+\infty} \|\chi_{-m}P_+(\chi_m f) - f\|_2 = 0.$$

Proof Even though the second relation is a special case of the first one, we verify it and then use it to prove the first one. To see how it works, for each $g \in L^2(\mathbb{T})$, put $g_m = \chi_{-m}P_+(\chi_m g)$, $m \in \mathbb{Z}$. Then, using the trivial fact that $\chi_m\chi_{-m} = 1$, write

$$\begin{aligned}
&\chi_{-m}T_\varphi(P_+(\chi_m f)) - \varphi f\\
&\quad = \chi_{-m}T_\varphi(\chi_m f_m) - \varphi f\\
&\quad = \chi_{-m}P_+(\varphi\chi_m f_m) - \varphi f\\
&\quad = \chi_{-m}P_+(\varphi\chi_m(f_m - f)) + \chi_{-m}P_+(\chi_m\varphi f) - \varphi f\\
&\quad = \chi_{-m}P_+(\varphi\chi_m(f_m - f)) + (\varphi f)_m - \varphi f.
\end{aligned}$$

Therefore, by (4.22), we have

$$\|\chi_{-m}T_\varphi(P_+(\chi_m f)) - \varphi f\|_2 \le \|\varphi\|_\infty\|f_m - f\|_2 + \|(\varphi f)_m - \varphi f\|_2.$$

So it remains to show that, for every $g \in L^2(\mathbb{T})$, we have $\|g_m - g\|_2 \longrightarrow 0$, as $m \longrightarrow +\infty$. Note that f and φf are both in $L^2(\mathbb{T})$.

Fix $g \in L^2(\mathbb{T})$. Then

$$\begin{aligned} g_m(e^{it}) &= e^{-imt} P_+\left(e^{imt} \sum_{k=-\infty}^{\infty} \hat{g}(k) e^{ikt} \right) \\ &= e^{-imt} P_+\left(\sum_{k=-\infty}^{\infty} \hat{g}(k) e^{i(k+m)t} \right) \\ &= e^{-imt} \sum_{k=-m}^{\infty} \hat{g}(k) e^{i(k+m)t} \\ &= \sum_{k=-m}^{\infty} \hat{g}(k) e^{ikt}, \end{aligned}$$

and thus, by Parseval's identity, we have

$$\|g_m - g\|_2^2 = \sum_{k=-\infty}^{-m-1} |\hat{g}(k)|^2.$$

Therefore, for each $g \in L^2(\mathbb{T})$,

$$\lim_{m \to +\infty} \|g_m - g\|_2 = 0. \qquad \square$$

We are now ready to show that $\|T_\varphi\| = \|\varphi\|_\infty$.

Theorem 12.2 *Let* $\varphi \in L^\infty(\mathbb{T})$. *Then*

$$\|T_\varphi\|_{\mathcal{L}(H^2(\mathbb{T}))} = \|\varphi\|_{L^\infty(\mathbb{T})}.$$

Proof We have already seen that $\|T_\varphi\| \le \|\varphi\|_\infty$. Lemma 12.1 and the inequality

$$\|T_\varphi g\|_2 \le \|T_\varphi\| \, \|g\|_2 \qquad (g \in H^2(\mathbb{T}))$$

help us to obtain a lower bound for $\|T_\varphi\|$.

Let $f \in L^2(\mathbb{T})$ and put $g = P_+(\chi_m f)$, $m \in \mathbb{Z}$. Since χ_m is unimodular and $\|g\|_2 \le \|f\|_2$, the last inequality implies that

$$\|\chi_{-m} T_\varphi(P_+(\chi_m f))\|_2 \le \|T_\varphi\| \, \|f\|_2 \qquad (f \in L^2(\mathbb{T})).$$

Now, we let $m \longrightarrow +\infty$. Then Lemma 12.1 implies that

$$\|\varphi f\|_2 \le \|T_\varphi\| \, \|f\|_2 \qquad (f \in L^2(\mathbb{T})).$$

In other words,

$$\|M_\varphi f\|_2 \le \|T_\varphi\| \, \|f\|_2 \qquad (f \in L^2(\mathbb{T})),$$

where M_φ is the multiplication operator on $L^2(\mathbb{T})$. The last inequality is equivalent to $\|M_\varphi\| \le \|T_\varphi\|$. But, by Theorem 2.20, we know that $\|M_\varphi\| = \|\varphi\|_\infty$. Hence, we also have $\|\varphi\|_\infty \le \|T_\varphi\|$. $\square$

Consider the mapping

$$\begin{array}{rcl} L^\infty(\mathbb{T}) & \longrightarrow & \mathcal{L}(H^2(\mathbb{T})) \\ \varphi & \longmapsto & T_\varphi. \end{array}$$

The definition of T_φ immediately implies that this mapping is linear. Moreover, Theorem 12.2 reveals that it is an injective mapping that preserves the norm. In other words, we have succeeded in finding an isometric isomorphic copy of $L^\infty(\mathbb{T})$ inside $\mathcal{L}(H^2(\mathbb{T}))$. In particular, we emphasize that

$$T_\varphi = T_\psi \iff \varphi = \psi. \tag{12.1}$$

It is not difficult to determine the adjoint of T_φ. In fact, by Lemma 4.8, for each $f, g \in H^2(\mathbb{T})$, we have

$$\begin{aligned} \langle T_\varphi f, g\rangle_{H^2(\mathbb{T})} &= \langle P_+(\varphi f), g\rangle_{H^2(\mathbb{T})} \\ &= \langle \varphi f, g\rangle_{L^2(\mathbb{T})} \\ &= \langle f, \bar{\varphi} g\rangle_{L^2(\mathbb{T})} \\ &= \langle f, P_+(\bar{\varphi} g)\rangle_{H^2(\mathbb{T})} \\ &= \langle f, T_{\bar{\varphi}} g\rangle_{H^2(\mathbb{T})}. \end{aligned}$$

Therefore, for each $\varphi \in L^\infty(\mathbb{T})$,

$$T_\varphi^* = T_{\bar{\varphi}}. \tag{12.2}$$

Two special Toeplitz operators with different names have been studied earlier. If $\varphi = \chi_1$, then the Toeplitz operator T_{χ_1} is exactly the unilateral forward shift operator S on H^2, which was introduced in Section 8.2. Moreover, by (8.14) and (12.2), we have $S^* = T_{\chi_{-1}}$.

In the following result, by saying that φ is real, we mean that $\varphi(e^{it}) \in \mathbb{R}$ for almost all $e^{it} \in \mathbb{T}$. Other statements, e.g. φ is positive, should be interpreted similarly.

Theorem 12.3 *Let $\varphi \in L^\infty(\mathbb{T})$. Then the following hold.*

(i) *The Toeplitz operator T_φ is self-adjoint if and only if φ is real.*
(ii) *The Toeplitz operator T_φ is positive if and only if φ is positive.*
(iii) *The Toeplitz operator T_φ is an orthogonal projection if and only if either $\varphi \equiv 1$ or $\varphi \equiv 0$. In this situation, T_φ is either the identity operator I or the zero operator 0.*

Proof (i) This is an immediate consequence of (12.1) and (12.2).

(ii) By Lemma 4.8, for each $f \in H^2(\mathbb{T})$, we have

$$\langle T_\varphi f, f\rangle_{H^2(\mathbb{T})} = \langle \varphi f, f\rangle_{L^2(\mathbb{T})} = \frac{1}{2\pi}\int_0^{2\pi} \varphi(e^{it})\, |f(e^{it})|^2\, dt.$$

If φ is positive, the last integral shows that T_φ is a positive operator.

Now suppose that T_φ is positive. This means that

$$\frac{1}{2\pi}\int_0^{2\pi} \varphi(e^{it})\,|f(e^{it})|^2\,dt \geq 0 \qquad (f \in H^2(\mathbb{T})).$$

We show that this inequality in fact holds for all $f \in L^2(\mathbb{T})$. Fix $f \in L^2(\mathbb{T})$, and let $m \geq 1$. Then $P_+(\chi_m f) \in H^2(\mathbb{T})$ and thus

$$\frac{1}{2\pi}\int_0^{2\pi} \varphi(e^{it})\,|P_+(\chi_m f)(e^{it})|^2\,dt \geq 0.$$

Since $|\chi_{-m}| = 1$, we have

$$\frac{1}{2\pi}\int_0^{2\pi} \varphi(e^{it})\,|\chi_{-m}(e^{it})P_+(\chi_m f)(e^{it})|^2\,dt \geq 0.$$

By Lemma 12.1, $\chi_{-m}P_+(\chi_m f) \longrightarrow f$ in $L^2(\mathbb{T})$ norm. Hence, for each $f \in L^2(\mathbb{T})$,

$$\frac{1}{2\pi}\int_0^{2\pi} \varphi(e^{it})\,|f(e^{it})|^2\,dt \geq 0.$$

Therefore, by taking f to be the characteristic function of an arbitrary measurable set, we conclude that $\varphi \geq 0$.

(iii) It is trivial that $T_1 = I$ and $T_0 = 0$ are orthogonal projections. Now, suppose that T_φ is an orthogonal projection. By (ii) and Theorem 12.2, we necessarily have $0 \leq \varphi \leq 1$. In light of Lemma 4.8, the property

$$\langle T_\varphi f, T_\varphi f\rangle_{H^2(\mathbb{T})} = \langle T_\varphi f, f\rangle_{H^2(\mathbb{T})} \qquad (f \in H^2(\mathbb{T}))$$

is written as

$$\langle \varphi f, T_\varphi f\rangle_{L^2(\mathbb{T})} = \langle \varphi f, f\rangle_{L^2(\mathbb{T})} \qquad (f \in H^2(\mathbb{T})).$$

Hence, for all $f \in H^2(\mathbb{T})$,

$$\frac{1}{2\pi}\int_0^{2\pi} \varphi(e^{it}) f(e^{it})\,\overline{P_+(\varphi f)(e^{it})}\,dt = \frac{1}{2\pi}\int_0^{2\pi} \varphi(e^{it})\,|f(e^{it})|^2\,dt.$$

In particular, for $f = \chi_m$, $m \geq 1$,

$$\frac{1}{2\pi}\int_0^{2\pi} \varphi(e^{it})\,\overline{\chi_{-m}(e^{it})P_+(\varphi\chi_m)(e^{it})}\,dt = \frac{1}{2\pi}\int_0^{2\pi} \varphi(e^{it})\,dt.$$

Again, by Lemma 12.1, $\chi_{-m}P_+(\chi_m\varphi) \longrightarrow \varphi$ in $L^2(\mathbb{T})$ norm. Thus, by letting $m \longrightarrow +\infty$, we get

$$\frac{1}{2\pi}\int_0^{2\pi} \varphi^2(e^{it})\,dt = \frac{1}{2\pi}\int_0^{2\pi} \varphi(e^{it})\,dt.$$

Since $0 \leq \varphi \leq 1$, the identity

$$\frac{1}{2\pi}\int_0^{2\pi} \varphi(e^{it})(1 - \varphi(e^{it}))\,dt = 0$$

holds if and only if

$$\varphi(e^{it})(1 - \varphi(e^{it})) = 0$$

for almost all $e^{it} \in \mathbb{T}$. This fact means that φ is the characteristic function of a measurable set.

Using the identity $\langle T_\varphi f, T_\varphi f\rangle_{H^2(\mathbb{T})} = \langle \varphi f, f\rangle_{L^2(\mathbb{T})}$, for each $f \in H^2(\mathbb{T})$, and the formulas

$$\begin{aligned}\langle \varphi f, f\rangle_{L^2(\mathbb{T})} &= \frac{1}{2\pi}\int_0^{2\pi} \varphi(e^{it})\,|f(e^{it})|^2\,dt\\ &= \frac{1}{2\pi}\int_0^{2\pi} |\varphi(e^{it})f(e^{it})|^2\,dt\end{aligned}$$

and

$$\langle T_\varphi f, T_\varphi f\rangle_{H^2(\mathbb{T})} = \frac{1}{2\pi}\int_0^{2\pi} |P_+(\varphi f)(e^{it})|^2\,dt,$$

we obtain

$$\|\varphi f\|_{L^2(\mathbb{T})} = \|P_+(\varphi f)\|_{L^2(\mathbb{T})} \qquad (f \in H^2(\mathbb{T})).$$

But this happens if and only if $\varphi f \in H^2(\mathbb{T})$, for all $f \in H^2(\mathbb{T})$. In particular, we must have $\varphi \in H^2(\mathbb{T})$. According to (4.12), the only real functions in $H^2(\mathbb{T})$ are real constants. Hence, either $\varphi \equiv 1$ or $\varphi \equiv 0$. □

Exercises

Exercise 12.1.1 Let $\varphi \in L^\infty(\mathbb{T})$. Show that the matrix of the Toeplitz operator $T_\varphi \in \mathcal{L}(H^2(\mathbb{T}))$ with respect to the orthonormal basis $(\chi_n)_{n\geq 0}$ is

$$\begin{bmatrix} \hat\varphi(0) & \hat\varphi(-1) & \hat\varphi(-2) & \hat\varphi(-3) & \ddots\\ \hat\varphi(1) & \hat\varphi(0) & \hat\varphi(-1) & \hat\varphi(-2) & \ddots\\ \hat\varphi(2) & \hat\varphi(1) & \hat\varphi(0) & \hat\varphi(-1) & \ddots\\ \hat\varphi(3) & \hat\varphi(2) & \hat\varphi(1) & \hat\varphi(0) & \ddots\\ \ddots & \ddots & \ddots & \ddots & \ddots \end{bmatrix}.$$

Hint: Apply Lemma 4.8 to get

$$\langle T_\varphi \chi_n, \chi_m\rangle = \hat\varphi(m-n).$$

Remark: Compare with (2.30).

Exercise 12.1.2 Let $A \in \mathcal{L}(H^2(\mathbb{T}))$. Show that the following are equivalent.

(i) $S^*AS = A$.
(ii) The matrix of A with respect to the standard basis $(\chi_n)_{n\geq 0}$ is a singly infinite Toeplitz matrix.
(iii) There is a function $\varphi \in L^\infty(\mathbb{T})$ such that $A = T_\varphi$.

Hint: The equivalence (i) $\Longleftrightarrow$ (ii) was established in Exercise 8.3.3. The implication (iii) $\Longrightarrow$ (ii) was discussed above in Exercise 12.1.1. Finally, to verify (ii) $\Longrightarrow$ (iii), define

$$A_n = Z^{*n} i_+ A P_+ Z^n \qquad (n \geq 1).$$

Then show that A_n strongly converges to a bounded operator on $L^2(\mathbb{T})$ whose matrix is a doubly infinite Toeplitz matrix. Then use Exercise 8.3.2.
Remark: Note that we cannot appeal to Exercise 8.3.3 and Theorem 8.14 to settle this exercise. This is because A fulfills $S^*AS = A$, and not necessarily $AS = SA$.

12.2 Composition of two Toeplitz operators

It is rather elementary to see that the multiplication operators on $L^2(\mathbb{T})$ commute and their composition is another multiplication operator, i.e. for each $\varphi, \psi \in L^\infty(\mathbb{T})$, $M_\varphi M_\psi = M_\psi M_\varphi = M_{\varphi\psi}$. However, the class of Toeplitz operators is not commutative. For instance, we can easily check that

$$T_{\bar{z}} T_z = I \quad \text{but} \quad T_z T_{\bar{z}} = I - \chi_0 \otimes \chi_0.$$

Nevertheless, the following theorem reveals a partial result of this type.

Theorem 12.4 *Let $\varphi, \psi \in L^\infty(\mathbb{T})$, and suppose that at least one of them is in $H^\infty(\mathbb{T})$. Then*

$$T_{\bar{\psi}} T_\varphi = T_{\bar{\psi}\varphi}.$$

Proof If $\varphi \in H^\infty(\mathbb{T})$, then, for each $f \in H^2(\mathbb{T})$,

$$T_{\bar{\psi}} T_\varphi f = T_{\bar{\psi}} P_+(\varphi f) = T_{\bar{\psi}}(\varphi f) = P_+(\bar{\psi}\varphi f) = T_{\bar{\psi}\varphi} f.$$

Hence $T_{\bar{\psi}} T_\varphi = T_{\bar{\psi}\varphi}$. If $\psi \in H^\infty(\mathbb{T})$, then, by what we just proved,

$$T_{\bar{\varphi}} T_\psi = T_{\bar{\varphi}\psi}.$$

Therefore, by (12.2), we have

$$T_{\bar{\psi}} T_\varphi = T_\psi^* T_{\bar{\varphi}}^* = (T_{\bar{\varphi}} T_\psi)^* = T_{\bar{\varphi}\psi}^* = T_{\varphi\bar{\psi}}. \qquad \square$$

Theorem 12.4 is actually reversible; see Exercise 12.2.1. Under the conditions of this theorem, we do not have $T_{\bar{\psi}} T_\varphi \neq T_\varphi T_{\bar{\psi}}$ unless one of the symbols is constant. However, if $\varphi, \psi \in H^\infty(\mathbb{T})$, we do have the elementary identities

$$T_\varphi T_\psi = T_\psi T_\varphi = T_{\varphi\psi} \quad \text{and} \quad T_{\bar\varphi} T_{\bar\psi} = T_{\bar\psi} T_{\bar\varphi} = T_{\bar\varphi\bar\psi}. \tag{12.3}$$

The first relation is trivial. The second is obtained by taking the conjugate of all sides of the first one. In fact, the two relations also follow from Theorem 12.4.

For an arbitrary $\varphi \in L^\infty(\mathbb{T})$, we know that $S^*T_\varphi - T_\varphi S^* \neq 0$. However, in a special case, this commutator is a rank-one operator.

Lemma 12.5 *Let $\varphi \in H^\infty$. Then we have*

$$S^*T_\varphi - T_\varphi S^* = S^*\varphi \otimes k_0.$$

Proof Let $f \in H^2$. Then, using Corollary 8.11, we have

$$\begin{aligned}(S^*T_\varphi - T_\varphi S^*)f &= S^*(\varphi f) - \varphi S^* f\\ &= f(0)S^*\varphi\\ &= \langle f, k_0\rangle_{H^2} S^*\varphi\\ &= (S^*\varphi \otimes k_0)(f).\end{aligned}$$

This completes the proof. □

If p is an analytic polynomial and $\varphi \in L^\infty(\mathbb{T})$, then $T_\varphi p$ is not necessarily a polynomial. However, if φ is conjugate analytic, then this is the case.

Theorem 12.6 *Let $\varphi \in H^\infty$, and let p be an analytic polynomial of degree n. Then $T_{\bar\varphi} p$ is also an analytic polynomial of degree at most n.*

Proof Since $\varphi \in H^\infty$, we can write

$$\varphi = \sum_{k=0}^{\infty} \hat\varphi(k)\chi_k$$

as a series in $L^2(\mathbb{T})$. Thus,

$$\bar\varphi\chi_n = \sum_{k=0}^{\infty} \overline{\hat\varphi(k)}\, \chi_{n-k},$$

which implies that

$$T_{\bar\varphi}\chi_n = \sum_{k=0}^{n} \overline{\hat\varphi(n-k)}\, \chi_k. \tag{12.4}$$

This identity reveals that the image of an analytic polynomial of degree n under $T_{\bar\varphi}$ is also an analytic polynomial of degree at most n. In fact, if $\hat\varphi(0) = \varphi(0) \neq 0$, the image has order n. □

Corollary 12.7 *Let $\varphi \in H^\infty$, $\varphi \not\equiv 0$. Then $\mathcal{R}(T_{\bar\varphi})$ is dense in H^2.*

Proof By Theorem 12.6, the range of $T_{\bar\varphi}$ contains all analytic polynomials. □

The following result provides a connection between Toeplitz and Hankel operators.

Lemma 12.8 *Let $\varphi, \psi \in L^\infty(\mathbb{T})$. Then*

$$T_{\varphi\psi} - T_\varphi T_\psi = H_{\bar{\varphi}}^* H_\psi.$$

Proof Let $f \in H^2$. Then according to Lemma 11.1, we have

$$\begin{aligned} H_{\bar{\varphi}}^* H_\psi f &= P_+(\varphi P_-(\psi f)) \\ &= P_+(\varphi\psi f) - P_+(\varphi P_+(\psi f)) \\ &= T_{\varphi\psi} f - T_\varphi T_\psi f, \end{aligned}$$

which gives the result. □

Corollary 12.9 *Let $\varphi, \psi \in L^\infty(\mathbb{T})$. Assume that $\varphi \in \mathcal{C}(\mathbb{T})$ or $\psi \in \mathcal{C}(\mathbb{T})$. Then the operator $T_{\varphi\psi} - T_\varphi T_\psi$ is compact.*

Proof If either φ or ψ is in $\mathcal{C}(\mathbb{T})$, then, according to Theorem 11.8, the operator $H_{\bar{\varphi}}^* H_\psi$ is compact. Then the result follows from Lemma 12.8. □

We finish this section with the following result, which compares $T_\varphi T_{\bar{\varphi}}$ and $T_{\bar{\varphi}} T_\varphi$.

Theorem 12.10 *Let $\varphi \in H^\infty(\mathbb{T})$. Then*

$$T_\varphi T_{\bar{\varphi}} \le T_{\bar{\varphi}} T_\varphi.$$

Moreover, the equality holds if and only if φ is a constant.

Proof For each $f \in H^2(\mathbb{T})$, by (4.22), we have

$$\langle T_\varphi T_{\bar{\varphi}} f, f\rangle_{H^2} = \langle T_{\bar{\varphi}} f, T_{\bar{\varphi}} f\rangle_{H^2} = \|P_+(\bar{\varphi} f)\|_{H^2}^2 \le \|\bar{\varphi} f\|_{L^2}^2.$$

But $\|\bar{\varphi} f\|_{L^2} = \|\varphi f\|_{L^2} = \|\varphi f\|_{H^2}$ and

$$\|\varphi f\|_{H^2}^2 = \|T_\varphi f\|_{H^2}^2 = \langle T_\varphi f, T_\varphi f\rangle_{H^2} = \langle T_{\bar{\varphi}} T_\varphi f, f\rangle_{H^2}.$$

Thus,

$$\langle T_\varphi T_{\bar{\varphi}} f, f\rangle_{H^2} \le \langle T_{\bar{\varphi}} T_\varphi f, f\rangle_{H^2} \qquad (f \in H^2(\mathbb{T})),$$

which means that $T_\varphi T_{\bar{\varphi}} \le T_{\bar{\varphi}} T_\varphi$.

In the above argument, there was just one place where we had an inequality, i.e.

$$\|P_+(\bar{\varphi} f)\|_{H^2}^2 \le \|\bar{\varphi} f\|_{L^2}^2 \qquad (f \in H^2).$$

Hence, the equality $T_\varphi T_{\bar{\varphi}} = T_{\bar{\varphi}} T_\varphi$ holds if and only if

$$\|P_+(\bar{\varphi} f)\|_{H^2}^2 = \|\bar{\varphi} f\|_{L^2}^2 \qquad (f \in H^2). \tag{12.5}$$

Take $f = 1$ to deduce that $\bar{\varphi} \in H^2$ and thus φ is a constant. Clearly, if φ is a constant, then (12.5) holds. □

Exercises

Exercise 12.2.1 Let $\varphi, \psi \in L^\infty(\mathbb{T})$. Show that $T_\varphi T_\psi$ is a Toeplitz operator if and only if $\bar{\varphi}$ or ψ belongs to H^∞.
Hint: Observe that, if $(c_{i,j})_{i,j \geq 0}$ is the matrix of the operator $T_\varphi T_\psi$ with respect to the standard basis $(z^n)_{n \geq 0}$ of H^2, then we have

$$c_{i+1,j+1} = c_{i,j} + \hat{\varphi}(i+1)\hat{\psi}(-j-1).$$

Exercise 12.2.2 Let $\varphi, \psi \in L^\infty(\mathbb{T})$. Show that $T_\varphi T_\psi = 0$ if and only if either $\varphi = 0$ or $\psi = 0$.
Hint: Use Exercise 12.2.1.

12.3 The spectrum of T_φ

An invertible operator is surely bounded below. The following Lemma, which is an easy consequence of Nehari's theorem, characterizes the Toeplitz operators that are bounded below. We recall that an operator is bounded below if and only if it is left-invertible. This fact will be exploited in applications of Lemma 12.11.

Lemma 12.11 *Let $\varphi \in L^\infty(\mathbb{T})$ and assume that $|\varphi| = 1$ a.e. on $\mathbb{T}$. Then T_φ is bounded below if and only if* $\operatorname{dist}(\varphi, H^\infty) < 1$.

Proof By definition, T_φ is bounded below if and only if there is a constant $c > 0$ such that

$$\|T_\varphi f\|_2 \geq c\|f\|_2 \tag{12.6}$$

for every $f \in H^2$. But, since φ is unimodular, we have

$$\|H_\varphi f\|_2^2 + \|T_\varphi f\|_2^2 = \|\varphi f\|_2^2 = \|f\|_2^2,$$

whence (12.6) is equivalent to

$$\|H_\varphi f\|_2^2 \leq (1 - c^2)\|f\|_2^2 \qquad (f \in H^2).$$

This inequality is equivalent to $\|H_\varphi\| < 1$ and it remains to apply Nehari's theorem to obtain the result. □

The precise relation between $\sigma(T_\varphi)$ and φ is not fully known yet. The best-known result states that $\sigma(T_\varphi)$ is in the convex hull of $\mathcal{R}_e(\varphi)$, the essential range of φ. To establish this result, we need to develop certain intermediate facts.

Lemma 12.12 *Let $\varphi \in L^\infty(\mathbb{T})$. Suppose that T_φ is lower bounded in $\mathcal{L}(H^2)$. Then φ is invertible in $L^\infty(\mathbb{T})$.*

Proof It is enough to show that M_φ is lower bounded in $\mathcal{L}(L^2)$ and then apply Lemma 2.22. Since T_φ is lower bounded, there is a constant $c > 0$ such that

$$\|T_\varphi f\|_2 \geq c\|f\|_2 \qquad (f \in H^2).$$

Replace f by $P_+(\chi_m f)$, where f is an arbitrary element of $L^2(\mathbb{T})$. Hence, since χ_m is unimodular, we have

$$\|\chi_{-m} T_\varphi P_+(\chi_m f)\|_2 \geq c\|\chi_{-m} P_+(\chi_m f)\|_2 \qquad (f \in L^2(\mathbb{T})).$$

Let $m \longrightarrow +\infty$. Thus, by Lemma 12.1, we obtain

$$\|\varphi f\|_2 \geq c\|f\|_2 \qquad (f \in L^2(\mathbb{T})).$$

This means that the operator M_φ is lower bounded. □

In a special case, the necessary condition presented in Lemma 12.12 becomes necessary and sufficient.

Corollary 12.13 *Let $\varphi \in H^\infty$. Then T_φ is invertible in $\mathcal{L}(H^2)$ if and only if $1/\varphi \in H^\infty$.*

Proof If $\varphi^{-1} \in H^\infty$, then, according to (12.3), we have

$$T_{\varphi^{-1}} T_\varphi = T_\varphi T_{\varphi^{-1}} = T_1 = I,$$

which proves that T_φ is invertible. Conversely, let us assume that T_φ is invertible. By Lemma 12.12, φ is invertible in $L^\infty(\mathbb{T})$. Moreover, T_φ is onto, and thus there is a $\psi \in H^2$ such that

$$\varphi\psi = 1 \qquad \text{(a.e. on } \mathbb{T}).$$

The uniqueness part of the canonical factorization theorem (Theorem 4.19) now shows either that φ and ψ are unimodular constants, or that they are bounded outer functions. In any case, $1/\varphi \in H^\infty$. □

Since A is invertible if and only if so is A^*, in the preceding result we can replace T_φ by $T_{\bar{\varphi}}$ to obtain the following corollary.

Corollary 12.14 *Let $\varphi \in H^\infty$ and assume that $\varphi^{-1} \notin H^\infty$. Then, for every $\varepsilon > 0$, there is an $f = f_\varepsilon \in H^2$ such that*

$$\|f\|_2 = 1$$

and

$$\|T_{\bar{\varphi}} f\|_2 \leq \varepsilon.$$

Proof By Corollary 12.13, $T_{\bar{\varphi}}$ is not invertible, and by Corollary 12.7, $\mathcal{R}(T_{\bar{\varphi}})$ is dense in H^2. Hence, in light of Corollary 1.31, $T_{\bar{\varphi}}$ cannot be a lower-bounded operator. □

The above results enable us to take the first step in recognizing the elements of $\sigma(T_\varphi)$. In the next two theorems, we presents two results that are somehow the best-known results in characterizing $\sigma(T_\varphi)$.

Theorem 12.15 *Let* $\varphi \in L^\infty(\mathbb{T})$. *Then*

$$\mathcal{R}_e(\varphi) \subset \sigma(T_\varphi).$$

Proof Suppose that we have $\lambda \notin \sigma(T_\varphi)$. This means that the operator $T_{\lambda-\varphi} = \lambda I - T_\varphi$ is invertible in $\mathcal{L}(H^2)$. Hence, by Lemma 12.12, $\lambda - \varphi$ is invertible in $L^\infty(\mathbb{T})$. Therefore, by Theorem 2.23, $\lambda \notin \mathcal{R}_e(\varphi)$. □

If $\varphi(z) = z$, then, on the one hand, $\mathcal{R}_e(\varphi) = \mathbb{T}$, and, on the other, $\sigma(T_\varphi) = \sigma(S) = \bar{\mathbb{D}}$. Hence, the inclusion $\mathcal{R}_e(\varphi) \subset \sigma(T_\varphi)$ can be proper. The *closed convex hull* of a set $E \in \mathbb{C}$ is the intersection of all closed convex sets that contain E. Using elementary plane geometry techniques, it is easy to see that the closed convex hull of E is precisely the intersection of all open half-planes that contain E.

Theorem 12.16 *Let* $\varphi \in L^\infty(\mathbb{T})$. *Then* $\sigma(T_\varphi)$ *is contained in the closed convex hull of* $\mathcal{R}_e(\varphi)$.

Proof It is enough to show that, if an open half-plane contains $\mathcal{R}_e(\varphi)$, it also includes $\sigma(T_\varphi)$. After a translation and rotation, we can assume that our half-plane is the open right half-plane $\{z : \Re(z) > 0\}$, and we proceed to show that $0 \notin \sigma(T_\varphi)$, i.e. T_φ is invertible.

Since $\mathcal{R}_e(\varphi)$ is a compact subset of the open right half-plane, there is an $\epsilon > 0$ such that

$$\epsilon\mathcal{R}_e(\varphi) = \{z : |z-1| < 1\}.$$

Thus, we have $\|1 - \epsilon\varphi\|_\infty < 1$, and, by Theorem 12.2, this implies $\|I - \epsilon T_\varphi\| < 1$. Therefore, ϵT_φ is invertible. □

Since the closed convex hull of $\mathbb{T}$ is $\bar{\mathbb{D}}$, and $\sigma(T_z) = \bar{\mathbb{D}}$, Theorem 12.16 is sharp. In light of Theorems 12.15 and 12.16, we can say that

$$\mathcal{R}_e(\varphi) \subset \sigma(T_\varphi) \subset \text{closed convex hull of } \mathcal{R}_e(\varphi)$$

for each $\varphi \in L^\infty(\mathbb{T})$. If φ is a real-valued bounded function, then the closed convex hull of φ is precisely the closed interval $[\operatorname{essinf}\varphi, \operatorname{esssup}\varphi]$. In this case, we can completely determine $\sigma(T_\varphi)$.

Corollary 12.17 *Let φ be a real-valued function in $L^\infty(\mathbb{T})$. Then*

$$\sigma(T_\varphi) = [\operatorname{essinf} \varphi, \operatorname{esssup} \varphi].$$

Proof If φ is a constant, the result is obvious. Hence, suppose that φ is a nonconstant bounded real function. By Theorem 12.16, we know that

$$\sigma(T_\varphi) \subset [\operatorname{essinf} \varphi, \operatorname{esssup} \varphi].$$

To show that equality holds, we prove that, if, for some $\lambda \in \mathbb{R}$, the operator $\lambda I - T_\varphi$ is invertible in $\mathcal{L}(H^2)$, then either $\lambda - \varphi \geq 0$ or $\lambda - \varphi \leq 0$ almost everywhere on $\mathbb{T}$. In other words, there are not two disjoint Borel subsets of $\mathbb{T}$ with positive Lebesgue measures such that $\lambda - \varphi$ is positive on one of them and negative on the other.

Hence, let $\lambda \in \mathbb{R}$, and suppose that $\lambda I - T_\varphi$ is invertible in $\mathcal{L}(H^2)$. Thus, by Lemma 12.12, $\lambda - \varphi$ is invertible in $L^\infty(\mathbb{T})$. Moreover, there is $f \in H^2$ such that

$$(\lambda I - T_\varphi)(f) = 1.$$

Clearly $f \neq 0$, and there is $g \in H^2_0$ such that

$$(\lambda - \varphi) f = 1 + \bar{g}.$$

Equivalently, we have $(\lambda - \varphi)\bar{f} = 1 + g \in H^2$, and, if we multiply both sides by $f \in H^2$, we deduce that

$$(\lambda - \varphi)\,|f|^2 = (1 + g) f \in H^1.$$

Therefore, by (4.12), this function is a nonzero real constant, say $\alpha \in \mathbb{R} \setminus \{0\}$, i.e.

$$(\lambda - \varphi)\,|f|^2 = \alpha.$$

By the uniqueness theorem for H^p classes (Lemma 4.30), $f \neq 0$ almost everywhere on $\mathbb{T}$. Hence, $\lambda - \varphi$ has the same sign as α. □

Exercises

Exercise 12.3.1 Suppose that T_φ is quasi-nilpotent. Show that $\varphi = 0$.
Hint: Use Theorem 12.15.
Remark: An operator $T \in \mathcal{L}(\mathcal{H})$ is called *quasi-nilpotent* if $\sigma(T) = 0$.

Exercise 12.3.2 Let $\varphi \in H^\infty$. Show that $\sigma(T_\varphi) = \operatorname{Clos}(\varphi(\mathbb{D}))$.
Hint: Use Corollary 12.13.

Exercise 12.3.3 Let $\varphi \in L^\infty(\mathbb{T})$. Show that $r(T_\varphi) = \|\varphi\|_\infty$.
Hint: Observe that, by Theorems 2.23 and 12.15,

$$\sigma(M_\varphi) \subset \sigma(T_\varphi).$$

This inclusion implies that

$$r(M_\varphi) \le r(T_\varphi).$$

Then use Theorem 2.17 to obtain $r(M_\varphi) = \|M_\varphi\|$ and hence

$$\|M_\varphi\| = r(M_\varphi) \le r(T_\varphi) \le \|T_\varphi\| \le \|\varphi\|_\infty.$$

Conclude by applying Theorem 2.20.

12.4 The kernel of T_φ

If $\varphi \in H^\infty(\mathbb{T})$, then, for each $f \in H^2(\mathbb{T})$, we have $\varphi f \in H^2(\mathbb{T})$ and thus

$$T_\varphi f = \varphi f.$$

In other words, φ is a multiplier of the reproducing kernel Hilbert space $H^2(\mathbb{T})$ and T_φ is nothing but the multiplier operator M_φ introduced in Section 9.1. In this case, if $\varphi \not\equiv 0$, we easily see that

$$\ker T_\varphi = \{0\}.$$

Note that the notation M_φ was also used for the multiplication operator on $L^2(\mathbb{T})$ in Section 7.2, but this will not cause any difficulty. Therefore, as a special case of Theorem 9.2 and by (12.2), for the multiplier T_φ we have

$$T_{\bar{\varphi}} k_z^{\mathcal{H}} = \overline{\varphi(z)} k_z^{\mathcal{H}} \qquad (z \in \mathbb{D}), \tag{12.7}$$

where $k_z^{\mathcal{H}}$ is the Cauchy kernel.

Theorem 12.18 *Let $\varphi \in H^\infty(\mathbb{T})$, $\varphi \not\equiv 0$. Then the following are equivalent:*

(i) *φ is inner;*
(ii) *T_φ is an isometry;*
(iii) *T_φ is a partial isometry.*

Proof (i) $\Longrightarrow$ (ii) Since φ is analytic, then $T_\varphi(f) = \varphi f$, $f \in H^2$, and, since φ is unimodular on $\mathbb{T}$, we have

$$\|T_\varphi(f)\|_{H^2} = \|f\|_{H^2}.$$

(ii) $\Longrightarrow$ (iii) This is trivial.

(iii) $\Longrightarrow$ (i) If T_φ is a partial isometry, then, by Theorem 7.22, $T_{|\varphi|^2} = T_\varphi^* T_\varphi$ is an orthogonal projection. By Theorem 12.3, we must have either $|\varphi| \equiv 1$ or $|\varphi| \equiv 0$. The second possibility is rolled out as an assumption. The first one means that φ is inner. □

Generally speaking, a Toeplitz operator is far from being one-to-one. In the following, we study the kernel of certain Toeplitz operators. We also explore the relation between the kernel and the model subspace K_Θ, which was introduced in Section 8.11.

Theorem 12.19 *Let $\varphi \in H^\infty(\mathbb{T})$, $\varphi \not\equiv 0$, and let Θ be the inner part of φ. Then the following hold.*

(i) *The Toeplitz operator T_φ is injective.*
(ii) *We have*

$$\ker T_{\bar{\varphi}} = K_\Theta.$$

In particular, if φ is an outer function in H^∞, then T_φ and $T_{\bar{\varphi}}$ are both injective.

Proof Part (i) follows immediately from Lemma 4.30. To prove (ii), first assume that φ is outer. Let f be in the kernel of $T_{\bar{\varphi}}$. By Lemma 4.10, this means that $\bar{\varphi} f \in \overline{H_0^2}$, which is equivalent to $\varphi \bar{f} \in H_0^2$. Hence, by Corollary 4.28, we deduce that $\bar{f} \in H_0^2$. Therefore, $f \in H^2 \cap \overline{H_0^2} = \{0\}$.

For the general case, let $\varphi = \Theta g$ be the inner–outer decomposition of φ. Then, according to (12.3) and Lemma 4.10, we have

$$\begin{aligned}\ker T_{\bar{\varphi}} &= \{f \in H^2 : T_{\bar{\varphi}} f = 0\}\\ &= \{f \in H^2 : T_{\bar{g}} T_{\bar{\Theta}} f = 0\}\\ &= \{f \in H^2 : T_{\bar{\Theta}} f = 0\}\\ &= \{f \in H^2 : \bar{\Theta} f \in \overline{H_0^2}\}\\ &= H^2 \cap \Theta \overline{H_0^2}\\ &= K_\Theta.\end{aligned}$$

This completes the proof. □

For the rest of this section, we consider the general case, i.e. we assume that $\varphi \in L^\infty(\mathbb{T})$. The following result shows that, for nonzero symbols φ, either T_φ or its adjoint is one-to-one.

Theorem 12.20 *Let $\varphi \in L^\infty(\mathbb{T})$, $\varphi \neq 0$. Then either $\ker T_\varphi = \{0\}$ or $\ker T_\varphi^* = \{0\}$.*

Proof Arguing by absurdity, assume that there are two functions h_1 and h_2 in H^2, not identically zero, such that $T_\varphi h_1 = 0$ and $T_\varphi^* h_2 = 0$. By Lemma 4.10, the first relation implies that $\varphi h_1 \in \overline{H_0^2}$ and then $\varphi h_1 \bar{h}_2 \in \overline{H_0^1}$. Since $T_\varphi^* = T_{\bar{\varphi}}$, the second relation gives that $\varphi \bar{h}_2 \in H_0^2$ and then $\varphi h_1 \bar{h}_2 \in H_0^1$. We thus obtain $\varphi h_1 \bar{h}_2 \in H_0^1 \cap \overline{H_0^1} = \{0\}$. But, since h_1 and h_2 are nonzero almost everywhere, the equality $\varphi h_1 \bar{h}_2 = 0$ forces φ to be zero almost everywhere, which is a contradiction. □

The following result guarantees that a nontrivial kernel of a Toeplitz operator always has an outer element.

Theorem 12.21 *Let $\varphi \in L^\infty(\mathbb{T})$ and assume that T_φ is not injective, i.e.*

$$\ker T_\varphi \neq \{0\}.$$

If $f \in \ker T_\varphi$, $f \neq 0$ and $f = \Theta h$, where Θ is an inner function and $h \in H^2$ (e.g. h can be the outer part of f), then $h \in \ker T_\varphi$. In particular, the outer part of f belongs to $\ker T_\varphi$.

Proof By Corollary 4.11, we have

$$T_\varphi h = T_\varphi(\bar{\Theta} f) = P_+(\varphi \bar{\Theta} f) = P_+(\bar{\Theta} P_+(\varphi f)) = P_+(\bar{\Theta} T_\varphi(f)) = 0.$$

Therefore, $h \in \ker T_\varphi$. □

Corollary 12.22 *Let $\varphi \in L^\infty(\mathbb{T})$. Assume that $\dim \ker T_\varphi = N < \infty$. Then*

$$\ker T_\varphi \cap S^N H^2 = \{0\}.$$

Proof Suppose, on the contrary, that there is a $g \in H^2$, $g \neq 0$, such that $g \in \ker T_\varphi \cap S^N H^2$. Write $g(z) = z^N h(z)$, $h \in H^2$, $h \neq 0$. Hence, by Theorem 12.21, we have

$$E = \{h, zh, z^2 h, \dots, z^N h\} \subset \ker T_\varphi.$$

Since $h \neq 0$, it is easy to see that E is a linearly independent set and

$$\dim \operatorname{span} E = N + 1.$$

But this is a contradiction to the assumption $\dim \ker T_\varphi = N$. □

We will need the following result in studying the nearly invariant subspaces of H^2 in Chapter 30.

Corollary 12.23 *Let $\varphi \in L^\infty(\mathbb{T})$. If $f \in \ker T_\varphi$ and $f(0) = 0$, then $S^* f \in \ker T_\varphi$.*

Proof Apply Theorem 12.21 with $\Theta(z) = z$. Note that, in this situation, we have $f = \Theta S^* f$. □

Theorem 12.24 *Let $f \in H^2$, $f \not\equiv 0$. Then $T_{f/\bar{f}}$ is injective.*

Proof Let $g \in H^2$ be such that $T_{f/\bar{f}}(g) = 0$. Then $P_+(fg/\bar{f}) = 0$, which implies that $fg/\bar{f} = \bar{h}$, where $h \in H^2_0$. Now, consider the function

$$\varphi = fg = \overline{fh}.$$

Then, on the one hand, the representation $\varphi = \overline{fh}$ shows that $\varphi \in \overline{H^1_0}$, and, on the other, by $\varphi = fg$, we have $\varphi \in H^1$. Therefore, $\varphi \in H^1 \cap \overline{H^1_0} = \{0\}$. Since, according to Lemma 4.30, $f \neq 0$ almost everywhere on $\mathbb{T}$, we deduce that $g \equiv 0$. This means that the operator $T_{f/\bar{f}}$ is injective. □

Let us highlight that the operator $T_{\bar{f}/f} \in \mathcal{L}(H^2)$ is not necessarily injective. The injectivity of this operator is characterized in Theorem 12.30. To establish that theorem we need the following result.

Theorem 12.25 *Let f be an outer function in H^2. Then*

$$\ker(T_{\bar{z}\bar{f}/f}) = \mathbb{C}f \oplus \ker T_{\bar{f}/f}.$$

Proof According to Theorem 12.4, we have $T_{\bar{z}\bar{f}/f} = S^* T_{\bar{f}/f}$. Thus, $g \in \ker(T_{\bar{z}\bar{f}/f})$ if and only if

$$T_{\bar{f}/f} g = \lambda,$$

where λ is a constant. It is easy to verify that $T_{\bar{f}/f} f = \overline{f(0)}$. First, since $f(0) \neq 0$, $f \notin \ker T_{\bar{f}/f}$. Second, we can write

$$T_{\bar{f}/f}(g - \lambda' f) = 0, \tag{12.8}$$

where $\lambda' = \lambda/\overline{f(0)}$. Therefore, we can say that $g \in \ker(T_{\bar{z}\bar{f}/f})$ if and only if there is a constant λ' such that (12.8) holds. The latter is equivalent to $g \in \mathbb{C}f + \ker T_{\bar{f}/f}$. □

Theorem 12.26 *Let f_1 and f_2 be two outer functions in H^1 such that*

$$|f_1| \le |f_2| \qquad (\text{a.e. on } \mathbb{T}).$$

Then

$$\frac{f_1}{f_2} \ker(T_{\bar{f}_2/f_2}) \subset \ker(T_{\bar{f}_1/f_1}).$$

Proof Let $g \in \ker(T_{\bar{f}_2/f_2})$. Put $h = gf_1/f_2$. By hypothesis, f_1/f_2 is in $L^\infty(\mathbb{T})$ and, since f_2 is outer, Corollary 4.28 implies that f_1/f_2 is in H^∞. Thus, the function h is in H^2. Moreover, by Corollary 4.11, we have

$$\begin{aligned} T_{\bar{f}_1/f_1}(h) &= P_+\left(\frac{\bar{f}_1}{f_1} h\right) \\ &= P_+\left(\frac{\bar{f}_1 g}{f_2}\right) \\ &= P_+\left(\frac{\bar{f}_1}{\bar{f}_2}\,\frac{\bar{f}_2 g}{f_2}\right) \\ &= P_+\left(\frac{\bar{f}_1}{\bar{f}_2}\, P_+\left(\frac{\bar{f}_2}{f_2} g\right)\right) \\ &= P_+\left(\frac{\bar{f}_1}{\bar{f}_2}\, T_{\bar{f}_2/f_2} g\right) = 0. \end{aligned}$$

Thus, $h \in \ker(T_{\bar{f}_1/f_1})$. □

If $\varphi \in H^\infty$, then the relation $T_z T_\varphi = T_\varphi T_z$, or equivalently $ST_\varphi = T_\varphi S$, is trivial. We use (12.7) to show that these are the only operators that commute with the forward shift operator S.

Theorem 12.27 *Let $A \in \mathcal{L}(H^2)$ be such that it commutes with S, i.e.*

$$AS = SA.$$

Then there exists $\varphi \in H^\infty$ such that $A = T_\varphi$.

Proof Using (12.7) and the commutation relation, we have

$$S^* A^* k_\lambda = A^* S^* k_\lambda = \bar{\lambda} A^* k_\lambda,$$

which shows that $A^* k_\lambda \in \ker(S^* - \bar{\lambda} I)$. But, according to Lemma 8.6, we know that $\ker(S^* - \bar{\lambda} I) = \mathbb{C} k_\lambda$, and thus we must have $A^* k_\lambda \in \mathbb{C} k_\lambda$. In other words, for every $\lambda \in \mathbb{D}$, the Cauchy kernel k_λ is an eigenvector for A^*. Hence, Theorem 9.3 implies that there exists $\varphi \in \mathfrak{Mult}(H^2)$ such that $A = M_\varphi$. But (9.13) says that $\mathfrak{Mult}(H^2) = H^\infty$, and thus $\varphi \in H^\infty$ and $A = M_\varphi = T_\varphi$. □

Exercises

Exercise 12.4.1 Let $\varphi \in L^\infty(\mathbb{T})$. Assume that $\dim \ker T_\varphi = N < \infty$. Show that no nonzero element of $\dim \ker T_\varphi$ can have a zero of order N at any $w \in \mathbb{D}$.
Hint: See the proof of Corollary 12.22.

Exercise 12.4.2 Let φ be a nonconstant function in $L^\infty(\mathbb{T})$. Show that

$$\sigma_p(T_\varphi) \cap \overline{\sigma_p(T_\varphi^*)} = \emptyset,$$

where $\overline{\sigma_p(T_\varphi^*)}$ denotes the set of complex conjugates of the eigenvalues of T_φ^*.
Hint: Use Theorem 12.20.

Exercise 12.4.3 Let φ be a real-valued nonconstant function in $L^\infty(\mathbb{T})$. Show that

$$\sigma_p(T_\varphi) = \emptyset.$$

Hint: By Theorem 12.3, T_φ is self-adjoint, and the spectrum of a self-adjoint operator is contained in the real line. Then apply Exercise 12.4.2.

Exercise 12.4.4 Let φ be a nonconstant function in H^∞. Show that the only invariant finite-dimensional subspace of T_φ is $\{0\}$.
Hint: Arguing by absurdity, assume that $\mathcal{M}$ is a finite-dimensional subspace of H^2 that is invariant under T_φ, and $\mathcal{M} \neq \{0\}$. Observe now that T_φ has necessarily an eigenvalue. Apply Theorem 12.19 to get a contradiction.

Exercise 12.4.5 Let φ be a nonconstant function in H^∞, and let K be a compact operator that commutes with T_φ. Show that K is nilpotent.
Hint: Let $\lambda \in \mathbb{C}$, $\lambda \neq 0$. Define $\mathcal{M}_\lambda = \ker(K - \lambda I)$ and observe that $\mathcal{M}_\lambda$ is a finite-dimensional subspace that is invariant for T_φ. Apply Exercise 12.4.4 to conclude that $\lambda \notin \sigma(K)$.

12.5 When is T_φ compact?

It is clear that $T_0 = 0$ is compact. In this section, we show that this is the only compact Toeplitz operator.

The Fourier coefficients of a function $\psi \in L^1(\mathbb{T})$ were defined by

$$\hat{\psi}(n) = \frac{1}{2\pi} \int_0^{2\pi} \psi(e^{it}) e^{-int}\, dt \qquad (n \in \mathbb{Z}).$$

By the Riemann–Lebesgue lemma,

$$\lim_{n \to \pm\infty} \hat{\psi}(n) = 0.$$

In particular, if $f, g \in L^2(\mathbb{T})$, then $f\bar{g} \in L^1(\mathbb{T})$, and the Riemann–Lebesgue lemma says that

$$\lim_{n \to \pm\infty} \frac{1}{2\pi} \int_0^{2\pi} f(e^{it}) \overline{g(e^{it})} e^{-int}\, dt = 0.$$

Using the inner product of $L^2(\mathbb{T})$, this fact can be rewritten as

$$\lim_{n \to \pm\infty} \langle \chi_n f, g \rangle = 0. \tag{12.9}$$

In technical language, this means that, for each fixed $f \in L^2(\mathbb{T})$, the sequence $(\chi_n f)_{n\in\mathbb{Z}}$ weakly converges to zero in $L^2(\mathbb{T})$ (see Section 1.2). As a special case, if $g \in H^2$, then, by Lemma 4.8, (12.9) implies that

$$\lim_{n \to +\infty} \langle P_+(\chi_n f), g \rangle = 0. \tag{12.10}$$

This means that, for each fixed $f \in L^2(\mathbb{T})$, the sequence $(P_+(\chi_n f))_{n\geq 1}$ weakly converges to zero in H^2.

Theorem 12.28 *Let $\varphi \in L^\infty(\mathbb{T})$. Then the Toeplitz operator T_φ is compact if and only if $\varphi \equiv 0$.*

Proof Clearly $T_0 = 0$ is compact. Now, suppose that K is an arbitrary compact operator on H^2. Then, by the triangle inequality and the fact that χ_n is unimodular on $\mathbb{T}$, we obtain

$$\begin{aligned} \|M_\varphi f\|_2 \leq\, & \|\varphi f - \chi_{-n} T_\varphi(P_+(\chi_n f))\|_2 \\ & + \|T_\varphi(P_+(\chi_n f)) - K(P_+(\chi_n f))\|_2 \\ & + \|K(P_+(\chi_n f))\|_2 \end{aligned}$$

for each $f \in L^2(\mathbb{T})$ and $n \in \mathbb{Z}$. By Lemma 12.1,

$$\lim_{n\to+\infty} \|\chi_{-n}T_\varphi(P_+(\chi_n f)) - \varphi f\|_2 = 0.$$

For the second term, we have

$$\begin{aligned}\|T_\varphi(P_+(\chi_n f)) - K(P_+(\chi_n f))\|_2 &\le \|T_\varphi - K\|\,\|P_+(\chi_n f)\|_2\\ &\le \|T_\varphi - K\|\,\|f\|_2.\end{aligned}$$

Since, by (12.10), $\chi_n f$ weakly converges to 0 and K is compact, then

$$\lim_{n\to+\infty} \|K(P_+(\chi_n f))\|_2 = 0.$$

Therefore, we obtain the estimation

$$\|M_\varphi f\|_2 \le \|T_\varphi - K\|\,\|f\|_2 \qquad (f \in L^2(\mathbb{T})).$$

By Theorem 2.20, we conclude that

$$\|\varphi\|_\infty \le \|T_\varphi - K\|$$

for all compact operators K on H^2. Hence, if T_φ is compact, we must have $\varphi \equiv 0$. □

Since an operator with a finite-dimensional rank is compact, we immediately obtain the following result.

Corollary 12.29 *Let $\varphi \in L^\infty(\mathbb{T})$, and suppose that $\varphi \not\equiv 0$. Then the range of T_φ is an infinite-dimensional subspace of H^2.*

12.6 Characterization of rigid functions

In Section 6.8, we introduced rigid functions and gave some sufficient conditions to create such elements (Theorems 6.22 and 6.23). In this section, we explore the relation between rigid functions and Toeplitz operators. This investigation enables us to find some new rigid functions that are not covered by the above-mentioned theorems.

Theorem 12.30 *Let f be an outer function in H^2. Then the following are equivalent:*

(i) *the function f^2 is rigid;*
(ii) *the operator $T_{\bar{f}/f} \in \mathcal{L}(H^2)$ is injective;*
(iii) $\ker(T_{\bar{z}\bar{f}/f})$ *is one-dimensional;*
(iv) $\ker(T_{\bar{z}\bar{f}/f}) = \mathbb{C}f$.

Proof (i) $\Longrightarrow$ (ii) Suppose, on the contrary, that the operator $T_{\bar{f}/f}$ is not injective. Hence, according to Theorem 12.21, there exists an outer function g in H^2 such that $T_{\bar{f}/f}g = 0$. This means that the function $g\bar{f}/f$ belongs to $\overline{H_0^2}$. Write

$$\frac{f\bar{g}}{\bar{f}} = h \in H_0^2.$$

Since $|h| = |g|$, by the canonical factorization theorem (Theorem 4.19), we can decompose h as $h = \Theta g$, where Θ is an inner function with $\Theta(0) = 0$. Therefore,

$$\frac{f^2}{|f|^2} = \frac{\Theta g^2}{|g|^2},$$

and this identity implies that

$$\arg(f^2) = \arg(\Theta g^2).$$

Since f^2 is rigid and $\Theta g^2 \in H^1$, there exists a positive constant λ such that $f^2 = \lambda \Theta g^2$. Remember that f and g are outer functions. Hence, by the uniqueness of canonical factorization, Θ has to be a unimodular constant. This is absurd since $\Theta(0) = 0$.

(ii) $\Longleftrightarrow$ (iii) $\Longleftrightarrow$ (iv) These follow from Theorem 12.25.

(iv) $\Longrightarrow$ (i) Suppose, on the contrary, that f^2 is not rigid. Hence, according to Lemma 6.17, there is an outer function $g_e \in H^1$ such that $\arg g_e = \arg f^2$ and $g_e \neq \lambda f^2$, for any positive constant λ. Consider the function $g = g_e^{1/2}$. Clearly, g is an outer function in H^2 that satisfies

$$\frac{\bar{g}}{g} = \frac{\bar{g}^2}{|g|^2} = \frac{\bar{g}_e}{|g_e|} = \frac{\bar{f}^2}{|f|^2} = \frac{\bar{f}}{f}, \tag{12.11}$$

almost everywhere on $\mathbb{T}$. From this last equality, one sees that

$$T_{\bar{z}\bar{f}/f}g = P_+\left(\bar{z}\frac{\bar{f}}{f}g\right) = P_+(\bar{z}\bar{g}) = 0,$$

whence $g \in \ker(T_{\bar{z}\bar{f}/f})$. Hence, by assumption, there is a constant $\alpha \in \mathbb{C}$ such that $g = \alpha f$. The equality (12.11) reveals that α is a real constant. Hence $g_e = g^2 = \alpha^2 f^2$, which is a contradiction. □

We exploit Theorem 12.30 to create an outer function $h \in H^1$ that is rigid, but at the same time $1/h \notin H^1$. This shows that the converse of Theorem 6.22 is not true.

Corollary 12.31 *The function $z \longmapsto 1 + z$ is rigid.*

Proof Let $f(z) = (1 + z)^{1/2}$, $z \in \mathbb{D}$, where the main branch is considered with $f(0) = 1$. By Corollary 4.25, f is outer. We claim that $T_{\bar{f}/f}$ is injective, i.e. $\ker(T_{\bar{f}/f}) = \{0\}$. This is simply based on the property

$$z\bar{f}^2 = f^2 \qquad (z \in \mathbb{T}). \tag{12.12}$$

Assume that $g \in \ker(T_{\bar{f}/f})$. Hence, $\bar{f}g/f = \bar{z}\bar{h}$ where $h \in H^2$. Square both sides and appeal to (12.12) to deduce that

$$zg^2 = \bar{h}^2 \qquad (z \in \mathbb{T}).$$

Since $zg^2 \in H_0^1$ and $\bar{h}^2 \in \overline{H^1}$, we immediately conclude that $g = 0$. □

Corollary 12.32 *Let h_1 and h_2 be two outer functions in H^1. Suppose that the function h_1 is rigid and*

$$|h_1| \le |h_2| \qquad (\textit{a.e. on } \mathbb{T}).$$

Then h_2 is also rigid.

Proof Let $f_k = h_k^{1/2}$, $k = 1, 2$. By Theorem 12.26,

$$\frac{f_1}{f_2}\ker(T_{\bar{f}_2/f_2}) \subset \ker(T_{\bar{f}_1/f_1}).$$

If h_1 is rigid, then by Theorem 12.30,

$$\ker(T_{\bar{f}_1/f_1}) = \{0\}.$$

Hence,

$$\ker(T_{\bar{f}_2/f_2}) = \{0\}$$

and, again by Theorem 12.30, we conclude that h_2 is rigid. □

We can also formulate Theorem 12.30 in terms of an exposed point of the unit ball of H^1.

Corollary 12.33 *Let f be an outer function in H^1, with $\|f\|_1 = 1$. Then the following are equivalent.*

(i) *The function f is an exposed point of the unit ball of H^1.*
(ii) *The operator $T_{\bar{f}/|f|}$ is injective.*
(iii) *The operator $T_{\bar{z}\bar{f}/|f|}$ has a one-dimensional kernel.*

Proof Since f is an outer function, the function $g = f^{1/2}$ is well defined and belongs to H^2. According to Theorem 6.15, f is an exposed point of the closed unit ball of H^1 if and only if f is rigid. But, by Theorem 12.30, this is equivalent to the injectivity of $T_{\bar{g}/g}$. It remains to note that

$$\frac{\bar{g}}{g} = \frac{\bar{g}^2}{|g|^2} = \frac{\bar{f}}{|f|}.$$ □

12.7 Toeplitz operators on $H^2(\mu)$

The relation (12.4) reveals that the image of an analytic polynomial under $T_{\bar{\varphi}}$ is again an analytic polynomial whenever $\varphi \in H^\infty$. This simple observation is the key point in defining Toeplitz operators on $H^2(\mu)$ spaces. The identity

$$T_{\bar{\varphi}}\chi_k = \sum_{j=0}^{k} \overline{\hat{\varphi}(k-j)}\, \chi_j$$

makes sense in $H^2(\mu)$ and thus $T_{\bar{\varphi}}$ is well defined on $\mathcal{P}_+$, the linear manifold of analytic polynomials. More specifically, we consider $\mathcal{P}_+ \subset H^2(\mu)$ and define

$$\begin{array}{rccc} T_{\bar{\varphi}}^{\mathcal{P}} : & \mathcal{P}_+ & \longrightarrow & \mathcal{P}_+ \\ & \displaystyle\sum_{k=0}^{n} a_k\chi_k & \longmapsto & \displaystyle\sum_{k=0}^{n} a_k\left(\sum_{j=0}^{k} \overline{\hat{\varphi}(k-j)}\, \chi_j\right). \end{array}$$

On the other hand, by Theorem 5.11, each $k_z \in H^2(\mu)$. Since, by (12.7), in the classic case we have

$$T_{\bar{\varphi}}k_w = \overline{\varphi(w)}k_w \qquad (w \in \mathbb{D}),$$

we also expect that the preceding equality holds when we extend the definition of $T_{\bar{\varphi}}$ from $\mathcal{P}_+$ to $H^2(\mu)$. In fact, one may equally well start with the definition

$$\begin{array}{rccc} T_{\bar{\varphi}}^{\mathcal{K}} : & \mathcal{K} & \longrightarrow & \mathcal{K} \\ & \displaystyle\sum_{j=0}^{n} a_j k_{w_j} & \longmapsto & \displaystyle\sum_{j=0}^{n} a_j\overline{\varphi(w_j)}k_{w_j}, \end{array}$$

where $\mathcal{K}$ stands for the linear manifold of all finite linear combinations of Cauchy kernels k_w, $w \in \mathbb{D}$, and then explore its possible extension to $H^2(\mu)$. These two approaches are shown to be equivalent in Theorem 12.36 below. To establish this connection, we need a technical lemma.

Lemma 12.34 *Let $\varphi \in H^\infty$, and let $w \in \mathbb{D}$. Then*

$$\overline{\varphi(w)}k_w(z) = \sum_{n=0}^{\infty} \bar{w}^n\left(\sum_{k=0}^{n} \overline{\hat{\varphi}(n-k)}\, \chi_k(z)\right) \qquad (z \in \mathbb{T}).$$

Moreover, the convergence is uniform with respect to $z \in \mathbb{T}$ and $w \in E$, where E is any compact subset of $\mathbb{D}$.

Proof We have

$$|w|^n \sum_{k=0}^{n} |\hat{\varphi}(n-k)|\, |z|^k \le (n+1)\, \|\varphi\|_\infty r^n$$

for all $z \in \mathbb{T}$ and $|w| \le r < 1$. Since $r < 1$, the series $\sum nr^n$ is convergent and thus the series

$$\sum_{n=0}^{\infty} \bar{w}^n \bigg(\sum_{k=0}^{n} \overline{\hat{\varphi}(n-k)}\, \chi_k(z) \bigg) \qquad (z \in \mathbb{T})$$

is absolutely and uniformly convergent for $z \in \mathbb{T}$ and $|w| \le r < 1$. Now using Fubini's theorem, we can write

$$\begin{aligned}
\sum_{n=0}^{\infty} \sum_{k=0}^{n} \overline{\hat{\varphi}(n-k)}\, \bar{w}^n z^k &= \sum_{k=0}^{\infty} \bigg(\sum_{n=k}^{\infty} \overline{\hat{\varphi}(n-k)}\, \bar{w}^n \bigg) z^k \\
&= \sum_{k=0}^{\infty} \bigg(\sum_{n=k}^{\infty} \overline{\hat{\varphi}(n-k)}\, \bar{w}^{n-k} \bigg) \bar{w}^k z^k \\
&= \sum_{k=0}^{\infty} \bigg(\sum_{n=0}^{\infty} \overline{\hat{\varphi}(n)}\, \bar{w}^n \bigg) \bar{w}^k z^k \\
&= \bigg(\sum_{k=0}^{\infty} \bar{w}^k z^k \bigg) \bigg(\sum_{n=0}^{\infty} \overline{\hat{\varphi}(n)}\, \bar{w}^n \bigg) \\
&= \frac{1}{1-\bar{w}z} \bigg(\sum_{n=0}^{\infty} \overline{\hat{\varphi}(n)}\, \bar{w}^n \bigg).
\end{aligned}$$

But, by (4.7), we have

$$\varphi(w) = \sum_{n=0}^{\infty} \frac{\varphi^{(n)}(0)}{n!} w^n = \sum_{n=0}^{\infty} \hat{\varphi}(n) w^n,$$

which implies that

$$\sum_{n=0}^{\infty} \sum_{k=0}^{n} \overline{\hat{\varphi}(n-k)}\, \bar{w}^n z^k = \frac{\overline{\varphi(w)}}{1-\bar{w}z} = \overline{\varphi(w)} k_w(z). \qquad \square$$

By (12.4),

$$T_{\bar{\varphi}}^{\mathcal{P}} \chi_n = \sum_{k=0}^{n} \overline{\hat{\varphi}(n-k)}\, \chi_k,$$

and we saw in the preceding lemma that the series

$$\sum_{n=0}^{\infty} \bar{w}^n \bigg(\sum_{k=0}^{n} \overline{\hat{\varphi}(n-k)}\, \chi_k(z) \bigg)$$

converges to $\overline{\varphi(w)} k_w(z)$ uniformly for $z \in \mathbb{T}$ and $|w| \le r < 1$. Hence, for each fixed $w \in \mathbb{D}$, the convergence also happens with respect to the norm of $H^2(\mu)$. We thus obtain the following result.

Corollary 12.35 *Let $\varphi \in H^\infty$, and let $w \in \mathbb{D}$. Then, in $H^2(\mu)$,*

$$\sum_{n=0}^{\infty} \bar{w}^n T_{\bar{\varphi}}^{\mathcal{P}} \chi_n = \overline{\varphi(w)} k_w.$$

Note that, up to here, no assumption on the boundedness of $T_{\bar{\varphi}}^{\mathcal{P}}$ or $T_{\bar{\varphi}}^{\mathcal{K}}$, as operators on $H^2(\mu)$, has been made. We now study this situation.

Theorem 12.36 *Let $\varphi \in H^\infty$. Then the mapping*

$$\begin{array}{rccc} T_{\bar{\varphi}}^{\mathcal{P}}: & \mathcal{P}_+ & \longrightarrow & \mathcal{P}_+ \\ & \displaystyle\sum_{k=0}^{n} a_k \chi_k & \longmapsto & \displaystyle\sum_{k=0}^{n} a_k \left(\sum_{j=0}^{k} \overline{\hat{\varphi}(k-j)}\, \chi_j \right) \end{array}$$

extends to a bounded operator on $H^2(\mu)$ if and only if so does the mapping

$$\begin{array}{rccc} T_{\bar{\varphi}}^{\mathcal{K}}: & \mathcal{K} & \longrightarrow & \mathcal{K} \\ & \displaystyle\sum_{j=0}^{n} a_j k_{w_j} & \longmapsto & \displaystyle\sum_{j=0}^{n} a_j \overline{\varphi(w_j)} k_{w_j}. \end{array}$$

Moreover, the extensions (if they exist) coincide.

Proof Suppose first that $T_{\bar{\varphi}}^{\mathcal{P}}$ extends to a bounded operator on $H^2(\mu)$. To make the distinction, temporarily denote this extension by $\mathbf{T}_{\bar{\varphi}}^{\mathcal{P}}$. Fix $w \in \mathbb{D}$. By Theorem 5.5(i), the analytic polynomials

$$\sum_{n=0}^{N} \bar{w}^n \chi_n \qquad (N \geq 0)$$

uniformly converge to k_w on $\mathbb{T}$. Hence, the convergence also holds in $H^2(\mu)$. Thus,

$$\mathbf{T}_{\bar{\varphi}}^{\mathcal{P}} k_w = \sum_{n=0}^{\infty} \bar{w}^n T_{\bar{\varphi}}^{\mathcal{P}} \chi_n,$$

with the series converging in $H^2(\mu)$. By Corollary 12.35, we know that this series converges to $\overline{\varphi(w)} k_w$ in $H^2(\mu)$. Therefore, we obtain

$$\mathbf{T}_{\bar{\varphi}}^{\mathcal{P}} k_w = \overline{\varphi(w)} k_w = T_{\bar{\varphi}}^{\mathcal{K}} k_w \qquad (w \in \mathbb{D}).$$

By Theorem 5.11, $\mathbf{T}_{\bar{\varphi}}^{\mathcal{P}}$ is precisely the extension of $T_{\bar{\varphi}}^{\mathcal{K}}$ to $H^2(\mu)$.

Now suppose that $T_{\bar{\varphi}}^{\mathcal{K}}$ extends to a bounded operator on $H^2(\mu)$, and denote its extension by $\mathbf{T}_{\bar{\varphi}}^{\mathcal{K}}$. As we saw above, the analytic polynomials

$$\sum_{n=0}^{N} \bar{w}^n \chi_n \qquad (N \geq 0)$$

converge to k_w in $H^2(\mu)$. Thus, we must have

$$T_{\bar{\varphi}}^{\mathcal{K}} k_w = \sum_{n=0}^{\infty} \bar{w}^n \mathbf{T}_{\bar{\varphi}}^{\mathcal{K}} \chi_n.$$

By definition, $T_{\bar{\varphi}}^{\mathcal{K}} k_w = \overline{\varphi(w)} k_w$. Therefore, by Corollary 12.35, for each $w \in \mathbb{D}$, the identity

$$\sum_{n=0}^{\infty} \bar{w}^n \mathbf{T}_{\bar{\varphi}}^{\mathcal{K}} \chi_n = \sum_{n=0}^{\infty} \bar{w}^n T_{\bar{\varphi}}^{\mathcal{P}} \chi_n$$

holds in $H^2(\mu)$. Thus, we deduce that

$$\mathbf{T}_{\bar{\varphi}}^{\mathcal{K}} \chi_n = T_{\bar{\varphi}}^{\mathcal{P}} \chi_n \qquad (n \geq 0).$$

This shows that $\mathbf{T}_{\bar{\varphi}}^{\mathcal{K}}$ is precisely the extension of $T_{\bar{\varphi}}^{\mathcal{P}}$ to $H^2(\mu)$.

The above argument shows that, if $\mathbf{T}_{\bar{\varphi}}^{\mathcal{P}}$ and $\mathbf{T}_{\bar{\varphi}}^{\mathcal{K}}$ exist, then

$$\mathbf{T}_{\bar{\varphi}}^{\mathcal{P}} = \mathbf{T}_{\bar{\varphi}}^{\mathcal{K}}.$$

□

The extension studied in Theorem 12.36, if it exists, will be denoted by $T_{\bar{\varphi}}$.

12.8 The Riesz projection on $L^2(\mu)$

In the classic case, the Toeplitz operator T_φ, with symbol $\varphi \in L^\infty(\mathbb{T})$, was defined on H^2 by the formula $T_\varphi f = P_+(\varphi f)$, where P_+ is the Riesz projection. We wish to generalize Toeplitz operators to generalized Hardy spaces $H^2(\mu)$, $\mu \in \mathcal{M}(\mathbb{T})$, by using the same formula. However, the difficulty is that the Riesz projection P_+ is not well defined on an arbitrary $L^2(\mu)$ space. However, at least P_+ is well defined on the dense manifold of trigonometric polynomials. In this section, we consider P_+ as such a densely defined operator and characterize its boundedness.

As before, let $\mathcal{P}$, $\mathcal{P}_+$ and $\mathcal{P}_-$, respectively, be the linear manifolds created by $\{z^n : n \in \mathbb{Z}\}$, $\{z^n : n \geq 0\}$ and $\{z^n : n \leq -1\}$. Using the notion of skew projection, which was developed in Section 1.12, we can say that

$$P_+ = P_{\mathcal{P}_+ \| \mathcal{P}_-} \qquad (\text{on } \mathcal{P}). \tag{12.13}$$

Lemma 12.37 *Let $\psi \in L^\infty(\mathbb{T})$ be such that $|\psi| = 1$ a.e. on $\mathbb{T}$. Then the following are equivalent.*

(i) *There is an outer function $h \in H^\infty$ such that $\|\psi - h\|_\infty < 1$.*
(ii) *There is an outer function $g \in H^\infty$ such that $\|\bar{\psi} - g\|_\infty < 1$.*
(iii) *There are real-valued bounded functions u and v on $\mathbb{T}$, with $\|u\|_\infty < \pi/2$, and a unimodular constant γ such that*

$$\psi = \gamma e^{i(u+\tilde{v})}.$$

Proof (i) $\Longrightarrow$ (iii) The assumption $\|\psi - h\|_\infty < 1$ implies that

$$0 < 1 - \|\psi - h\|_\infty \leq |h| \leq \|h\|_\infty \qquad \text{(a.e. on } \mathbb{T}).$$

Hence, $\log|h|$ is a bounded function. Write

$$\psi = \frac{\psi\bar{h}}{|\psi\bar{h}|} \times \frac{h}{|h|} \qquad \text{(a.e. on } \mathbb{T}).$$

By (4.53), we have

$$\frac{h}{|h|} = \gamma \exp(i\, \widetilde{\log|h|}) \qquad \text{(a.e. on } \mathbb{T}),$$

where γ is a unimodular constant. We also have

$$\frac{\psi\bar{h}}{|\psi\bar{h}|} = e^{i\arg(\psi\bar{h})},$$

and the condition $\|\psi - h\|_\infty < 1$ ensures that $|\arg(\psi\bar{h})| < \pi/2$. Thus, we can take $u = \arg(\psi\bar{h})$ and $v = \log|h|$.

(iii) $\Longrightarrow$ (i) Put $f = \gamma\exp(v + i\tilde{v})$. Then $f \in H^\infty$ and $1/f \in H^\infty$, and thus, by Corollary 4.24(iii), f is an outer function. Since

$$\bar{\psi} f = \exp(v - iu)$$

we have

$$\exp(-\|v\|_\infty) \leq |\bar{\psi} f| \leq \exp(\|v\|_\infty) \qquad \text{(a.e. on } \mathbb{T})$$

and

$$|\arg(\bar{\psi} f)| \leq \|u\|_\infty < \frac{\pi}{2} \qquad \text{(a.e. on } \mathbb{T}).$$

The above two inequalities show that there are $r, R > 0$, with $r < R$, such that

$$|\bar{\psi} f - R| \leq r \qquad \text{(a.e. on } \mathbb{T}).$$

Therefore, we have

$$\left|\psi - \frac{f}{R}\right| \leq \frac{r}{R} < 1 \qquad \text{(a.e. on } \mathbb{T}).$$

(ii) $\Longleftrightarrow$ (iii) It is trivial that

$$\psi = \gamma e^{i(u+\tilde{v})} \qquad \Longleftrightarrow \qquad \bar{\psi} = \bar{\gamma} e^{i(u'+\tilde{v}')},$$

where $u' = -u$ and $v' = -v$. Now apply the previously established equivalence (i) $\Longleftrightarrow$ (iii) to the function $\bar{\psi}$. □

Before stating the theorem, refer to Section 1.12 for the definition of the angle between two linear manifolds.

Lemma 12.38 *Let $d\mu = |h|^2\, dm$, where $h \in H^2$ is an outer function. Then*

$$\cos\langle \mathcal{P}_+, \mathcal{P}_- \rangle_{L^2(\mu)} = \mathrm{dist}_{L^\infty(\mathbb{T})}(\bar{h}/h, H^\infty).$$

Proof For each $f \in \mathcal{P}_+$ and $g \in \mathcal{P}_-$, by Lemma 4.9, we have

$$\begin{aligned}
\langle f, g \rangle_{L^2(\mu)} &= \int_{\mathbb{T}} f\bar{g}\, d\mu \\
&= \int_{\mathbb{T}} f\bar{g}|h|^2\, dm \\
&= \int_{\mathbb{T}} \frac{\bar{h}}{h} f h\, \bar{g}h\, dm \\
&= \left\langle \frac{\bar{h}}{h} fh,\, g\bar{h} \right\rangle_{L^2(\mathbb{T})} \\
&= \left\langle P_- \left(\frac{\bar{h}}{h} fh \right),\, g\bar{h} \right\rangle_{L^2(\mathbb{T})} \\
&= \langle H_{\bar{h}/h}(fh),\, g\bar{h} \rangle_{L^2(\mathbb{T})}.
\end{aligned}$$

Since h is an outer function, by Corollary 8.17, we have $\mathrm{Clos}_{L^2(\mathbb{T})} h\mathcal{P}_+ = H^2$ and $\mathrm{Clos}_{L^2(\mathbb{T})} \bar{h}\mathcal{P}_- = \overline{H_0^2} = (H^2)^\perp$. Moreover, $\|fh\|_{H^2} = \|f\|_{L^2(\mu)}$, for all $f \in \mathcal{P}_+$, and similarly $\|g\bar{h}\|_{H^2} = \|g\|_{L^2(\mu)}$, for all $g \in \mathcal{P}_-$. Therefore, if we take the supremum of

$$\frac{|\langle f, g \rangle_{L^2(\mu)}|}{\|f\|_{L^2(\mu)} \|g\|_{L^2(\mu)}} = \frac{|\langle H_{\bar{h}/h}(fh),\, g\bar{h} \rangle_{L^2(\mathbb{T})}|}{\|fh\|_{H^2} \|g\bar{h}\|_{H^2}}$$

with respect to all $f \in \mathcal{P}_+$, $f \neq 0$, and $g \in \mathcal{P}_-$, $g \neq 0$, we obtain

$$\cos\langle \mathcal{P}_+, \mathcal{P}_- \rangle = \|H_{\bar{h}/h}\|.$$

Thus, by Corollary 11.4,

$$\cos\langle \mathcal{P}_+, \mathcal{P}_- \rangle = \mathrm{dist}_{L^\infty(\mathbb{T})}(\bar{h}/h, H^\infty). \qquad \square$$

Theorem 12.39 *Let $\mu \in \mathcal{M}(\mathbb{T})$. Then the following are equivalent.*

(i) *The family $(z^n)_{n\in\mathbb{Z}}$ is a nonsymmetric basis of $L^2(\mu)$.*
(ii) *The family $(z^n)_{n\in\mathbb{Z}}$ is a symmetric basis of $L^2(\mu)$.*
(iii) *The Riesz projection P_+, as an operator from $L^2(\mu)$ onto $H^2(\mu)$, is well defined and bounded.*
(iv) $\langle \mathcal{P}_+, \mathcal{P}_- \rangle_{L^2(\mu)} > 0$.
(v) *The measure μ is absolutely continuous with respect to the Lebesgue measure with $d\mu = |h|^2\, dm$, where $h \in H^2$ is an outer function such that*

$$\mathrm{dist}_{L^\infty(\mathbb{T})}(\bar{h}/h, H^\infty) < 1.$$

(vi) *The measure* μ *is absolutely continuous with respect to the Lebesgue measure and* $d\mu = \omega\, dm$, *where*

$$\omega = e^{u+\tilde{v}},$$

and u, v *are real bounded functions with* $\|v\|_\infty < \pi/2$ ($\tilde{v}$ *is the Hilbert transform of* v).

Proof Our plan is to show that (i) $\Longrightarrow$ (ii) $\Longrightarrow$ (iii) $\Longrightarrow$ (i), (iii) $\Longleftrightarrow$ (iv), (i) + (iv) $\Longrightarrow$ (v), (v) $\Longrightarrow$ (iv) and finally (v) $\Longleftrightarrow$ (vi).

(i) $\Longrightarrow$ (ii) This is trivial.

(ii) $\Longrightarrow$ (iii) Fix any trigonometric polynomial p. Then, for each $n > \deg(p)$, we have

$$P_+(p) = z^n S_{-n,n}(z^{-n}p).$$

See Section 10.4 for the definition of $S_{m,n}$. Therefore,

$$\|P_+(p)\|_{L^2(\mu)} \le \left(\sup_{n\ge1} \|S_{-n,n}\| \right) \|p\|_{L^2(\mu)},$$

and thus, by Theorem 10.12, P_+ is bounded on $L^2(\mu)$.

(iii) $\Longrightarrow$ (i) Fix any trigonometric polynomial p. Then we have

$$S_{m,n}(p) = z^{n+1}(I - P_+)z^{-(n+m+1)}P_+(z^m p).$$

An easy way to verify this identity is to consider $p = z^k$ and treat the three cases $k < m$, $k > n$ and $m \le k \le n$. This identity implies that

$$\begin{aligned}
\|S_{m,n}(p)\|_{L^2(\mu)} &= \|(I - P_+)z^{-(n+m+1)}P_+(z^m p)\|_{L^2(\mu)} \\
&\le \|I - P_+\|\,\|P_+(z^m p)\|_{L^2(\mu)} \\
&\le \|I - P_+\|\,\|P_+\|\,\|p\|_{L^2(\mu)}.
\end{aligned}$$

This implies that

$$\sup_{m,n\in\mathbb{Z}} \|S_{m,n}\| < \infty.$$

Thus, by Theorem 10.11, $(z^n)_{n\in\mathbb{Z}}$ is a nonsymmetric basis of $L^2(\mu)$.

(iii) $\Longleftrightarrow$ (iv) This equivalence follows from Corollary 1.45 and (12.13).

(i) + (iv) $\Longrightarrow$ (v) Since $(z^n)_{n\in\mathbb{Z}}$ is a basis, we have

$$\bigcap_{n\ge0} z^n H^2(\mu) = \{0\}.$$

In fact, let $(\Lambda_n)_{n\in\mathbb{Z}}$ denote the biorthogonal sequence corresponding to $(z^n)_{n\in\mathbb{Z}}$. Then, if $f \in \bigcap_{n\ge0} z^n H^2(\mu)$, we have $\Lambda_n(f) = 0$ for all $n \in \mathbb{Z}$, which implies $f = 0$. By Lemma 8.19, we have $H^2(\mu) = H^2(\mu_a) \oplus L^2(\mu_s)$. In particular, $L^2(\mu_s) \subset H^2(\mu)$. Hence,

$$L^2(\mu_s) = \bigcap_{n\geq 0} z^n L^2(\mu_s) \subset \bigcap_{n\geq 0} z^n H^2(\mu) = \{0\}.$$

Therefore, μ is absolutely continuous with respect to the Lebesgue measure. Write $d\mu = \omega\, dm$, with $\omega \in L^1(\mathbb{T})$, $\omega \geq 0$.

Since $(z^n)_{n\in\mathbb{Z}}$ is minimal, $1 \notin H_0^2(\mu)$. Thus, by Theorem 8.22, $\log\omega \in L^1(\mathbb{T})$. Let h be the outer function such that $|h|^2 = \omega$ a.e. on $\mathbb{T}$. Then, by Lemma 12.38, we have

$$\operatorname{dist}_{L^\infty(\mathbb{T})}(\bar{h}/h, H^\infty) = \cos\langle \mathcal{P}_+, \mathcal{P}_- \rangle_{L^2(\mu)} < 1.$$

(v) $\Longrightarrow$ (iv) This follows immediately from Lemma 12.38.

(v) $\Longrightarrow$ (vi) Put $\omega = |h|^2 \in L^1(\mathbb{T})$. By assumption, there is a function $g \in H^\infty$ such that $\|\bar{h}/h - g\|_\infty < 1$. In other words, there is $\delta > 0$ such that

$$|\bar{h}/h - g| < 1 - \delta \qquad \text{(a.e. on } \mathbb{T}). \tag{12.14}$$

Multiply both sides by $|h|^2$ to get

$$|gh^2 - |h|^2| < (1-\delta)|h|^2 \qquad \text{(a.e. on } \mathbb{T}). \tag{12.15}$$

All disks with center $a > 0$ and radius $(1-\delta)a$ are contained in the open sector

$$\Omega_\delta = \{z : |\arg(z)| < \arcsin(1-\delta)\}.$$

(Note that $\arcsin(1-\delta) < \pi/2$.) Hence, by (12.15), we have

$$gh^2 \in \Omega_\delta \qquad \text{(a.e. on } \mathbb{T}).$$

The Poisson integral formula immediately implies that

$$g(z)h^2(z) \in \Omega_\delta \qquad (z \in \mathbb{D}).$$

Hence, by Corollary 4.25, gh^2 is outer and $\log(gh^2)$ is a well-defined analytic function on $\mathbb{D}$. We can thus write

$$\log(gh^2) = \log|gh^2| + i\,\widetilde{\log|gh^2|} \qquad \text{(a.e. on } \mathbb{T}), \tag{12.16}$$

with $\left|\widetilde{\log|gh^2|}\right| \leq \arcsin(1-\delta)$. Put

$$v = -\widetilde{\log|gh^2|} \qquad \text{(a.e. on } \mathbb{T}).$$

Since gh^2 is a nonzero outer function in H^1, by (4.44), $\log|gh^2| \in L^1(\mathbb{T})$. Moreover,

$$|v| \leq \arcsin(1-\delta) < \pi/2 \qquad \text{(a.e. on } \mathbb{T}),$$

and thus, by Theorem 3.14, we have

$$\tilde{v} = \log |gh^2| + c,$$

where c is a constant. Therefore, we can rewrite (12.16) as

$$gh^2 = \exp(\tilde{v} - iv - c) \qquad \text{(a.e. on } \mathbb{T}).$$

Taking the absolute values of both sides gives

$$|h|^2 = \exp(-\log |g| - c + \tilde{v}) \qquad \text{(a.e. on } \mathbb{T}).$$

The relation (12.14) implies that

$$\delta \le |g| \le 2 - \delta \qquad \text{(a.e. on } \mathbb{T}).$$

Hence, $u = -\log |g| - c \in L^\infty(\mathbb{T})$ and

$$|h|^2 = \exp(u + \tilde{v}) \qquad \text{(a.e. on } \mathbb{T}).$$

(vi) $\Longrightarrow$ (v) We have $\log \omega = u + \tilde{v}$, and, by Theorem 3.15, $\omega \in L^1(\mathbb{T})$. Hence, there is an outer function $h \in H^2$ such that $|h|^2 = \omega$. Since $\log |h|^2 = u + \tilde{v}$, by Theorem 3.14 and (4.53),

$$\log h^2 = (u + \tilde{v}) + i\widetilde{(u + \tilde{v})} = u + \tilde{v} + i\tilde{u} - iv - ic,$$

where c is a real constant. Then

$$\bar{h}/h = |h|^2/h^2 = \exp(i(v + c - \tilde{u})).$$

Therefore, by Lemma 12.37, there is an outer function $g \in H^\infty$ such that $\|\bar{h}/h - g\|_\infty < 1$.

This completes the proof of Theorem 12.39. □

We will say that a positive function g on $\mathbb{T}$ satisfies the *Helson–Szegő condition*, and write $g \in (HS)$, if there exist two bounded real functions u and v on $\mathbb{T}$, with $\|v\|_\infty < \pi/2$, such that

$$g = \exp(u + \tilde{v}).$$

12.9 Characterization of invertibility

In this section, we give a criterion for the invertibility of Toeplitz operators. For the following result, we recall that $[\varphi]$ stands for the outer function created by φ (see (4.51)).

Lemma 12.40 *Let $\varphi \in L^\infty(\mathbb{T})$ be such that $1/\varphi \in L^\infty(\mathbb{T})$. Then the following are equivalent:*

(i) T_φ *is invertible;*
(ii) $T_{\varphi/|\varphi|}$ *is invertible;*
(iii) $T_{\varphi/[\varphi]}$ *is invertible.*

Proof (i) $\Longleftrightarrow$ (ii) Put $h = [|\varphi|^{-1/2}]$. The outer function h satisfies the identity $|h|^2 = |\varphi|^{-1}$ on the boundary and thus it is bounded from above and below on $\mathbb{D}$. Hence, T_h is invertible and, in fact, $T_h^{-1} = T_{1/h}$. Moreover, by Theorem 12.4,

$$T_{\varphi/|\varphi|} = T_{\varphi|h|^2} = T_{\bar{h}\varphi h} = T_h^* T_\varphi T_h.$$

This identity shows that T_φ is invertible if and only if so is $T_{\varphi/|\varphi|}$.

(i) $\Longleftrightarrow$ (iii) Since the outer functions $[\varphi]$ and $1/[\varphi]$ are well defined and bounded, we have $T_{[\varphi]}^{-1} = T_{1/[\varphi]}$. Hence, again by Theorem 12.4, the identity

$$T_{\varphi/[\varphi]} = T_\varphi T_{1/[\varphi]}$$

reveals that $T_{\varphi/[\varphi]}$ is invertible if and only if so is T_φ. □

Lemma 12.41 *Let $\psi \in L^\infty(\mathbb{T})$ be such that $|\psi| = 1$ a.e. on $\mathbb{T}$. Then the following are equivalent:*

(i) T_ψ *is invertible;*
(ii) $\operatorname{dist}(\psi, H^\infty) < 1$ *and* $\operatorname{dist}(\bar{\psi}, H^\infty) < 1$*;*
(iii) *there is an outer function $h \in H^\infty$ such that $\|\psi - h\|_\infty < 1$;*
(iv) *there is an outer function $g \in H^\infty$ such that $\|\bar{\psi} - g\|_\infty < 1$;*
(v) *there are real-valued bounded functions u and v on $\mathbb{T}$, with $\|u\|_\infty < \pi/2$, and a unimodular constant γ such that*

$$\psi = \gamma e^{i(u+\tilde{v})}.$$

Proof (i) $\Longleftrightarrow$ (ii) According to Lemma 12.11, (ii) is equivalent to saying that T_ψ and T_ψ^* are bounded below. Then, by Corollary 1.31, the latter is equivalent to T_ψ being invertible.

(ii) $\Longrightarrow$ (iii) + (iv) By assumption, there are functions $g, h \in H^\infty$ such that $\|\psi - h\|_\infty < 1$ and $\|\bar{\psi} - g\|_\infty < 1$. We show that g and h are necessarily outer functions.

Write the last two inequalities as $\|1 - \bar{\psi}h\|_\infty < 1$ and $\|1 - \psi g\|_\infty < 1$. These two inequalities imply that the image of $\mathbb{T}$ under either of the mappings $\bar{\psi}h$ and ψg lies in the curvilinear rectangle

$$R = \{z : 1 - r \le |z| \le 1 + r,\ |\arg z| \le \arcsin(r)\},$$

where $r = \max\{\|1 - \bar{\psi}h\|_\infty,\ \|1 - \psi g\|_\infty\}$. Note that $0 \le r < 1$ and thus $0 \notin R$ and, moreover, $\arcsin(r) < \pi/2$. Hence, $K = R^2 = \{z^2 : z \in R\}$ is a compact set such that $\mathbb{C} \setminus K$ is connected and $0 \notin K$. Since $gh = \psi g \times \bar{\psi}h$

maps $\mathbb{T}$ to K, Theorem 4.5 ensures that gh maps the whole unit disk into K. In particular,

$$\inf_{z\in\mathbb{D}} |g(z)h(z)| > 0.$$

By Corollary 4.24(iii), gh is an outer function. By part (ii) of the same result, we deduce that g and h are outer functions.

(iii) + (iv) $\Longrightarrow$ (ii) This is trivial.

(iii) $\Longleftrightarrow$ (iv) $\Longleftrightarrow$ (v) These are proved in Lemma 12.37.

This completes the proof. □

Theorem 12.42 *Let $\varphi \in L^\infty(\mathbb{T})$. Then the following are equivalent:*

(i) *T_φ is invertible;*

(ii) *$\varphi^{-1} \in L^\infty(\mathbb{T})$ and*

$$\operatorname{dist}(\varphi/|\varphi|, H^\infty) < 1 \quad \textit{and} \quad \operatorname{dist}(\bar{\varphi}/|\varphi|, H^\infty) < 1;$$

(iii) *$\varphi^{-1} \in L^\infty(\mathbb{T})$ and there exists an outer function $h \in H^\infty$ such that*

$$\left\| \frac{\varphi}{|\varphi|} - h \right\|_\infty < 1;$$

(iv) *$\varphi^{-1} \in L^\infty(\mathbb{T})$ and there are real-valued bounded functions u and v on $\mathbb{T}$, with $\|u\|_\infty < \pi/2$, and a unimodular constant γ such that*

$$\frac{\varphi}{|\varphi|} = \gamma e^{i(u+\tilde{v})};$$

(v) *$\varphi^{-1} \in L^\infty(\mathbb{T})$ and there exists an outer function $h \in H^2$, with $|h|^2 \in (HS)$, such that*

$$\frac{\varphi}{|\varphi|} = \gamma \frac{h}{\bar{h}},$$

where γ is a unimodular constant.

Proof By Lemma 12.12, $\varphi^{-1} \in L^\infty(\mathbb{T})$ is a necessary condition for the invertibility of T_φ. Then the equivalence of (i), (ii), (iii) and (iv) follows immediately from Lemmas 12.40 and 12.41.

(iv) $\Longrightarrow$ (v) Define the outer function h such that

$$\log|h| = (v - \tilde{u})/2 \qquad \text{(a.e. on } \mathbb{T}).$$

Note that, according to Corollary 3.16, $\tilde{u} \in L^1(\mathbb{T})$, and thus h is well defined. Moreover,

$$|h|^2 = \exp(v - \tilde{u}) \in (HS).$$

Then, by Theorem 3.14,

$$2\,\widetilde{\log|h|} = u + \tilde{v} - \hat{u}(0) \qquad \text{(a.e. on } \mathbb{T}),$$

and note that $\hat{u}(0)$ is a real constant. Since, by (4.53),

$$h/\bar{h} = \exp\big(2i\,\widetilde{\log|h|}\big),$$

the preceding identity is equivalent to $\varphi/|\varphi| = \gamma h/\bar{h}$, where γ is a unimodular constant.

(v) $\Longrightarrow$ (iv) The preceding argument is almost reversible. We assume that $|h|^2 \in (HS)$, i.e.

$$|h|^2 = \exp(v - \tilde{u}),$$

where u, v are real and bounded on $\mathbb{T}$ and $\|u\|_\infty < \pi/2$. Hence, by Theorem 3.14,

$$2\,\widetilde{\log|h|} = \tilde{v} + u - \hat{u}(0).$$

Since $\varphi/|\varphi| = \gamma h/\bar{h} = \gamma \exp\big(2i\,\widetilde{\log|h|}\big)$, the result follows. □

Corollary 12.43 *Let h be an outer function in H^2. Then the following are equivalent:*

(i) *$T_{h/\bar{h}}$ is invertible;*
(ii) *$h^{-1} \in H^2$ and $\operatorname{dist}(h/\bar{h}, H^\infty) < 1$;*
(iii) *$\operatorname{dist}(\bar{h}/h, H^\infty) < 1$;*
(iv) *$|h|^2 \in (HS)$.*

Proof We will proceed as follows: (i) $\Longrightarrow$ (iii) $\Longrightarrow$ (iv) $\Longrightarrow$ (i) and then (iv) $\Longrightarrow$ (ii) $\Longrightarrow$ (iii).

(i) $\Longrightarrow$ (iii) Apply Theorem 12.42 to $\varphi = h/\bar{h}$, which is in $L^\infty(\mathbb{T})$ and $|\varphi| = 1$ almost everywhere on $\mathbb{T}$.

(iii) $\Longrightarrow$ (iv) Apply Theorem 12.39 to $d\mu = |h|^2\,dm$.

(iv) $\Longrightarrow$ (i) Apply Theorem 12.42 to $\varphi = h/\bar{h}$.

(iv) $\Longrightarrow$ (ii) Assume that $|h|^2 \in (HS)$. Hence, there exist two bounded real functions u and v on $\mathbb{T}$, $\|v\|_\infty < \pi/2$, such

$$|h|^2 = \exp(u + \tilde{v}).$$

This implies that

$$|h|^{-2} = \exp(-u - \tilde{v}) \in (HS).$$

We now apply Theorem 12.39 to $d\mu = |h|^{-2}\,dm$ to deduce that there exists an outer function $g \in H^2$ such that $|h|^{-2} = |g|^2$ and

$$\operatorname{dist}(\bar{g}/g, H^\infty) < 1.$$

Note that, since $|h|^{-2} = |g|^2$, we get that $1/h \in L^2(\mathbb{T})$ and, since h is an outer function, Corollary 4.28(iii) implies that $1/h$ is an outer function in H^2.

But, since $|h|^{-1} = |g|$, there exists a constant λ of modulus one such that $h^{-1} = \lambda g$. Hence

$$\frac{h}{\bar{h}} = \frac{\overline{h^{-1}}}{h^{-1}} = \frac{\bar{\lambda}\,\bar{g}}{\lambda\,g},$$

and we get $\text{dist}(h/\bar{h}, H^\infty) = \text{dist}(\bar{g}/g, H^\infty) < 1$.

(ii) $\Longrightarrow$ (iii) Assume that h is an outer function such that $h, h^{-1} \in H^2$ and $\text{dist}(h/\bar{h}, H^\infty) < 1$. By Corollary 4.24(iii), h^{-1} is outer. Since

$$\text{dist}(\overline{h^{-1}}/h^{-1}, H^\infty) = \text{dist}(h/\bar{h}, H^\infty) < 1,$$

we can apply the implication (vi) $\Longrightarrow$ (ii) (already proved) to h^{-1} to get

$$\text{dist}(\bar{h}/h, H^\infty) = \text{dist}(h^{-1}/\overline{h^{-1}}, H^\infty) < 1. \qquad \square$$

Corollary 12.44 *Let h be an outer function in H^2 such that $|h|^2 \in (HS)$. Then h^2 is a rigid function.*

Proof According to Corollary 12.43, the operator $T_{\bar{h}/h}$ is invertible. In particular, $\ker T_{\bar{h}/h} = \{0\}$. Hence, Theorem 12.30 implies that h^2 is rigid. □

12.10 Fredholm Toeplitz operators

In this section, we study some characterizations of Fredholm Toeplitz operators.

Lemma 12.45 *Let $\varphi \in L^\infty(\mathbb{T})$. Then the following are equivalent.*

(i) *The operator T_φ is invertible.*
(ii) *The operator T_φ is a Fredholm operator of index* 0.

Proof The implication (i) $\Longrightarrow$ (ii) is trivial (and valid for any operator).

Let us prove the reverse implication. Assume that the operator T_φ is a Fredholm operator of index 0. By Theorem 12.20, either T_φ or T_φ^* has a trivial kernel. But since

$$0 = \text{ind}T_\varphi = \dim\ker T_\varphi - \dim\ker T_\varphi^*,$$

we thus conclude that both kernels are trivial. Therefore, T_φ and T_φ^* are both bounded below and then T_φ is invertible. □

Theorem 12.46 *Let $\varphi \in L^\infty(\mathbb{T})$ be such that $|\varphi| = 1$ a.e. on $\mathbb{T}$. Then the following are equivalent:*

(ii) *T_φ is a left semi-Fredholm operator;*
(ii) $\text{dist}(\varphi, H^\infty + \mathcal{C}) < 1$.

Proof (i) $\Longrightarrow$ (ii) Assume that T_φ is a left semi-Fredholm operator. According to Theorem 12.20, we see that T_φ is either left-invertible (in the case $\ker T_\varphi = \{0\}$) or Fredholm (in the case $\ker T_\varphi^* = \{0\}$). In the former case, we know by Lemma 12.11 that $\operatorname{dist}(\varphi, H^\infty) < 1$ and then $\operatorname{dist}(\varphi, H^\infty + \mathcal{C}) < 1$. In the latter case, let

$$\operatorname{ind} T_\varphi = \dim \ker T_\varphi - \dim \ker T_\varphi^* = n - 0 = n.$$

Then we claim that $T_{z^n\varphi}$ is a Fredholm operator of index 0. Indeed, by Lemma 8.7, we know that $T_{z^n} = S^n$ is a Fredholm operator of index $-n$, and thus Theorem 7.39 implies that $T_{z^n}T_\varphi$ is a Fredholm operator of index

$$\operatorname{ind}(T_{z^n}T_\varphi) = \operatorname{ind} T_{z^n} + \operatorname{ind} T_\varphi = -n + n = 0.$$

Now, according to Corollary 12.9, the operator $T_{z^n\varphi} - T_{z^n}T_\varphi$ is compact, and it remains to apply Corollary 7.41 to derive the claim. By Lemma 12.45, the operator $T_{z^n\varphi}$ is thus invertible and so, using Lemma 12.11, we conclude that

$$\operatorname{dist}(z^n\varphi, H^\infty) < 1.$$

This means that there exists a function $h \in H^\infty$ such that $\|\varphi - \overline{z^n}h\|_\infty < 1$. Let p be the analytic polynomial of degree n such that

$$p^{(j)}(0) = h^{(j)}(0) \qquad (0 \le j \le n-1).$$

Hence, $h = p + z^n h_n$, with $h_n \in H^\infty$. Then $\|\varphi - \overline{z^n}p - h_n\|_\infty < 1$, which implies that

$$\operatorname{dist}(\varphi, H^\infty + \mathcal{C}) < 1.$$

(ii) $\Longrightarrow$ (i) Assume that $\operatorname{dist}(\varphi, H^\infty + \mathcal{C}) < 1$. Then there exist $h \in H^\infty$ and $g \in \mathcal{C}(\mathbb{T})$ such that $\|\varphi - (h+g)\|_\infty < 1$. Let $\varepsilon > 0$ be such that

$$\|\varphi - (h+g)\|_\infty + \varepsilon < 1.$$

By the Weierstrass approximation theorem, we know that there exists a trigonometric polynomial p such that $\|g - p\|_\infty < \varepsilon$, and then, for any integer n,

$$\begin{aligned}\|z^n\varphi - (z^n h + z^n p)\|_\infty &= \|\varphi - (h+p)\|_\infty\\ &\le \|\varphi - (h+g)\|_\infty + \|g - p\|_\infty\\ &\le \|\varphi - (h+g)\|_\infty + \varepsilon < 1.\end{aligned}$$

Choosing n sufficiently large so that $z^n p$ is an analytic polynomial, and in particular belonging to H^∞, we thus get

$$\operatorname{dist}(z^n\varphi, H^\infty) < 1.$$

By Lemma 12.11, it then follows that $T_{z^n\varphi}$ is left-invertible. In particular, there exists an operator $R \in \mathcal{L}(H^2)$ such that $RT_{z^n\varphi} = I$. Using Theorem 12.4, we then get

$$RT_\varphi T_{z^n} T_{\overline{z^n}} = T_{\overline{z^n}} . \tag{12.17}$$

But Corollary 12.9 implies that $I - T_{z^n} T_{\overline{z^n}}$ is compact, and then so is $RT_\varphi(I - T_{z^n} T_{\overline{z^n}})$. Using (12.17) and Theorem 12.4 once more, we finally get that the operator $T_{z^n} RT_\varphi$ is of the form $I + K$ with K compact. Finally, arguing as in the proof of Theorem 7.34, we deduce that T_φ is a left semi-Fredholm operator. □

By applying the above result, we immediately get the following characterization.

Corollary 12.47 *Let $\varphi \in L^\infty(\mathbb{T})$ be such that $|\varphi| = 1$ a.e. on $\mathbb{T}$. Then the following are equivalent:*

(i) *the operator T_φ is a Fredholm operator;*
(ii) $\text{dist}(\varphi, H^\infty + \mathcal{C}) < 1$ *and* $\text{dist}(\bar{\varphi}, H^\infty + \mathcal{C}) < 1$.

Exercises

Exercise 12.10.1 Let $\varphi \in L^\infty(\mathbb{T})$. Show that

$$\mathcal{R}_e(\varphi) \subset \sigma_{ess}(T_\varphi).$$

Hint: Assume that T_φ is a Fredholm operator and use Theorem 12.20 and Lemma 12.12 to conclude that φ is invertible in $L^\infty(\mathbb{T})$.

Exercise 12.10.2 Let $\varphi \in H^\infty + \mathcal{C}(\mathbb{T})$. The aim of this exercise is to show that the following are equivalent:

(i) T_φ is Fredholm;
(ii) φ is invertible in $H^\infty + \mathcal{C}(\mathbb{T})$.

To do so, take the following steps.

Step I: Assume that there exists $\psi \in H^\infty + \mathcal{C}(\mathbb{T})$ such that $\varphi\psi = 1$.

(a) Show that $I - T_\psi T_\varphi$ and $I - T_\varphi T_\psi$ are compact.
Hint: Use Lemma 12.8 and Theorem 11.8.
(b) Deduce that (ii) $\Longrightarrow$ (i).

Step II: Assume that T_φ is Fredholm.

(a) Show that φ is invertible in $L^\infty(\mathbb{T})$.
Hint: Use Exercise 12.10.1.
(b) Write $\varphi = u\eta$, where η is an outer function invertible in H^∞ and u is a unimodular function. Show that T_u is Fredholm. Let $n = \text{ind}\, T_\varphi$.

(c) Show that $T_{z^n u}$ is Fredholm with index 0.
Hint: Separate the two cases $n \geq 0$ and $n < 0$ and then use Lemma 8.7.
(d) Conclude that $T_{z^n u}$ is invertible.
(e) Show that there is an outer function $h \in H^\infty$ such that

$$\|1 - z^n uh\|_\infty = \|\overline{z^n u} - h\|_\infty < 1.$$

(f) Conclude that $z^n uh$ is invertible in $H^\infty + \mathcal{C}(\mathbb{T})$.
(g) Deduce that (i) $\Longrightarrow$ (ii).

Exercise 12.10.3 Let $\varphi \in \mathcal{C}(\mathbb{T})$. Show that

$$\sigma_{ess}(T_\varphi) = \varphi(\mathbb{T}).$$

Hint: Use Exercise 12.10.2

12.11 Characterization of surjectivity

Lemma 12.12 enables us to limit our attention, in problems of surjectivity, invertibility, left- or right-invertibility, to Toeplitz operators with unimodular symbols. Indeed, if T_φ is surjective, then $T_{\bar{\varphi}}$ is left-invertible, hence bounded below. By Lemma 12.12, the function φ is then invertible in $L^\infty(\mathbb{T})$. Assuming this to be the case, we consider h to be the outer function whose modulus on $\mathbb{T}$ equals $|\varphi|$ and is positive at the origin, and set $\psi = \varphi/\bar{h}$. Thus, ψ is unimodular and Theorem 12.4 gives $T_\varphi = T_{\bar{h}} T_\psi$. Since $1/h$ belongs to H^∞, we deduce that $T_{\bar{h}}$ is invertible. Hence, T_φ is invertible if and only if so is T_ψ. In general, the two operators have the same kernel.

Theorem 12.48 *Let $\varphi \in L^\infty(\mathbb{T})$, with $|\varphi| = 1$ on $\mathbb{T}$. Then T_φ is surjective if and only if there is an inner function Θ and an outer function h with $|h|^2 \in (HS)$ such that*

$$\varphi = \bar{\Theta}\frac{\bar{h}}{h}.$$

Proof Assume that $\varphi = \bar{\Theta}\bar{h}/h$, for some inner function Θ and an outer function h such that $|h|^2 \in (HS)$. Then, by Corollary 12.43, the operator $T_{\varphi\Theta} = T_{\bar{h}/h}$ is invertible. But, according to Theorem 12.4, we have $T_\varphi = T_{\bar{\Theta}} T_{\varphi\Theta}$ and then

$$T_\varphi (T_{\varphi\Theta})^{-1} T_\Theta = T_{\bar{\Theta}} T_{\varphi\Theta} (T_{\varphi\Theta})^{-1} T_\Theta = T_{\bar{\Theta}} T_\Theta = I,$$

which proves that T_φ is right-invertible (or surjective).

Conversely, assume that T_φ is surjective. Then it follows from Corollary 1.34 that $T_{\bar{\varphi}}$ is left-invertible, and Lemma 12.11 implies that $\mathrm{dist}(\bar{\varphi}, H^\infty) < 1$.

Thus, there is a function $g \in H^\infty$ such that $\|\bar{\varphi} - g\|_\infty < 1$. Write $g = \Theta g_o$, with Θ inner and g_o outer. Hence, $\|\bar{\varphi}\bar{\Theta} - g_o\|_\infty < 1$, which, by Theorem 12.42, means that the operator $T_{\bar{\varphi}\bar{\Theta}}$ is invertible and there exists an outer function h in H^2 such that $|h|^2 \in (HS)$ and $\bar{\varphi}\bar{\Theta} = h/\bar{h}$. Thus $\varphi = \bar{\Theta}\bar{h}/h$. □

If T_φ is surjective, according to Theorem 12.48, there exists an inner function Θ and an outer function h such that $|h|^2 \in (HS)$ and $\varphi = \bar{\Theta}\bar{h}/h$. Furthermore, if T_φ is invertible, then the inner function Θ is necessarily constant and the function h is unique (up to multiplication by a nonzero real number). Indeed, according to Corollary 12.43, the operator $T_{\bar{h}/h}$ is invertible and, since $T_\varphi = T_{\bar{\Theta}} T_{\bar{h}/h}$, the operator $T_{\bar{\Theta}}$ is also invertible. But, Theorem 12.19 implies that $\ker T_{\bar{\Theta}} = K_\Theta$, and thus $K_\Theta = \{0\}$, which means that Θ is constant. For the uniqueness of h, assume that there exists another outer function $h_1 \in H^2$ such that

$$\frac{\bar{h}}{h} = \frac{\bar{h}_1}{h_1} \qquad \text{(a.e. on } \mathbb{T}).$$

Then

$$\frac{h^2}{|h|^2} = \frac{h_1^2}{|h_1|^2} \qquad \text{(a.e. on } \mathbb{T}).$$

In particular, we get $\arg(h^2) = \arg(h_1^2)$. Since $|h|^2 \in (HS)$, Corollary 12.44 implies that h^2 is rigid. Thus, $h_1^2 = \lambda h^2$, with $\lambda > 0$. This implies that $h_1 = \mu h$ for some real constant μ.

On the contrary, if T_φ is surjective but not invertible, the functions Θ and h that appear in Theorem 12.48 are not unique. Indeed, we know that there is an outer function h_0 such that $\|\bar{\Theta}\bar{\varphi} - h_0\|_\infty < 1$. Let g be any function in H^∞ such that

$$\|g - h_0\|_\infty < 1 - \|\bar{\Theta}\bar{\varphi} - h_0\|_\infty.$$

Then $\|\bar{\Theta}\bar{\varphi} - g\|_\infty < 1$. If $g = \Theta_0 g_0$, with Θ_0 inner and g_0 outer, then

$$\|\overline{\Theta_0}\,\bar{\Theta}\bar{\varphi} - g_0\|_\infty.$$

Now, Theorem 12.42 implies that $T_{\overline{\Theta_0}\,\bar{\Theta}\bar{\varphi}}$ is invertible and there is an outer function $h_1 \in H^2$ such that $|h_1|^2 \in (HS)$ and

$$\Theta_0 \Theta \varphi = \frac{\bar{h}_1}{h_1}.$$

That is, $\varphi = \overline{\Theta_0}\,\bar{\Theta}\bar{h}_1/h_1$.

In the case where the Toeplitz operator T_φ is not invertible, since we do have uniqueness in the representation given in Theorem 12.48, the conditions given in Theorem 12.48 are rather difficult to use in practice. In Section 30.6, using the language of $\mathcal{H}(b)$ spaces, we give another characterization of surjectivity.

12.12 The operator $X_{\mathcal{H}}$ and its invariant subspaces

In Chapter 9, we introduced the forward shift operator $S_{\mathcal{H}}$ on abstract reproducing kernel Hilbert spaces $\mathcal{H}$ and we discussed its invariant subspaces. In this section, we discuss parallel results for the backward shift operator $X_{\mathcal{H}}$ on an abstract reproducing kernel Hilbert space $\mathcal{H}$. Here, we assume that $\mathcal{H}$ is an analytic reproducing kernel Hilbert space on the open unit disk $\mathbb{D}$ satisfying the following three additional conditions:

(H4) The space $\mathcal{H} \subset H^2$.

(H5) The space $\mathcal{H}$ is invariant under S^*. Hence, using the closed graph theorem (Corollary 1.18), the mapping

$$\begin{array}{rccc} X_{\mathcal{H}}: & \mathcal{H} & \longrightarrow & \mathcal{H} \\ & f & \longmapsto & S^* f \end{array}$$

is continuous.

(H6) The operator $X_{\mathcal{H}}$ is polynomially bounded, that is, there is a universal constant $c > 0$ such that

$$\|p(X_{\mathcal{H}})\|_{\mathcal{L}(\mathcal{H})} \le c\|p\|_\infty$$

for any analytic polynomial p.

If $\mathcal{H}$ is an analytic reproducing kernel space satisfying (H4), then, by the closed graph theorem (Corollary 1.18), that $\mathcal{H}$ is indeed boundedly contained in H^2. More explicitly, there is a constant $\kappa > 0$ such that

$$\|f\|_2 \le \kappa\|f\|_{\mathcal{H}} \qquad (f \in \mathcal{H}).$$

Theorem 12.49 *Let $\mathcal{H}$ be an analytic reproducing kernel Hilbert space on the open unit disk $\mathbb{D}$ satisfying (H4), (H5) and (H6). Let $\varphi \in H^\infty$. Then $\mathcal{H}$ is invariant under $T_{\bar{\varphi}}$. Moreover, if*

$$\begin{array}{rccc} T_{\bar{\varphi},\mathcal{H}}: & \mathcal{H} & \longrightarrow & \mathcal{H} \\ & f & \longmapsto & T_{\bar{\varphi}} f, \end{array}$$

then $T_{\bar{\varphi},\mathcal{H}}$ is bounded and

$$\|T_{\bar{\varphi},\mathcal{H}}\|_{\mathcal{L}(\mathcal{H})} \le c\|\varphi\|_\infty.$$

Proof First note that, if p is a polynomial, say $p(z) = \sum_{k=0}^N a_k z^k$, and if p^* is the polynomial associated with p by $p^*(z) = \overline{p(\bar{z})}$, then

$$T_{\bar{p}} f = p^*(X_{\mathcal{H}}) f$$

for all $f \in \mathcal{H}$. Indeed, we have

$$\begin{aligned} T_{\bar{p}} f &= \sum_{k=0}^{N} \bar{a}_k T_{\bar{z}^k} f \\ &= \sum_{k=0}^{N} \bar{a}_k S^{*k} f \\ &= \sum_{k=0}^{N} \bar{a}_k X_{\mathcal{H}}^k f \\ &= p^*(X_{\mathcal{H}}) f. \end{aligned}$$

In particular, we get from (H5) and (H6) that $\mathcal{H}$ is invariant under $T_{\bar{p}}$ and

$$\|T_{\bar{p},\mathcal{H}}\|_{\mathcal{L}(\mathcal{H})} \le c\|p\|_\infty. \tag{12.18}$$

Now, we should extend (12.18) for an arbitrary function $\varphi \in H^\infty$. For that purpose, we use a standard approximation argument used several times in Chapter 9. So, let $\varphi \in H^\infty$ and let $(p_n)_{n\ge1}$ be the sequence of Fejér means of its Fourier sums, that is,

$$p_n = \frac{1}{n+1}(s_0 + s_1 + \cdots + s_n),$$

with $s_j(e^{i\theta}) = \sum_{k=0}^{j} \hat{\varphi}(k) e^{ik\theta}$. Recall that $\|p_n\|_\infty \le \|\varphi\|_\infty$ and $(p_n)_{n\ge1}$ converges to φ in the weak-star topology of $L^\infty(\mathbb{T})$. Thus, according to (12.18),

$$\|T_{\bar{p}_n} f\|_{\mathcal{H}} \le c\|\varphi\|_\infty \|f\|_{\mathcal{H}}$$

for any $f \in \mathcal{H}$. Hence, by passing to a subsequence if needed, there exists a $g \in \mathcal{H}$ such that $T_{\bar{p}_n} f$ converges weakly to g and $\|g\|_{\mathcal{H}} \le c\|\varphi\|_\infty \|f\|_{\mathcal{H}}$. The argument finishes if we show that $g = T_{\bar{\varphi}} f$.

Fix $z \in \mathbb{D}$, and let $k_k^{\mathcal{H}}$ denote the reproducing kernel of $\mathcal{H}$. Then, on the one hand, we have

$$\lim_{n\to\infty} (T_{\bar{p}_n} f)(z) = \lim_{n\to\infty} \langle T_{\bar{p}_n} f, k_z^{\mathcal{H}} \rangle_{\mathcal{H}} = \langle g, k_z^{\mathcal{H}} \rangle_{\mathcal{H}} = g(z).$$

On the other hand, since $\mathcal{H} \subset H^2$, we can write

$$\begin{aligned} (T_{\bar{p}_n} f)(z) &= \langle T_{\bar{p}_n} f, k_z \rangle_2 = \langle \bar{p}_n f, k_z \rangle_2 \\ &= \frac{1}{2\pi} \int_0^{2\pi} \overline{p_n(e^{i\theta})} f(e^{i\theta}) \overline{k_z(e^{i\theta})}\, d\theta. \end{aligned}$$

Since $f\overline{k_z} \in L^1(\mathbb{T})$ and $(p_n)_{n\ge1}$ converges to φ in the weak-star topology of $L^\infty(\mathbb{T})$, we have

$$\begin{aligned} \lim_{n\to\infty} (T_{\bar{p}_n} f)(z) &= \frac{1}{2\pi} \int_0^{2\pi} \overline{\varphi(e^{i\theta})} f(e^{i\theta}) \overline{k_z(e^{i\theta})}\, d\theta \\ &= \langle \bar{\varphi} f, k_z \rangle_2 = (T_{\bar{\varphi}} f)(z). \end{aligned}$$

Therefore, by the uniqueness of the limit, we have $g = T_{\bar{\varphi}} f \in \mathcal{H}$. □

Corollary 12.50 *Let $\mathcal{H}$ be an analytic reproducing kernel Hilbert space on the open unit disk $\mathbb{D}$ satisfying (H4), (H5) and (H6). Let $\mathcal{E}$ be a closed subspace of $\mathcal{H}$ that is invariant under $X_{\mathcal{H}}$. Then, for every $\varphi \in H^\infty$, the subspace $\mathcal{E}$ is also invariant under $T_{\bar{\varphi},\mathcal{H}}$.*

Proof Let $\varphi \in H^\infty$ and let $(p_n)_{n\geq 1}$ be the sequence of Fejér means of its Fourier sums. According to (12.18), for any $f \in \mathcal{H}$, the sequence $(T_{\bar{p}_n} f)$ is bounded in the norm of $\mathcal{H}$, and, as we saw in the proof of Theorem 12.49,

$$\lim_{n\to\infty} (T_{\bar{p}_n} f)(z) \to (T_{\bar{\varphi}} f)(z) \qquad (z \in \mathbb{D}).$$

Since $T_{\bar{p}_n} f = p^*(X_{\mathcal{H}})f \in \mathcal{E}$, using Lemma 9.1, we conclude that $T_{\bar{\varphi}} f \in \mathcal{E}$. □

Notes on Chapter 12

In an early paper [523], Toeplitz investigated finite matrices that are constant on diagonals and their relation to the corresponding one- and two-sided infinite matrices. Then Toeplitz operators were studied by Wintner [555], by Hartman and Wintner [260, 261] and by Grenander and Szegő [244]. The first systematic study of Toeplitz operators emphasizing the mapping $\varphi \longmapsto T_\varphi$ was made by Brown and Halmos in the paper [119], which profoundly influenced the development of the theory. As already mentioned, together with Hankel operators, Toeplitz operators constitute a very important class of operators on spaces of analytic functions. For further results on Toeplitz operators, we refer to the following books and survey articles: Douglas [176, 177], Böttcher and Silbermann [113], Nikolskii [386, 387], Peller [408], Peller and Khrushchëv [409] and Litvinchuk and Spitkovskii [339].

Section 12.1

Theorem 12.2 is due to Brown and Halmos [119]. The result proved in Exercise 12.1.2 appeared also in that paper.

Section 12.2

In [494], Stampfli observed that a proof of Coburn in [146] actually yields Corollary 12.9. The results proved in Exercises 12.2.1 and 12.2.2 appeared in Brown and Halmos [119]. A necessary and sufficient condition that the semicommutator $T_{\varphi\psi} - T_\varphi T_\psi$ be compact (for general φ, ψ in $L^\infty(\mathbb{T})$) has been obtained by Axler, Chang and Sarason [54] and Volberg [533].

Section 12.3

The spectral inclusion theorem (Theorem 12.15) is due to Hartman and Wintner [260]. They were also the first to determine the spectra of self-adjoint Toeplitz operators; see Corollary 12.17. The proof of Theorem 12.15 given here, which is based on Lemma 12.12, first occurs in a paper of Widom [548]. Theorem 12.16 was obtained by Brown and Halmos [119]. Lemma 12.11 was given independently by Widom [549] and Devinatz [170].

Two important further results should be mentioned here. Douglas proved in [177, theorem 7.45] that the essential spectrum of a Toeplitz operator is a connected subset of the complex plane. This result was suggested by Abrahamse. Moreover, the spectrum of a Toeplitz operator is also a connected subset of $\mathbb{C}$. This was proved by Widom [550, 552] and answered a question posed by Halmos [252].

Section 12.4

Theorem 12.20 appeared in Coburn [145]. Despite its simplicity, this result has numerous consequences. The results of Exercises 12.4.4 and 12.4.5 come from Cowen [162]. Theorem 12.27 is due to Brown and Halmos [119].

The problem of the point spectrum of a Toeplitz operator has attracted quite a lot of attention. The reason is the link with many other important questions in analysis. In particular, the completeness of reproducing kernels in model spaces is linked with the injectivity of certain Toeplitz operators; see Exercise 31.8.3. The kernels $\ker T_\varphi$ are also related to exposed points of the unit ball of H^1, as shown in Section 12.6. In Chapter 30 we will give a parameterization for $\ker T_\varphi$ in the case when $\ker T_\varphi \neq \{0\}$.

Section 12.5

Theorem 12.28 was obtained by Brown and Halmos [119].

Section 12.6

Theorem 12.30 is due to Bloomfield, Jewell and Hayashi [101]. The relation between the problem of rigid functions (or exposed points) in H^1 and the problem of the injectivity of Toeplitz operators was the source of motivation for many papers. It also permits some new examples of rigid functions to be given.

Section 12.7

Theorem 12.36 was established by Lotto [343].

Section 12.8

The problem of the basis of exponentials in $L^2(\mu)$ goes back to the work of Kolmogorov [318]. The fundamental theorem Theorem 12.39 is due to Helson and Szegő [283]. The implication that, if $(z^n)^{n\in\mathbb{N}}$ is a basis of $L^2(\mu)$, then μ is absolutely continuous is due to Kolmogorov. Another approach to weighted estimates of the Riesz projection, and hence to the exponential bases problem, is given by Hunt, Muckenhoupt and Wheeden in [294, 365]. They introduced the so-called Hunt–Muckenhoupt–Wheeden condition

$$\sup_{I\subset\mathbb{T}}\left(\frac{1}{|I|}\int_I w\,dm\right)\left(\frac{1}{|I|}\int_I \frac{1}{w}\,dm\right) < +\infty, \tag{A_2}$$

where I runs over all subarcs of $\mathbb{T}$ and $|I|$ is the Lebesgue measure of I. Then they proved that w satisfies condition (A_2) if and only if $(z^n)_{n\geq 0}$ is a basis of the weighted space $L^2(w\,dm)$. In particular, we see that the Hunt–Muckenhoupt–Wheeden condition (A_2) is equivalent to the Helson–Szegő condition (HS). However, it should be noted that no direct proof of this equivalence is known.

Section 12.9

Lemma 12.41 was given independently by Widom [549] and Devinatz [170]. The fundamental invertibility criterion, Theorem 12.42, was also found by Widom and Devinatz.

Several further developments and additional topics should be mentioned here. The invertibility problem for symbols in various algebras has been considered. The study of Toeplitz operators with symbols in $H^\infty + C(\mathbb{T})$ was carried out by Douglas [174] with complements given by Lee and Sarason [335] and Stampfli [494]. The study of the invertibility of Toeplitz operators for continuous symbols has been done by several authors, including Kreĭn [328], Calderón, Spitzer and Widom [121], Widom [548], Devinatz [170] and Stampfli [494]. The analysis of the smallest closed subalgebra of $\mathcal{L}(H^2)$ containing $\{T_\varphi : \varphi \in \mathcal{C}(\mathbb{T})\}$ was made by Coburn [146, 147], while its applicability to the invertibility problem for Toeplitz operators with continuous symbols was studied by Douglas [175]. The invertibility problem has also been considered for symbols in the algebra of piecewise continuous functions by Widom [548] and Devinatz [170], and from the algebra point of view by Gohberg and Krupnik [243]. The algebra of almost periodic functions was considered by Coburn and Douglas [148] and by Gohberg and Feldman [239, 240]. Lastly, Lee and Sarason [335] and Douglas and Sarason [179] have considered the invertibility problem for symbols of the form $\varphi\bar{\psi}$, where φ and ψ are inner functions.

Section 12.11

Theorem 12.48 is due to Nakazi [371].

Section 12.10

Theorem 12.46 is due to Douglas and Sarason [178]. The result proved in Exercise 12.10.2 is due to Douglas [174]; see also Peller [408, p. 97]. The result proved in Exercise 12.10.1 can be found in Douglas [177, corollary 7.14.].

13

Cauchy transform and Clark measures

There is no doubt that an integral representation is a precious tool in analysis. We have already seen several such representations, e.g. for functions in the Hardy space H^p. In the first part of this chapter, we consider the more general concept of the Cauchy transform C_μ of a complex Borel measure μ on the unit circle $\mathbb{T}$. It is an analytic function on an open set, which, in particular, contains the open unit disk $\mathbb{D}$ and the exterior disk $\mathbb{D}_e$. We obtain series representations of C_μ on $\mathbb{D}$ and $\mathbb{D}_e$, determine its inclusion in H^p, and study its boundary behavior at $\mathbb{T}$. Then we define the important transformation K_μ. In fact, in most standard textbooks, the Cauchy transform is denoted by K_μ or $K\mu$. However, we keep this notation for this new and related object, which is essential in our studies of $\mathcal{H}(b)$ spaces. Among other things, we show that K_φ maps $L^2(\varphi)$ into H^2, and its kernel is $(H^2(\varphi))^\perp$. Moreover, its adjoint is the natural inclusion J_φ of H^2 into $L^2(\varphi)$. We also obtain the adjoint of the forward shift operator S_φ, present a functional calculus for this operator and define $\psi(S_\varphi)$. We use K_φ to define Toeplitz operators with L^2 symbols. The second part is devoted to Clark measures. We define the Herglotz transformation and then use it to define the Clark measure μ_α. We determine the Cauchy transform of μ_α, and its series representations. At the end, we define the function ρ and discuss its properties. This function will be used in studying $\mathcal{H}(b)$ spaces, where b is an extreme point of H^∞.

13.1 The space $\mathfrak{K}(\mathbb{D})$

The *Cauchy transform* of a measure $\mu \in \mathcal{M}(\mathbb{T})$ is defined by the integral formula

$$C_\mu(z) = \int_{\mathbb{T}} \frac{d\mu(\zeta)}{1 - \bar{\zeta} z}. \tag{13.1}$$

It represents an analytic function that is defined on the open set

$$\Omega_\mu = \mathbb{C} \setminus \operatorname{supp} \mu.$$

It is even analytic at ∞ with $C_\mu(\infty) = 0$. If $d\mu(\zeta) = \varphi(\zeta)\,dm(\zeta)$, where $\varphi \in L^1(\mathbb{T})$ and m is the normalized Lebesgue measure on $\mathbb{T}$, then, instead of C_μ, we write C_φ.

The set Ω_μ certainly contains $\mathbb{D} \cup \mathbb{D}_e$, and, if $\operatorname{supp}\mu \neq \mathbb{T}$, then there is at least one open arc on $\mathbb{T}$ that acts as a channel between $\mathbb{D}$ and $\mathbb{D}_e$. In this case, Ω_μ becomes a domain. We first obtain the power series representations of C_μ in terms of the Fourier coefficients

$$\hat{\mu}(n) = \int_{\mathbb{T}} e^{-in\theta}\,d\mu(e^{i\theta}) \qquad (n \in \mathbb{Z})$$

on $\mathbb{D}$ and $\mathbb{D}_e$.

Theorem 13.1 *Let $\mu \in \mathcal{M}(\mathbb{T})$. Then*

$$C_\mu(z) = \sum_{n=0}^{\infty} \hat{\mu}(n) z^n \qquad (z \in \mathbb{D})$$

and

$$C_\mu(z) = -\sum_{n=1}^{\infty} \frac{\hat{\mu}(-n)}{z^n} \qquad (z \in \mathbb{D}_e).$$

Proof The two series representations

$$\frac{1}{1-\bar{\zeta}z} = \sum_{n=0}^{\infty} \bar{\zeta}^n z^n \qquad (\zeta \in \mathbb{T},\ z \in \mathbb{D})$$

and

$$\frac{1}{1-\bar{\zeta}z} = -\sum_{n=1}^{\infty} \frac{\zeta^n}{z^n} \qquad (\zeta \in \mathbb{T},\ z \in \mathbb{D}_e)$$

easily follow from the geometric series $1/(1-w) = \sum_{n=0}^{\infty} w^n$, $w \in \mathbb{D}$. Moreover, for a fixed z, both series are boundedly and uniformly convergent on $\mathbb{T}$. Hence, we can change the order of summation and integration and the result follows. □

Since

$$|\hat{\mu}(n)| \leq \|\mu\| \qquad (n \in \mathbb{Z}), \tag{13.2}$$

Theorem 13.1 implicitly says that the Cauchy transform of a Borel measure has uniformly bounded Taylor coefficients. Moreover, if $\varphi \in L^2(\mathbb{T})$, the first formula in Theorem 13.1 is written as

$$C_\varphi(z) = \sum_{n=0}^{\infty} \hat{\varphi}(n) z^n \qquad (z \in \mathbb{D}). \tag{13.3}$$

Since $\hat{\varphi} \in \ell^2(\mathbb{Z})$, we deduce that

$$\varphi \in L^2(\mathbb{T}) \implies C_\varphi \in H^2(\mathbb{D}) \tag{13.4}$$

and, moreover, that

$$\|C_\varphi\|_{H^2} \le \|\varphi\|_2. \tag{13.5}$$

As usual, we also use C_φ to denote the corresponding boundary value function on $\mathbb{T}$ that is an element of $H^2(\mathbb{T})$. With this *convention*, C_φ becomes a creature that also lives on $\mathbb{T}$. Keeping this fact in mind, in light of (13.3) and (13.4), we can say that

$$C_\varphi = P_+\varphi \qquad (\varphi \in L^2(\mathbb{T})), \tag{13.6}$$

where P_+ is the Riesz projection from $L^2(\mathbb{T})$ onto $H^2(\mathbb{T})$.

Corollary 13.2 *Let $\mu \in \mathcal{M}(\mathbb{T})$. Then*

$$|C_\mu(z)| \le \frac{\|\mu\|}{1-|z|} \qquad (z \in \mathbb{D}).$$

Proof By the first formula in Theorem 13.1 and (13.2), we have

$$\begin{aligned}
|C_\mu(z)| &\le \sum_{n=0}^{\infty} |\hat{\mu}(n)|\,|z|^n \\
&\le \|\mu\| \sum_{n=0}^{\infty} |z|^n \\
&\le \frac{\|\mu\|}{1-|z|}.
\end{aligned}$$

A more direct proof can also be obtained via the formula (13.1). □

According to (8.14), for each $f \in H^2$, we have

$$(S^*f)(z) = \frac{f(z)-f(0)}{z} \qquad (z \in \mathbb{D}).$$

Hence, finding a representation for the combination $(f(z)-f(0))/z$ should be rewarding.

Theorem 13.3 *Let $\mu \in \mathcal{M}(\mathbb{T})$. Then*

$$\frac{C_\mu(z)-C_\mu(0)}{z} = \int_{\mathbb{T}} \frac{d\mu(\zeta)}{\zeta - z} \qquad (z \in \Omega_\mu).$$

In particular, if $\varphi \in L^2(\mathbb{T})$, then

$$(S^*C_\varphi)(z) = \int_{\mathbb{T}} \frac{\varphi(\zeta)}{\zeta - z}\, dm(\zeta) \qquad (z \in \mathbb{D}).$$

Proof For each $z \in \Omega_\mu$, we have

$$\frac{1}{\zeta - z} = \frac{\bar{\zeta}}{1-\bar{\zeta}z} = \frac{1}{z}\left(\frac{1}{1-\bar{\zeta}z} - 1\right)$$

and thus

$$\int_{\mathbb{T}} \frac{d\mu(\zeta)}{\zeta - z} = \frac{1}{z}\bigg(\int_{\mathbb{T}} \frac{d\mu(\zeta)}{1-\bar{\zeta}z} - \int_{\mathbb{T}} d\mu(\zeta) \bigg).$$

The result follows from the definition (13.1). □

The collection of all functions on $\mathbb{D}$ that are the restriction of the Cauchy transform of a measure in $\mathcal{M}(\mathbb{T})$ to $\mathbb{D}$ is denoted by $\mathfrak{K}$. More specifically, we define

$$\mathfrak{K}(\mathbb{D}) = \{f : \exists\, \mu \in \mathcal{M}(\mathbb{T}) \text{ such that } f(z) = C_\mu(z),\ z \in \mathbb{D}\}.$$

Note that each function in $\mathfrak{K}(\mathbb{D})$ is represented by infinitely many measures. In fact, if $f \in \mathfrak{K}$ has the representation $f = C_\mu$, for some $\mu \in \mathcal{M}(\mathbb{T})$, and we pick any arbitrary $\varphi \in H_0^1$ and define ν by

$$d\nu = d\mu + \bar{\varphi}\, dm, \tag{13.7}$$

then

$$\hat{\nu}(n) = \hat{\mu}(n) \qquad (n \geq 0),$$

and thus, by the first formula in Theorem 13.1, we also have $f = C_\nu$. In fact, if $\nu, \mu \in \mathcal{M}(\mathbb{T})$ fulfill the identity $C_\nu = C_\mu$ on $\mathbb{D}$, then there is a $\varphi \in H_0^1$ such that (13.7) holds (see Exercise 13.1.2). In particular,

$$C_\mu \equiv 0 \text{ on } \mathbb{D} \quad \Longleftrightarrow \quad \exists\, \varphi \in H_0^1 \text{ such that } d\mu = \bar{\varphi}\, dm. \tag{13.8}$$

Corollary 13.2 indicates that the Cauchy transform does not have a very rapid growth as we approach the boundary points. This observation is further supported by the following result.

Theorem 13.4 *We have*

$$\mathfrak{K}(\mathbb{D}) \subset \bigcap_{0<p<1} H^p(\mathbb{D}). \tag{13.9}$$

Moreover, for each $\mu \in \mathcal{M}(\mathbb{T})$,

$$\|C_\mu\|_p \leq c_p \|\mu\|,$$

where c_p *is a constant satisfying*

$$c_p = O\bigg(\frac{1}{1-p}\bigg) \qquad (\textit{as } p \longrightarrow 1).$$

Proof Fix $\mu \in \mathcal{M}(\mathbb{T})$. Let

$$\mu = (\mu_1 - \mu_2) + i(\mu_3 - \mu_4)$$

be the Jordan decomposition of μ. Since

$$C_\mu = (C_{\mu_1} - C_{\mu_2}) + i(C_{\mu_3} - C_{\mu_4})$$

and

$$C_{\mu_k}(0) = \|\mu_k\| \le \|\mu\| \qquad (1 \le k \le 4),$$

it is enough to prove the result for a positive Borel measure. But, in this case, we have

$$\Re(C_{\mu_k}(z)) = \int_{\mathbb{T}} \frac{1 - \Re(\bar{\zeta} z)}{|1 - \bar{\zeta} z|^2}\, d\mu_k(\zeta) > 0 \qquad (z \in \mathbb{D}).$$

The result now follows from Corollary 4.25. □

Appealing to the boundary value properties of Hardy spaces, Theorem 13.4 says that each $f \in \mathfrak{K}(\mathbb{D})$ has a finite nontangential limit almost everywhere on $\mathbb{T}$. As in the Hardy spaces, this allows us to define $\mathfrak{K}(\mathbb{T})$. However, in the following, we simply write $\mathfrak{K}$ for both spaces. According to (4.13), the representation

$$f(z) = \int_0^{2\pi} \frac{f(\zeta)}{1 - \bar{\zeta} z}\, dm(\zeta) \qquad (z \in \mathbb{D})$$

holds for each $f \in H^1$. In our new language, this means that $f = C_{fdm}$. Note that, for this subclass, we have $C_{fdm} \equiv 0$ on $\mathbb{D}_e$. Hence, considering (13.9), we can say that

$$H^1 \subset \mathfrak{K} \subset \bigcap_{0<p<1} H^p. \tag{13.10}$$

Both inclusions are proper (see Exercise 13.1.6). With a little work, we can distinguish more elements of $\mathfrak{K}$. In light of Theorem 13.4, the following result can be considered as a slight improvement of Corollary 4.25.

Theorem 13.5 *Let f be analytic on $\mathbb{D}$ with*

$$\Re(f(z)) > 0 \qquad (z \in \mathbb{D}).$$

Then $f \in \mathfrak{K}$.

Proof Since $u(z) = \Re(f(z))$ is a positive harmonic function on $\mathbb{D}$, by Herglotz's result (Corollary 3.7), there is a positive Borel measure μ on $\mathbb{T}$ such that

$$u(z) = \int_{\mathbb{T}} \frac{1 - |z|^2}{|\zeta - z|^2}\, d\mu(\zeta) \qquad (z \in \mathbb{D}).$$

This equality is rewritten as

$$\Re(f(z)) = \Re\bigg(\int_{\mathbb{T}} \frac{\zeta + z}{\zeta - z}\, d\mu(\zeta)\bigg) \qquad (z \in \mathbb{D}).$$

Therefore, by the uniqueness theorem for analytic functions, we have

$$f(z) = i\,\Im(f(0)) + \int_{\mathbb{T}} \frac{\zeta + z}{\zeta - z}\, d\mu(\zeta) \qquad (z \in \mathbb{D}). \tag{13.11}$$

Since

$$\frac{\zeta + z}{\zeta - z} = \frac{2}{1 - \bar{\zeta} z} - 1$$

and

$$\int_{\mathbb{T}} \frac{1}{1 - \bar{\zeta} z} \, dm(\zeta) = 1 \qquad (z \in \mathbb{D}),$$

the representation (13.11) becomes

$$f(z) = \int_{\mathbb{T}} \frac{1}{1 - \bar{\zeta} z} \, d\mu(\zeta) \qquad (z \in \mathbb{D}),$$

where $\mu = 2\mu + (-\|\mu\| + i\,\Im(f(0)))m$. □

Exercises

Exercise 13.1.1 Let $\mu, \nu \in \mathcal{M}(\mathbb{T})$ be such that

$$C_\mu(z) = C_\nu(z) \qquad (z \in \mathbb{D} \cup \mathbb{D}_e).$$

Show that $\mu = \nu$.
Hint: Use Theorem 13.1 and the uniqueness of the Fourier coefficients of L^1 functions.

Exercise 13.1.2 Let $\mu, \nu \in \mathcal{M}(\mathbb{T})$ be such that

$$C_\mu(z) = C_\nu(z) \qquad (z \in \mathbb{D}).$$

Show that there is a $\varphi \in H_0^1$ such that

$$d\mu = d\nu + \bar{\varphi}\, dm.$$

Hint: Use Theorems 4.4 and 13.1.
Remark: Compare with Exercise 13.1.1.

Exercise 13.1.3 Let $\mu, \nu \in \mathcal{M}(\mathbb{T})$ be such that

$$C_\mu(z) = C_\nu(z) \qquad (z \in \mathbb{D}).$$

Show that $\mu_s = \nu_s$, i.e. μ and ν have the same singular parts.
Hint: Use Exercise 13.1.2.

Exercise 13.1.4 Let

$$f(z) = \frac{1}{1-z} \log\left(\frac{1}{1-z}\right) \qquad (z \in \mathbb{D}).$$

Show that

$$f \notin \mathfrak{K}.$$

Hint: By Exercise 4.2.3, f does not have uniformly bounded coefficients. Now, apply (13.2). Another proof is obtained via Corollary 13.2.

Exercise 13.1.5 Let

$$f(z) = \frac{1}{1-z} \qquad (z \in \mathbb{D}).$$

Show that $f \in \mathfrak{K}$, but $f \notin H^1$.
Hint: Use $f = C_{\delta_1}$ and

$$|f(e^{i\theta})| \sim \frac{1}{|\theta|} \qquad (\text{as } \theta \longrightarrow 0).$$

Exercise 13.1.6 Show that

$$H^1 \subsetneqq \mathfrak{K} \subsetneqq \bigcap_{0<p<1} H^p.$$

Hint: Use (13.9), (13.10) and Exercises 4.2.3, 13.1.4 and 13.1.5.

Exercise 13.1.7 Let $\varphi \in L^1(\mathbb{T})$. Show that

$$\|C_\varphi\|_p = o\left(\frac{1}{1-p}\right) \qquad (\text{as } p \longrightarrow 1^-).$$

Hint: Write $\varphi = \psi + f$, where f is a trigonometric polynomial and $\psi \in L^1(\mathbb{T})$ with $\|\psi\|_1 \leq \varepsilon$. Then apply Theorem 13.4.

Exercise 13.1.8 Let f be analytic on $\mathbb{D}$, and assume that

$$f(\zeta) = \lim_{r \to 1} f(r\zeta)$$

exists for almost all $\zeta \in \mathbb{T}$, $f \in L^1(\mathbb{T})$, and, moreover, that

$$f(z) = \int_{\mathbb{T}} \frac{f(\zeta)}{1-\bar{\zeta}z}\, dm(\zeta) \qquad (z \in \mathbb{D}).$$

Show that $f \in H^1(\mathbb{D})$.
Hint: Use Theorem 13.4 and Corollary 4.28.

13.2 Boundary behavior of C_μ

Since C_μ is defined at least on $\mathbb{D}\cup\mathbb{D}_e$, it makes sense to question the existence of the two limits

$$\lim_{r\to 1^-} C_\mu(r\zeta) \quad \text{and} \quad \lim_{R\to 1^+} C_\mu(R\zeta)$$

for each $\zeta\in\mathbb{T}$. According to Theorem 13.4, $\mathfrak{K}\subset\bigcap_{0<p<1} H^p(\mathbb{D})$, and thus

$$\lim_{r\to 1^-} C_\mu(r\zeta)$$

exists and is finite almost everywhere on $\mathbb{T}$. By the same token, we can consider the function $z\longmapsto C_\mu(1/z)$, $z\in\mathbb{D}$, and show that it belongs to $\bigcap_{0<p<1} H^p(\mathbb{D})$. Therefore, the radial limit

$$\lim_{R\to 1^+} C_\mu(R\zeta)$$

also exists and is finite almost everywhere on $\mathbb{T}$. The identities

$$C_\mu(r\zeta) - C_\mu(\zeta/r) = P\mu(r\zeta)$$

and

$$C_\mu(r\zeta) + C_\mu(\zeta/r) = \mu(\mathbb{T}) + iQ\mu(r\zeta)$$

are easy to establish and, in light of (3.24) and Theorem 3.2, they are used to obtain the following result.

Theorem 13.6 *Let $\mu\in\mathcal{M}(\mathbb{T})$. Then, for almost all $\zeta\in\mathbb{T}$, we have*

$$\begin{aligned}
\lim_{r\to 1^-}\Big(C_\mu(r\zeta) - C_\mu(\zeta/r)\Big) &= D\mu(\zeta),\\
\lim_{r\to 1^-}\Big(C_\mu(r\zeta) + C_\mu(\zeta/r)\Big) &= \mu(\mathbb{T}) + i\tilde{\mu}(\zeta)\\
&= P.V.\int_{\mathbb{T}} \frac{2}{1-\bar{\omega}\zeta}\,d\mu(\omega).
\end{aligned}$$

(Here, as before, P.V. indicates the principal value of the integral.)

As an immediate consequence, we deduce

$$\lim_{r\to 1^-} C_\mu(r\zeta) = \tfrac{1}{2}(\mu(\mathbb{T}) + D\mu(\zeta) + i\tilde{\mu}(\zeta))$$

and

$$\lim_{R\to 1^+} C_\mu(r\zeta) = \tfrac{1}{2}(\mu(\mathbb{T}) - D\mu(\zeta) + i\tilde{\mu}(\zeta))$$

for almost all $\zeta\in\mathbb{T}$.

We end this section with an elementary observation about the boundary behavior of C_μ at points where μ has a Dirac mass.

Theorem 13.7 *Let $\mu \in \mathcal{M}(\mathbb{T})$, and let $\zeta \in \mathbb{T}$. Then*

$$\lim_{r \to 1^-} (1-r) C_\mu(r\zeta) = \mu(\{\zeta\}).$$

Proof By (13.1),

$$(1-r) C_\mu(r\zeta) = \int_{\mathbb{T}} \frac{1-r}{1-r\zeta\bar{\omega}} \, d\mu(\omega).$$

The integrands

$$\varphi_r(\omega) = \frac{1-r}{1-r\zeta\bar{\omega}} \qquad (0 \le r < 1, \ \omega \in \mathbb{T})$$

are uniformly bounded by 1 and, moreover,

$$\lim_{r \to 1} \varphi_r(\omega) = \begin{cases} 1 & \text{if} \quad \omega = \zeta, \\ 0 & \text{if} \quad \omega \neq \zeta. \end{cases}$$

Hence, the result immediately follows from the dominated convergence theorem. □

Corollary 13.8 *Let $\mu \in \mathcal{M}(\mathbb{T})$, and let $\zeta \in \mathbb{T}$ be such that $\mu(\{\zeta\}) \neq 0$. Then*

$$\lim_{r \to 1^-} |C_\mu(r\zeta)| = +\infty.$$

Exercise

Exercise 13.2.1 Let $(\zeta_n)_{n \ge 1}$ be a dense subset of $\mathbb{T}$, and let $(c_n)_{n \ge 1} \in \ell^1$. Define

$$f(z) = \sum_{n=1}^{\infty} \frac{c_n}{1-\bar{\zeta}_n z} \qquad (z \in \mathbb{D}).$$

Show that $\mathbb{T}$ is a natural boundary for f, i.e. f does not have an analytic continuation along any open arc of $\mathbb{T}$.
Hint: Use Corollary 13.8.

13.3 The mapping K_μ

When we write $f \in H^2$, we might consider f as a measurable function that lives on $\mathbb{T}$, or as an analytic function on $\mathbb{D}$. The passage from $\mathbb{T}$ to $\mathbb{D}$ can be made via the Poisson or Cauchy integral formulas. We defined $H^2(\mu)$ as a subclass of $L^2(\mu)$. Hence, it is a family of functions that live on $\mathbb{T}$. By the same token, we may wonder if there is an analytic continuation of such functions into the open unit disk $\mathbb{D}$. In this section, we define the mapping K_μ and show that the right object to consider is $K_\mu f$.

For each $f \in L^1(\mu)$, where $\mu \in \mathcal{M}(\mathbb{T})$, $f\,d\mu$ is a well-defined complex Borel measure on $\mathbb{T}$. Keeping this fact in mind, we define the mapping

$$K_\mu f = C_{f\,d\mu}. \tag{13.12}$$

In other words, $K_\mu f$ is given by the more explicit formula

$$(K_\mu f)(z) = \int_{\mathbb{T}} \frac{f(\zeta)}{1-\bar{\zeta}z}\,d\mu(\zeta). \tag{13.13}$$

Clearly, $K_\mu f$ is an analytic function on Ω_μ. However, it might have analytic continuation on some open arcs of $\mathbb{T}$ that intersect $\operatorname{supp}\mu$.

Note that, with this terminology, $K_\mu 1$ is precisely the Cauchy transform of μ. If $d\mu = \varphi\,dm$, where $\varphi \in L^1(\mathbb{T})$ and m is the normalized Lebesgue measure on $\mathbb{T}$, then, instead of $K_{\varphi\,dm}$, we write K_φ. Thus, for $\varphi \in L^1(\mathbb{T})$, we have

$$(K_\varphi f)(z) = \int_{\mathbb{T}} \frac{f(\zeta)\varphi(\zeta)}{1-\bar{\zeta}z}\,dm(\zeta) \qquad (f \in L^1(\varphi)),$$

and K_φ is considered as a mapping from $L^1(\varphi)$ into $\mathrm{Hol}(\mathbb{C}\setminus\mathbb{T})$. As another convention, if $\mu = m$, then, instead of the original notation K_m, we simply write K. In other words, K is the mapping

$$(Kf)(z) = \int_{\mathbb{T}} \frac{f(\zeta)}{1-\bar{\zeta}z}\,dm(\zeta) \qquad (f \in L^1(\mathbb{T})).$$

If $f \in H^1$, then by (4.13), we have

$$(Kf)(z) = \int_{\mathbb{T}} \frac{f(\zeta)}{1-\bar{\zeta}z}\,dm(\zeta) = f(z) \qquad (z \in \mathbb{D}). \tag{13.14}$$

On the contrary, if $f \in \overline{H_0^1}$, then, by (13.8),

$$(Kf)(z) = 0 \qquad (z \in \mathbb{D}). \tag{13.15}$$

For the following result, we recall that the space $\mathrm{Hol}(\Omega)$, the collection of all analytic functions on Ω, is endowed with the topology of uniform convergence on compact subsets of Ω.

Lemma 13.9 *Let $\mu \in \mathcal{M}(\mathbb{T})$. Then the linear map $K_\mu : L^1(\mu) \longrightarrow \mathrm{Hol}(\Omega_\mu)$ is continuous.*

Proof Let $f \in L^1(\mu)$ and let E be a compact subset of Ω_μ. Then, for each $z \in E$, we have

$$|(K_\mu f)(z)| = \left| \int_{\mathbb{T}} \frac{f(\zeta)}{1-\bar{\zeta}z}\,d\mu(\zeta) \right| \le \frac{\|f\|_{L^1(\mu)}}{\mathrm{dist}(z, \operatorname{supp}\mu)}.$$

Since $E \subset \Omega_\mu$ is compact and $E \cap \operatorname{supp}\mu = \emptyset$, we have

$$\delta = \inf_{z\in E} \mathrm{dist}(z, \operatorname{supp}\mu) > 0,$$

and thus the above estimation gives

$$\sup_{z\in E} |(K_\mu f)(z)| \le \frac{\|f\|_{L^1(\mu)}}{\delta}. \tag{13.16}$$

Now, let $(f_n)_{n\ge1}$ be a sequence in $L^1(\mu)$ that converges to $f \in L^1(\mu)$, i.e.

$$\lim_{n\to\infty} \|f_n - f\|_{L^1(\mu)} = 0.$$

Then it follows from (13.16) that $(K_\mu f_n)_{n\ge1}$ converges uniformly to $K_\mu f$ on the compact set E. This precisely means that K_μ is a continuous mapping from $L^1(\mu)$ into the topological space $\mathrm{Hol}(\Omega_\mu)$. □

Since $L^2(\mu) \subset L^1(\mu)$, we may restrict the domain of definition of K_μ to $L^2(\mu)$. The range of K_μ when f runs over $L^2(\mu)$ is denoted by $K^2(\mu)$. Hence, for the time being, $K^2(\mu)$ is simply an aggregate of analytic functions that are defined on Ω_μ. In short, we have defined the surjective map

$$\begin{array}{rccc} K_\mu: & L^2(\mu) & \longrightarrow & K^2(\mu) \\ & f & \longmapsto & K_\mu f. \end{array}$$

For fixed $z \in \mathbb{D}$, the application $k_z(\zeta) = 1/(1-\bar{z}\,\zeta)$, which is the reproducing kernel for the Hardy space H^2, is bounded on $\mathbb{T}$ and thus it belongs to $L^2(\mu)$. Therefore, if $f \in L^2(\mu)$, then the relation (13.13), restricted to the open unit disk, can be rewritten as

$$(K_\mu f)(z) = \langle f, k_z\rangle_{L^2(\mu)} \qquad (z \in \mathbb{D}). \tag{13.17}$$

This identity along with (5.18) immediately imply the following result.

Theorem 13.10 *Let $\mu \in \mathcal{M}(\mathbb{T})$, and let $f \in L^2(\mu)$. Then*

$$K_\mu f|_{\mathbb{D}} \equiv 0 \quad\Longleftrightarrow\quad f \in H^2(\mu)^\perp = L^2(\mu) \ominus H^2(\mu).$$

In particular, the mapping

$$\begin{array}{ccc} H^2(\mu) & \longrightarrow & \mathrm{Hol}(\mathbb{D}) \\ f & \longmapsto & K_\mu f \end{array}$$

is injective.

The elements of $L^2(\mu)$ live on $\mathbb{T}$, and so do the members of $H^2(\mu)$. The preceding result indicates that the right analytic continuation of f to $\mathbb{D}$ is the restriction of $K_\mu f$ to $\mathbb{D}$. This belief is strengthened when we further observe that, in the classic case, according to (13.14), K maps an element of $H^2(\mathbb{T})$ into its corresponding analytic extension in $H^2(\mathbb{D})$. In this situation, the identity (13.17) will play the role of the evaluation functional at z for elements of $H^2(\mu)$.

The mapping $K : L^1(\mathbb{T}) \longrightarrow H(\mathbb{D} \cup \mathbb{D}_e)$ was defined by the formula

$$Kf(z) = \int_{\mathbb{T}} \frac{f(\zeta)}{1 - \bar{\zeta} z}\, dm(\zeta).$$

By Theorem 13.4, we know that

$$Kf \in \bigcap_{0<p<1} H^p(\mathbb{D}).$$

In the following, we put in place a more restrictive condition than $f \in L^1(\mathbb{T})$, e.g. $f \in L^p(\mathbb{T})$ with $p > 1$, and then study its consequence on Kf. Since

$$\frac{2}{1 - \bar{\zeta} z} = 1 + \Re\left(\frac{\zeta + z}{\zeta - z}\right) + i\,\Im\left(\frac{\zeta + z}{\zeta - z}\right),$$

we immediately deduce that

$$2Kf(z) = \hat{f}(0) + Pf(z) + iQf(z), \tag{13.18}$$

at least for $z \in \mathbb{D} \cup \mathbb{D}_e$. Hence, the study of Kf is intrinsically related to the three quantities $\hat{f}(0)$, Pf and Qf. The first two usually behave well. However, the conjugate Poisson kernel sometimes shows a different behavior and this makes the difference. By the same token, according to (3.24) and Theorem 3.2, or by using Theorem 13.6, the identity (13.18) implies that

$$2Kf(\zeta) = \hat{f}(0) + f(\zeta) + i\tilde{f}(\zeta) \tag{13.19}$$

for almost all $\zeta \in \mathbb{T}$. Therefore, it is no wonder that the behavior of Kf on $\mathbb{T}$ is closely related to the Hilbert transform of f. The following results are a direct consequence of the corresponding results in Sections 3.4 and 3.6.

Theorem 13.11 *Let $f \in L^p(\mathbb{T})$, $1 < p < \infty$. Then $Kf \in H^p$ and there is a constant c_p such that*

$$\|Kf\|_p \le c_p \|f\|_p.$$

Proof Clearly we have

$$|\hat{f}(0)| \le \|f\|_1 \le \|f\|_p.$$

By Theorem 3.8, $\|(Pf)_r\|_p \le \|f\|_p$. Moreover, by M. Riesz's theorem (Theorem 3.10), we have

$$\|(Qf)_r\|_p \le c_p \|f\|_p,$$

where c_p is a constant. Hence, by (13.18), $Kf \in H^p$ and its norm is controlled by $\|f\|_p$. □

Since $Kf = f$, $f \in H^p$ (see (13.14)), Theorem 13.11 shows that the mapping

$$\begin{array}{rcl} L^p(\mathbb{T}) & \longrightarrow & H^p(\mathbb{T}) \\ f & \longmapsto & Kf \end{array}$$

is a bounded surjective operator whenever $1 < p < \infty$. However, Theorem 13.11 is not valid if $p = 1$ or $p = \infty$. These two problematic cases are treated below.

Theorem 13.12 *Suppose that f satisfies*

$$\int_{\mathbb{T}} |f| \log^+ |f| \, dm < \infty.$$

Then $Kf \in H^1$.

Proof Clearly we have

$$|\hat{f}(0)| \leq \|f\|_1$$

and, by Theorem 3.6, $\|(Pf)_r\|_1 \leq \|f\|_1$. Moreover, by Zygmund's theorem (Theorem 3.12), we have

$$\sup_{0 \leq r < 1} \|(Qf)_r\|_1 < \infty.$$

Hence, by (13.18), $Kf \in H^1$. □

If we do not have the strong $L \log L$ condition of Theorem 13.12, and we only know that $f \in L^1(\mathbb{T})$, then Kolmogorov's theorem (Theorem 3.13) applies and we get the following weaker result. Note that, since $Kf \in \bigcap_{0<p<1} H^p(\mathbb{D})$, we can look at Kf as a measurable function on $\mathbb{T}$.

Theorem 13.13 *Let $f \in L^1(\mathbb{T})$. Then*

$$m(|Kf| > t) \leq C \frac{\|f\|_1}{t} \qquad (t > 0),$$

where C is a universal constant.

To treat the case $f \in L^\infty$, we define $BMOA$, the space of *analytic functions of bounded mean oscillation*, by

$$BMOA = BMO \cap H^1.$$

Theorem 13.14 *Let $f \in L^\infty(\mathbb{T})$. Then $Kf \in BMOA$. Moreover, there is a universal constant c such that*

$$\|Kf\|_{BMOA} \leq c \|f\|_\infty.$$

Proof Clearly we have

$$|\hat{f}(0)| \leq \|f\|_1 \leq \|f\|_\infty.$$

Moreover, by Stein's theorem (Theorem 3.17), we have

$$\|\tilde{f}\|_{BMO} \leq c\|f\|_\infty.$$

Hence, by (13.19), $Kf \in BMOA$ and its norm is controlled by $\|f\|_\infty$. □

Exercises

Exercise 13.3.1 Let $\varphi \in L^1(\mathbb{T})$. Show that

$$(K_\varphi \chi_n)(z) = \sum_{k=0}^{\infty} \hat{\varphi}(k-n) z^k \qquad (z \in \mathbb{D}).$$

Hint: Use Theorem 13.1 and (13.12).

Exercise 13.3.2 Let $\varphi \in H^2$. Show that

$$K_\varphi \chi_{-1} = S^* \varphi.$$

What can you say about $K_\varphi \chi_1$?
Hint: Use Exercise 13.3.1.

Exercise 13.3.3 Let $f \in H^1$. Show that

$$\tilde{f} = -if + i\hat{f}(0).$$

Hint: Use (13.14) and (13.19).

Exercise 13.3.4 Let $(a_n)_{n\geq 0}$ be a real sequence satisfying the following properties:

(i) $a_n \longrightarrow 0$, as $n \longrightarrow \infty$;
(ii) $(a_n)_{n\geq 0}$ is convex, which means that $\Delta^2 a_n \geq 0$, where $\Delta a_n = a_{n+1} - a_n$.

Our goal is to prove that the series

$$\frac{a_0}{2} + \sum_{n=1}^{\infty} a_n \cos(nt) \tag{13.20}$$

converges for all $t \in [-\pi, 0) \cup (0, \pi]$ to a nonnegative integrable function $\varphi(t)$ whose Fourier series is given by (13.20).

To do so, take the following steps.

(a) Let

$$S_N(t) = \frac{a_0}{2} + \sum_{n=1}^{N} a_n \cos(nt).$$

We recall that the Dirichlet kernel and the Fejér kernel are respectively given by

$$D_N(t) = \frac{1}{2} + \sum_{n=1}^{N} \cos(nt) = \frac{\sin((N+\frac{1}{2})t)}{2\sin(\frac{1}{2}t)}$$

and

$$K_N(t) = \frac{1}{N+1}\sum_{n=0}^{N} D_n(t) = \frac{\sin^2(\frac{1}{2}(N+1)t)}{2(N+1)\sin^2(\frac{1}{2}t)}.$$

Using Abel's summation method twice successively, show that

$$\begin{aligned} S_N(t) = &\sum_{n=0}^{N} (\Delta^2 a_n)(n+1)K_n(t) \\ &- (\Delta a_{N+1})(N+1)K_N(t) + a_{N+1}D_N(t). \end{aligned}$$

(b) Show that the series $\sum_n (\Delta^2 a_n)(n+1)K_n(t)$ converges to a nonnegative function $\varphi(t)$, which is continuous for $t \in [-\pi, \pi]$, $t \neq 0$.
Hint: Observe that $(n+1)K_n(t)$ is uniformly bounded in each set $0 < \delta \leq |t| \leq \pi$.

(c) Show that

$$S_N(t) \longrightarrow \varphi(t) = \sum_{n=0}^{\infty} (\Delta^2 a_n)(n+1)K_n(t).$$

(d) Show that, for any integer $m \geq 0$, we have

$$\frac{1}{\pi}\int_{-\pi}^{\pi} \varphi(t)\cos(mt)\,dt = \sum_{n=m}^{\infty} (\Delta^2 a_n)(n-m+1).$$

(e) Conclude that

$$\frac{1}{\pi}\int_{-\pi}^{\pi} \varphi(t)\cos(mt)\,dt = a_m.$$

Hint: Use Exercise 1.1.12 to deduce that $k\Delta a_k \longrightarrow 0$ as $k \longrightarrow \infty$.

Exercise 13.3.5 This exercise shows that the $L \log L$ condition in Theorem 13.12 cannot be replaced by the weaker assumption $f \in L^1(\mathbb{T})$.

(i) Show that there is a function $f \in L^1(\mathbb{T})$ whose Fourier series is given by

$$\sum_{n=2}^{\infty} \frac{\cos(nt)}{\log n}, \qquad t \in [-\pi, \pi],\ t \neq 0.$$

Hint: Use Exercise 13.3.4.

(ii) Show that the Cauchy transform of f is given by

$$(Kf)(z) = \frac{1}{2}\sum_{n=2}^{\infty} \frac{z^n}{\log n}.$$

(iii) Show that $Kf \notin H^1$.
Hint: Argue by absurdity and use Exercise 11.3.4.

13.4 The operator $K_\varphi : L^2(\varphi) \longrightarrow H^2$

If $\varphi \in L^\infty(\mathbb{T})$ and $f \in L^2(\varphi)$, then

$$\|\varphi f\|_2 \leq \|\varphi\|_\infty^{1/2} \|f\|_{L^2(\varphi)}.$$

This means that $\varphi f \in L^2(\mathbb{T})$, and thus, by (13.4) or (13.6),

$$K_\varphi f = C_{\varphi f} = P_+(\varphi f) \in H^2. \tag{13.21}$$

Moreover, by (13.5),

$$\|K_\varphi f\|_2 = \|C_{\varphi f}\|_2 \leq \|\varphi f\|_2 \leq \|\varphi\|_\infty^{1/2} \|f\|_{L^2(\varphi)}.$$

Therefore, the mapping

$$\begin{array}{rccc} K_\varphi : & L^2(\varphi) & \longrightarrow & H^2 \\ & f & \longmapsto & C_{\varphi f} \end{array}$$

is a well-defined operator whose norm is at most $\|\varphi\|_\infty^{1/2}$. By Theorem 13.10, we have

$$\ker K_\varphi = H^2(\varphi)^\perp. \tag{13.22}$$

Hence,

$$K_\varphi P_{H^2(\varphi)} = K_\varphi, \tag{13.23}$$

where $P_{H^2(\varphi)} \in \mathcal{L}(L^2(\varphi))$ is the orthogonal projection on the closed subspace $H^2(\varphi)$ (see Figure 13.1).

The relation (13.22) also motivates us to define the injective operator

$$\begin{array}{rccc} \mathbf{K}_\varphi : & H^2(\varphi) & \longrightarrow & H^2 \\ & f & \longmapsto & K_\varphi f = C_{\varphi f}. \end{array}$$

In other words, using the operator i_φ that was defined at the end of Section 4.3, the operator $\mathbf{K}_\varphi$ is given by

$$\mathbf{K}_\varphi = K_\varphi i_\varphi \tag{13.24}$$

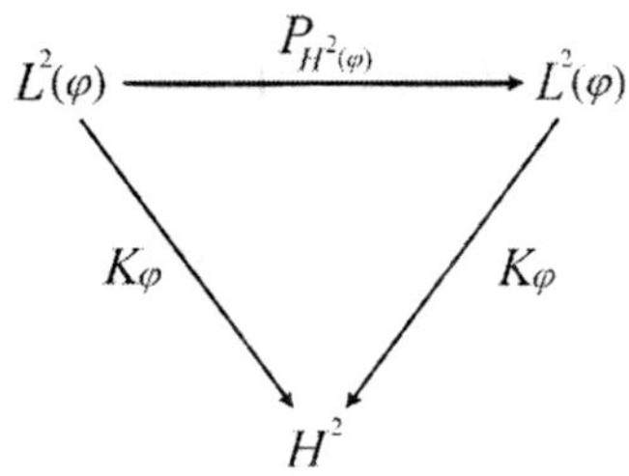

Figure 13.1 $H^2(\varphi)^\perp$ is the kernel of K_φ.

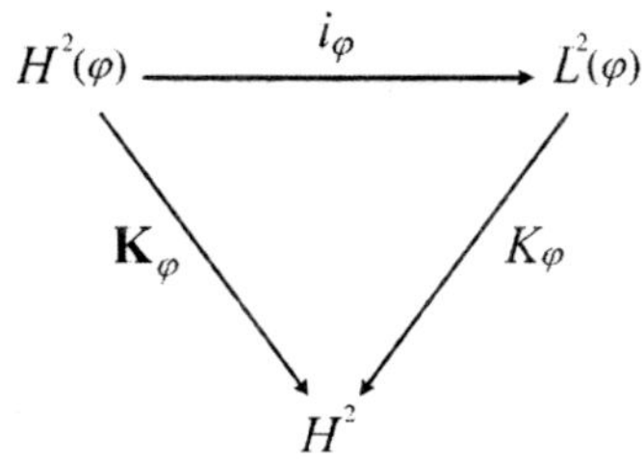

Figure 13.2 The relation between K_φ and $\mathbf{K}_\varphi$.

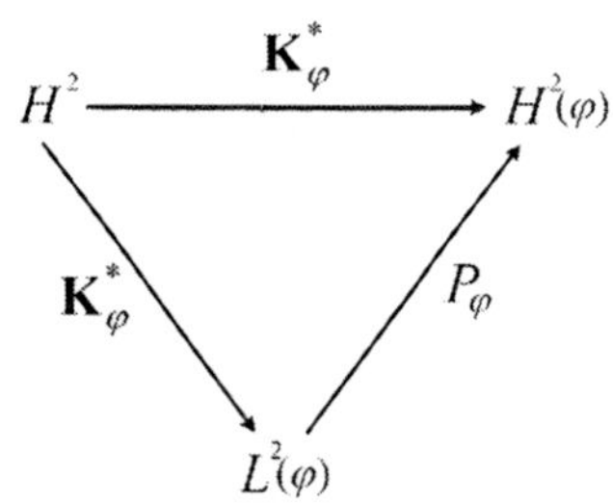

Figure 13.3 A decomposition of $\mathbf{K}_\varphi^*$.

(see Figure 13.2). Hence, by Theorem 13.15, (1.37) and (5.22), we have

$$\mathbf{K}_\varphi^* = P_\varphi \circ K_\varphi^* : H^2 \longrightarrow H^2(\varphi) \tag{13.25}$$

(see Figure 13.3). To have an explicit formula for K_φ^* we need to introduce another inclusion operator.

Since φ is bounded on $\mathbb{T}$, the inclusions

$$H^2(\mathbb{T}) \subset L^2(\mathbb{T}) \subset L^2(\varphi)$$

are trivial. Moreover, we have

$$\|f\|_{L^2(\varphi)} \le \|\varphi\|_\infty^{1/2} \|f\|_2 \qquad (f \in H^2).$$

Thus, the injection mapping from H^2 into $L^2(\varphi)$ is well defined and bounded. For further reference, we denote this operator by J_φ, i.e.

$$\begin{array}{rccc} J_\varphi : & H^2(\mathbb{T}) & \longrightarrow & L^2(\varphi) \\ & f & \longmapsto & f. \end{array} \tag{13.26}$$

We also need the function

$$\operatorname{sgn}(z) = \begin{cases} z/|z| & \text{if} \quad z \neq 0, \\ 1 & \text{if} \quad z = 0. \end{cases}$$

Thus, for each measurable function φ, $\operatorname{sgn}(\varphi)$ is a unimodular measurable function such that $\varphi = |\varphi| \times \operatorname{sgn}(\varphi)$. Therefore, the multiplication operator $M_{\operatorname{sgn}(\varphi)}$ is well defined on $L^2(\varphi)$.

Theorem 13.15 *Let* $\varphi \in L^\infty(\mathbb{T})$. *Then*

$$K_\varphi^* = M_{\operatorname{sgn}(\bar{\varphi})} J_\varphi.$$

In particular, if φ *is a nonnegative function, then* $K_\varphi^* = J_\varphi$.

Proof For simplicity, write $\Phi = \operatorname{sgn}(\bar{\varphi})$. Then, by (13.21), we have

$$\langle K_\varphi f, g \rangle_{H^2} = \langle P_+(\varphi f), g \rangle_{H^2}$$

for each $f \in L^2(\varphi)$ and $g \in H^2$. Hence, by Lemma 4.8,

$$\langle K_\varphi f, g \rangle_{H^2} = \langle \varphi f, g \rangle_{L^2} = \langle f, \bar{\varphi} g \rangle_{L^2} = \langle f, |\varphi| \Phi g \rangle_{L^2}.$$

But, according to the definition of inner product in $L^2(\varphi)$,

$$\langle f, |\varphi| \Phi g \rangle_{L^2} = \langle f, \Phi g \rangle_{L^2(\varphi)}.$$

Hence, we can write

$$\langle K_\varphi f, g \rangle_{H^2} = \langle f, M_\Phi J_\varphi g \rangle_{L^2(\varphi)}.$$

This identity shows that $K_\varphi^* = M_\Phi J_\varphi$. If $\varphi \geq 0$, then M_Φ is the identity operator on $L^2(\varphi)$, and thus we obtain $K_\varphi^* = J_\varphi$. □

According to Theorem 13.15, (13.25) is rewritten as

$$\mathbf{K}_\varphi^* = J_\varphi \tag{13.27}$$

whenever $\varphi \geq 0$. Based on Theorem 13.15, Figure 13.4 provides a decomposition of K_φ^* as the product of an inclusion operator and a multiplication operator.

The following result reveals the relation between the operator K_φ and the Toeplitz operator T_φ.

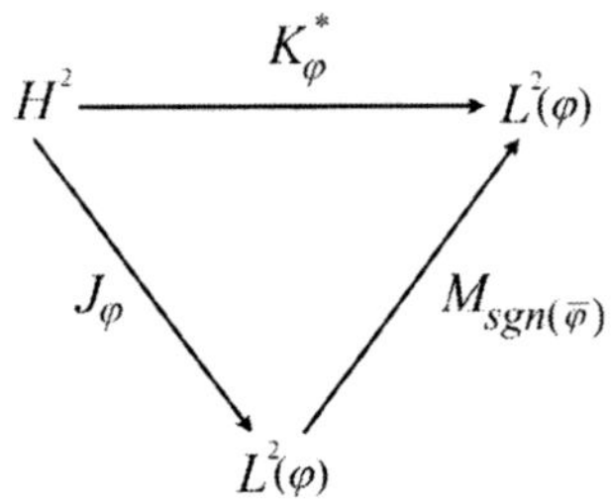

Figure 13.4 A decomposition of K_φ^*.

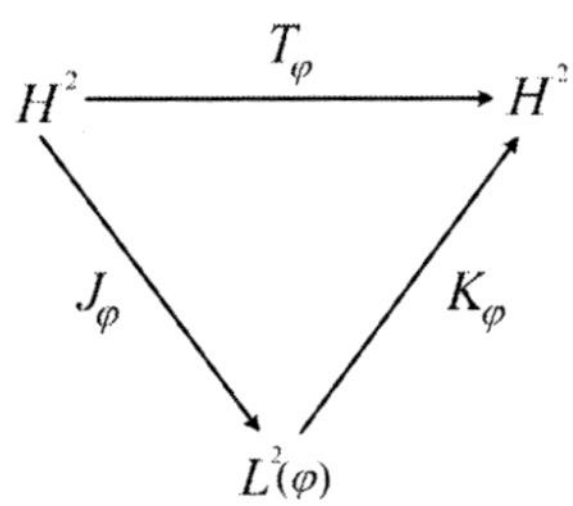

Figure 13.5 A decomposition of T_φ.

Corollary 13.16 *Let $\varphi \in L^\infty(\mathbb{T})$. Then*

$$K_\varphi J_\varphi = T_\varphi.$$

In particular, if φ is a nonnegative function, then

$$K_\varphi K_\varphi^* = J_\varphi^* J_\varphi = T_\varphi.$$

Proof Let $f \in H^2$. Then, by (13.21),

$$K_\varphi f = C_{\varphi f} = P_+(\varphi f) = T_\varphi f.$$

The rest follows from Theorem 13.15. □

Based on Corollary 13.16, Figure 13.5 provides a decomposition of T_φ.

Exercises

Exercise 13.4.1 Let $\varphi \in L^2(\mathbb{T})$. Show that

$$K_\varphi \chi_n = \sum_{k=0}^{\infty} \hat{\varphi}(k-n)\chi_k,$$

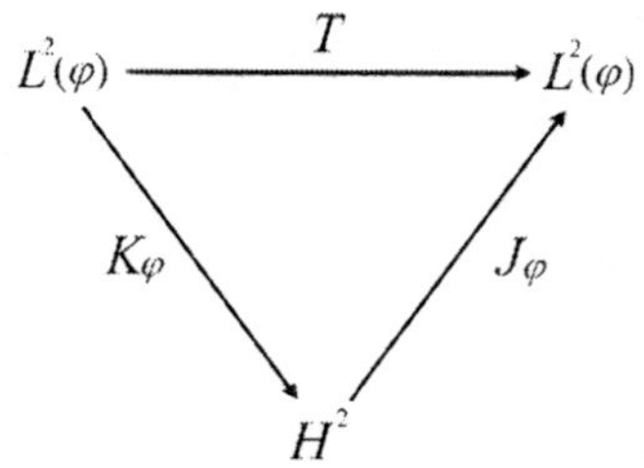

Figure 13.6 A decomposition of T.

where we interpret the identity in $H^2(\mathbb{T})$.
Remark: The above formula is a special case of the formula given in Exercise 13.3.1.

Exercise 13.4.2 Show that $J_\varphi^* = K_{|\varphi|}$.
Hint: Use $J_\varphi = J_{|\varphi|}$.

Exercise 13.4.3 Let $\varphi \in L^\infty(\mathbb{T})$, and let

$$T = J_\varphi K_\varphi.$$

Show that

$$TJ_\varphi = J_\varphi T_\varphi$$

and

$$K_\varphi T = T_\varphi K_\varphi$$

(see Figure 13.6).
Hint: Use Corollary 13.16.

13.5 Functional calculus for S_φ

The operator $S_\mu = S_{H^2(\mu)}$ was defined in Section 8.2 by the formula

$$\begin{array}{rccc} S_\mu : & H^2(\mu) & \longrightarrow & H^2(\mu) \\ & f & \longmapsto & \chi_1 f. \end{array}$$

We now have enough tools to study further properties of S_μ. First, we establish the following connection between K_μ and S_μ. The following result is a generalization of (8.14).

Lemma 13.17 *Let μ be a complex Borel measure on $\mathbb{T}$, and let $f \in H^2(\mu)$. Then*

$$(K_\mu S_\mu^* f)(z) = \frac{(K_\mu f)(z) - (K_\mu f)(0)}{z} \qquad (z \in \mathbb{D},\ z \neq 0).$$

Proof Fix $z \in \mathbb{D} \setminus \{0\}$. Then, according to (13.17), we have

$$(K_\mu S_\mu^* f)(z) = \langle S_\mu^* f, k_z \rangle_{H^2(\mu)}.$$

But, by definition,

$$\langle S_\mu^* f, k_z \rangle_{H^2(\mu)} = \langle f, S_\mu k_z \rangle_{H^2(\mu)}.$$

By a direct verification, we see that

$$S_\mu k_z = \frac{k_z - k_0}{\bar{z}}.$$

Now, use (13.17) again. □

If μ is the normalized Lebesgue measure m on $\mathbb{T}$, then, by (4.13), K_m is the identity operator on H^2. Moreover, S_μ is precisely the forward shift operator S on the classic Hardy space H^2. In this case, Lemma 13.17 reduces to the well-known formula (8.14).

The special case $\varphi \in L^\infty(\mathbb{T})$ leads to more interesting conclusions. In this situation, we have seen that K_φ is a bounded operator from $L^2(\varphi)$ into H^2. Hence, by (13.24), Lemma 13.17 is written as

$$\mathbf{K}_\varphi S_\varphi^* = S^* \mathbf{K}_\varphi \tag{13.28}$$

(see Figures 13.8–13.10).

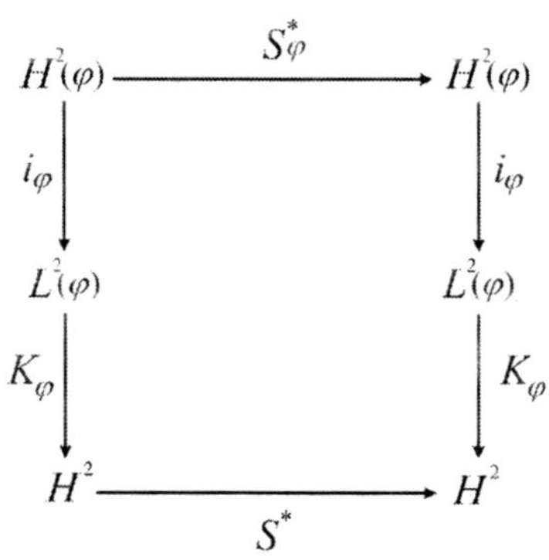

Figure 13.7 $\mathbf{K}_\varphi$ intertwines with S^* and S_μ^*.

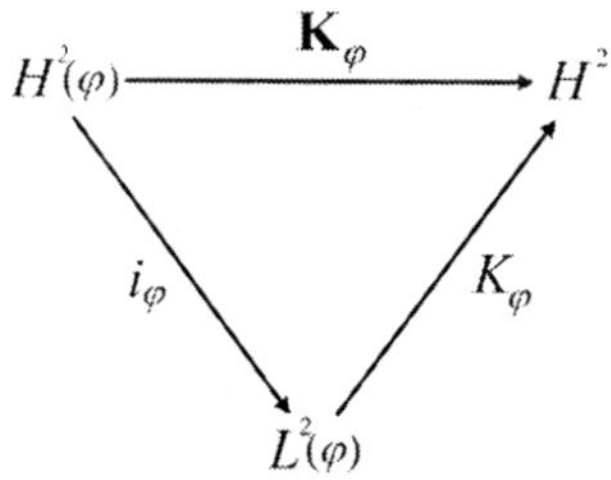

Figure 13.8 The definition of $\mathbf{K}_\varphi$.

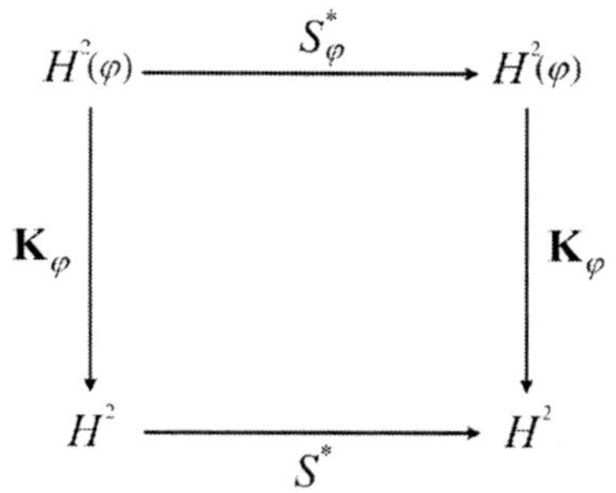

Figure 13.9 The relation between S^* and S_φ^*.

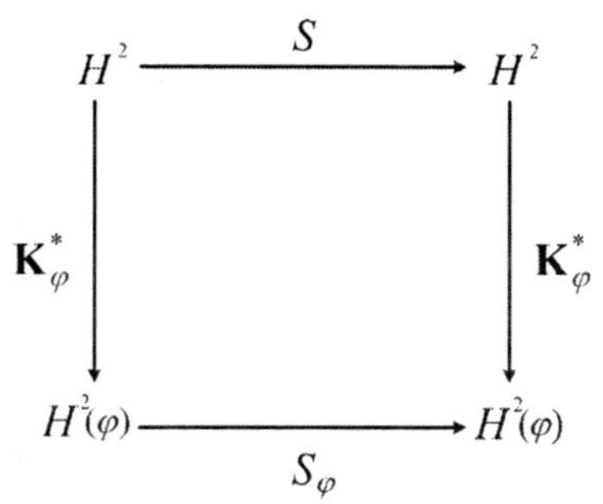

Figure 13.10 The relation between S and S_φ.

Corollary 13.18 *Let $\varphi \in L^\infty(\mathbb{T})$, $\varphi \geq 0$, be such that*

$$\int_0^{2\pi} \log \varphi(e^{it})\, dt = -\infty.$$

Then

$$K_\varphi Z_\varphi^* = S^* K_\varphi.$$

Proof By Corollary 8.23, we have $L^2(\varphi) = H^2(\varphi)$. Hence, the inclusion i_φ is the identity operator on $L^2(\varphi)$ and also we have $S_\varphi = Z_\varphi$. The result now follows from (13.28). □

Since, by (8.13),

$$S^* = P_+ Z^* i = P_+ M_{\bar{z}} i,$$

we also expect to have a similar formula for S_φ^*. To obtain such a formula, note that

$$\begin{aligned}
\langle S_\varphi^{*n} f, g\rangle_{H^2(\varphi)} &= \langle f, S_\varphi^n g\rangle_{H^2(\varphi)} \\
&= \langle f, z^n g\rangle_{H^2(\varphi)} \\
&= \langle i_\varphi f, Z_\varphi^n i_\varphi g\rangle_{L^2(\varphi)} \\
&= \langle i_\varphi^* Z_\varphi^{*n} i_\varphi f, g\rangle_{L^2(\varphi)} \\
&= \langle P_\varphi Z_\varphi^{*n} i_\varphi f, g\rangle_{L^2(\varphi)}.
\end{aligned}$$

Hence, we have

$$S_\varphi^{*n} = P_\varphi Z_\varphi^{*n} i_\varphi = P_\varphi M_{\bar{z}^n} i_\varphi \qquad (n \geq 0). \tag{13.29}$$

These observations are generalized in the following corollary.

Corollary 13.19 *Let $\varphi \in L^\infty(\mathbb{T})$, and let q be any analytic polynomial. Then the following assertions hold:*

(i) $\mathbf{K}_\varphi q(S_\varphi^*) = q(S^*)\mathbf{K}_\varphi$;
(ii) $q(S_\varphi)^* = P_\varphi q(Z_\varphi)^* i_\varphi = P_\varphi M_{\bar{q}}\, i_\varphi$.

Proof (i) By (13.28) and by induction, we have

$$\mathbf{K}_\varphi S_\varphi^{*n} = S^{*n}\mathbf{K}_\varphi \qquad (n \geq 0).$$

Then, by taking a finite linear combination, we obtain the result.

The statement (ii) follows similarly by using (13.29). □

Note that, in the same manner, we can show $q(S_\varphi^*) = P_\varphi q(Z_\varphi^*) i_\varphi$. However, if the coefficients of q are not real numbers, this operator is not the same as $P_\varphi M_{\bar{q}}\, i_\varphi$.

If q is an analytic polynomial and $f \in H^2(\mu)$, then, as a direct consequence of the definition of S_φ, we have

$$q(S_\varphi) f = q f.$$

Hence, given $\psi \in H^\infty$, it is natural to define the operator

$$\begin{array}{rccc}
\psi(S_\varphi): & H^2(\varphi) & \longrightarrow & H^2(\varphi) \\
& f & \longmapsto & \psi f.
\end{array}$$

But, as the first step, we need to justify that $\psi(S_\varphi)$ is well defined and bounded.

Lemma 13.20 *Let $\varphi \in L^\infty(\mathbb{T})$ and $\psi \in H^\infty$. Then $\psi(S_\varphi)$ is a bounded operator on $H^2(\varphi)$.*

Proof The multiplication operator

$$\begin{array}{rccc} M_\psi : & L^2(\varphi) & \longrightarrow & L^2(\varphi) \\ & f & \longmapsto & \psi f \end{array}$$

is well defined and bounded, and so is

$$\begin{array}{rccc} M_\psi i_\varphi : & H^2(\varphi) & \longrightarrow & L^2(\varphi) \\ & f & \longmapsto & \psi f. \end{array}$$

The statement will be proved if we can show that the range of this operator is $H^2(\varphi)$. Equivalently, we need to prove that $M_\psi H^2(\varphi) \subset H^2(\varphi)$. We first show that M_ψ maps analytic polynomials into $H^2(\varphi)$, and then exploit this special case to establish the general case.

Hence, fix an analytic polynomial q. Define ψ_n by

$$\psi_n(\zeta) = \sum_{k=0}^{n} \hat{\psi}(k)\zeta^k.$$

Then $(\psi_n)_{n\geq 1}$ converges to ψ in L^2 norm. Therefore, from

$$\|\psi_n q - \psi q\|_{L^2(\varphi)} \leq \|q\|_\infty \|\varphi\|_\infty^{1/2} \|\psi_n - \psi\|_2,$$

we deduce that $(\psi_n q)_{n\geq 1}$ converges to ψq in $L^2(\varphi)$. Since $\psi_n q$ is a polynomial, we conclude that $M_\psi q = \varphi q \in H^2(\varphi)$.

Now, let $f \in H^2(\varphi)$. Then, by definition, there exists a sequence of analytic polynomials $(q_n)_{n\geq 1}$ such that

$$\|q_n - f\|_{L^2(\varphi)} \longrightarrow 0.$$

Since M_ψ is bounded on $L^2(\varphi)$, we also have

$$\|\psi q_n - \psi f\|_{L^2(\varphi)} \longrightarrow 0.$$

But, we have just seen that $\psi q_n \in H^2(\varphi)$, $n \geq 1$, and thus $\psi f \in H^2(\varphi)$. □

The proof of Lemma 13.20 implicitly shows that

$$\psi(S_\varphi) = P_\varphi M_\psi i_\varphi. \tag{13.30}$$

Now, we explore the relations between $\psi(S_\varphi)$, on the one hand, and T_ψ, K_φ and I_φ, on the other.

Theorem 13.21 *Let $\varphi \in L^\infty(\mathbb{T})$ and $\psi \in H^\infty$. Then the following hold:*

(i) $\psi(S_\varphi)^* = P_\varphi M_{\bar{\psi}} i_\varphi$;
(ii) $K_\varphi M_{\bar{\psi}} = T_{\bar{\psi}} K_\varphi$;
(iii) $\mathbf{K}_\varphi \psi(S_\varphi)^* = T_{\bar{\psi}} \mathbf{K}_\varphi$.

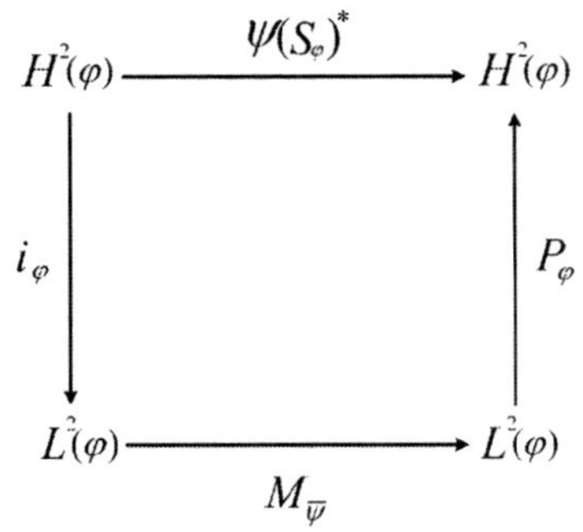

Figure 13.11 The relation between $\psi(S_\varphi)^*$ and $M_{\bar{\psi}}$.

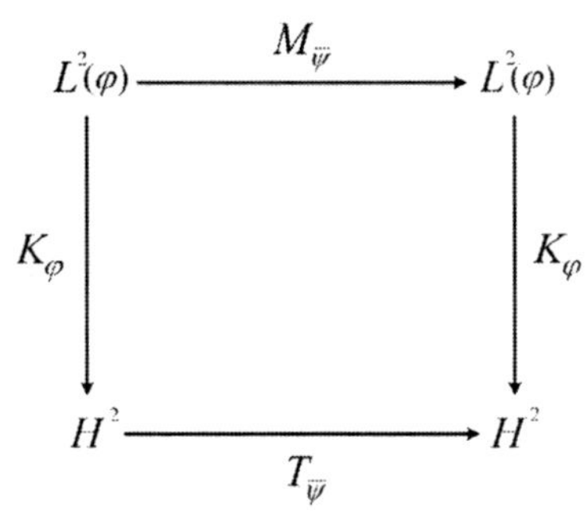

Figure 13.12 K_φ intertwines $T_{\bar{\psi}}$ and $M_{\bar{\psi}}$.

Proof (i) This follows from (13.30) and the fact that $i_\varphi^* = P_\varphi$ (see Figure 13.11).

(ii) Let $f \in L^2(\varphi)$. Then, according to Corollary 4.11 and (13.21), we have

$$T_{\bar{\psi}} K_\varphi f = P_+(\bar{\psi} P_+(\varphi f)) = P_+(\bar{\psi}\varphi f) = P_+(\varphi M_{\bar{\psi}} f) = K_\varphi M_{\bar{\psi}} f$$

(see Figure 13.12).

(iii) According to (i), (ii), (5.20) and (13.23), we have

$$\begin{aligned}
\mathbf{K}_\varphi \psi(S_\varphi)^* &= K_\varphi i_\varphi P_\varphi M_{\bar{\psi}} i_\varphi \\
&= K_\varphi P_{H^2(\varphi)} M_{\bar{\psi}} i_\varphi \\
&= K_\varphi M_{\bar{\psi}} i_\varphi \\
&= T_{\bar{\psi}} K_\varphi i_\varphi \\
&= T_{\bar{\psi}} \mathbf{K}_\varphi
\end{aligned}$$

(see Figure 13.13). □

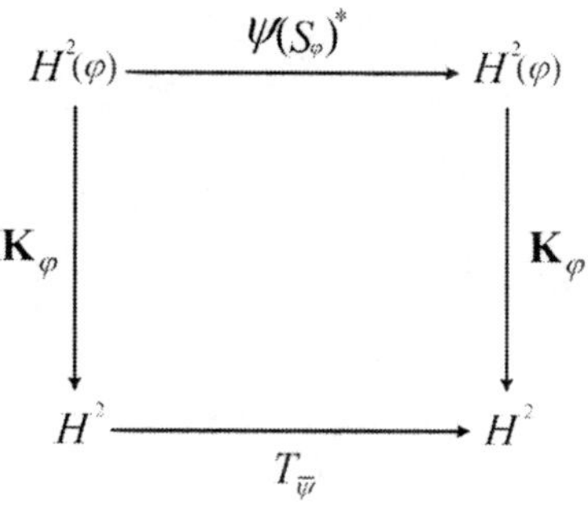

Figure 13.13 $\mathbf{K}_\varphi$ intertwines $T_{\bar\psi}$ and $\psi(S_\varphi)^*$.

Exercises

Exercise 13.5.1 Let $\varphi \in L^\infty(\mathbb{T})$. Show that

$$S_\varphi P_\varphi J_\varphi = P_\varphi J_\varphi S.$$

Hint: Use (13.27) and (13.28).

Exercise 13.5.2 Let $\varphi \in L^\infty(\mathbb{T})$. Show that

$$S^* C_\varphi = K_\varphi S_\varphi^* \chi_0.$$

Hint: Use Theorem 13.3 and Lemma 13.17.

13.6 Toeplitz operators with symbols in $L^2(\mathbb{T})$

In the classic definition of Toeplitz operators, the symbol φ in T_φ belongs to $L^\infty(\mathbb{T})$. In this section, we define Toeplitz operators with symbols in $L^2(\mathbb{T})$ and study their continuity as applications from H^2 into $H(\mathbb{D})$.

Let $\varphi \in L^2(\mathbb{T})$. In Lemma 13.9, we saw that the linear map $K_\varphi : L^1(\varphi) \longrightarrow H(\mathbb{D})$ is continuous. Using the Cauchy–Schwarz inequality, it is easy to see that the injection $i : H^2 \longrightarrow L^1(\varphi)$ is well defined and continuous, and thus so is the mapping

$$\begin{array}{ccccc} H^2 & \longrightarrow & L^1(\varphi) & \longrightarrow & H(\mathbb{D}) \\ f & \longmapsto & f & \longmapsto & K_\varphi f. \end{array} \tag{13.31}$$

Corollary 13.16 says that, if we have the stronger assumption $\varphi \in L^\infty(\mathbb{T})$, then $K_\varphi J_\varphi = T_\varphi$. Hence, this result provides the motivation to denote the above mapping by T_φ. Therefore, we define the continuous mapping T_φ from H^2 into $H(\mathbb{D})$ by

$$\begin{array}{rccc} T_\varphi: & H^2 & \longrightarrow & H(\mathbb{D}) \\ & f & \longmapsto & K_\varphi f. \end{array}$$

If φ and f are such that $\varphi f \in L^2(\mathbb{T})$, then, by (13.6), we have

$$T_\varphi(f) = K_\varphi(f) = K(\varphi f) = P_+(\varphi f) \qquad (f \in H^2). \tag{13.32}$$

For example, if $\varphi \in L^\infty(\mathbb{T})$, then surely $\varphi f \in L^2(\mathbb{T})$ and thus our new definition coincides with the old one. In this situation, T_φ maps H^2 into itself and, moreover, it is a bounded operator on H^2. In fact, as shown in the following result, this is indeed the only case where T_φ is a bounded operator on H^2.

Theorem 13.22 *Let $\varphi \in L^2(\mathbb{T})$. Then T_φ belongs to $\mathcal{L}(H^2)$ if and only if $\varphi \in L^\infty(\mathbb{T})$.*

Proof If $\varphi \in L^\infty(\mathbb{T})$, then we know that T_φ is a bounded operator on H^2 whose norm is equal to $\|\varphi\|_\infty$; see Section 12.1. To treat the other direction, assume that T_φ is bounded on H^2. Let p be any trigonometric polynomial. Then, for each fixed $m \geq \deg(p)$, $\chi_m p$ is an analytic polynomial. Since we assumed that T_φ is bounded on H^2, we have

$$\|\chi_{-m} T_\varphi(\chi_m p)\|_2 = \|T_\varphi(\chi_m p)\|_2 \leq \|T_\varphi\| \times \|\chi_m p\|_2 = \|T_\varphi\| \times \|p\|_2.$$

But, by (13.32),

$$\chi_{-m} T_\varphi(\chi_m p) = \chi_{-m} P_+(\chi_m \varphi p)$$

and, by Lemma 12.1,

$$\lim_{m \to \infty} \|\chi_{-m} P_+(\chi_m \varphi p) - \varphi p\|_2 = 0.$$

Hence, we have

$$\|\varphi p\|_2 \leq \|T_\varphi\| \times \|p\|_2. \tag{13.33}$$

An expert in measure theory would immediately deduce that $\varphi \in L^\infty(\mathbb{T})$ with $\|\varphi\|_\infty \leq \|T_\varphi\|$. This conclusion is explained in detail below.

For each positive integer n, consider

$$E_n = \{\zeta \in \mathbb{T} : |\varphi(\zeta)| \leq n\}$$

and define

$$\varphi_n(\zeta) = \begin{cases} \varphi(\zeta) & \text{if} \quad \zeta \in E_n, \\ 0 & \text{if} \quad \zeta \notin E_n. \end{cases}$$

It is clear that $\varphi_n \in L^\infty(\mathbb{T})$, $\|\varphi_n\|_\infty \longrightarrow \|\varphi\|_\infty$, as $n \longrightarrow \infty$, and also that $|\varphi_n(\zeta)|$ increases monotonically to $|\varphi(\zeta)|$ almost everywhere on $\mathbb{T}$. Hence, for each fixed n, (13.33) implies that the inequality

$$\|\varphi_n p\|_2 \leq \|T_\varphi\| \times \|p\|_2$$

holds for all trigonometric polynomials p. Since φ_n is bounded and the trigonometric polynomials are dense in $L^2(\mathbb{T})$, we deduce that

$$\|\varphi_n f\|_2 \leq \|T_\varphi\| \times \|f\|_2 \qquad (f \in L^2(\mathbb{T})).$$

Writing the above identity as

$$\|M_{\varphi_n} f\|_2 \leq \|T_\varphi\| \times \|f\|_2 \qquad (f \in L^2(\mathbb{T})),$$

shows that

$$\|M_{\varphi_n}\|_{\mathcal{L}(L^2(\mathbb{T}))} \leq \|T_\varphi\|.$$

But, by Theorem 2.20, $\|M_{\varphi_n}\|_{\mathcal{L}(L^2(\mathbb{T}))} = \|\varphi_n\|_\infty$. Therefore, for each $n \geq 1$,

$$\|\varphi_n\|_\infty \leq \|T_\varphi\|.$$

Let $n \longrightarrow \infty$ to get

$$\|\varphi\|_\infty \leq \|T_\varphi\|. \qquad \square$$

For $\varphi \in L^2(\mathbb{T})$ and $\psi \in H^2$, the product $T_\psi T_\varphi$ can be interpreted as ψT_φ. With this convention, $T_\psi T_\varphi$ becomes a bounded operator from H^2 into $H(\mathbb{D})$. It is an interesting and difficult problem to determine if, for a given $\varphi \in L^2(\mathbb{T})$ and $\psi \in H^2$, the product of the two unbounded operators T_ψ and T_φ on H^2 is bounded on H^2. In Chapter 23, we will see some partial results in this direction. With the above convention, we can generalize Theorem 12.4 to the case where one of the symbols is in $L^2(\mathbb{T})$.

Theorem 13.23 *Let $\varphi \in H^2$ and $\psi \in L^\infty(\mathbb{T})$. Then*

$$T_{\bar{\varphi}} T_\psi = T_{\bar{\varphi}\psi}.$$

Proof Let $f \in H^2$. Then we have

$$T_{\bar{\varphi}} T_\psi f = K_{\bar{\varphi}} T_\psi f = K(\bar{\varphi} P_+(\psi f)).$$

But, by (13.15), we have $K(\bar{\varphi} P_-(\psi f)) = 0$ and thus

$$T_{\bar{\varphi}} T_\psi f = K(\bar{\varphi}\psi f) = K_{\bar{\varphi}\psi}(f) = T_{\bar{\varphi}\psi} f. \qquad \square$$

We denote by $X : H(\mathbb{D}) \longrightarrow H(\mathbb{D})$ the linear map defined by

$$(Xf)(z) = \frac{f(z) - f(0)}{z} \qquad (z \in \mathbb{D}).$$

Note that, if $f \in H^2$, then $Xf = S^* f = T_{\bar{z}} f$ and thus X is an extension of the adjoint of the shift operator. The following result is a generalization of (12.3).

Corollary 13.24 *Let $\varphi \in H^2$. Then we have*

$$T_{\bar{\varphi}} S^* = T_{\bar{z}\bar{\varphi}} = X T_{\bar{\varphi}}.$$

Proof Using Theorem 13.23 and the fact that $S^* = T_{\bar{z}}$, we immediately get $T_{\bar{\varphi}}S^* = T_{\bar{z}\bar{\varphi}}$. To establish the second identity, fix any function $f \in H^2$ and $z \in \mathbb{D}$, $z \neq 0$. Then

$$
\begin{aligned}
(XT_{\bar{\varphi}}f)(z) &= \frac{T_{\bar{\varphi}}f(z) - T_{\bar{\varphi}}f(0)}{z} \\
&= \frac{K_{\bar{\varphi}}f(z) - K_{\bar{\varphi}}f(0)}{z} \\
&= \frac{1}{z}\left(\int_{\mathbb{T}} \frac{\overline{\varphi(\zeta)}f(\zeta)}{1-\bar{\zeta}z}\, dm(\zeta) - \int_{\mathbb{T}} \overline{\varphi(\zeta)}f(\zeta)\, dm(\zeta)\right) \\
&= \frac{1}{z}\int_{\mathbb{T}} \overline{\varphi(\zeta)}f(\zeta)\left(\frac{1}{1-\bar{\zeta}z} - 1\right) dm(\zeta) \\
&= \int_{\mathbb{T}} \frac{\overline{\zeta\varphi(\zeta)}f(\zeta)}{1-\bar{\zeta}z}\, dm(\zeta) \\
&= (T_{\bar{z}\bar{\varphi}}f)(z),
\end{aligned}
$$

which proves the second equality. □

When φ is in H^2, then we extend T_φ to $H(\mathbb{D})$ by defining $T_\varphi h = \varphi h$, $h \in H(\mathbb{D})$. Then the following result is a generalization of Lemma 12.5.

Lemma 13.25 *Let $\varphi \in H^2$ and $h \in H(\mathbb{D})$. Then we have*

$$(XT_\varphi - T_\varphi X)h = h(0)S^*\varphi.$$

Proof Fix $z \in \mathbb{D}$. Then

$$
\begin{aligned}
((XT_\varphi - T_\varphi X)h)(z) &= (X(\varphi h))(z) - \varphi(z)(Xh)(z) \\
&= \frac{\varphi(z)h(z) - \varphi(0)h(0)}{z} - \varphi(z)\frac{h(z) - h(0)}{z} \\
&= h(0)\frac{\varphi(z) - \varphi(0)}{z} = h(0)X\varphi(z).
\end{aligned}
$$

Since φ is in H^2, we have $X\varphi = S^*\varphi$ and hence we get the result. □

Lemma 13.26 *Let $\varphi \in H^2$ and let $w \in \mathbb{D}$. Then we have*

$$T_{\bar{\varphi}}k_w = \overline{\varphi(w)}k_w.$$

Proof First note that since $k_w \in H^\infty$, we have $\bar{\varphi}k_w \in L^2(\mathbb{T})$ and thus $T_{\bar{\varphi}}k_w = P_+(\bar{\varphi}k_w) \in H^2$. Now let $g \in H^\infty$. Then we have

$$\begin{aligned}\langle T_{\bar\varphi} k_w, g\rangle_{H^2} &= \langle P_+(\bar\varphi k_w), g\rangle_{H^2}\\ &= \langle \bar\varphi k_w, g\rangle_{L^2}\\ &= \langle k_w, \varphi g\rangle_{L^2}\\ &= \langle k_w, \varphi g\rangle_{H^2}\\ &= \overline{\varphi(w) g(w)}\\ &= \overline{\varphi(w)}\langle k_w, g\rangle_{H^2}.\end{aligned}$$

Since this equality is valid for all $g \in H^\infty$, and H^∞ is dense in H^2, we get the result. □

Exercise

Exercise 13.6.1 Let $\varphi, \psi \in H^2$. Show that

$$(XT_\varphi - T_\varphi X)T_{\bar\psi} = S^*\varphi \otimes \psi.$$

Hint: Use Lemma 13.25.

13.7 Clark measures μ_α

If μ is a positive Borel measure on $\mathbb{T}$ and we define the function u by the formula

$$u(z) = \int_{\mathbb{T}} \frac{1-|z|^2}{|\zeta - z|^2}\, d\mu(\zeta) \qquad (z \in \mathbb{D}), \tag{13.34}$$

then it is easy to verify directly that u is a positive harmonic function on $\mathbb{D}$. Conversely, if a positive harmonic function u on $\mathbb{D}$ is given, then, by Herglotz's result (Corollary 3.7), there is a unique positive Borel measure μ on $\mathbb{T}$ such that (13.34) holds. Since

$$\frac{1-|z|^2}{|\zeta - z|^2} = \Re\left(\frac{\zeta + z}{\zeta - z}\right),$$

then u is the real part of the analytic function

$$f(z) = \int_{\mathbb{T}} \frac{\zeta + z}{\zeta - z}\, d\mu(\zeta) + ic \qquad (z \in \mathbb{D}), \tag{13.35}$$

where c is any arbitrary real constant. That is why the mapping

$$(H\mu)(z) = \int_{\mathbb{T}} \frac{\zeta + z}{\zeta - z}\, d\mu(\zeta) \qquad (z \in \mathbb{D}),$$

which sends a positive Borel measure on $\mathbb{T}$ to an analytic function on $\mathbb{D}$ with a positive real part, is called the *Herglotz transformation*.

Since μ is a positive measure, we can apply Corollaries 3.3 and 3.5 to study the radial behavior of u. On the one hand, according to Corollary 3.3 and (13.34), we have

$$u(\zeta) = \lim_{r\to 1} u(r\zeta) = \frac{d\mu}{dm}(\zeta) \tag{13.36}$$

for almost all $\zeta \in \mathbb{T}$ and, moreover, $u \in L^1(\mathbb{T})$. Hence, the Lebesgue decomposition of μ is written as

$$d\mu = u\,dm + d\mu_s, \tag{13.37}$$

where μ_s is a positive singular measure. On the other hand, by Corollary 3.5 and (13.34), we also have

$$u(\zeta) = \lim_{r\to 1} u(r\zeta) = +\infty \tag{13.38}$$

for almost all $\zeta \in \mathbb{T}$ with respect to the measure μ_s. In particular, (13.38) holds at all points $\zeta \in \mathbb{T}$, where μ has a strictly positive mass. The following result is a direct consequence of (13.36).

Lemma 13.27 *Let u be a positive harmonic function on $\mathbb{D}$. Then the following are equivalent:*

(i) *for almost all $\zeta \in \mathbb{T}$,*

$$u(\zeta) = \lim_{r\to 1} u(r\zeta) = 0;$$

(ii) *u is defined by* (13.34)*, where μ is a positive singular Borel measure on $\mathbb{T}$.*

Let b be a nonconstant function in the closed unit ball of H^∞. Then, for each $\alpha \in \mathbb{T}$, define

$$u_\alpha(z) = \Re\left(\frac{\alpha + b(z)}{\alpha - b(z)}\right) \qquad (z \in \mathbb{D}). \tag{13.39}$$

By direct calculation, we see that u_α is also given by the formula

$$u_\alpha(z) = \frac{1 - |b(z)|^2}{|\alpha - b(z)|^2} \qquad (z \in \mathbb{D}). \tag{13.40}$$

The first representation reveals that u_α is harmonic, while the second ensures that it is positive on $\mathbb{D}$. Hence, by Herglotz's result (Corollary 3.7), there is a unique positive Borel measure μ_α on $\mathbb{T}$, such that

$$\frac{1 - |b(z)|^2}{|\alpha - b(z)|^2} = \int_{\mathbb{T}} \frac{1 - |z|^2}{|\zeta - z|^2}\, d\mu_\alpha(\zeta) \qquad (z \in \mathbb{D}). \tag{13.41}$$

The collection $(\mu_\alpha)_{\alpha\in\mathbb{T}}$ is the family of *Clark measures* associated with the function b. In particular, instead of μ_1, we simply write μ and address it as the Clark measure associated with b. With this convention, we can also say that

μ_α is the Clark measure associated with the function $\bar{\alpha}b$. We emphasize that (13.41) is the defining identity for μ_α, and all its properties are deduced from this equation. Some of these properties are discussed below.

Since b is not a constant function, μ_α is not a multiple of the Lebesgue measure m. In particular, $\mu_\alpha \neq 0$. If we put $z = 0$ in (13.39), (13.40) and (13.41), we obtain

$$\hat{\mu}_\alpha(0) = \|\mu_\alpha\| = \mu_\alpha(\mathbb{T}) = \Re\left(\frac{\alpha + b(0)}{\alpha - b(0)}\right) = \frac{1 - |b(0)|^2}{|\alpha - b(0)|^2}. \tag{13.42}$$

This relation shows that the family $(\mu_\alpha)_{\alpha\in\mathbb{T}}$ consists only of probability measures if and only if $b(0) = 0$.

By (13.39) and (13.41), the uniqueness theorem for analytic functions implies that

$$\frac{\alpha + b(z)}{\alpha - b(z)} = \int_{\mathbb{T}} \frac{\zeta + z}{\zeta - z}\, d\mu_\alpha(\zeta) + i\,\Im\left(\frac{\alpha + b(0)}{\alpha - b(0)}\right). \tag{13.43}$$

In other words,

$$(H\mu_\alpha)(z) = \int_{\mathbb{T}} \frac{\zeta + z}{\zeta - z}\, d\mu_\alpha(\zeta) = \frac{\alpha + b(z)}{\alpha - b(z)} - i\,\Im\left(\frac{\alpha + b(0)}{\alpha - b(0)}\right). \tag{13.44}$$

Since

$$\Im\left(\frac{\alpha + w}{\alpha - w}\right) = \frac{2\,\Im(\bar{\alpha}w)}{|\alpha - w|^2},$$

the identity (13.43) can also be rewritten as

$$\frac{\alpha + b(z)}{\alpha - b(z)} = \int_{\mathbb{T}} \frac{\zeta + z}{\zeta - z}\, d\mu_\alpha(\zeta) + 2i\,\frac{\Im(\bar{\alpha}b(0))}{|\alpha - b(0)|^2}. \tag{13.45}$$

Let us study a very specific case in which we explicitly obtain the Clark measure of a given function b. Let $\alpha \in \mathbb{T}$. Since δ_α is the unit Dirac mass at the point $\alpha \in \mathbb{T}$, for each $z \in \mathbb{D}$, we have

$$\frac{1 + \Theta(z)}{1 - \Theta(z)} = \frac{1 + \bar{\alpha}z}{1 - \bar{\alpha}z} = \frac{\alpha + z}{\alpha - z} = \int_{\mathbb{T}} \frac{\zeta + z}{\zeta - z}\, d\delta_\alpha(\zeta). \tag{13.46}$$

Hence, by the uniqueness of the Clark measure, we conclude that the Clark measure associated with Θ is the Dirac measure δ_α. This observation also says that the family of Clark measures associated with the inner function $\Theta(z) = z$ is precisely the family of Dirac measures $(\delta_\alpha)_{\alpha\in\mathbb{T}}$.

According to Exercise 4.2.2, for each $\alpha \in \mathbb{T}$, the set

$$\{b = \alpha\} = \left\{\zeta \in \mathbb{T} : \lim_{r\to 1} b(r\zeta) = \alpha\right\}$$

is a well-defined Borel subset of $\mathbb{T}$. By the same token, the set

$$\{u_\alpha = +\infty\} = \left\{\zeta \in \mathbb{T} : \lim_{r\to 1} u_\alpha(r\zeta) = +\infty\right\}$$

is a Borel subset of $\mathbb{T}$ (see Exercise 13.7.2). Moreover, by (13.40), we have

$$\{u_\alpha = +\infty\} \subset \{b = \alpha\}. \tag{13.47}$$

Now, we show that $\{b = \alpha\}$ is a carrier of μ_α.

Theorem 13.28 *Let b be a nonconstant function in the closed unit ball of H^∞, and let $(\mu_\alpha)_{\alpha\in\mathbb{T}}$ denote the associated Clark measures. Then the Lebesgue decomposition of μ_α is*

$$d\mu_\alpha(\zeta) = \frac{1 - |b(\zeta)|^2}{|\alpha - b(\zeta)|^2}\, dm(\zeta) + d\sigma_\alpha(\zeta), \tag{13.48}$$

where σ_α is a positive Borel singular measure on $\mathbb{T}$ carried on the set $\{b = \alpha\}$.

Proof According to (13.41), we have

$$\frac{1 - |b(z)|^2}{|\alpha - b(z)|^2} = \int_{\mathbb{T}} \frac{1 - |z|^2}{|\zeta - z|^2}\, d\mu_\alpha(\zeta) \qquad (z \in \mathbb{D}),$$

and, applying Fatou's theorem, this formula shows that

$$D\mu_\alpha(\zeta) = \frac{1 - |b(z)|^2}{|\alpha - b(z)|^2} \qquad \text{(a.e. on } \mathbb{T}). \tag{13.49}$$

Therefore, (13.37) reduces to (13.48).

By (3.13), σ_α is carried on the set $\{\zeta \in \mathbb{T} : D\mu_\alpha(\zeta) = +\infty\}$. But, by Theorem 3.2, at such points we certainly have $u_\alpha(\zeta) = +\infty$. Therefore, (13.47) reveals that σ_α is also carried on the set $\{b = \alpha\}$. □

A byproduct of Theorem 13.28 is that $(1 - |b|^2)/(|\alpha - b|^2) \in L^1(\mathbb{T})$, for each $\alpha \in \mathbb{T}$. In fact, we are even able to give an upper estimate for its norm. Since

$$\int_{\mathbb{T}} \frac{1 - |b(\zeta)|^2}{|\alpha - b(\zeta)|^2}\, dm(\zeta) \le \int_{\mathbb{T}} d\mu_\alpha(\zeta) = \|\mu_\alpha\|$$

and, by (13.42),

$$\|\mu_\alpha\| = \frac{1 - |b(0)|^2}{|\alpha - b(0)|^2} \le \frac{1 + |b(0)|}{1 - |b(0)|},$$

we have

$$\int_{\mathbb{T}} \frac{1 - |b(\zeta)|^2}{|\alpha - b(\zeta)|^2}\, dm(\zeta) \le \frac{1 + |b(0)|}{1 - |b(0)|}. \tag{13.50}$$

The identity (13.49) immediately implies the following result, which is interesting in its own right.

Corollary 13.29 *Let b be a nonconstant function in the closed unit ball of H^∞, and let $(\mu_\alpha)_{\alpha\in\mathbb{T}}$ denote the associated family of Clark measures. Then the following are equivalent.*

(i) *The function b is an inner function.*
(ii) *For each $\alpha \in \mathbb{T}$, the Clark measure μ_α is singular.*
(iii) *There is an $\alpha_0 \in \mathbb{T}$ such that the Clark measure μ_{α_0} is singular.*

In this situation, μ_α is carried on the set $\{b = \alpha\}$.

As a very special inner function, assume that b is a finite Blaschke product. Then b has an analytic continuation across $\mathbb{T}$ and, moreover, $|b| \equiv 1$ on $\mathbb{T}$. Hence, by Corollary 13.29, μ_α is a singular measure that is carried on the finite set $\{\zeta \in \mathbb{T} : b(\zeta) = \alpha\}$. In fact, this set is precisely the support of μ_α. A more general version of this fact is studied below.

Corollary 13.30 *Let b be a nonconstant function in the closed unit ball of H^∞, and let*

$$d\mu(\zeta) = \frac{1 - |b(\zeta)|^2}{|1 - b(\zeta)|^2}\, dm(\zeta) + d\sigma(\zeta)$$

be the corresponding Clark measure. Suppose that I is an open subarc of $\mathbb{T}$ such that b has an analytic continuation across I and $|b| \equiv 1$ on I. Then, on I, the measure μ is a discrete measure consisting of Dirac masses anchored precisely at the points of

$$\{\zeta \in I : b(\zeta) = 1\}.$$

Moreover, if the above set has infinite cardinality, its points can only cluster at the end points of I.

Proof Put

$$E = \{\zeta \in I : b(\zeta) = 1\}.$$

The set E is countable and its (possible) accumulations points can only be the end points of I, since, otherwise, E would have an accumulation point inside the arc I, and then the uniqueness principle for analytic functions would imply that $b \equiv 1$ on $\mathbb{D}$, which is absurd.

By hypothesis, $|b| \equiv 1$ on I and $b = 1$ on a countable subset of I. Thus, by Theorem 13.28, on I, $\mu = \sigma$ is a countable collection of Dirac measures anchored precisely at (some or all of) the points of E. However, at each point of E, we must have a nonzero Dirac mass. This is because, on the one hand, by (13.43), we have

$$\frac{1 + b(z)}{1 - b(z)} = \int_{\mathbb{T}} \frac{\zeta + z}{\zeta - z}\, d\mu(\zeta) + ic,$$

and the right-hand side of this representation shows that the function $(1 + b)/(1 - b)$ is analytic at all points of I except at those points where there is a Dirac mass, and, on the other hand, $(1 + b)/(1 - b)$ has a pole at points of I where $b = 1$. □

Corollary 13.31 *Let b be a function in the unit ball of H^∞ and let $\alpha \in \mathbb{T}$. Assume that the function*

$$\frac{\alpha + b}{\alpha - b}$$

is in H^1. Then the measure μ_α is absolutely continuous.

Proof Using (4.8), we have

$$\frac{\alpha + b(z)}{\alpha - b(z)} = \int_{\mathbb{T}} \frac{1 - |z|^2}{|\zeta - z|^2} \frac{\alpha + b(\zeta)}{\alpha - b(\zeta)} \, dm(\zeta),$$

that is, taking the real part,

$$\frac{1 - |b(z)|^2}{|\alpha - b(z)|^2} = \int_{\mathbb{T}} \frac{1 - |z|^2}{|\zeta - z|^2} \frac{1 - |b(\zeta)|^2}{|\alpha - b(\zeta)|^2} \, dm(\zeta).$$

By uniqueness, (13.41) that implies that

$$d\mu_\alpha = \frac{1 - |b|^2}{|\alpha - b|^2} \, dm,$$

which means that μ_α is absolutely continuous. □

As a special case of Corollary 13.31, if b is a function in the closed unit ball of H^∞ such that $\|b\|_\infty < 1$, then, for any $\alpha \in \mathbb{T}$, we even have

$$\frac{\alpha + b}{\alpha - b} \in H^\infty, \tag{13.51}$$

and thus each Clark measure μ_α is absolutely continuous.

Exercises

Exercise 13.7.1 Given a positive Borel measure μ on $\mathbb{T}$, μ not a constant multiple of the Lebesgue measure, define the analytic function b by the formula

$$\frac{1 + b(z)}{1 - b(z)} = \int_{\mathbb{T}} \frac{\zeta + z}{\zeta - z} \, d\mu(\zeta) \qquad (z \in \mathbb{D}). \tag{13.52}$$

Show that b is a nonconstant element in the closed unit ball of $H^\infty(\mathbb{D})$ with $-1 < b(0) < 1$.
Hint: Take the real part of both sides of (13.52) to show that (13.41) holds.

Exercise 13.7.2 Fix $\alpha \in \mathbb{T}$. Show that the set of all $\zeta \in \mathbb{T}$ such that

$$\lim_{r \to \infty} u_\alpha(r\zeta) = \lim_{r \to \infty} \frac{1 - |b(r\zeta)|^2}{|\alpha - b(r\zeta)|^2} = +\infty$$

is an $F_{\sigma\delta}$ set.
Hint: Put

$$E(m,n) = \{\zeta \in \mathbb{T} : u_\alpha(r\zeta) \geq m \text{ for all } r \geq 1 - 1/n\}$$

and consider the $F_{\sigma\delta}$ set

$$E = \bigcap_{m=1}^{\infty} \bigcup_{n=1}^{\infty} E(m,n).$$

Exercise 13.7.3 Let μ be a positive measure on $\mathbb{T}$ and assume that μ is singular and $\|\mu\| = 1$. Show that $H\mu \in H^p$, for all $0 < p < 1$, and in particular $H\mu$ has nontangential boundary values m-almost everywhere. The purpose of this exercise is now to prove the identity

$$m(\{\zeta \in \mathbb{T} : |(H\mu)(\zeta)| > t\}) = \frac{2}{\pi} \arctan\left(\frac{1}{t}\right). \tag{13.53}$$

(i) Let $\phi(z) = ((H\mu)(z) - 1)/((H\mu)(z) + 1)$, $z \in \mathbb{D}$. Show that ϕ is an analytic function that maps the disk into itself.
Hint: Observe that $H\mu$ maps the disk to the domain $\{z : \Re(z) > 0\}$ and $z \longmapsto (z-1)/(z+1)$ maps $\{z : \Re(z) > 0\}$ to $\mathbb{D}$.

(ii) Verify that $\phi(0) = 0$ and show that ϕ is an inner function.

(iii) Establish the claim.
Hint: Write

$$m(\{\zeta \in \mathbb{T} : |(H\mu)(\zeta)| > t\}) = m\left(\left\{\zeta \in \mathbb{T} : \left|\frac{1+\phi(\zeta)}{1-\phi(\zeta)}\right| > t\right\}\right)$$

and use the fact that ϕ is measure-preserving (see Exercise 4.10.4).

Exercise 13.7.4 Let $b \in H^\infty$, $\|b\|_\infty \leq 1$, and let $(\mu_\alpha)_{\alpha \in \mathbb{T}}$ denote the associated family of Clark measures.

(i) Show that, for any $z \in \mathbb{D}$, we have

$$\int_{\mathbb{T}} \left(\int_{\mathbb{T}} P_z(\zeta)\, d\mu_\alpha(\zeta) \right) dm(\alpha) = \int_{\mathbb{T}} P_z(\zeta)\, dm(\zeta),$$

where $P_z(\zeta)$ denotes the Poisson kernel,

$$P_z(\zeta) = \frac{1 - |z|^2}{|z - \zeta|^2}.$$

(ii) Show that, for any $g \in \mathcal{C}(\mathbb{T})$, we have

$$\int_{\mathbb{T}} \left(\int_{\mathbb{T}} g(\zeta)\, d\mu_\alpha(\zeta) \right) dm(\alpha) = \int_{\mathbb{T}} g(\zeta)\, dm(\zeta). \tag{13.54}$$

Hint: Use the fact that the linear span of $\{P_z : z \in \mathbb{D}\}$ is dense in $\mathcal{C}(\mathbb{T})$.

13.8 The Cauchy transform of μ_α

The identities (13.43) and (13.45) enable us to easily find the Cauchy transform of μ_α.

Theorem 13.32 *Let b be a nonconstant function in the closed unit ball of H^∞, and let $(\mu_\alpha)_{\alpha\in\mathbb{T}}$ denote the associated Clark measures. Then*

$$C_{\mu_\alpha}(z) = \int_{\mathbb{T}} \frac{d\mu_\alpha(\zeta)}{1-\bar{\zeta}z} = \frac{1}{1-\bar{\alpha}b(z)} + \frac{\alpha\overline{b(0)}}{1-\alpha\overline{b(0)}} \qquad (z\in\mathbb{D}).$$

In particular, if $b(0)=0$, we have

$$C_{\mu_\alpha}(z) = \frac{1}{1-\bar{\alpha}b(z)} \qquad (z\in\mathbb{D}).$$

Proof It is enough to observe that

$$\frac{2}{1-\bar{\zeta}z} = \frac{\zeta+z}{\zeta-z} + 1,$$

and then apply (13.43) and (13.42) to obtain

$$\begin{aligned}
2C_{\mu_\alpha}(z) &= \int_{\mathbb{T}} \frac{2}{1-\bar{\zeta}z}\, d\mu_\alpha(\zeta) \\
&= \int_{\mathbb{T}} \frac{\zeta+z}{\zeta-z}\, d\mu_\alpha(\zeta) + \int_{\mathbb{T}} d\mu_\alpha(\zeta) \\
&= \frac{\alpha+b(z)}{\alpha-b(z)} - i\,\Im\left(\frac{\alpha+b(0)}{\alpha-b(0)}\right) + \Re\left(\frac{\alpha+b(0)}{\alpha-b(0)}\right) \\
&= \frac{\alpha+b(z)}{\alpha-b(z)} + \frac{\bar{\alpha}+\overline{b(0)}}{\bar{\alpha}-\overline{b(0)}} \\
&= \frac{1+\bar{\alpha}b(z)}{1-\bar{\alpha}b(z)} + \frac{1+\alpha\overline{b(0)}}{1-\alpha\overline{b(0)}} \\
&= \frac{2}{1-\bar{\alpha}b(z)} + \frac{2\alpha\overline{b(0)}}{1-\alpha\overline{b(0)}}.
\end{aligned}$$

This completes the proof. □

The identity (13.42) can be rewritten as

$$\hat{\mu}_\alpha(0) = \Re\left(\frac{\alpha+b(0)}{\alpha-b(0)}\right) = \frac{1-|b(0)|^2}{|\alpha-b(0)|^2}.$$

Since μ is a real measure, we also have

$$\hat{\mu}_\alpha(-n) = \overline{\hat{\mu}_\alpha(n)} \qquad (n\geq 1).$$

To evaluate the positive part of spectrum, we use Theorem 13.32.

Corollary 13.33 *Let b be a nonconstant function in the closed unit ball of H^∞, and let $(\mu_\alpha)_{\alpha\in\mathbb{T}}$ denote the associated Clark measures. Then*

$$\hat{\mu}_\alpha(n) = \sum_{k=1}^{\infty} \bar{\alpha}^k \widehat{b^k}(n) \qquad (n \geq 1).$$

In particular, if $b(0) = 0$, then we have

$$\hat{\mu}_\alpha(n) = \sum_{k=1}^{n} \bar{\alpha}^k \widehat{b^k}(n) \qquad (n \geq 1).$$

Proof By the first formula in Theorem 13.1, $\hat{\mu}_\alpha(n)$ is the coefficient of z^n in the Taylor expansion of C_{μ_α} on $\mathbb{D}$. But, by Theorem 13.32, the series representation

$$\begin{aligned} C_{\mu_\alpha}(z) &= \frac{1}{1-\bar{\alpha}b(z)} + \frac{\alpha\overline{b(0)}}{1-\alpha\overline{b(0)}} \\ &= \frac{\alpha\overline{b(0)}}{1-\alpha\overline{b(0)}} + \sum_{k=0}^{\infty} \bar{\alpha}^k b^k(z) \\ &= \frac{1}{1-\alpha\overline{b(0)}} + \sum_{k=1}^{\infty} \bar{\alpha}^k b^k(z) \end{aligned}$$

holds on the open unit disk $\mathbb{D}$. The last identity immediately shows that the coefficient of z^n, $n \geq 1$, in the Taylor expansion of C_{μ_α} is

$$\sum_{k=1}^{\infty} \bar{\alpha}^k \widehat{b^k}(n) \qquad (n \geq 1).$$

This proves our first formula. If $b(0) = 0$, then the coefficients of $1, \dots, z^{k-1}$ in the expansion of b^k are zero. Hence, for $k > n$, the coefficient of z^n in b^k is zero, and we obtain the second formula. □

13.9 The function ρ

For a nonconstant function b in the closed unit ball of H^∞, we define

$$\rho = 1 - |b|^2.$$

This function plays an essential role in the study of $\mathcal{H}(b)$ spaces. In this section, we just treat a variant of Theorem 8.22 and Corollary 8.23.

Corollary 13.34 *Let b be a nonconstant function in the closed unit ball of H^∞, and let μ denote the associated Clark measure. Then the following are equivalent:*

(i) *b is an extreme point of H^∞;*

(ii) $\displaystyle\int_{\mathbb{T}} \log \rho(\zeta)\, dm(\zeta) = -\infty$;

(iii) $\displaystyle\int_{\mathbb{T}} \log\left(\frac{d\mu}{dm}(\zeta)\right) dm(\zeta) = -\infty$;

(iv) $1 \in H_0^2(\rho)$;

(v) $1 \in H_0^2(\mu)$;

(vi) $H^2(\rho) = L^2(\rho)$;

(vii) $H^2(\mu) = L^2(\mu)$.

Proof (i) $\Longleftrightarrow$ (ii) By Theorem 6.7, b is an extreme point of H^∞ if and only if

$$\int_{\mathbb{T}} \log(1 - |b(\zeta)|)\, dm(\zeta) = -\infty,$$

and, since $1 - |b| \le \rho \le 2(1 - |b|)$, this is equivalent to

$$\int_{\mathbb{T}} \log \rho(\zeta)\, dm(\zeta) = -\infty.$$

(ii) $\Longleftrightarrow$ (iii) By (13.49), we have

$$\varphi(\zeta) = \frac{d\mu}{dm}(\zeta) = \frac{1 - |b(\zeta)|^2}{|1 - b(\zeta)|^2} = \frac{\rho(\zeta)}{|1 - b(\zeta)|^2} \qquad (a.e.\text{ on } \mathbb{T}),$$

and thus

$$\int_{\mathbb{T}} \log \varphi(\zeta)\, dm(\zeta) = -\infty \quad \Longleftrightarrow \quad \int_{\mathbb{T}} \log \rho(\zeta)\, dm(\zeta) = -\infty.$$

Note that

$$\int_{\mathbb{T}} \log |1 - b(\zeta)|^2\, dm(\zeta) = \log |1 - b(0)|^2.$$

(ii) $\Longleftrightarrow$ (iv) $\Longleftrightarrow$ (vi) and (iii) $\Longleftrightarrow$ (v) $\Longleftrightarrow$ (vii) These were established in Corollary 8.23. □

Notes on Chapter 13

The study of Cauchy transforms of measures supported in the plane is an old and vast subject. It dates back to the mid-1800s with the classic Cauchy integral formula. In this chapter, we restrict ourselves to Cauchy transforms of measures supported on the unit circle, and, even in that case, the subject has deep connections to complex analysis, functional analysis, distribution theory, perturbation theory and mathematical physics. We have developed here just the material we need for our study of $\mathcal{H}(b)$ spaces; for a detailed treatment of Cauchy transforms, the reader should consult the main reference [141].

The Clark measures appeared in the work of Clark [144] on perturbation theory of certain linear operators. Since then, they have played an important key role in several topics, e.g. spectral theory and mathematical physics, composition operators, rigid functions, generalized factorizations, value distribution theory of self-maps of $\mathbb{D}$, and the theory of Cauchy transforms. For more on these, see Clark [144], Aleksandrov [14–17], Sarason [455–458], Shapiro [479], Simon and Wolff [484], Makarov and Poltoratski [350] and many others. We also refer the interested reader to the excellent survey of Poltoratski and Sarason [417].

Section 13.1

Theorem 13.4 is due to Smirnov [489].

Section 13.2

The first formula in Theorem 13.6 is due to Fatou [210] and is known as *Fatou's jump theorem*. The second formula in Theorem 13.6 is due to several authors. Sokhotski [492] in 1873 was the first to prove this formula for $d\mu = f\,dm$, where f is a Lipschitz function. Plemelj [413] refined this result in 1908. The version given in this text is due to Privalov [418]. Presently, this formula is known as the *Plemelj formula* or the *Sokhotski–Plemelj formula*.

Section 13.3

Theorem 13.12 is due to Zygmund [571]. Theorem 13.13 follows from Kolmogorov's theorem (Theorem 3.13). Similarly, Theorem 13.14 is a consequence of Stein's theorem (Theorem 3.17) [495]. The result proved in Exercise 13.3.4 is contained in Zygmund [572], where a similar result may be found. The example of an $L^1(\mathbb{T})$ function such that its Cauchy transform is not in H^1 (constructed in Exercise 13.3.5) is taken from [141].

Section 13.7

The family of measures $(\mu_\alpha)_{\alpha\in\mathbb{T}}$ was studied by Clark in a seminal paper [144], which was concerned with the case when the function b is an inner function. For the general case of a function in the unit ball of H^∞, the theory was developed by Aleksandrov. Note that, in [144], Clark introduced these measures in the study of rank-one unitary perturbations of S_Θ, the compression of the shift to a model space $(\Theta H^2)^\perp$. More precisely, he proved that, if Θ is an inner function such that $\Theta(0) = 0$, then the rank-one unitary perturbations of S_Θ are of the form

$$U_\alpha = S_\Theta + \bar{\alpha}1 \otimes S^*\Theta$$

for some $\alpha \in \mathbb{T}$. Moreover, the spectral measure associated with U_α is precisely the Clark measure μ_α, that is, U_α is unitarily equivalent to multiplication by the independent variable ζ on $L^2(\mu_\alpha)$. See Section 31.2 for an extension of this result to the general case of a function in the unit ball of H^∞.

The formula (13.54) proved in Exercise 13.7.4 is due to Aleksandrov and is known as *Aleksandrov's disintegration formula*. See [141] for further details.

14

Model subspaces K_Θ

Our main object in these studies, i.e. the $\mathcal{H}(b)$ space, is a linear manifold in the Hardy space H^2. More precisely, $\mathcal{H}(b)$ is not necessarily a closed subspace with respect to the norm of H^2. In the special case where $\mathcal{H}(b)$ is a closed subspace of H^2, the function b is necessarily inner and $\mathcal{H}(b)$ is precisely the model subspace K_b that has been introduced before. In this chapter, we study more properties of this special, but ultra-important, case.

We start by studying the structure of inner functions. Then we obtain a generator set for K_Θ, in particular, an orthonormal basis when Θ is a Blaschke product, and also show that it is invariant under any Toeplitz operator with an antianalytic symbol. In particular, each model subspace is invariant under the unilateral backward shift operator S^*. We give an explicit formula for the projection of $L^2(\mathbb{T})$ onto K_Θ, and then define the Crofoot transform, a conjugation operator and an involution. The minimal sequences of reproducing kernels is studied next. The restriction of S^* to K_Θ is denoted by M_Θ. The properties of this operator are studied in detail, e.g. we obtain its spectrum, develop a functional calculus and eventually present a commutant lifting theorem.

14.1 The arithmetic of inner functions

Let Θ_1 and Θ_2 be two inner functions. We say that Θ_1 *divides* Θ_2 if there exists an inner function Θ such that $\Theta_2 = \Theta\Theta_1$. As a consequence of the uniqueness of the inner–outer factorization, this is equivalent to saying that $\Theta_2/\Theta_1 \in H^p$, for some $0 < p \le \infty$.

Theorem 14.1 *Let Θ_1 and Θ_2 be two inner functions. Then the following are equivalent:*

(i) Θ_1 *divides* Θ_2*;*
(ii) $\Theta_2 H^2 \subset \Theta_1 H^2$*;*
(iii) $K_{\Theta_1} \subset K_{\Theta_2}$.

Proof (i) $\Longrightarrow$ (ii) Assume that $\Theta_2 = \Theta_1\Theta$, for some inner function Θ. Then we have

$$\Theta_2 H^2 = \Theta_1 \Theta H^2 \subset \Theta_1 H^2.$$

(ii) $\Longrightarrow$ (i) Assume that $\Theta_2 H^2 \subset \Theta_1 H^2$. Since $\Theta_2 \in \Theta_2 H^2$, we also have $\Theta_2 \in \Theta_1 H^2$. This means that there exists a function $\Theta \in H^2$ such that $\Theta_2 = \Theta_1\Theta$. In particular, we must have $|\Theta| = 1$ almost everywhere on $\mathbb{T}$, which implies that Θ is an inner function and thus Θ_1 divides Θ_2.

(ii) $\Longleftrightarrow$ (iii) This follows from the fact that $K_\Theta = (\Theta H^2)^\perp$, which was established in Theorem 8.44. □

The above result helps to shed more light on the lattice of inner functions and to define the notions of the greatest common divisor and the least common multiple.

Corollary 14.2 *Let* Θ_1 *and* Θ_2 *be two inner functions. Then there exists an inner function* Θ *such that*

$$\Theta_1 H^2 \cap \Theta_2 H^2 = \Theta H^2.$$

Moreover, the function Θ *is unique (up to a constant of modulus one) and satisfies the following properties:*

(i) *both* Θ_1 *and* Θ_2 *divide* Θ*;*
(ii) *if* Θ_1 *and* Θ_2 *divide an inner function* Θ_3*, then* Θ *also divides* Θ_3*.*

In other words, the function Θ *is the least common multiple of* Θ_1 *and* Θ_2*, and is denoted by*

$$\Theta = LCM(\Theta_1, \Theta_2).$$

Proof The subspace $\Theta_1 H^2 \cap \Theta_2 H^2$ is a closed S-invariant subspace of H^2, which is not reduced to $\{0\}$, because it contains $\Theta_1\Theta_2$. Hence, according to Beurling's theorem (Theorem 8.16), there exists a unique inner function Θ such that

$$\Theta_1 H^2 \cap \Theta_2 H^2 = \Theta H^2.$$

The property (i) follows immediately from Theorem 14.1. Now, for (ii), if both Θ_1 and Θ_2 divide an inner function Θ_3, then we have

$$\Theta_3 H^2 \subset \Theta_1 H^2 \cap \Theta_2 H^2 = \Theta H^2,$$

which, by the same theorem, implies that Θ divides Θ_3. □

In a similar way, we can also prove the existence of the greatest common divisor of two inner functions.

Corollary 14.3 *Let Θ_1 and Θ_2 be two inner functions. Then there exists an inner function Θ such that*

$$K_{\Theta_1} \cap K_{\Theta_2} = K_\Theta.$$

Moreover, the function Θ is unique (up to a constant of modulus one) and satisfies the following properties:

(i) *Θ divides both Θ_1 and Θ_2;*
(ii) *if Θ_3 divides both functions Θ_1 and Θ_2, then Θ_3 also divides Θ.*

In other words, the function Θ is the greatest common divisor of the functions Θ_1 and Θ_2, and is denoted by

$$\Theta = GCD(\Theta_1, \Theta_2).$$

Proof The subspace $K_{\Theta_1} \cap K_{\Theta_2}$ is a proper closed subspace of H^2, which is invariant under S^*. Thus, by Corollary 8.33, there exists a unique inner function Θ such that

$$K_{\Theta_1} \cap K_{\Theta_2} = K_\Theta.$$

The property (i) follows immediately from Theorem 14.1. Now, for (ii), if Θ_3 divides both Θ_1 and Θ_2, then we have

$$K_{\Theta_3} \subset K_{\Theta_1} \cap K_{\Theta_2} = K_\Theta,$$

which, by the same theorem, implies that Θ_3 divides Θ. □

Exercises

Exercise 14.1.1 Let Θ_1 and Θ_2 be two inner functions and put

$$\theta = GCD(\Theta_1, \Theta_2) \quad \text{and} \quad \Theta = LCM(\Theta_1, \Theta_2).$$

Show that:

(i) $\vee(\Theta_1 H^2, \Theta_2 H^2) = \theta H^2$;
(ii) $\vee(K_{\Theta_1}, K_{\Theta_2}) = K_\Theta$.

Hint: Use Corollaries 14.2 and 14.3 and the fact that

$$\vee(M, N) = (M^\perp \cap N^\perp)^\perp.$$

Remark: We remind the reader that $\vee(M, N)$ represents the smallest closed subspace that contains both M and N.

Exercise 14.1.2 Let $1 \le p < \infty$, let q be the exponent conjugate of p, and let Θ_1 and Θ_2 be two inner functions. Show that the following are equivalent:

(i) Θ_1 divides Θ_2;
(ii) $\Theta_2 H^q \subset \Theta_1 H^q$;
(iii) $K^p_{\Theta_1} \subset K^p_{\Theta_2}$.

Hint: Use Exercise 8.11.11. See also the proof of Theorem 14.1.

14.2 A generator for K_Θ

In Section 8.11, we defined the model subspace K_Θ by the formula

$$K_\Theta = H^2 \cap (\Theta \overline{H^2_0})$$

and showed that

$$K_\Theta = (\Theta H^2)^\perp.$$

This property can be equivalently written as

$$H^2 = \Theta H^2 \oplus K_\Theta. \tag{14.1}$$

Given an operator T on a space X, we say that $x \in M \subset X$ generates the subspace M if M is the smallest closed T-invariant subspace of X that contains x. This is equivalent to saying that M is the closure of the linear span of $x, Tx, T^2x, \dots$, and we write

$$M = \langle x \rangle = \text{Span}\{x, Tx, T^2x, \dots\}.$$

The following result shows that $S^*\Theta$ is a generator for K_Θ.

Theorem 14.4 *Let Θ be an inner function. Then*

$$K_\Theta = \text{Span}\{S^*\Theta, S^{*2}\Theta, S^{*3}\Theta, \dots\}.$$

Proof For simplicity of notation, put

$$M = \text{Span}\{S^*\Theta, S^{*2}\Theta, S^{*3}\Theta, \dots\}.$$

Then, for each $n \ge 1$ and each $f \in H^2$, we have

$$\begin{aligned} \langle S^{*n}\Theta, \Theta f\rangle_{H^2} &= \langle \Theta, S^n\Theta f\rangle_{H^2} \\ &= \langle \Theta, z^n\Theta f\rangle_{H^2} \\ &= \langle 1, z^n f\rangle_{H^2} = 0. \end{aligned}$$

In the third identity, we used the fact that Θ is unimodular. Hence, we have $S^{*n}\Theta \perp \Theta H^2$, which, by Theorem 8.44, implies that

$$M \subset K_\Theta.$$

To establish equality, it is enough to show that $f \in K_\Theta$ and $f \perp M$ implies that $f = 0$. Hence, assume that $f \in K_\Theta$ and $f \perp M$. This means that, on the one hand,

$$\langle f, \Theta g\rangle_{H^2} = 0 \qquad (g \in H^2)$$

and, on the other,

$$\langle f, S^{*n}\Theta\rangle_{H^2} = 0 \qquad (n \geq 1).$$

The latter assumption implies that

$$0 = \langle S^n f, \Theta\rangle_{H^2} = \langle z^n f, \Theta\rangle_{H^2} = \langle z^n, \Theta\bar{f}\rangle_{L^2} \qquad (n \geq 1).$$

Taking $g = z^n$, $n \geq 0$, the former assumption also gives

$$\langle z^{-n}, \Theta\bar{f}\rangle_{L^2} = 0 \qquad (n \geq 0).$$

Hence, by the uniqueness theorem for the Fourier coefficients, we conclude that $\Theta\bar{f} = 0$. Once again, since Θ is unimodular, this implies that $f = 0$. □

Theorem 14.4 shows that K_Θ is the model subspace that is generated by $S^*\Theta$. Naturally, one may wonder about the model subspace that is generated by Θ itself. But the answer is immediate if we note that

$$S^*(z\Theta) = \Theta,$$

and thus $S^{*n}(z\Theta) = S^{*(n-1)}(\Theta)$, $n \geq 1$. Thus, we obtain the following result.

Corollary 14.5 *The model subspace generated by the inner function Θ is precisely $K_{z\Theta}$.*

Theorem 14.4 can be applied to obtain all the elements of some particular model subspaces. As the first example, suppose that Θ is a simple Blaschke factor, i.e.

$$\Theta(z) = \gamma\frac{\lambda - z}{1 - \bar{\lambda}z},$$

where $\gamma \in \mathbb{T}$ and $\lambda \in \mathbb{D}$. Then, by direct verification, we have

$$(S^*\Theta)(z) = \frac{\Theta(z) - \Theta(0)}{z} = \frac{-\gamma(1 - |\lambda|^2)}{1 - \bar{\lambda}z} = -\gamma(1 - |\lambda|^2)k_\lambda(z)$$

and, by induction,

$$S^{*n}k_\lambda = \bar{\lambda}^n k_\lambda \qquad (n \geq 1).$$

Therefore, we conclude that

$$\Theta(z) = \gamma\frac{\lambda - z}{1 - \bar{\lambda}z} \quad \Longrightarrow \quad K_\Theta = \mathbb{C}k_\lambda. \tag{14.2}$$

Moreover, instead of k_λ as a generator for this special model subspace, we can also take

$$f(z) = \frac{\sqrt{1-|\lambda|^2}}{1-\bar{\lambda}z}. \tag{14.3}$$

The advantage of this choice as a generator is that $\|f\|_2 = 1$. Now, we proceed to generalize this result for an arbitrary Blaschke product.

Lemma 14.6 *Let Θ_1 and Θ_2 be inner functions. Then*

$$K_{\Theta_1\Theta_2} = K_{\Theta_1} \oplus \Theta_1 K_{\Theta_2}.$$

Proof First, since Θ_1 divides $\Theta_1\Theta_2$, we have $K_{\Theta_1} \subset K_{\Theta_1\Theta_2}$, by Theorem 14.1. Second, since $\Theta_1 K_{\Theta_2} \subset \Theta_1 H^2$ and, by (14.1), $K_{\Theta_1} \perp \Theta_1 H^2$, we surely have $K_{\Theta_1} \perp \Theta_1 K_{\Theta_2}$. Therefore, $K_{\Theta_1} \oplus \Theta_1 K_{\Theta_2}$ is an orthogonal direct sum. Third, we show that $\Theta_1 K_{\Theta_2} \subset K_{\Theta_1\Theta_2}$. Hence, we conclude that

$$K_{\Theta_1} \oplus \Theta_1 K_{\Theta_2} \subset K_{\Theta_1\Theta_2}.$$

To verify that $\Theta_1 K_{\Theta_2} \subset K_{\Theta_1\Theta_2}$, take $f \in \Theta_1 K_{\Theta_2}$. Hence, $f = \Theta_1 g$, where $g \in K_{\Theta_2}$. Therefore, for each $h \in H^2$, we have

$$\begin{aligned}\langle f, \Theta_1\Theta_2 h\rangle_{H^2} &= \langle \Theta_1 g, \Theta_1\Theta_2 h\rangle_{H^2}\\ &= \langle g, \Theta_2 h\rangle_{H^2} = 0.\end{aligned}$$

The last equality is a consequence of the assumption $g \in K_{\Theta_2}$. Hence, we have $f \perp \Theta_1\Theta_2 H^2$, which means that $f \in K_{\Theta_1\Theta_2}$.

To establish the inverse inclusion

$$K_{\Theta_1\Theta_2} \subset K_{\Theta_1} \oplus \Theta_1 K_{\Theta_2},$$

take any $f \in K_{\Theta_1\Theta_2}$. Hence, for all $g \in H^2$,

$$\langle f, \Theta_1\Theta_2 g\rangle_{H^2} = 0.$$

But, by (14.1), there are functions $h \in H^2$ and $k \in K_{\Theta_1}$ such that

$$f = \Theta_1 h + k.$$

Therefore, for each $g \in H^2$,

$$\begin{aligned}0 &= \langle f, \Theta_1\Theta_2 g\rangle_{H^2}\\ &= \langle \Theta_1 h, \Theta_1\Theta_2 g\rangle_{H^2} + \langle k, \Theta_1\Theta_2 g\rangle_{H^2}\\ &= \langle h, \Theta_2 g\rangle_{H^2} + \langle k, \Theta_1(\Theta_2 g)\rangle_{H^2}\\ &= \langle h, \Theta_2 g\rangle_{H^2}.\end{aligned}$$

This means that $h \in K_{\Theta_2}$. In other words, $f \in K_{\Theta_1} \oplus \Theta_1 K_{\Theta_2}$. □

Let $(\lambda_k)_{1\le k\le n}$ be a finite sequence in $\mathbb{D}$, in which repetition is allowed. Write

$$b_\lambda(z) = \frac{\lambda - z}{1-\bar{\lambda}z},$$

and consider the finite Blaschke product

$$B = \prod_{k=1}^{n} b_{\lambda_k}.$$

By repeated application of Lemma 14.6, we have

$$K_B = K_{b_{\lambda_1}} \oplus b_{\lambda_1} K_{b_{\lambda_2}} \oplus b_{\lambda_1} b_{\lambda_2} K_{b_{\lambda_3}} \oplus \cdots \oplus (b_{\lambda_1} b_{\lambda_2} \cdots b_{\lambda_{n-1}}) K_{b_{\lambda_n}}.$$

By (14.2), each K_{b_λ} is one-dimensional and its unit vector is given by (14.3). Hence, based on the above orthogonal decomposition and the fact that b_{λ_k} are unimodular, we immediately deduce that K_B is an n-dimensional subspace of H^2 with the orthonormal basis $\{f_1, f_2, \dots, f_n\}$, where

$$f_j(z) = \left(\prod_{k=1}^{j-1} \frac{\lambda_k - z}{1-\bar{\lambda}_k z}\right) \frac{\sqrt{1-|\lambda_j|^2}}{1-\bar{\lambda}_j z} \qquad (1 \le j \le n). \tag{14.4}$$

This result can easily be generalized to infinite Blaschke products.

Theorem 14.7 *Let $(\lambda_k)_{k\ge1}$ be a Blaschke sequence, and denote the corresponding Blaschke product by B. Define f_j, $j \ge 1$, by* (14.4)*. Then $\{f_j : j \ge 1\}$ is an orthonormal basis for K_B.*

Proof Based on the above discussion, it is clear that $\{f_j : j \ge 1\}$ is at least an orthonormal set in K_B. To show that it is a basis, we take $f \in K_B$ such that $f \perp \{f_j : j \ge 1\}$ and show that $f = 0$.

Hence, suppose that $g \in K_B$ and $g \perp \{f_j : j \ge 1\}$. The latter, for $j = 1$, means that

$$0 = \langle g, f_1\rangle_{H^2} = \sqrt{1-|\lambda_1|^2}\, \langle g, k_{\lambda_1}\rangle_{H^2} = \sqrt{1-|\lambda_1|^2}\, g(\lambda_1).$$

Therefore, we can write $g = b_{\lambda_1} g_1$, where $g_1 \in H^2$. Since $g \perp f_2$, we also have

$$0 = \langle g, f_2\rangle_{H^2} = \sqrt{1-|\lambda_2|^2}\, \langle b_{\lambda_1} g_1, b_{\lambda_1} k_{\lambda_2}\rangle_{H^2} = \sqrt{1-|\lambda_2|^2}\, g_1(\lambda_2).$$

Therefore, we can write $g_1 = b_{\lambda_2} g_2$, and thus $g = b_{\lambda_1} b_{\lambda_2} g_2$, where $g_2 \in H^2$. Note that this reasoning works even if $\lambda_1 = \lambda_2$. By induction, we deduce that, for each $n \ge 1$, the Blaschke product $b_{\lambda_1} b_{\lambda_2} \cdots b_{\lambda_n}$ is a divisor of g. By passing to the limit, we conclude that B divides g, i.e. $g \in BH^2$. But, at the same time, $g \in K_B$. Hence, $g = 0$. □

If there is no repetition in the sequence $(\lambda_k)_{k\ge1}$, then we easily see that

$$f_j \in \operatorname{Span}\{k_{\lambda_1}, k_{\lambda_2}, \dots, k_{\lambda_j}\}$$

and conversely

$$k_{\lambda_j} \in \operatorname{Span}\{f_1, f_2, \dots, f_j\}.$$

Therefore, we obtain a weaker version of Theorem 14.7.

Corollary 14.8 *Let $(\lambda_k)_{k\ge1}$ be a Blaschke sequence of distinct points in $\mathbb{D}$, and denote the corresponding Blaschke product by B. Then we have*

$$K_B = \operatorname{Span}_{H^2}\{k_{\lambda_j} : j \ge 1\}.$$

Proof By Theorem 14.7, $\{f_j : j \ge 1\}$, where f_j is defined by (14.4), is an orthonormal basis for K_B. Since the points of $(\lambda_j)_{j\ge1}$ are distinct, it is easy to see that

$$\operatorname{Span}\{f_1, f_2, \dots, f_n\} = \operatorname{Span}\{k_{\lambda_1}, k_{\lambda_2}, \dots, k_{\lambda_n}\}$$

for each $n \ge 1$. Hence, the result follows immediately. □

As a special case of Corollary 14.8, if $\lambda_1, \dots, \lambda_n$ are n distinct points in $\mathbb{D}$, and B denotes the finite Blaschke product formed with these zeros, then K_B is an n-dimensional vector space spanned by $\{k_{\lambda_1}, k_{\lambda_2}, \dots, k_{\lambda_n}\}$.

Even though it is implicit in the definition of K_Θ, Theorem 14.4 is also a manifestation of the fact that K_Θ is invariant under S^*, i.e.

$$S^* K_\Theta \subset K_\Theta.$$

We recall that $S^* = T_{\bar{z}}$, i.e. S^* is a Toeplitz operator with an antianalytic symbol. Looking at S^* from this angle, we now generalize the preceding observation.

Lemma 14.9 *Let Θ be any inner function, and let $\varphi \in H^\infty$. Then*

$$T_{\bar{\varphi}} K_\Theta \subset K_\Theta.$$

Proof Let $f \in K_\Theta = (\Theta H^2)^\perp$. It is trivial that $T_{\bar{\varphi}} f \in H^2$. Hence, we need to show that $T_{\bar{\varphi}} f \perp \Theta g$, for all $g \in H^2$, in order to conclude that $T_{\bar{\varphi}} f \in K_\Theta$. But, by (12.2), we have

$$\begin{aligned}
\langle T_{\bar{\varphi}} f, \Theta g\rangle_{H^2} &= \langle T_\varphi^* f, \Theta g\rangle_{H^2} \\
&= \langle f, T_\varphi \Theta g\rangle_{L^2} \\
&= \langle f, \varphi\Theta g\rangle_{L^2} \\
&= \langle f, \Theta(\varphi g)\rangle_{H^2} = 0,
\end{aligned}$$

since $f \perp \Theta H^2$ and $\varphi g \in H^2$. Therefore, $T_{\bar{\varphi}} f \in K_\Theta$. □

The preceding Lemma ensures that the operator

$$\begin{array}{ccc} K_\Theta & \longrightarrow & K_\Theta \\ f & \longmapsto & T_{\bar{\varphi}} f \end{array}$$

is well defined, and since it is the restriction of $T_{\bar{\varphi}} \in \mathcal{L}(H^2)$ to K_Θ, and K_Θ inherits the topological structure of H^2, the norm of the restricted operator does not exceed $\|\varphi\|_\infty$ (see Theorem 12.2). However, the norm might be smaller than $\|\varphi\|_\infty$. For example, by Theorem 12.19(ii), if $\varphi = \Theta$, then

$$T_{\bar{\Theta}} f = 0 \qquad (f \in K_\Theta), \tag{14.5}$$

i.e. the above restricted operator is identically zero. We will discuss this result again in the study of $\mathcal{H}(b)$ spaces.

Theorem 14.10 *Let $f \in H^2$, and let $\varphi \in H^\infty$, $\varphi \not\equiv 0$. Then f is a cyclic vector for S^* if and only if so is $T_{\bar{\varphi}} f$.*

Proof Since $S^{*n} = T_{\bar{z}^n}$, $n \geq 0$, by (12.3), we have

$$S^{*n} T_{\bar{\varphi}} = T_{\bar{\varphi}} S^{*n}. \tag{14.6}$$

Assume first that $T_{\bar{\varphi}} f$ is not cyclic for S^*, i.e.

$$\text{Span}\{S^{*n} T_{\bar{\varphi}} f : n \geq 0\} \subsetneqq H^2.$$

Then, by (14.6),

$$\text{Span}\{T_{\bar{\varphi}} S^{*n} f : n \geq 0\} \subsetneqq H^2.$$

Since

$$T_{\bar{\varphi}}(\text{Span}\{S^{*n} f : n \geq 0\}) \subset \text{Span}\{T_{\bar{\varphi}} S^{*n} f : n \geq 0\},$$

and $T_{\bar{\varphi}}$ has a dense range in H^2 (Corollary 12.7), we must have

$$\text{Span}\{S^{*n} f : n \geq 0\} \subsetneqq H^2.$$

This means that f is not a cyclic vector for S^*.

Conversely, assume that f is not cyclic for S^*, i.e.

$$M = \text{Span}\{S^{*n} f : n \geq 0\} \subsetneqq H^2.$$

Since M is a closed invariant subspace of S^*, by Corollary 8.33, there is an inner function Θ such that $M = K_\Theta$. Lemma 14.9 now ensures that $T_{\bar{\varphi}} M \subset M$. In particular, we have $T_{\bar{\varphi}} f \in M$. Therefore, $T_{\bar{\varphi}} f$ is not cyclic for S^*. □

Exercise

Exercise 14.2.1 Let $n \geq 0$, and let $a_0, a_1, \dots, a_n$ be complex numbers. Consider $B(z) = z^{n+1}$ and $K_B = H^2 \ominus BH^2$.

(i) Show that $K_B = \mathrm{Lin}\{z^k : 0 \leq k \leq n\}$.

(ii) Let P_B be the orthogonal projection of H^2 onto K_B. Show that

$$P_B f = \sum_{k=0}^{n} \widehat{f}(k) z^k \qquad (f \in H^2).$$

(iii) Show that the matrix of the model operator M_B, where $M_B f = P_B(zf)$, relative to the basis $(z^k)_{0 \leq k \leq n}$, has the form

$$[M_B] = \begin{pmatrix} 0 & 0 & 0 & 0 & \cdots & 0 & 0 \\ 1 & 0 & 0 & 0 & \cdots & 0 & 0 \\ 0 & 1 & 0 & 0 & \cdots & 0 & 0 \\ 0 & 0 & 1 & 0 & \cdots & 0 & 0 \\ \vdots & \vdots & \vdots & \vdots & \ddots & \vdots & \vdots \\ 0 & 0 & 0 & 0 & \cdots & 0 & 0 \\ 0 & 0 & 0 & 0 & \cdots & 1 & 0 \end{pmatrix}.$$

14.3 The orthogonal projection P_Θ

Since K_Θ is a closed subspace of H^2, the orthogonal projection $P_{K_\Theta} \in \mathcal{L}(H^2)$, as defined in Section 1.2, is well defined. The following result reveals the connection to Toeplitz operators.

Theorem 14.11 *Let Θ be an inner function. Then*

$$P_{\Theta H^2} = T_\Theta T_{\bar{\Theta}}$$

and

$$P_{K_\Theta} = I - T_\Theta T_{\bar{\Theta}}.$$

Proof Since $I - P_M = P_{M^\perp}$, we just need to establish the first identity. To do so, we show that both sides of the identity behave the same on ΘH^2 and its orthogonal complement K_Θ.

Let $f \in \Theta H^2$, i.e. $f = \Theta g$ for some $g \in H^2$. Then, by definition, $P_{\Theta H^2} f = f$, and, since Θ is unimodular, $T_{\bar{\Theta}} f = P_+(\bar{\Theta} f) = P_+ g = g$. Thus $T_\Theta T_{\bar{\Theta}} f = T_\Theta g = P_+(\Theta g) = P_+ f = f$. Therefore,

$$T_\Theta T_{\bar{\Theta}} f = P_{\Theta H^2} f \qquad (f \in \Theta H^2).$$

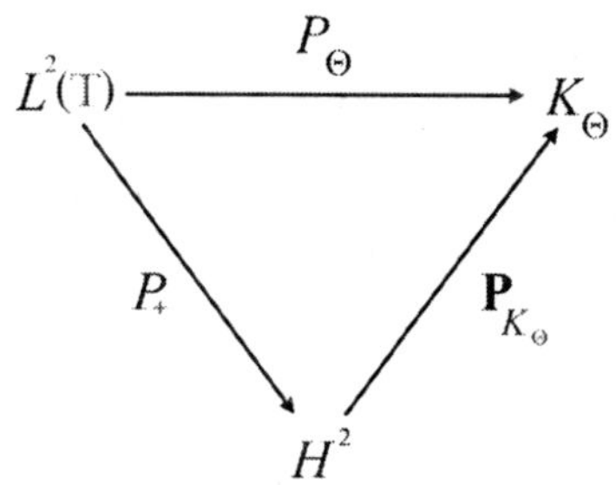

Figure 14.1 The relation between P_Θ and $\mathbf{P}_\Theta$.

Then suppose that $f \in (\Theta H^2)^\perp = K_\Theta$. By (14.5), $T_{\bar\Theta} f = 0$, and, by definition, $P_{\Theta H^2} f = 0$. Therefore, we also have

$$T_\Theta T_{\bar\Theta} f = P_{\Theta H^2} f = 0 \qquad (f \in K_\Theta). \qquad \square$$

We recall that $P_{K_\Theta} \in \mathcal{L}(H^2)$, i.e. it is an operator defined on H^2 with its values in H^2. However, we are mostly concerned with the restricted projection $\mathbf{P}_\Theta \in \mathcal{L}(H^2, K_\Theta)$, which is defined by

$$\begin{array}{rccl} \mathbf{P}_\Theta : & H^2 & \longrightarrow & K_\Theta \\ & f & \longmapsto & P_{K_\Theta} f, \end{array} \tag{14.7}$$

and the projection $P_\Theta \in \mathcal{L}(L^2(\mathbb{T}), K_\Theta)$, given by

$$P_\Theta = \mathbf{P}_\Theta \circ P_+ \tag{14.8}$$

(see Figure 14.1).

Since K_Θ is a closed subspace of H^2, and H^2 has a reproducing kernel, then so does K_Θ. Using the operator $\mathbf{P}_\Theta$ and the Cauchy kernel for H^2, we can find the kernel of K_Θ.

Corollary 14.12 *Let Θ be an inner function. Then, for each $\lambda \in \mathbb{D}$,*

$$k_\lambda^\Theta(z) = \frac{1 - \overline{\Theta(\lambda)}\Theta(z)}{1 - \bar\lambda z} \in K_\Theta$$

and, moreover,

$$f(\lambda) = \langle f, k_\lambda^\Theta \rangle_2 \qquad (f \in K_\Theta).$$

Proof We know that $k_\lambda \in H^2$. Hence, $P_{K_\Theta} k_\lambda \in K_\Theta$. But, by (12.7) and Theorem 14.11, we have

$$P_{K_\Theta} k_\lambda = k_\lambda - T_\Theta T_{\bar\Theta} k_\lambda = k_\lambda - T_\Theta(\overline{\Theta(\lambda)} k_\lambda) = (1 - \overline{\Theta(\lambda)}\Theta) k_\lambda = k_\lambda^\Theta.$$

Moreover, for each $f \in K_\Theta$, since $f = P_{K_\Theta} f$, then

$$f(\lambda) = \langle f, k_\lambda \rangle_2 = \langle P_{K_\Theta} f, k_\lambda \rangle_2 = \langle f, P_{K_\Theta} k_\lambda \rangle_2 = \langle f, k_\lambda^\Theta \rangle_2. \qquad \square$$

We now give a more explicit formula for P_Θ.

Corollary 14.13 *Let Θ be an inner function and let P_Θ be the orthogonal projection of $L^2(\mathbb{T})$ onto K_Θ; see* (14.8). *Then we have*

$$P_\Theta f = P_+ f - \Theta P_+(\bar{\Theta} f) \qquad (f \in L^2(\mathbb{T})).$$

In particular, if $f \in H^2$, then we have

$$\mathbf{P}_\Theta f = P_\Theta f = f - \Theta P_+(\bar{\Theta} f) = \Theta P_-(\bar{\Theta} f).$$

Proof Fix any $f \in L^2(\mathbb{T})$. By Theorem 14.11, (14.7) and (14.8), we have

$$P_\Theta f = P_{K_\Theta} P_+ f = P_+ f - T_\Theta T_{\bar{\Theta}} P_+ f.$$

Then, by Corollary (Theorem 4.11),

$$T_{\bar{\Theta}} P_+ f = P_+(\bar{\Theta} P_+ f) = P_+(\bar{\Theta} f).$$

Hence, we deduce that

$$P_\Theta f = P_+ f - \Theta P_+(\bar{\Theta} f).$$

Moreover, if $f \in H^2$, then, using the fact that $|\Theta| = 1$ a.e. on $\mathbb{T}$, we get

$$P_\Theta f = f - \Theta P_+(\bar{\Theta} f) = \Theta(I - P_+)(\bar{\Theta} f) = \Theta P_-(\bar{\Theta} f). \qquad \square$$

From Corollary 14.13, we can also deduce that

$$\begin{aligned} \Theta P_+(\bar{\Theta} f) &= f \qquad (f \in \Theta H^2), \\ \Theta P_-(\bar{\Theta} f) &= f \qquad (f \in K_\Theta). \end{aligned}$$

Surely, the above relations can also be proved directly.

Exercises

Exercise 14.3.1 Consider the operator

$$\begin{array}{rccc} V: & L^2(\mathbb{T}) & \longrightarrow & L^2(\mathbb{T}) \\ & f & \longmapsto & \Theta P_+ f. \end{array}$$

Show that V is a partial isometry with the initial space H^2 and final space ΘH^2. Moreover,

$$V^* f = P_+(\bar{\Theta} f) \qquad (f \in L^2(\mathbb{T})).$$

Conclude that

$$V V^* f = \Theta P_+(\bar{\Theta} f) \qquad (f \in L^2(\mathbb{T})),$$

and then give another proof for the formula for P_Θ that was given in Corollary 14.13.

Exercise 14.3.2 Let Θ be an inner function, and consider ΘH^2 as a closed subspace of $L^2(\mathbb{T})$. Find an explicit formula for the orthogonal projection from $L^2(\mathbb{T})$ onto ΘH^2.
Hint: Use Exercise 14.3.1.

14.4 The conjugation Ω_Θ

A mapping $A : \mathcal{H} \longrightarrow \mathcal{H}$, on the Hilbert space $\mathcal{H}$, is called a *conjugation* if it is antilinear, isometric and involutive. Briefly speaking, all these terms mean that $A^2 = I$ and

$$\langle Ax, Ay\rangle = \langle y, x\rangle \qquad (x, y \in \mathcal{H}).$$

Given the inner function Θ, there exists a natural conjugation Ω_Θ on the model subspace K_Θ defined by

$$\Omega_\Theta(f) = \Theta\bar{z}\bar{f} \qquad (f \in K_\Theta).$$

The verification of this fact depends heavily on the defining property $K_\Theta = H^2 \cap \Theta\overline{H_0^2}$ and the fact that $|\Theta| = 1$ a.e. on $\mathbb{T}$. In fact, the representation ensures that Ω_Θ maps K_Θ into itself. Moreover,

$$\Omega_\Theta^2(f) = \Omega_\Theta(\Theta\overline{zf}) = \Theta z\overline{\Theta\overline{zf}} = f$$

and

$$\langle \Omega_\Theta(f), \Omega_\Theta(g)\rangle = \langle \Theta\overline{zf}, \Theta\overline{zg}\rangle = \langle \bar{f}, \bar{g}\rangle = \langle g, f\rangle.$$

Thus, Ω_Θ is a conjugation on K_Θ.

For future applications, we determine the action of Ω_Θ on the reproducing kernels of K_Θ. For this purpose, given $\lambda \in \mathbb{D}$, define

$$\hat{k}_\lambda^\Theta = Q_\lambda\Theta.$$

In other words, the value $\hat{k}_\lambda^\Theta(z)$ is given by the formula

$$\hat{k}_\lambda^\Theta(z) = \frac{\Theta(z) - \Theta(\lambda)}{z - \lambda} \qquad (z \in \mathbb{D}).$$

Note that $\hat{k}_\lambda^\Theta \in H^\infty$.

Lemma 14.14 *Let Θ be an inner function and let $\lambda \in \mathbb{D}$. Then*

$$\Omega_\Theta(k_\lambda^\Theta) = \hat{k}_\lambda^\Theta.$$

Proof Since $|\Theta(z)| = 1$, for almost all $z \in \mathbb{T}$, we get

$$\begin{aligned}(\Omega_\Theta(k_\lambda^\Theta))(z) &= \bar{z}\Theta(z)\frac{1-\Theta(\lambda)\overline{\Theta(z)}}{1-\lambda\bar{z}}\\ &= \frac{\Theta(z)-\Theta(\lambda)}{z-\lambda} = \hat{k}_\lambda^\Theta(z).\end{aligned}$$

This completes the proof. □

The family $(\hat{k}_\lambda^\Theta)_{\lambda\in\mathbb{D}}$ is known as the family of *difference quotients*. Since the family of reproducing kernels is complete in K_Θ and the map Ω_Θ is isometric and onto, then we immediately obtain the following result.

Corollary 14.15 *Let Θ be an inner function. Then*

$$\text{Span}\{\hat{k}_\lambda^\Theta : \lambda \in \mathbb{D}\} = K_\Theta.$$

14.5 Minimal sequences of reproducing kernels in K_B

Let $(\lambda_n)_{n\geq 1}$ be a Blaschke sequence of distinct points in the open unit disk, and denote the corresponding Blaschke product by B. Then, by Corollary 14.8,

$$K_B = \text{Span}_{H^2}\{k_{\lambda_n} : n \geq 1\}.$$

In other words, the sequence $(k_{\lambda_n})_{n\geq 1}$ is complete in K_B. In the following result, we explore the connection to minimality.

Lemma 14.16 *Let $(\lambda_n)_{n\geq 1}$ be a sequence of distinct points in the open unit disk $\mathbb{D}$. Then the following assertions hold.*

(i) *If $(\lambda_n)_{n\geq 1}$ is not a Blaschke sequence, then $(k_{\lambda_n})_{n\geq 1}$ is complete in H^2, i.e.*

$$\text{Span}_{H^2}\{k_{\lambda_n} : n \geq 1\} = H^2,$$

but $(k_{\lambda_n})_{n\geq 1}$ is not minimal in H^2.

(ii) *If $(\lambda_n)_{n\geq 1}$ is a Blaschke sequence and if B denotes the corresponding Blaschke product, then $(k_{\lambda_n})_{n\geq 1}$ is minimal and complete in K_B.*

Proof (i) Assume that $(\lambda_n)_{n\geq 1}$ is not a Blaschke sequence. The completeness of $(k_{\lambda_n})_{n\geq 1}$ in H^2 easily follows from the fact that the zero sequence of a nonnull H^2 function must satisfy the Blaschke condition. We implicitly used the fact that these points are distinct. Otherwise, we should also consider the derivatives of f.

Now, assume that $(k_{\lambda_n})_{n\geq 1}$ is minimal. Then, according to Lemma 10.1, there exists a function $f \in H^2$ such that $f(\lambda_1) = 1$ and $f(\lambda_n) = 0$, $n \geq 2$. But, since $f \not\equiv 0$, we must have

$$\sum_{n\geq 2}(1 - |\lambda_n|) < \infty,$$

which contradicts the hypothesis. Thus, we conclude that $(k_{\lambda_n})_{n\geq 1}$ is not minimal. Another way to observe this conclusion is that, by the above paragraph, the span of $(\lambda_n)_{n\geq k}$, where $k \geq 2$ is fixed but arbitrary, is also H^2. Hence, $(k_{\lambda_n})_{n\geq 1}$ is not minimal in H^2.

(ii) Assume now that $(\lambda_n)_{n\geq 1}$ is a Blaschke sequence, and let B be the corresponding Blaschke product. Denote by B_n the Blaschke product corresponding to the sequence $(\lambda_k)_{k\neq n}$, that is,

$$B_n = \frac{B}{b_{\lambda_n}} = \prod_{k\neq n} b_{\lambda_k},$$

where we recall that

$$b_\lambda(z) = \frac{|\lambda|}{\lambda}\frac{\lambda - z}{1 - \bar{\lambda}z} \qquad (z, \lambda \in \mathbb{D}).$$

Clearly, for each $n \geq 1$, the function B_n belongs to H^2 and $B_n(\lambda_n) \neq 0$. Moreover, for each $n, p \geq 1$, we have

$$\left\langle \frac{B_n}{B_n(\lambda_n)}, k_{\lambda_p} \right\rangle_{H^2} = \frac{B_n(\lambda_p)}{B_n(\lambda_n)} = \delta_{n,p}. \tag{14.9}$$

Again, this is the place where we need the points λ_n to be distinct. Hence, $(B_n/B_n(\lambda_n))_{n\geq 1}$ is a biorthogonal sequence of $(k_{\lambda_n})_{n\geq 1}$. Therefore, by Lemma 10.1, we conclude that $(k_{\lambda_n})_{n\geq 1}$ is minimal in H^2. The completeness of $(k_{\lambda_n})_{n\geq 1}$ in K_B was already shown in Corollary 14.8. Hence, $(k_{\lambda_n})_{n\geq 1}$ is in fact minimal in K_B. □

It follows from Lemma 14.16 that a distinct sequence $(k_{\lambda_n})_{n\geq 1}$ of reproducing kernels in H^2 is minimal if and only if $\Lambda = (\lambda_n)_{n\geq 1}$ is a Blaschke sequence in the open unit disk $\mathbb{D}$. Moreover, in this case, the sequence $(k_{\lambda_n})_{n\geq 1}$ is complete in K_B. Hence, it follows from Lemma 10.1 that $(k_{\lambda_n})_{n\geq 1}$ has a unique biorthogonal sequence in K_B, say $(k^*_{\lambda_n})_{n\geq 1}$. See also the comments preceding Corollary 10.15. It will be important for us to have a formula for this biorthogonal. This is done below.

Lemma 14.17 *Let $\Lambda = (\lambda_n)_{n\geq 1}$ be a Blaschke sequence of distinct points in $\mathbb{D}$ and let B be the associated Blaschke product. Then the unique biorthogonal sequence associated with $(k_{\lambda_n})_{n\geq 1}$ in K_B is $(k^*_{\lambda_n})_{n\geq 1}$, given by*

$$k^*_{\lambda_n} = \frac{(1-|\lambda_n|^2)}{B_n(\lambda_n)} B_n k_{\lambda_n} = \frac{(1-|\lambda_n|^2)}{B_n(\lambda_n)} \frac{\lambda_n}{|\lambda_n|} \frac{B}{z-\lambda_n},$$

where $B_n = B/b_{\lambda_n}$.

Proof We recall that $\mathbf{P}_B$ is the orthogonal projection of H^2 onto K_B; see (14.7). Since $k_{\lambda_p} \in K_B$, we have $\mathbf{P}_B k_{\lambda_p} = k_{\lambda_p}$ and thus, by (14.9),

$$\left\langle \frac{\mathbf{P}_B B_n}{B_n(\lambda_n)}, k_{\lambda_p} \right\rangle_{H^2} = \left\langle \frac{B_n}{B_n(\lambda_n)}, \mathbf{P}_B k_{\lambda_p} \right\rangle_{H^2} = \delta_{n,p}.$$

Hence, we have

$$k^*_{\lambda_n} = \frac{1}{B_n(\lambda_n)} \mathbf{P}_B B_n.$$

Then, using Corollary 14.13, we have

$$\mathbf{P}_B B_n = BP_-(\bar{B} B_n) = BP_- \bar{b}_{\lambda_n}$$

and

$$P_- \bar{b}_{\lambda_n} = \bar{b}_{\lambda_n} - P_+ \bar{b}_{\lambda_n} = \bar{b}_{\lambda_n} - \overline{b_{\lambda_n}(0)} = \bar{b}_{\lambda_n} - |\lambda_n|.$$

Since

$$b_{\lambda_n} - |\lambda_n| = \frac{|\lambda_n|}{\lambda_n}\left(\frac{\lambda_n - z}{1-\bar{\lambda}_n z} - \lambda_n \right) = -\frac{|\lambda_n|}{\lambda_n} \frac{(1-|\lambda_n|^2)z}{1-\bar{\lambda}_n z},$$

we deduce that

$$(P_- \bar{b}_{\lambda_n})(z) = \frac{|\lambda_n|}{\bar{\lambda}_n} \frac{1-|\lambda_n|^2}{\lambda_n - z} \qquad (z \in \mathbb{T}).$$

Hence, we have

$$\begin{aligned} \mathbf{P}_B B_n &= \frac{|\lambda_n|}{\bar{\lambda}_n}(1-|\lambda_n|^2)\frac{B}{\lambda_n - z} \\ &= (1-|\lambda_n|^2)\frac{B_n}{1-\bar{\lambda}_n z} = (1-|\lambda_n|^2) B_n k_{\lambda_n}. \end{aligned}$$

This completes the proof. □

If $\mathfrak{X} = (x_n)_{n\geq 1}$ is a minimal and complete sequence in a Banach space $\mathcal{X}$, then $\mathfrak{X}$ has a unique biorthogonal sequence $\mathfrak{X}^* = (x^*_n)_{n\geq 1}$. But, in general, the sequence $\mathfrak{X}^*$ is not complete in $\mathcal{X}$. However, it is rather surprising that such a result is true for sequences of reproducing kernels in H^2.

Theorem 14.18 *Let $\Lambda = (\lambda_n)_{n\geq 1}$ be a Blaschke sequence of distinct points in $\mathbb{D}$, let B be the associated Blaschke product, and let $(k^*_{\lambda_n})_{n\geq 1}$ be the unique biorthogonal sequence associated with $(k_{\lambda_n})_{n\geq 1}$ in K_B. Then the sequence $(k^*_{\lambda_n})_{n\geq 1}$ is complete in K_B, that is,*

$$\mathrm{Span}_{H^2}\{k^*_{\lambda_n} : n \geq 1\} = K_B.$$

Proof According to Lemma 14.17, we have

$$k^*_{\lambda_n} = \frac{(1-|\lambda_n|^2)}{B_n(\lambda_n)} B_n k_{\lambda_n}.$$

Remember that Ω_B denotes the conjugation on K_B. Then, by Lemma 14.14,

$$\Omega_B(k_{\lambda_n}) = \Omega_B(k^B_{\lambda_n}) = \frac{B}{z-\lambda_n} = c_n k^*_{\lambda_n},$$

where

$$c_n = -\frac{|\lambda_n|}{\lambda_n}\frac{B_n(\lambda_n)}{1-|\lambda_n|^2} \neq 0.$$

The precise value of c_n is irrelevant here. Using the fact that Ω_B is surjective and an isometry on K_B, we conclude that

$$\begin{aligned}\mathrm{Span}_{H^2}\{k^*_{\lambda_n} : n \geq 1\} &= \mathrm{Span}_{H^2}\{c_n k^*_{\lambda_n} : n \geq 1\}\\ &= \mathrm{Span}_{H^2}\{\Omega_B k_{\lambda_n} : n \geq 1\}\\ &= \Omega_B\, \mathrm{Span}_{H^2}\{k_{\lambda_n} : n \geq 1\}\\ &= \Omega_B K_B = K_B.\end{aligned}$$

This completes the proof. □

14.6 The operators J and $\mathbf{M}_\Theta$

On the one hand, we have the orthogonal decomposition

$$L^2(\mathbb{T}) = H^2_- \oplus H^2 = H^2_- \oplus K_\Theta \oplus \Theta H^2$$

and, on the other, since multiplication by Θ acts as a unitary operator on L^2, the orthogonal decomposition $\bar{\Theta}L^2(\mathbb{T}) = L^2(\mathbb{T}) = H^2_- \oplus H^2$ is rewritten as

$$L^2(\mathbb{T}) = \Theta H^2_- \oplus \Theta H^2.$$

Hence, comparing the previous identities, we obtain the orthogonal decomposition

$$\Theta H^2_- = H^2_- \oplus K_\Theta. \tag{14.10}$$

For the following result, let $J : L^2(\mathbb{T}) \longrightarrow L^2(\mathbb{T})$ be the mapping defined by

$$Jf = \bar{z}\bar{f} \qquad (f \in L^2(\mathbb{T})).$$

Two easily verified properties of J are

$$J^2 = I \quad \text{and} \quad JP_+J = P_-. \tag{14.11}$$

Hence, we have

$$JP_+ = P_-J \quad \text{and} \quad P_+J = JP_-. \tag{14.12}$$

Finally, if $\varphi \in L^\infty(\mathbb{T})$ and M_φ denotes multiplication by φ on $L^2(\mathbb{T})$, then

$$M_\varphi J = JM_{\bar{\varphi}} \tag{14.13}$$

(see Exercise 14.6.1). This identity is usually written as $\varphi J = J\bar{\varphi}$. It implies the following two special cases:

$$J\bar{\Theta}J = \Theta \tag{14.14}$$

and

$$\Theta J\Theta = J, \tag{14.15}$$

which are needed in the proof of the next lemma.

Lemma 14.19 *Let Θ_1 and Θ_2 be two inner functions. Then we have*

$$(\Theta_1 JT_{\Theta_1\overline{\Theta_2}}J\overline{\Theta_2})f = f \qquad (f \in H^2_-)$$

and

$$(\Theta_1 JT_{\Theta_1\overline{\Theta_2}}J\overline{\Theta_2})f = P_{\Theta_1}f \qquad (f \in K_{\Theta_2}).$$

Proof Fix $f \in H^2_-$. Put $g = Jf$. Then, by (14.11), $f = Jg$ and $g \in H^2$. Hence, by (14.14) and (14.15), we have

$$\begin{aligned}
(\Theta_1 JT_{\Theta_1\overline{\Theta_2}}J\overline{\Theta_2})f &= (\Theta_1 JT_{\Theta_1\overline{\Theta_2}}J\overline{\Theta_2})(Jg)\\
&= (\Theta_1 JT_{\Theta_1\overline{\Theta_2}})(J\overline{\Theta_2}Jg)\\
&= (\Theta_1 JT_{\Theta_1\overline{\Theta_2}})(\Theta_2 g)\\
&= (\Theta_1 J)(P_+\Theta_1 g)\\
&= (\Theta_1 J)(\Theta_1 g)\\
&= Jg = f,
\end{aligned}$$

which reveals that $\Theta_1 JT_{\Theta_1\overline{\Theta_2}}J\overline{\Theta_2}$ acts as the identity on H^2_-.

Now let $f \in K_{\Theta_2}$ and write $f = \Theta_2 \bar{z}\bar{g} = \Theta_2 J g$, with $g \in H^2$. Then, by (14.12),

$$\begin{aligned}(\Theta_1 J T_{\Theta_1\overline{\Theta_2}} J \overline{\Theta_2})(f) &= (\Theta_1 J T_{\Theta_1\overline{\Theta_2}} J \overline{\Theta_2})(\Theta_2 J g)\\ &= (\Theta_1 J T_{\Theta_1\overline{\Theta_2}} J^2)(g)\\ &= (\Theta_1 J T_{\Theta_1\overline{\Theta_2}})(g)\\ &= (\Theta_1 J P_+)(\Theta_1\overline{\Theta_2} g)\\ &= (\Theta_1 P_- J)(\Theta_1\overline{\Theta_2} g)\\ &= (\Theta_1 P_-)(\bar{z}\overline{\Theta_1}\Theta_2\bar{g})\\ &= (\Theta_1 P_- \overline{\Theta_1})(f).\end{aligned}$$

Hence, by Corollary 14.13,

$$(\Theta_1 J T_{\Theta_1\overline{\Theta_2}} J \overline{\Theta_2})(f) = P_{\Theta_1}(f). \qquad \square$$

In light of the decomposition (14.10), Lemma 14.19 can be rewritten as

$$\Theta_1 J T_{\Theta_1\overline{\Theta_2}} J \overline{\Theta_2}_{|\Theta_2 H^2_-} = I_{H^2_-} \oplus P_{\Theta_1|K_{\Theta_2}}. \tag{14.16}$$

If Θ is inner, for any fixed $w \in \mathbb{D}$, the function

$$\Theta_w(z) = \frac{\Theta(w) - \Theta(z)}{1 - \overline{\Theta(w)}\Theta(z)} \qquad (z \in \mathbb{D})$$

is also inner. In fact, a result of Frostman says that Θ_w is a Blaschke product for all $w \in \mathbb{D}$, except possibly on a set of logarithmic capacity zero. But, we do not need this result here. The elementary fact that Θ_w is inner will be enough for us.

Since, for each $z \in \mathbb{D}$,

$$2 \ge |1 - \overline{\Theta(w)}\Theta(z)| \ge 1 - |\Theta(w)| > 0,$$

the formula

$$h_w(z) = \frac{(1 - |\Theta(w)|^2)^{1/2}}{1 - \overline{\Theta(w)}\Theta(z)}$$

gives us an outer function with $h_w, 1/h_w \in H^\infty$.

Theorem 14.20 *Let Θ be an inner function, and let $w \in \mathbb{D}$. Then*

$$\begin{array}{rccc} T_{h_w} : & K_\Theta & \longrightarrow & K_{\Theta_w} \\ & f & \longmapsto & h_w f \end{array}$$

is a unitary operator, i.e. T_{h_w} acts as an isometry from K_Θ onto K_{Θ_w}.

Proof For almost all $\zeta \in \mathbb{T}$, we have

$$\begin{aligned}\Theta_w(\zeta)\overline{h_w(\zeta)} &= \frac{\Theta(w)-\Theta(\zeta)}{1-\overline{\Theta(w)}\Theta(\zeta)}\,\frac{(1-|\Theta(w)|^2)^{1/2}}{1-\Theta(w)\overline{\Theta(\zeta)}}\\ &= \frac{\Theta(w)-\Theta(\zeta)}{1-\overline{\Theta(w)}\Theta(\zeta)}\,\frac{(1-|\Theta(w)|^2)^{1/2}}{\Theta(\zeta)-\Theta(w)}\Theta(\zeta)\\ &= -\Theta(\zeta)h_w(\zeta).\end{aligned}$$

Hence, we have

$$\Theta_w\overline{h_w} = -\Theta h_w. \tag{14.17}$$

To verify that $T_{h_w}K_\Theta \subset K_{\Theta_w}$, take any function $f \in K_\Theta$. Since $h_w \in H^\infty$, it is trivial that $h_w f \in H^2$. If we write $f = \Theta\bar{z}\bar{f}_1$, then, using (14.17), we get

$$h_w f = h_w\Theta\bar{z}\overline{f_1} = \Theta_w\overline{f_2},$$

with $f_2 = -zh_w f_1$. Since $f_1 \in H^2$ and $h_w \in H^\infty$, we have $f_2 \in H^2_0$ and thus

$$h_w f \in H^2 \cap \Theta_w\overline{H^2_0} = K_{\Theta_w}.$$

Moreover, we have

$$\|h_w f\|_2 = (1-|\Theta(w)|^2)^{1/2}\left\|\frac{f}{1-\overline{\Theta(w)}\Theta}\right\|_2.$$

But, now write

$$\begin{aligned}\frac{f}{1-\overline{\Theta(w)}\Theta} &= \frac{1-\overline{\Theta(w)}\Theta+\overline{\Theta(w)}\Theta}{1-\overline{\Theta(w)}\Theta}f\\ &= f+\overline{\Theta(w)}\frac{\Theta f}{1-\overline{\Theta(w)}\Theta},\end{aligned}$$

and note that $\Theta f/(1-\overline{\Theta(w)}\Theta)$ belongs to ΘH^2, and $f \in K_\Theta$. Thus,

$$\left\|\frac{f}{1-\overline{\Theta(w)}\Theta}\right\|_2^2 = \|f\|_2^2 + |\Theta(w)|^2 \times \left\|\frac{f}{1-\overline{\Theta(w)}\Theta}\right\|_2^2,$$

which gives

$$(1-|\Theta(w)|^2)\left\|\frac{f}{1-\overline{\Theta(w)}\Theta}\right\|_2^2 = \|f\|_2^2.$$

Thus $\|h_w f\|_2 = \|f\|_2$.

It remains to prove that the operator is surjective. Let $f \in K_{\Theta_w}$. Then $f/h_w \in H^2$ (remember, $1/h_w \in H^\infty$). Moreover, if we write $f = \Theta_w\bar{z}\bar{f}_1$, with $f_1 \in H^2$, then, using (14.17), we have

$$\frac{f}{h_w} = \frac{\Theta_w}{h_w}\bar{z}\bar{f}_1 = -\Theta\bar{f}_2, \quad \text{where } f_2 = z\frac{f_1}{h_w} \in H^2_0.$$

This shows that $f/h_w \in H^2 \cap \Theta\overline{H^2_0} = K_\Theta$. Clearly, $T_{h_w}(f/h_w) = f$. □

The mapping T_{h_w} is called the *Crofoot transform.*

As we have seen in Corollary 8.33, the subspace K_Θ is invariant under the unilateral backward shift S^*. In particular, we can consider S^* as an operator from K_Θ into itself, and we denote this restriction by X_Θ. In other words,

$$\begin{array}{rccl} X_\Theta : & K_\Theta & \longrightarrow & K_\Theta \\ & f & \longmapsto & S^* f. \end{array}$$

Note that we can write

$$X_\Theta = \mathbf{P}_\Theta S^* \mathbf{i}_{K_\Theta}, \tag{14.18}$$

where $\mathbf{P}_\Theta \in \mathcal{L}(H^2, K_\Theta)$ is the restricted orthogonal projection of H^2 onto K_Θ and $\mathbf{i}_{K_\Theta} \in \mathcal{L}(K_\Theta, L^2)$ is the inclusion mapping. Hence,

$$X_\Theta^* = \mathbf{P}_\Theta S \mathbf{i}_{K_\Theta}. \tag{14.19}$$

This is the so-called *compression of the shift operator*. This terminology is justified by Corollary 1.42. Since $K_\Theta = H^2 \ominus \Theta H^2$, this theorem implies that

$$p(\mathbf{P}_\Theta S \mathbf{i}_{K_\Theta}) = \mathbf{P}_\Theta p(S) \mathbf{i}_{K_\Theta}, \tag{14.20}$$

where p is any analytic polynomial.

Since S is a multiplication operator and X_Θ^* is its compression to K_Θ, in the following we also denote X_Θ^* by $\mathbf{M}_\Theta$. More explicitly, we write

$$\mathbf{M}_\Theta = \mathbf{P}_\Theta S \mathbf{i}_{K_\Theta} \tag{14.21}$$

Thus, the relation (14.20) becomes

$$p(\mathbf{M}_\Theta) = \mathbf{P}_\Theta p(S) \mathbf{i}_{K_\Theta} = p(S)_{|K_\Theta}. \tag{14.22}$$

Since

$$\mathbf{P}_\Theta f = P_\Theta f \qquad (f \in H^2),$$

the identity (14.22) can also be rewritten as

$$p(\mathbf{M}_\Theta) f = P_\Theta (pf) \qquad (f \in H^2). \tag{14.23}$$

Let us treat a simple example. If $\Theta(z) = z^{n+1}$, then

$$K_\Theta = \mathrm{Lin}\{z^k : 0 \le k \le n\},$$

and, since $(z^k)_{0 \le k \le n}$ is an orthonormal basis for K_Θ, we have

$$\mathbf{P}_\Theta f = \sum_{k=0}^{n} \hat{f}(k) z^k \qquad (f \in H^2).$$

The action of $\mathbf{M}_\Theta$ on $(z^k)_{0 \le k \le n}$ is thus given by

$$\mathbf{M}_\Theta z^k = \mathbf{P}_\Theta S \mathbf{i}_{K_\Theta} z^k = \mathbf{P}_\Theta z^{k+1} = \begin{cases} z^{k+1} & \text{if} \quad 0 \le k < n, \\ 0 & \text{if} \quad k = n. \end{cases}$$

Therefore, the matrix of $\mathbf{M}_\Theta$ with respect to this basis is the Jordan matrix

$$[\mathbf{M}_\Theta] = \begin{pmatrix} 0 & 0 & 0 & 0 & \cdots & 0 & 0 \\ 1 & 0 & 0 & 0 & \cdots & 0 & 0 \\ 0 & 1 & 0 & 0 & \cdots & 0 & 0 \\ 0 & 0 & 1 & 0 & \cdots & 0 & 0 \\ \vdots & \vdots & \vdots & \vdots & \ddots & \vdots & \vdots \\ 0 & 0 & 0 & 0 & \cdots & 0 & 0 \\ 0 & 0 & 0 & 0 & \cdots & 1 & 0 \end{pmatrix}.$$

For the description of the spectrum of $\mathbf{M}_\Theta$, the following result will be needed.

Lemma 14.21 *Let Θ be an inner function. Then, for each $f \in K_\Theta$, we have*

$$\lim_{n\to\infty} \|X_\Theta^n f\|_2 = 0$$

and

$$\lim_{n\to\infty} \|\mathbf{M}_\Theta^n f\|_2 = 0.$$

Proof The assertion for X_Θ is rather trivial since X_Θ is the restriction of S^* to K_Θ. Hence, by (8.16), for each $f \in K_\Theta$, we have

$$\|X_\Theta^n f\|_2^2 = \|S^{*n} f\|_2^2 = \sum_{k\geq n} |\hat{f}(k)|^2 \longrightarrow 0.$$

For the second assertion, note that, by (14.23), we have

$$\mathbf{M}_\Theta^n f = P_\Theta(z^n f) \qquad (f \in K_\Theta).$$

Thus, by Corollary 14.13,

$$\mathbf{M}_\Theta^n f = z^n f - \Theta P_+(\bar{\Theta} z^n f),$$

which implies that

$$\|\mathbf{M}_\Theta^n f\|_2 = \|\bar{\Theta} f - \bar{z}^n P_+(z^n \bar{\Theta} f)\|_2.$$

The result now follows from Lemma 12.1. □

Lemma 14.22 *Let Θ be an inner function, and let $\lambda \in \mathbb{D}$. Then we have*

$$k_\lambda^\Theta = (I - \bar{\lambda}\mathbf{M}_\Theta)^{-1} k_0^\Theta.$$

Proof By (8.19), for each $f \in H^2$, we have

$$f(\lambda) = \langle (I - \lambda S^*)^{-1} f, k_0 \rangle_2.$$

Thus, if $f \in K_\Theta$, we can write

$$\begin{aligned} f(\lambda) &= \langle (I - \lambda X_\Theta)^{-1} f, k_0 \rangle_2 \\ &= \langle (I - \lambda X_\Theta)^{-1} f, P_\Theta k_0 \rangle_2 \\ &= \langle (I - \lambda X_\Theta)^{-1} f, k_0^\Theta \rangle_2 \\ &= \langle f, (I - \bar{\lambda} X_\Theta^*)^{-1} k_0^\Theta \rangle_2. \end{aligned}$$

Hence, we obtain

$$k_\lambda^\Theta = (I - \bar{\lambda} X_\Theta^*)^{-1} k_0^\Theta = (I - \bar{\lambda} \mathbf{M}_\Theta)^{-1} k_0^\Theta. \qquad \square$$

Exercises

Exercise 14.6.1 Let $\varphi \in L^\infty(\mathbb{T})$. Remember that we denote the operator of multiplication by φ on $L^2(\mathbb{T})$ by M_φ. Show that

$$M_\varphi J = J M_{\bar{\varphi}}.$$

Deduce (14.14) and (14.15).

Exercise 14.6.2 Let Θ be an inner function. Recall that K_Θ is the closed S^*-invariant subspace of H^2 defined by $K_\Theta = (\Theta H^2)^\perp$. For simplicity, we denote X_{K_Θ} by X_Θ. Let $\mathcal{E}$ be a closed subspace of K_Θ. Show that $\mathcal{E}$ is invariant under X_Θ if and only if there is an inner function Θ_1 dividing Θ such that $\mathcal{E} = K_{\Theta_1}$.
Hint: Use Corollary 8.33 and Theorem 14.1.

14.7 Functional calculus for $\mathbf{M}_\Theta$

The following result, a simple consequence of Theorem 1.41, is the main step in the construction of H^∞ functional calculus for the operator $\mathbf{M}_\Theta$.

Lemma 14.23 *Let Θ be an inner function, and let p be an analytic polynomial. Then*

$$p(\mathbf{M}_\Theta) = \mathbf{P}_\Theta p(S) \mathbf{i}_{K_\Theta} = \mathbf{P}_\Theta T_p \mathbf{i}_{K_\Theta}$$

and

$$\|p(\mathbf{M}_\Theta)\|_{\mathcal{L}(K_\Theta)} \le \|p\|_{H^\infty}.$$

Proof The first identity was established before; see (14.20). The second formula follows at once from (8.23). Hence, by Theorem 12.2,

$$\|p(\mathbf{M}_\Theta)\|_{\mathcal{L}(K_\Theta)} = \|\mathbf{P}_\Theta T_p \mathbf{i}_{K_\Theta}\|_{\mathcal{L}(K_\Theta)} \le \|T_p\|_{\mathcal{L}(H^2)} = \|p\|_\infty. \qquad \square$$

In light of the identity $p(\mathbf{M}_\Theta) = \mathbf{P}_\Theta T_p \mathbf{i}_{K_\Theta}$, which was established in Lemma 14.23, we define

$$\varphi(\mathbf{M}_\Theta) = \mathbf{P}_\Theta T_\varphi \mathbf{i}_{K_\Theta}, \tag{14.24}$$

for any $\varphi \in H^\infty$. More explicitly, $\varphi(\mathbf{M}_\Theta)$ is given by the formula

$$\varphi(\mathbf{M}_\Theta)(f) = \mathbf{P}_\Theta(\varphi f) = P_\Theta(\varphi f) \qquad (f \in K_\Theta). \tag{14.25}$$

First, it is clear that $\varphi(\mathbf{M}_\Theta)$ is a linear map from K_Θ into itself. Second,

$$\|\varphi(\mathbf{M}_\Theta)\|_{\mathcal{L}(K_\Theta)} = \|\mathbf{P}_\Theta T_\varphi \mathbf{i}_{K_\Theta}\|_{\mathcal{L}(K_\Theta)} \le \|T_\varphi\|_{\mathcal{L}(H^2)} = \|\varphi\|_{H^\infty}.$$

The mapping

$$\begin{array}{rccc} \Lambda: & H^\infty & \longrightarrow & \mathcal{L}(K_\Theta) \\ & \varphi & \longmapsto & \varphi(\mathbf{M}_\Theta) \end{array}$$

is called the H^∞ *functional calculus* for the operator $\mathbf{M}_\Theta$.

Theorem 14.24 *Let Θ be an inner function. Then the mapping Λ is linear, multiplicative and contractive.*

Proof Considering the discussion above, the only fact that remains to be proved is that the mapping is multiplicative. To verify this fact, let $\varphi, \psi \in H^\infty$ and let $f \in K_\Theta$. Then we have

$$\begin{aligned} \Lambda(\varphi\psi)f &= (\varphi\psi)(\mathbf{M}_\Theta)f \\ &= \mathbf{P}_\Theta(\varphi\psi f) \\ &= \mathbf{P}_\Theta(\varphi \mathbf{P}_\Theta(\psi f)) + \mathbf{P}_\Theta(\varphi(I - \mathbf{P}_\Theta)(\psi f)). \end{aligned}$$

But, $\varphi(I - \mathbf{P}_\Theta)(\psi f) \in \varphi\Theta H^2 \subset \Theta H^2 = K_\Theta^\perp$, whence

$$\Lambda(\varphi\psi)f = \mathbf{P}_\Theta(\varphi \mathbf{P}_\Theta(\psi f)) = \varphi(\mathbf{M}_\Theta)\psi(\mathbf{M}_\Theta)f = \Lambda(\varphi)\Lambda(\psi)f. \qquad \square$$

In the following results, we gather the main properties of this functional calculus.

Theorem 14.25 *Let Θ be an inner function, and let $\varphi \in H^\infty$. Then the following assertions hold.*

(i) *For each $f \in K_\Theta$, we have*

$$\varphi(\mathbf{M}_\Theta)^* f = P_+(\bar{\varphi} f) = T_{\bar{\varphi}} f.$$

(ii) *If* $\sum_{n\geq 0} |\hat{\varphi}(n)| < \infty$, *then*

$$\varphi(\mathbf{M}_\Theta) = \sum_{n=0}^{\infty} \hat{\varphi}(n)\mathbf{M}_\Theta^n.$$

The series converges in $\mathcal{L}(K_\Theta)$.

(iii) *We have* $\varphi(\mathbf{M}_\Theta) = 0$ *if and only if* $\varphi \in \Theta H^\infty$.

Proof (i) By (14.24),

$$\varphi(\mathbf{M}_\Theta)^* = \mathbf{P}_\Theta T_\varphi^* \mathbf{i}_{K_\Theta} = \mathbf{P}_\Theta T_{\bar{\varphi}} \mathbf{i}_{K_\Theta}.$$

Note that $\mathbf{i}_{K_\Theta}^* = \mathbf{P}_\Theta$. Hence,

$$\varphi(\mathbf{M}_\Theta)^* f = \mathbf{P}_\Theta T_{\bar{\varphi}} f.$$

But, by Lemma 14.9, $T_{\bar{\varphi}} f \in K_\Theta$. Therefore,

$$\varphi(\mathbf{M}_\Theta)^* f = T_{\bar{\varphi}} f.$$

(ii) Put

$$\varphi_N(z) = \sum_{n=0}^{N} \widehat{\varphi}(n) z^n \qquad (N \in \mathbb{N}).$$

Then we have

$$\varphi_N(\mathbf{M}_\Theta) = \sum_{n=0}^{N} \widehat{\varphi}(n)\mathbf{M}_\Theta^n$$

and, using Theorem 14.24, we get

$$\begin{aligned}\left\| \sum_{n=0}^{N} \widehat{\varphi}(n)\mathbf{M}_\Theta^n - \varphi(\mathbf{M}_\Theta) \right\|_{\mathcal{L}(K_\Theta)} &= \|\varphi_N(\mathbf{M}_\Theta) - \varphi(\mathbf{M}_\Theta)\|_{\mathcal{L}(K_\Theta)} \\ &= \|\Lambda(\varphi_N - \varphi)\|_{\mathcal{L}(K_\Theta)} \\ &\leq \|\varphi_N - \varphi\|_\infty.\end{aligned}$$

Since we have assumed that $\sum_{n\geq 0} |\widehat{\varphi}(n)| < \infty$, then $\|\varphi_N - \varphi\|_\infty \longrightarrow 0$, as $N \longrightarrow \infty$. Thus, $\sum_{n=0}^{N} \widehat{\varphi}(n)\mathbf{M}_\Theta^n$ converges to $\varphi(\mathbf{M}_\Theta)$ as $N \longrightarrow \infty$.

(iii) Assume that $\varphi = \Theta g$ for some $g \in H^\infty$. Then, for each $f \in K_\Theta$, we have

$$\varphi(\mathbf{M}_\Theta) f = P_\Theta(\varphi f) = P_\Theta(\Theta f g) = 0,$$

because $\Theta f g \in \Theta H^2 \subset (K_\Theta)^\perp$. Conversely, assume that $\varphi(\mathbf{M}_\Theta) = 0$. Then, for each $f \in H^2$, we have

$$P_\Theta(\varphi f) = P_\Theta(\varphi P_\Theta f) + P_\Theta(\varphi P_{\Theta H^2} f) = \varphi(\mathbf{M}_\Theta) P_\Theta f = 0,$$

which implies that $\varphi H^2 \subset \Theta H^2$. Hence, $\varphi = g\Theta$, for some $g \in H^2$. But taking the absolute values of both sides shows that $g \in H^\infty$. Therefore, $\varphi \in \Theta H^\infty$. □

Theorem 14.26 *Let Θ be an inner function, and let $\varphi \in H^\infty$. Let $(\varphi_n)_{n\geq 1}$ be a sequence of functions in H^∞ such that*

$$\sup_{n\geq 1} \|\varphi_n\|_\infty < \infty.$$

Then the following assertions hold.

(i) *If*

$$\lim_{n\to\infty} \varphi_n(\zeta) = \varphi(\zeta) \qquad (\textit{a.e. on } \mathbb{T}),$$

then $\varphi_n(\mathbf{M}_\Theta)$ tends to $\varphi(\mathbf{M}_\Theta)$ with respect to the strong operator topology.

(ii) *If*

$$\lim_{n\to\infty} \varphi_n(\lambda) = \varphi(\lambda) \qquad (\lambda \in \mathbb{D}),$$

then $\varphi_n(\mathbf{M}_\Theta)$ tends to $\varphi(\mathbf{M}_\Theta)$ with respect to the weak operator topology.

Proof (i) Fix $f \in K_\Theta$. Then we have

$$\|\varphi_n(\mathbf{M}_\Theta)f - \varphi(\mathbf{M}_\Theta)f\|_2 = \|P_\Theta(\varphi_n f) - P_\Theta(\varphi f)\|_2 \leq \|\varphi_n f - \varphi f\|_2$$

and

$$\|\varphi_n f - \varphi f\|_2^2 = \frac{1}{2\pi}\int_0^{2\pi} |\varphi_n(e^{i\theta}) - \varphi(e^{i\theta})|^2 |f(e^{i\theta})|^2 \, d\theta.$$

Since $\varphi_n(e^{i\theta}) - \varphi(e^{i\theta}) \longrightarrow 0$, a.e. on $\mathbb{T}$, and

$$|\varphi_n(e^{i\theta}) - \varphi(e^{i\theta})|\,|f(e^{i\theta})| \leq (\|\varphi_n\|_\infty + \|\varphi\|_\infty)\,|f(e^{i\theta})| \leq C|f(e^{i\theta})|,$$

where C is a constant independent of n, then Lebesgue's dominated convergence theorem implies that $\lim_{n\to\infty} \|\varphi_n f - \varphi f\|_2 = 0$. Thus, we obtain $\varphi_n(\mathbf{M}_\Theta)f \longrightarrow \varphi(\mathbf{M}_\Theta)f$ in norm.

(ii) Fix $f, g \in K_\Theta$. Then we have

$$\begin{aligned}
\langle \varphi_n(\mathbf{M}_\Theta)f - \varphi(\mathbf{M}_\Theta)f, g\rangle &= \langle P_\Theta(\varphi_n f) - P_\Theta(\varphi f), g\rangle \\
&= \langle \varphi_n f - \varphi f, g\rangle \\
&= \frac{1}{2\pi}\int_0^{2\pi} (\varphi_n(e^{i\theta}) - \varphi(e^{i\theta}))f(e^{i\theta})\overline{g(e^{i\theta})}\, d\theta.
\end{aligned}$$

Since $f(e^{i\theta})\overline{g(e^{i\theta})} \in L^1(\mathbb{T})$, it is sufficient to apply Theorem 4.16 to conclude. □

Exercises

Exercise 14.7.1 Let Θ be an inner function, and let $\varphi \in H^\infty$. Show that

$$P_\Theta(\varphi f) = \varphi(\mathbf{M}_\Theta)(P_\Theta f) \qquad (f \in H^2).$$

Remark: Compare with (14.25).

Exercise 14.7.2 Let Θ be an inner function. Show that

$$\Theta(\mathbf{M}_\Theta) = 0.$$

Hint: Apply (14.24).

14.8 Spectrum of $\mathbf{M}_\Theta$ and $\varphi(\mathbf{M}_\Theta)$

The following result establishes a useful connection between the spectral properties of the operator $\mathbf{M}_\Theta$ and the properties of the inner function Θ. The spectrum of Θ was defined in Section 5.2. We first need the following technical lemma. This result, in a more general form, will appear again in our discussion.

Lemma 14.27 *Let Θ be an inner function, and let $\zeta \in \mathbb{T} \setminus \sigma(\mathbf{M}_\Theta)$. Then each $f \in K_\Theta$ admits an analytic continuation across the point ζ.*

Proof Since $\sigma(\mathbf{M}_\Theta)$ is closed, there exists a neighborhood of ζ on $\mathbb{T}$, say J_ζ, such that $J_\zeta \subset \mathbb{T} \setminus \sigma(\mathbf{M}_\Theta)$. We show that f can be analytically continued across J_ζ. Fix $\zeta_0 \in J_\zeta$. Write

$$I - \bar{z}\mathbf{M}_\Theta = I - \bar{\zeta}_0\mathbf{M}_\Theta - (\bar{z} - \bar{\zeta}_0)\mathbf{M}_\Theta,$$

and, since $I - \bar{\zeta}_0\mathbf{M}_\Theta = \bar{\zeta}_0(\zeta_0 I - \mathbf{M}_\Theta)$ is invertible, we get

$$I - \bar{z}\mathbf{M}_\Theta = (I - \bar{\zeta}_0\mathbf{M}_\Theta)(I - (\bar{z} - \bar{\zeta}_0)(I - \bar{\zeta}_0\mathbf{M}_\Theta)^{-1}\mathbf{M}_\Theta). \tag{14.26}$$

Put $r_0 = \|(I - \bar{\zeta}_0\mathbf{M}_\Theta)^{-1}\mathbf{M}_\Theta\|^{-1}$. Then, for $z \in D(\zeta_0, r_0)$,

$$\|(\bar{z} - \bar{\zeta}_0)(I - \bar{\zeta}_0\mathbf{M}_\Theta)^{-1}\mathbf{M}_\Theta\| \le |z - \zeta_0| r_0^{-1} < 1.$$

Thus, the operator $I - (\bar{z} - \bar{\zeta}_0)(I - \bar{\zeta}_0\mathbf{M}_\Theta)^{-1}\mathbf{M}_\Theta$ is invertible and its inverse is given by

$$(I - (\bar{z} - \bar{\zeta}_0)(I - \bar{\zeta}_0\mathbf{M}_\Theta)^{-1}\mathbf{M}_\Theta)^{-1} = \sum_{n=0}^{\infty} (\bar{z} - \bar{\zeta}_0)^n (I - \bar{\zeta}_0\mathbf{M}_\Theta)^{-n}\mathbf{M}_\Theta^n,$$

where the series is convergent in the operator norm. We deduce from (14.26) that $I - \bar{z}\mathbf{M}_\Theta$ is invertible and

$$(I - \bar{z}\mathbf{M}_\Theta)^{-1} = \sum_{n=0}^{\infty} (\bar{z} - \bar{\zeta}_0)^n (I - \bar{\zeta}_0 \mathbf{M}_\Theta)^{-(n+1)} \mathbf{M}_\Theta^n,$$

where the series is still convergent in operator norm. Therefore, for $z \in D(\zeta_0, r_0)$, we have

$$\langle f, (I - \bar{z}\mathbf{M}_\Theta)^{-1} k_0^\Theta \rangle_2 = \sum_{n=0}^{\infty} (z - \zeta_0)^n \langle f, (I - \bar{\zeta}_0 \mathbf{M}_\Theta)^{-(n+1)} \mathbf{M}_\Theta^n k_0^\Theta \rangle_2.$$

The last equality reveals that the function $z \longmapsto \langle f, (I - \bar{z}\mathbf{M}_\Theta)^{-1} k_0^\Theta \rangle_2$ is analytic on the disk $D(\zeta_0, r_0)$. But, by Lemma 14.22, we have

$$(I - \bar{z}\mathbf{M}_\Theta)^{-1} k_0^\Theta = k_z^\Theta \qquad (z \in \mathbb{D})$$

and thus

$$\langle f, (I - \bar{z}\mathbf{M}_\Theta)^{-1} k_0^\Theta \rangle_2 = \langle f, k_z^\Theta \rangle_2 = f(z)$$

for each point $z \in D(\zeta_0, r_0) \cap \mathbb{D}$. This implies that the function f can be analytically continued in a neighborhood of ζ_0. Since ζ_0 is any point of J_ζ, we deduce that the function f in K_Θ has an analytic continuation across J_ζ. □

Theorem 14.28 *Let Θ be an inner function. Then*

$$\sigma(\mathbf{M}_\Theta) = \sigma(\Theta)$$

and

$$\sigma_p(\mathbf{M}_\Theta) = \sigma(\Theta) \cap \mathbb{D} = \{\lambda \in \mathbb{D} : \Theta(\lambda) = 0\}.$$

Proof Let us first verify that $\sigma(\mathbf{M}_\Theta) \subset \sigma(\Theta)$. As $\|\mathbf{M}_\Theta\| \le 1$, we have at least $\sigma(\mathbf{M}_\Theta) \subset \bar{\mathbb{D}}$. Fix $\lambda \in \bar{\mathbb{D}} \setminus \sigma(\Theta)$. Then Θ is analytic in a neighborhood of λ and $\Theta(\lambda) \neq 0$. Therefore, the function u, defined by

$$u(z) = \frac{\Theta(z) - \Theta(\lambda)}{z - \lambda} \qquad (z \in \mathbb{D}),$$

belongs to H^∞ and, by Theorems 14.24 and 14.25(iii), we have

$$\begin{aligned}(\lambda I - \mathbf{M}_\Theta) u(\mathbf{M}_\Theta) &= u(\mathbf{M}_\Theta)(\lambda I - \mathbf{M}_\Theta)\\ &= \Theta(\lambda) I - \Theta(\mathbf{M}_\Theta)\\ &= \Theta(\lambda) I.\end{aligned} \tag{14.27}$$

This identity ensures that $\lambda \notin \sigma(\mathbf{M}_\Theta)$ and

$$(\lambda I - \mathbf{M}_\Theta)^{-1} = \Theta(\lambda)^{-1} u(\mathbf{M}_\Theta).$$

We thus deduce that $\sigma(\mathbf{M}_\Theta) \subset \sigma(\Theta)$.

We proceed now to prove that $\sigma_p(\mathbf{M}_\Theta) = \sigma(\Theta) \cap \mathbb{D}$. Since, by Lemma 14.21,

$$\|\mathbf{M}_\Theta^n f\|_2 \longrightarrow 0 \qquad (f \in K_\Theta),$$

as $n \longrightarrow \infty$, we must have

$$\sigma_p(\mathbf{M}_\Theta) \subset \mathbb{D}.$$

Since $\sigma_p(\mathbf{M}_\Theta) \subset \sigma(\mathbf{M}_\Theta) \subset \sigma(\Theta)$, we deduce that

$$\sigma_p(\mathbf{M}_\Theta) \subset \sigma(\Theta) \cap \mathbb{D}.$$

According to the definition of the spectrum of an inner function, if $\lambda \in \sigma(\Theta) \cap \mathbb{D}$, then $\Theta(\lambda) = 0$ and, using the H^∞ function u once more, by (14.27), we have

$$(\lambda I - \mathbf{M}_\Theta) u(\mathbf{M}_\Theta) = 0. \tag{14.28}$$

By Lemma 14.14, $u = \hat{k}_\lambda^\Theta \in K_\Theta$. Since $u \not\equiv 0$ and $K_\Theta \cap \Theta H^\infty = \{0\}$, surely $u \notin \Theta H^\infty$ and thus, by Theorem 14.25(iii), $u(\mathbf{M}_\Theta) \neq 0$. This means that there is $f \in K_\Theta$ such that $g = u(\mathbf{M}_\Theta) f \in K_\Theta$ and $g \neq 0$. The identity (14.28), applied to f, now reveals that $\lambda \in \sigma_p(\mathbf{M}_\Theta)$.

To finish the proof, it remains to show that $\sigma(\Theta) \cap \mathbb{T} \subset \sigma(\mathbf{M}_\Theta)$. Assume that $\zeta \in \mathbb{T} \setminus \sigma(\mathbf{M}_\Theta)$. By Lemma 14.27, every function in K_Θ admits an analytic continuation across the point ζ. Fix any $\lambda \in \mathbb{D}$ such that $\Theta(\lambda) \neq 0$. Then the function $f = k_\lambda^\Theta$ belongs to K_Θ and thus admits an analytic continuation across the point ζ. But, since

$$\Theta(z) = \frac{1 - (1 - \bar{\lambda} z) k_\lambda^\Theta(z)}{\overline{\Theta(\lambda)}} \qquad (z \in \mathbb{D}),$$

the function Θ also admits an analytic continuation across the point ζ. In other words, $\zeta \in \mathbb{T} \setminus \sigma(\Theta)$. This is equivalent to saying that

$$\mathbb{T} \setminus \sigma(\mathbf{M}_\Theta) \subset \mathbb{T} \setminus \sigma(\Theta),$$

and thus $\sigma(\Theta) \cap \mathbb{T} \subset \sigma(\mathbf{M}_\Theta)$. □

The following result is an immediate consequence of Lemma 14.27 and Theorem 14.28.

Corollary 14.29 *Let Θ be an inner function and let $\zeta \in \mathbb{T}$. The function Θ admits an analytic continuation across a neighborhood of ζ if and only if every function $f \in K_\Theta$ has this property.*

The next result is needed to show that $\mathbf{M}_{\Theta^*}$ and X_Θ are unitarily equivalent, where Θ^* is the inner function

$$\Theta^*(z) = \overline{\Theta(\bar{z})} \qquad (z \in \mathbb{D}). \tag{14.29}$$

Note that $\Theta^{**} = \Theta$.

Lemma 14.30 *Let $f \in K_\Theta$ and define*

$$(V_\Theta f)(z) = \bar{z}\Theta^*(z)f(\bar{z}) \qquad (z \in \mathbb{T}).$$

Then V_Θ is a unitary operator from K_Θ onto K_{Θ^*} and

$$V_\Theta^* = V_\Theta^{-1} = V_{\Theta^*}.$$

Proof Let $f \in L^2(\mathbb{T})$. Then, by definition (see Section 8.11),

$$\begin{aligned} f \in K_\Theta \quad &\Longleftrightarrow \quad (\, f \in H^2 \text{ and } f \in \bar{z}\Theta\overline{H^2}\,) \\ &\Longleftrightarrow \quad (\, f \in H^2 \text{ and } zf(z)\overline{\Theta(z)} \in \overline{H^2}\,) \\ &\Longleftrightarrow \quad (\, f \in H^2 \text{ and } \bar{z}f(\bar{z})\Theta^*(z) \in H^2\,). \end{aligned}$$

Applying the preceding argument to $g = V_\Theta f$ gives us

$$g \in K_{\Theta^*} \quad \Longleftrightarrow \quad (\, g \in H^2 \text{ and } \bar{z}g(\bar{z})\Theta^{**}(z) \in H^2\,).$$

But $g(z) = \bar{z}\Theta^*(z)f(\bar{z})$ and $\Theta^{**} = \Theta$, which gives

$$\bar{z}g(\bar{z})\Theta^{**}(z) = \bar{z}\Theta(z)z\Theta^*(\bar{z})f(z) = f(z)$$

for almost all $z \in \mathbb{T}$. We rewrite the last identity as

$$V_{\Theta^*}g = f, \tag{14.30}$$

and hence

$$g \in K_{\Theta^*} \quad \Longleftrightarrow \quad (\, \bar{z}f(\bar{z})\Theta^*(z) \in H^2 \text{ and } f \in H^2\,).$$

Therefore, we deduce that $f \in K_\Theta$ if and only if $g \in K_{\Theta^*}$. The operator V_Θ is an isometry because Θ is unimodular a.e. on $\mathbb{T}$. Moreover, according to (14.30), we have $V_\Theta^{-1} = V_{\Theta^*}$. □

Theorem 14.31 *Let Θ be an inner function. Then X_Θ and $\mathbf{M}_{\Theta^*}$ are unitarily equivalent. More explicitly, we have*

$$V_\Theta^* \mathbf{M}_{\Theta^*} V_\Theta = X_\Theta.$$

Proof Let $f \in K_\Theta$, and let $g = f^*$. Thus, $g \in H^2$ and

$$P_+(\bar{g}) = \overline{g(0)} = f(0). \tag{14.31}$$

Let $z \in \mathbb{T}$. The following discussion is valid for almost all such values of z. Remember that $|\Theta^*(z)| = |\Theta(z)| = 1$. By definition

$$(V_\Theta f)(z) = \bar{z}\Theta^*(z)f(\bar{z}).$$

By Lemma 14.30, we know that $V_\Theta f \in K_{\Theta^*}$. Hence,

$$(S\mathbf{i}_{K_{\Theta^*}} V_\Theta f)(z) = \Theta^*(z)f(\bar{z}) = \Theta^*(z)\overline{g(z)}.$$

This shows $\Theta^*\bar{g} \in H^2$. By (14.21), applied to Θ^*, we have $\mathbf{M}_{\Theta^*} = \mathbf{P}_{\Theta^*} S \mathbf{i}_{K_{\Theta^*}}$, and thus

$$\mathbf{M}_{\Theta^*} V_\Theta f = P_{\Theta^*}(\Theta^* \bar{g}).$$

By Corollary 14.13, again applied to Θ^*, we have $P_{\Theta^*}h = h - \Theta^* P_+(\overline{\Theta^*}h)$, $h \in H^2$, and hence, by (14.31),

$$\mathbf{M}_{\Theta^*} V_\Theta f = \Theta^* \bar{g} - \Theta^* P_+(\bar{g}) = \Theta^*(\bar{g} - f(0)).$$

Finally, by Lemma 14.30, $V_\Theta^* = V_{\Theta^*}$, and thus

$$\begin{aligned}
(V_\Theta^* \mathbf{M}_{\Theta^*} V_\Theta f)(z) &= \bar{z}\Theta(z)\Theta^*(\bar{z})(\overline{g(\bar{z})} - f(0)) \\
&= \bar{z}\Theta(z)\overline{\Theta(z)}(g^*(z) - f(0)) \\
&= \bar{z}(f(z) - f(0)) \\
&= (S^* f)(z) \\
&= (X_\Theta f)(z).
\end{aligned}$$

This completes the proof. □

We know from Theorem 14.28 that

$$\sigma_p(\mathbf{M}_\Theta) = \sigma(\Theta) \cap \mathbb{D} = \{\lambda \in \mathbb{D} : \Theta(\lambda) = 0\}. \tag{14.32}$$

Hence, using Theorem 14.31, we get

$$\sigma_p(X_\Theta) = \sigma_p(\mathbf{M}_{\Theta^*}) = \overline{\{\lambda \in \mathbb{D} : \Theta(\lambda) = 0\}}, \tag{14.33}$$

where the bar means the complex conjugate. We now have a complete description of the eigensubspaces of X_Θ and $\mathbf{M}_\Theta$. In particular, we show that they are of dimension one.

Theorem 14.32 *Let Θ be an inner function, and let $\lambda \in \mathbb{D}$ be such that $\Theta(\lambda) = 0$. Then*

$$\ker(X_\Theta - \bar{\lambda} I) = \mathbb{C} k_\lambda$$

and

$$\ker(\mathbf{M}_\Theta - \lambda I) = \mathbb{C} \hat{k}_\lambda^\Theta.$$

Proof The first identity is rather trivial. Since $\Theta(\lambda) = 0$, then $k_\lambda = k_\lambda^\Theta \in K_\Theta$. Moreover,

$$\ker(X_\Theta - \bar{\lambda} I) = \ker(S^* - \bar{\lambda} I) \cap K_\Theta.$$

But, by Lemma 8.6, $\ker(S^* - \bar{\lambda} I) = \mathbb{C} k_\lambda$, whence

$$\ker(X_\Theta - \bar{\lambda} I) = \mathbb{C} k_\lambda.$$

In the same spirit, for the second equality, according to Theorem 14.31, we have

$$\begin{aligned}\ker(\mathbf{M}_\Theta - \lambda I) &= \ker(V_{\Theta^*} X_{\Theta^*} V_{\Theta^*}^* - \lambda I) \\ &= V_{\Theta^*} \ker(X_{\Theta^*} - \lambda I) = \mathbb{C} V_{\Theta^*} k_{\bar{\lambda}}.\end{aligned}$$

But, for almost all $z \in \mathbb{T}$,

$$\begin{aligned}(V_{\Theta^*} k_{\bar{\lambda}})(z) &= \bar{z}\Theta(z) k_{\bar{\lambda}}(\bar{z}) \\ &= \frac{\bar{z}\Theta(z)}{1-\lambda\bar{z}} \\ &= \frac{\Theta(z)}{z-\lambda} \\ &= \hat{k}_\lambda^\Theta(z).\end{aligned}$$

This completes the proof. □

The following result describes the spectrum of the operator $\varphi(\mathbf{M}_\Theta)$ where $\varphi \in H^\infty$. There are some classes of operators and some classes of functional calculus for which we have the spectral mapping theorem (Theorem 1.22), which means that

$$\sigma(f(T)) = f(\sigma(T)).$$

In our situation, we must pay attention to the meaning of the right-hand side, because for functions in H^∞ the values at points of $\mathbb{T}$ are not defined everywhere. To overcome this difficulty, instead of taking into account the value $\varphi(\lambda)$, $\lambda \in \mathbb{T}$, which is rather ambiguous, the idea is to consider the set of limiting values of the function φ at λ, that is, the set

$$\left\{\zeta \in \mathbb{C} : \liminf_{z\to\lambda,\, z\in\mathbb{D}} |\varphi(z) - \zeta| = 0\right\}.$$

Theorem 14.33 *Let Θ be an inner function, and let $\varphi \in H^\infty$. Then*

$$\sigma(\varphi(\mathbf{M}_\Theta)) = \left\{\zeta \in \mathbb{C} : \inf_{z\in\mathbb{D}} (|\Theta(z)| + |\varphi(z) - \zeta|) = 0\right\}.$$

In particular, if φ has a continuous extension to $\bar{\mathbb{D}}$, then we have

$$\sigma(\varphi(\mathbf{M}_\Theta)) = \varphi(\sigma(\mathbf{M}_\Theta)) = \varphi(\sigma(\Theta)). \tag{14.34}$$

Proof First, let us prove the inclusion

$$\left\{\zeta \in \mathbb{C} : \inf_{z\in\mathbb{D}} (|\Theta(z)| + |\varphi(z) - \zeta|) = 0\right\} \subset \sigma(\varphi(\mathbf{M}_\Theta)).$$

If $\psi \in H^\infty$, then, according to Theorem 14.25, we have

$$\psi(\mathbf{M}_\Theta)^* k_\lambda^\Theta = P_+(\bar{\psi} k_\lambda^\Theta) = P_+(\bar{\psi} k_\lambda) - \overline{\Theta(\lambda)} P_+(\bar{\psi}\Theta k_\lambda)$$

for each $\lambda \in \mathbb{D}$. But $P_+\bar{\psi}k_\lambda = T_{\psi^*}k_\lambda = \overline{\psi(\lambda)}k_\lambda$, and thus

$$\|\psi(\mathbf{M}_\Theta)^* k_\lambda^\Theta\|_2 \le |\psi(\lambda)|\,\|k_\lambda\|_2 + |\Theta(\lambda)|\,\|\psi\|_\infty \|k_\lambda\|_2. \tag{14.35}$$

Let $\zeta \in \mathbb{C}$ be such that $\inf_{z\in\mathbb{D}}(|\Theta(z)| + |\varphi(z) - \zeta|) = 0$, and take $\psi = \varphi - \zeta$. Then, using (14.35), we have

$$\begin{aligned}
\left\|(\varphi(\mathbf{M}_\Theta)^* - \bar{\zeta}I)\left(\frac{k_\lambda^\Theta}{\|k_\lambda^\Theta\|}\right)\right\|_2 &= \left\|\psi(\mathbf{M}_\Theta)^*\left(\frac{k_\lambda^\Theta}{\|k_\lambda^\Theta\|}\right)\right\|_2 \\
&\le (|\psi(\lambda)| + |\Theta(\lambda)|\,\|\psi\|_\infty)\frac{\|k_\lambda\|_2}{\|k_\lambda^\Theta\|_2} \\
&= \frac{|\varphi(\lambda) - \zeta| + |\Theta(\lambda)|\,\|\varphi - \zeta\|_\infty}{(1 - |\Theta(\lambda)|^2)^{1/2}} \\
&\le \frac{1 + \|\varphi\|_\infty}{(1 - |\Theta(\lambda)|^2)^{1/2}}(|\varphi(\lambda) - \zeta| + |\Theta(\lambda)|).
\end{aligned}$$

This implies that

$$\inf\{\|(\varphi(\mathbf{M}_\Theta)^* - \bar{\zeta}I)f\| : f \in K_\Theta,\ \|f\|_2 = 1\} = 0.$$

In other words, $\bar{\zeta} \in \sigma_a(\varphi(\mathbf{M}_\Theta)^*) \subset \sigma(\varphi(\mathbf{M}_\Theta)^*) = \overline{\sigma(\varphi(\mathbf{M}_\Theta))}$.

The reverse inclusion is deduced from the corona theorem. Indeed, if

$$\inf_{z\in\mathbb{D}}(|\Theta(z)| + |\varphi(z) - \zeta|) > 0,$$

then there exist functions u and v in H^∞ such that

$$\Theta(z)u(z) + v(z)(\varphi(z) - \zeta) = 1 \qquad (z \in \mathbb{D}).$$

Then, by Theorem 14.24, we get

$$\Theta(\mathbf{M}_\Theta)u(\mathbf{M}_\Theta) + v(\mathbf{M}_\Theta)(\varphi(\mathbf{M}_\Theta) - \zeta I) = I.$$

Since $\Theta(\mathbf{M}_\Theta) = 0$, the above identity simplifies to

$$v(\mathbf{M}_\Theta)(\varphi(\mathbf{M}_\Theta) - \zeta I) = I.$$

Therefore, $\zeta \in \sigma(\varphi(\mathbf{M}_\Theta))$.

It remains to prove (14.34) when φ has a continuous extension to $\bar{\mathbb{D}}$. In that case we have $\inf_{z\in\mathbb{D}}(|\Theta(z)| + |\varphi(z) - \zeta|) = 0$ if and only if there exists a sequence $(\lambda_n)_{n\ge1}$ of points of $\mathbb{D}$ such that

$$\Theta(\lambda_n) \longrightarrow 0 \quad \text{and} \quad \varphi(\lambda_n) \longrightarrow \zeta,$$

as $n \longrightarrow \infty$. There exists a subsequence $(\lambda_{n_p})_{p\ge1}$ that converges to some point $\lambda \in \bar{\mathbb{D}}$. According to Theorem 5.4, we necessarily have $\lambda \in \sigma(\Theta)$, and, since φ is continuous on $\bar{\mathbb{D}}$, $\varphi(\lambda_{n_p}) \longrightarrow \varphi(\lambda)$ as $p \longrightarrow \infty$. Hence,

$\zeta = \varphi(\lambda) \in \varphi(\sigma(\mathbf{M}_\Theta))$. Conversely, let $\zeta = \varphi(\lambda)$ for some $\lambda \in \sigma(\mathbf{M}_\Theta)$. Then, using Theorem 5.4 once again, there exists $(\lambda_n)_{n\geq 1} \subset \mathbb{D}$ such that

$$\Theta(\lambda_n) \longrightarrow 0 \quad \text{and} \quad \lambda_n \longrightarrow \lambda,$$

as $n \longrightarrow \infty$. By continuity, we have $\varphi(\lambda_n) \longrightarrow \varphi(\lambda) = \zeta$. Hence,

$$|\Theta(\lambda_n)| + |\varphi(\lambda_n) - \zeta| \longrightarrow 0 \qquad (n \longrightarrow \infty),$$

which gives

$$\inf_{z\in\mathbb{D}} (|\Theta(z)| + |\varphi(z) - \zeta|) = 0. \qquad \square$$

The following result improves Theorem 14.33 by giving a complete description of the point spectrum of $\varphi(\mathbf{M}_\Theta)$.

Theorem 14.34 *Let Θ be an inner function, and let $\varphi \in H^\infty$. Fix any $\lambda \in \mathbb{C}$, and put*

$$\vartheta = GCD((\varphi - \lambda)_i, \Theta),$$

where $(\varphi - \lambda)_i$ is the inner part of $\varphi - \lambda$. Then

$$\ker(\varphi(\mathbf{M}_\Theta) - \lambda I) = \frac{\Theta}{\vartheta} K_\vartheta \tag{14.36}$$

and

$$\ker(\varphi(\mathbf{M}_\Theta)^* - \bar{\lambda} I) = K_\vartheta. \tag{14.37}$$

In particular, the following are equivalent:

(i) $\lambda \in \sigma_p(\varphi(\mathbf{M}_\Theta))$;
(ii) $\bar{\lambda} \in \sigma_p(\varphi(\mathbf{M}_\Theta)^*)$;
(iii) *ϑ is not constant.*

Proof Let $\psi = \varphi - \lambda$, and let ψ_i be the inner part of ψ. Then we have $\vartheta = GCD(\psi_i, \Theta)$. To prove the first assertion, assume that $f \in K_\Theta$. Then $f \in \ker(\varphi(\mathbf{M}_\Theta) - \lambda I)$ if and only if $\psi(\mathbf{M}_\Theta) f = P_\Theta(\psi f) = 0$. But the latter is equivalent to the condition that Θ divides the product $\psi_i f_i$, where f_i is the inner part of f. Since $GCD(\Theta, \psi_i) = \vartheta$, we necessarily have $GCD(\Theta/\vartheta, \psi/\vartheta) = 1$ and thus Θ divides $\psi_i f_i$ if and only if Θ/ϑ divides f_i, which is equivalent to $f \in (\Theta/\vartheta)H^2$. Therefore,

$$\ker(\varphi(\mathbf{M}_\Theta) - \lambda I) = K_\Theta \cap \frac{\Theta}{\vartheta} H^2 = \left(K_{\Theta/\vartheta} \oplus \frac{\Theta}{\vartheta} K_\vartheta\right) \cap \frac{\Theta}{\vartheta} H^2 = \frac{\Theta}{\vartheta} K_\vartheta.$$

For the second assertion, according to Theorem 14.25, we have

$$\ker(\varphi(\mathbf{M}_\Theta)^* - \bar{\lambda} I) = \ker(T_{\bar{\varphi}|K_\Theta} - \bar{\lambda} I) = \ker(T_{\bar{\varphi} - \bar{\lambda}}) \cap K_\Theta.$$

Now, according to Theorem 12.19, we have $\ker(T_{\bar{\varphi}-\bar{\lambda}}) = K_{\psi_i}$, whence

$$\ker(\varphi(\mathbf{M}_\Theta)^* - \lambda I) = K_{\psi_i} \cap K_\Theta.$$

It remains to apply Lemma 14.3 to conclude that

$$\ker(\varphi(\mathbf{M}_\Theta)^* - \lambda I) = K_\vartheta.$$

The equivalence between the three assertions follows easily from (14.36) and (14.37). □

Note that, using Theorem 14.34, we can recover Theorem 14.32. To this end, pick up the function $\varphi(z) = z$. On the one hand, if $|\lambda| \geq 1$, then the function $\varphi - \lambda$ is outer (for instance, note that $\Re(\varphi(z) - \lambda) = \Re(z - \lambda) \geq 0$ for all $z \in \mathbb{D}$). Hence, $GCD((\varphi - \lambda)_i, \Theta)$ is constant. On the other hand, if $\lambda \in \mathbb{D}$, then write

$$z - \lambda = \frac{z - \lambda}{1 - \bar{\lambda} z}(1 - \bar{\lambda} z),$$

and, since the function $z \longmapsto 1 - \bar{\lambda} z$ is outer, we deduce that the inner part of $\varphi - \lambda$ is the Blaschke factor b_λ. Thus $GCD((\varphi - \lambda)_i, \Theta)$ is not constant if and only if $\Theta(\lambda) = 0$. Moreover, in this case, we have

$$\vartheta = GCD(\varphi - \lambda, \Theta) = b_\lambda.$$

Hence, $K_\vartheta = \mathbb{C} k_\lambda$ and, according to Theorem 14.34, we thus get

$$\ker(\mathbf{M}_\Theta - \lambda I) = \mathbb{C}\frac{\Theta}{b_\lambda} k_\lambda = \mathbb{C}\frac{\Theta}{z - \lambda}$$

and

$$\ker(X_\Theta - \bar{\lambda} I) = \mathbb{C} k_\lambda.$$

Corollary 14.35 *Let B be a Blaschke product associated with the Blaschke sequence Λ in the unit disk. Let $\varphi \in H^\infty$. Then $\lambda_0 \in \sigma_p(\varphi(\mathbf{M}_B))$ if and only if $\Lambda' = \varphi^{-1}(\{\lambda_0\}) \cap \Lambda \neq \emptyset$. Moreover, we have*

$$\ker(\varphi(\mathbf{M}_B) - \lambda_0) = B_{\Lambda \setminus \Lambda'} K_{B_{\Lambda'}} \tag{14.38}$$

and

$$\ker(\varphi(\mathbf{M}_B)^* - \bar{\lambda}_0) = K_{B_{\Lambda'}}, \tag{14.39}$$

where $B_{\Lambda'} = \prod_{\lambda' \in \Lambda'} b_{\lambda'}$ and $B_{\Lambda \setminus \Lambda'} = B / B_{\Lambda'}$. In particular, we have

$$\varphi(\mathbf{M}_B)\left(\frac{B}{z - \lambda}\right) = \varphi(\lambda)\frac{B}{z - \lambda} \tag{14.40}$$

and

$$\varphi(\mathbf{M}_B)^* k_\lambda = \overline{\varphi(\lambda)} k_\lambda, \tag{14.41}$$

for each $\lambda \in \Lambda$.

Proof Let $\vartheta = GCD((\varphi - \lambda_0)_i, B)$. Since ϑ should divide the Blaschke product B, we necessarily must have $\vartheta = B_{\Lambda^\sharp}$, where $\Lambda^\sharp \subset \Lambda$. Moreover, for any $\lambda \in \Lambda^\sharp$, we have $\varphi(\lambda) = \lambda_0$. This proves that $\Lambda^\sharp = \Lambda'$. Hence, $GCD((\varphi - \lambda_0)_i, B) = B_{\Lambda'}$. According to Theorem 14.34, we have $\lambda_0 \in \sigma_p(\varphi(\mathbf{M}_B))$ if and only if $B_{\Lambda'}$ is not constant, that is, $\Lambda' \neq \emptyset$.

The equations (14.38) and (14.39) follow immediately from (14.36) and (14.37). Now, for $\lambda \in \Lambda$, if $\lambda_0 = \varphi(\lambda)$, we have $\lambda \in \Lambda'$. Thus both functions k_λ and $B_{\Lambda'}/(z-\lambda)$ belong to $K_{B_{\Lambda'}}$, which gives (14.40) and (14.41). □

Exercises

Exercise 14.8.1 Let Θ be an inner function, let $\varphi \in H^\infty$ and let $\lambda \in \mathbb{D}$ be such that $\Theta(\lambda) = 0$. Show that $\varphi(\lambda) \in \sigma_p(\varphi(\mathbf{M}_\Theta))$ and $\overline{\varphi(\lambda)} \in \sigma_p(\varphi(\mathbf{M}_\Theta)^*)$.

Exercise 14.8.2 Let Θ be an inner function, let $\varphi \in H^\infty$ and let $\lambda \in \mathbb{D}$ be such that $\Theta(\lambda) = 0$. Assume that φ is one-to-one. Show that

$$\ker(\varphi(\mathbf{M}_\Theta) - \varphi(\lambda)) = \mathbb{C}\frac{\Theta}{z-\lambda}$$

and

$$\ker(\varphi(\mathbf{M}_\Theta)^* - \bar{\varphi}(\lambda)) = \mathbb{C}k_\lambda.$$

14.9 The commutant lifting theorem for $\mathbf{M}_\Theta$

Let $A : \mathcal{H} \longrightarrow \mathcal{H}$ be a linear and bounded operator on a Hilbert space $\mathcal{H}$. Recall that the *commutant* of A, denoted by $\{A\}'$, is the subset of $\mathcal{L}(\mathcal{H})$ defined by

$$\{A\}' = \{B \in \mathcal{L}(\mathcal{H}) : AB = BA\}.$$

The purpose of this section is to describe the commutant of $\mathbf{M}_\Theta$. First, using the properties of the functional calculus Λ for the operator $\mathbf{M}_\Theta$, it is clear that, for each $f \in H^\infty$, we have

$$f(\mathbf{M}_\Theta)\mathbf{M}_\Theta = \Lambda(f)\Lambda(z) = \Lambda(fz) = \Lambda(zf) = \Lambda(z)\Lambda(f) = \mathbf{M}_\Theta f(\mathbf{M}_\Theta).$$

Hence, $f(\mathbf{M}_\Theta) \in \{\mathbf{M}_\Theta\}'$. The converse is also correct and is called the *commutant lifting theorem* for $\mathbf{M}_\Theta$. The terminology comes from the fact that it lifts the equation $A\mathbf{M}_\Theta = \mathbf{M}_\Theta A$ up to the level of the isometric dilation $S : H^2 \to H^2$ of $\mathbf{M}_\Theta$. We will give two proofs of this important result. The first proof is based on Nehari's theorem (Theorem 11.3) and uses a link between the commutant of $\mathbf{M}_\Theta$ and Hankel operators. This link is precisely stated in the following lemma.

Lemma 14.36 *Let Θ be an inner function, and let $A \in \mathcal{L}(K_\Theta)$. Define*

$$\begin{array}{rccc} A_* : & H^2 & \longrightarrow & \overline{H_0^2} \\ & f & \longmapsto & \bar{\Theta} A P_\Theta f. \end{array}$$

Then the following assertions are equivalent:

(i) $A\mathbf{M}_\Theta = \mathbf{M}_\Theta A$;
(ii) A_* *is a Hankel operator.*

Proof First, let us verify that A_* is a well-defined operator from H^2 into $\overline{H_0^2}$. Indeed, since $K_\Theta = H^2 \cap \Theta \overline{H_0^2}$, we have $AP_\Theta H^2 \subset \Theta \overline{H_0^2}$, which implies that $\bar{\Theta} A P_\Theta H^2 \subset \overline{H_0^2}$.

Moreover,

$$\begin{aligned} A\mathbf{M}_\Theta = \mathbf{M}_\Theta A \quad &\Longleftrightarrow \quad AP_\Theta S = P_\Theta S A \qquad (\text{on } K_\Theta), \\ &\Longleftrightarrow \quad AP_\Theta S P_\Theta = P_\Theta S A P_\Theta \qquad (\text{on } H^2). \end{aligned}$$

Since $z\Theta H^2 \subset \Theta H^2$, we have $P_\Theta S(I - P_\Theta) = 0$, whence $P_\Theta S P_\Theta = P_\Theta S$ and

$$AP_\Theta S P_\Theta = AP_\Theta S \qquad (\text{on } H^2).$$

According to Corollary 14.13, $P_\Theta = \Theta P_- \bar{\Theta}$. Thus, on H^2,

$$\begin{aligned} A\mathbf{M}_\Theta = \mathbf{M}_\Theta A \quad &\Longleftrightarrow \quad AP_\Theta S = P_\Theta S A P_\Theta \\ &\Longleftrightarrow \quad AP_\Theta S = \Theta P_- \bar{\Theta} S A P_\Theta \\ &\Longleftrightarrow \quad \bar{\Theta} A P_\Theta S = P_- Z \bar{\Theta} A P_\Theta \\ &\Longleftrightarrow \quad A_* S = P_- Z A_*. \end{aligned}$$

The latter condition exactly means that A^* is a Hankel operator; see Exercise 11.1.3. □

The second proof of the commutant lifting theorem for the operator $\mathbf{M}_\Theta$ is based on the theory of isometric dilations developed in Sections 7.10 and 7.11. For this approach, the following lemma is used.

Lemma 14.37 *Let Θ be a nonconstant inner function. Then the shift operator $S : H^2 \longrightarrow H^2$ is the minimal isometric dilation of $\mathbf{M}_\Theta$.*

Proof According to Lemma 14.23, we already know that S is an isometric dilation of $\mathbf{M}_\Theta$. It remains to check that S is minimal, and, according to Lemma 7.44, this is equivalent to saying that

$$H^2 = \bigvee_{n \geq 0} S^n K_\Theta.$$

Let $f \in H^2$, $f \perp S^n K_\Theta$, $n \geq 0$. We should prove that $f \equiv 0$. First, note that $f \perp K_\Theta$, and thus $f = \Theta g_1$, for some $g_1 \in H^2$. Now, since $S^*\Theta \in K_\Theta$, we have

$$\langle f, S^n S^* \Theta \rangle_2 = 0 \qquad (n \geq 1).$$

But

$$\begin{aligned} \langle f, S^n S^* \Theta \rangle_2 &= \langle \Theta g_1, S^{n-1}(\Theta - \Theta(0)) \rangle_2 \\ &= \langle g_1, z^{n-1} \rangle_2 - \overline{\Theta(0)} \langle \Theta g_1, z^{n-1} \rangle_2 \\ &= g_1^{(n-1)}(0) - \overline{\Theta(0)} (\Theta g_1)^{(n-1)}(0) \end{aligned}$$

and hence, for $n = 1$, we obtain

$$(1 - |\Theta(0)|^2) g_1(0) = 0,$$

which gives $g_1(0) = 0$. For $n \geq 2$, by Leibniz's rule, we get

$$\begin{aligned} 0 &= g_1^{(n-1)}(0) - \overline{\Theta(0)} \sum_{k=0}^{n-1} \binom{n-1}{k} \Theta^{(k)}(0) g_1^{(n-1-k)}(0) \\ &= (1 - |\Theta(0)|^2) g_1^{(n-1)}(0) - \overline{\Theta(0)} \sum_{k=1}^{n-1} \binom{n-1}{k} \Theta^{(k)}(0) g_1^{(n-1-k)}(0). \end{aligned}$$

Therefore, by induction, $g_1^{(k)}(0) = 0$ for any $k \geq 0$, and thus $g_1 \equiv 0$. □

We are now ready for the commutant lifting theorem for the operator $\mathbf{M}_\Theta$.

Theorem 14.38 *Let Θ be an inner function, and let $A \in \{\mathbf{M}_\Theta\}'$, i.e. $A \in \mathcal{L}(K_\Theta)$ and $A\mathbf{M}_\Theta = \mathbf{M}_\Theta A$. Then there exists $\varphi \in H^\infty$ such that $A = \varphi(\mathbf{M}_\Theta)$. Moreover, the following assertions hold.*

(i) *For any representation $A = \varphi(\mathbf{M}_\Theta)$, with $\varphi \in H^\infty$, we have*

$$\|A\| = \text{dist}(\varphi, \Theta H^\infty).$$

(ii) $$\|A\| = \inf\{\|\varphi\|_\infty : A = \varphi(\mathbf{M}_\Theta) \text{ with } \varphi \in H^\infty\}.$$

(iii) *There exists a particular choice $\varphi \in H^\infty$ such that*

$$A = \varphi(\mathbf{M}_\Theta) \quad \text{and} \quad \|A\| = \|\varphi\|_\infty.$$

We now give the two proofs of the theorem separately.

First proof – based on Nehari's theorem According to Lemma 14.36, the operator $A_* = \bar{\Theta} A P_\Theta$ is a Hankel operator. Since A is bounded, we have

$$\|A_* f\|_2 = \|\bar{\Theta} A P_\Theta f\|_2 = \|A P_\Theta f\|_2 \leq \|A\| \, \|f\|_2 \qquad (f \in H^2),$$

which implies that A_* is bounded and $\|A_*\| \leq \|A\|$. Furthermore,

$$\|Af\|_2 = \|A P_\Theta f\|_2 = \|\bar{\Theta} A P_\Theta f\|_2 = \|A_* f\|_2 \leq \|A_*\| \, \|f\|_2 \qquad (f \in K_\Theta),$$

whence $\|A\| \le \|A_*\|$. Therefore,

$$\|A\|_{\mathcal{L}(K_\Theta)} = \|A_*\|_{\mathcal{L}(H^2,\overline{H_0^2})}.$$

Nehari's theorem (Theorem 11.3) implies that there exists a function (symbol) $\eta \in L^\infty(\mathbb{T})$ such that $A_* = H_\eta$ and $\|A_*\| = \|\eta\|_\infty = \text{dist}(\eta, H^\infty)$. Hence,

$$\|A\| = \|A_*\| = \|\eta\|_\infty = \text{dist}(\eta, H^\infty). \tag{14.42}$$

Since $\Theta\eta \in L^2(\mathbb{T}) = H^2 \oplus \overline{H_0^2}$, we can write

$$\Theta\eta = \varphi + \bar{\psi},$$

where $\varphi \in H^2$ and $\psi \in H_0^2$. But,

$$P_-\Theta\eta = H_\eta\Theta = A_*\Theta = \bar{\Theta}AP_\Theta\Theta = 0,$$

which means that $\bar{\psi} = P_-\Theta\eta = 0$. Hence,

$$\Theta\eta = \varphi \in H^2 \cap L^\infty = H^\infty.$$

Rewrite the last identity as $\eta = \bar{\Theta}\varphi$, with $\varphi \in H^\infty$. This is a rewarding representation. In fact, for $f \in K_\Theta$, we have

$$\begin{aligned} Af &= \Theta(\bar{\Theta}AP_\Theta f) \\ &= \Theta A_* f \\ &= \Theta H_{\bar{\Theta}\varphi} f \\ &= \Theta P_- \bar{\Theta}\varphi f \\ &= P_\Theta \varphi f \\ &= \varphi(\mathbf{M}_\Theta) f, \end{aligned}$$

whence $A = \varphi(\mathbf{M}_\Theta)$. Moreover, according to (14.42) and using the fact that $|\Theta| = 1$ a.e. on $\mathbb{T}$, we have

$$\|A\| = \|\varphi\|_\infty = \text{dist}(\varphi, \Theta H^\infty). \tag{14.43}$$

Since $\varphi(\mathbf{M}_\Theta) = A$, we have

$$\|A\| = \|\varphi\|_\infty \ge \inf\{\|h\|_\infty : A = h(\mathbf{M}_\Theta), \text{ with } h \in H^\infty\}.$$

If $h \in H^\infty$ is such that $h(\mathbf{M}_\Theta) = A = \varphi(\mathbf{M}_\Theta)$, then according to Theorem 14.25(iii), we have $h - \varphi \in \Theta H^\infty$. In other words, there exists $g \in H^\infty$ such that $h = \varphi + \Theta g$. Thus,

$$\text{dist}(h, \Theta H^\infty) = \text{dist}(\varphi, \Theta H^\infty) = \|A\|$$

and

$$\|h\|_\infty = \|\varphi + \Theta g\| \ge \text{dist}(\varphi, \Theta H^\infty) = \|A\|.$$

Hence, we have

$$\inf\{\|h\|_\infty : A = h(\mathbf{M}_\Theta), \text{ with } h \in H^\infty\} \geq \|A\|,$$

which proves that

$$\|A\| = \inf\{\|h\|_\infty : A = h(\mathbf{M}_\Theta), \text{ with } h \in H^\infty\} = \operatorname{dist}(\varphi, \Theta H^\infty).$$

Finally, (14.43) shows that the infimum is attained, which ends the first proof. □

Second proof – based on the theory of isometric dilations By Lemma 14.37, we know that S is the minimal isometric dilation of $\mathbf{M}_\Theta$. Hence, Theorem 7.50 implies that there exists $Y \in \mathcal{L}(H^2)$ such that

$$YS = SY, \qquad \|A\| = \|Y\|, \qquad A = P_\Theta Y|K_\Theta.$$

Now, Theorem 8.14 says that there exists $\varphi \in H^\infty$ such that $Y = M_\varphi$ and $\|Y\| = \|\varphi\|_\infty$. Thus $A = P_\Theta M_\varphi|K_\Theta = \varphi(\mathbf{M}_\Theta)$ and $\|A\| = \|\varphi\|_\infty$. This ends the second proof. □

According to the commutant lifting theorem, if $A \in \mathcal{L}(K_\Theta)$ is such that $A\mathbf{M}_\Theta = \mathbf{M}_\Theta A$, then there exists a function $\varphi \in H^\infty$ such that the diagram in Figure 14.2 commutes. We use $S_\varphi : H^2 \to H^2$ to denote the operator $S_\varphi f = \varphi f$, $f \in H^2$.

Exercises

Exercise 14.9.1 Let Θ be an inner function, and let φ be an H^∞ function. Prove that

$$\varphi(\mathbf{M}_\Theta) = \mathbf{M}_\Theta H_{\bar{\Theta}\varphi|K_\Theta}. \tag{14.44}$$

Exercise 14.9.2 Let $\psi \in L^\infty(\mathbb{T})$. The purpose of this exercise is to prove that the Hankel operator H_ψ has a closed range in H^2_- if and only if $\psi = \bar{\Theta}\varphi$, where $\varphi \in H^\infty$ and Θ is an inner function such that $(\Theta, \varphi) \in (HCR)$.

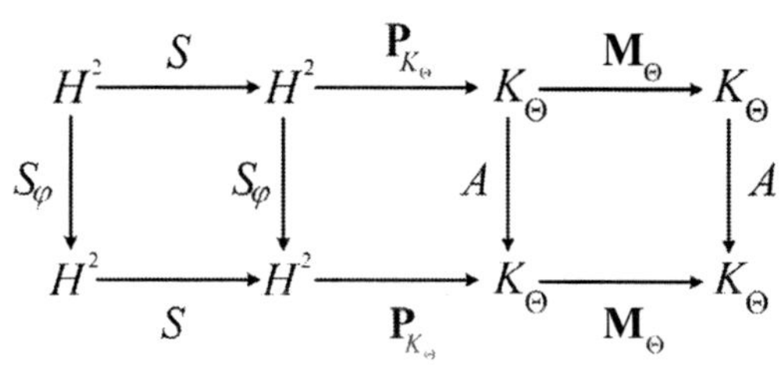

Figure 14.2 The commutant diagram.

Remark: Recall that $(\Theta, \varphi) \in (HCR)$ means that (Θ, φ) satisfies the corona hypothesis, that is,

$$\inf_{z\in\mathbb{D}} (|\Theta(z)| + |\varphi(z)|) > 0.$$

The proof is split into two main parts:

(I) Assume that H_φ has a closed range in H^2_-.

(i) Prove that $\ker H_\psi \neq \{0\}$.
Hint: Argue by absurdity. Hence, suppose that there exists $\delta > 0$ such that $\|H_\psi f\|_2 \geq c\|f\|_2$ for any $f \in H^2$. Apply this inequality to $f(z) = z^n$ and observe that $\|H_\psi z^n\|_2 \longrightarrow 0$ as $n \longrightarrow \infty$.

(ii) Show that there is an inner function Θ and a function φ in H^∞ such that $\psi = \bar{\Theta}\varphi$.

(iii) Show that we may assume that Θ and φ are such that $GCD(\Theta, \varphi_i) = 1$, where φ_i is the inner factor of φ.

(iv) Deduce that $\varphi(\mathbf{M}_\Theta)$ is one-to-one.

(v) Deduce that $H_{\bar{\Theta}\varphi}$ is one-to-one from K_Θ onto $H^2_- \ominus \bar{\Theta}H^2_-$.
Hint: Use Exercise 14.9.1.

(vi) Deduce that $\varphi(\mathbf{M}_\Theta)$ is onto and conclude that $(\Theta, \psi) \in (HCR)$.

(II) Assume that $\psi = \bar{\Theta}\varphi$, where $\varphi \in H^\infty$ and Θ is an inner function such that $(\Theta, \psi) \in (HCR)$.

(i) Show that $\varphi(\mathbf{M}_\Theta)$ is invertible.

(ii) Deduce from Exercise 14.9.1 that $H_{\bar{\Theta}\varphi}$ has a closed range in H^2_-.

Exercise 14.9.3 Let Θ be an inner function and let φ be an H^∞ function. Prove that $\varphi(\mathbf{M}_\Theta)$ is a compact operator if and only if $\bar{\Theta} \in H^\infty + \mathcal{C}(\mathbb{T})$.
Hint: Use Exercise 14.9.1 and Theorem 11.8.

14.10 Multipliers of K_Θ

In this section, we will show that K_Θ has no nonconstant multipliers.

Lemma 14.39 *Let Θ be a nonconstant inner function and let φ be a nonconstant function in H^∞. Then $\varphi S^*\Theta \notin K_\Theta$.*

Proof Assume on the contrary that $\varphi S^*\Theta \in K_\Theta$. Then, for all $n \geq 0$,

$$\begin{aligned}
0 &= \langle \varphi S^*\Theta, \Theta z^n \rangle_2 \\
&= \langle \varphi(\Theta - \Theta(0)), \Theta z^{n+1} \rangle_2 \\
&= \langle \varphi, z^{n+1} \rangle_2 - \langle \Theta(0)\bar{\Theta}\varphi, z^{n+1} \rangle_2 \\
&= \langle \varphi - \Theta(0)T_{\bar{\Theta}}\varphi, z^{n+1} \rangle_2.
\end{aligned}$$

Therefore, $\varphi - \Theta(0)T_{\bar{\Theta}}\varphi \perp z^{n+1}$, $n \geq 0$. But, since $\varphi - \Theta(0)T_{\bar{\Theta}}\varphi$ belongs to H^2, this function should be a constant. In particular, $S^*(\varphi - \Theta(0)T_{\bar{\Theta}}\varphi) = 0$. If $\Theta(0) = 0$, this means that φ is a constant, which is a contradiction. So $\Theta(0) \neq 0$. Hence, we can write $S^*\varphi = \Theta(0)S^*T_{\bar{\Theta}}\varphi$, which according to (12.3) implies that

$$S^*\varphi = \Theta(0)T_{\bar{\Theta}}S^*\varphi.$$

In particular, using the fact that $\|T_{\bar{\Theta}}\| = \|T_\Theta\| = \|\Theta\|_\infty = 1$, we have

$$\|S^*\varphi\|_2 \leq |\Theta(0)|\,\|S^*\varphi\|_2,$$

whence $|\Theta(0)| \geq 1$ (note that $\|S^*\varphi\|_2$ is not zero because φ is not constant). But, by the maximum principle, Θ must be a constant, which is again a contradiction. □

We recall that χ_0 is the function identically equal to 1.

Theorem 14.40 *Let Θ be a nonconstant inner function. Then*

$$\mathfrak{Mult}(K_\Theta) = \mathbb{C}\chi_0.$$

Proof Using Theorem 14.4, we see that $S^*\Theta \in K_\Theta$. Assume that $\varphi \in \mathfrak{Mult}(K_\Theta)$. According to Corollary 9.7, we have $\varphi \in H^\infty$. Also we should have $\varphi S^*\Theta \in K_\Theta$. Then Lemma 14.39 implies that φ is a constant. This proves that $\mathfrak{Mult}(K_\Theta) \subset \mathbb{C}\chi_0$. The reverse inclusion is trivial. □

Notes on Chapter 14

The model spaces K_Θ are of considerable importance in operator theory and function theory. One main reason is that they serve in modeling Hilbert space contractions. More explicitly, through Sz.-Nagy–Foiaş theory, we know that, if T is a Hilbert space contraction such that $\partial_T = \partial_{T^*} = 1$ and $T^n \longrightarrow 0$, $n \longrightarrow \infty$, in the strong operator topology, then there exists an inner function Θ such that T is unitarily equivalent to $S^* : K_\Theta \longrightarrow K_\Theta$. Since the work of Beurling [97] and Sz.-Nagy and Foiaş [505], we have discovered that subspaces K_Θ also have a key role to play in several other questions. As we already saw (see Chapter 8), they appeared naturally in questions concerning pseudocontinuation and cyclicity through the work of Douglas, Shapiro and Shields [180]. But they are also important in the theory of Cauchy transforms, in the theory of interpolation, in rational approximation and in harmonic analysis. The theory is very rich, and we refer the interested reader to the monograph of Nikolskii [386] and the following papers: Ahern and Clark [7–9], Aleksandrov [14–17], Baranov [69–72], Baranov and Belov [73], Baranov, Belov and Borichev [74], Baranov, Bessonov and Kapustin [75], Baranov, Chalendar, Fricain, Mashreghi and Timotin [76], Baranov and Dyakonov [77], Baranov and Fedorovskiĭ [78], Baranov, Fricain and Mashreghi [79], Baranov and Havin [80], Belov [85–87], Chalendar, Fricain and Partington [131], Chalendar, Fricain and Timotin [133], Cima, Garcia, Ross and Wogen [138, 143], Cima and Matheson [140], Clark [144], W. S. Cohn [150–157], Dyakonov [190–203], Fricain [221, 222], Garcia, Poore and Ross [231], Garcia, Ross and Wogen [232], Hartmann and Ross [263], Havin and Mashreghi [265, 266, 356], Hruščëv, Nikolskii and Pavlov [293], Nazarov and Volberg [372], Nikolskii [382, 383], Nikolskii and Pavlov [390–392], Nikolskii and Vasyunin [394, 395], Poltoratski and Sarason [417], Sarason [446, 447, 465, 466], Sedlock [473], Shapiro [479], Treil [525, 526], Volberg [532, 533] and Volberg and Treil [534].

Section 14.2

The standard text [386] refers to Theorem 14.7 as the Malmquist–Walsh lemma (and it can be found in the book of Walsh [541]), but this basis appeared as early as 1925 in a paper of Takenaka. The functions f_j of that theorem are then usually called the Malmquist–Walsh functions or the Malmquist–Walsh–Takenaka functions. They are often employed in the theory of interpolation; see Walsh [541] and Džrbašjan [204] for instance.

Section 14.5

The results of this section appeared in the standard book of Nikolskii [386].

Section 14.6

The formula (14.16) is due to Nikolskii [385, 386] and will be used to reduce the problem of the invertibility of $P_{\Theta_1|K_{\Theta_2}}$ to the invertibility of the Toeplitz operator $T_{\Theta_1\overline{\Theta_2}}$; see Theorem 31.19. Theorem 14.20 is due to Crofoot [163].

Section 14.7

The inequality $\|p(\mathbf{M}_\Theta)\| \le \|p\|_\infty$ that appears in Lemma 14.23 is a particular case of the von Neumann inequality (because $\mathbf{M}_\Theta$ is a contraction); see Exercise 2.5.1. This lemma can be found in Sarason [445]. The H^∞ functional calculus for $\mathbf{M}_\Theta$ has been constructed by Sz.-Nagy and Foiaş in the vector-valued case [505].

Section 14.8

Lemma 14.27 and Theorem 14.28 are due to Livšic [340] and Moeller [363]. Their analog for general completely nonunitary contractions T (in place of

M_Θ) is proved by Sz.-Nagy and Foiaş [505] and Helson [279]; see also [386]. Theorem 14.33 is due to Fuhrmann [225, 226]. Theorems 14.32 and 14.34 can be found in the book of Nikolskii [386].

Section 14.9

Theorem 14.38 is due to Sarason [447] and plays an important role in interpolation theory. It is a special case of the commutant lifting theory of Sz.-Nagy and Foiaş [506, 507]; see Theorem 7.50. The approach based on Hankel operators and the Nehari theorem was suggested by Nikolskii [381]. The formula (14.44) proved in Exercise 14.9.1 is due to Nikolskii [381]. The characterization of Hankel operators that have a closed range proved in Exercise 14.9.2 is a result of Clark [144].

15

Bases of reproducing kernels and interpolation

This chapter is devoted to the bases of K_B that are formed with a sequence of reproducing kernels, and the interpolation problem. We start by characterizing the uniformly minimal sequences, which leads to the definition of Carleson sequence and Carleson constant. Then we study the Carleson–Newman condition and eventually the Riesz basis of reproducing kernels for the model space K_B. The Nevanlinna–Pick interpolation problem is introduced next. This is a very vast subject. We just have a short glimpse and merely give a matrix characterization of the problem, and then determine the H^∞- and H^2-interpolating sequences. In the last section, we treat the asymptotically orthonormal sequences of K_B.

15.1 Uniform minimality of $(k_{\lambda_n})_{n\geq 1}$

Let $\Lambda = (\lambda_n)_{n\geq 1}$ be a sequence of distinct points of the open unit disk $\mathbb{D}$. Then we recall that the sequence $(k_{\lambda_n})_{n\geq 1}$ of reproducing kernels is minimal if and only if Λ is a Blaschke sequence (see Lemma 14.16). Moreover, in that case, if B denotes the associated Blaschke product, the sequence $(k_{\lambda_n})_{n\geq 1}$ is complete in $K_B = H^2 \ominus BH^2$ and the unique biorthogonal sequence in K_B associated with $(k_{\lambda_n})_{n\geq 1}$ is given by

$$k^*_{\lambda_n}(z) = \frac{1 - |\lambda_n|^2}{B_n(\lambda_n)} \frac{B_n(z)}{1 - \bar{\lambda}_n z} \qquad (z \in \mathbb{D}).$$

The next result gives a characterization of uniform minimal sequences.

Theorem 15.1 *Let $\Lambda = (\lambda_n)_{n\geq 1}$ be a Blaschke sequence of distinct points of the unit disk $\mathbb{D}$, and let B be the associated Blaschke product. Then the sequence $(k_{\lambda_n})_{n\geq 1}$ is uniformly minimal if and only if*

$$\delta(\Lambda) = \inf_{n\geq 1} |B_n(\lambda_n)| > 0, \tag{15.1}$$

where $B_n = B/b_{\lambda_n}$.

Proof According to Lemma 10.1, $(k_{\lambda_n})_{n\geq 1}$ is uniformly minimal if and only if

$$\sup_{n\geq 1} \|k_{\lambda_n}\|_2 \|k^*_{\lambda_n}\|_2 < \infty.$$

But

$$\begin{aligned} \|k^*_{\lambda_n}\|_2 &= \frac{1-|\lambda_n|^2}{|B_n(\lambda_n)|} \left\| \frac{B_n}{1-\bar{\lambda}_n z} \right\|_2 \\ &= \frac{1-|\lambda_n|^2}{|B_n(\lambda_n)|} \left\| \frac{1}{1-\bar{\lambda}_n z} \right\|_2 \\ &= \frac{1-|\lambda_n|^2}{|B_n(\lambda_n)|} \|k_{\lambda_n}\|_2 \end{aligned}$$

and hence

$$\|k_{\lambda_n}\|_2 \|k^*_{\lambda_n}\|_2 = \frac{1-|\lambda_n|^2}{|B_n(\lambda_n)|} \|k_{\lambda_n}\|_2^2 = \frac{1}{|B_n(\lambda_n)|}, \tag{15.2}$$

which ends the proof. □

A sequence $\Lambda = (\lambda_n)_{n\geq 1}$ that satisfies (15.1) is said to be a *Carleson sequence* and we write $\Lambda \in (C)$. The constant $\delta(\Lambda)$ is called the *Carleson constant* of Λ. Recall that, according to (10.1), the constant $\delta(\mathfrak{X})$ of uniform minimality of a sequence $\mathfrak{X} = (k_{\lambda_n})_{n\geq 1}$ satisfies

$$\delta(\mathfrak{X})^{-1} = \sup_{n\geq 1} \|k_{\lambda_n}\|_2 \|k^*_{\lambda_n}\|_2.$$

Hence, it follows from (15.2) that

$$\delta(\mathfrak{X}) = \delta(\Lambda).$$

Note that $\delta(\lambda) \leq 1$.

15.2 The Carleson–Newman condition

Let λ be a point of the unit disk. We denote by $\tilde{k}_\lambda$ the associated normalized reproducing kernel, that is,

$$\tilde{k}_\lambda(z) = \frac{k_\lambda(z)}{\|k_\lambda\|_2} = \frac{(1-|\lambda|^2)^{1/2}}{1-\bar{\lambda}z} \qquad (z \in \mathbb{D}).$$

Furthermore, if $\Lambda = (\lambda_n)_{n\geq 1}$ is a Blaschke sequence, we introduce the positive and finite discrete measure μ_Λ on the unit disk by

$$\mu_\Lambda = \sum_{n=1}^{\infty} (1-|\lambda_n|^2)\delta_{\{\lambda_n\}}.$$

In other words, for any Borel subset $A \subset \mathbb{D}$, we have

$$\mu_\Lambda(A) = \sum_{\lambda_n \in A} (1-|\lambda_n|^2).$$

Theorem 15.2 *Let $\Lambda = (\lambda_n)_{n\geq 1}$ and assume that Λ is a Carleson sequence. Then the measure μ_Λ is a Carleson measure. More precisely, we have*

$$\sum_{n=1}^{\infty}(1-|\lambda_n|^2)|f(\lambda_n)|^2 \leq 2e\left(1+2\log\frac{1}{\delta(\Lambda)}\right)\|f\|_2^2$$

for all functions $f \in H^2$.

Proof Fix any $p \geq 1$. Then

$$\begin{aligned}(1-|\lambda_p|^2)\|k_{\lambda_p}\|^2_{L^2(\mu_\Lambda)} &= (1-|\lambda_p|^2)\int_{\mathbb{D}}\frac{1}{|1-\bar{\lambda}_p z|^2}\,d\mu_\Lambda(z)\\ &= (1-|\lambda_p|^2)\sum_{n\geq 1}\frac{1-|\lambda_n|^2}{|1-\bar{\lambda}_p\lambda_n|^2}\\ &= 1+\sum_{n\neq p}\frac{(1-|\lambda_n|^2)(1-|\lambda_p|^2)}{|1-\bar{\lambda}_p\lambda_n|^2}.\end{aligned}$$

According to (4.41), we can write

$$\frac{(1-|\lambda_n|^2)(1-|\lambda_p|^2)}{|1-\bar{\lambda}_p\lambda_n|^2} = 1-|b_{\lambda_p}(\lambda_n)|^2 \in (0,1).$$

Hence, using the simple inequality $x \leq -\log(1-x)$, $x \in (0,1)$, we obtain

$$\begin{aligned}(1-|\lambda_p|^2)\|k_{\lambda_p}\|^2_{L^2(\mu_\Lambda)} &\leq 1-\sum_{n\neq p}\log\left(1-\frac{(1-|\lambda_n|^2)(1-|\lambda_p|^2)}{|1-\bar{\lambda}_p\lambda_n|^2}\right)\\ &= 1-\sum_{n\neq p}\log|b_{\lambda_p}(\lambda_n)|^2\\ &= 1-\sum_{n\neq p}\log|b_{\lambda_n}(\lambda_p)|^2\\ &= 1-\log\prod_{n\neq p}|b_{\lambda_n}(\lambda_p)|^2\\ &= 1-\log|B_p(\lambda_p)|^2.\end{aligned}$$

Since $\Lambda = (\lambda_n)_{n\geq 1} \in (C)$, we have $\delta(\Lambda) = \inf_{p\geq 1}|B_p(\lambda_p)| > 0$, whence

$$(1-|\lambda_p|^2)\|k_{\lambda_p}\|^2_{L^2(\mu_\Lambda)} \leq 1-2\log(\delta(\Lambda)) = 1+2\log\frac{1}{\delta(\Lambda)}.$$

Finally, we get

$$K'(\mu_\Lambda) = \sup_{p\geq 1}(1-|\lambda_p|^2)\|k_{\lambda_p}\|^2_{L^2(\mu_\Lambda)} \leq 1+2\log\frac{1}{\delta(\Lambda)},$$

and Theorem 5.15 implies that μ_Λ is a Carleson measure such that

$$\sum_{n=1}^{\infty}(1-|\lambda_n|^2)|f(\lambda_n)|^2 = \|f\|^2_{L^2(\mu_\Lambda)} \leq 2e\left(1+2\log\frac{1}{\delta(\Lambda)}\right)\|f\|^2_{H^2}. \quad \square$$

Using our technical language of Chapter 10, Theorem 15.2 says that, if $(\tilde{k}_{\lambda_n})_{n\geq 1}$ is uniformly minimal, then it is a Bessel sequence.

Theorem 15.2 inspires the following definition. Let $\Lambda = (\lambda_n)_{n\geq 1}$ be a sequence in the unit disk. We say that Λ satisfies the *Carleson–Newman condition*, and we write $\Lambda \in (CN)$, if there exists a constant $C > 0$ such that

$$\sum_{n=1}^{\infty}(1 - |\lambda_n|^2)|f(\lambda_n)|^2 \leq C\|f\|_2^2, \tag{CN}$$

for all functions $f \in H^2$. Furthermore, we say that Λ is *separated*, and we write $\Lambda \in (R)$, if there exists a constant $\delta > 0$ such that

$$\inf_{n\neq p} |b_{\lambda_n}(\lambda_p)| \geq \delta. \tag{R}$$

It is immediate to see that $\Lambda \in (CN)$ if and only if the normalized sequence of reproducing kernels $(\tilde{k}_{\lambda_n})_{n\geq 1}$ is a Bessel sequence. Moreover, since we have

$$|\langle \tilde{k}_{\lambda_n}, \tilde{k}_{\lambda_p}\rangle|^2 = \frac{(1 - |\lambda_n|^2)(1 - |\lambda_p|^2)}{|1 - \bar{\lambda}_n\lambda_p|^2} = 1 - |b_{\lambda_n}(\lambda_p)|^2,$$

we see that $\Lambda \in (R)$ if and only if

$$\sup_{n\neq p} |\langle \tilde{k}_{\lambda_n}, \tilde{k}_{\lambda_p}\rangle| < 1,$$

which is equivalent to saying that $(\tilde{k}_{\lambda_n})_{n\geq 1}$ is separated in the sense introduced in Section 10.8.

It is clear that, if $\Lambda \in (C)$, then $\Lambda \in (R)$. Furthermore, according to Theorem 15.2, if $\Lambda \in (C)$, then $\Lambda \in (CN)$. Thus

$$(C) \Longrightarrow (R) + (CN).$$

The following result shows that the converse is also true.

Theorem 15.3 *Let $\Lambda = (\lambda_n)_{n\geq 1}$ be a sequence in the open unit disk $\mathbb{D}$. Then Λ satisfies (C) if and only if Λ satisfies (R) and (CN).*

Proof We already saw that, if Λ satisfies (C), then it satisfies (R) and (CN). So it remains to prove the converse. Fix $p \geq 1$. We rewrite the identity (4.41) as

$$\left|\frac{1 - \bar{\lambda}\mu}{\lambda - \mu}\right|^2 = 1 + \frac{(1 - |\lambda|^2)(1 - |\mu|^2)}{|\lambda - \mu|^2} \qquad (\lambda \neq \mu).$$

Then, since $\log(1+x) \leq x$, $x \geq 0$, we have

$$\log \prod_{n \neq p} \left| \frac{1 - \bar{\lambda}_n \lambda_p}{\lambda_n - \lambda_p} \right|^2 = \sum_{n \neq p} \log \left(1 + \frac{(1-|\lambda_n|^2)(1-|\lambda_p|^2)}{|\lambda_n - \lambda_p|^2} \right)$$
$$\leq \sum_{n \neq p} \frac{(1-|\lambda_n|^2)(1-|\lambda_p|^2)}{|\lambda_n - \lambda_p|^2}$$
$$= (1-|\lambda_p|^2) \sum_{n \neq p} \frac{1-|\lambda_n|^2}{|1-\bar{\lambda}_n \lambda_p|^2} |b_{\lambda_n}(\lambda_p)|^{-2}.$$

Since Λ is separated, there exists a constant $\delta > 0$ such that

$$|b_{\lambda_n}(\lambda_p)|^{-2} \leq \delta^{-2} \qquad (n \neq p),$$

and it follows from condition (CN), applied to $f = k_{\lambda_p}$, that there exists $C > 0$ such that

$$\sum_{n \neq p} \frac{1-|\lambda_n|^2}{|1-\bar{\lambda}_n \lambda_p|^2} \leq \frac{C^2}{1-|\lambda_p|^2}.$$

Hence, we have

$$\log \prod_{n \neq p} \left| \frac{1 - \bar{\lambda}_n \lambda_p}{\lambda_n - \lambda_p} \right|^2 \leq \frac{C^2}{\delta^2},$$

which gives

$$|B_{\lambda_p}(\lambda_p)| \geq e^{-\frac{1}{2}C^2/\delta^2}$$

for all $p \geq 1$. Thus, Λ satisfies the Carleson condition. □

Lemma 15.4 *Let $(\lambda_n)_{n \geq 1}$ be a Blaschke sequence of $\mathbb{D}$ such that $0 < \alpha < \lambda_n < 1$. Then*

$$\prod_{n \geq 1} \left(\frac{\lambda_n - \alpha}{1 - \alpha \lambda_n} \right) \geq \frac{\prod_{n \geq 1} \lambda_n - \alpha}{1 - \alpha \prod_{n \geq 1} \lambda_n}. \tag{15.3}$$

Proof Let B be the Blaschke product with zeros λ_n, $n \geq 1$. The inequality (15.3) can be rewritten as

$$B(\alpha) \geq \frac{B(0) - \alpha}{1 - \alpha B(0)}. \tag{15.4}$$

Now, by (4.1) applied to B, we have

$$\rho(B(\alpha), B(0)) \leq \alpha,$$

where we recall that ρ denotes the hyperbolic distance between two points z and w in $\mathbb{D}$ and is defined by

$$\rho(z,w) = \left| \frac{z - w}{1 - \bar{w} z} \right|.$$

Hence, $B(\alpha)$ belongs to the hyperbolic disk $D_{hyp}(B(0),\alpha)$. So, by (4.3), we get

$$B(\alpha) = |B(\alpha)| \geq \frac{|B(0)| - \alpha}{1 - \alpha|B(0)|} = \frac{B(0) - \alpha}{1 - \alpha B(0)},$$

which gives (15.4) and thus (15.3) holds. □

Lemma 15.5 *Let $(\lambda_n)_{n\geq 1}$ be a Blaschke sequence in $\mathbb{D}$, and let $(\mu_n)_{n\geq 1}$ be a sequence in $\mathbb{D}$ satisfying*

$$\sup_{n\geq 1} \left| \frac{\lambda_n - \mu_n}{1 - \bar{\lambda}_n \mu_n} \right| \leq \lambda < 1.$$

Then $(\mu_n)_{n\geq 1}$ is also a Blaschke sequence.

Proof Applying inequality (4.41), we have

$$\frac{(1 - |\lambda_n|^2)(1 - |\mu_n|^2)}{|1 - \bar{\lambda}_n \mu_n|^2} = 1 - |b_{\lambda_n}(\mu_n)|^2 \geq 1 - \lambda^2 > 0.$$

Hence, $(1 - |\lambda_n|^2)(1 - |\mu_n|^2) \geq (1 - |\lambda|^2)(1 - |\mu_n|)^2$, which gives

$$1 - |\mu_n| \leq \frac{4}{1 - |\lambda|^2}(1 - |\lambda_n|).$$

This inequality proves that the series $\sum_n (1 - |\mu_n|)$ is convergent, which means that $(\mu_n)_{n\geq 1}$ is a Blaschke sequence. □

Theorem 15.6 *Let $\Lambda = (\lambda_n)_{n\geq 1}$ be a Carleson sequence in the open unit disk, and let $\delta = \delta(\Lambda)$ be its Carleson constant. Let $M = (\mu_n)_{n\geq 1}$ be a sequence in $\mathbb{D}$ satisfying*

$$\sup_{n\geq 1} \left| \frac{\lambda_n - \mu_n}{1 - \bar{\lambda}_n \mu_n} \right| \leq \lambda,$$

where λ is any real number such that

$$0 < \lambda < 1 \quad \textit{and} \quad \frac{2\lambda}{1 + \lambda^2} < \delta.$$

Then M is a Carleson sequence and

$$\delta(M) \geq \frac{\delta - 2\lambda/(1 + \lambda^2)}{1 - 2\lambda\delta/(1 + \lambda^2)}.$$

Proof Using the inequality (4.4) twice, we have, for $j \neq k$,

$$\rho(\mu_k, \mu_j) \geq \frac{\rho(\mu_k, \lambda_j) - \rho(\lambda_j, \mu_j)}{1 - \rho(\mu_k, \lambda_j)\rho(\lambda_j, \mu_j)} \geq \frac{\rho(\mu_k, \lambda_j) - \lambda}{1 - \lambda\rho(\mu_k, \lambda_j)}$$

and

$$\rho(\mu_k, \lambda_j) = \rho(\lambda_j, \mu_k) \geq \frac{\rho(\lambda_j, \lambda_k) - \rho(\lambda_k, \mu_k)}{1 - \rho(\lambda_j, \lambda_k)\rho(\lambda_k, \mu_k)} \geq \frac{\rho(\lambda_j, \lambda_k) - \lambda}{1 - \lambda\rho(\lambda_j, \lambda_k)}.$$

Put $\alpha = 2\lambda/(1+\lambda^2)$. Then we obtain

$$\rho(\mu_k, \mu_j) \geq \frac{\dfrac{\rho(\lambda_j, \lambda_k) - \lambda}{1 - \lambda\rho(\lambda_j, \lambda_k)} - \lambda}{1 - \lambda\dfrac{\rho(\lambda_j, \lambda_k) - \lambda}{1 - \lambda\rho(\lambda_j, \lambda_k)}} = \frac{\rho(\lambda_j, \lambda_k) - \alpha}{1 - \alpha\rho(\lambda_j, \lambda_k)}.$$

Finally, Lemma 15.4 implies that

$$\begin{aligned}\prod_{k \neq j} \rho(\mu_k, \mu_j) \geq \prod_{k \neq j} \left(\frac{\rho(\lambda_j, \lambda_k) - \alpha}{1 - \alpha\rho(\lambda_j, \lambda_k)} \right) &\geq \frac{\prod_{k \neq j} \rho(\lambda_k, \lambda_j) - \alpha}{1 - \alpha \prod_{k \neq j} \rho(\lambda_k, \lambda_j)} \\ &\geq \frac{\delta - \alpha}{1 - \alpha\delta},\end{aligned}$$

which exactly means that M is a Carleson sequence with the desired estimate on $\delta(M)$. □

Exercises

Exercise 15.2.1 Let $\lambda, \mu \in \mathbb{D}$. Show that

$$|b_\lambda(\mu)|^2 = 1 - \frac{(1 - |\lambda|^2)(1 - |\mu|^2)}{|1 - \bar{\lambda}\mu|^2}.$$

Then deduce that

$$|b_\lambda(\mu)| \geq |b_{|\lambda|}(|\mu|)|.$$

Exercise 15.2.2 Let $(\lambda_n)_{n \geq 1}$ be a Blaschke sequence such that

$$|\lambda_n| \leq |\lambda_{n+1}| \qquad (n \geq 1).$$

Assume that

$$\gamma = \lim_{n \to +\infty} \frac{1 - |\lambda_{n+1}|}{1 - |\lambda_n|} < 1.$$

We proceed to show that $(\lambda_n)_{n \geq 1} \in (C)$.

(i) Show that there exist $d < 1$ and $n_0 \in \mathbb{N}$ such that

$$n > k \geq n_0 \quad \Longrightarrow \quad (1 - |\lambda_n|) \leq d^{n-k}(1 - |\lambda_k|).$$

(ii) Deduce that, for each $n > k \geq n_0$, we have

$$|\lambda_n| - |\lambda_k| \geq (1 - d^{n-k})(1 - |\lambda_k|)$$

and

$$1 - |\lambda_n \lambda_k| \leq (1 + d^{n-k})(1 - |\lambda_k|).$$

(iii) Show that, for all $n, k \geq n_0$, $n \neq k$, we have

$$|b_{\lambda_k}(\lambda_n)| \geq \frac{1 - d^{|n-k|}}{1 + d^{|n-k|}}.$$

Hint: Apply Exercise 15.2.1.

(iv) Show that the infinite product

$$\prod_{p \geq 1} \frac{1 - d^p}{1 + d^p}$$

converges and that

$$|B_n(\lambda_n)| \geq \prod_{k \leq n_0} |b_{\lambda_k}(\lambda_n)| \bigg(\prod_{p \geq 1} \frac{1 - d^p}{1 + d^p} \bigg)^2 \qquad (n \geq n_0).$$

(v) Deduce that $\inf_{n \geq 1} |B_n(\lambda_n)| > 0$.

Exercise 15.2.3 Show that $\lambda_n := 1 - q^n$, where $0 < q < 1$, is a Carleson sequence.
Hint: Apply Exercise 15.2.2.

Exercise 15.2.4 Let $\Lambda = (\lambda_n)_{n \geq 1}$ be a real sequence of $[0, 1]$, and assume that $\lambda_n < \lambda_{n+1}$ and $\Lambda \in (C)$. Show that

$$\gamma = \lim_{n \to +\infty} \frac{1 - \lambda_{n+1}}{1 - \lambda_n} < 1.$$

Hint: Use

$$1 - \frac{1 - \lambda_{n+1}}{1 - \lambda_n} = \frac{\lambda_{n+1} - \lambda_n}{1 - \lambda_n} \geq \frac{\lambda_{n+1} - \lambda_n}{1 - \lambda_n \lambda_{n+1}} = b_{\lambda_{n+1}}(\lambda_n) \geq |B_{\lambda_n}(\lambda_n)|.$$

15.3 Riesz basis of reproducing kernels

Lemma 14.16 implies that, if $(k_{\lambda_n})_{n \geq 1}$ is minimal, then the sequence $(\lambda_n)_{n \geq 1}$ is a Blaschke sequence. In particular, we have $|\lambda_n| \longrightarrow 1$ as $n \longrightarrow \infty$. Thus,

$$\|k_{\lambda_n}\|_2 = (1 - |\lambda_n|^2)^{-1/2} \longrightarrow \infty$$

as $n \longrightarrow \infty$. This means that the sequence $(k_{\lambda_n})_{n \geq 1}$ is never a Riesz sequence. Therefore, we need to normalize the sequence of reproducing kernels if we wish to obtain a Riesz sequence. We recall that

$$\tilde{k}_\lambda(z) = \frac{k_\lambda(z)}{\|k_\lambda\|_2} = \frac{(1 - |\lambda|^2)^{1/2}}{1 - \bar{\lambda} z} \qquad (z \in \mathbb{D}).$$

For the notions used in the following result, see Chapter 10.

Theorem 15.7 *Let $\Lambda = (\lambda_n)_{n\geq 1}$ be a Blaschke sequence of distinct points of the unit disk $\mathbb{D}$, let B be the Blaschke product associated with Λ, and let*

$$\mathfrak{X} = (\tilde{k}_{\lambda_n})_{n\geq 1}$$

be the associated sequence of normalized reproducing kernels. Then the following are equivalent:

(i) *$\mathfrak{X}$ is a Riesz basis of K_B;*
(ii) *$\mathfrak{X}$ is uniformly minimal;*
(iii) *$\mathcal{R}(J_{\mathfrak{X}}) = \ell^2$, where*

$$J_{\mathfrak{X}} f = ((1-|\lambda_n|^2)^{1/2} f(\lambda_n))_{n\geq 1} \qquad (f \in H^2);$$

(iv) *Λ is a Carleson sequence.*

Proof (i) $\Longleftrightarrow$ (iii) This follows immediately from Theorem 10.21.

(i) $\Longrightarrow$ (ii) According to Theorem 10.16, this is true for any sequence in a Hilbert space.

(ii) $\Longrightarrow$ (iv) This follows from Theorem 15.1.

(iv) $\Longrightarrow$ (i) First note that, since the sequence $\mathfrak{X}$ is minimal and complete in K_B, Theorem 10.21 implies that $\mathfrak{X}$ is a Riesz basis of K_B if and only if $\mathcal{R}(J_{\mathfrak{X}}) \subset \ell^2$ and $\mathcal{R}(J_{\mathfrak{X}^*}) \subset \ell^2$, where

$$J_{\mathfrak{X}} f = ((1-|\lambda_n|^2)^{1/2} f(\lambda_n))_{n\geq 1} \qquad (f \in H^2)$$

and

$$J_{\mathfrak{X}^*} f = (\langle f, \tilde{k}^*_{\lambda_n}\rangle_2)_{n\geq 1} \qquad (f \in H^2),$$

with $(\tilde{k}^*_{\lambda_n})_{n\geq 1}$ be the unique biorthogonal sequence associated with $(\tilde{k}_{\lambda_n})_{n\geq 1}$ in K_B. Using Lemma 14.17, we see that

$$\begin{aligned}
\tilde{k}^*_{\lambda_n} &= \|k_{\lambda_n}\|_2 k^*_{\lambda_n} \\
&= \frac{(1-|\lambda_n|^2)^{1/2}}{B_n(\lambda_n)} \frac{\lambda_n}{|\lambda_n|} \frac{B}{z-\lambda_n} \\
&= \frac{(1-|\lambda_n|^2)^{1/2}}{B_n(\lambda_n)} \frac{\lambda_n}{|\lambda_n|} \Omega_B(k^B_{\lambda_n}).
\end{aligned}$$

Hence, by Lemma 14.14, we have

$$\begin{aligned}
|\langle f, \tilde{k}^*_{\lambda_n}\rangle_2|^2 &= \frac{1-|\lambda_n|^2}{|B_n(\lambda_n)|^2} |\langle f, \Omega_B(k^B_{\lambda_n})\rangle_2|^2 \\
&= \frac{1-|\lambda_n|^2}{|B_n(\lambda_n)|^2} |\langle \Omega_B(f), k^B_{\lambda_n}\rangle_2|^2 \\
&= \frac{1-|\lambda_n|^2}{|B_n(\lambda_n)|^2} |(\Omega_B f)(\lambda_n)|^2 \qquad (f \in K_B).
\end{aligned}$$

Thus, $(\tilde{k}_{\lambda_n})_{n\geq 1}$ is a Riesz basis of K_B if and only if

$$\sum_{n\geq 1}(1-|\lambda_n|^2)|f(\lambda_n)|^2 < \infty$$

and

$$\sum_{n\geq 1}\frac{1-|\lambda_n|^2}{|B_n(\lambda_n)|^2}|(\Omega_B f)(\lambda_n)|^2 < \infty.$$

But, according to Theorem 15.2 and the fact that $\inf_n |B_n(\lambda_n)| > 0$, these two inequalities hold. Therefore, $\mathfrak{X}$ is a Riesz basis of K_B. □

Corollary 15.8 *Let $\Lambda = (\lambda_n)_{n\geq 1}$ be a sequence in the open unit disk. Then $(\tilde{k}_{\lambda_n})_{n\geq 1}$ satisfies the Feichtinger conjecture.*

Proof According to Corollary 10.28, it is sufficient to prove that $(\tilde{k}_{\lambda_n})_{n\geq 1}$ satisfies the weaker Feichtinger conjecture (WFC). But, if $(\tilde{k}_{\lambda_n})_{n\geq 1}$ is a separated Bessel sequence, then it means that Λ satisfies (CN) and (R). Then it follows from Theorem 15.3 that Λ satisfies the Carleson condition (C). This implies, by Theorem 15.7, that $(\tilde{k}_{\lambda_n})_{n\geq 1}$ is a Riesz basis. Therefore, $(\tilde{k}_{\lambda_n})_{n\geq 1}$ satisfies WFC and thus FC. □

Corollary 15.8 can be rephrased as follows.

Corollary 15.9 *Let $\Lambda = (\lambda_n)_{n\geq 1}$ be a sequence of the unit disk. Assume that $\Lambda \in (CN)$. Then Λ can be partitioned into a finite union of Carleson sequences.*

Exercises

Exercise 15.3.1 Let $(\lambda_n)_{n\geq 1}$ be a Blaschke sequence of distinct points in the open unit disk $\mathbb{D}$, and let B be the associated Blaschke product. Let $(\varphi_n)_{n\geq 1}$ be the biorthogonal sequence associated with (k_{λ_n}) in K_B.

(i) Put

$$B^{(k)} = \prod_{n\geq k} b_{\lambda_n}.$$

Show that, for each $f \in K_B$, we have

$$B^{(k)}(\mathbf{M}_B)^* f = \sum_{n=1}^{k-1} \overline{B^{(k)}(\lambda_n)}\langle f, \varphi_n\rangle k_{\lambda_n}.$$

(ii) Deduce that $(k_{\lambda_n})_{n\geq 1}$ is a summation basis of K_B, that is, there exists an infinite matrix $A = (a_{n,k})_{n,k\geq 1}$ such that, for any $f \in K_B$, we have

$$x = \lim_{k\to\infty} \sum_{n\geq 1} a_{n,k} \langle f, \varphi_n \rangle k_{\lambda_n}.$$

(iii) Deduce that

$$\text{Span}\{\varphi_n : n \geq 1\} = K_B.$$

Exercise 15.3.2 Let $\varphi \in H^\infty$, let $(\lambda_n)_{n\geq 1}$ be a Carleson sequence, and let B be the associated Blaschke product. Show that $\varphi(\mathbf{M}_B)$ is compact if and only if $\varphi(\lambda_n) \longrightarrow 0$, as $n \longrightarrow \infty$.
Hint: Use (14.41) and Theorem 15.7.

15.4 Nevanlinna–Pick interpolation problem

Given n points $\lambda_1, \lambda_2, \ldots, \lambda_n$ in the open unit disk $\mathbb{D}$, and n points $\omega_1, \omega_2, \ldots, \omega_n$ in the complex plane, we would like to know if there exists a function f in the closed unit ball of H^∞ interpolating the points λ_i to the points ω_i, that is,

$$f(\lambda_i) = \omega_i \qquad (1 \leq i \leq n).$$

The following result answers this question.

Theorem 15.10 *Let $(\lambda_i)_{1\leq i\leq n}$ be n distinct points in $\mathbb{D}$, and let $(\omega_i)_{1\leq i\leq n}$ be complex numbers. Then the following are equivalent.*

(i) *There exists a function f in H^∞ such that*

$$f(\lambda_i) = \omega_i \qquad (1 \leq i \leq n)$$

and, moreover, $\|f\|_\infty \leq 1$.

(ii) *The matrix $Q = (Q_{j,k})_{1\leq j,\, k\leq n}$, where*

$$Q_{j,k} = \frac{1 - \overline{\omega_j}\omega_k}{1 - \overline{\lambda_j}\lambda_k} \qquad (j, k = 1, \ldots, n),$$

is nonnegative.

Proof We start by introducing the operator T, which is exploited in the proof of equivalence. Let B be the (finite) Blaschke product associated with the sequence $(\lambda_i)_{1\leq i\leq n}$. We know from Theorem 5.5(iii) and Lemma 14.16 that the sequence of reproducing kernels $(k_{\lambda_i})_{1\leq i\leq n}$ forms a basis of K_B. Note that K_B is finite-dimensional. Hence we can consider the linear bounded operator T from K_B into itself defined by

$$Tk_{\lambda_i} = \overline{\omega_i} k_{\lambda_i} \qquad (1 \leq i \leq n). \tag{15.5}$$

By Theorem 14.32, we have

$$T\mathbf{M}_B^* k_{\lambda_i} = \mathbf{M}_B^* T k_{\lambda_i} = \overline{\lambda_i}\overline{\omega_i} k_{\lambda_i} \qquad (1 \le i \le n).$$

Since $(k_{\lambda_i})_{1\le i\le n}$ is a basis of K_B, we deduce that $T\mathbf{M}_B^* = \mathbf{M}_B^* T$. Hence, Theorem 14.38 ensures that there exists $g \in H^\infty$ such that $T = g(\mathbf{M}_B)^*$ and

$$\|T\| = \min\{\|g\|_\infty : T = g(\mathbf{M}_B)^*\}. \tag{15.6}$$

For any function g that fulfills $T = g(\mathbf{M}_B)^*$, we have

$$Tk_{\lambda_i} = g(\mathbf{M}_B)^* k_{\lambda_i} = \overline{g(\lambda_i)} k_{\lambda_i} \qquad (1 \le i \le n)$$

and therefore

$$g(\lambda_i) = \omega_i \qquad (1 \le i \le n). \tag{15.7}$$

To establish the connection between Q and T, let $a_i \in \mathbb{C}$, $1 \le i \le n$. Then, using the operator $\mathbb{T}$, we can write

$$\begin{aligned}\sum_{1\le i,j\le n} a_i \overline{a_j} \frac{1-\overline{\omega_i}\omega_j}{1-\overline{\lambda_i}\lambda_j} &= \sum_{1\le i,j\le n} a_i\overline{a_j}(\langle k_{\lambda_i}, k_{\lambda_j}\rangle - \langle Tk_{\lambda_i}, Tk_{\lambda_j}\rangle)\\ &= \sum_{1\le i,j\le n} a_i\overline{a_j}\langle (I - T^*T)k_{\lambda_i}, k_{\lambda_j}\rangle\\ &= \langle (I-T^*T)f, f\rangle,\end{aligned}$$

where

$$f = \sum_{i=1}^n a_i k_{\lambda_i}.$$

Hence, the matrix Q is nonnegative if and only if the operator $I - T^*T$ is positive, which, by (2.13), is equivalent to saying that T is a contraction.

(i) $\Longrightarrow$ (ii) Assume that there exists $f \in H^\infty$, $\|f\|_\infty \le 1$, such that $f(\lambda_i) = \omega_i$, $1 \le i \le n$. Hence, using (15.5) and (14.41), we have

$$Tk_{\lambda_i} = \overline{\omega_i} k_{\lambda_i} = \overline{f(\lambda_i)} k_{\lambda_i} = f(\mathbf{M}_B)^* k_{\lambda_i} \qquad (1 \le i \le n).$$

Thus we get $T = f(\mathbf{M}_B)^*$. Now, (15.6) implies that

$$\|T\| \le \|f\|_\infty \le 1,$$

i.e. T is a contraction, and thus Q is nonnegative.

(ii) $\Longrightarrow$ (i) Assume that Q is nonnegative. Thus, T is a contraction. By (15.6), we know that there exists a function $f \in H^\infty$, $\|f\|_\infty \le 1$, such that $T = f(\mathbf{M}_B)^*$. Thus, by (15.7), $f(\lambda_i) = \omega_i$, $1 \le i \le n$, which gives (i). □

Exercise

Exercise 15.4.1 Let A be the operator on K_B whose matrix relative to the basis $(z^k)_{0\le k\le n}$ is

$$[A] = \begin{pmatrix} a_0 & 0 & 0 & \cdots & 0 & 0 \\ a_1 & a_0 & 0 & \cdots & 0 & 0 \\ a_2 & a_1 & a_0 & \cdots & 0 & 0 \\ \vdots & \vdots & \vdots & \ddots & \vdots & \vdots \\ a_{n-1} & a_{n-2} & a_{n-3} & \cdots & a_0 & 0 \\ a_n & a_{n-1} & a_{n-2} & \cdots & a_1 & a_0 \end{pmatrix}.$$

Show that the following are equivalent.

(i) There exists a function $f \in H^\infty$ such that

$$\widehat{f}(k) = a_k \quad (0 \le k \le n) \quad \text{and} \quad \|f\|_\infty \le 1.$$

(ii) $\|A\| \le 1$.

15.5 H^∞-interpolating sequences

Let $\Lambda = (\lambda_n)_{n\ge1}$ be a sequence in the open unit disk. We say that Λ is an H^∞*-interpolating* sequence if, for each sequence $(\omega_n)_{n\ge1}$ in ℓ^∞, there exists a function $f \in H^\infty$ such that $f(\lambda_n) = \omega_n$, $n \ge 1$. The following result characterizes such sequences.

Theorem 15.11 *Let $\Lambda = (\lambda_n)_{n\ge1}$ be a sequence of distinct points in the open unit disk $\mathbb{D}$. Then the following assertions are equivalent:*

(i) *$(\lambda_n)_{n\ge1}$ is an H^∞-interpolating sequence;*
(ii) *$(\lambda_n)_{n\ge1}$ is a Carleson sequence.*

Proof If Λ is an H^∞-interpolating sequence, then Λ is a Blaschke sequence. Indeed, taking $\omega_1 = 1$ and $\omega_n = 0$, $n \ge 2$, our assumption says that there exists a function $f \in H^\infty$ such that $f(\lambda_1) = 1$ and $f(\lambda_n) = 0$, $n \ge 2$. Hence, the sequence $(\lambda_n)_{n\ge2}$ is the zero sequence of a nonzero function in H^∞ and thus it must be a Blaschke sequence.

We will now denote by $\mathfrak{X} = (\tilde{k}_{\lambda_n})_{n\ge1}$ the sequence of normalized reproducing kernels and by B the Blaschke product associated with Λ. Recall that $\mathfrak{Mult}(\mathfrak{X})$ denotes the space of multipliers of $\mathfrak{X}$ in K_B, that is,

$$\mathfrak{Mult}(\mathfrak{X}) = \{(u_n)_{n\ge1} : u_n \in \mathbb{C}, \text{ and } \exists\, T \in \mathcal{L}(K_B),\ T\tilde{k}_n = u_n\tilde{k}_n,\ n \ge 1\}.$$

Let us show that

$$\mathfrak{Mult}(\mathfrak{X}) = \overline{H^\infty_{|\Lambda}}, \tag{15.8}$$

where

$$H^\infty_{|\Lambda} = \{(f(\lambda_n))_{n\geq 1} : f \in H^\infty\}.$$

The overbar in (15.8) stands for complex conjugation. To prove (15.8), first note that, if $f \in H^\infty$, then according to (14.41), we have

$$f(\mathbf{M}_B)^* \tilde{k}_{\lambda_n} = \overline{f(\lambda_n)} \tilde{k}_{\lambda_n}.$$

Hence, the sequence $(\overline{f(\lambda_n)})_{n\geq 1}$ is a multiplier of $\mathfrak{X}$, and this proves the inclusion $\overline{H^\infty_{|\Lambda}} \subset \mathfrak{Mult}(\mathfrak{X})$. For the reverse inclusion, if $(u_n)_{n\geq 1}$ is a multiplier of $\mathfrak{X}$, then, by definition, there exists $T \in \mathcal{L}(K_B)$ such that $T\tilde{k}_{\lambda_n} = u_n \tilde{k}_{\lambda_n}$. Hence

$$T\mathbf{M}_B^* \tilde{k}_{\lambda_n} = \mathbf{M}_B^* T \tilde{k}_{\lambda_n} = \overline{\lambda_n} u_n \tilde{k}_{\lambda_n}.$$

Since the sequence $(\tilde{k}_{\lambda_n})_{n\geq 1}$ is complete in K_B, we deduce that $T\mathbf{M}_B^* = \mathbf{M}_B^* T$. But, according to Theorem 14.38, the latter equation implies the existence of a function $f \in H^\infty$ such that $T = f(\mathbf{M}_B)^*$. Therefore,

$$u_n \tilde{k}_{\lambda_n} = T\tilde{k}_{\lambda_n} = f(\mathbf{M}_B)^* \tilde{k}_{\lambda_n} = \overline{f(\lambda_n)} \tilde{k}_{\lambda_n},$$

which gives that $\overline{f(\lambda_n)} = u_n$. That proves (15.8).

Theorems 15.7, 10.30 and 10.14 together show that

$$(\lambda_n)_{n\geq 1} \in (C) \quad \Longleftrightarrow \quad \mathfrak{Mult}(\mathfrak{X}) = \ell^\infty.$$

Hence, in light of (15.8), we can say that

$$(\lambda_n)_{n\geq 1} \in (C) \quad \Longleftrightarrow \quad \overline{H^\infty_{|\Lambda}} = \ell^\infty \quad \Longleftrightarrow \quad H^\infty_{|\Lambda} = \ell^\infty.$$

The last identity precisely means that $(\lambda_n)_{n\geq 1}$ is an H^∞-interpolating sequence. □

15.6 H^2-interpolating sequences

Let $(\lambda_n)_{n\geq 1}$ be a sequence of points in $\mathbb{D}$. We say that $(\lambda_n)_{n\geq 1}$ is an H^2-*interpolating* sequence if, for each sequence $(a_n)_{n\geq 1}$ in ℓ^2, there exists a function $f \in H^2$ such that

$$f(\lambda_n) = a_n(1 - |\lambda_n|^2)^{-1/2} \qquad (n \geq 1). \tag{15.9}$$

The following result shows that, if $(\lambda_n)_{n\geq 1}$ is an H^2-interpolating sequence, then not only does the interpolation problem (15.9) have a solution but also we can somehow control the norm of the interpolating function f, which is particularly important in applications (for instance, in control theory).

Lemma 15.12 *Let $\Lambda = (\lambda_n)_{n\geq 1} \subset \mathbb{D}$. If $(\lambda_n)_{n\geq 1}$ is an H^2-interpolating sequence, then there exists a constant $C > 0$ such that, for any sequence $a = (a_n)_{n\geq 1}$ in ℓ^2, there exists a function f in H^2 satisfying*

$$f(\lambda_n) = a_n(1 - |\lambda_n|^2)^{-1/2} \qquad (n \geq 1)$$

and

$$\|f\|_{H^2} \leq C\|a\|_{\ell^2}.$$

Proof If $(\lambda_n)_{n\geq 1}$ is an H^2-interpolating sequence, then it is in particular a Blaschke sequence. Indeed, similar to the proof of Theorem 15.11, it is sufficient to solve the interpolating problem with the sequence $a = (a_n)_{n\geq 1}$ defined by $a_1 = 1$ and $a_n = 0$, $n \geq 2$.

We denote by B the Blaschke product associated with Λ. Let us show that, for any sequence $a = (a_n)_{n\geq 1}$ in ℓ^2, there exists a unique function $f_a \in K_B$ such that $f_a(\lambda_n) = a_n(1 - |\lambda_n|^2)^{-1/2}$, $n \geq 1$. Indeed, since $(\lambda_n)_{n\geq 1}$ is an H^2-interpolating sequence, there exists a function $f \in H^2$ satisfying the interpolation problem (15.9). Writing f as $f = f_a + Bg$, with $f_a \in K_B$ and $g \in H^2$, we have

$$a_n(1 - |\lambda_n|^2)^{-1/2} = f(\lambda_n) = f_a(\lambda_n) + B(\lambda_n)g(\lambda_n) = f_a(\lambda_n),$$

which proves the existence of f_a. For the uniqueness, let $g \in K_B$ such that $g(\lambda_n) = a_n(1 - |\lambda_n|^2)^{-1/2}$. Then the function $f_a - g$ vanishes at $(\lambda_n)_{n\geq 1}$, which implies that it belongs to BH^2. But, since it also belongs to K_B, we must have $g = f_a$.

Based on the above paragraph, the mapping

$$\begin{array}{rccc} \Lambda: & \ell^2 & \longrightarrow & K_B \\ & a & \longmapsto & f_a \end{array}$$

is well defined. It is elementary to verify that Λ is linear. To check that Λ is continuous, we use the closed graph theorem (Corollary 1.18). Hence, let $(a^{(m)})_{m\geq 1}$ in ℓ^2 and assume that $(a^{(m)})_{m\geq 1}$ converges to a in ℓ^2 and $(f_{a^{(m)}})_{m\geq 1}$ converges to f in K_B. We have to prove that $f_a = f$. But, for each fixed $n \geq 1$, we have

$$|f_{a^{(m)}}(\lambda_n) - f(\lambda_n)| \leq \|f_{a^{(m)}} - f\|_2 \|k_{\lambda_n}\|_2$$

and

$$|a_n^{(m)} - a_n| \leq \|a^{(m)} - a\|_2,$$

which implies that

$$\lim_{m\to\infty} f_{a^{(m)}}(\lambda_n) = f(\lambda_n) \quad \text{and} \quad \lim_{m\to\infty} a_n^{(m)} = a_n.$$

But $f_{a^{(m)}}(\lambda_n) = (1-|\lambda_n|^2)^{-1/2} a_n^{(m)}$, $n \geq 1$, from which we deduce that

$$f(\lambda_n) = (1-|\lambda_n|^2)^{-1/2} a_n \qquad (n \geq 1).$$

By definition, the last identity means that $f = f_a$. Therefore, Λ is continuous, which means that there exists a constant $C > 0$ such that

$$\|f_a\|_2 \leq C\|a\|_2. \qquad \square$$

The previous lemma paves the way to the characterization of H^2-interpolating sequences. It is rather surprising that we have precisely the same characterization as in Theorem 15.11.

Theorem 15.13 *Let $(\lambda_n)_{n\geq 1}$ be a sequence of distinct points in $\mathbb{D}$. Then the following assertions are equivalent:*

(i) *$(\lambda_n)_{n\geq 1}$ is an H^2-interpolating sequence;*
(ii) *$(\lambda_n)_{n\geq 1}$ is Carleson sequence.*

Proof (i) $\Longrightarrow$ (ii) Let $a^{(m)} = (\delta_{m,n})_{n\geq 1}$. Then, according to Lemma 15.12, there exists $f_{a^{(m)}} \in K_B$ such that

$$f_{a^{(m)}}(\lambda_n) = (1-|\lambda_n|^2)^{-1/2}\delta_{m,n} \qquad (n \geq 1)$$

and

$$|f_{a^{(m)}}\|_2 \leq C\|a^{(m)}\|_2 = C,$$

for some constant $C > 0$ (depending only on Λ). Set $g_m = (1-|\lambda_m|^2)^{1/2} f_{a^{(m)}}$. We have $g_m(\lambda_n) = \delta_{m,n}$ and

$$\|g_m\|_2\|k_{\lambda_m}\|_2 = \|f_{a^{(m)}}\|_2 \leq C.$$

Therefore, $(k_{\lambda_m})_{m\geq 1}$ is uniformly minimal and thus, according to Theorem 15.1, the sequence $(\lambda_n)_{n\geq 1}$ is a Carleson sequence. Let $\mathfrak{X} = (x_{\lambda_n})_{n\geq 1}$ be the sequence of normalized reproducing kernels and let B be the Blaschke product associated with Λ.

(ii) $\Longrightarrow$ (i) If $(\lambda_n)_{n\geq 1}$ is a Carleson sequence, then it follows from Theorem 15.7 that $\mathcal{R}(J_{\mathfrak{X}}) = \ell^2$, where

$$J_{\mathfrak{X}} f = ((1-|\lambda_n|^2)^{1/2} f(\lambda_n))_{n\geq 1} \qquad (f \in H^2).$$

Hence, if $(a_n)_{n\geq 1}$ is in ℓ^2, then there exists a function $f \in H^2$ such that

$$(1-|\lambda_n|^2)^{1/2} f(\lambda_n) = a_n \qquad (n \geq 1).$$

This means that $(\lambda_n)_{n\geq 1}$ is an H^2-interpolating sequence. $\square$

Exercises

Exercise 15.6.1 Let $\Lambda = (\lambda_n)_{n\geq 1}$ be a sequence in the open unit disk satisfying the Carleson condition. Denote by B the Blaschke product associated with Λ and let $(\omega_n)_{n\geq 1} \in \ell^2$.

(i) Set

$$g_k(z) = \sum_{n=1}^{k} \omega_n \frac{(1-|\lambda_n|^2)^{1/2}}{B_n(\lambda_n)} \frac{B_n(z)}{1-\bar{\lambda}_n z} \qquad (z \in \mathbb{D}),$$

where $B_n = B/b_{\lambda_n}$. Show that the sequence $(g_k)_{k\geq 1}$ converges in H^2 to a function, which we denote by g.

(ii) Show that $g \in K_B$ and satisfies the interpolation problem

$$g(\lambda_n) = \omega_n(1-|\lambda_n|^2)^{-1/2} \qquad (n \geq 1).$$

(iii) Show that

$$\|g\|_2 = \inf\{\|f\|_2 : f \in H^2,\ f(\lambda_n) = \omega_n(1-|\lambda_n|^2)^{-1/2},\ n \geq 1\}.$$

15.7 Asymptotically orthonormal sequences

Suppose $\Lambda = (\lambda_n)_{n\geq 1}$ is a Blaschke sequence of distinct points in $\mathbb{D}$. Since the reproducing kernels $(k_{\lambda_n})_{n\geq 1}$ are complete and minimal in K_B, if $(\tilde{k}_{\lambda_n})_{n\geq 1}$ is an asymptotically orthonormal sequence (AOS), then it is also a complete asymptotically orthonormal basic sequence (AOB) in K_B; see Section 10.10 for the relevant definitions. Such sequences are characterized in Theorem 15.18. We need some preliminary results to prove that characterization.

Lemma 15.14 *Let B be a Blaschke product formed with zeros $(\lambda_n)_{n\geq 1}$, and let*

$$\delta = \inf_{n\geq 1} |B_n(\lambda_n)|.$$

Then, for every decomposition $B = B_1 B_2$, where B_1 and B_2 are Blaschke subproducts, we have

$$\max\{|B_1(z)|, |B_2(z)|\} \geq \frac{\delta^2}{(1+\sqrt{1-\delta^2})^2} \qquad (z \in \mathbb{D}).$$

Proof For simplicity, put

$$\eta = \frac{\delta}{1+\sqrt{1-\delta^2}}.$$

Fix any $\zeta \in D_{hyp}(\lambda_n, \eta)$ and $n \geq 1$. The inequality (4.1) implies that, if $\zeta \in D_{hyp}(\lambda_n, \eta)$, then $B_n(\zeta) \in D_{hyp}(B_n(\lambda_n), \eta)$. Hence, by (4.3),

$$|B_n(\zeta)| \geq \frac{|B_n(\lambda_n)| - \eta}{1 - \eta|B_n(\lambda_n)|}.$$

The real function $t \longmapsto (t - \eta)/(1 - \eta t)$, $0 < t < 1$, is increasing, and, since $|B_n(\lambda_n)| > \delta$, we obtain

$$|B_n(\zeta)| \geq \frac{\delta - \eta}{1 - \eta\delta} = \eta.$$

In particular, using this inequality, we show that

$$D_{hyp}(\lambda_n, \eta) \cap D_{hyp}(\lambda_m, \eta) = \emptyset \qquad (n \neq m).$$

Indeed, assume on the contrary that there is a point $\zeta \in D_{hyp}(\lambda_n, \eta) \cap D_{hyp}(\lambda_m, \eta)$ for some $n \neq m$. Then we have

$$\eta \leq |B_n(\zeta)| \leq |b_{\lambda_m}(\zeta)| < \eta,$$

which gives a contradiction.

Now let $B = B_1 B_2$ be a decomposition, where B_1 and B_2 are Blaschke subproducts. For $\zeta \in \partial D_{hyp}(\lambda_n, \eta)$, we have

$$|B(\zeta)| = |B_n(\zeta)|\,|b_{\lambda_n}(\zeta)| = \eta|B_n(\zeta)| \geq \eta^2.$$

Hence $|B(\zeta)| \geq \eta^2$ for any $\zeta \in \bigcup_{n\geq 1} \partial D_{hyp}(\lambda_n, \eta)$. Let us now denote by $B^{(k)}$ the finite Blaschke subproduct formed with zeros $(\lambda_n)_{1\leq n\leq k}$. Then we have

$$|B^{(k)}(\zeta)| \leq |B(\zeta)| \geq \eta^2,$$

for any $\zeta \in \bigcup_{n\geq 1} \partial D_{hyp}(\lambda_n, \eta)$ and

$$|B^{(k)}(\zeta)| = 1 \qquad (\zeta \in \mathbb{T}).$$

By the maximum principle applied to $1/B^{(k)}$ on $\mathbb{D} \setminus \bigcup_{n\geq 1} \partial D_{hyp}(\lambda_n, \eta)$, we get that

$$|B^{(k)}(\zeta)| \geq \eta^2,$$

for any $\zeta \notin \bigcup_{n\geq 1} \partial D_{hyp}(\lambda_n, \eta)$. Hence letting $k \to +\infty$, we obtain

$$|B(z)| \geq \eta^2 \qquad \left(z \notin \bigcup_{n\geq 1} \partial D_{hyp}(\lambda_n, \eta)\right).$$

Thus for any $z \notin \bigcup_{n\geq 1} \partial D_{hyp}(\lambda_n, \eta)$, we have

$$\max\{|B_1(z)|, |B_2(z)|\} \geq \eta^2.$$

Suppose now that $z \in D_{hyp}(\lambda_n, \eta))$. We may assume that λ_n is a zero of B_1. Then

$$|B_2(z)| \geq |B_n(z)| \geq \eta.$$

Hence, for any $z \in \bigcup_{n\geq 1} \partial D_{hyp}(\lambda_n, \eta)$, we also have

$$\max\{|B_1(z)|, |B_2(z)|\} \geq \eta^2. \qquad \square$$

The proof of the following lemma will use the so-called *Rademacher system*. Recall that the Rademacher functions r_n, $n \geq 0$, are the functions defined on $\mathbb{R}$ by the conditions:

(i) r_0 is periodic with period 1, and $r_0(t) = 1$, $0 < t < 1/2$, $r_0(t) = -1$, $1/2 < t < 1$, $r_0(0) = r_0(1) = 0$;
(ii) for $n \geq 1$ and all $t \in \mathbb{R}$, $r_n(t) = r_0(2^n t)$.

Thus, in the open intervals $]0, 1/2^{n+1}[$, $]1/2^{n+1}, 2/2^{n+1}[$, $]2/2^{n+1}, 3/2^{n+1}[$, ..., the function r_n takes alternately the values $+1$ and -1. It is familiar that the Rademacher functions form an orthonormal system in $L^2(0,1)$.

Lemma 15.15 *Let $\mathfrak{X} = (x_n)_{n\geq 1}$ be a normalized Riesz basis of the Hilbert space $\mathcal{H}$. For $I \subset \mathbb{N}$, let $\mathcal{M}_I = \operatorname{Span}\{x_n : n \in I\}$ and let $P_{\mathcal{M}_I}$ denote the orthogonal projection of $\mathcal{H}$ onto $\mathcal{M}_I$. Suppose that there is a constant $\varepsilon > 0$ such that, for each set $I \subset \mathbb{N}$, we have*

$$\|P_{\mathcal{M}_{\mathbb{N}\setminus I}} P_{\mathcal{M}_I}\| < \varepsilon.$$

Then

$$\|\Gamma_{\mathfrak{X}} - I\| \leq 4\varepsilon \|V_{\mathfrak{X}}\|^2.$$

Proof Let $U_{\mathfrak{X}}$ be the orthogonalizer of $\mathfrak{X}$. Remember that $V_{\mathfrak{X}} = U_{\mathfrak{X}}^{-1}$. Let $\mathfrak{e} = \sum_{k=1}^{N} c_k \mathfrak{e}_k$, $x = V_{\mathfrak{X}} \mathfrak{e} = \sum_{k=1}^{N} c_k x_k$ and

$$K = \|V_{\mathfrak{X}} \mathfrak{e}\|^2 = \left\| \sum_{k=1}^{N} c_k x_k \right\|^2,$$

and, for $t \in [0, 1]$, put

$$\mathcal{L}(t) = \left\| \sum_{k=1}^{N} c_k r_k(t) x_k \right\|^2,$$

where $(r_k)_{k\geq 0}$ is the Rademacher system. Note that

$$\mathcal{L}(t) = \sum_{1\leq k,j\leq N} c_k \bar{c}_j \langle x_k, x_j \rangle r_k(t) r_j(t).$$

Hence

$$\int_0^1 \mathcal{L}(t)\,dt = \sum_{1\le k,j\le N} c_k\bar{c}_j\langle x_k, x_j\rangle \int_0^1 r_k(t)r_j(t)\,dt.$$

Since $(r_n)_{n\ge 1}$ forms an orthonormal system in $L^2(0,1)$, we get

$$\int_0^1 \mathcal{L}(t)\,dt = \sum_{k=1}^N |c_k|^2 = \|\mathfrak{c}\|^2.$$

Now, for every fixed number $t \in [0,1]$, put

$$\sigma_1 = \{k \in \mathbb{N} : r_k(t) = 1\} \quad \text{and} \quad \sigma_2 = \{k \in \mathbb{N} : r_k(t) = -1\}.$$

By Lemma 1.44, we have

$$\cos\langle \mathcal{M}_{\sigma_1}, \mathcal{M}_{\sigma_2}\rangle = \|P_{\mathcal{M}_{\sigma_2}} P_{\mathcal{M}_{\sigma_1}}\| = \|P_{\mathcal{M}_{\mathbb{N}\setminus\sigma_1}} P_{\mathcal{M}_{\sigma_1}}\| < \varepsilon.$$

Moreover, if we denote by $\mathcal{P}_1$ the projection onto $\mathcal{M}_{\sigma_1}$ parallel to $\mathcal{M}_{\sigma_2}$ and $\mathcal{P}_2 = I - \mathcal{P}_1$, then we have

$$\mathcal{P}_1 x = \sum_{\substack{k\in\sigma_1\\ 1\le k\le N}} c_k x_k = \sum_{\substack{k\in\sigma_1\\ 1\le k\le N}} c_k r_k(t) x_k$$

and

$$\mathcal{P}_2 x = \sum_{\substack{k\in\sigma_2\\ 1\le k\le N}} c_k x_k = -\sum_{\substack{k\in\sigma_2\\ 1\le k\le N}} c_k r_k(t) x_k.$$

Thus,

$$\begin{aligned}\mathcal{P}_1 x - \mathcal{P}_2 x &= \sum_{\substack{k\in\sigma_1\\ 1\le k\le N}} c_k r_k(t) x_k + \sum_{\substack{k\in\sigma_2\\ 1\le k\le N}} c_k r_k(t) x_k\\ &= \sum_{k=1}^N c_k r_k(t) x_k = \mathcal{L}(t),\end{aligned}$$

and, since $\mathcal{P}_1 x + \mathcal{P}_2 x = x$, we get

$$K - \mathcal{L}(t) = \|\mathcal{P}_1 x + \mathcal{P}_2 x\|^2 - \|\mathcal{P}_1 x - \mathcal{P}_2 x\|^2 = 4\Re(\langle \mathcal{P}_1 x, \mathcal{P}_2 x\rangle).$$

Hence,

$$\begin{aligned}|K - \mathcal{L}(t)| &\le 4|\langle \mathcal{P}_1 x, \mathcal{P}_2 x\rangle| \le 4\|\mathcal{P}_1 x\|\,\|\mathcal{P}_2 x\| \cos\langle \mathcal{M}_{\sigma_1}, \mathcal{M}_{\sigma_2}\rangle\\ &\le 4\varepsilon \|\mathcal{P}_1 x\|\,\|\mathcal{P}_2 x\|.\end{aligned}$$

But,

$$\begin{aligned}\|\mathcal{P}_1 x\| &= \Big\| \sum_{\substack{k\in\sigma_1\\ 1\le k\le N}} c_k x_k \Big\| \\ &= \Big\| \sum_{\substack{k\in\sigma_1\\ 1\le k\le N}} c_k V_{\mathfrak{X}} \mathfrak{e}_k \Big\| \\ &\le \|V_{\mathfrak{X}}\| \Big\| \sum_{\substack{k\in\sigma_1\\ 1\le k\le N}} c_k \mathfrak{e}_k \Big\| \\ &\le \|V_{\mathfrak{X}}\| \left(\sum_{k=1}^{N} |c_k|^2 \right)^{1/2}.\end{aligned}$$

We have the same estimate for $\|\mathcal{P}_2 x\|$, which gives

$$|K - \mathcal{L}(t)| \le 4\varepsilon \|V_{\mathfrak{X}}\|^2 \sum_{k=1}^{N} |c_k|^2 = 4\varepsilon \|V_{\mathfrak{X}}\|^2 \|\mathfrak{e}\|^2.$$

Thus,

$$(1 - 4\varepsilon\|V_{\mathfrak{X}}\|^2)\, \|\mathfrak{e}\|^2 \le \|V_{\mathfrak{X}}\mathfrak{e}\|^2 \le (1 + 4\varepsilon\|V_{\mathfrak{X}}\|^2)\, \|\mathfrak{e}\|^2.$$

To finish, just note that

$$\langle \Gamma_{\mathfrak{X}} \mathfrak{e}, \mathfrak{e} \rangle_{\ell^2} = \|V_{\mathfrak{X}}\mathfrak{e}\|^2 \qquad (\mathfrak{e} \in \ell^2).$$

Hence, the last two inequalities can be rewritten as

$$|\langle (\Gamma_{\mathfrak{X}} - I)\mathfrak{e}, \mathfrak{e} \rangle_{\ell^2}| \le 4\varepsilon \|V_{\mathfrak{X}}\|^2 \|\mathfrak{e}\|^2 \qquad (\mathfrak{e} \in \ell^2),$$

and thus the result follows. □

Lemma 15.16 *Let Θ_1 and Θ_2 be inner functions. Then*

$$\|P_{\Theta_1} P_{\Theta_2}\| = \|H_{\bar{\Theta}_1} H^*_{\bar{\Theta}_2}\| = \|H^*_{\bar{\Theta}_2} H_{\bar{\Theta}_1}\|.$$

Proof Let

$$\begin{array}{rccc} V: & H^2 & \longrightarrow & H^2_- \\ & f & \longmapsto & P_- \bar{\Theta}_2 f \end{array}$$

and

$$\begin{array}{rccc} U: & L^2 & \longrightarrow & L^2 \\ & f & \longmapsto & \Theta_1 f. \end{array}$$

Clearly, U is an isometry and V is a partial isometry with initial space K_{Θ_2}. Then, for each $f \in H^2$,

$$\begin{aligned} P_{\Theta_1} P_{\Theta_2} f &= \Theta_1 P_- \bar{\Theta}_1 \Theta_2 P_- \bar{\Theta}_2 f \\ &= \Theta_1 P_- \bar{\Theta}_1 P_+ \Theta_2 P_- \bar{\Theta}_2 f \\ &= \Theta_1 H_{\bar{\Theta}_1} H^*_{\bar{\Theta}_2} P_- \bar{\Theta}_2 f \\ &= U H_{\bar{\Theta}_1} H^*_{\bar{\Theta}_2} V f. \end{aligned}$$

Therefore, we immediately deduce that

$$\|P_{\Theta_1} P_{\Theta_2}\| = \|H_{\bar{\Theta}_1} H^*_{\bar{\Theta}_2}\|.$$

To obtain the second identity, consider the operator J defined in Section 14.6. Then, by (14.12), (14.14) and (14.15), we have

$$\begin{aligned} J H_{\bar{\Theta}_1} H^*_{\bar{\Theta}_2} J f &= J P_- \bar{\Theta}_1 P_+ \Theta_2 J f \\ &= P_+ J \bar{\Theta}_1 P_+ \Theta_2 J f \\ &= P_+ \Theta_1 J P_+ \Theta_2 J f \\ &= P_+ \Theta_1 P_- J \Theta_2 J f \\ &= P_+ \Theta_1 P_- \bar{\Theta}_2 f \\ &= H^*_{\bar{\Theta}_1} H_{\bar{\Theta}_2} f \\ &= (H^*_{\bar{\Theta}_2} H_{\bar{\Theta}_1})^* f \qquad (f \in H^2). \end{aligned}$$

This completes the proof. □

Lemma 15.17 *Let B_1 and B_2 be two Blaschke products satisfying*

$$\inf_{z \in \mathbb{D}} \max\{|B_1(z)|, |B_2(z)|\} \geq \delta > 0.$$

Then there are two constants C_1 and C_2 such that

$$\|H^*_{\bar{B}_1} H^*_{B_2}\| \geq C_1 (1 - \delta)^{C_2}.$$

Theorem 15.18 *The sequence $(\tilde{k}_{\lambda_n})_{n \geq 1}$ is a complete AOB in K_B if and only if*

$$\lim_{n \to \infty} |B_n(\lambda_n)| = 1. \tag{15.10}$$

Proof Suppose that $(\tilde{k}_{\lambda_n})_{n \geq 1}$ is a complete AOB in K_B. Denote its biorthogonal sequence by $(\tilde{k}^*_{\lambda_n})_{n \geq 1}$. Then, by Corollary 10.33, we have

$$\lim_{n \to \infty} \|\tilde{k}^*_{\lambda_n}\| = 1.$$

But, $\tilde{k}^*_{\lambda_n} = \|k_{\lambda_n}\| k^*_{\lambda_n}$, and thus by (15.2), we deduce that (15.10) holds.

For the inverse, we use several results that were developed just before the theorem. If (15.10) holds, then at least

$$\inf_{n\geq 1} |B_n(\lambda_n)| > 0,$$

and thus, by Theorem 15.7, $(\tilde{k}_{\lambda_n})_{n\geq 1}$ is a Riesz basis for K_B. Hence, by Theorem 10.21, the Gram matrix Γ of $(\tilde{k}_{\lambda_n})_{n\geq 1}$ is bounded and invertible on ℓ^2. According to Theorem 10.32, to show that $(\tilde{k}_{\lambda_n})_{n\geq 1}$ is an AOB, it is enough to show that $K = \Gamma - I$ is compact.

Let Q_k denote the orthogonal projection of ℓ^2 onto $\mathrm{Span}\{\mathfrak{e}_j : j > k\}$. Write

$$\begin{aligned}\Gamma - I &= (\Gamma - I)Q_k + (\Gamma - I)(I - Q_k)\\ &= Q_k(\Gamma - I)Q_k + (I - Q_k)(\Gamma - I)Q_k + (\Gamma - I)(I - Q_k).\end{aligned}$$

Since $(I - Q_k)(\Gamma - I)Q_k + (\Gamma - I)(I - Q_k)$ is a finite-rank operator, it is enough to show that $Q_k(\Gamma - I)Q_k \longrightarrow 0$.

Given $E \subset \mathbb{N}$, we write B_E for the Blaschke product formed with all zeros λ_j, where $j \in E$. Fix $k \geq 1$. Put

$$E_k = \{n \in \mathbb{N} : n \geq k\}$$

and

$$\delta_k = \inf_{n\geq k} |B_{E_k\setminus\{n\}}(\lambda_n)|.$$

Note that our main hypothesis implies that $\delta_k \longrightarrow 1$ as $k \longrightarrow \infty$. Fix any subset $F \subset E_k$. According to Lemma 15.14,

$$\max\{|B_{E_k\setminus F}(z)|, |B_F(z)|\} \geq \frac{\delta_k^2}{(1+\sqrt{1-\delta_k^2})^2}.$$

Thus, by Lemma 15.17,

$$\|H^*_{\overline{B_{E_k\setminus F}}} H_{\overline{B_F}}\| \leq c_1\left(1 - \frac{\delta_k^2}{(1+\sqrt{1-\delta_k^2})^2}\right)^{c_2}.$$

Then, by Lemma 15.16, we deduce that

$$\|P_{B_{E_k\setminus F}} P_{B_F}\| \leq c_1\left(1 - \frac{\delta_k^2}{(1+\sqrt{1-\delta_k^2})^2}\right)^{c_2}.$$

Finally, by Lemma 15.15, we get

$$\|Q_k(\Gamma - I)Q_k\| \leq c_3\left(1 - \frac{\delta_k^2}{(1+\sqrt{1-\delta_k^2})^2}\right)^{c_2}.$$

Therefore, as $k \longrightarrow \infty$, we see that $\|Q_k(\Gamma - I)Q_k\| \longrightarrow 0$. □

Henceforth, we will call $(\lambda_n)_{n\geq 1}$ a *thin sequence*, and write $(\lambda_n) \in (T)$, if it satisfies the equivalent conditions of Theorem 15.18. In other words, a thin sequence is the asymptotically orthonormal sequence of normalized reproducing kernels. A different characterization of these objects can be stated by appealing to the Gram matrix.

Theorem 15.19 *Let $\Lambda = (\lambda_n)_{n\geq 1}$ be a Blaschke sequence of distinct points in $\mathbb{D}$, and let B be the associated Blaschke product. Let Γ denote the Gram matrix of $(\tilde{k}_{\lambda_n})_{n\geq 1}$. Then the following are equivalent:*

(i) $(\tilde{k}_{\lambda_n})_{n\geq 1}$ *is thin in* K_B;
(ii) $(\Gamma - I)e_n \longrightarrow 0$.

Proof (i) $\Longrightarrow$ (ii) By Lemma 10.32(iii), it follows that $\Gamma = I + K$, where K is compact. Since $Ke_n \longrightarrow 0$, we deduce that $(\Gamma - I)e_n \longrightarrow 0$.

(ii) $\Longrightarrow$ (i) By hypothesis $(\Gamma - I)e_n \longrightarrow 0$. But, by (4.41),

$$\begin{aligned}\|(\Gamma - I)e_n\|^2 &= \sum_{p\neq n,\, p\geq 1} |\Gamma_{n,p}|^2 \\ &= \sum_{p\neq n,\, p\geq 1} \frac{(1-|\lambda_n|^2)(1-|\lambda_p|^2)}{|1-\bar{\lambda}_n\lambda_p|^2} \\ &= \sum_{p\neq n,\, p\geq 1} (1-|b_{\lambda_p}(\lambda_n)|^2).\end{aligned}$$

In particular, there is an integer N such that, for all $n \geq N$, we have $\|(\Gamma - I)e_n\|^2 < 1/2$, and therefore $|b_{\lambda_p}(\lambda_n)|^2 > 1/2$ if p or n is larger than N. Since the points λ_n are distinct, the whole sequence Λ is separated, and there exists $\varepsilon > 0$ such that $|b_{\lambda_p}(\lambda_n)| \geq \varepsilon$ for all $n \neq p$. Therefore,

$$1 - |b_{\lambda_p}(\lambda_n)|^2 \geq -c \log |b_{\lambda_p}(\lambda_n)| \geq 0$$

for some suitable constant $c > 0$. It follows that

$$\|(\Gamma - I)e_n\|^2 \geq -c \log |B_n(\lambda_n)|,$$

whence $|B_n(\lambda_n)| \longrightarrow 1$. By Theorem 15.19, it follows that $(\tilde{k}_{\lambda_n})_{n\geq 1}$ is a thin (complete AOB) in K_B. □

We now give a practical method to construct thin sequences. This result reveals that the family of thin sequences is somehow much smaller than the family of geometric sequences, i.e. a sequence $(\lambda_n)_{n\geq 1}$ in $\mathbb{D}$ that satisfies the growth restriction

$$1 - |\lambda_{n+1}| \leq c(1 - |\lambda_n|) \qquad (n \geq 1),$$

for some constant $0 \leq c < 1$.

Theorem 15.20 *Suppose that $\Lambda = (\lambda_n)_{n\geq 1}$ is a sequence in $\mathbb{D}$ such that*

$$\lim_{n\to\infty} |\lambda_n| = 1$$

and $|\lambda_n| \leq |\lambda_{n+1}|$ for all $n \geq 1$. Put

$$\gamma = \lim_{k\to\infty} \frac{1-|\lambda_{k+1}|}{1-|\lambda_k|}.$$

If $\gamma = 0$, then $\Lambda \in (T)$. If, moreover, $\Lambda \subset [0,1)$, then $\Lambda \in (T)$ if and only if $\gamma = 0$.

Proof Fix $\varepsilon \in (0,1)$. Since the sequence $|\lambda_n|$ is increasing, there exists $N = N(\varepsilon)$ such that

$$\frac{1-|\lambda_n|}{1-|\lambda_k|} < \varepsilon \qquad (n > k \geq N).$$

In fact, using induction, we get

$$\frac{1-|\lambda_n|}{1-|\lambda_k|} < \varepsilon^{n-k} \qquad (n > k \geq N).$$

Hence, it follows that

$$\begin{aligned} |b_{\lambda_k}(\lambda_n)| &\geq |b_{|\lambda_k|}(|\lambda_n|)| \\ &= \frac{|\lambda_n| - |\lambda_k|}{1 - |\lambda_k\lambda_n|} \\ &\geq \frac{1-\varepsilon^{n-k}}{1+\varepsilon^{n-k}} \qquad (n, k \geq N,\ n \neq k). \end{aligned}$$

For $n > N$, write

$$|B_n(\lambda_n)| = \prod_{k\leq N} |b_{\lambda_k}(\lambda_n)| \times \prod_{k=N+1}^{n-1} |b_{\lambda_k}(\lambda_n)| \times \prod_{k=n+1}^{\infty} |b_{\lambda_k}(\lambda_n)|.$$

It then follows that

$$|B_n(\lambda_n)| \geq \prod_{k\leq N} |b_{\lambda_k}(\lambda_n)| \left(\prod_{p\geq 1} \frac{1-\varepsilon^p}{1+\varepsilon^p} \right)^2$$

and therefore

$$\begin{aligned} \log|B_n(\lambda_n)| &\geq \sum_{1\leq k\leq N} \log|b_{\lambda_k}(\lambda_n)| + 2\sum_{p\geq 1} \log\left(1 - \frac{2\varepsilon^p}{1+\varepsilon^p}\right) \\ &\geq \sum_{1\leq k\leq N} \log|b_{\lambda_k}(\lambda_n)| + 2\sum_{p\geq 1} \log(1-2\varepsilon^p). \end{aligned}$$

But, if $\varepsilon \in (0, \frac{1}{4})$, then $\log(1 - 2\varepsilon^p) \geq -4\varepsilon^p$ and

$$2\sum_{p\geq 1} \log(1 - 2\varepsilon^p) \geq -\frac{8\varepsilon}{1-\varepsilon}.$$

Since $\lim_{n\to\infty} |\lambda_n| = 1$, it follows that $\lim_{n\to\infty} |b_{\lambda_k}(\lambda_n)| = 1$, for any $1 \leq k \leq N$. Thus, if N is sufficiently large, there exists $C > 0$ such that, for $n \geq N$, $\log |B_n(\lambda_n)| \geq -C\varepsilon$. Consequently, $\lim_{n\to\infty} |B_n(\lambda_n)| = 1$.

If $\lambda \in (0, 1)$, then

$$1 - \frac{1-\lambda_{k+1}}{1-\lambda_k} = \frac{\lambda_{k+1} - \lambda_k}{1-\lambda_k} \geq \frac{\lambda_{k+1} - \lambda_k}{1-\lambda_k\lambda_{k+1}} = b_{\lambda_k}(\lambda_{k+1}) \geq |B_k(\lambda_k)|.$$

Therefore, if $\lim_{k\to\infty} |B_k(\lambda_k)| = 1$, then

$$\lim_{k\to\infty} \frac{1-\lambda_{k+1}}{1-\lambda_k} = 0. \qquad \square$$

In the next result, we show that, by changing the arguments, we can transform any Blaschke sequence to a thin sequence.

Theorem 15.21 *Suppose that $(r_n)_{n\geq 1}$ is a sequence of distinct positive numbers, $0 < r_n < 1$, such that $\sum_{n\geq 1}(1 - r_n) < \infty$. Then there exist $\theta_n \geq 0$ such that $(r_n e^{i\theta_n})_{n\geq 1} \in (T)$.*

Proof For simplicity, we will denote $a_n = 1 - r_n$. If needed, we change the indexing so that a_n is a decreasing sequence. Put $A = \sum_{n\geq 1} a_n < \infty$. Define

$$n_k = \min\left\{ n : \sum_{p=1}^{n-1} a_p \geq A\left(1 - \frac{1}{2^k}\right)\right\}.$$

Then $n_k \uparrow \infty$, as $k \to \infty$. We put $b_1 = 1$ and $b_{n+1} = k a_n$ if $n_k \leq n < n_{k+1}$. Thus,

$$\begin{aligned}
\sum_{n\geq n_1} b_{n+1} &= \sum_{k\geq 1} \sum_{n_k \leq n < n_{k+1}} b_{n+1} \\
&= \sum_{k\geq 1} k\Bigg(\sum_{n_k \leq n < n_{k+1}} a_n \Bigg) \\
&\leq \sum_{k\geq 1} k\Bigg(\sum_{n_k \leq n} a_n \Bigg) \\
&\leq \sum_{k\geq 1} k\frac{A}{2^k},
\end{aligned}$$

where the last inequality is a consequence of the definition of n_k. Therefore, $\sum_{n\geq 1} b_n < \infty$, and $a_n/b_{n+1} \longrightarrow 0$. Since a_n is decreasing, we also have $a_{n+1}/b_{n+1} \longrightarrow 0$

Since the property of being a thin sequence is not changed by adding or deleting a finite number of distinct points, we may suppose that $\sum_{n\geq 1} b_n < \pi/2$. We now define $\theta_n = \sum_{k=1}^{n} b_k$, and $\lambda_n = r_n e^{i\theta_n}$.

In order to show that $(\lambda_n)_{n\geq 1} \in (T)$, we have to show, by Theorem 15.19, that

$$S_n = \sum_{k\neq n} \frac{(1-|\lambda_k|^2)(1-|\lambda_n|^2)}{|1-\bar{\lambda}_k\lambda_n|^2} \longrightarrow 0$$

as $n \longrightarrow \infty$. Since

$$(1-|\lambda_k|^2)(1-|\lambda_n|^2) \leq 4a_k a_n,$$

$$(1-|\lambda_k\lambda_n|)^2 = (a_n + a_k - a_n a_k)^2 \geq a_n^2$$

and

$$4|\lambda_k\lambda_n| \sin^2 \frac{\theta_k - \theta_n}{2} \geq c(\theta_k - \theta_n)^2,$$

where $c > 0$ is a constant, formula (4.42) implies that, for some $C > 0$,

$$\frac{(1-|\lambda_k|^2)(1-|\lambda_n|^2)}{|1-\bar{\lambda}_k\lambda_n|^2} \leq C \frac{a_n a_k}{a_n^2 + (\theta_n - \theta_k)^2}.$$

Therefore, we have

$$\begin{aligned} S_n &\leq C \sum_{k\neq n} \frac{a_n a_k}{a_n^2 + (\theta_n - \theta_k)^2} \\ &= C\bigg(\sum_{k<n} \frac{a_n a_k}{a_n^2 + (\theta_n - \theta_k)^2} + \sum_{k>n} \frac{a_n a_k}{a_n^2 + (\theta_n - \theta_k)^2} \bigg) \\ &= C\bigg(\sum_{k<n} \frac{a_n b_{k+1}}{a_n^2 + (\theta_n - \theta_k)^2} + \sum_{k>n} \frac{a_n b_k}{a_n^2 + (\theta_n - \theta_k)^2} \bigg). \end{aligned}$$

Define

$$f(x) = \frac{a_n}{a_n^2 + (x - \theta_n)^2}.$$

Then we can write the last inequality as

$$S_n \leq C\bigg(\sum_{k<n} (\theta_{k+1} - \theta_k) f(\theta_k) + \sum_{k>n} (\theta_k - \theta_{k-1}) f(\theta_k) \bigg).$$

A glimpse at the graph of f shows that this last quantity is majorized by

$$\int_{-\infty}^{\theta_{n-1}} f(x)\,dx + \int_{\theta_{n+1}}^{\infty} f(x)\,dx + b_n f(\theta_{n-1}) + b_{n+1} f(\theta_{n+1}).$$

Since

$$b_n f(\theta_{n-1}) = \frac{b_n a_n}{a_n^2 + b_n^2},$$

$$b_{n+1} f(\theta_{n+1}) = \frac{b_{n+1} a_n}{a_n^2 + b_{n+1}^2},$$

$$\int_{-\infty}^{\theta_{n-1}} f(x)\, dx = \frac{\pi}{2} - \arctan\left(\frac{b_n}{a_n}\right)$$

and finally

$$\int_{\theta_{n+1}}^{\infty} f(x)\, dx = \frac{\pi}{2} - \arctan\left(\frac{b_{n+1}}{a_n}\right),$$

all these quantities tend to 0 as $n \longrightarrow \infty$. Thus, $S_n \longrightarrow 0$. □

Lemma 15.22 *Let* $\Lambda = (\lambda_n)_{n\geq 1}$ *and* $\Lambda' = (\lambda'_n)_{n\geq 1}$ *be two sequences in* $\mathbb{D}$. *If* $\sup_{n\geq 1} |b_{\lambda_n}(\lambda'_n)| < 1$, *then* $\Lambda \in (T)$ *if and only if* $\Lambda' \in (T)$.

Proof Put $\lambda = \sup_{n\geq 1} |b_{\lambda_n}(\lambda'_n)|$ and $\delta_n = |B_n(\lambda_n)|$. By Theorem 15.18, we know that $\delta_n \longrightarrow 1$, as $n \longrightarrow \infty$. Hence, there exists $N \in \mathbb{N}$ such that

$$\alpha = \frac{2\lambda}{1+\lambda^2} < \inf_{n\geq N} \delta_n.$$

By Theorem 15.6, we have, for all $k \geq N$,

$$\prod_{j\neq k} |b_{\lambda'_j}(\lambda'_k)| \geq \frac{\prod_{j\neq k} |b_{\lambda_j}(\lambda_k)| - \alpha}{1 - \alpha \prod_{j\neq k} |b_{\lambda_j}(\lambda_k)|} = \frac{\delta_k - \alpha}{1 - \alpha\delta_k}.$$

Hence, we get

$$\lim_{k\to\infty} \prod_{j\neq k} |b_{\lambda'_j}(\lambda'_k)| \geq \frac{1-\alpha}{1-\alpha} = 1.$$

Using Theorem 15.18, we conclude that $\Lambda' \in (T)$. □

Notes on Chapter 15

The problem of the convergence of series

$$\sum_n a_n k_{\lambda_n}$$

of reproducing kernels of a reproducing kernel Hilbert space $\mathcal{H}$ on a set Ω is a dual form of describing the restriction spaces $H_{|\Lambda}$ onto discrete subsets $\Lambda = (\lambda_n)_{n\geq 1}$ of Ω. For a few general facts about this duality, we refer to Nikolskii [379, 384, 386]. Note that using the Fourier transform (and Paley–Wiener theorem) and the conformal mapping, which sends the unit disk onto the upper half-plane, one can see that the study of the geometry of sequences of reproducing kernels $(k_{\lambda_n})_{n\geq 1}$ in H^2 is equivalent to the study of exponential systems $(e^{i\mu_n t})_{n\geq 1}$ in $L^2(\mathbb{R}_+)$, $\Im(\mu_n) > 0$. See [293, 386, 388] for more information concerning this link. In particular, all the results given in this chapter for the sequence $(k_{\lambda_n})_{n\geq 1}$ can be translated into results concerning exponential systems in $L^2(\mathbb{R}_+)$. This can be viewed as a motivation to study the sequences of reproducing kernels of H^2, since exponential systems appear naturally in many applications such as control theory.

Section 15.1

The Carleson condition (15.1) plays a prominent role in the theory of interpolation and basis. It was introduced independently by Newman [376] and Carleson [125].

Section 15.2

Theorem 15.2 is due to Carleson [125]. The results proved in Exercises 15.2.2 and Exercise 15.2.4 are due to Kabaĭla [305] and Newman [376].

Section 15.3

The equivalence (iii) $\Longleftrightarrow$ (iv) of Theorem 15.7 was established by Shapiro and Shields [477, 478]. The equivalence (i) $\Longleftrightarrow$ (iv), together with an explicit statement of the connection between the problems of bases of rational functions and interpolation, is due to Nikolskii and Pavlov [389, 392].

Section 15.4

Theorem 15.10 is originally due to F. and R. Nevanlinna [375] and Pick [412]. The approach given in this text, which is based on the commutant lifting theorem, is contained in the paper of Sarason [447]. The result proved in Exercise 15.4.1 is known as the Carathéodory–Fejér theorem.

Section 15.5

The problem of describing the interpolating sequences for H^∞ was proposed by R. C. Buck. After partial solutions by Hayman [273], Newman [376] proved that a sequence $\Lambda = (\lambda_n)_{n\geq 1}$ is an H^∞-interpolating sequence if and only if Λ satisfies the conditions (C) and (CN). Then Carleson [125] got the final and beautiful solution through Theorem 15.11. Numerous comments and problems connected with it can be found in Duren [188] and in Vinogradov and

Havin [529, 530]. The proof of Theorem 15.11 given in this text is proposed in Nikolskii [380]. The same approach can be adapted in other situations, as was discussed in [386].

Section 15.6

Theorem 15.13 is due to Shapiro and Shields [477]. In fact, they proved the theorem in H^p, $1 \leq p < \infty$. There exists an extension to $0 < p < 1$ due to Kabaĭla [306, 307].

Section 15.7

Lemma 15.17 is taken from [533, lemma 5]. Its proof is rather long and takes us way beyond the scope of this book. Hence, it was not included in the text. Theorem 15.18 is due to Volberg [533]. Theorem 15.19 is due to Chalendar, Fricain and Timotin [133]. Theorem 15.20 comes also from [133]. It is a generalization of a result of Kabaĭla [305] and Newman [376]; see Exercise 15.2.2. See also the book of Nikolskii [386] and the paper of Vinogradov and Havin [529], where various geometric conditions imposed on the sequence $(\lambda_n)_{n\geq 1}$ are analyzed through the point of view of interpolation. Theorem 15.21 is also taken from [133]. It is a generalization of a result of Naftalevič concerning Carleson sequences [367].

References

[1] Abramovich, Y. A., and Aliprantis, C. D. *An invitation to operator theory*, vol. 50 of *Graduate Studies in Mathematics*. American Mathematical Society, Providence, RI, 2002.

[2] Adamjan, V. M., Arov, D. Z., and Kreĭn, M. G. Infinite Hankel matrices and generalized Carathéodory–Fejér and I. Schur problems. *Funkcional. Anal. Priložen. 2*, 4 (1968), 1–17.

[3] Adamjan, V. M., Arov, D. Z., and Kreĭn, M. G. Infinite Hankel matrices and generalized problems of Carathéodory–Fejér and F. Riesz Problems. *Funkcional. Anal. Priložen. 2*, 1 (1968), 1–19.

[4] Adamjan, V. M., Arov, D. Z., and Kreĭn, M. G. Analytic properties of the Schmidt pairs of a Hankel operator and the generalized Schur–Takagi problem. *Mat. Sbornik (N.S.) 86*, 128 (1971), 34–75.

[5] Adamjan, V. M., Arov, D. Z., and Kreĭn, M. G. Infinite Hankel block matrices and related problems of extension. *Izv. Akad. Nauk Armjan. SSR Ser. Mat. 6*, 2–3 (1971), 87–112.

[6] Agler, J., and McCarthy, J. E. *Pick interpolation and Hilbert function spaces*, vol. 44 of *Graduate Studies in Mathematics*. American Mathematical Society, Providence, RI, 2002.

[7] Ahern, P. R., and Clark, D. N. On functions orthogonal to invariant subspaces. *Acta Math. 124* (1970), 191–204.

[8] Ahern, P. R., and Clark, D. N. On star-invariant subspaces. *Bull. Amer. Math. Soc. 76* (1970), 629–632.

[9] Ahern, P. R., and Clark, D. N. Radial limits and invariant subspaces. *Amer. J. Math. 92* (1970), 332–342.

[10] Akhiezer, N. I. *The classical moment problem and some related questions in analysis*. Translated by N. Kemmer. Hafner, New York, 1965.

[11] Akhiezer, N. I., and Glazman, I. M. *Theory of linear operators in Hilbert space*. Translated from Russian and with preface by M. Nestell; reprint of 1961 and 1963 translations. Dover, New York, 1993.

[12] Albiac, F., and Kalton, N. J. *Topics in Banach space theory*, vol. 233 of *Graduate Texts in Mathematics*. Springer, New York, 2006.

[13] ALEKSANDROV, A. B. Invariant subspaces of the backward shift operator in the space H^p $(p \in (0, 1))$. *Zap. Nauchn. Sem. Leningrad. Otdel. Mat. Inst. Steklov. (LOMI) 92, Issled. Linein. Oper. Teorii Funktsii. 9* (1979), 7–29, 318.

[14] ALEKSANDROV, A. B. Multiplicity of boundary values of inner functions. *Izv. Akad. Nauk Armyan. SSR Ser. Mat. 22*, 5 (1987), 490–503, 515.

[15] ALEKSANDROV, A. B. Inner functions and related spaces of pseudocontinuable functions. *Zap. Nauchn. Sem. Leningrad. Otdel. Mat. Inst. Steklov. (LOMI) 170, Issled. Linein. Oper. Teorii Funktsii. 17* (1989), 7–33, 321.

[16] ALEKSANDROV, A. B. On the existence of angular boundary values of pseudocontinuable functions. *Zap. Nauchn. Sem. St.-Peterburg. Otdel. Mat. Inst. Steklov. (POMI) 222, Issled. Linein. Oper. Teorii Funktsii. 23* (1995), 5–17, 307.

[17] ALEKSANDROV, A. B. Isometric embeddings of co-invariant subspaces of the shift operator. *Zap. Nauchn. Sem. St.-Peterburg. Otdel. Mat. Inst. Steklov. (POMI) 232, Issled. Linein. Oper. Teorii Funktsii. 24* (1996), 5–15, 213.

[18] ALEMAN, A., RICHTER, S., AND ROSS, W. T. Pseudocontinuations and the backward shift. *Indiana Univ. Math. J. 47*, 1 (1998), 223–276.

[19] ALPAY, D. *Algorithme de Schur, espaces à noyau reproduisant et théorie des systèmes*, vol. 6 of *Panoramas et Synthèses*. Société Mathématique de France, Paris, 1998.

[20] ALPAY, D. (Ed.) *Reproducing kernel spaces and applications*, vol. 143 of *Operator Theory: Advances and Applications*. Birkhäuser, Basel, 2003.

[21] ALPAY, D., ATTIA, H., AND VOLOK, D. Realization theorems for stationary multiscale systems. *Linear Algebra Appl. 412* (2–3) (2006), 326–347.

[22] ALPAY, D., BALL, J. A., AND PERETZ, Y. System theory, operator models and scattering: the time-varying case. *J. Operator Theory 47*, 2 (2002), 245–286.

[23] ALPAY, D., BOLOTNIKOV, V., DIJKSMA, A., ROVNYAK, J., AND SADOSKY, C. Espaces de Hilbert inclus contractivement dans l'espace de Hardy du bi-disque. *C. R. Acad. Sci. Paris, Sér. I Math. 326*, 12 (1998), 1365–1370.

[24] ALPAY, D., BOLOTNIKOV, V., DIJKSMA, A., AND SADOSKY, C. Hilbert spaces contractively included in the Hardy space of the bidisk. *Positivity 5*, 1 (2001), 25–50.

[25] ALPAY, D., CONSTANTINESCU, T., DIJKSMA, A., AND ROVNYAK, J. Notes on interpolation in the generalized Schur class. II. Nudelman's problem. *Trans. Amer. Math. Soc. 355*, 2 (2003), 813–836 (electronic).

[26] ALPAY, D., DIJKSMA, A., AND LANGER, H. The Schur transformation for Nevanlinna functions: operator representations, resolvent matrices, and orthogonal polynomials. In vol. 1 of *Operator Theory: Advances and Applications*. Birkhäuser, Basel, 2008, pp. 1–37.

[27] ALPAY, D., DIJKSMA, A., AND PERETZ, Y. Nonstationary analogs of the Herglotz representation theorem: the discrete case. *J. Funct. Anal. 166*, 1 (1999), 85–129.

[28] ALPAY, D., DIJKSMA, A., ROVNYAK, J., AND DE SNOO, H. *Schur functions, operator colligations, and reproducing kernel Pontryagin spaces*, vol. 96 of *Operator Theory: Advances and Applications*. Birkhäuser, Basel, 1997.

[29] ALPAY, D., DIJKSMA, A., ROVNYAK, J., AND DE SNOO, H. S. V. Reproducing kernel Pontryagin spaces. In *Holomorphic spaces (Berkeley, CA, 1995)*, vol. 33 of *Mathematical Sciences Research Institute Publications*. Cambridge University Press, Cambridge, UK, 1998, pp. 425–444.

[30] ALPAY, D., AND DUBI, C. Boundary interpolation in the ball. *Linear Algebra Appl. 340* (2002), 33–54.

[31] ALPAY, D., AND DUBI, C. Backward shift operator and finite dimensional de Branges–Rovnyak spaces in the ball. *Linear Algebra Appl. 371* (2003), 277–285.

[32] ALPAY, D., AND DUBI, C. Carathéodory–Fejér interpolation in the ball. *Linear Algebra Appl. 359* (2003), 1–19.

[33] ALPAY, D., AND DYM, H. Hilbert spaces of analytic functions, inverse scattering and operator models. I. *Integral Equations Operator Theory 7*, 5 (1984), 589–641.

[34] ALPAY, D., AND DYM, H. Hilbert spaces of analytic functions, inverse scattering and operator models. II. *Integral Equations Operator Theory 8*, 2 (1985), 145–180.

[35] ALPAY, D., AND DYM, H. On a new class of structured reproducing kernel spaces. *J. Funct. Anal. 111*, 1 (1993), 1–28.

[36] ALPAY, D., AND KAPTANOĞLU, H. T. Some finite-dimensional backward-shift-invariant subspaces in the ball and a related interpolation problem. *Integral Equations Operator Theory 42*, 1 (2002), 1–21.

[37] ALPAY, D., AND PERETZ, Y. Realizations for Schur upper triangular operators centered at an arbitrary point. *Integral Equations Operator Theory 37*, 3 (2000), 251–323.

[38] ALPAY, D., REICH, S., AND SHOIKHET, D. Rigidity, boundary interpolation and reproducing kernels. arXiv:0709.2504v1, 2007.

[39] ALPAY, D., SHAPIRO, M., AND VOLOK, D. Espaces de Branges–Rovnyak et fonctions de Schur: le cas hyper-analytique. *C. R. Math. Acad. Sci. Paris 338*, 6 (2004), 437–442.

[40] ALPAY, D., TIMOSHENKO, O., AND VOLOK, D. Carathéodory functions in the Banach space setting. *Linear Algebra Appl. 425*, 2–3 (2007), 700–713.

[41] AMAR, E., AND LEDERER, A. Points exposés de la boule unité de $H^\infty(D)$. *C.R. Acad. Sci. Paris, Sér. A, 272* (1971), 1449–1452.

[42] AMAR, E., AND MATHERON, E. *Analyse complexe*. In *Enseignement des mathématiques*. Cassini, 2003.

[43] ANDERSON, J. M., DRITSCHEL, M. A., AND ROVNYAK, J. Schwarz–Pick inequalities for the Schur–Agler class on the polydisk and unit ball. *Comput. Methods Funct. Theory 8*, 1–2 (2008), 339–361.

[44] ANDERSON, J. M., AND ROVNYAK, J. On generalized Schwarz–Pick estimates. *Mathematika 53*, 1 (2006), 161–168 (2007).

[45] ARONSZAJN, N. Theory of reproducing kernels. *Trans. Amer. Math. Soc. 68* (1950), 337–404.

[46] AROV, D. Z., AND DYM, H. J-inner matrix functions, interpolation and inverse problems for canonical systems. I. Foundations. *Integral Equations Operator Theory 29*, 4 (1997), 373–454.

[47] ARSOVE, M. G. Similar bases and isomorphisms in Fréchet spaces. *Math. Ann. 135* (1958), 283–293.

[48] ARSOVE, M. G. The Paley–Wiener theorem in metric linear spaces. *Pacific J. Math. 10* (1960), 365–379.

[49] ARSOVE, M. G., AND EDWARDS, R. E. Generalized bases in topological linear spaces. *Studia Math. 19* (1960), 95–113.

[50] ATKINSON, F. V. The normal solubility of linear equations in normed spaces. *Mat. Sbornik (N.S.) 28*, 70 (1951), 3–14.

[51] Avdonin, S. A., and Ivanov, S. A. *Families of exponentials: the method of moments in controllability problems for distributed parameter systems.* Translated from Russian and revised by authors. Cambridge University Press, Cambridge, UK, 1995.

[52] Axler, S., Berg, I. D., Jewell, N., and Shields, A. Approximation by compact operators and the space $H^\infty + C$. *Ann. of Math. (2) 109*, 3 (1979), 601–612.

[53] Axler, S., Bourdon, P., and Ramey, W. *Harmonic function theory*, 2nd edn, vol. 137 of *Graduate Texts in Mathematics*. Springer, New York, 2001.

[54] Axler, S., Chang, S.-Y. A., and Sarason, D. Products of Toeplitz operators. *Integral Equations Operator Theory 1*, 3 (1978), 285–309.

[55] Bachman, G., and Narici, L. *Functional analysis*. Reprint of 1966 original. Dover, Mineola, NY, 2000.

[56] Baire, R. Sur les fonctions de variables réelles. *Ann. Mat. Ser. 3*, 3 (1899), 1–123.

[57] Ball, J. A. Models for noncontractions. *J. Math. Anal. Appl. 52*, 2 (1975), 235–254.

[58] Ball, J. A., Bolotnikov, V., and Fang, Q. Multivariable backward-shift-invariant subspaces and observability operators. arXiv:0610634v1, 2006.

[59] Ball, J. A., Bolotnikov, V., and Fang, Q. Schur-class multipliers on the Fock space: De Branges–Rovnyak reproducing kernel spaces and transfer-function realizations. arXiv:0610635v1, 2006.

[60] Ball, J. A., Bolotnikov, V., and Fang, Q. Transfer-function realization for multipliers of the Arveson space. *J. Math. Anal. Appl. 333*, 1 (2007), 68–92.

[61] Ball, J. A., Bolotnikov, V., and Fang, Q. Schur-class multipliers on the Arveson space: De Branges–Rovnyak reproducing kernel spaces and commutative transfer-function realizations. *J. Math. Anal. Appl. 341* (2008), 519–539.

[62] Ball, J. A., and Kriete, III, T. L. Operator-valued Nevanlinna–Pick kernels and the functional models for contraction operators. *Integral Equations Operator Theory 10*, 1 (1987), 17–61.

[63] Ball, J. A., and Lubin, A. On a class of contractive perturbations of restricted shifts. *Pacific J. Math. 63*, 2 (1976), 309–323.

[64] Ball, J. A., Sadosky, C., and Vinnikov, V. Scattering systems with several evolutions and multidimensional input/state/output systems. *Integral Equations Operator Theory 52*, 3 (2005), 323–393.

[65] Ball, J. A., Trent, T. T., and Vinnikov, V. Interpolation and commutant lifting for multipliers on reproducing kernel Hilbert spaces. In *Operator theory and analysis (Amsterdam, 1997)*, vol. 122 of *Operator Theory: Advances and Applications*. Birkhäuser, Basel, 2001, pp. 89–138.

[66] Banach, S. Sur les fonctionelles linéaires I, II. *Studia Math. 1* (1929), 211–216, 223–239.

[67] Banach, S. *Théorie des opérations linéaires*. Reprint of the 1932 original. Éditions Jacques Gabay, Sceaux, 1993.

[68] Banach, S., and Steinhaus, H. Sur le principe de la condensation de singularités. *Fund. Math. 9* (1927), 50–61.

[69] Baranov, A. Stability of bases and frames of reproducing kernels in model spaces. *Ann. Inst. Fourier (Grenoble) 55*, 7 (2005), 2399–2422.

[70] BARANOV, A. D. Bernstein-type inequalities for shift-coinvariant subspaces and their applications to Carleson embeddings. *J. Funct. Anal. 223*, 1 (2005), 116–146.

[71] BARANOV, A. Completeness and Riesz bases of reproducing kernels in model subspaces. *Int. Math. Res. Not. 2006* (2006), 81530.

[72] BARANOV, A. D. Embeddings of model subspaces of the Hardy class: compactness and Schatten–von Neumann ideals. *Izv. Ross. Akad. Nauk Ser. Mat. 73*, 6 (2009), 3–28.

[73] BARANOV, A., AND BELOV, Y. Systems of reproducing kernels and their biorthogonal: completeness or incompleteness? *Int. Math. Res. Not. 22* (2011), 5076–5108.

[74] BARANOV, A., BELOV, Y., AND BORICHEV, A. A restricted shift completeness problem. *J. Funct. Anal. 263*, 7 (2012), 1887–1893.

[75] BARANOV, A., BESSONOV, R., AND KAPUSTIN, V. Symbols of truncated Toeplitz operators. *J. Funct. Anal. 261*, 12 (2011), 3437–3456.

[76] BARANOV, A., CHALENDAR, I., FRICAIN, E., MASHREGHI, J., AND TIMOTIN, D. Bounded symbols and reproducing kernel thesis for truncated Toeplitz operators. *J. Funct. Anal. 259*, 10 (2010), 2673–2701.

[77] BARANOV, A., AND DYAKONOV, K. The Feichtinger conjecture for reproducing kernels in model subspaces. *J. Geom. Anal. 21*, 2 (2011), 276–287.

[78] BARANOV, A. D., AND FEDOROVSKIĬ, K. Y. Boundary regularity of Nevanlinna domains and univalent functions in model subspaces. *Mat. Sbornik 202*, 12 (2011), 3–22.

[79] BARANOV, A., FRICAIN, E., AND MASHREGHI, J. Weighted norm inequalities for de Branges–Rovnyak spaces and their applications. *Amer. J. Math. 132*, 1 (2010), 125–155.

[80] BARANOV, A. D., AND HAVIN, V. P. Admissible majorants for model subspaces and arguments of inner functions. *Funktsional. Anal. Prilozhen. 40*, 4 (2006), 3–21, 111.

[81] BARI, N. K. Sur les bases dans l'espace de Hilbert. *Dokl. Akad. Nauk. SSSR 54* (1946), 379–382.

[82] BARI, N. K. Biorthogonal systems and bases in Hilbert space. *Moskov. Gos. Univ. Učenye Zap. Mat. 148*, 4 (1951), 69–107.

[83] BEAUZAMY, B. *Introduction to Banach spaces and their geometry*, 2nd edn, vol. 68 of *North-Holland Mathematics Studies, Notas de Matemática*. North-Holland, Amsterdam, 1985.

[84] BEAUZAMY, B. Un opérateur sans sous-espace invariant: simplification de l'exemple de P. Enflo. *Integral Equations Operator Theory 8*, 3 (1985), 314–384.

[85] BELOV, Y. S. Criteria for the admissibility of majorants for model subspaces with a rapidly increasing argument of the generating inner function. *Zap. Nauchn. Sem. St.-Peterburg. Otdel. Mat. Inst. Steklov. (POMI) 345, Issled. Linein. Oper. Teor. Funkts. 34* (2007), 55–84, 141.

[86] BELOV, Y. S. Model functions with an almost prescribed modulus. *Algebra Analiz 20*, 2 (2008), 3–18.

[87] BELOV, Y. S. Necessary conditions for the admissibility of majorants for some model subspace. *Algebra Analiz 20*, 4 (2008), 1–26.

[88] BENEKER, P. Strongly exposed points, Helson–Szegő weights and Toeplitz operators. *Integral Equations Operator Theory 31*, 3 (1998), 299–306.

[89] Beneker, P., and Wiegerinck, J. Strongly exposed points in uniform algebras. *Proc. Amer. Math. Soc. 127*, 5 (1999), 1567–1570.

[90] Bercovici, H. *Operator theory and arithmetic in H^∞*, vol. 26 of *Mathematical Surveys and Monographs*. American Mathematical Society, Providence, RI, 1988.

[91] Berenstein, C. A., and Gay, R. *Complex variables: an introduction*, vol. 125 of *Graduate Texts in Mathematics*. Springer, New York, 1991.

[92] Bergman, S. *The kernel function and conformal mapping*, 2nd (rev.) edn, *Mathematical Surveys and Monographs, No. 5*. American Mathematical Society, Providence, RI, 1970.

[93] Bergman, S. Über die Entwicklung der harmonischen Funktionen der Ebene und des Raumes nach Orthogonalfunktionen. *Math. Ann. 86*, 3–4 (1922), 238–271.

[94] Bessaga, C., and Pełczyński, A. On bases and unconditional convergence of series in Banach spaces. *Studia Math. 17* (1958), 151–164.

[95] Bessaga, C., and Pełczyński, A. Properties of bases in spaces of type B_0. *Prace Mat. 3* (1959), 123–142.

[96] Beurling, A. Sur les intégrales de Fourier absolument convergentes et leur application à une transformation fonctionnelle. *Neuvième Congrès Math. Scand. Helsinki* (1938), 345–366.

[97] Beurling, A. On two problems concerning linear transformations in Hilbert space. *Acta Math. 81* (1948), 17.

[98] Bini, D., and Pan, V. Y. Polynomial and matrix computations. Vol. 1. Fundamental algorithms. In *Progress in Theoretical Computer Science*. Birkhäuser, Boston, MA, 1994.

[99] Birkhoff, G. *Lattice theory*. American Mathematical Society, New York, 1940.

[100] Blaschke, W. Eine Erweiterung des Satzes von Vitali über Folgen analytischer Funktionen. *Math.-Phys. Kl., Sächs. Gesell. Wiss. Leipzig 67* (1915), 194–200.

[101] Bloomfield, P., Jewell, N., and Hayashi, E. Characterizations of completely nondeterministic processes. *Pacific J. Math. 107* (1983), 307–317.

[102] Boas, Jr., R. P. Expansions of analytic functions. *Trans. Amer. Math. Soc. 48* (1940), 467–487.

[103] Boas, Jr., R. P. A general moment problem. *Amer. J. Math. 63* (1941), 361–370.

[104] Bochner, S. Über orthogonale Systeme analytischer Funktionen. *Math. Z. 14*, 1 (1922), 180–207.

[105] Bočkarev, S. V. Existence of a basis in the space of functions analytic in the disc, and some properties of Franklin's system. *Mat. Sbornik (N.S.) 95*, 137 (1974), 3–18, 159.

[106] Bolotnikov, V. A boundary Nevanlinna–Pick problem for a class of analytic matrix-valued functions in the unit ball. *Linear Algebra Appl. 346* (2002), 239–260.

[107] Bolotnikov, V. Interpolation for multipliers on reproducing kernel Hilbert spaces. *Proc. Amer. Math. Soc. 131*, 5 (2003), 1373–1383 (electronic).

[108] Bolotnikov, V., and Kheifets, A. Boundary Nevanlinna–Pick interpolation problems for generalized Schur functions. In *Interpolation, Schur functions and moment problems*, vol. 165 of *Operator Theory: Advances and Applications*. Birkhäuser, Basel, 2006, pp. 67–119.

[109] BOLOTNIKOV, V., AND KHEIFETS, A. A higher order analogue of the Carathéodory–Julia theorem. *J. Funct. Anal. 237*, 1 (2006), 350–371.

[110] BOLOTNIKOV, V. AND KHEIFETS, A. The higher order Carathéodory–Julia theorem and related boundary interpolation problems. arXiv:0701106v1, 2007.

[111] BONSALL, F. F., AND GILLESPIE, T. A. Hankel operators with PC symbols and the space $H^\infty + PC$. *Proc. R. Soc. Edinburgh Sect. A 89*, 1–2 (1981), 17–24.

[112] BÖTTCHER, A., DIJKSMA, A., LANGER, H., DRITSCHEL, M. A., ROVNYAK, J., AND KAASHOEK, M. A. *Lectures on operator theory and its applications*, vol. 3 of *Fields Institute Monographs*. Edited by P. Lancaster. American Mathematical Society, Providence, RI, 1996.

[113] BÖTTCHER, A., AND SILBERMANN, B. *Analysis of Toeplitz operators*, 2nd edn. Prepared jointly with A. Karlovich. In *Springer Monographs in Mathematics*. Springer, Berlin, 2006.

[114] BOURGAIN, J. Strongly exposed points in weakly compact convex sets in Banach spaces. *Proc. Amer. Math. Soc. 58* (1976), 197–200.

[115] BRAY, H. Elementary properties of the Stieltjes integral. *Ann. of Math. (2) 20*, 3 (1919), 177–186.

[116] BRENNAN, J. E. Invariant subspaces and subnormal operators. In *Harmonic analysis in Euclidean spaces (Proc. Sympos. Pure Math. XXXV, Williams Coll., Williamstown, MA, 1978), Part 1*. American Mathematical Society, Providence, RI, 1979, pp. 303–309.

[117] BRENNAN, J. E. Point evaluations, invariant subspaces and approximation in the mean by polynomials. *J. Funct. Anal. 34*, 3 (1979), 407–420.

[118] BREZIS, H. *Functional analysis, Sobolev spaces and partial differential equations*. Universitext, Springer, New York, 2011.

[119] BROWN, A., AND HALMOS, P. R. Algebraic properties of Toeplitz operators. *J. Reine Angew. Math. 213* (1963/1964), 89–102.

[120] BULTHEEL, A., AND VAN BAREL, M. *Linear algebra, rational approximation and orthogonal polynomials*, vol. 6 of *Studies in Computational Mathematics*. North-Holland, Amsterdam, 1997.

[121] CALDERÓN, A., SPITZER, F., AND WIDOM, H. Inversion of Toeplitz matrices. *Illinois J. Math. 3* (1959), 490–498.

[122] CARATHÉODORY, C. Über den Variabilitätsbereich der Koeffizienten von Potenzreflien, die gegebene Werte nicht annehmen. *Math. Ann. 64* (1907), 95–115.

[123] CARATHÉODORY, C. Über den Variabilitätsbereich der Fourier'schen Konstanten von positiven harmonischen Funktionen. *Rend. Circ. Mat. Palermo 32* (1911), 193–217.

[124] CARATHÉODORY, C., AND FEJÉR, L. Über den Zusammenhang der extremen von harmonischen Funktionen mit ihren Koeffizienten und über den Picard–Landau'schen Satz. *Rend. Circ. Mat. Palermo 32* (1911), 218–239.

[125] CARLESON, L. An interpolation problem for bounded analytic functions. *Amer. J. Math. 80* (1958), 921–930.

[126] CARLESON, L. Interpolations by bounded analytic functions and the corona problem. *Ann. of Math. (2) 76* (1962), 547–559.

[127] CASAZZA, P. G., CHRISTENSEN, O., LINDNER, A. M., AND VERSHYNIN, R. Frames and the Feichtinger conjecture. *Proc. Amer. Math. Soc. 133*, 4 (2005), 1025–1033 (electronic).

[128] Casazza, P. G., and Tremain, J. C. The Kadison–Singer problem in mathematics and engineering. *Proc. Natl. Acad. Sci. USA 103*, 7 (2006), 2032–2039 (electronic).

[129] Cauer, W. The Poisson integral for functions with positive real part. *Bull. Amer. Math. Soc. 38*, 10 (1932), 713–717.

[130] Chalendar, I., and Esterle, J. Le problème du sous-espace invariant. In *Development of mathematics 1950–2000*. Birkhäuser, Basel, 2000, pp. 235–267.

[131] Chalendar, I., Fricain, E., and Partington, J. R. Overcompleteness of sequences of reproducing kernels in model spaces. *Integral Equations Operator Theory 56*, 1 (2006), 45–56.

[132] Chalendar, I., Fricain, E., and Timotin, D. A short note on the Feichtinger conjecture. arXiv:1106.3408, 2011.

[133] Chalendar, I., Fricain, E., and Timotin, D. Functional models and asymptotically orthonormal sequences. *Ann. Inst. Fourier (Grenoble) 53*, 5 (2003), 1527–1549.

[134] Chalendar, I., and Partington, J. R. *Modern approaches to the invariant-subspace problem*, vol. 188 of *Cambridge Tracts in Mathematics*. Cambridge University Press, Cambridge, UK, 2011.

[135] Chevrot, N., Fricain, E., and Timotin, D. On certain Riesz families in vector-valued de Branges–Rovnyak spaces. *J. Math. Anal. Appl. 355*, 1 (2009), 110–125.

[136] Choimet, D., and Queffélec, H. *Analyse mathématique – Grands théorèmes du vingtième siècle*. Éditeur Calvage et Mounet, 2009.

[137] Christensen, O. *Frames and bases: An introductory course*. In *Applied and Numerical Harmonic Analysis*. Birkhäuser, Boston, MA, 2008.

[138] Cima, J. A., Garcia, S. R., Ross, W. T., and Wogen, W. R. Truncated Toeplitz operators: spatial isomorphism, unitary equivalence, and similarity. *Indiana Univ. Math. J. 59*, 2 (2010), 595–620.

[139] Cima, J. A., and Matheson, A. L. Essential norms of composition operators and Aleksandrov measures. *Pacific J. Math. 179*, 1 (1997), 59–64.

[140] Cima, J. A., and Matheson, A. L. On Carleson embeddings of star-invariant supspaces. *Quaest. Math. 26*, 3 (2003), 279–288.

[141] Cima, J. A., Matheson, A. L., and Ross, W. T. *The Cauchy transform*, vol. 125 of *Mathematical Surveys and Monographs*. American Mathematical Society, Providence, RI, 2006.

[142] Cima, J. A., and Ross, W. T. *The backward shift on the Hardy space*, vol. 79 of *Mathematical Surveys and Monographs*. American Mathematical Society, Providence, RI, 2000.

[143] Cima, J. A., Ross, W. T., and Wogen, W. R. Truncated Toeplitz operators on finite dimensional spaces. *Oper. Matrices 2*, 3 (2008), 357–369.

[144] Clark, D. N. One dimensional perturbations of restricted shifts. *J. Analyse Math. 25* (1972), 169–191.

[145] Coburn, L. A. Weyl's theorem for nonnormal operators. *Michigan Math. J. 13* (1966), 285–288.

[146] Coburn, L. A. The C^*-algebra generated by an isometry. I. *Bull. Amer. Math. Soc. 73* (1967), 722–726.

[147] Coburn, L. A. The C^*-algebra generated by an isometry. II. *Trans. Amer. Math. Soc. 137* (1969), 211–217.

[148] Coburn, L. A., and Douglas, R. G. Translation operators on the half-line. *Proc. Natl. Acad. Sci. USA 62* (1969), 1010–1013.

[149] Cohen, P. J. A note on constructive methods in Banach algebras. *Proc. Amer. Math. Soc. 12* (1961), 159–163.

[150] Cohn, B. Carleson measures for functions orthogonal to invariant subspaces. *Pacific J. Math. 103*, 2 (1982), 347–364.

[151] Cohn, W. S. Orthogonal complements to invariant subspaces. *Indiana Univ. Math. J. 29*, 5 (1980), 683–692.

[152] Cohn, W. S. Unitary equivalence of restricted shifts. *J. Operator Theory 5*, 1 (1981), 17–28.

[153] Cohn, W. S. On the H^p classes of derivatives of functions orthogonal to invariant subspaces. *Michigan Math. J. 30*, 2 (1983), 221–229.

[154] Cohn, W. S. Carleson measures and operators on star-invariant subspaces. *J. Operator Theory 15*, 1 (1986), 181–202.

[155] Cohn, W. S. Radial limits and star invariant subspaces of bounded mean oscillation. *Amer. J. Math. 108*, 3 (1986), 719–749.

[156] Cohn, W. S. On fractional derivatives and star invariant subspaces. *Michigan Math. J. 34*, 3 (1987), 391–406.

[157] Cohn, W. S. A maximum principle for star invariant subspaces. *Houston J. Math. 14*, 1 (1988), 23–37.

[158] Coifman, R. R., Rochberg, R., and Weiss, G. Factorization theorems for Hardy spaces in several variables. *Ann. of Math. (2) 103*, 3 (1976), 611–635.

[159] Conway, J. B. *A course in functional analysis*, 2nd edn, vol. 96 of *Graduate Texts in Mathematics*. Springer, New York, 1990.

[160] Conway, J. B. *The theory of subnormal operators*, vol. 36 of *Mathematical Surveys and Monographs*. American Mathematical Society, Providence, RI, 1991.

[161] Corach, G., Maestripieri, A., and Stojanoff, D. Projections in operator ranges. *Proc. Amer. Math. Soc. 134*, 3 (2006), 765–778 (electronic).

[162] Cowen, C. C. The commutant of an analytic Toeplitz operator. *Trans. Amer. Math. Soc. 239* (1978), 1–31.

[163] Crofoot, R. B. Multipliers between invariant subspaces of the backward shift. *Pacific J. Math. 166*, 2 (1994), 225–246.

[164] Davies, E. B. *Linear operators and their spectra*, vol. 106 of *Cambridge Studies in Advanced Mathematics*. Cambridge University Press, Cambridge, UK, 2007.

[165] Davis, B. M., and McCarthy, J. E. Multipliers of de Branges spaces. *Michigan Math. J. 38*, 2 (1991), 225–240.

[166] de Branges, L. A construction of Kreĭn spaces of analytic functions. *J. Funct. Anal. 98*, 1 (1991), 1–41.

[167] de Branges, L., and Rovnyak, J. Canonical models in quantum scattering theory. In *Perturbation theory and its applications in quantum mechanics (Proc. Adv. Sem. Math. Res. Center, US Army, Theor. Chem. Inst., Univ. Wisconsin, Madison, WI, 1965)*. Wiley, New York, 1966, pp. 295–392.

[168] de Branges, L., and Rovnyak, J. *Square summable power series*. Holt, Rinehart and Winston, New York, 1966.

[169] de Leeuw, K., and Rudin, W. Extreme points and extremum problems in H_1. *Pacific J. Math. 8* (1958), 467–485.

[170] Devinatz, A. Toeplitz operators on H^2 spaces. *Trans. Amer. Math. Soc. 112* (1964), 304–317.

[171] DIEUDONNÉ, J. Sur les homomorphismes d'espaces normés. *Bull. Sci. Math. (2) 67* (1943), 72–84.

[172] DIEUDONNÉ, J. A. The index of operators in Banach spaces. *Integral Equations Operator Theory 8*, 5 (1985), 580–589.

[173] DOUGLAS, R. G. On majorization, factorization, and range inclusion of operators on Hilbert space. *Proc. Amer. Math. Soc. 17* (1966), 413–415.

[174] DOUGLAS, R. G. Toeplitz and Wiener–Hopf operators in $H^\infty + C$. *Bull. Amer. Math. Soc. 74* (1968), 895–899.

[175] DOUGLAS, R. G. On the spectrum of a class of Toeplitz operators. *J. Math. Mech. 18* (1968/1969), 433–435.

[176] DOUGLAS, R. G. *Banach algebra techniques in the theory of Toeplitz operators*. In *Conference Board of the Mathematical Sciences, Regional Conference Series in Mathematics, No. 15*. American Mathematical Society, Providence, RI, 1973.

[177] DOUGLAS, R. G. *Banach algebra techniques in operator theory*, 2nd edn, vol. 179 of *Graduate Texts in Mathematics*. Springer, New York, 1998.

[178] DOUGLAS, R. G., AND SARASON, D. Fredholm Toeplitz operators. *Proc. Amer. Math. Soc. 26* (1970), 117–120.

[179] DOUGLAS, R. G., AND SARASON, D. A class of Toeplitz operators. In *Proceedings of an International Symposium on Operator Theory (Indiana Univ., Bloomington, IN, 1970)* (1971), vol. 20, pp. 891–895.

[180] DOUGLAS, R. G., SHAPIRO, H. S., AND SHIELDS, A. L. Cyclic vectors and invariant subspaces for the backward shift operator. *Ann. Inst. Fourier (Grenoble) 20*, 1 (1970), 37–76.

[181] DRITSCHEL, M. A., AND ROVNYAK, J. Extension theorems for contraction operators on Krein spaces. In *Extension and interpolation of linear operators and matrix functions*, vol. 47 of *Operator Theory: Advances and Applications*. Birkhäuser, Basel, 1990, pp. 221–305.

[182] DRITSCHEL, M. A., AND ROVNYAK, J. Julia operators and complementation in Kreĭn spaces. *Indiana Univ. Math. J. 40*, 3 (1991), 885–901.

[183] DUFFIN, R. J., AND EACHUS, J. J. Some notes on an expansion theorem of Paley and Wiener. *Bull. Amer. Math. Soc. 48* (1942), 850–855.

[184] DUNFORD, N. A mean ergodic theorem. *Duke Math. J. 5* (1939), 635–646.

[185] DUNFORD, N., AND PETTIS, B. J. Linear operations on summable functions. *Trans. Amer. Math. Soc. 47* (1940), 323–392.

[186] DUNFORD, N., AND SCHWARTZ, J. T. *Linear operators. Part I. General theory*. With the assistance of W. G. Bade and R. G. Bartle. Reprint of the 1958 original. Wiley, New York, 1988.

[187] DUNFORD, N., AND SCHWARTZ, J. T. *Linear operators. Part II. Spectral theory*. With the assistance of W. G. Bade and R. G. Bartle. Reprint of the 1963 original. Wiley, New York, 1988.

[188] DUREN, P. L. *Theory of H^p spaces*, vol. 38 of *Pure and Applied Mathematics*. Academic Press, New York, 1970.

[189] DUREN, P. L., ROMBERG, B. W., AND SHIELDS, A. L. Linear functionals on H^p spaces with $0 < p < 1$. *J. Reine Angew. Math. 238* (1969), 32–60.

[190] DYAKONOV, K. M. The geometry of the unit ball in the space K^1_θ. In *Geometric problems of the theory of functions and sets*. Kalinin. Gos. Univ., Kalinin, 1987, pp. 52–54 (in Russian).

[191] DYAKONOV, K. M. An interpolation problem and equivalent norms in the spaces K^p_θ. *Vestnik Leningrad. Univ. Mat. Mekh. Astron. 4* (1988), 104–105, 115.

[192] DYAKONOV, K. M. Moduli and arguments of analytic functions from subspaces in H^p that are invariant under the backward shift operator. *Sibirsk. Mat. Zh. 31*, 6 (1990), 64–79.

[193] DYAKONOV, K. M. Entire functions of exponential type and model subspaces in H^p. *Zap. Nauchn. Sem. Leningrad. Otdel. Mat. Inst. Steklov. (LOMI) 190, Issled. Linein. Oper. Teor. Funktsii. 19* (1991), 81–100, 186.

[194] DYAKONOV, K. M. Interpolating functions of minimal norm, star-invariant subspaces and kernels of Toeplitz operators. *Proc. Amer. Math. Soc. 116*, 4 (1992), 1007–1013.

[195] DYAKONOV, K. M. Kernels of Toeplitz operators, smooth functions and Bernstein-type inequalities. *Zap. Nauchn. Sem. St.-Peterburg. Otdel. Mat. Inst. Steklov. (POMI) 201, Issled. Linein. Oper. Teor. Funktsii. 20* (1992), 5–21, 190.

[196] DYAKONOV, K. M. Smooth functions and co-invariant subspaces of the shift operator. *Algebra Analiz 4*, 5 (1992), 117–147.

[197] DYAKONOV, K. M. Division and multiplication by inner functions and embedding theorems for star-invariant subspaces. *Amer. J. Math. 115*, 4 (1993), 881–902.

[198] DYAKONOV, K. M. Generalized Hardy inequalities and pseudocontinuable functions. *Ark. Mat. 34*, 2 (1996), 231–244.

[199] DYAKONOV, K. M. Embedding theorems for star-invariant subspaces generated by smooth inner functions. *J. Funct. Anal. 157*, 2 (1998), 588–598.

[200] DYAKONOV, K. M. Kernels of Toeplitz operators via Bourgain's factorization theorem. *J. Funct. Anal. 170*, 1 (2000), 93–106.

[201] DYAKONOV, K. M. Differentiation in star-invariant subspaces. I. Boundedness and compactness. *J. Funct. Anal. 192*, 2 (2002), 364–386.

[202] DYAKONOV, K. M. Differentiation in star-invariant subspaces. II. Schatten class criteria. *J. Funct. Anal. 192*, 2 (2002), 387–409.

[203] DYAKONOV, K. M. Zero sets and multiplier theorems for star-invariant subspaces. *J. Anal. Math. 86* (2002), 247–269.

[204] DŽRBAŠJAN, M. M. Biorthogonal systems, and the solution of an interpolation problem with nodes of bounded multiplicity in the class H_2. *Izv. Akad. Nauk Armjan. SSR Ser. Mat. 9*, 5 (1974), 339–373, 428.

[205] EBERLEIN, W. F. Weak compactness in Banach spaces. I. *Proc. Natl. Acad. Sci. USA 33* (1947), 51–53.

[206] ENFLO, P. A counterexample to the approximation problem in Banach spaces. *Acta Math. 130* (1973), 309–317.

[207] ENFLO, P. On the invariant subspace problem for Banach spaces. *Acta Math. 158*, 3–4 (1987), 213–313.

[208] EVANS, L. C., AND GARIEPY, R. F. *Measure theory and fine properties of functions*. In *Studies in Advanced Mathematics*. CRC Press, Boca Raton, FL, 1992.

[209] FABIAN, M., HABALA, P., HÁJEK, P., MONTESINOS, V., AND ZIZLER, V. *Banach space theory: The basis for linear and nonlinear analysis*. In *CMS Books in Mathematics/Ouvrages de Mathématiques de la SMC*. Springer, New York, 2011.

[210] FATOU, P. Séries trigonométriques et séries de Taylor. *Acta Math. 30*, 1 (1906), 335–400.

[211] FEJÉR, L. Untersuchungen über Fouriersche Reihen. *Math. Ann. 58* (1904), 501–569.

[212] FEJÉR, L. Über trigonometrische Polynome. *J. Reine Angew. Math. 146* (1916), 53–82.

[213] Fejér, L., and Riesz, F. Über einige funktionentheoretische Ungleichungen. *Math. Z. 11*, 3–4 (1921), 305–314.

[214] Ferguson, S. H. Backward shift invariant operator ranges. *J. Funct. Anal. 150*, 2 (1997), 526–543.

[215] Fillmore, P. A., and Williams, J. P. On operator ranges. *Adv. Math. 7* (1971), 254–281.

[216] Fischer, E. Über das Carathéodory'sche Problem, Potenzreihen mit positiven reellen Teil betreffend. *Rend. Circ. Mat. Palermo 32* (1911), 240–256.

[217] Fisher, S. Exposed points in spaces of bounded analytic functions. *Duke Math. J. 36* (1969), 479–484.

[218] Foiaş, C. Sur certains théorèmes de J. von Neumann concernant les ensembles spectraux. *Acta Sci. Math. Szeged 18* (1957), 15–20.

[219] Fonf, V. P., Lindenstrauss, J., and Phelps, R. R. Infinite dimensional convexity. In *Handbook of the geometry of Banach spaces, Vol. I*. North-Holland, Amsterdam, 2001, pp. 599–670.

[220] Francis, B. A. *A course in H_∞ control theory*, vol. 88 of *Lecture Notes in Control and Information Sciences*. Springer, Berlin, 1987.

[221] Fricain, E. Bases of reproducing kernels in model spaces. *J. Operator Theory 46*, 3, suppl. (2001), 517–543.

[222] Fricain, E. Complétude des noyaux reproduisants dans les espaces modèles. *Ann. Inst. Fourier (Grenoble) 52*, 2 (2002), 661–686.

[223] Fricain, E. Bases of reproducing kernels in de Branges spaces. *J. Funct. Anal. 226*, 2 (2005), 373–405.

[224] Frostman, O. Potentiel d'équilibre et capacité des ensembles avec quelques applications à la théorie des fonctions. *Meddel. Lunds Univ. Mat. Sem. 3* (1935), 1–118.

[225] Fuhrmann, P. A. On the corona theorem and its application to spectral problems in Hilbert space. *Trans. Amer. Math. Soc. 132* (1968), 55–66.

[226] Fuhrmann, P. A. Exact controllability and observability and realization theory in Hilbert space. *J. Math. Anal. Appl. 53*, 2 (1976), 377–392.

[227] Fuhrmann, P. A. *Linear systems and operators in Hilbert space*. McGraw-Hill, New York, 1981.

[228] Fuhrmann, P. A. *A polynomial approach to linear algebra*. Universitext, Springer, New York, 1996.

[229] Gamelin, T. W. Wolff's proof of the corona theorem. *Israel J. Math. 37*, 1–2 (1980), 113–119.

[230] Gamelin, T. W. *Uniform algebras*. In *Cambridge Mathematical Library*. American Mathematical Society, Cambridge, MA, 2005.

[231] Garcia, S. R., Poore, D. E., and Ross, W. T. Unitary equivalence to a truncated Toeplitz operator: analytic symbols. *Proc. Amer. Math. Soc. 140*, 4 (2012), 1281–1295.

[232] Garcia, S. R., Ross, W. T., and Wogen, W. R. Spatial isomorphisms of algebras of truncated Toeplitz operators. *Indiana Univ. Math. J. 59*, 6 (2010), 1971–2000.

[233] Garnett, J. B. *Bounded analytic functions*, 1st edn, vol. 236 of *Graduate Texts in Mathematics*. Springer, New York, 2007.

[234] Gelfand, I. Normierte Ringe. *Rec. Math. [Mat. Sbornik] N.S. 9 (51)* (1941), 3–24.

[235] GELFAND, I. M. Remark on the work of N. K. Bari, "Biorthogonal systems and bases in Hilbert space." *Moskov. Gos. Univ. Učenye Zap. Mat. 148(4)* (1951), 224–225.

[236] GHOSECHOWDHURY, S. A local expansion theorem for the Löwner equation. *Integral Equations Operator Theory 48*, 1 (2004), 1–25.

[237] GLOVER, K. All optimal Hankel-norm approximations of linear multivariable systems and their L^∞-error bounds. *Int. J. Control 39*, 6 (1984), 1115–1193.

[238] GOHBERG, I. C. On linear equations in normed spaces. *Dokl. Akad. Nauk SSSR (N.S.) 76* (1951), 477–480.

[239] GOHBERG, I. C., AND FELDMAN, I. A. Wiener–Hopf integro-difference equations. *Dokl. Akad. Nauk SSSR 183* (1968), 25–28.

[240] GOHBERG, I. C., AND FELDMAN, I. A. Integro-difference Wiener–Hopf equations. *Acta Sci. Math. (Szeged) 30* (1969), 199–224.

[241] GOHBERG, I. C., AND KREĬN, M. G. Fundamental aspects of defect numbers, root numbers and indexes of linear operators. *Uspekhi Mat. Nauk (N.S.) 12*, 2(74) (1957), 43–118.

[242] GOHBERG, I. C., AND KREĬN, M. G. *Introduction to the theory of linear nonselfadjoint operators*, vol. 18 of *Translations of Mathematical Monographs*. Translated from the Russian by A. Feinstein. American Mathematical Society, Providence, RI, 1969.

[243] GOHBERG, I. C., AND KRUPNIK, N. J. The algebra generated by the Toeplitz matrices. *Funkcional. Anal. Prilozen. 3*, 2 (1969), 46–56 (in Russian); *Functional Anal. Appl. 3* (1969), 119–127.

[244] GRENANDER, U., AND SZEGÖ, G. *Toeplitz forms and their applications*. In *California Monographs in Mathematical Sciences*. University of California Press, Berkeley, CA, 1958.

[245] GRINBLYUM, M. Certains théorèmes sur la base dans un espace de type (B). *Dokl. Akad. Nauk. SSSR 31* (1941), 428–432.

[246] GUREVICH, L. Sur une propriété de la base dans l'espace de Hilbert. *Dokl. Akad. Nauk. SSSR 30* (1941), 289–291.

[247] HAHN, H. Über lineare Gleichungssystem in linearen Räumen. *J. Reine Angew. Math. 157* (1926), 214–229.

[248] HALMOS, P. R. Normal dilations and extensions of operators. *Summa Brasil. Math. 2* (1950), 125–134.

[249] HALMOS, P. R. *Introduction to Hilbert space and the theory of spectral multiplicity*. Chelsea, New York, 1951.

[250] HALMOS, P. R. *Finite-dimensional vector spaces*, 2nd edn. In *University Series in Undergraduate Mathematics*. Van Nostrand, Princeton, NJ, 1958.

[251] HALMOS, P. R. Shifts on Hilbert spaces. *J. Reine Angew. Math. 208* (1961), 102–112.

[252] HALMOS, P. R. A glimpse into Hilbert space. In *Lectures on Modern Mathematics, Vol. I*. Wiley, New York, 1963, pp. 1–22.

[253] HALMOS, P. R. *A Hilbert space problem book*, 2nd edn, vol. 19 of *Graduate Texts in Mathematics*. Springer, New York, 1982.

[254] HALMOS, P. R. *I want to be a mathematician: An automathography in three parts*. In *MAA Spectrum Series*. Mathematical Association of America, Washington, DC, 1985.

[255] HAMBURGER, H. Über eine Erweiterung des Stieltjesschen Momentenproblems. *Math. Ann. 81*, 2–4 (1920), 235–319.

[256] Hankel, H. Über eine besondere Classe der symmetrischen Determinanten. PhD Thesis, Göttingen, 1861.

[257] Hardy, G. H., and Littlewood, J. E. Some new properties of Fourier constants. *Math. Ann. 97*, 1 (1927), 159–209.

[258] Hardy, G. H., Littlewood, J. E., and Pólya, G. *Inequalities*. In *Cambridge Mathematical Library*. Reprint of the 1952 edition. Cambridge University Press, Cambridge, UK, 1988.

[259] Hartman, P. On completely continuous Hankel matrices. *Proc. Amer. Math. Soc. 9* (1958), 862–866.

[260] Hartman, P., and Wintner, A. On the spectra of Toeplitz's matrices. *Amer. J. Math. 72* (1950), 359–366.

[261] Hartman, P., and Wintner, A. The spectra of Toeplitz's matrices. *Amer. J. Math. 76* (1954), 867–882.

[262] Hartmann, A., Massaneda, X., Nicolau, A., and Ortega-Cerdà, J. Reverse Carleson measures in Hardy spaces. *Collect. Math.* to appear.

[263] Hartmann, A., and Ross, W. T. Boundary values in range spaces of co-analytic truncated Toeplitz operators. *Publ. Mat. 56*, 1 (2012), 191–223.

[264] Hartmann, A., Sarason, D., and Seip, K. Surjective Toeplitz operators. *Acta Sci. Math. (Szeged) 70*, 3–4 (2004), 609–621.

[265] Havin, V., and Mashreghi, J. Admissible majorants for model subspaces of H^2. I. Slow winding of the generating inner function. *Canad. J. Math. 55*, 6 (2003), 1231–1263.

[266] Havin, V., and Mashreghi, J. Admissible majorants for model subspaces of H^2. II. Fast winding of the generating inner function. *Canad. J. Math. 55*, 6 (2003), 1264–1301.

[267] Havin, V. P., Hruščëv, S. V., and Nikolskii, N. K. (Eds.) *Linear and complex analysis problem book: 199 research problems*, vol. 1043 of *Lecture Notes in Mathematics*. Springer, Berlin, 1984.

[268] Havin, V. P., and Nikolski, N. K. Stanislav Aleksandrovich Vinogradov, his life and mathematics. In *Complex analysis, operators, and related topics*, vol. 113 of *Operator Theory: Advances and Applications*. Birkhäuser, Basel, 2000, pp. 1–18.

[269] Havinson, S. Ya. On some extremal problems of the theory of analytic functions. *Moskov. Gos. Univ. Učenye Zap. Mat. 148(4)* (1951), 133–143.

[270] Hayashi, E. The solution of extremal problems in H^1. *Proc. Amer. Math. Soc. 93* (1985), 690–696.

[271] Hayashi, E. The kernel of a Toeplitz operator. *Integral Equations Operator Theory 9*, 4 (1986), 588–591.

[272] Hayashi, E. Classification of nearly invariant subspaces of the backward shift. *Proc. Amer. Math. Soc. 110*, 2 (1990), 441–448.

[273] Hayman, W. Interpolation by bounded functions. *Ann. Inst. Fourier (Grenoble) 8* (1958), 277–290.

[274] Hayman, W. K., and Kennedy, P. B. *Subharmonic functions. Vol. I.* In *London Mathematical Society Monographs, No. 9*. Academic Press, London, 1976.

[275] Heinz, E. Beiträge zur Störungstheorie der Spektralzerlegung. *Math. Ann. 123* (1951), 415–438.

[276] Heinz, E. Ein v. Neumannscher Satz über beschränkte Operatoren im Hilbertschen Raum. *Nachr. Akad. Wiss. Göttingen. Math.-Phys. Kl. IIa, 1952* (1952), 5–6.

[277] HELLY, E. Über lineare Funktionaloperationen. *Sitzungsber. Math. Naturwiss. Kl. Akad. Wiss. 121* (1912), 265–291.

[278] HELLY, E. Über Systeme linearer Gleichungen mit unendlich vielen Unbekannten. *Monatsh. Math. Phys. 31*, 1 (1921), 60–91.

[279] HELSON, H. *Lectures on invariant subspaces*. Academic Press, New York, 1964.

[280] HELSON, H. Large analytic functions. II. In *Analysis and partial differential equations*, vol. 122 of *Lecture Notes in Pure and Applied Mathematics*. Dekker, New York, 1990, pp. 217–220.

[281] HELSON, H., AND LOWDENSLAGER, D. Invariant subspaces. In *Proc. Int. Symp. on Linear Spaces (Jerusalem, 1960)*. Jerusalem Academic Press, Jerusalem, 1961, pp. 251–262.

[282] HELSON, H., AND SARASON, D. Past and future. *Math. Scand 21* (1967), 5–16 (1968).

[283] HELSON, H., AND SZEGŐ, G. A problem in prediction theory. *Ann. Mat. Pura Appl. (4) 51* (1960), 107–138.

[284] HERGLOTZ, G. Über Potenzreihen mit positivem reellen Teil im Einheitskreis. *Ber. Ver. Ges. Wiss. Leipzig 63* (1911), 501–511.

[285] HILBERT, D. Grundzüge einer allgemeinen Theorie der linearen Integralgleichung. *Nach. Akad. Wiss. Göttingen Math. Phys.* (1904), 49–91.

[286] HILBERT, D. Grundzüge einer allgemeinen Theorie der linearen integralgleichung. *Nach. Akad. Wiss. Göttingen Math. Phys.* (1906), 157–227.

[287] HILBERT, D. *Grundzüge einer allgemeinen Theorie der linearen Integralgleichungen*. Fortschritte der Mathematischen Wissenschaften. Monographien Herausgegeben von Otto Blumenthal, Leipzig and Berlin, 1912.

[288] HILDEBRANDT, T. H. Convergence of sequences of linear operations. *Bull. Amer. Math. Soc. 28*, 1–2 (1922), 53–58.

[289] HILDEBRANDT, T. H. Über vollstetige lineare Transformationen. *Acta Math. 51*, 1 (1928), 311–318.

[290] HITT, D. Invariant subspaces of H^2 of an annulus. *Pacific J. Math. 134*, 1 (1988), 101–120.

[291] HOFFMAN, K. *Banach spaces of analytic functions*. Reprint of 1962 original. Dover, New York, 1988.

[292] HORN, R. A., AND JOHNSON, C. R. *Matrix analysis*, 2nd edn. Cambridge University Press, Cambridge, UK, 2013.

[293] HRUŠČĖV, S. V., NIKOLSKII, N. K., AND PAVLOV, B. S. Unconditional bases of exponentials and of reproducing kernels. In *Complex analysis and spectral theory (Leningrad, 1979/1980)*, vol. 864 of *Lecture Notes in Mathematics*. Springer, Berlin, 1981, pp. 214–335.

[294] HUNT, R., MUCKENHOUPT, B., AND WHEEDEN, R. Weighted norm inequalities for the conjugate function and Hilbert transform. *Trans. Amer. Math. Soc. 176* (1973), 227–251.

[295] INGHAM, A. E. Some trigonometrical inequalities with applications to the theory of series. *Math. Z. 41*, 1 (1936), 367–379.

[296] INOUE, J. An example of a nonexposed extreme function in the unit ball of H^1. *Proc. Edinburgh Math. Soc. (2) 37*, 1 (1994), 47–51.

[297] INOUE, J., AND NAKAZI, T. Polynomials of an inner function which are exposed points in H^1. *Proc. Amer. Math. Soc. 100*, 3 (1987), 454–456.

[298] Iohvidov, I. S. *Hankel and Toeplitz matrices and forms: Algebraic theory.* Translated from Russian by G. Philip and A. Thijsse, with an introduction by I. Gohberg. Birkhäuser, Boston, MA, 1982.

[299] Johnson, W. B., and Lindenstrauss, J. Basic concepts in the geometry of Banach spaces. In *Handbook of the geometry of Banach spaces, Vol. I.* North-Holland, Amsterdam, 2001, pp. 1–84.

[300] Johnson, W. B., Rosenthal, H. P., and Zippin, M. On bases, finite dimensional decompositions and weaker structures in Banach spaces. *Israel J. Math. 9* (1971), 488–506.

[301] Julia, G. Sur la représentation analytique des opérateurs bornés ou fermés de l'espace hilbertien. *C. R. Acad. Sci. Paris 219* (1944), 225–227.

[302] Jury, M. T. An improved Julia–Carathéodory theorem for Schur–Agler mappings of the unit ball. arXiv:0707.3423v1, 2007.

[303] Jury, M. T. Norms and spectral radii of linear fractional composition operators on the ball. *J. Funct. Anal. 254* (2008), 2387–2400.

[304] Jury, M. T. Reproducing kernels, de Branges–Rovnyak spaces, and norms of weighted composition operators. *Proc. Amer. Math. Soc. 135*, 11 (2007), 3669–3675 (electronic).

[305] Kabaĭla, V. On interpolation of functions of class H_δ. *Uspekhi Mat. Nauk (N.S.) 13*, 1(79) (1958), 181–188.

[306] Kabaĭla, V. Sharpening some theorems on interpolation in the class H_δ. *Vilniaus Valst. Univ. Mokslo Darbai Mat. Fiz. 9* (1960), 15–19.

[307] Kabaĭla, V. Interpolation sequences for the H_p classes in the case $p < 1$. *Litovsk. Mat. Sbornik 3*, 1 (1963), 141–147.

[308] Kaczmarz, S., and Steinhaus, H. *Theorie der Orthogonalreihen*, vol. VI in *Monografje Matematyczne*. Seminar. Matemat. Uniwers., Warsaw, 1935.

[309] Kadec, M. Ĭ. Bases and their spaces of coefficients. *Dopovidi Akad. Nauk Ukraïn. RSR 1964* (1964), 1139–1141.

[310] Kadec, M. Ĭ. The exact value of the Paley–Wiener constant. *Dokl. Akad. Nauk SSSR 155* (1964), 1253–1254.

[311] Kapustin, V. V. Real functions in weighted Hardy spaces. *Zap. Nauchn. Sem. St.-Peterburg. Otdel. Mat. Inst. Steklov. (POMI) 262, Issled. Linein. Oper. Teor. Funkts. 27* (1999), 138–146, 233–234.

[312] Kapustin, V., and Poltoratski, A. Boundary convergence of vector-valued pseudocontinuable functions. *J. Funct. Anal. 238*, 1 (2006), 313–326.

[313] Karlin, S. Bases in Banach spaces. *Duke Math. J. 15* (1948), 971–985.

[314] Kato, T. *Perturbation theory for linear operators*, vol. 132 of *Die Grundlehren der mathematischen Wissenschaften*. Springer, New York, 1966.

[315] Kellogg, C. N. Pseudo-uniform convexity in H^1. *Proc. Amer. Math. Soc. 23* (1969), 190–192.

[316] Kheifets, A. Hamburger moment problem: Parseval equality and A-singularity. *J. Funct. Anal. 141*, 2 (1996), 374–420.

[317] Klee, Jr., V. L. Extremal structure of convex sets. II. *Math. Z. 69* (1958), 90–104.

[318] Kolmogorov, A. Stationary sequences in Hilbert space. *Moscow Univ. Math. Bull. 2* (1941), 1–40.

[319] Kolmogorov, A. N. Sur les fonctions harmoniques conjugées et les séries de Fourier. *Fund. Math. 7* (1925), 24–29.

[320] KOOSIS, P. *Introduction to H_p spaces*, 2nd edn, vol. 115 of *Cambridge Tracts in Mathematics*. With two appendices by V. P. Havin. Cambridge University Press, Cambridge, UK, 1998.

[321] KÖTHE, G., AND TOEPLITZ, O. Lineare Räume mit unendlich vielen Koordinaten und Ringe unendlicher Matrizen. *J. Reine Angew. Math. 171* (1934), 251–270.

[322] KOVALISHINA, I. V. The Carathéodory–Julia theorem for matrix-functions. *Teor. Funktsiĭ Funktsional. Anal. Prilozhen.*, 43 (1985), 70–82.

[323] KOVALISHINA, I. V. Theory of a multiple j-elementary matrix-function with a pole on the boundary of the unit disk. *Teor. Funktsiĭ Funktsional. Anal. Prilozhen.*, 50 (1988), 62–74, iii.

[324] KREĬN, M. On a generalization of some investigations of G. Szegő, V. Smirnoff and A. Kolmogoroff. *C. R. (Doklady) Acad. Sci. URSS (N.S.) 46* (1945), 91–94.

[325] KREĬN, M., AND LIUSTERNIK, L. Functional analysis. In "Thirty years of mathematics". *Trans. Amer. Math. Soc. 48* (1940), 467–487.

[326] KREĬN, M., AND MILMAN, D. On extreme points of regular convex sets. *Studia Math. 9* (1940), 133–138.

[327] KREĬN, M. G. On Bari bases of a Hilbert space. *Uspekhi Mat. Nauk (N.S.) 12*, 3(75) (1957), 333–341.

[328] KREĬN, M. G. Integral equations on the half-line with a kernel depending on the difference of the arguments. *Uspekhi Mat. Nauk 13*, 5 (83) (1958), 3–120. (in Russian); *Amer. Math. Soc. Transl.* 22 (2), 163–288 (1962).

[329] KRIETE, T., AND MOORHOUSE, J. Linear relations in the Calkin algebra for composition operators. *Trans. Amer. Math. Soc. 359*, 6 (2007), 2915–2944 (electronic).

[330] KRONECKER, L. Zur Theorie der Elimination einer Variablen aus zwei algebraischen Gleichungen. *Monatsber. Königl. Preuss. Akad. Wiss. (Berlin)* (1881), 535–600.

[331] KUBRUSLY, C. S. *Hilbert space operators. A problem solving approach.* Birkhäuser, Boston, MA, 2003.

[332] KURATOWSKI, C. La propriété de baire dans les espaces métriques. *Fund. Math. 16* (1930), 390–394.

[333] KURATOWSKI, C. Théorèmes sur les ensembles de première catégorie. *Fund. Math. 16* (1930), 395–398.

[334] LANG, S. *Real and functional analysis*, 3rd edn, vol. 142 of *Graduate Texts in Mathematics*. Springer, New York, 1993.

[335] LEE, M., AND SARASON, D. The spectra of some Toeplitz operators. *J. Math. Anal. Appl. 33* (1971), 529–543.

[336] LEFÈVRE, P., LI, D., QUEFFÉLEC, H., AND RODRÍGUEZ-PIAZZA, L. Some revisited results about composition operators on Hardy spaces. *Rev. Mat. Iberoamer. 28*, 1 (2012), 57–76.

[337] LINDENSTRAUSS, J. On operators which attain their norm. *Israel J. Math. 1* (1963), 139–148.

[338] LINDENSTRAUSS, J., AND TZAFRIRI, L. *Classical Banach spaces. I. Sequence spaces*, vol. 92 of *Ergebnisse der Mathematik und ihrer Grenzgebiete*. Springer, Berlin, 1977.

[339] LITVINCHUK, G. S., AND SPITKOVSKII, I. M. *Factorization of measurable matrix functions*, vol. 25 of *Operator Theory: Advances and Applications*. Translated from Russian by B. Luderer. Birkhäuser, Basel, 1987.

[340] LIVŠIC, M. S. On a class of linear operators in Hilbert space. *Mat. Sbornik N.S. 19* (1946), 239–262.

[341] LORCH, E. R. Bicontinuous linear transformations in certain vector spaces. *Bull. Amer. Math. Soc. 45* (1939), 564–569.

[342] LOTTO, B. A. Inner multipliers of de Branges's spaces. *Integral Equations Operator Theory 13*, 2 (1990), 216–230.

[343] LOTTO, B. A. Toeplitz operators on weighted Hardy spaces. In *Function spaces (Edwardsville, IL, 1990)*, vol. 136 of *Lecture Notes in Pure and Applied Mathematics*. Dekker, New York, 1992, pp. 295–300.

[344] LOTTO, B. A., AND SARASON, D. Multiplicative structure of de Branges's spaces. *Rev. Mat. Iberoamer.7*, 2 (1991), 183–220.

[345] LOTTO, B. A., AND SARASON, D. Multipliers of de Branges–Rovnyak spaces. *Indiana Univ. Math. J. 42*, 3 (1993), 907–920.

[346] LOTTO, B. A., AND SARASON, D. Multipliers of de Branges–Rovnyak spaces. II. In *Harmonic analysis and hypergroups (Delhi, 1995)*, in *Trends in Mathematics*. Birkhäuser, Boston, MA, 1998, pp. 51–58.

[347] LÖWNER, K. Über monotone Matrixfunktionen. *Math. Z. 38*, 1 (1934), 177–216.

[348] LUECKING, D. H. The compact Hankel operators form an M-ideal in the space of Hankel operators. *Proc. Amer. Math. Soc. 79*, 2 (1980), 222–224.

[349] LUSIN, N., AND PRIWALOFF, J. Sur l'unicité et la multiplicité des fonctions analytiques. *Ann. Sci. École Norm. Sup. (3) 42* (1925), 143–191.

[350] MAKAROV, N., AND POLTORATSKI, A. Meromorphic inner functions, Toeplitz kernels and the uncertainty principle. In *Perspectives in analysis*, vol. 27 of *Mathematical Physics Studies*. Springer, Berlin, 2005, pp. 185–252.

[351] MARKUSHEVICH, A. Sur les bases (au sens large) dans les espaces linéaires. *C. R. (Doklady) Acad. Sci. URSS (N.S.) 41* (1943), 227–229.

[352] MARTI, J. T. *Introduction to the theory of bases*. vol. 18 of *Springer Tracts in Natural Philosophy*. Springer, New York, 1969.

[353] MARTÍNEZ-AVENDAÑO, R. A., AND ROSENTHAL, P. *An introduction to operators on the Hardy–Hilbert space*, vol. 237 of *Graduate Texts in Mathematics*. Springer, New York, 2007.

[354] MASHREGHI, J. *Representation theorems in Hardy spaces*, vol. 74 of *London Mathematical Society Student Texts*. Cambridge University Press, Cambridge, UK, 2009.

[355] MASHREGHI, J. On a family of oute functions. In *Complex analysis and potential theory*, vol. 55 of *CRM Proc. Lecture Notes*. American Mathematical Society, Providence, R1, 2012, pp. 193–199.

[356] MASHREGHI, J., NAZAROV, F. L., AND HAVIN, V. P. The Beurling–Malliavin multiplier theorem: the seventh proof. *Algebra Analiz 17*, 5 (2005), 3–68.

[357] MAUREY, B. Isomorphismes entre espaces H_1. *Acta Math. 145*, 1–2 (1980), 79–120.

[358] MCCARTHY, J. E. Boundary values and Cowen–Douglas curvature. *J. Funct. Anal. 137*, 1 (1996), 1–18.

[359] MCGIVNEY, R. J., AND RUCKLE, W. Multiplier algebras of biorthogonal systems. *Pacific J. Math. 29* (1969), 375–387.

[360] MERCER, J. Functions of positive and negative type and their connection with the theory of integral equations. *Phil. Trans. R. Soc. London 209* (1909), 415–446.

[361] MILLER, T. L., SMITH, W., AND YANG, L. Bounded point evaluations for certain $P^t(\mu)$ spaces. *Illinois J. Math. 43*, 1 (1999), 131–150.

[362] MINKOWSKI, H. *Gesammelte Abhandlungen.* B. G. Teubner, Leipzig, 1911.

[363] MOELLER, J. W. On the spectra of some translation invariant spaces. *J. Math. Anal. Appl. 4* (1962), 276–296.

[364] MOORE, E. *General analysis.* In *Memoirs of the American Philosophical Society.* American Philosophical Society, Philadelphia, 1935.

[365] MUCKENHOUPT, B. Weighted norm inequalities for the Hardy maximal function. *Trans. Amer. Math. Soc. 165* (1972), 207–226.

[366] MURRAY, F. J., AND VON NEUMANN, J. On rings of operators. *Ann. of Math. (2) 37*, 1 (1936), 116–229.

[367] NAFTALEVIČ, A. G. On interpolation by functions of bounded characteristic. *Vilniaus Valst. Univ. Mokslų Darbai. Mat. Fiz. Chem. Mokslų Ser. 5* (1956), 5–27.

[368] NAKAMURA, Y. One-dimensional perturbations of the shift. *Integral Equations Operator Theory 17*, 3 (1993), 373–403.

[369] NAKAZI, T. Exposed points and extremal problems in H^1. *J. Funct. Anal. 53* (1983), 224–230.

[370] NAKAZI, T. Exposed points and extremal problems in H^1, II. *Tôhoku Math. J. 37* (1985), 265–269.

[371] NAKAZI, T. Toeplitz operators and weighted norm inequalities. *Acta Sci. Math. (Szeged) 58*, 1–4 (1993), 443–452.

[372] NAZAROV, F., AND VOLBERG, A. The Bellman function, the two-weight Hilbert transform, and embeddings of the model spaces K_θ. *J. Anal. Math. 87* (2002), 385–414.

[373] NEHARI, Z. On bounded bilinear forms. *Ann. of Math. (2) 65* (1957), 153–162.

[374] NEUWIRTH, J., AND NEWMAN, D. J. Positive $H^{1/2}$ functions are constants. *Proc. Amer. Math. Soc. 18* (1967), 958.

[375] NEVANLINNA, F., AND NEVANLINNA, R. Über die Eigenschaften analytischer Funktionen in der Umgebung einer singulären Stelle oder Linie. *Acta Soc. Sci. Fennicae 50* (1922), 1–46.

[376] NEWMAN, D. J. Interpolation in H^∞. *Trans. Amer. Math. Soc. 92* (1959), 501–507.

[377] NEWMAN, D. J. Pseudo-uniform convexity in H^1. *Proc. Amer. Math. Soc. 14* (1963), 676–679.

[378] NEWNS, W. F. On the representation of analytic functions by infinite series. *Phil. Trans. R. Soc. London. Ser. A. 245* (1953), 429–468.

[379] NIKOLSKII, N. K. Invariant subspaces in operator theory and function theory. In *Mathematical analysis, Vol. 12* (in Russian). *Akad. Nauk SSSR Vsesojuz. Inst. Naučn. Tehn. Informacii, Moscow*, 1974, pp. 199–412, 468.

[380] NIKOLSKII, N. K. Lectures on the shift operator. II. *Zap. Naučn. Sem. Leningrad. Otdel. Mat. Inst. Steklov. (LOMI) 47, Issled. Linein. Oper. Teor. Funktsii. 5* (1974), 90–119, 186–187, 192.

[381] NIKOLSKII, N. K. *Selected problems of weighted approximation and spectral analysis.* Translated from Russian by F. A. Cezus. American Mathematical Society, Providence, RI, 1976.

[382] NIKOLSKII, N. K. Bases of invariant subspaces and operator interpolation. Spectral theory of functions and operators. *Trudy Mat. Inst. Steklov. 130* (1978), 50–123, 223.

[383] NIKOLSKII, N. K. Bases of exponentials and values of reproducing kernels. *Dokl. Akad. Nauk SSSR 252*, 6 (1980), 1316–1320.

[384] Nikolskii, N. K. What problems do spectral theory and complex analysis solve for each other? In *Proceedings of the International Congress of Mathematicians (Helsinki, 1978)*. Acad. Sci. Fennica, Helsinki, 1980, pp. 631–638.

[385] Nikolskii, N. K. Ha-plitz operators: a survey of some recent results. In *Operators and function theory (Lancaster, 1984)*, vol. 153 of *NATO Adv. Sci. Inst. Ser. C Math. Phys. Sci.* Reidel, Dordrecht, 1985, pp. 87–137.

[386] Nikolskii, N. K. *Treatise on the shift operator. Spectral function theory*, vol. 273 of *Grundlehren der Mathematischen Wissenschaften*. With appendix by S. V. Hruščev and V. V. Peller. Translated from Russian by J. Peetre. Springer, Berlin, 1986.

[387] Nikolskii, N. K. *Operators, functions, and systems: an easy reading. Vol. 1, Hardy, Hankel and Toeplitz*, vol. 92 of *Mathematical Surveys and Monographs*. Translated from French by A. Hartmann and revised by the author. American Mathematical Society, Providence, RI, 2002.

[388] Nikolskii, N. K. *Operators, functions, and systems: an easy reading. Vol. 2, Model operators and systems*, vol. 93 of *Mathematical Surveys and Monographs*. Translated from French by A. Hartmann and revised by the author. American Mathematical Society, Providence, RI, 2002.

[389] Nikolskii, N. K., and Pavlov, B. S. Eigenvector expansions of nonunitary operators, and the characteristic function. *Zap. Naučn. Sem. Leningrad. Otdel. Mat. Inst. Steklov. (LOMI) 11* (1968), 150–203.

[390] Nikolskii, N. K., and Pavlov, B. S. Bases of eigenvectors, characteristic function and interpolation problems in the Hardy space H^2. *Dokl. Akad. Nauk SSSR 184* (1969), 550–553.

[391] Nikolskii, N. K., and Pavlov, B. S. Bases from the eigenvectors of totally nonunitary contractions. *Dokl. Akad. Nauk SSSR 184* (1969), 778–781.

[392] Nikolskii, N. K., and Pavlov, B. S. Bases of eigenvectors of completely nonunitary contractions, and the characteristic function. *Izv. Akad. Nauk SSSR Ser. Mat. 34* (1970), 90–133.

[393] Nikolskii, N. K., and Vasyunin, V. I. Notes on two function models. In *The Bieberbach conjecture (West Lafayette, IN, 1985)*, vol. 21 of *Mathematical Surveys and Monographs*. American Mathematical Society, Providence, RI, 1986, pp. 113–141.

[394] Nikolskii, N. K., and Vasyunin, V. I. Quasi-orthogonal Hilbert space decompositions and estimates of univalent functions. I. In *Functional analysis and related topics (Sapporo, 1990)*. World Scientific, River Edge, NJ, 1991, pp. 19–28.

[395] Nikolskii, N. K., and Vasyunin, V. I. Quasi-orthogonal Hilbert space decompositions and estimates of univalent functions. II. In *Progress in approximation theory (Tampa, FL, 1990)*, vol. 19 of *Springer Series in Computational Mathematics*. Springer, New York, 1992, pp. 315–331.

[396] Nikolskii, V. N. Some questions of best approximation in a function space. *Uč. Zap. Kalininsk. Pedagog. Inst. 16* (1953), 119–160 (1954).

[397] Noether, F. Über eine Klasse singulärer Integralgleichungen. *Math. Ann. 82*, 1–2 (1920), 42–63.

[398] Orlicz, W. Beiträge zur Theorie der Orthogonalentwicklungen, I. *Studia Math. 1* (1929), 1–39.

[399] Orlicz, W. Über unbedingte Konvergenz in Funktionenräumen. *Studia Math. 2* (1933), 41–47.

[400] Page, L. B. Bounded and compact vectorial Hankel operators. *Trans. Amer. Math. Soc. 150* (1970), 529–539.

[401] Paley, R. E. A. C., and Wiener, N. *Fourier transforms in the complex domain*, vol. 19 of *American Mathematical Society Colloquium Publications*. Reprint of the 1934 original. American Mathematical Society, Providence, RI, 1987.

[402] Parrott, S. On a quotient norm and the Sz.-Nagy–Foiaş lifting theorem. *J. Funct. Anal. 30*, 3 (1978), 311–328.

[403] Paulsen, V. I., and Singh, D. A Helson–Lowdenslager–de Branges theorem in L^2. *Proc. Amer. Math. Soc. 129*, 4 (2001), 1097–1103 (electronic).

[404] Pełczyński, A. All separable Banach spaces admit for every $\varepsilon > 0$ fundamental total and bounded by $1 + \varepsilon$ biorthogonal sequences. *Studia Math. 55*, 3 (1976), 295–304.

[405] Peller, V. V. Hankel operators of class $\mathfrak{S}_p$ and their applications (rational approximation, Gaussian processes, the problem of majorization of operators). *Mat. Sbornik (N.S.) 113(155)*, 4(12) (1980), 538–581, 637.

[406] Peller, V. V. Description of Hankel operators of the class $\mathfrak{S}_p$ for $p > 0$, investigation of the rate of rational approximation and other applications. *Mat. Sbornik (N.S.) 122(164)*, 4 (1983), 481–510.

[407] Peller, V. V. Hankel operators in the theory of perturbations of unitary and selfadjoint operators. *Funktsional. Anal. Prilozhen. 19*, 2 (1985), 37–51, 96.

[408] Peller, V. V. *Hankel operators and their applications*. In *Springer Monographs in Mathematics*. Springer, New York, 2003.

[409] Peller, V. V., and Khrushchëv, S. V. Hankel operators, best approximations and stationary Gaussian processes. *Uspekhi Mat. Nauk 37*, 1(223) (1982), 53–124, 176.

[410] Petermichl, S., Treil, S., and Wick, B. D. Carleson potentials and the reproducing kernel thesis for embedding theorems. *Illinois J. Math. 51*, 4 (2007), 1249–1263.

[411] Phelps, R. R. Dentability and extreme points in Banach spaces. *J. Funct. Anal. 17* (1974), 78–90.

[412] Pick, G. Über die Beschränkungen analytische Funktionen, welche durch vorgegebene Funktionswerte bewirkt werden. *Math. Ann. 77* (1916), 7–23.

[413] Plemelj, J. Ein Ergänzungssatz zur Cauchyschen Integraldarstellung analytischer Funktionen, Randwerte betreffend. *Monatsh. Math. Phys. 19*, 1 (1908), 205–210.

[414] Poltoratski, A. G. Boundary behavior of pseudocontinuable functions. *Algebra Analiz 5*, 2 (1993), 189–210.

[415] Poltoratski, A. G. Properties of exposed points in the unit ball of H^1. *Indiana Univ. Math. J. 50*, 4 (2001), 1789–1806.

[416] Poltoratski, A. Integral representations and uniqueness sets for star-invariant subspaces. In *Systems, approximation, singular integral operators, and related topics (Bordeaux, 2000)*, vol. 129 of *Operator Theory: Advances and Applications*. Birkhäuser, Basel, 2001, pp. 425–443.

[417] Poltoratski, A., and Sarason, D. Aleksandrov–Clark measures. In *Recent advances in operator-related function theory*, vol. 393 of *Contemporary Mathematics*. American Mathematical Society, Providence, RI, 2006, pp. 1–14.

[418] Privalov, I. Intégrale de Cauchy. PhD Thesis, Saratov, 1919.

[419] Read, C. J. A solution to the invariant subspace problem. *Bull. London Math. Soc. 16*, 4 (1984), 337–401.

[420] READ, C. J. A solution to the invariant subspace problem on the space l_1. *Bull. London Math. Soc. 17*, 4 (1985), 305–317.

[421] REED, M., AND SIMON, B. *Methods of modern mathematical physics. I. Functional analysis*, 2nd edn. Academic Press, New York, 1980.

[422] RETHERFORD, J. R. *Hilbert space: compact operators and the trace theorem*, vol. 27 of *London Mathematical Society Student Texts*. Cambridge University Press, Cambridge, UK, 1993.

[423] RIESZ, F. Sur certains systèmes d'équations intégrales. *Ann. École Norm. Sup. 28*, 3 (1911), 33–62.

[424] RIESZ, F. *Les systèmes d'équations linéaires à une infinité d'inconnues*. Gauthier Villars, Paris, 1913.

[425] RIESZ, F. Über ein Problem des Herrn Carathéodory. *J. Reine Angew. Math. 146* (1916), 83–87.

[426] RIESZ, F. Über lineare Funktionalgleichungen. *Acta Math. 41*, 1 (1916), 71–98.

[427] RIESZ, F. Über die Randwerte einer analytischen Funktion. *Math. Z. 18* (1923), 87–95.

[428] RIESZ, F. Sur la décomposition des opérateurs linéaires. *Atti Congr. Int. Mat., Bologna, 3*, 1928.

[429] RIESZ, F. Über die linearen Transformationen des komplexen Hilbertschen Raumes. *Acta. Sci. Math. Szeged 5*, 1 (1930), 23–54.

[430] RIESZ, F., AND RIESZ, M. Über die Randwerte einer analytischen Funktion. *Quatrième Congrés des Mathématiciens Scandinaves, Stockholm*, 1916, pp. 27–44.

[431] RIESZ, F., AND SZ.-NAGY, B. *Functional analysis*. In *Dover Books on Advanced Mathematics*. Translated from French 2nd edn by L. F. Boron. Reprint of 1955 original. Dover, New York, 1990.

[432] RIESZ, M. Sur le problème des moments, troisième note. *Ark. Mat. Astr. Fys. 17*, 16 (1923), 62pp.

[433] RIESZ, M. Sur les fonctions conjuguées. *Math. Z. 27*, 1 (1928), 218–244.

[434] RINGROSE, J. *Compact non-self-adjoint operators*, vol. 35 of *Van Nostrand Reinhold Mathematical Studies*. Van Nostrand Reinhold, London, 1971.

[435] ROGOSINSKI, W. W., AND SHAPIRO, H. S. On certain extremum problems for analytic functions. *Acta Math. 90* (1953), 287–318.

[436] ROMBERG, B. The H_p spaces with $0 < p < 1$. PhD Thesis, University of Rochester, NY, 1960.

[437] ROSENBLUM, M., AND ROVNYAK, J. *Hardy classes and operator theory*. Corrected reprint of 1985 original. Dover, Mineola, NY, 1997.

[438] ROVNYAK, J. Characterization of spaces $H(m)$. Unpublished paper, 1968.

[439] ROVNYAK, J. An extension problem for the coefficients of Riemann mappings. *Operator Theory Seminar*, University of Virginia, 1991.

[440] ROVNYAK, J. Methods of Kreĭn space operator theory. In *Interpolation theory, systems theory and related topics (Tel Aviv/Rehovot, 1999)*, vol. 134 of *Operator Theory: Advances and Applications*. Birkhäuser, Basel, 2002, pp. 31–66.

[441] RUDIN, W. *Functional analysis*. In *McGraw-Hill Series in Higher Mathematics*. McGraw-Hill, New York, 1973.

[442] RUDIN, W. *Real and complex analysis*, 2nd edn. In *McGraw-Hill Series in Higher Mathematics*. McGraw-Hill, New York, 1974.

[443] SAND, M. Operator ranges and non-cyclic vectors for the backward shift. *Integral Equations Operator Theory 22*, 2 (1995), 212–231.

[444] Sarason, D. The H^p spaces of an annulus. *Mem. Amer. Math. Soc. 56* (1965), 78.

[445] Sarason, D. On spectral sets having connected complement. *Acta Sci. Math. (Szeged) 26* (1965), 289–299.

[446] Sarason, D. A remark on the Volterra operator. *J. Math. Anal. Appl. 12* (1965), 244–246.

[447] Sarason, D. Generalized interpolation in H^∞. *Trans. Amer. Math. Soc. 127* (1967), 179–203.

[448] Sarason, D. Weak-star density of polynomials. *J. Reine Angew. Math. 252* (1972), 1–15.

[449] Sarason, D. Functions of vanishing mean oscillation. *Trans. Amer. Math. Soc. 207* (1975), 391–405.

[450] Sarason, D. *Function theory on the unit circle*. Notes for lectures given at Department of Mathematics, Virginia Polytechnic Institute and State University, Blacksburg, VA, June 19–23, 1978.

[451] Sarason, D. Doubly shift-invariant spaces in H^2. *J. Operator Theory 16*, 1 (1986), 75–97.

[452] Sarason, D. Shift-invariant spaces from the Brangesian point of view. In *The Bieberbach conjecture (West Lafayette, IN, 1985)*, vol. 21 of *Mathematical Surveys and Monographs*. American Mathematical Society, Providence, RI, 1986, pp. 153–166.

[453] Sarason, D. Angular derivatives via Hilbert space. *Complex Variables Theory Appl. 10*, 1 (1988), 1–10.

[454] Sarason, D. Nearly invariant subspaces of the backward shift. In *Contributions to operator theory and its applications (Mesa, AZ, 1987)*, vol. 35 of *Operator Theory: Advances and Applications*. Birkhäuser, Basel, 1988, pp. 481–493.

[455] Sarason, D. Exposed points in H^1. I. In *The Gohberg anniversary collection, Vol. II (Calgary, AB, 1988)*, vol. 41 of *Operator Theory: Advances and Applications*. Birkhäuser, Basel, 1989, pp. 485–496.

[456] Sarason, D. Exposed points in H^1. II. In *Topics in operator theory: Ernst D. Hellinger memorial volume*, vol. 48 of *Operator Theory: Advances and Applications*. Birkhäuser, Basel, 1990, pp. 333–347.

[457] Sarason, D. Composition operators as integral operators. In *Analysis and partial differential equations*, vol. 122 of *Lecture Notes in Pure and Applied Mathematics*. Dekker, New York, 1990, pp. 545–565.

[458] Sarason, D. Weak compactness of holomorphic composition operators on H^1. In *Functional analysis and operator theory (New Delhi, 1990)*, vol. 1511 of *Lecture Notes in Mathematics*. Springer, Berlin, 1992, pp. 75–79.

[459] Sarason, D. Kernels of Toeplitz operators. In *Toeplitz operators and related topics (Santa Cruz, CA, 1992)*, vol. 71 of *Operator Theory: Advances and Applications*. Birkhäuser, Basel, 1994, pp. 153–164.

[460] Sarason, D. *Sub-Hardy Hilbert spaces in the unit disk*, no. 10 in *University of Arkansas Lecture Notes in the Mathematical Sciences*. Wiley-Interscience, New York, 1994.

[461] Sarason, D. Local Dirichlet spaces as de Branges–Rovnyak spaces. *Proc. Amer. Math. Soc. 125*, 7 (1997), 2133–2139.

[462] Sarason, D. Harmonically weighted Dirichlet spaces associated with finitely atomic measures. *Integral Equations Operator Theory 31*, 2 (1998), 186–213.

[463] SARASON, D. Errata: "Harmonically weighted Dirichlet spaces associated with finitely atomic measures". *Integral Equations Operator Theory 36*, 4 (2000), 499–504.

[464] SARASON, D. Nevanlinna–Pick interpolation with boundary data. *Integral Equations Operator Theory 30*, 2 (1998), 231–250.

[465] SARASON, D. Algebraic properties of truncated Toeplitz operators. *Oper. Matrices 1*, 4 (2007), 491–526.

[466] SARASON, D. Unbounded operators commuting with restricted backward shifts. *Oper. Matrices 2*, 4 (2008), 583–601.

[467] SCHAUDER, J. Zur Theorie stetiger Abbildungen in Funktionalräumen. *Math. Z. 26*, 1 (1927), 47–65.

[468] SCHAUDER, J. Über lineare vollstetige Funktionaloperationen. *Studia Math. 2* (1930), 183–196.

[469] SCHMIDT, E. Entwicklung willkürlicher Funktionen nach Systemen vorgeschriebener. *Math. Ann. 63* (1907), 433–476.

[470] SCHMIDT, E. Über die Auflösung linearen Gleichungen mit unendlich vielen Unbekanten. *Rend. Circ. Mat. Palermo 25* (1908), 53–77.

[471] SCHUR, I. Über Potenzreihen, die im Innern des Einheitskreises beschränkt sind, I. *J. Reine Angew. Math. 147* (1917), 205–232.

[472] SCHUR, I. Über Potenzreihen, die im Innern des Einheitskreises beschränkt sind, II. *J. Reine Angew. Math. 148* (1918), 122–145.

[473] SEDLOCK, N. A. Algebras of truncated Toeplitz operators. *Oper. Matrices 5*, 2 (2011), 309–326.

[474] SEGAL, I. E. Irreducible representations of operator algebras. *Bull. Amer. Math. Soc. 53* (1947), 73–88.

[475] SHAPIRO, H. S. Weakly invertible elements in certain function spaces, and generators in ℓ_1. *Michigan Math. J. 11* (1964), 161–165.

[476] SHAPIRO, H. S. Generalized analytic continuation. In *Symposia on Theoretical Physics and Mathematics, Vol. 8 (Madras, 1967)*. Plenum, New York, 1968, pp. 151–163.

[477] SHAPIRO, H. S., AND SHIELDS, A. L. On some interpolation problems for analytic functions. *Amer. J. Math. 83* (1961), 513–532.

[478] SHAPIRO, H. S., AND SHIELDS, A. L. Interpolation in Hilbert spaces of analytic functions. *Studia Math. (Ser. Specjalna) Zeszyt 1* (1963), 109–110.

[479] SHAPIRO, J. E. Aleksandrov measures used in essential norm inequalities for composition operators. *J. Operator Theory 40*, 1 (1998), 133–146.

[480] SHAPIRO, J. E. Relative angular derivatives. *J. Operator Theory 46*, 2 (2001), 265–280.

[481] SHAPIRO, J. E. More relative angular derivatives. *J. Operator Theory 49*, 1 (2003), 85–97.

[482] SHAPIRO, J. H., AND SUNDBERG, C. Compact composition operators on L^1. *Proc. Amer. Math. Soc. 108*, 2 (1990), 443–449.

[483] SHIELDS, A. L., AND WALLEN, L. J. The commutants of certain Hilbert space operators. *Indiana Univ. Math. J. 20* (1970/1971), 777–788.

[484] SIMON, B., AND WOLFF, T. Singular continuous spectrum under rank one perturbations and localization for random Hamiltonians. *Comm. Pure Appl. Math. 39*, 1 (1986), 75–90.

[485] SINGER, I. Weak* bases in conjugate Banach spaces. *Studia Math. 21* (1961/1962), 75–81.

[486] SINGER, I. *Bases in Banach spaces. I*, vol. 154 of *Die Grundlehren der Mathematischen Wissenschaften*. Springer, New York, 1970.

[487] SINGH, D., AND THUKRAL, V. Multiplication by finite Blaschke factors on de Branges spaces. *J. Operator Theory 37*, 2 (1997), 223–245.

[488] SMIRNOV, V. Sur la théorie des polynômes orthogonaux à une variable complexe. *J. Leningrad Fiz.-Mat. Obsch.* 2, 1 (1928), 155–179.

[489] SMIRNOV, V. Sur les valeurs limites des fonctions régulières à l'intérieur d'un cercle. *J. Leningrad Fiz.-Mat. Obsch. 2*, 2 (1928), 22–37.

[490] SMIRNOV, V. Sur les formules de Cauchy et Green et quelques problèmes qui s'y rattachent. *Izv. Akad. Nauk SSSR, Ser. Fiz.-Mat. 3* (1932), 338–372.

[491] ŠMULIAN, V. Über lineare topologische Räume. *Rec. Math. [Mat. Sbornik] N.S. 7 (49)* (1940), 425–448.

[492] SOKHOTSKI, Y. On definite integrals and functions using series expansions. PhD Thesis, St. Petersburg, 1873.

[493] SRINIVASAN, T. P., AND WANG, J.-K. On closed ideals of analytic functions. *Proc. Amer. Math. Soc. 16* (1965), 49–52.

[494] STAMPFLI, J. G. On hyponormal and Toeplitz operators. *Math. Ann. 183* (1969), 328–336.

[495] STEIN, E. M. Singular integrals, harmonic functions, and differentiability properties of functions of several variables. In *Singular integrals (Proc. Symp. Pure Math., Chicago, IL, 1966)*. American Mathematical Society, Providence, RI, 1967, pp. 316–335.

[496] STONE, M. *Linear transformations in Hilbert space and their applications to analysis*, vol. 15 of *American Mathematical Society Colloquium Publications*. American Mathematical Society, Providence, RI, 1932.

[497] STRASZEWICZ, S. Über exponierte Punkte abgeschlossener Punktmengen. *Fund. Math. 24* (1935), 139–143.

[498] STRAUSS, Y. Sz.-Nagy–Foias theory and Lax–Phillips type semigroups in the description of quantum mechanical resonances. *J. Math. Phys. 46*, 3 (2005), 1–25.

[499] SUÁREZ, D. Multipliers of de Branges–Rovnyak spaces in H^2. *Rev. Mat. Iberoamer. 11*, 2 (1995), 375–415.

[500] SUÁREZ, D. Backward shift invariant spaces in H^2. *Indiana Univ. Math. J. 46*, 2 (1997), 593–619.

[501] SUÁREZ, D. Closed commutants of the backward shift operator. *Pacific J. Math. 179*, 2 (1997), 371–396.

[502] SULTANIC, S. Sub-Bergman Hilbert spaces. *J. Math. Anal. Appl. 324*, 1 (2006), 639–649.

[503] SZ.-NAGY, B. On semi-groups of self-adjoint transformations in Hilbert space. *Proc. Natl. Acad. Sci. USA 24* (1938), 559–560.

[504] SZ.-NAGY, B. *Spektraldarstellung linearer Transformationen des Hilbertschen Raumes*, no. 5 of *Ergebnisse der Mathematik und ihrer Grenzgebiete*. Springer, Berlin, 1942.

[505] SZ.-NAGY, B. Sur les contractions de l'espace de Hilbert. *Acta Sci. Math. (Szeged) 15* (1953), 87–92.

[506] SZ.-NAGY, B., AND FOIAŞ, C. Commutants de certains opérateurs. *Acta Sci. Math. (Szeged) 29* (1968), 1–17.

[507] SZ.-NAGY, B., AND FOIAŞ, C. Dilatation des commutants d'opérateurs. *C. R. Acad. Sci. Paris, Sér. A–B 266* (1968), A493–A495.

[508] Sz.-Nagy, B., and Foiaş, C. *Harmonic analysis of operators on Hilbert space*. Translated from French and revised. North-Holland, Amsterdam, 1970.

[509] Szegő, G. Über die Entwickelung einer analytischen Funktion nach den Polynomen eines Orthogonalsystems. *Math. Ann. 82*, 3–4 (1921), 188–212.

[510] Szegő, G. Über orthogonale Polynome, die zu einer gegebenen Kurve der komplexen Ebene gehören. *Math. Z. 9*, 3–4 (1921), 218–270.

[511] Szegő, G. Beitrage zur Theorie der Toeplitzchen Formen, I. *Math. Z. 6* (1920), 167–202.

[512] Szegő, G. Über die Randwerte einer analytischen Funktion. *Math. Ann. 84*, 3–4 (1921), 232–244.

[513] Takenaka, S. On the orthonormal functions and a new formula of interpolation. *Jpn. J. Math. 2* (1925), 129–145.

[514] Taussky, O. A remark concerning the characteristic roots of the finite segments of the Hilbert matrix. *Q. J. Math., Oxford 20* (1949), 80–83.

[515] Taylor, A. E. The resolvent of a closed transformation. *Bull. Amer. Math. Soc. 44*, 2 (1938), 70–74.

[516] Taylor, A. E. Weak convergence in the spaces H^p. *Duke Math. J. 17* (1950), 409–418.

[517] Taylor, A. E. Banach spaces of functions analytic in the unit circle. I. *Studia Math. 11* (1950), 145–170.

[518] Taylor, A. E. Banach spaces of functions analytic in the unit circle. II. *Studia Math. 12* (1951), 25–50.

[519] Taylor, G. D. Contributions to the theory of multipliers. PhD Thesis, University of Michigan, Ann Arbor, MI, 1965.

[520] Temme, D., and Wiegerinck, J. Extremal properties of the unit ball in H^1. *Indag. Math. (N.S.) 3*, 1 (1992), 119–127.

[521] Thomson, J. E. Approximation in the mean by polynomials. *Ann. of Math. (2) 133*, 3 (1991), 477–507.

[522] Toeplitz, O. Über die Fouriersche Entwickelung positiver Funktionen. *Rend. Circ. Mat. Palermo 32* (1911), 191–192.

[523] Toeplitz, O. Zur Theorie der quadratischen und bilinearen Formen von unendlichvielen Veränderlichen. *Math. Ann. 70*, 3 (1911), 351–376.

[524] Tolokonnikov, V. A. Estimates in the Carleson corona theorem, ideals of the algebra H^∞, a problem of Sz.-Nagy. *Zap. Nauchn. Sem. Leningrad. Otdel. Mat. Inst. Steklov. (LOMI) 113, Issled. Linein. Oper. Teor. Funktsii. 11* (1981), 178–198, 267.

[525] Treil, S. R. Angles between co-invariant subspaces, and the operator corona problem. The Szőkefalvi-Nagy problem. *Dokl. Akad. Nauk SSSR 302*, 5 (1988), 1063–1068.

[526] Treil, S. R. Hankel operators, embedding theorems and bases of co-invariant subspaces of the multiple shift operator. *Algebra Analiz 1*, 6 (1989), 200–234.

[527] Troyanski, S. L. On locally uniformly convex and differentiable norms in certain non-separable Banach spaces. *Studia Math. 37* (1970/71), 173–180.

[528] Vigier, J. P. Etude sur les suites infinies d'opérateurs hermitiens. PhD Thesis, Geneva, 1946.

[529] Vinogradov, S. A., and Havin, V. P. Free interpolation in H^∞ and in certain other classes of functions. I. *Zap. Naučn. Sem. Leningrad. Otdel. Mat. Inst. Steklov. (LOMI) 47, Issled. Linein. Oper. Teor. Funktsii. 5* (1974), 15–54, 184–185, 191.

[530] Vinogradov, S. A., and Havin, V. P. Free interpolation in H^∞ and in certain other classes of functions. II. *Zap. Naučn. Sem. Leningrad. Otdel. Mat. Inst. Steklov. (LOMI) 56, Issled. Linein. Oper. Teor. Funktsii. 6* (1976), 12–58, 195.

[531] Visser, C. Note on linear operators. *Proc. Acad. Amsterdam 40* (1937), 270–272.

[532] Volberg, A. L. Thin and thick families of rational fractions. In *Complex analysis and spectral theory (Leningrad, 1979/1980)*, vol. 864 of *Lecture Notes in Mathematics*. Springer, Berlin, 1981, pp. 440–480.

[533] Volberg, A. L. Two remarks concerning the theorem of S. Axler, S.-Y. A. Chang and D. Sarason. *J. Operator Theory 7*, 2 (1982), 209–218.

[534] Volberg, A. L., and Treil, S. R. Embedding theorems for invariant subspaces of the inverse shift operator. *Zap. Nauchn. Sem. Leningrad. Otdel. Mat. Inst. Steklov. (LOMI) 149, Issled. Linein. Teor. Funktsii. 15* (1986), 38–51, 186–187.

[535] von Neumann, J. Allgemeine Eigenwerttheorie Hermitescher Funktionaloperatoren. *Math. Ann. 102*, 1 (1930), 49–131.

[536] von Neumann, J. Zur Algebra der Funktionaloperationen und Theorie der normalen Operatoren. *Math. Ann. 102*, 1 (1930), 370–427.

[537] von Neumann, J. Über adjungierte Funktionaloperatoren. *Ann. of Math. (2) 33*, 2 (1932), 294–310.

[538] von Neumann, J. On a certain topology for rings of operators. *Ann. of Math. (2) 37*, 1 (1936), 111–115.

[539] von Neumann, J. On rings of operators, III. *Ann. of Math. (2) 41* (1940), 94–161.

[540] von Neumann, J. Eine Spektraltheorie für allgemeine Operatoren eines unitären Raumes. *Math. Nachr. 4* (1951), 258–281.

[541] Walsh, J. L. *Interpolation and approximation by rational functions in the complex domain*, 4th edn, vol. 20 of *American Mathematical Society Colloquium Publications*. American Mathematical Society, Providence, RI, 1965.

[542] Walters, S. S. The space H^p with $0 < p < 1$. *Proc. Amer. Math. Soc. 1* (1950), 800–805.

[543] Walters, S. S. Remarks on the space H^p. *Pacific J. Math. 1* (1951), 455–471.

[544] Wecken, F. J. Zur Theorie linearer Operatoren. *Math. Ann. 110*, 1 (1935), 722–725.

[545] Wermer, J. On algebras of continuous functions. *Proc. Amer. Math. Soc. 4* (1953), 866–869.

[546] Weyl, H. Über gewöhnliche Differentialgleichungen mit Singularitäten und die zugehörigen Entwicklungen willkürlicher Funktionen. *Math. Ann. 68*, 2 (1910), 220–269.

[547] Whitley, R. An elementary proof of the Eberlein–Šmulian theorem. *Math. Ann. 172* (1967), 116–118.

[548] Widom, H. Inversion of Toeplitz matrices. II. *Illinois J. Math. 4* (1960), 88–99.

[549] Widom, H. Inversion of Toeplitz matrices. III. *Notices Amer. Math. Soc. 7* (1960), 63.

[550] Widom, H. On the spectrum of a Toeplitz operator. *Pacific J. Math. 14* (1964), 365–375.

[551] Widom, H. Hankel matrices. *Trans. Amer. Math. Soc. 121* (1966), 1–35.

[552] Widom, H. Toeplitz operators on H_p. *Pacific J. Math. 19* (1966), 573–582.

[553] Wiegerinck, J. A characterization of strongly exposed points of the unit ball of H^1. *Indag. Math. (N.S.) 4*, 4 (1993), 509–519.

[554] WIENER, N. *The Fourier integral and certain of its applications*. In *Cambridge Mathematical Library*. Reprint of 1933 edition. Cambridge University Press, Cambridge, UK, 1988.

[555] WINTNER, A. Zur Theorie der beschränkten Bilinearformen. *Math. Z. 30*, 1 (1929), 228–281.

[556] WOJTASZCZYK, P. The Franklin system is an unconditional basis in H_1. *Ark. Mat. 20*, 2 (1982), 293–300.

[557] WOJTASZCZYK, P. *Banach spaces for analysts*, vol. 25 of *Cambridge Studies in Advanced Mathematics*. Cambridge University Press, Cambridge, UK, 1991.

[558] WOLD, H. *A study in the analysis of stationary time series*, 2nd edn, with an appendix by P. Whittle. Almqvist and Wiksell, Stockholm, 1954.

[559] YABUTA, K. Remarks on extremum problems in H^1. *Tohoku Math. J. 23* (1971), 129–137.

[560] YABUTA, K. Unicity of the extremum problems in $H^1(U^n)$. *Proc. Amer. Math. Soc. 28* (1971), 181–184.

[561] YABUTA, K. Some uniqueness theorems for $H^p(U^n)$ functions. *Tohoku Math. J. 24* (1972), 353–357.

[562] YAMAZAKI, S. Normed rings and bases in Banach spaces. *Sci. Papers College Gen. Educ. Univ. Tokyo 15* (1965), 1–13.

[563] YOOD, B. Properties of linear transformations preserved under addition of a completely continuous transformation. *Duke Math. J. 18* (1951), 599–612.

[564] YOSIDA, K. *Functional analysis*. In *Classics in Mathematics*. Reprint of 6th (1980) edition. Springer, Berlin, 1995.

[565] YOUNG, N. *An introduction to Hilbert space*. In *Cambridge Mathematical Textbooks*. Cambridge University Press, Cambridge, UK, 1988.

[566] YOUNG, R. M. *An introduction to nonharmonic Fourier series*, vol. 93 of *Pure and Applied Mathematics*. Academic Press, New York, 1980.

[567] ZAREMBA, S. L'équation biharmonique et une classe remarquable de fonctions fondamentales harmoniques. *Bull. Int. Acad. Sci. Cracovie* (1907), 147–196.

[568] ZHU, K. Sub-Bergman Hilbert spaces on the unit disk. *Indiana Univ. Math. J. 45*, 1 (1996), 165–176.

[569] ZHU, K. Sub-Bergman Hilbert spaces in the unit disk. II. *J. Funct. Anal. 202*, 2 (2003), 327–341.

[570] ZHU, K. *Operator theory in function spaces*, 2nd edn, vol. 138 of *Mathematical Surveys and Monographs*. American Mathematical Society, Providence, RI, 2007.

[571] ZYGMUND, A. Sur les fonctions conjugées. *Fund. Math. 13* (1929), 284–303.

[572] ZYGMUND, A. *Trigonometric series. Vols. I, II*, 2nd edn. Cambridge University Press, New York, 1959.

Symbol index

Author index

Subject index

CPSIA information can be obtained
at www.ICGtesting.com
Printed in the USA
LVOW12*1106270516
490276LV00006B/11/P

9 781107 027770